AF704202

$19.95C

R.E. RIDSDALE
BRITISH COLUMBIA
INSTITUTE OF TECHNOLOGY

ELECTRIC CIRCUITS

SECOND EDITION

McGraw-Hill Book Company

New York Atlanta Dallas St. Louis San Francisco
Auckland Bogotá Guatemala Hamburg Johannesburg
Lisbon London Madrid Mexico Montreal New Delhi
Panama Paris San Juan São Paulo Singapore
Sydney Tokyo Toronto

Sponsoring Editor: George Z. Kuredjian
Editing Supervisor: Patricia H. Nolan
Design and Art Supervisor: Caryl Valerie Spinka
Production Supervisor: Frank Bellantoni

Interior Design: Sharkey Design
Cover Photographer: Ken Karp
Technical Studio: Technical Graphics

Library of Congress Cataloging in Publication Data

Ridsdale, R. E.
Electric circuits.

Rev. ed. of: Electric circuits for engineering technology. 1976.
Includes bibliographical references and index.
1. Electric circuits. 2. Electric networks.
I. Title.
TK454.R52 1983 621.319′2 83-939
ISBN 0-07-052948-5

ELECTRIC CIRCUITS, Second Edition

Copyright © 1984, 1976 by McGraw-Hill, Inc. All rights reserved. Printed in the United States of America. Except as permitted under the United States Copyright Act of 1976, no part of this publication may be reproduced or distributed in any form or by any means, or stored in a data base or retrieval system, without the prior written permission of the publisher.

The first edition was published under the title, *Electric Circuits for Engineering Technology*

1 2 3 4 5 6 7 8 9 0 VNHVNH 8 9 0 9 8 7 6 5 4 3

ISBN 0-07-052948-5

CONTENTS

PREFACE

This book is designed as a basic text in electric circuits for students of electrical and electronics technology.

Electric circuits is the backbone of the study of what is usually identified as electricity—an identity used primarily to distinguish it from the more recognizable subject of electronics. If any bias is detected in the choice of problems and applications it is toward electronics. The fact is that it is impossible to work in any medium concerned with electricity without being involved in electronics. Those former discrete disciplines of power technology and electronics are now so interrelated and interdependent that to still hold to the boundary lines of the past is unproductive and unreal. But to admit to this bias nevertheless does not detract from the treatment of electrical principles which after all forms the strength of this text.

The second edition maintains the balance between theory and practice that proved so successful in the previous edition. Comments from both students and instructors have resulted in changes in organization and content that strengthen the basic pedagogical elements of the first edition. For example:

- In addition to Exercise Problems, Review Questions are provided to check the student's understanding of the text material. These Questions and Problems now follow immediately after blocks of text material within a chapter.
- New Chapter Questions and Chapter Problems are provided at the end of each chapter. These are more general, more difficult, or more practical than those problem sets following the individual text blocks. Chapter ending materials emphasize specific major points of the chapter and tie together interrelated concepts.
- The order in which topics now appear reflects curriculum trends as indicated by comments from instructors. For example, the discussion of capacitance and inductance appears early in the text to form a three-part group with the discussion of resistance. This unites the three most important circuit characteristics and helps dispel the false idea that capacitance and inductance are somehow difficult subjects not directly associated with resistance. Related to this is the introduction of resistors, capacitors, and inductors as practical devices that exhibit the three main circuit characteristics.
- Mathematical tables have given way to the recognition of the availability and use of the ubiquitous scientific calculator by all technology students. By working accurately and quickly, it enables more problems to be solved correctly in a given amount of time. This is a definite aid to learning since it focuses attention on circuits rather than mathematics.
- The use of the calculator also facilitates the use of practical commercial

values of components. The internationally standardized approved values are used in most of the examples and problems. While simple, tidy problem solutions could be obtained by selecting appropriate data, continual avoidance of general but awkward situations gives a false impression of the ease of obtaining real solutions to problems that might occur on the job.

For the most part technical terminology and symbols approved by international authority are used throughout the text. The units are SI (MKSA); quantity and unit symbols conform to various IEEE and ANSI standards.

Unit symbols are used for all quantities in calculations, even though such a routine is too laborious for use in practical electrical work. This total adoption of unit symbols results in a good learning situation, because it provides a rationale for the units in which the answers are given.

From a realistic viewpoint, accuracy in calculating needs to be no greater than that of the available data. Tolerances in electric components and measurements are rarely better than 1 percent, so three significant figures are considered adequate for most electrical situations. But for learning it is often desirable to solve a given problem several times using different techniques. Identical answers encourage confidence in the validity of the techniques. The use of three significant figures is sometimes inadequate to ensure that answers are identical, whereas four significant figures are appreciably better in this regard. This learning advantage is accepted as the justification for the adoption of four significant figures (including angles to 0.01°) throughout this text.

No attempt will be made to suggest to teachers how they should schedule the text material. Many factors will determine the sections to be selected; the order of presentation and the timing will be chosen to suit the purposes of the complete learning program.

I wish to thank my friends and colleagues for their invaluable assistance in preparing this second edition and to express my appreciation to the many readers who have offered suggestions for its successful completion.

R. E. Ridsdale

INTRODUCTION

Electric circuits—what are they? They are systems of conductors and elements connected together so that electric current can exist.

What is a system? What is a conductor? What is an element? What is current? What is needed for the existence of current?

The original question produced five further questions. This text will help you to answer these questions and others that have not yet arisen.

Electrical technology is very closely associated with physical science and mathematics. The nature of electricity has its roots in physics, and the laws governing circuit behavior are physical laws. These laws are usually stated in the language of mathematics. The development of electrical concepts from physical laws depends on manipulation of the mathematical relationships between quantities. While it is certainly not the aim of this book to teach physics and mathematics, it is necessary that some physical and mathematical ideas be dealt with as an introduction to the behavior of electric circuits. Thus we start by introducing some of the terms, symbols, equations, and methods you will meet frequently in succeeding chapters. For whatever time and effort you spend in assimilating this material, you will be amply repaid by the ease with which you will be able to handle circuit principles and applications.

Because the terms "parameter" and "variable" are related to the basic elements of which electric circuits are fashioned, it is logical that we start our investigation of circuits by looking at these quantities. This investigation is followed by an introduction to some familiar basic elements and the relationships between them.

1

SOME FUNDAMENTALS

Principles and techniques used extensively in electrical work were originated by Maxwell, Laplace, and other mathematicians. Mathematical methods are used to determine the quantities involved in electric circuits. It is not surprising, therefore, to find that mathematical terms such as "equation" and "square root," and symbols such as = and $\sqrt{\ }$, are common in electrical language.

Many of the relationships between electrical quantities can be expressed in the form of simple equations of the type $y = kx$ (one quantity equals the product of two others). Applications of this equation simply use different symbols for different quantities. The most common electrical version of this equation is $V = IR$, relating voltage, current, and resistance (Ohm's law).

Usually, in the Ohm's law equation two of the quantities (voltage and current) are variables; that is, their values can vary over a wide range. The other quantity (resistance) is a parameter of nominally fixed value.

Variables are always specified in terms of units. Parameters too are usually associated with units, but sometimes they are dimensionless numbers (numbers without units).

In your study of electric circuits, you will become familiar with many variables, parameters, and units. An understanding of electric circuits depends on this familiarity.

1-1 VARIABLES

A *variable* is a quantity which can readily be made to change value.

Continuing to use Ohm's law as a convenient, frequently used equation, without concern for its electrical meaning, let us look at voltage and current.

In any circuit, it is possible to have a wide range of voltage and current values. Thus, in the Ohm's law equation, V (voltage) and I (current) are variables.

It is common practice to use a graphical representation of a range of voltage and current values, as in Fig. 1-1.

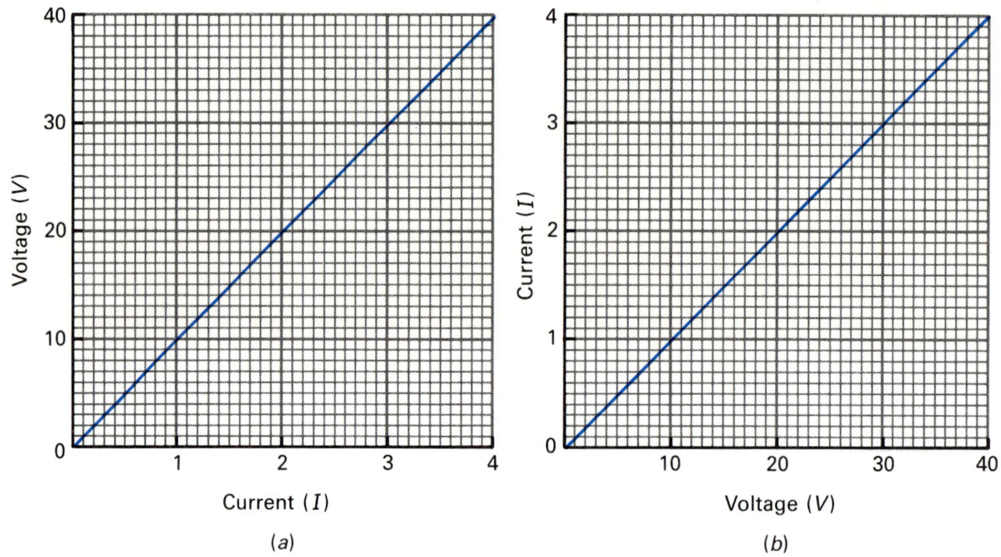

Fig. 1-1. Linear voltage-current graphs.

In Fig. 1-1a, the values of current are plotted horizontally (on the X axis). By convention this means that current is the independent variable, its values having been deliberately selected. The values of voltage are plotted vertically (on the Y axis), implying that this quantity is the dependent variable, its values being dependent on those selected for current.

The graph in Fig. 1-1b shows voltage as the independent variable and current as the dependent variable.

Both graphs in Fig. 1-1 depict the same electrical situation. The choice of presentation will be determined by the purpose of the graph.

Voltage and current are the most common electrical variables. Some others are power, energy, charge, and time.

REVIEW QUESTIONS FOR SECTION 1-1

1. What is meant by the term "variable"?
2. What is a convenient method of displaying a relationship between two variables?
3. What is the name given to the variable whose values are deliberately selected?
4. What is the name given to the variable whose values rely on those of another variable?
5. In the equation $a = kb$ (where k is a constant), which is the dependent variable and which is the independent variable?

1-2 PARAMETERS

A *parameter* is a constant, the value of which shows the relationship between two variables.

In electrical terms, a parameter is usually regarded as a circuit characteristic which describes the circuit's behavior.

Perhaps the word constant should have been placed in quotes. Constant means unvarying, but some circuit parameters can change value under operating conditions. Figure 1-1 relates to truly constant parameters in linear circuit operation (the ''curves'' of the graphs are straight lines). In any such situation the value of the parameter is given by the slope of the curve (the angle of the graph line from the horizontal). The numerical value of the parameter is determined by the ratio of any chosen value of the dependent variable to the value of the related independent variable.

In Fig. 1-1*a*, the parameter has a value of 10, any value of voltage V being 10 times the corresponding value of current I. The equation for this graph is $V = IR$, where R is the parameter resistance, a circuit characteristic.

The equation for Fig. 1-1*b* is $I = VG$, where G is conductance, another circuit parameter. The value of G in this case is $\frac{1}{10}$ or 0.1.

Resistance (voltage divided by current) and conductance (current divided by voltage) are different parameters even when, as in Fig. 1-1, the voltage and current values are the same. Some other common circuit parameters are capacitance, impedance, and phase angle.

REVIEW QUESTIONS FOR SECTION 1-2

1. What is meant by the term ''parameter''?
2. Can a parameter for one situation be a variable for another?

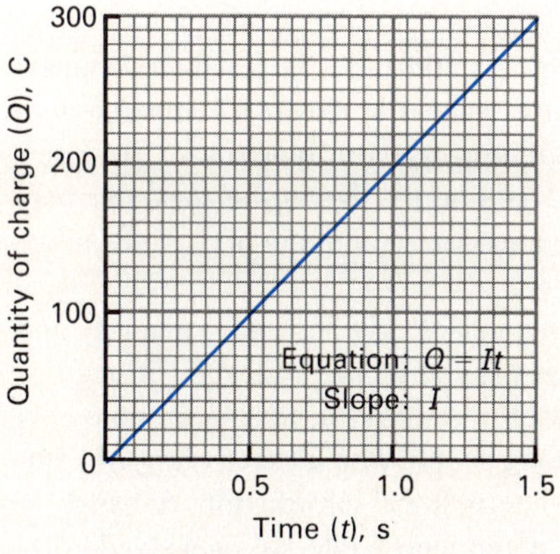

Fig. 1-2. Review question 3: Graph.

3. For the graph in Fig. 1-2,
   a. What is the term applied to:
      i. I *ii*. Q *iii*. t
   b. What is the value of I?

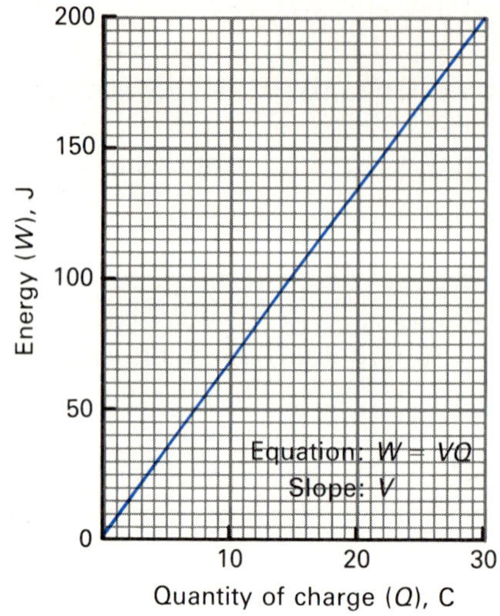

Fig. 1-3. Review question 4: Graph.

4. For the graph of Fig. 1-3,
a. What is the term applied to:
i. V *ii*. Q *iii*. W
b. What is the value of V?

1-3 UNITS

Units reveal the magnitudes, or sizes, of quantities. Most numbers associated with electric circuits have no meaning unless they are accompanied by units.

A *unit* is a specific value of a quantity with which other quantities of the same type can be compared.

Is 100 larger than 2? Obviously it is. But is 100 miles larger than 2 hours? Clearly it is not. The mile, a unit of distance, cannot be directly compared with the hour, a unit of time; they represent different types of quantities. However, there can be a relationship between miles and hours. For instance, a car may travel 100 miles in 2 hours; but this merely introduces miles per hour, the unit for another quantity, velocity or speed.

In technical work, it is important to distinguish clearly between a quantity and the unit in which it is specified. Most of us use the word ''mileage'' instead of the correct term ''distance'' to refer to how far a car travels. Such terminology is harmless enough in social conversation among people who all speak English. But for technological uses, which may involve international information exchange, it is unacceptable to add ''age'' to the end of a unit name. International standardizing authorities have made a great effort to eliminate the confusion which can arise as a result of basing the name of a quantity on the term used for its unit.

Of course, there seems to be an exception to every rule. The term ''voltage'' is so firmly entrenched in electrical language that it has won official approval. But ''amperage'' for ''current'' and ''wattage'' for ''power'' are not acceptable, even though they are used in the industry.

Many units bear the names of internationally famous scientists. Thus you will encounter such unit names as ampere, hertz, ohm, joule, and newton. Volt is an abbreviation of Volta, another scientist's name. The mho, the unit for conductance for many years, has been officially changed to the siemens (named after the scientist von Siemens). Therefore, in this book we will refer to the siemens as the unit of conductance.

REVIEW QUESTIONS FOR SECTION 1-3

1. What is meant by the term "unit"?
2. Distinguish between the terms "quantity" and "unit." Give four examples of quantities and their units.

1-4 SYSTEMS OF UNITS

There are several different unit systems in existence. A quantity can be represented in more than one unit system and can have several different units. To illustrate this, let us consider some units of length or distance:

Statute mile	1609.35 meters
Nautical mile	1852 meters
Foot	0.3048 meter
Rod	5.029 meters

There are other distance units in use throughout the world for various purposes. To reduce the confusion in modern communication caused by such a proliferation of different units for the same quantity, the meter has been agreed upon as the international standard, although we know that it has not yet displaced other units.

The reasons for the existence of various unit systems are historical or geographical, going back to the early need to measure. The meter was originally based on a calculation of one-ten-millionth of the surface distance from pole to equator. The nautical mile originated as 1 minute of longitude measured at the equator. A foot was the length of a man's foot.

Despite attempts to encourage the exclusive use of MKS units, there are three unit systems in use in electrical work. Each of them is named in terms of its units for length, mass, and time.

MKS	meter-kilogram-second
FPS (British)	foot-pound-second
CGS	centimeter-gram-second

The FPS and CGS systems have not been approved for electrical purposes. But many charts and tables have been compiled over the years with these units, and people have become so familiar with them that there is a natural reluctance to change, and so it is necessary that we recognize their existence.

REVIEW QUESTIONS FOR SECTION 1-4

1. Can a quantity be expressed in more than one type of unit?
2. What is the international standard unit for distance?
3. What are the three unit systems found in electrical work?
4. Which unit system has been universally approved for use in electrical work?

1-5 MKSA UNITS

The MKS unit system is extended, for electrical purposes, to include the ampere. It is then known as the MKSA system. With the units for distance (or length), mass, time, and current defined, it is possible to define all other electrical units. Because of its widespread international use and its decimal base, the MKSA unit system was agreed on by the Conference Générale des Poids et Mesures (General Conference on Weights and Measures, CGPM) in 1960. It was designated SI, Système International d'Unités (International System of Units). Since that time most, if not all, scientific and engineering organizations throughout the world have endorsed the adoption of SI units. The Institute of Electrical and Electronics Engineers has recommended the universal adoption of SI units in electrical technology.

All the variables and parameters in this text will be associated with the relevant SI units.

If you already have some electrical knowledge, you are well on the way to learning the SI unit system. Most of the traditional electrical units, including the volt, ampere, ohm, farad, and henry, are MKSA units and are SI units.

REVIEW QUESTIONS FOR SECTION 1-5

1. The system of units used for electrical purposes has four basic units. What are they?
2. What are the quantities expressed by the four basic units?
3. What is meant by SI units?

1-6 DEFINITIONS OF MKSA UNITS

Recent advances in technology are shown by the changes which have taken place in the definitions of basic units. As recently as 1960, the meter was defined in terms of the length of a platinum-iridium bar kept at the International Bureau of Weights and Measures. But measuring between two marks on a piece of metal, however stable, is not good enough for many present-day purposes.

The modern definition of the meter refers length to atomic radiation and results in a defined meter almost exactly the same as that defined by the platinum-iridium bar. Any organization which can justify the elaborate equipment needed is able to set up the modern standard and be sure that it complies with the definition of the meter.

Mass is not so easy to define by direct measurement. Advantage is taken of the fact that, at any place on earth, two bodies which weigh the same on a balance have the same mass. A standard kilogram weight is maintained at the International Bureau of Weights and Measures. Other weights can be compared with it.

The precise definition of the second, the SI basic unit of time, has still not been agreed on. In the meantime, most authorities accept the use of instruments, such as a cesium clock, for time standardization. These instruments use molecular resonance, a phenomenon of accurately known time characteristics.

Two current-carrying conductors exert a force on each other. The magnitude of this force depends on the magnitudes of the currents. This principle forms the basis of the modern electromagnetic definition of the SI basic unit of current, the ampere.

Two additional SI units are of interest in electric circuits, the kelvin and the radian.

Temperature is a variable which can affect the behavior of electrical materials and components. The SI basic unit of temperature is the kelvin. The degree Celsius, which has the same unit value as the kelvin, is commonly used. Kelvin temperatures can be converted to degrees Celsius[1] by subtracting 273.15. A temperature of 0 degrees Celsius is the same as 273.15 kelvins.

Plane angles (angles between lines) are often important in electric circuits. For instance, phase relationships in ac circuits are usually specified in terms of angles. The radian is the SI unit for plane angle. There are 2π radians in one complete revolution. The degree is frequently used in preference to the radian. There are 360 degrees in one revolution. Thus, 2π radians = 360 degrees.

EXAMPLE 1-1

The relationship between degrees Fahrenheit and degrees Celsius is:

Temperature in degrees Celsius = (temperature in degrees Fahrenheit − 32) × $\frac{5}{9}$

Express normal room temperature. 68 degrees Fahrenheit, in degrees Celsius and in kelvins.

Solution

$$\text{Temperature in degrees Celsius} = (68 - 32) \times \tfrac{5}{9} = 36 \times \tfrac{5}{9} = \mathbf{20\ degrees\ Celsius}$$

$$\text{Temperature in kelvins} = \text{temperature in degrees Celsius} + 273.15 = 20 + 273.15 = \mathbf{293.15\ kelvins}$$

Therefore 68 degrees Fahrenheit is the same as 20 degrees Celsius or 293.15 kelvins.

[1] The degree Celsius has superseded the degree centigrade. They are identical in value, the name change being for purposes of international standardization.

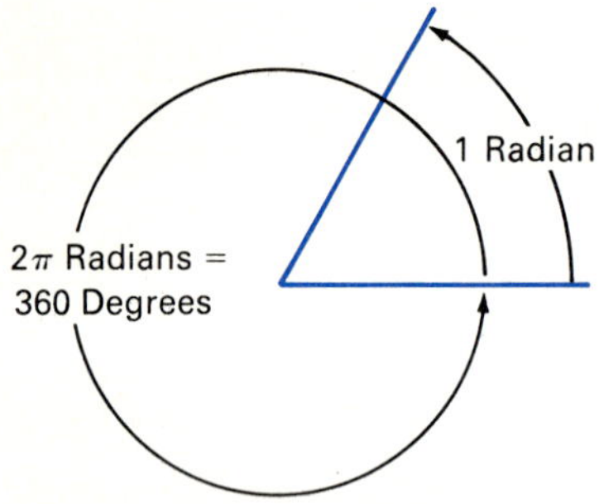

Fig. 1-4. Example 1-2: The radian.

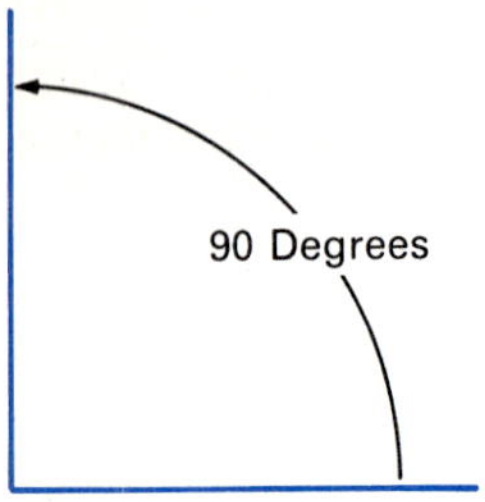

Fig. 1-5. Example 1-3: The right angle.

EXAMPLE 1-2 How many degrees equal 1 radian?

Solution (See Fig. 1-4.)

$$2\pi \text{ radians} = 360 \text{ degrees}$$
$$1 \text{ radian} = \frac{360}{2\pi} \text{ degrees}$$
$$1 \text{ radian} = \mathbf{57.3\ degrees}$$

EXAMPLE 1-3 How many radians are there in a right angle (90 degrees)?

Solution (See Fig. 1-5.)

$$360 \text{ degrees} = 2\pi \text{ radians}$$
$$90 \text{ degrees} = \frac{360}{4} \text{ degrees} = \frac{2\pi}{4} \text{ radians}$$
$$90 \text{ degrees} = \frac{\pi}{2} \textbf{ radians}$$

Often the answer is left in the form $\pi/2$ radians, although it can be written as 1.571 radians.

REVIEW QUESTIONS FOR SECTION 1-6

1. What is the SI unit for temperature?
2. How does the degree Celsius compare in magnitude with the kelvin?
3. To obtain Celsius values, what numerical adjustment must be made to kelvin values?
4. What is the SI unit of angle?
5. How many degrees are required to approximately represent the SI unit of angle?

EXERCISE PROBLEMS FOR SECTION 1-6

1. Express the following temperatures in kelvins.
   a. 0°C (*ans.* 273.15 K) *e.* 212°F (*ans.* 373.15 K)
   b. 0°F (*ans.* 255.4 K) *f.* 150°F (*ans.* 338.7 K)
   c. 100°C (*ans.* 373.15 K) *g.* −40°C (*ans.* 233.15 K)
   d. 32°F (*ans.* 273.15 K) *h.* −40°F (*ans.* 233.15 K)
2. Express the following temperatures in degrees Celsius.
   a. −60°F *c.* 283 K
   b. 120°F *d.* 390 K
3. Express the following temperatures in degrees Fahrenheit.
   a. 0 K (*ans.* −459.7°F) *c.* 25°C (*ans.* 77°F)
   b. 300 K (*ans.* 80.33°F) *d.* −25°C (*ans.* −13°F)

4. Express the following angles in degrees.
   a. 0.5 rad   *c*. 0.785 rad
   b. 2.1 rad   *d*. 4.71 rad
5. Express the following angles in radians.
   a. 150° (*ans*. 2.618 rad)   *c*. 990° (*ans*. 17.28 rad)
   b. 30° (*ans*. 0.5235 rad)   *d*. 1° (*ans*. 0.01745 rad)

1-7 SYMBOLS

It is laborious to write words repeatedly, particularly long ones. Language difficulties make words undesirable for information exchange. Symbols are short and international.

To illustrate the effectiveness of symbols, Examples 1-1 and 1-2 are repeated in the more usual symbolic form.

EXAMPLE 1-1 The relationship between degrees Fahrenheit (θ°F) and degrees Celsius (t°C) is

$$t°\text{C} = (\theta°\text{F} - 32) \times \tfrac{5}{9} \qquad \textbf{(1-1)}$$

Express normal room temperature, 68°F, in degrees Celsius and in kelvins.

Solution

$$t = (68°\text{F} - 32) \times \tfrac{5}{9} = 36 \times \tfrac{5}{9} = \mathbf{20°C}$$

The temperature in degrees Celsius (t°C) and in kelvins (T K) is given by

$$T\ \text{K} = t°\text{C} + 273.15 \qquad \textbf{(1-2)}$$

$$T = 20°\text{C} + 273.15 = \mathbf{293.15\ K}$$

EXAMPLE 1-2 How many degrees (°) equal 1 radian (1 rad)?

Solution

$$2\pi\ \text{rad} = 360° \qquad \textbf{(1-3)}$$

$$1\ \text{rad} = \frac{360°}{2\pi}$$

$$1\ \text{rad} = \mathbf{57.3°}$$

A quantity symbol is a single letter, in some cases Greek, with a subscript added when required. Some examples are

I = current
v = instantaneous voltage
I_o = output current
ω = angular velocity or angular frequency
P_{av} = average power
ϕ_1 = phase angle number one

A unit symbol is a letter, a group of letters, or a special sign. A few typical examples are

K = kelvin
cm = centimeter
P = pascal
Ω · m = ohm-meter
°C = degree Celsius

There is a total of 100 capital and lowercase Roman (ordinary) and Greek letters. There are many more scientific and engineering quantities and units that require symbols. To minimize confusion, *italic* (sloping) letters are used for quantities, and roman (upright) letters for units. It is therefore important to distinguish between, say, *m* for mass and m for meter.

The symbols for the SI quantities and units referred to in this chapter are listed in Table 1-1.

Other units which have appeared in this chapter but which are not SI units are listed in Table 1-2.

The Greek letters used in these tables are

ρ = lowercase rho
ϕ = lowercase phi
θ = lowercase theta
ω = lowercase omega
Ω = capital omega

TABLE 1-1 Symbols for SI Quantities and Units

Quantity	Symbol	Unit	Symbol
Angular velocity or angular frequency	ω	radian per second	rad/s
Conductance	G	siemens	S
Current*	I	ampere	A
Distance†	s	meter	m
Mass	m	kilogram	kg
Plane angle‡	ϕ	radian	rad
Power*	P	watt	W
Pressure	p	pascal	P
Resistance†	R	ohm	Ω
Resistivity	ρ	ohm-meter	Ω · m
Temperature	T	kelvin	K
Time	t	second	s
Voltage*	V	volt	V

*Lowercase i, p, or v may be used as the symbols for current, power, or voltage, respectively, to indicate instantaneous values, or magnitudes at specified instants of time. Subscripts may be used to identify specific quantities (for example, P_o may mean power at the output terminals).

†Multiples and fractions of SI units are frequently used. Centimeter (cm) and kilohm (kΩ) are examples.

‡Other Greek letters may be used as symbols for plane angle. ϕ is shown because it is commonly used for phase angle.

TABLE 1-2 Symbols in Chap. 1 Which Are Not SI

Quantity	Symbol	Unit	Symbol
plane angle*	ϕ	degree	°
temperature	t	degree Celsius	°C
temperature	θ	degree Fahrenheit	°F

*Other Greek letters may be used as symbols for plane angle. ϕ is shown because it is commonly used for phase angle.

REVIEW QUESTIONS FOR SECTION 1-7

1. Why are symbols used for quantities and units?
2. How can quantity symbols be distinguished from unit symbols?

1-8 SUPER- AND SUBUNITS

Primary SI or MKSA units are not always of a convenient size. For example, the meter is a satisfactory unit in which to express the length of a table, but for specifying distances between cities the kilometer (1000 m) is more useful. Units larger or smaller than the primary unit may be convenient for use in electric circuit work.

A superunit is greater than the primary unit, the kilometer being an example. A subunit is smaller than the primary unit, for instance, the millisecond = $\frac{1}{1000}$ s. It has been found convenient to place super- and subunits on a scale such that adjacent values are spaced by a ratio of 1000. A list with the standard prefixes is given in Table 1-3.

TABLE 1-3 Prefixes for Use with SI Units

Prefix	Symbol	Relationship to Primary Unit*
tera	T	unit × 10^{12} or × 1 000 000 000 000
giga	G	unit × 10^{9} or × 1 000 000 000
mega	M	unit × 10^{6} or × 1 000 000
kilo	k	unit × 10^{3} or × 1 000
unit		unit × 10^{0} or × 1
milli	m	unit × 10^{-3} or × 0.001
micro	μ	unit × 10^{-6} or × 0.000 001
nano	n	unit × 10^{-9} or × 0.000 000 001
pico	p	unit × 10^{-12} or × 0.000 000 000 001
femto	f	unit × 10^{-15} or × 0.000 000 000 000 001
atto	a	unit × 10^{-18} or × 0.000 000 000 000 000 001

*For numbers with more than four digits on either side of the decimal point (greater than 9999 or smaller than 0.0001), it is recommended by international standard that spaces be used to separate *triads,* groups of three digits, commencing at the decimal point. Examples: 12 345 or 0.123 45. This use of spaces supersedes the use of commas.

Typical examples of the use of Table 1-3 are

$$\begin{aligned}\text{10 microseconds} &= 10 \times 10^{-6}\text{ s} = 10\ \mu\text{s}\\ \text{50 kilovolts} &= 50 \times 10^{3}\text{ V} = 50\text{ kV}\\ \text{6.8 megohms} &= 6.8 \times 10^{6}\ \Omega = 6.8\text{ M}\Omega\\ \text{200 picofarads} &= 200 \times 10^{-12}\text{ F} = 200\text{ pF}\\ \text{4 gigahertz} &= 4 \times 10^{9}\text{ Hz} = 4\text{ GHz}\end{aligned}$$

The centimeter, cm, is a direct decimal submultiple of the meter ($\frac{1}{100}$ m). However, it is preferable to call 1 cm either 10^{-2} m or 0.01 m.

Arithmetic Using Super- and Subunits

All calculations involving units can be carried out in scientific notation or standard form. This is the expression of any number as a number between 1 and 10 multiplied by a power of 10 (for example, $123.4 = 1.234 \times 10^2$). This notation has been widely used, as it simplifies writing and calculation methods and helps to reduce errors.[1] A further simplification can be achieved by using super- and subunits and applying arithmetic methods directly to them. The following examples demonstrate the technique.

1. mega × milli = kilo

$$4\text{ M}\Omega \times 5\text{ mA} = 20\text{ kV} \qquad (4 \times 10^{6}\ \Omega \times 5 \times 10^{-3}\text{ A} = 20 \times 10^{3}\text{ V})$$

2. unit/micro = mega

$$\frac{10\text{ V}}{5\ \mu\text{A}} = 2\text{ M}\Omega \qquad [10\text{ V} \div (5 \times 10^{-6})\text{ A} = 2 \times 10^{6}\ \Omega]$$

3. milli squared/kilo = nano

$$\frac{(5\text{ mV})^2}{5\text{ k}\Omega} = 5\text{ nW} \qquad [(5 \times 10^{-3}\text{ V})^2 \div (5 \times 10^{3})\ \Omega = 5 \times 10^{-9}\text{ W}]$$

4. Square root of mega = kilo

$$\sqrt{1\text{ W} \times 1\text{ M}\Omega} = 1\text{ kV} \qquad (\sqrt{10^{0}\text{ W} \times 10^{6}\ \Omega} = 10^{3}\text{ V})$$

5. Reciprocal of mega = micro

$$\frac{1}{1\text{ MHz}} = 1\ \mu\text{s} \qquad (1 \div 10^{6}\text{ Hz} = 10^{-6}\text{ s})$$

Note that the result should be 1 μs/Hz, but the form given (1 μs) is that most commonly used.

EXERCISE PROBLEMS FOR SECTION 1-8

1. Express the following in kilounits and megaunits.
   a. 150 000 V (*ans*. 150 kV, 0.15 MV)

[1] For a review of arithmetic using powers of 10 there are many good mathematics books (for example, N. M. Cooke, H. F. R. Adams, and P. B. Dell, *Basic Mathematics for Electronics*, 5th ed., McGraw-Hill Book Company, New York, 1982).

b. 5 000 000 W (*ans*. 5000 kW, 5 MW)
c. 680 000 Ω (*ans*. 680 kΩ, 0.68 MΩ)
d. 400 000 000 Hz (*ans*. 400 000 kHz, 400 MHz)

2. Express the following in milliunits and microunits.
a. 0.0001 V
b. 0.025 A
c. 0.001 25 S
d. 0.000 01 W

3. Express the following as numbers between 1 and 1000 followed by super- or subunits.
a. 2 800 000 000 Hz (*ans*. 2.8 GHz)
b. 0.000 005 5 V (*ans*. 5.5 μV)
c. 0.000 000 000 01 F (*ans*. 10 pF)
d. 0.000 000 1 s (*ans*. 100 ns)
e. 0.000 02 H (*ans*. 20 μH)
f. 300 000 000 Ω (*ans*. 300 MΩ)

4. Express the values given in Probs. 1, 2, and 3 in scientific notation. (*ans*. Problem 1: *a*. 1.5×10^5 V; *b*. 5×10^6 W; *c*. 6.8×10^5 Ω; *d*. 4×10^8 Hz. Problem 2: *a*. 1×10^{-4} V; *b*. 2.5×10^{-2} A; *c*. 1.25×10^{-3} S; *d*. 1×10^{-5} W. Problem 3: *a*. 2.8×10^9 Hz; *b*. 5.5×10^{-6} V; *c*. 1×10^{-11} F; *d*. 1×10^{-7} s; *e*. 2×10^{-5} H; *f*. 3×10^8 Ω)

SUMMARY OF EQUATIONS

$t°\text{C} = (\theta°\text{F} - 32) \times \frac{5}{9}$ (1-1)

$T\ \text{K} = t°\text{C} + 273.15$ (1-2)

$2\pi\ \text{rad} = 360°$ (1-3)

CHAPTER QUESTIONS

1-1. In electrical equations, or formulas, relating three quantities, two of the quantities are *variables*. What is the third quantity called? Is it usually the operating characteristic of the circuit or a part of it?

1-2. Why are equations and graphs important in describing the operation of electrical equipment and circuits?

1-3. The *distance* between two cities is 200 *miles*. The time taken to drive between them at an average *speed* of 50 *miles per hour* is 4 *hours*. Which of these terms refers to a *quantity* and which to a *unit*?

1-4. Why has the *unit of frequency*, the *cycle per second* (c/s), had its name changed to *hertz* (Hz)?

1-5. What is the difference between *mass* and *weight*? Is the difference often of importance?

1-6. Why are Greek letters often used as symbols in electrical work?

1-7. There is a common ratio between adjacent lines in Table 1-3. What is this ratio?

CHAPTER PROBLEMS

1-1. The temperature of a certain furnace for processing silicon for semiconductors is 2400°F. What is this temperature in degrees Celsius? (*ans*. 1316°C)

1-2. The temperature of a certain type of transistor rises from 25 to 150°C during operation. Express this temperature rise in

a. Kelvins

b. Degrees Fahrenheit

1-3. A generator shaft turns at 600 revolutions per minute. What is this rate in degrees per second and radians per second? (*ans*. 3600°/s, 62.83 rad/s)

1-4. The *phase angle* of an electric circuit is given as 0.6 radians. How many degrees is this?

1-5. Electricity in American homes has a frequency of 60 hertz, or 60 cycles per second. If one cycle is equivalent to one complete revolution, what is this frequency in radians per second? (*ans*. 377 rad/s)

1-6. What is the angle, in degrees and radians, traversed by a watch hand in 1 hour?

a. The hour hand

b. The minute hand

c. The second hand

1-7. The frequency of the light from one type of laser is 280 THz. What is this frequency in *hertz* and *gigahertz*? (*ans*. 280 000 000 000 000 Hz, 280 000 GHz)

1-8. A radio transmitter has an output power of 100 kilowatts. Express this power in watts and megawatts.

1-9. The variable capacitor tuning a radio set has a capacitance range of 0.000 05 to 0.000 35 μF. What is this range in picofarads? (*ans*. 50 to 350 pF)

1-10. The inductor that operates in conjunction with the capacitor of Prob. 1-9 has an inductance of 200 μH. Express this inductance in henrys.

2

ELECTRICAL QUANTITIES AND RELATIONSHIPS

A *system* is an orderly arrangement of elements into a useful whole.

All electric systems have certain features in common:

1. There must be an interchange of energy in order for the system to be of practical use.
2. There must be a part of the system where energy can be put to use.
3. There must be a part of the system where energy can be introduced, because it cannot be taken out for use unless it is first put in.

For a complete electric circuit three subsystems are necessary: the *generator* (or *source*), where energy enters the system; the *load,* where energy is usefully employed; and the *network,* which conveys the energy from the generator to the load.

It must be possible to specify, predict, and measure energy levels within the system. Many quantities are involved in this process. This chapter is concerned with some of these quantities and the relationships between them, as a foundation for the investigation of electric circuits. We will start with energy as that is what electricity is all about.

2-1 ENERGY AND WORK

Electric circuits exist for the purpose of transferring energy and doing work. The electrical utility which supplies your home sells you electrical energy. It is able to do so because it converts other forms of energy to electrical energy. When you use electricity, you put electrical energy to work, converting it to some other form of energy.

Figure 2-1 shows a hydroelectric system.

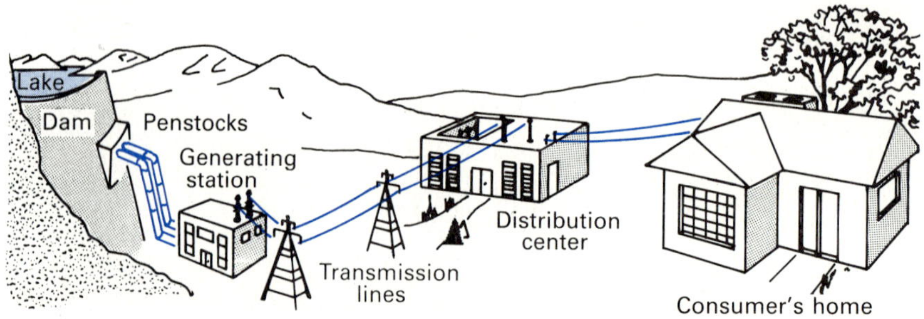

Fig. 2-1. Hydroelectric system.

A lake holds a large volume of water, maintained high above sea level by a dam. The water thus stored possesses *potential energy,* the energy of position. The water is directed down into the generating station through pipes called penstocks. The falling water has *kinetic energy,* the energy of motion. The generators produce electrical energy when the kinetic energy of the water causes the shafts of the generators to turn. Energy is converted from potential energy to kinetic energy to electrical energy. None of these energy forms is visible or otherwise capable of direct perception by our senses. But all are evident through what they can do. Perhaps this best explains what is meant by energy: It causes things to happen; we are aware of the existence of energy through what it does.

Energy is sometimes defined as the ability to do work. This is just another way of saying that energy causes things to happen, because work in the physical sense is a quantity by which action is measured. The electrical energy from the generating station is conducted by transmission lines to centers which then distribute it to houses, factories, hospitals, farms, etc. Here the electrical energy is used—put to work. A heater converts electrical energy to heat energy. A light bulb converts electrical energy to light energy. An electric motor converts electrical energy to mechanical energy. Energy is converted; work is done.

Figure 2-2 shows this energy and work system in block diagram form.[1]

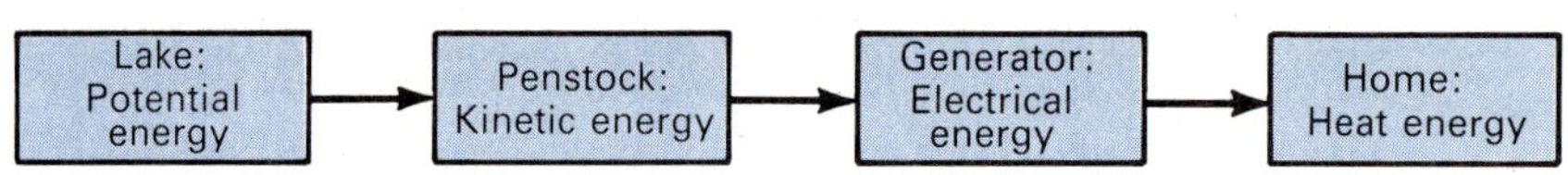

Fig. 2-2. Energy, from lake to home.

Energy and work are closely allied. *Energy is the ability to do work. Work is the result of converting energy*. The amount of energy expended is the amount of work done.

Both energy and work have the same symbol, W, and the same SI unit, the joule. The symbol for joule is J.

[1] A block diagram is a drawing showing the major components of a system as blocks of various shapes; it depicts the overall action in a simple manner.

When a force is applied to an object, causing the object to move, work is done. The amount of work done is the force multiplied by the distance moved:

$$W = Fs \tag{2-1}$$

where W = energy, J*
F = force, N
s = distance, m

Writing Eq. (2-1) in unit form gives

$$\mathrm{J} = \mathrm{N} \cdot \mathrm{m} \tag{2-2}$$

In Eq. (2-2) all the terms are units: J is joule, N is newton, and m is meter.

Energy Loss

Whenever energy is converted, all of it is converted to other forms, but not all to the intended form. An electric motor converts electrical energy to mechanical energy; this is the intent. But the motor becomes warm; some heat energy is produced. A stove element converts electrical energy to heat energy, but you can see the element glow; some light energy is produced. Energy converted to an unintended form represents an energy loss. The amount of energy loss W_L is the difference between the total amount of energy W_T and the amount of energy converted as intended, the useful energy W_U:

$$W_L = W_T - W_U \tag{2-3}$$

The total energy is sometimes called the input energy W_i, the useful energy being the output energy W_o:

$$W_L = W_i - W_o \tag{2-4}$$

Efficiency

The ratio of the output energy to the input energy is frequently of interest. This ratio is called the *efficiency* and is given by the equation

$$\eta = \frac{W_o}{W_i} \tag{2-5}$$

In Eqs. (2-3) to (2-5), W_L is energy loss, W_T and W_i are total and input energies, and W_U and W_o are useful and output energies. All the energy terms are in joules. The symbol η for efficiency is the lowercase Greek letter eta. As efficiency is the ratio of two similar quantities, η has no unit; it is a dimensionless number; it is always less than 1.

REVIEW QUESTIONS FOR SECTION 2-1

1. What is the purpose of an electric circuit?
2. When electrical energy is put to work, what is being done?
3. How can energy be defined?

*W is energy expressed in joules, J. Statements of this type will be abbreviated as shown.

4. What is the unit of:
   a. Work?
   b. Energy?
5. What is the standard symbol for the quantity:
   a. Work?
   b. Energy?
6. When electrical energy is converted to other forms, it is not always converted to the desired form. What unintentional forms of energy may be present at:
   a. An electric motor?
   b. An electric light bulb?
   c. An electric stove element?
7. Express the efficiency η of an electrical system as a ratio.

2-2 GENERATOR AND LOAD

Electrical energy enters the circuit because of the conversion of energy from some other form to the electrical form. The potential energy of stored water, the caloric energy of oil or gas, the nuclear energy of atoms, and the chemical energy of acids are all used for conversion to electrical energy. A *generator* is an item of equipment or a system in which this conversion of energy to the electrical form is carried out. It may be large and elaborate, as in a nuclear power station. It may be as small and simple as a flashlight battery. Energy *source* is another name for a generator.

The light energy of a television tube, the sound energy of a radio, the chemical energy of a plating bath, and the mechanical energy of an electric hammer are the results of conversion from the electrical form. This conversion of electrical energy to some other form takes place in a *load*. (See Fig. 2-3.)

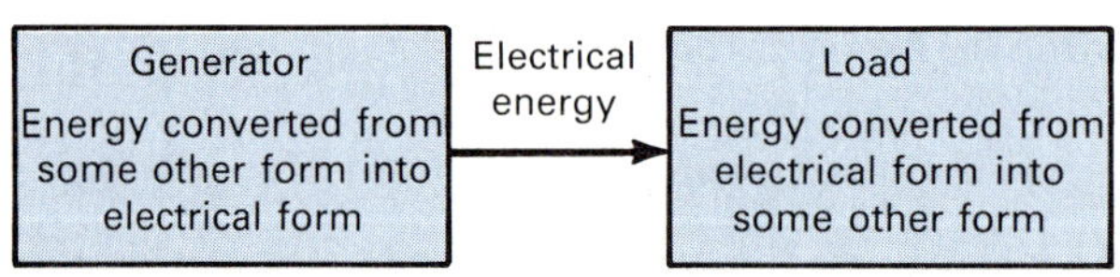

Fig. 2-3. Energy conversions in generator and load.

Application of the term "load" can be confusing. Everything connected to a generator constitutes a load on the generator because all the electrical energy generated is converted to other forms of energy. The total circuit consists basically of three sections: generator, network, and load. (See Fig. 2-4.)

Fig. 2-4. Basic electric system.

In this case the network, which connects the generator to the load, may convert some of the electrical energy to heat, leaving the load to convert less energy than was produced by the generator. It is perhaps best to view this situation in two ways.

In Fig. 2-5*a* the network and the load are together considered to be the load on the generator, taking all the energy output. This situation might be described by saying that, from the generator's viewpoint, the load includes the network.

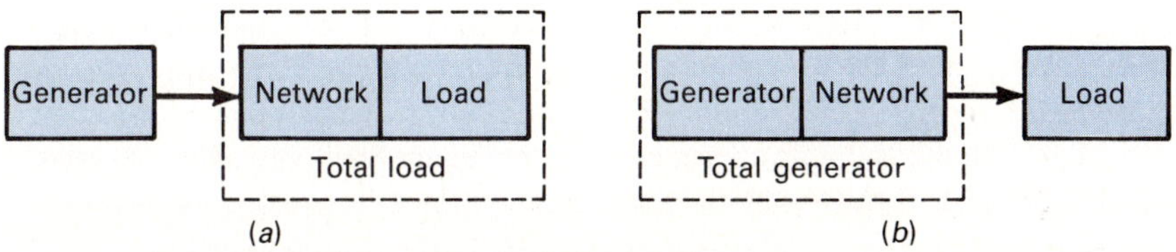

Fig. 2-5. *(a)* Network + load = total generator load; *(b)* generator + network = total generator.

In Fig. 2-5*b* the generator and the network are together considered as the total generator, with the load taking all the energy output. Another way of describing this concept is to say that the load sees the generator and the network as if the combination were a single energy source.

REVIEW QUESTIONS FOR SECTION 2-2

1. Name four forms of energy which can be converted to electrical energy.
2. What is the primary function of an electric generator?
3. What is the general term used for the device in which electrical energy is usefully employed?
4. In an electric circuit, what is the function of the "network"?

2-3 CHARGE

Charge is the medium through which electrical energy is conducted by the atoms of a substance. The amount of charge held by an electron is usually considered the fundamental quantity of charge. The charge of an electron has been measured and found to be extremely small—too small for practical use as an electrical unit. A much larger unit, the coulomb, symbol C, is the SI unit of charge.[1] The charge of a single electron is $1.602\ 03 \times 10^{-19}$ C, so that one coulomb represents the total charge of approximately 6.242×10^{18} electrons. To more clearly visualize what a large number this is, think of it as about $6\frac{1}{4}$ million million million electron charges.

How did these numbers arise? Why is the coulomb not specified as, say, 10^{18} electron charges? Like some other SI units, the coulomb was a unit of known magnitude before the charge of the electron was accepted as a basis of electricity. The number of electron charges in a coulomb was determined fairly recently

[1] The symbol C (roman) for the unit coulomb should not be confused with the symbol *C* (italic) for the quantity capacitance.

using modern measuring techniques. The other SI units to which the coulomb is related would have to be changed if the value of the coulomb were to be changed.

Charge is sometimes called quantity of electricity, giving rise to its standard symbol Q.

As an illustration of the availability of electrical charge, consider the metal copper, a good material for electrical conduction. Copper contains approximately 8.54×10^{28} free electrons per cubic meter. This means that there is about 1.37×10^{10} C (13.7 thousand million coulombs) charge available in a cubic meter of copper. If this number surprises you, remember that a cubic meter is a large volume; a cubic meter of copper weighs 8890 kg (about 10 tons).

EXAMPLE 2-1 How many free electrons exist in a copper rod of $\frac{1}{4}$-in diameter, 6 in long? What charge does this number of free electrons represent?

Solution (See Fig. 2-6.)

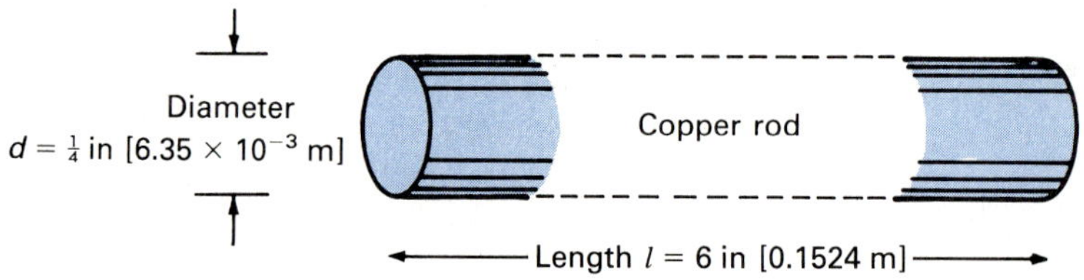

Fig. 2-6. Example 2-1: Dimensions of a copper rod.

The cross-sectional area of the rod is

$$A = \pi r^2 = \pi(3.175 \times 10^{-3}\ \text{m})^2 = 3.167 \times 10^{-5}\ \text{m}^2$$

The volume of copper in the rod is

$$V = Al = 3.167 \times 10^{-5}\ \text{m}^2 \times 0.1524\ \text{m} = 4.827 \times 10^{-6}\ \text{m}^3$$

The number of free electrons equals the number of free electrons per unit volume times the volume. There are 8.54×10^{28} free electrons per cubic meter of copper, and so the number of free electrons in the copper rod is

$$8.54 \times 10^{28} \times V = 8.54 \times 10^{28}\ \frac{\text{electrons}}{\text{m}^3} \times 4.827 \times 10^{-6}\ \text{m}^3$$

The number of free electrons in the rod = $\mathbf{4.122 \times 10^{23}}$.

The charge in coulombs equals the number of free electrons divided by the number of electrons per coulomb:

$$Q = \frac{4.122 \times 10^{23}\ \text{electrons}}{6.242 \times 10^{18}\ \text{electrons/C}}$$

$$Q = \mathbf{6.604 \times 10^{4}\ C}$$

The copper rod in Example 2-1, about the size of an ordinary pencil, contains close to 1 million million million million free electrons, the equivalent of approximately 66 000 C of charge. The very large number of electrons involved in familiar situations is illustrated by considering that, when an ordinary 100-watt light bulb is lit steadily for 1 day (24 hours), the number of electrons which pass through it is about the same as the number of electrons in the copper rod.

REVIEW QUESTIONS FOR SECTION 2-3

1. What is the SI unit of charge?
2. What is the symbol for charge?
3. What is the symbol for the unit of charge?

2-4 VOLTAGE

Whenever energy is converted to or from the electrical state, movement of charge is involved. Voltage is the factor relating the electrical energy and the charge.

Figure 2-7 shows a basic electric circuit consisting of a generator connected to a load by two conductors. Energy enters the circuit at the generator, is transferred to the load by the conductors, and leaves the circuit at the load.

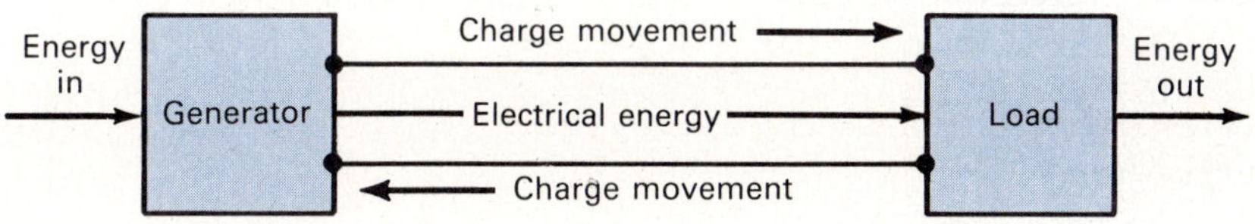

Fig. 2-7. Electrical energy associated with charge movement.

Electromotive Force

At the generator external energy is converted to electrical energy. This conversion is associated with a voltage which causes charge movement to take place. The voltage of the generator is sometimes called *electromotive force,* abbreviated emf. "Electromotive force" is a useful term because it conveys the idea of a force which causes electrons to move. It is important to realize, though, that voltage is not a true force, since its unit does not have the same MKSA dimensions as the newton, the SI unit of force.

Potential Difference

At the load the charge movement causes electrical energy to be converted to other forms (heat, light, mechanical, etc.). This energy conversion at a load sometimes results in energy storage, as when the load is a battery under charge. Usually, however, the energy is dissipated or released from the circuit at the load, mainly in the form of heat. A voltage across a load is sometimes referred to as *voltage drop, potential difference* (pd), or just *voltage*. (See Fig. 2-8.)

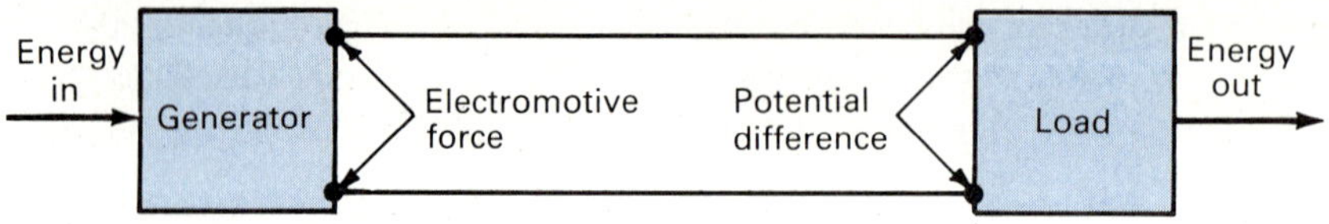

Fig. 2-8. Electrical energy associated with voltage.

The two interpretations of voltage can be summarized as follows.

Electromotive force is voltage associated with *generated energy*. *Potential difference* is voltage associated with *dissipated energy*.

The symbol for voltage used in this book is V. If it is necessary to distinguish between emf, the voltage of generation (source voltage), and pd, the voltage of dissipation (load voltage), the symbols E and V may be used. However, the use of two different symbols for voltage can cause confusion.

The unit for either type of voltage, emf or pd, is the volt, symbol V. The volt is defined by the statement: If between two points in a circuit one joule of energy conversion is associated with the movement of one coulomb of charge, one volt exists between those two points.

This definition gives rise to the equation

$$W = VQ \tag{2-6}$$

where W = energy or work, J
V = voltage (emf or pd), V
Q = charge, C

Inserting unit symbols into Eq. (2-6) gives

$$\mathrm{J} = \mathrm{V} \cdot \mathrm{C} \tag{2-7}$$

where J = joules
V = volts
C = coulombs

It can be seen from Eq. (2-7) that one volt is one joule per second.

EXAMPLE 2-2 A typical 12-V car battery can store a charge of 2×10^5 C. What amount of energy does this charge represent?

Solution

$$W = VQ \tag{2-6}$$

$$W = 12\ \mathrm{V} \times 2 \times 10^5\ \mathrm{C} = \mathbf{2.4 \times 10^6\ J}$$

The automobile battery of Example 2-2 can store about $2\frac{1}{2}$ million J of energy for the operation of the car's electrical equipment.

EXAMPLE 2-3 What would be the emf of a battery which is capable of storing 2.4×10^6 J of energy (as in Example 2-2) but does so with a stored charge of 4×10^5 C, double that of the charge in Example 2-2?

Solution

$$W = VQ \tag{2-6}$$

$$V = \frac{W}{Q} = \frac{2.4 \times 10^6 \text{ J}}{4 \times 10^5 \text{ C}} = \mathbf{6\ V}$$

The result of Example 2-3 shows that, to store the same amount of energy as a 12-V battery, a 6-V battery must store twice as much charge.

Closed and Open Circuits

Energy can be converted to and from the electrical state only if charge actually moves. To accomplish this charge movement, a complete or *closed circuit* is necessary, as in Fig. 2-9.

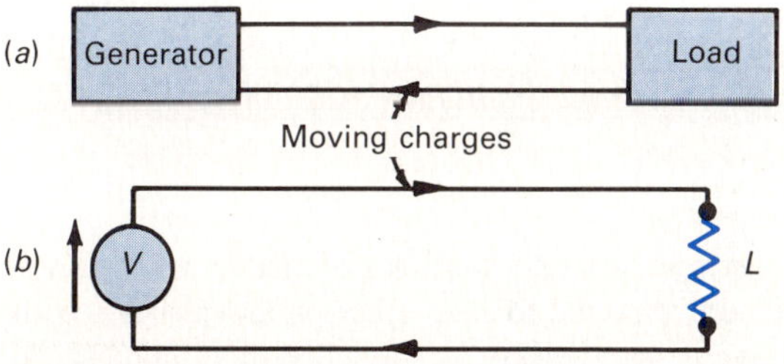

Fig. 2-9. Circuit complete for charge movement: *(a)* Block diagram; *(b)* schematic diagram.

Figure 2-9*a* is a block diagram of a generator connected to a load. Figure 2-9*b* is a schematic diagram[1] showing the generator as a voltage source V connected to the load L.

In Fig. 2-10*a* an open switch results in an *open circuit*. Charge movement cannot take place through an open switch, and hence there cannot be any energy conversion. The generator emf is present, ready to cause charge movement, but there is no load voltage.

In Fig. 2-10*b* the switch is closed, producing a closed circuit in which charge can move. The generator emf is still present, and now the load voltage also exists.

Emf and pd are sometimes defined so as to emphasize this difference between generator and load voltages:

If between two points in a circuit there exists the *capability for converting energy* from some other form to the electrical form, an *emf* exists between those two points.

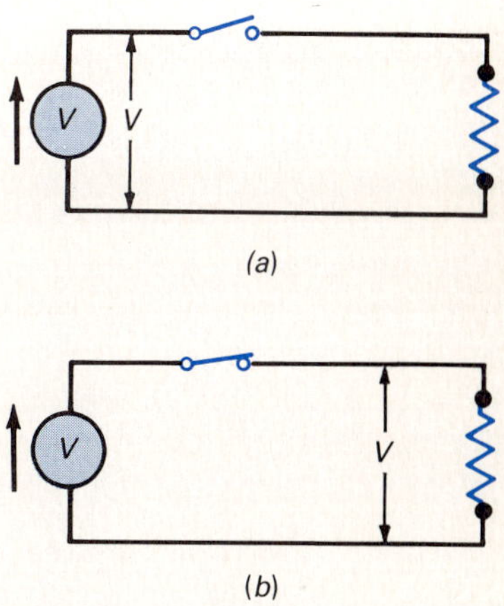

Fig. 2-10. Using a switch in a circuit: *(a)* Open circuit; *(b)* closed circuit.

[1] A schematic diagram is a drawing that shows the circuit elements and connections in a standard universal form, each element having a distinct symbol.

If between two points in a circuit *energy is converted* from the electrical form to some other form, a *pd* exists between those two points.

The fact that a generator emf can exist when the circuit it feeds is open leads to the often used terms "open-circuit generator voltage" and "open-circuit emf."

Polarity

The terminals of a generator may be assigned the polarities positive and negative (+ and −). Negative electron charges move from the negative terminal around the circuit to the positive terminal, as depicted in Fig. 2-11.

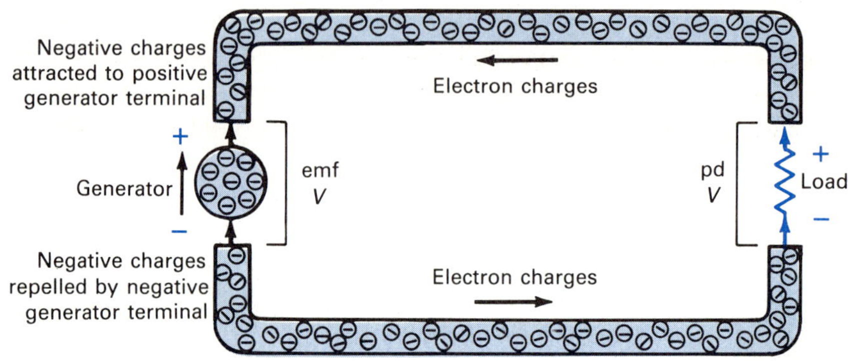

Fig. 2-11. Charges move within a circuit.

Figure 2-11 illustrates the principle whereby particles of matter with polarities of opposite sign (+ and −) are attracted to each other, while particles with polarities of the same sign (+ and +, or − and −) are repelled by each other. In Fig. 2-11 the negative electron charges are attracted to the positive terminal of the generator and are repelled by the negative generator terminal. In this case the combined action results in the charges moving around the circuit in a counterclockwise direction.

When the generator polarity is reversed, as in Fig. 2-12, the attracting and repelling forces cause the charge movement to become clockwise around the

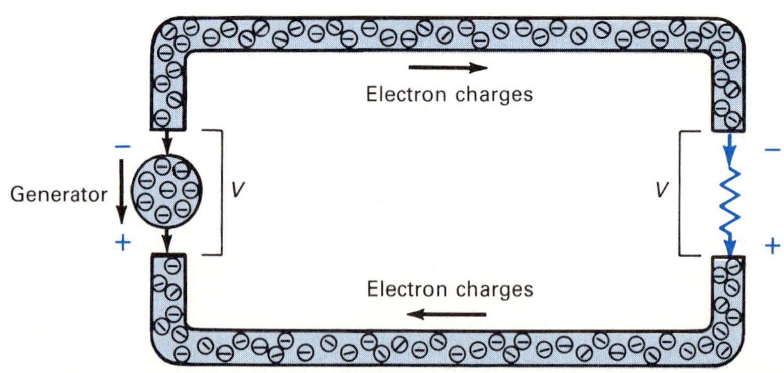

Fig. 2-12. Polarity reversal causes current direction reversal.

circuit. This means that a reversal of generator polarity results in a reversal of the direction of charge movement.

Arrows (↑) next to generator symbols can be used to indicate polarity, the arrowhead pointing in the direction of the positive terminal. It is sometimes said that the arrows point in the direction of potential rise.

In semiconductor technology (the basis of transistor operation) it is said that conduction may take place as a result of the movement of negative-charge carriers (electrons) or positive-charge carriers (holes). If this concept is accepted, no problem is created, since all circuit relationships are the same for either type of carrier. The only distinction is that negative and positive charges will move in opposite directions, although this movement direction is rarely of interest.

In Figs. 2-11 and 2-12 the load voltage is shown as having the same polarity as the generator voltage.

If the distance between the generator and the load is gradually reduced, they will eventually be directly joined, as in Fig. 2-13. Here it can clearly be seen that the touching terminals of the generator and the load must have the same polarity.

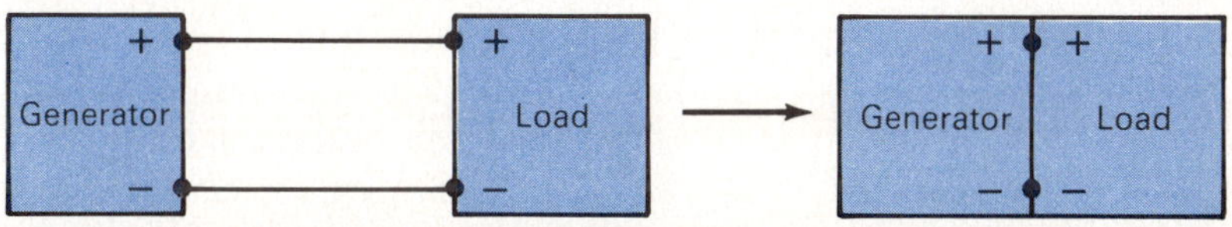

Fig. 2-13. The polarities of generator and load are the same.

REVIEW QUESTIONS FOR SECTION 2-4

1. What is the specific symbol for generator voltage?
2. The movement of charge is required to convert energy to or from the electrical state. What form must the circuit take to allow charge movement?
3. An electric circuit contains a switch with its contacts not touching. What term might be used to describe such a circuit?
4. What are the polarities of a generator's terminals usually called?

EXERCISE PROBLEMS FOR SECTION 2-4

1. How many coulombs of charge must pass through a load while it is converting 17 280 J of electrical energy to heat energy if the supply emf is 24 V? (*ans*. 720 C)
2. How many joules of energy are dissipated by a 6-V lamp bulb while passing 1500 C of charge?
3. What is the voltage developed across a load if 52 J of energy is dissipated while a movement of 130 C of charge occurs? (*ans*. 0.4 V)
4. If it requires 90 kJ of electrical energy to operate a 120-V toaster for

$1\frac{1}{2}$ min, how many coulombs of charge must pass through the toaster during this period?

5. Find the amount of energy converted to heat by a stove element during a period in which 3×10^4 C of charge pass through the element when the supply voltage is 240 V. (*ans*. 7.2 MJ)
6. If the dissipation of energy in Prob. 3 occurred in 1 h, what amount of energy was dissipated per second?
7. A television set with a line voltage of 120 V dissipates, mainly as heat, 630 kJ of energy each hour.
   a. How many coulombs of charge must pass through the set during this period? (*ans*. 5250 C)
   b. How many coulombs of charge must pass through the set per second? (*ans*. 1.458 C)
   c. How many joules of energy are dissipated per second? (*ans*. 175 J)

2-5 CURRENT

Charges available for movement are called *mobile charges* or *charge carriers*. Their presence in a material depends on its nature and its environment. Metals, which are normally good conductors, have a significant quantity of charge carriers available under ordinary environmental conditions. Semiconductors, which also are metals, need special conditions (such as the presence of strong light, heat, or nuclear radiation) or special chemical treatment before they can readily have charge carriers available for movement. Materials such as plastics or ceramics are insulators and ordinarily do not possess many mobile charges.

Mobile charges can be either negative or positive. This text is mainly concerned with conductor operation in which negative electron charges may drift or move.

In the absence of an applied voltage, these charges move within the conductor as a result of the forces existing between atoms.

As depicted in Fig. 2-14, the residual charge movement has no net direction; it is a random movement.

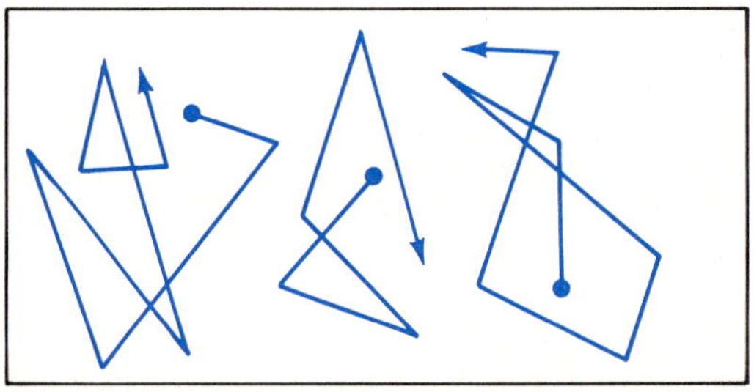

Fig. 2-14. Random movement of charges results in no current.

When a voltage is applied across the conductor, the attraction and repulsion of the negative charges results in a net direction of charge movement; a tendency for the charges to move in one general direction. Figure 2-15 shows this as a net movement from left to right.

This drift of charges in one net direction is called an electric current.

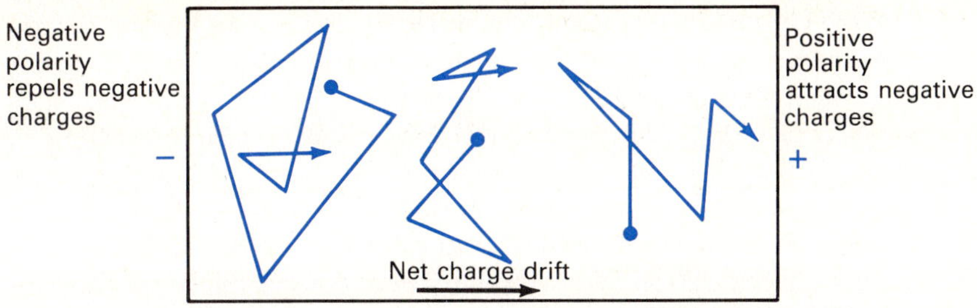

Fig. 2-15. Net drift of charges in one direction results in a current.

It is necessary to distinguish between "a current" and "current."

A current is a phenomenon. It is the *existence* of a net drift of charges through a circuit.[1]

Current is a quantity. It is the *magnitude* of the net drift of charges through a circuit.

The symbol for current is I, the initial letter of the word "intensity"—intensity of charge drift.

Current is the rate of charge drift—the average amount of charge moving past a given point in a circuit each second. From this description we obtain the equation

$$I = \frac{Q}{t} \tag{2-8}$$

where I = current, A
Q = charge, C
t = time, s

Equation (2-8) is frequently rearranged into product form for ease of memorization:

$$Q = It \tag{2-9}$$

The Ampere

The SI unit of current is the ampere, symbol A.*

Substituting unit symbols into Eq. (2-8),

$$\mathrm{A} = \frac{\mathrm{C}}{\mathrm{s}} \tag{2-10}$$

One ampere is one coulomb per second.

Current Direction

The direction of a current is defined as the direction in which *positive*-charge carriers drift through a circuit.

[1]This duality is not uncommon. For example, "a voltage" is a phenomenon, while "voltage" is a quantity. But, for some obscure reason, it is only "current" that seems to cause a difficulty.
*The ampere is one of the four basic MKSA units.

Before the electron theory of current conduction became popular, it was believed that current was always due to the motion of positively charged particles of matter. With the advent of the belief that current is due to the movement of negative electron charges, a determined effort was made to change the definition of current direction. The controversy was so great that there was a period during which the electrical industry was split into two factions. One of them preferred the conventional definition of current direction (positive-charge carriers), while the other group adopted a definition involving electron-flow direction (negative-charge carriers). To prevent the difficulties presented by having two definitions of current direction, the conventional definition was made the standard; it is used throughout this text.

Figure 2-16 shows what is meant by the conventional definition of current direction.

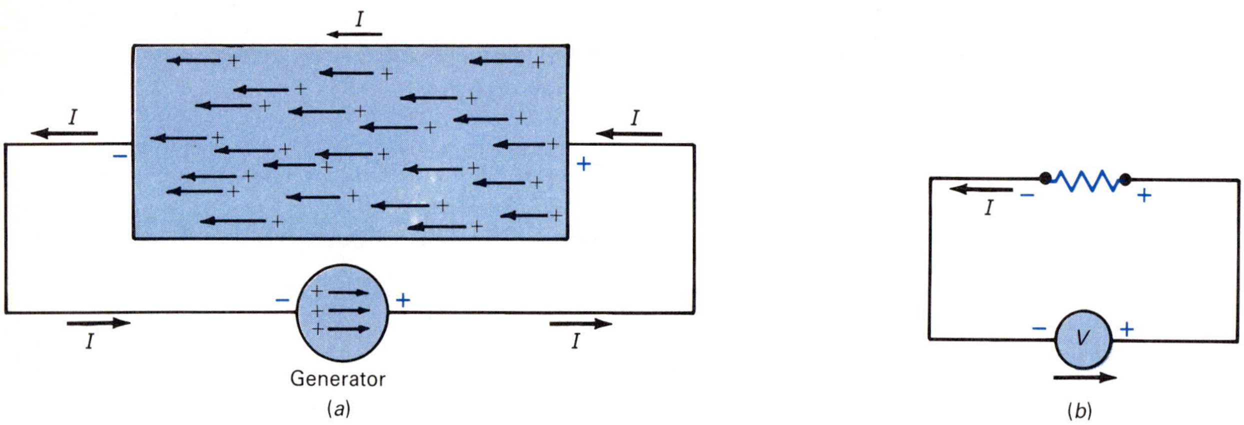

Fig. 2-16. Conventional current direction diagrams: *(a)* Pictorial; *(b)* schematic.

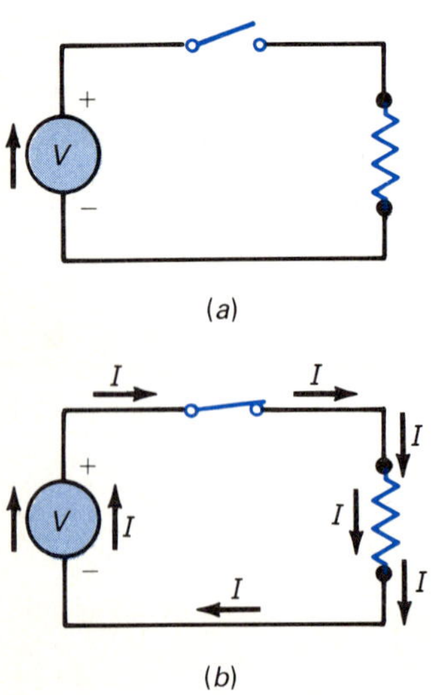

Fig. 2-17. Using a switch for current: *(a)* Open circuit, no current; *(b)* closed circuit, current.

Positive charges are attracted toward the negative side of the applied emf and repelled away from the positive side. The current direction is from the positive generator terminal, through the circuit, to the negative generator terminal. The current completes its path around the circuit inside the generator, from its negative terminal to its positive terminal.

Figure 2-16*b* is the schematic equivalent of Fig. 2-16*a*. The arrow below the generator symbol indicates the direction of the resulting current.

Current Velocity

Just as it was necessary to distinguish between "a current" and "current," so it is necessary to distinguish between "velocity of propagation of electrical energy" and "velocity of an electric current."

When a circuit is open, as in Fig. 2-17*a*, there is no current. After the circuit is closed, as in Fig. 2-17*b*, there will be current *I* throughout the circuit.

The speed at which the effect of the emf is experienced at all parts of the circuit, resulting in the existence of a current, is the *velocity of propagation of electrical energy*.[1] This is a high velocity, approximately 3×10^8 (300 million) m/s. It is independent of the value of the current and voltage.

The speed at which the charges drift is the *velocity of the current*. This is quite low, typically fractions of a meter per second. This velocity is influenced by the value of the current and other characteristics of the circuit.

EXAMPLE 2-4 If 1 A of current exists in a copper conductor 1 cm^2 in cross-sectional area and 10 km long, what is the velocity of the current? (See Fig. 2-18.)

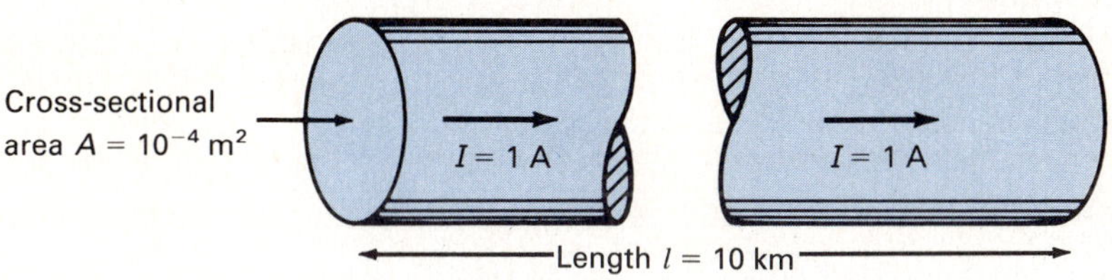

Fig. 2-18. Example 2-4: Dimensions of conductor.

Solution The volume of the copper conductor is 1 m^3. This volume of copper contains approximately 8.54×10^{28} free electrons, which is equivalent to 1.37×10^{10} C of available charge carriers.

A current of 1 A is 1 C of charge per second. At this rate, 1.37×10^{10} C represents a time of 1.37×10^{10} s. In that time, all the available charge will have drifted through the conductor. As the charge drifts through, it is replaced by charge moving out of the energy source. Therefore the total charge of 1.37×10^{10} C will, in effect, travel the full 10-km length of the conductor in 1.37×10^{10} s. Velocity is distance per unit time. The velocity of drift can therefore be calculated:

$$v = \frac{l}{t}$$

$$= \frac{10^4 \text{ m}}{1.37 \times 10^{10} \text{ s}}$$

$$\mathbf{v = 7.3 \times 10^{-7} \text{ m/s} = 0.73\ \mu\text{m/s}}$$

The velocity 0.73 μm/s is very low, roughly 30 millionths of an inch per second. It only takes 33 millionths of a second for the current to start to move at all points along the entire length of the conductor in Example 2-4, but it would take 13.7 thousand million seconds, about 400 years, for charge to travel from one end of the conductor to the other, a distance of about 6 miles.

A conductor of 1 cm^2 cross-sectional area would normally be used to carry currents closer to 200 A than to 1 A. With 200 A the velocity would be increased 200-fold to 146 μm/s. This is still a very low speed.

[1] The velocity of propagation of electrical energy, or electromagnetic waves, is the "speed of light." This velocity has the symbol c and is 2.99793×10^8 m/s.

REVIEW QUESTIONS FOR SECTION 2-5

1. In the absence of an applied generator voltage, is there any movement of electron charges around a circuit?
2. When a generator voltage is applied to a circuit, what is the term applied to the drift of electric charges, *current* or *a current?*
3. What is the symbol for current?
4. Express the relationship between current and charge in equation form.
5. What is the SI unit of current? Is this unit the same as that used by electricians in the past?
6. At what velocity is electrical energy propagated? Is this the speed at which charges move around a circuit?
7. The "conventional" current direction used in electrical and electronic work is opposite to the direction of electron charge movement. True?

EXERCISE PROBLEMS FOR SECTION 2-5

1. What current is present when 5 kC of charge pass a point in a circuit in 10 s? (*ans*. 500 A)
2. What length of time is required to accumulate a charge of 0.8 mC if the average current is 25 mA?
3. What quantity of charge passes a point in a circuit in 1 min if the average current is 75 μA? (*ans*. 4.5 mC)
4. In a 120-V light bulb, a charge of 0.5 C passes a given point each second. What current does the bulb draw?
5. A 120-V heating element has a current of 8.33 A.
   a. How many coulombs of charge pass a given point in the heater each second? (*ans*. 8.33 C)
   b. How many coulombs pass a given point each hour? (*ans*. 30 kC)
6. A 12-V car battery storing a charge of 200 kC was completely discharged in 24 h. If the rate of discharge was constant, what current was drawn from the battery?
7. A 6-V battery storing 2.4 MJ of energy was completely discharged in 12 h. If the rate of discharge was constant, what current was drawn from the battery? (*ans*. 9.259 A)
8. A 120-V heating element converts energy at the rate of 1500 J/s. What is the current in the element?

2-6 POWER

Power is the rate of converting energy or the rate of doing work. The symbol for power is *P*. The SI unit for power is the watt, symbol W.*

*It is easy to confuse W, the symbol for the unit watt, with *W*, the symbol for the quantity energy. To avoid this confusion, do not use the word "wattage" when you mean "power." Use *P* for power, not *W* for wattage.

From the definition of power the following equation can be derived:

$$P = \frac{W}{t} \tag{2-11}$$

where P = power, W
W = energy or work, J
t = time, s

Equation (2-11) is often expressed in product form:

$$W = Pt \tag{2-12}$$

From Eq. (2-11) it can be seen that one watt is one joule per second.

EXAMPLE 2-5 An electric heater is rated at 1.5 kW. How much heat energy does it give out each hour (3600 s)?

Solution

$$W = Pt \tag{2-12}$$
$$= 1.5 \text{ kW} \times 3.6 \text{ ks}$$
$$W = \mathbf{5.4\ MJ}$$

Energy, Power, and Time

It is sometimes convenient to express energy in terms of current and time. Combining Eqs. (2-6) and (2-9),

$$W = VQ \tag{2-6}$$
$$Q = It \tag{2-9}$$

Therefore,

$$W = VIt \tag{2-13}$$

where W = energy, J
V = voltage, V
I = current, A
t = time, s

By combining Eqs. (2-11) and (2-13),

$$P = \frac{W}{t} \tag{2-11}$$
$$W = VIt \tag{2-13}$$

we can obtain one of the best known formulas in electrical technology:

$$P = VI \tag{2-14}$$

Almost everyone who has studied or worked with electric circuits knows that *watts are volts times amperes*. This is a statement of Eq. (2-14) in unit form.

There are two energy units, based on the watt, which are extensions of the joule. They are the watthour (Wh) and the kilowatthour (kWh). A joule is a wattsecond, so

$$1\ \text{Wh} = 1\ \text{W} \times 3600\ \text{s}$$

$$1\ \text{Wh} = 3600\ \text{J} = 3.6\ \text{kJ} \tag{2-15}$$

$$1\ \text{kWh} = 1000\ \text{Wh}$$

$$1\ \text{kWh} = 3.6 \times 10^6\ \text{J} = 3.6\ \text{MJ} \tag{2-16}$$

EXAMPLE 2-6 How much current is drawn by a 120-V 100-W light bulb when it is lit?

Solution

$$P = VI \tag{2-14}$$

$$I = \frac{P}{V}$$

$$= \frac{100\ \text{W}}{120\ \text{V}}$$

$$I = \mathbf{0.8333\ A}$$

EXAMPLE 2-7 A house is heated by an oil furnace with a heat energy output of 30 000 J/s. The furnace is to be replaced by electric heaters. What current must the electrical service to the house be capable of supplying for these heaters alone if the supply voltage is 240 V?

Solution

$$W = VIt \tag{2-13}$$

$$I = \frac{W}{Vt}$$

$$= \frac{3 \times 10^4\ \text{J}}{240\ \text{V} \times 1\ \text{s}}$$

$$I = \mathbf{125\ A}$$

British Thermal Unit and Horsepower

The British thermal unit (Btu) is an FPS unit of heat energy. As it is not an MKSA unit, it is currently not recommended for use. But it is still used extensively for many applications, including the rating of oil furnaces. One Btu is equivalent to 1055 J, so that the furnace in Example 2-7 has a heat energy output of approximately 100 000 Btu/h, a typical value for a three-bedroom house. Thus, to replace a 100 000-Btu furnace with electric heat operating from an ordinary domestic 240-V supply requires 125 A of current.

The horsepower (hp) is another FPS unit which, although not approved for modern use, is still widely used. It will probably be many years before it becomes common to speak of an automobile of "75 kilowatts" instead of "100 horsepower."

$$1\ \text{hp} = 746\ \text{W} \tag{2-17}$$

EXAMPLE 2-8 A 2-hp electric motor has an efficiency of 80 percent. What is (*a*) its electric power input in watts, (*b*) its energy consumption per day in watthours and in kilowatthours, and (*c*) the current-carrying capacity of its connecting wire, if it is to operate from a supply of 120 V?

Solution *a.*

$$1 \text{ hp} = 746 \text{ W} \tag{2-17}$$
$$2 \text{ hp} = 1492 \text{ W}$$

The rated power of an electric motor is the mechanical power its rotating shaft is capable of producing. From Eqs. (2-5) and (2-11),

$$\eta = \frac{W_o}{W_i} = \frac{P_o}{P_i}$$

where η = efficiency (0.8 in this example)
P_o = output power (2 hp or 1492 W)
P_i = electric input power, W

$$P_i = \frac{P_o}{\eta} = \frac{1492 \text{ W}}{0.8}$$
$$P_i = \mathbf{1865\ W}$$

b. Applying Eq. (2-12):

$$W_i = Pt \tag{2-12}$$
$$W_i = 1865 \text{ W} \times 24 \text{ h} = \mathbf{44\ 760\ Wh}$$
or
$$W_i = 1.865 \text{ kW} \times 24 \text{ h} = \mathbf{44.76\ kWh}$$

c. Applying Eq. (2-14):

$$P_i = VI \tag{2-14}$$
$$I = \frac{P_i}{V}$$
$$= \frac{1865 \text{ W}}{120 \text{ V}}$$
$$I = \mathbf{15.54\ A}$$

Thus this 2-hp motor will require an input power of 1865 W, will utilize 44 760 Wh or 44.76 kWh of electrical energy every day (if it is operated continuously), and will need to be connected to the electric system by wire capable of continuously carrying 15.54 A.

EXAMPLE 2-9 A portable transistor radio has a 9-V battery with a usable capacity of 120 mA · h. The average current required by the radio is 20 mA. (*a*) How long can the radio be operated? (*b*) How much power does the radio require? (*c*) How much energy is the battery capable of supplying? (*d*) What charge is initially available from the battery?

Solution *a*. The capacity of a battery is a product of current and time. Thus capacity is the same as quantity (Q).

$$Q = It \qquad (2\text{-}9)$$

$$t = \frac{Q}{I} = \frac{120 \text{ mA} \cdot \text{h}}{20 \text{ mA}} = \mathbf{6\ h}$$

or $t = 6 \text{ h} = \mathbf{2.16 \times 10^4\ s}$ or **21.6 ks**

b.

$$P = VI \qquad (2\text{-}14)$$

$$P = 9 \text{ V} \times 20 \text{ mA} = \mathbf{180\ mW} \quad \text{or} \quad \mathbf{0.18\ W}$$

c.

$$W = Pt \qquad (2\text{-}12)$$

$$W = 180 \text{ mW} \times 21.6 \text{ ks} = \mathbf{3888\ J}$$

d. Capacity is the same as quantity or charge:

$$Q = 120 \text{ mA} \cdot \text{h} = 120 \text{ mA} \times 3.6 \text{ ks} = \mathbf{432\ C}$$

This miniature battery will operate the radio for 6 h at an average current of 20 mA. In doing so, it will supply the radio with 3888 J of energy (1.08 Wh) while discharging 432 C of charge. While the radio is being operated, 0.18 W of power will be drawn from the battery. A small portion will be given off as audio (sound) power, but most of it will be given off in the form of heat.

REVIEW QUESTIONS FOR SECTION 2-6

1. Define the term "power."
2. What is the SI unit of power, and what is its symbol?
3. Express power in terms of voltage and current.
4. Express power in terms of energy.

EXERCISE PROBLEMS FOR SECTION 2-6

1. How much time is required for a device with a rating of 600 W to do 2.7×10^5 J of work? (*ans*. 450 s)
2. How much time is required for a transistor dissipating 50 mW to convert 1000 J of energy?
3. What is the power of a motor which does 1.44 MJ of work per hour? (*ans*. 400 W)
4. What is the output power rating of a radio transmitter which radiates 108 kJ of energy per hour?
5. A 120-V light bulb is rated at 200 W. How much energy does it convert in 24 h? State your answer in:
   a. Joules (*ans*. 17.28 MJ)
   b. Watthours (*ans*. 4800 Wh)
   c. Kilowatthours (*ans*. 4.8 kWh)

6. A 12-V battery has 7 A drawn from it for 30 min. How much energy, in watthours, has been delivered by the battery?
7. The heating element of an electric dryer is rated at 3 kW and is supplied from a 240-V source.
   a. What current is drawn by the dryer? (*ans*. 12.5 A)
   b. How many joules of energy are converted to heat each hour? (*ans*. 10.8 MJ)
8. How long will it take a 120-V electric motor drawing 1.5 A to consume 100 kWh of electrical energy?
9. A 120-V motor operating at 70 percent efficiency has a rated output of 1 hp. How long will it take this motor to do 10^6 J of work? (*ans*. 1340 s)

2-7 SUMMARY OF QUANTITIES AND UNITS

TABLE 2-1 Fundamental Quantities and Units Introduced in Chaps. 1 and 2

Quantity	Symbol	Unit	Symbol
Energy or work	W	joule	J
emf, pd or voltage drop	V	volt	V
Charge or quantity	Q	coulomb	C
Current	I	ampere	A
Force	F	newton	N
Resistance*	R	ohm	Ω
Conductance*	G	siemens	S
Power	P	watt	W
Distance or length	l	meter	m
Mass	m	kilogram	kg
Time	t	second	s
Velocity	v	meters per second	m/s
Temperature	T	kelvin	K
	t	degree Celsius	°C
Plane angle	$\phi, \theta, \ldots$	radian	rad
		degree	°
Efficiency	η		
Capacitance*	C	farad	F
Inductance*	L	henry	H

*Resistance, conductance, capacitance and inductance, together with their units, the ohm, siemens, farad, and henry, are included in this table for the sake of completeness, although these topics are not covered in the text until later.

SUMMARY OF EQUATIONS

$W = Fs$ (2-1)

$J = N \cdot m$ (2-2)

$W_L = W_T - W_U$ (2-3)

$W_L = W_i - W_o$ (2-4)

$\eta = W_o/W_i$ (2-5)

$W = VQ$ (V = emf or pd) (2-6)

$J = V \cdot C$ (2-7)

$I = Q/t$ (2-8)

$Q = It$ (2-9)

$A = C/s$ (2-10)

$P = W/t$ (2-11)

$W = Pt$ (2-12)

$W = VIt$ (2-13)

$P = VI$ (2-14)

1 Wh = 3.6 kJ (2-15)

1 kWh = 3.6 MJ (2-16)

1 hp = 746 W (2-17)

CHAPTER QUESTIONS

2-1. What are the three subsystems needed for a complete electric circuit?

2-2. For the following devices or systems, say which you think are *generator, load,* or *network:*

a. Light bulb
b. Battery
c. Microphone
d. TV cable
e. Loudspeaker
f. Telephone exchange or office

2-3. What is the difference between *energy* and *work?*

2-4. When a circuit produces *energy loss,* what happens to the "lost" energy?

2-5. When a car battery is "charged," how does this relate to the electrical quantity *charge?*

2-6. Why are *electromotive force* and *potential difference* both regarded as *voltage?*

2-7. Referring to Fig. 2-10, what is the voltage across the switch, *V* or zero?

a. In Fig. 2-10*a*
b. In Fig. 2-10*b*

2-8. In Fig. 2-16, the current is shown as moving from the negative to the positive in the generator but from positive to negative in the remainder of the circuit. Why is this correct?

2-9. It may be said that an electric circuit only uses energy when the circuit is *closed*. Why is this so?

2-10. What conditions are required for an electric current to exist?

2-11. How important is it to know

a. The direction of a current?
b. The speed of a current?

2-12. What is the difference between *energy* and *power?*

2-13. Why is *time* regarded as important to electric circuits?

CHAPTER PROBLEMS

2-1. A portable tape recorder operates from a 3-V battery (two cells of 1.5 V each) that holds 3000 J of energy when it is new.

a. What charge does the battery hold when new? (*ans*. 1000 C)

b. How long would it take for all the battery's energy to be dissipated by the recorder if it were

i. Recording at a steady current drain of 25 mA? (*ans*. 40 000 s or 11.11 h)

ii. Playing back at a steady current of 10 mA? (*ans*. 100 000 s or 27.78 h)

2-2. What power would the battery in Prob. 2-1 dissipate for each of the situations, recording and playback?

2-3. A heat-treatment furnace used in metallurgy takes a power of 30 kW from a 240-V power point. What is the furnace current? (*ans*. 125 A)

2-4. In how much time will the furnace of Prob. 2-3 consume 1 MWh of energy?

2-5. The electric motor in a hand drill is rated at 0.25 hp. What current does this drill take, at this power, from a 120-V outlet? (*ans*. 3.108 A)

2-6. A stereo amplifier can deliver a total audio (sound) power of 50 W. When doing this, the amplifier takes 1.2 A of current from a 120-V power outlet. What is the efficiency of the amplifier?

2-7. A pair of headphones can handle up to 100 mW of power. At this power, the current supplied to the phones is 50 mA. What voltage does this represent? (*ans*. 2 V)

2-8. A fully discharged 12-V car battery is charged at 10 A for 8 h. What amount of charge is put into the battery?

2-9. For how long can the battery of Prob. 2-8 supply a 60-W headlamp load, if the voltage remains at 12 V? (*ans*. 16 h)

2-10. The rooftop antenna feeding a home television set delivers 75 μV to the TV. At this voltage level, the set draws a signal current of 250 nA. How much power does the antenna feed to the TV set?

3

RESISTANCE AND RESISTORS

Energy, power, current, and voltage are among the most important electrical quantities, but there are others. Resistance is one. The nature of a circuit or device, its size, shape, and what it is made of, controls its current. Something about it *resists* current flow; we call this effect *resistance*.

Any electrical or electronic circuit or equipment dissipates power, usually by generating heat. Resistance may be just one of the characteristics of the circuit, but only the resistance dissipates power. Power dissipation may be useful; for example, we want an electric heater to get hot. We do not want a transistor to get hot, so when it does, this means that energy is being wasted. Depending on whether or not power dissipation (or energy conversion) is wanted, resistance may be wanted or not wanted.

Resistance can be used to limit current to a definite value, to select a desired level of voltage, or to perform any of a number of other functions. Often these jobs are done by components called *resistors,* specially made to have a known amount of resistance.

3-1 RESISTANCE AND CONDUCTANCE

An emf applied to a closed circuit produces a current. But the same emf does not always produce the same current.

In Fig. 3-1*a* the 100-V emf causes a current of 1 A, while in Fig. 3-1*b* the same 100-V emf produces a current of 10 A. The circuit in Fig. 3-1*a* has the greater ability to oppose current. To describe this ability, we say that this circuit has the higher *resistance*.

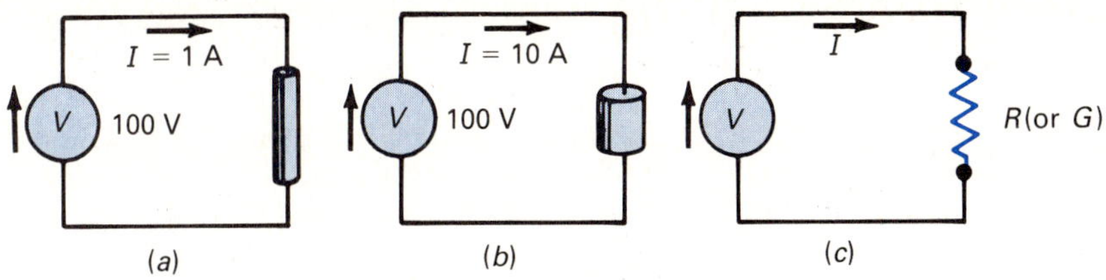

Fig. 3-1. *(a)* Low conductance, high resistance; *(b)* high conductance, low resistance; *(c)* schematic diagram of a resistive circuit.

Resistance is the ability of a circuit to oppose current. The symbol for resistance is R. The SI unit of resistance is the ohm, symbol Ω (capital Greek letter omega).

The circuit in Fig. 3-1*b* has the greater tendency to admit current. It is said to have the higher *conductance*.

Conductance is the ability of a circuit to permit a current to flow. The symbol for conductance is G. The SI unit of conductance is the siemens, symbol S.*

The circuits in Fig. 3-1*a* and *b* can both be represented by the schematic diagram in Fig. 3-1*c*. The schematic symbol for a circuit element gives no indication of the value of the element. The schematic symbol for resistance (or conductance, which has the same symbol) does not suggest its resistance value, just as the generator symbol gives no information about its emf.

If a circuit has *high resistance*, it must have *low conductance*, and vice versa; that is,

$$G = \frac{1}{R} \quad \text{or} \quad R = \frac{1}{G} \qquad \textbf{(3-1)}$$

EXAMPLE 3-1 The cable connecting a television set to its antenna has a resistance of 10 Ω/100 m length. If the cable is 20 m long, what is its conductance?

Solution The resistance of the cable equals the resistance per unit length multiplied by the length:

$$R = \frac{10\ \Omega}{100\ \text{m}} \times 20\ \text{m}$$

$$R = 2\ \Omega$$

$$G = \frac{1}{R} \qquad (3\text{-}1)$$

$$= \frac{1}{2\ \Omega}$$

$$G = \mathbf{0.5\ S}$$

*A unit of conductance still used is the mho (symbol ℧). The mho is numerically the same as the siemens.

EXAMPLE 3-2 An electric heater has a conductance of 50 mS when hot. What is its resistance?

Solution

$$R = \frac{1}{G} \tag{3-1}$$

$$= \frac{1}{50 \text{ mS}} = \frac{1}{5 \times 10^{-2} \text{ S}}$$

$$R = \mathbf{20\ \Omega}$$

REVIEW QUESTIONS FOR SECTION 3-1

1. What is the symbol for conductance?
2. What is the SI unit of conductance, and what is its symbol?
3. What is the symbol for resistance?
4. What is the SI unit of resistance, and what is its symbol?
5. What is the relationship between conductance and resistance?

EXERCISE PROBLEMS FOR SECTION 3-1

1. What is the conductance of the following resistors?
   a. 18 Ω (*ans*. 55.56 mS)
   b. 470 Ω (*ans*. 2.128 mS)
   c. 1200 Ω (*ans*. 0.8333 mS)
   d. 56 kΩ (*ans*. 17.86 μS)
   e. 2.2 MΩ (*ans*. 0.4545 μS)
   f. 3 mΩ (*ans*. 333.3 S)
2. What is the resistance of devices with the following conductances?
   a. 1.2 mS
   b. 2.5 S
   c. 33.3 μS
   d. 470 S
   e. 8000 μS
   f. 0.06 S
3. An electronic instrument has a conductance of 25 μS. What is its resistance? (*ans*. 40 kΩ)
4. No. 2 copper wire has a resistance of 0.513 Ω/km. What is the conductance of 12 km of this wire?
5. Find the conductance of a resistor in a transistor radio which has a current of 220 μA and a voltage of 5.2 V. (*ans*. 42.3 μS)
6. The input resistance of an oscilloscope is 1 MΩ. What is its input conductance?
7. Find the resistance of a device which has a voltage of 300 V and a conductance of 4.55 mS. (*ans*. 219.8 Ω)

3-2 OHM'S LAW

The most widely known principle in electrical technology is *Ohm's law,* which for steady direct current is usually stated:

In any electric circuit, the current is directly proportional to the applied voltage and inversely proportional to the total circuit resistance.

This may be said more simply as:

The effect (current *I*) is the cause (voltage *V*) divided by the opposition (resistance *R*).

This statement can be written in the form of an equation usually known as Ohm's law:

$$I = \frac{V}{R} \qquad \text{or} \qquad V = IR \qquad \text{or} \qquad R = \frac{V}{I} \tag{3-2}$$

The Ohm's law equation can also be expressed in terms of conductance as

$$I = VG \qquad \text{or} \qquad V = \frac{I}{G} \qquad \text{or} \qquad G = \frac{I}{V} \tag{3-3}$$

In Eqs. (3-2) and (3-3),

V = voltage (emf or pd), V
I = current, A
R = resistance, Ω
G = conductance, S

The statements in Sec. 3-1 represent a physical viewpoint of resistance and conductance—opposing and admitting current. We must look to Ohm's law to obtain a circuit view of these quantities and their units:

$$R = \frac{V}{I} \tag{3-2}$$

Resistance is the ratio of voltage to current; one ohm is one volt per ampere.

$$G = \frac{I}{V} \tag{3-3}$$

Conductance is the ratio of current to voltage; one siemens is one ampere per volt.

In unit form Eqs. (3-2) and (3-3) are

$$\Omega = \frac{\text{V}}{\text{A}} \tag{3-4}$$

$$\text{S} = \frac{\text{A}}{\text{V}} \tag{3-5}$$

EXAMPLE 3-3 An emf of 100 V is applied across a resistance of 20 Ω. What is the current?

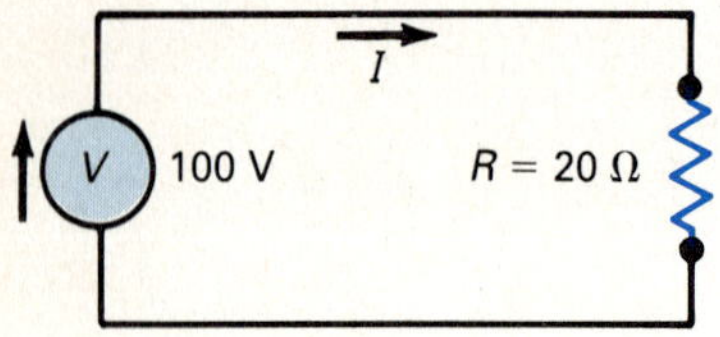

Fig. 3-2. Example 3-3: Circuit.

Solution (See Fig. 3-2.)

$$I = \frac{V}{R} \qquad (3\text{-}2)$$

$$I = \frac{100\ \text{V}}{20\ \Omega} = \mathbf{5\ A}$$

EXAMPLE 3-4 A current of 100 mA exists in a 5-kΩ resistance. What is the voltage across the resistance?

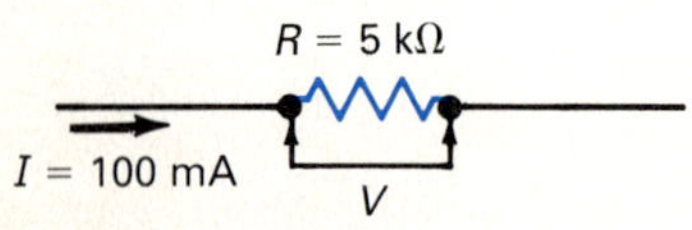

Fig. 3-3. Example 3-4: Conditions associated with the resistance.

Solution (See Fig. 3-3.)

$$V = IR \qquad (3\text{-}2)$$

$$V = 100\ \text{mA} \times 5\ \text{k}\Omega = \mathbf{500\ V}$$

EXAMPLE 3-5 If 3 mA is drawn from a 9-V battery, what is the circuit resistance?

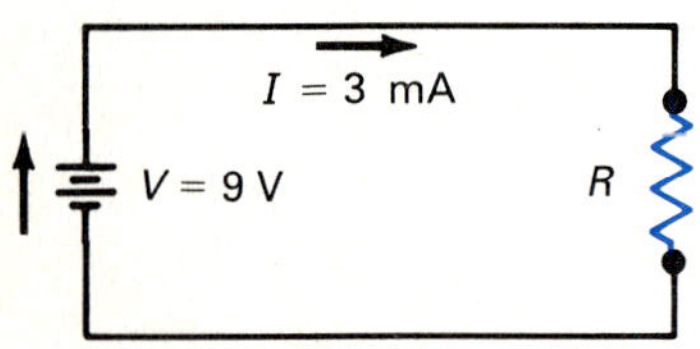

Fig. 3-4. Example 3-5: Circuit.

Solution (See Fig. 3-4.)

$$R = \frac{V}{I} \qquad (3\text{-}2)$$

$$R = \frac{9\ \text{V}}{3\ \text{mA}} = \mathbf{3\ k\Omega}$$

EXAMPLE 3-6 An electric motor armature has a conductance of 5 mS. What current will be produced in it by a 125-V emf?

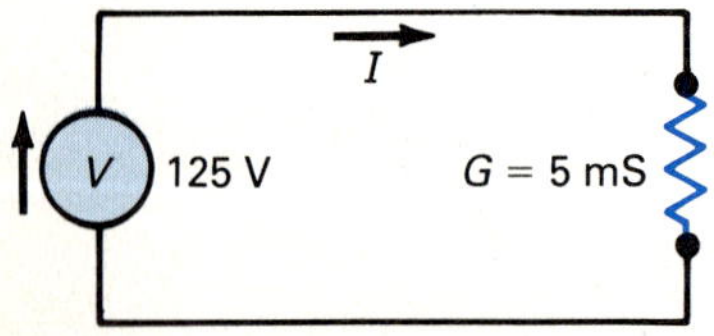

Fig. 3-5. Example 3-6: Circuit.

Solution (See Fig. 3-5.)

$$I = VG \qquad (3\text{-}3)$$

$$I = 125\ \text{V} \times 5\ \text{mS} = \mathbf{625\ mA} \quad \text{or} \quad \mathbf{0.625\ A}$$

EXAMPLE 3-7 What is the conductance of a 100-W 120-V light bulb when it is lit?

Solution

$$P = VI \qquad (2\text{-}14)$$

$$I = \frac{P}{V} = \frac{100\ \text{W}}{120\ \text{V}} = 0.8333\ \text{A}$$

$$G = \frac{I}{V} \qquad (3\text{-}3)$$

$$= \frac{0.8333\ \text{A}}{120\ \text{V}} = \mathbf{6.944\ mS}$$

REVIEW QUESTIONS FOR SECTION 3-2

1. Give the three forms of the Ohm's law equation in terms of resistance.
2. Give the product form of the Ohm's law equation in terms of conductance.

EXERCISE PROBLEMS FOR SECTION 3-2

1. Find the voltage across an 18-kΩ resistor which has a current of 0.5 mA. (*ans*. 9 V)
2. Determine the value of resistance in the following examples:
   a. $V = 300$ V; $I = 2$ A
   b. $V = 24$ V; $I = 0.3$ A
   c. $V = 10$ V; $I = 260$ A
   d. $V = 3.4$ V; $I = 1.9$ mA
   e. $V = 12$ V; $I = 30.8$ μA
3. Determine the values of conductance for Prob. 2*a* to *e*.
   (*ans*. *a*. 6.667 mS; *b*. 12.5 mS; *c*. 26 S; *d*. 0.5588 mS; *e*. 2.567 μS)
4. Determine the voltage in the following examples:
   a. $I = 3$ A; $R = 27$ Ω
   b. $I = 15$ mA; $R = 27$ kΩ
   c. $I = 30$ mA; $G = 350$ μS
   d. $R = 82$ Ω; $I = 8$ mA
   e. $G = 45$ μS; $I = 75$ μA
5. Determine the current in the following examples:
   a. $G = 25$ mS; $V = 80$ V (*ans*. 2 A)
   b. $G = 10$ S; $V = 6$ V (*ans*. 60 A)
   c. $R = 180$ Ω; $V = 1.57$ V (*ans*. 8.722 mA)
   d. $V = 240$ V; $R = 1500$ Ω (*ans*. 0.16 A)
   e. $V = 9$ V; $G = 0.05$ S (*ans*. 0.45 A)
6. Find the current in a circuit consisting of a 12-V battery and a 100-Ω load.
7. The input resistance of an amplifier is 50 Ω. If the input voltage is 75 μV, what current is developed? (*ans*. 1.5 μA)
8. A current of 20 mA is required in a circuit which has a supply voltage of 15 V. What value of resistance must be used?
9. How much greater than 100 Ω must a resistor be to result in a current of 0.75 A in a circuit being supplied from a 120-V source? (*ans*. 60 Ω)

3-3 POWER IN TERMS OF RESISTANCE

Combining Ohm's law [Eq. (3-2)] and the formula for power calculation [Eq. (2-14)] leads to two other very useful formulas.

Substituting for V from Eq. (3-2) into Eq. (2-14),

$$V = IR \qquad (3\text{-}2)$$

$$P = VI \qquad (2\text{-}14)$$

$$= IR \times I$$

$$P = I^2R \qquad \textbf{(3-6)}$$

Substituting for I from Eq. (3-2) into Eq. (2-14),

$$I = \frac{V}{R} \qquad (3\text{-}2)$$

$$P = VI \qquad (2\text{-}14)$$

$$= V \times \frac{V}{R}$$

$$P = \frac{V^2}{R} \qquad \mathbf{(3\text{-}7)}$$

EXAMPLE 3-8 What minimum power rating is required for a resistor which is to operate in the collector circuit of a transistor, if its resistance is to be 1 kΩ and the collector current is 50 mA?

Solution

$$P = I^2R \qquad (3\text{-}6)$$

$$P = (50 \text{ mA})^2 \times 1 \text{ k}\Omega = \mathbf{2.5\ W}$$

The resistor must be capable of dissipating a power of at least 2.5 W.

EXAMPLE 3-9 What is the conductance of a 100-W 120-V light bulb? (This is the same question as that in Example 3-7.)

Solution

$$P = \frac{V^2}{R} \qquad (3\text{-}7)$$

$$R = \frac{V^2}{P}$$

Therefore, as $G = 1/R$,

$$G = \frac{P}{V^2}$$

$$= \frac{100 \text{ W}}{(120 \text{ V})^2}$$

$$G = \mathbf{6.944\ mS}$$

REVIEW QUESTIONS FOR SECTION 3-3

1. What is the equation expressing power in terms of current and resistance?
2. What is the equation expressing power in terms of voltage and resistance?

3-4 GRAPHICAL VIEW OF OHM'S LAW

Plotting values of voltage against current gives graphs of resistance and conductance.

If the current is the independent variable and the voltage the dependent variable, the slope of the curve will represent resistance.

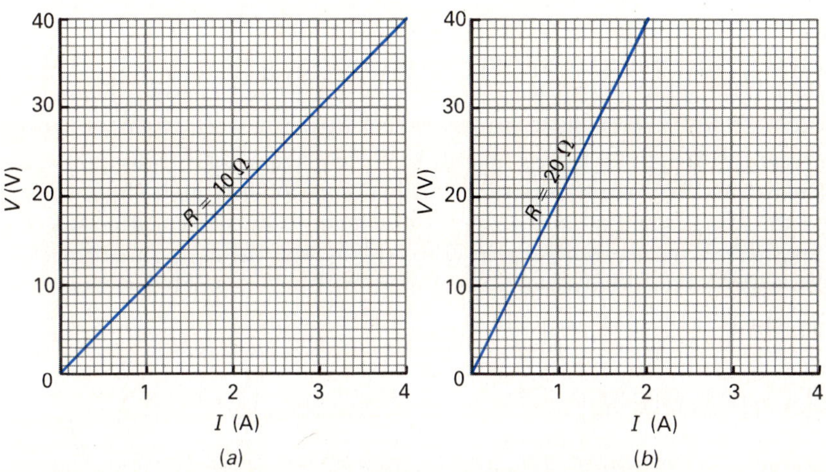

Fig. 3-6. Graphs showing resistance: *(a)* 10 Ω; *(b)* 20 Ω.

In Fig. 3-6, at all points on curve *a* the ratio *V*/*I* is 10, indicating a resistance of 10 Ω. Curve *b* represents a resistance of 20 Ω. The resistance is proportional to the slope of the curve.

If voltage is the independent variable and current the dependent variable, the slope of the curve will represent the conductance.

In Fig. 3-7, at all points on curve *a* the ratio *I*/*V* is 1:10 (for example, when $V = 20$ V, $I = 2$ A). Therefore curve *a* represents a conductance of 0.1 S. Similarly, curve *b* represents a conductance of 0.2 S (when $V = 20$ V, $I = 4$ A). The slope of the curve (its steepness) indicates the magnitude of the conductance.

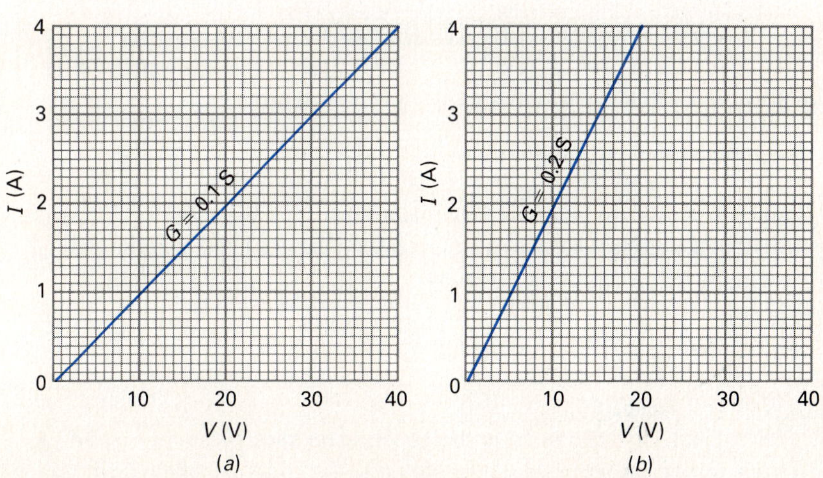

Fig. 3-7. Graphs showing conductance: *(a)* 0.1 S; *(b)* 0.2 S.

The curves in Figs. 3-6*a* and 3-7*a* both represent the same circuit element, since a conductance of 10 Ω is the same as a resistance of 0.1 S. The curves in Figs. 3-6*b* and 3-7*b* both have slopes greater than those of their *a* counterparts. They do *not* represent the same circuit element, because higher resistance means lower conductance (Fig. 3-6*b*), and higher conductance means lower resistance (Fig. 3-7*b*).

The curves in Figs. 3-6 and 3-7 represent linear resistances and conductances. In each case the current and voltage vary in direct proportion to each other. In this text we are concerned only with such linear elements. However, it should be realized that nonlinear elements are quite common. In fact, resistors, which are devices designed to provide the quantity resistance, are slightly nonlinear. When a resistor passes current, it heats up. The rise in temperature causes the resistance to change. Well-designed resistors, correctly chosen for their application, do not show this effect to a significant degree. We are quite justified in considering that, in practical circuits, resistors can be dealt with as constant-value resistance elements and thus as linear devices.

3-5 RESISTORS

A *resistor* is a component designed to have a known value of resistance.

In its simplest form, a resistor is just a length of some material. Most commercial resistors are made of a compound of carbon and plastic "filler." The proportion of filler to carbon determines the resistance of a resistor of a certain size and shape.

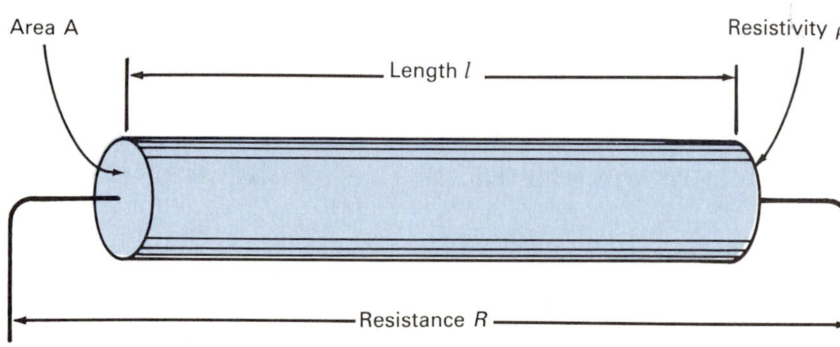

Fig. 3-8. Factors controlling resistance.

Figure 3-8 shows the main factors that control the resistance of a resistor. Provided that the resistor has the same cross-sectional area and material throughout its length, its resistance is given by

$$R = \frac{\rho l}{A} \tag{3-8}$$

where R = resistance, Ω
ρ = resistivity, Ω · m (ρ is the Greek letter rho)
l = length, m
A = area of cross-section, m^2

Resistivity

Resistors of the same length and area will have different resistances if they are made of different materials. For example, a resistor with a high plastic filler content will have more resistance than one of the same size made of pure carbon. A length of metal wire will have much less resistance than a carbon filament of the same size. This resistance difference between materials is called *resistivity* or *specific resistance*.

The resistivity of a material is the resistance of a piece with a standard length and area. In SI units, the standard size is one meter long and one square meter in area. Using this size, the unit of resistivity is the ohm-meter, $\Omega \cdot$ m.

TABLE 3-1 Resistivities of Materials

Material	Resistivity ρ in n$\Omega \cdot$m
Silver	16
Copper	17
Lead	22
Gold	24
Aluminum	27
Iron	100
Platinum	100
Manganin	440
Carbon	35 000

The resistivity values in Table 3-1 are given in nano-ohm-meters. To get the values in ohm-meters, multiply the values in the table by 10^{-9}. For instance, the resistivity of silver is 16 n$\Omega \cdot$ m or 1.6×10^{-8} $\Omega \cdot$ m.

Quoting the resistivity values in n$\Omega \cdot$ m results in simple numbers that make it easy to compare materials from a resistance viewpoint. You can see that most of the metals have similar resistivities, around 20 n$\Omega \cdot$ m. Manganin, a metal alloy of manganese, copper, and nickel formed into wires, is used for making high-precision resistors. Its resistivity is 26 times that of copper. For a given resistance, you would need 26 times the length of copper wire as of manganin wire.

Except for carbon, the materials in Table 3-1 are all metals. It is clear that carbon has 80 to 2000 times the resistivity of any of them. Therefore a carbon resistor can be quite small; it can be much smaller still, if the carbon is mixed with a plastic filler.

The 35 000-n$\Omega \cdot$ m resistivity of carbon may appear to be so high that it would effectively stop current flow. Compared with the insulating materials, such as porcelain, rubber, mica, and polystyrene, which have resistivities from 10^{18} to 10^{26} n$\Omega \cdot$ m, even carbon looks like a pretty good conductor.

EXAMPLE 3-10 What is the resistance of a carbon resistor $\frac{1}{2}$ in (1.27 cm) long with a diameter of $\frac{1}{16}$ in (1.59 mm)?

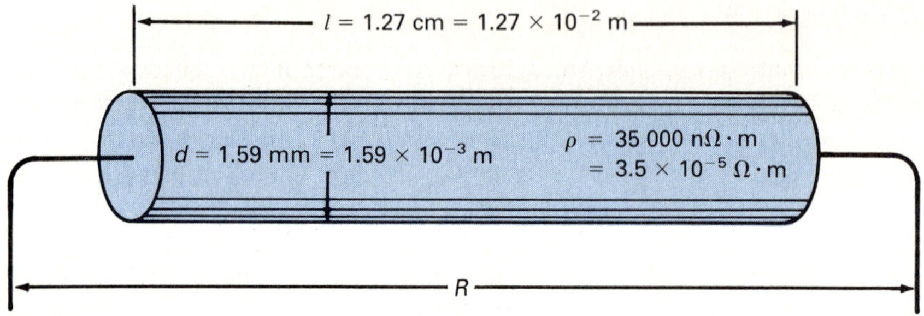

Fig. 3-9. Example 3-10: Data.

Solution (See Fig. 3-9.)

$$\text{Area of cross section } A = \pi r^2$$

$$A = \pi \times \left[\frac{1.59 \times 10^{-3}\ \text{m}}{2}\right]^2$$

$$= 1.986 \times 10^{-6}\ \text{m}^2$$

$$R = \frac{\rho l}{A} \qquad (3\text{-}8)$$

$$= \frac{3.5 \times 10^{-5}\ \Omega \cdot \text{m} \times 1.27 \times 10^{-2}\ \text{m}}{1.986 \times 10^{-6}\ \text{m}^2}$$

$$R = \mathbf{0.2238\ \Omega} \quad \text{or} \quad \mathbf{223.8\ m\Omega}$$

EXAMPLE 3-11 If a resistor of the same physical size as that in Example 3-10 has a resistance of 1 kΩ, what would be the resistivity of its material?

Solution

$$\rho = \frac{RA}{l} \qquad (3\text{-}8)$$

$$= \frac{1\ \text{k}\Omega \times 1.986 \times 10^{-6}\ \text{m}^2}{1.27 \times 10^{-2}\ \text{m}}$$

$$\rho = \mathbf{0.1564\ \Omega \cdot m}$$

The size and shape of the resistors in Examples 3-10 and 3-11 are about the same as those of a commercial $\frac{1}{2}$-W resistor. The 224 mΩ of Example 3-10 suggests that pure carbon is not much use as a resistor material. The 1 kΩ of Example 3-11 is a useful value. To get the resistivity up from the $3.5 \times 10^{-5}\ \Omega \cdot \text{m}$ of Example 3-10 to the $0.16\ \Omega \cdot \text{m}$ of Example 3-11, the carbon would need enough plastic filler to raise the resistivity by a factor of 4500. This is not as difficult as it sounds. For example, the resistivity of polystyrene is $10^{17}\ \Omega \cdot \text{m}$, 3×10^{21} times that of carbon. So a relatively small amount of this plastic mixed with carbon can give a high value of resistivity.

Resistor Types

Carbon (or *carbon-composition*) *resistors* are popular for general-purpose use. They are small, low-cost, reliable, and available in a wide range of resistance values. A carbon resistor consists of a rod of filler-mixed carbon with connecting wires at each end.

Figure 3-10 shows the main features of a typical carbon resistor. Its resistive element is a rod of carbon compound with wire connections at each end. These connections may be molded directly into the carbon, as shown in Fig. 3-10*b*, or attached to metal end caps, as in Fig. 3-10*c*.

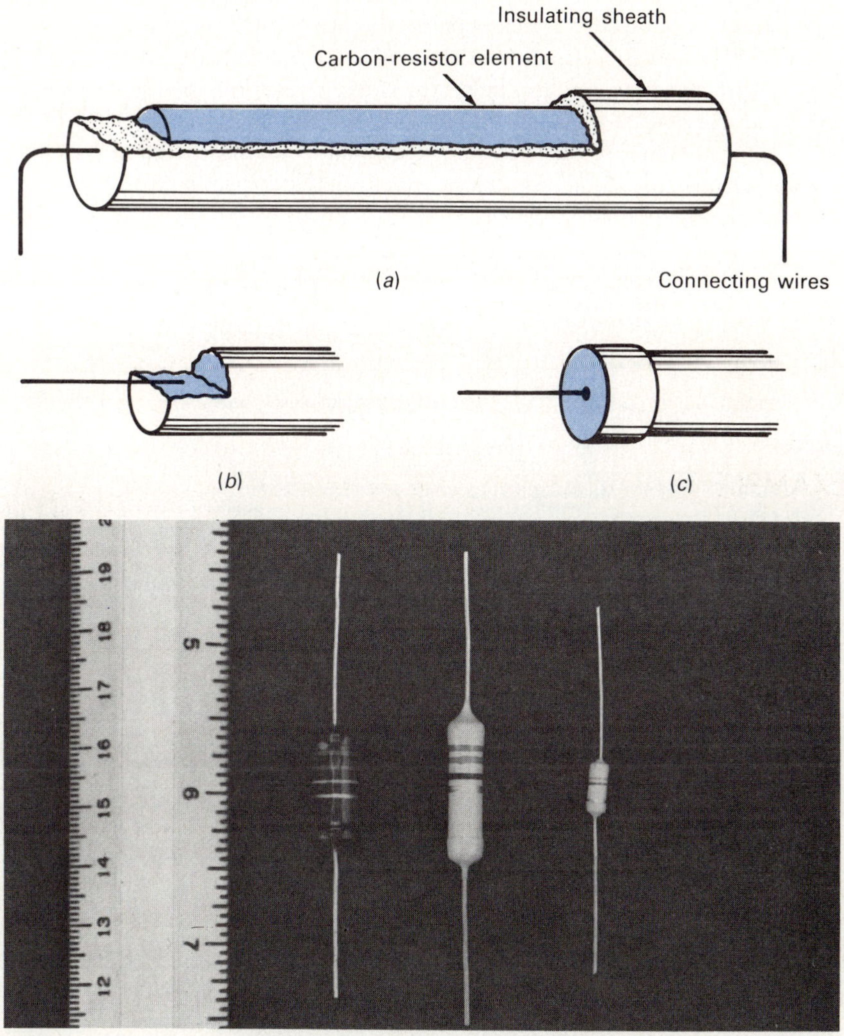

(*d*)

Fig. 3-10. Carbon resistor. *(a)* Cutaway view of the resistor; *(b)* connecting wires embedded in the carbon composition; *(c)* connecting wires attached to end caps. *(d)* *(Left to right)* Photo shows carbon resistors of 3 W, 2 W, and ½ W power ratings.

The carbon element in Fig. 3-10*a* is encased in a sheath of plastic or some other insulating material. This sheath serves two purposes: It protects against contact between the resistor and another part of the equipment, and it spreads the resistor's heat over the element surface, limiting the formation of hot spots that might cause resistor failure.

Carbon resistors are also made in film form. Instead of a solid carbon rod, a *carbon-film resistor* (Fig. 3-11*a*) consists of a ceramic rod on which is deposited a film of carbon composition. The connecting wires are attached to end caps. This type of resistor has higher stability of resistance value than the solid carbon type.

For high power dissipation, *wirewound resistors* are commonly used. (See Fig. 3-11*b*.) A wirewound resistor consists of a coil of nickel-alloy wire wound on an insulating former. The former is usually made of a ceramic material, such as porcelain. The wire coil may be coated with a layer of ceramic-based enamel to insulate the resistor element and prevent oxidation of the metal. This use of metal wire as a resistor element and its ceramic core and coating produces a resistor that can operate at high temperatures without breaking down or permanently changing its basic resistance.

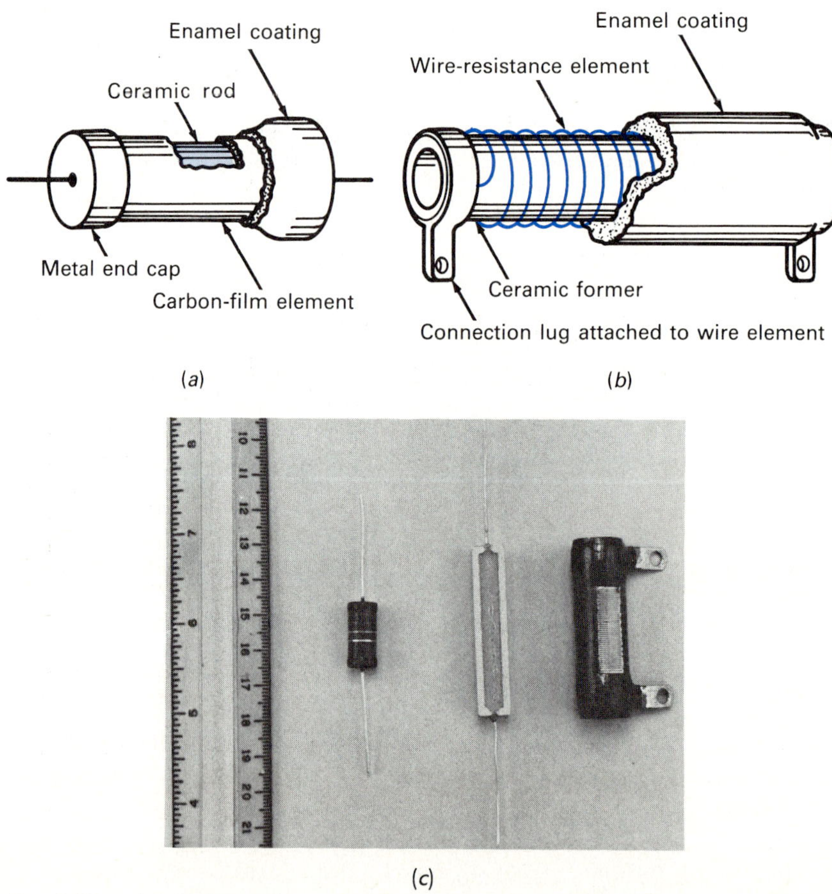

Fig. 3-11. *(a)* Carbon-film and *(b)* wirewound resistors. *(c) (Left to right)* Photo of 3-W carbon film, 10-W wirewound, and 50-W wirewound resistors.

Resistors wound with manganin wire are used in precision test equipment, such as the Wheatstone bridge. The manganin wire has a low *temperature coefficient of resistance,*[1] so that a resistor made from it maintains its resistance value over a wide range of operating temperatures.

Tolerance

The actual resistance of any practical resistor will lie within a range of values on either side of its *nominal* or specified value. The degree to which any single resistor is assured to be close to its nominal value is called its *tolerance*. For example, the resistance of a 100-Ω, 10 percent tolerance resistor will be within 100 Ω ± 10 percent, or between 90 Ω and 110 Ω. The most readily available tolerances are 10 percent and 5 percent, although tolerances of 2 percent, 1 percent, and even higher are obtainable in some resistor types. You should note that *high* tolerance means a *small* tolerance number; 5 percent tolerance is higher than 10 percent.

Color Code

Most carbon resistors, solid or film type, and some wirewound resistors are made with colored bands to show the nominal resistance value and tolerance. The method of color coding is outlined in Fig. 3-12.

Bands 1, 2, and 3 give the nominal value of resistance to two significant figures. Bands 1 and 2 give the first and second digits. Band 3 gives the order of magnitude as an exponent of 10 that follows the two digits. The colors used for bands 1, 2, and 3 and the numbers they represent are:

black	0	green	5
brown	1	blue	6
red	2	violet	7
orange	3	gray	8
yellow	4	white	9

Band 3 may have one of two other colors; gold for −1 or silver for −2. Band 4 gives the tolerance. For 10% and 5% tolerance, the colors are silver and gold, respectively.

Fig. 3-12. Resistor color code.

[1] *Temperature coefficient of resistance* is the amount by which resistance changes when temperature changes by one degree Celsius.

The nominal values for 10 percent and 5 percent tolerance resistors are listed in Table 3-2.

TABLE 3-2 Resistor Values*

5% Tolerance	10% Tolerance	Color Bands 1	Color Bands 2
10	10	Brown	Black
11			Brown
12	12		Red
13			Orange
15	15		Green
16			Blue
18	18		Gray
20		Red	Black
22	22		Red
24			Yellow
27	27		Violet
30		Orange	Black
33	33		Orange
36			Blue
39	39		White
43		Yellow	Orange
47	47		Violet
51		Green	Brown
56	56		Blue
62		Blue	Red
68	68		Gray
75		Violet	Green
82	82	Gray	Red
91		White	Brown

*The table gives the first two digits of the nominal value of resistance and the colors of the first two bands of the color code, as in Fig. 3-12.

Power Rating

Resistors of any resistance can be obtained in various *power ratings*. For carbon resistors, the common power ratings are $\frac{1}{10}$, $\frac{1}{4}$, $\frac{1}{2}$, 1, and 2 W. The range of powers for wirewound resistors goes from $\frac{1}{4}$ W to over 200 W.

In each case, resistor power rating refers to the maximum amount of power a resistor can handle without exceeding a safe working temperature, under conditions specified by the manufacturer or some commercial or military standard.

The power rating of a resistor of a given type is determined mainly by its physical size; a high-power resistor needs to be larger than one for low-power use.

EXAMPLE 3-12

What are the band colors on the following resistors: (*a*) 4.7 Ω, 10 percent tolerance; (*b*) 68 Ω, 5 percent tolerance; (*c*) 56 kΩ, 5 percent tolerance; and (*d*) 3.3 MΩ, 10 percent tolerance?

Solution

Converting each of the resistance values into a two-digit number and a power of 10, we get

a. $4.7\ \Omega = (47 \times 10^{-1})\ \Omega$
b. $68\ \Omega = (68 \times 10^{0})\ \Omega$
c. $56\ \text{k}\Omega = (56 \times 10^{3})\ \Omega$
d. $3.3\ \text{M}\Omega = (33 \times 10^{5})\ \Omega$

Using Fig. 3-12 and Table 3-2, the following list can be drawn up:

	Band 1	Band 2	Band 3	Band 4
a. 4.7 Ω, 10%	= Yellow	Violet	Gold	Silver
b. 68 Ω, 5%	= Blue	Gray	Black	Gold
c. 56 kΩ, 5%	= Green	Blue	Orange	Gold
d. 3.3 MΩ, 10%	= Orange	Orange	Green	Silver

The resistance values of Example 3-12 have been expressed in *powers-of-10* form. When working with the resistor color code, it is common to think in terms of "the number of zeros following the second digit." For this approach, 68 Ω has no zeros following the 8, so the third band is colored black for "zero zeros." The 56 kΩ is 56 000 Ω or "3 zeros following 56." As you can see, this gives the same colors: green, blue, and orange.

EXAMPLE 3-13

In an electronic circuit, a resistor is connected to the collector of a transistor. Under certain conditions, the 20-V supply appears across the resistor. For proper circuit operation, the current through the resistor must not exceed 5 mA. What is the best available 10 percent or 5 percent tolerance resistor to ensure that the resistor's current is close to, but no higher than, 5 mA?

Solution

$$R = \frac{V}{I} \tag{3-2}$$

$$R = \frac{20\ \text{V}}{5\ \text{mA}} = 4\ \text{k}\Omega$$

To limit the current to 5 mA, the resistance must be no *lower* than 4 kΩ.

From Table 3-2, the resistance value will be 4.7 kΩ at 10 percent tolerance or 4.3 kΩ at 5 percent.

In Example 3-13, the 4.7-kΩ, 10 percent resistor's actual value will lie between 4.23 kΩ and 5.17 kΩ, a safe margin above 4 kΩ. The 4.3-kΩ, 5 percent resistor's value will be between 4.09 kΩ and 4.52 kΩ. You might consider the 4.09 kΩ to be too close to the limit of 4 kΩ to be safe.

EXAMPLE 3-14 The resistor of Example 3-13 has the nominal value of 4.7 kΩ. Which of the power ratings given in the text is the smallest able to carry the current when the 20-V supply is developed across the resistor?

Solution

$$P = \frac{V^2}{R} \qquad (3\text{-}7)$$

$$P = \frac{(20\ \text{V})^2}{4.7\ \text{k}\Omega} = 85.11\ \text{mW}$$

Any of the standard power ratings would be suitable for handling 85 mW. The $\frac{1}{10}$-W (or 100-mW) size would be the smallest of those given in the text.

EXAMPLE 3-15 If the resistor of Examples 3-13 and 3-14 were required to carry 20 mA in a larger transistor, would the $\frac{1}{10}$-W resistor still do the job?

Solution

$$P = I^2R \qquad (3\text{-}6)$$

$$P = (20\ \text{mA})^2 \times 4.7\ \text{k}\Omega = 1.88\ \text{W}$$

To handle 1.88 W of power, the resistor should have a power rating of at least 2 W. The $\frac{1}{10}$-W resistor would not be suitable.

Tapped and Variable Resistors

The resistors discussed up to this point have each had a single resistance value; they are *fixed resistors*. Resistors may have more than one value, either more or less fixed or continuously variable.

Resistors may have two or more values obtained by *tapping,* or making connections somewhere between the element ends. A wirewound resistor similar to that of Fig. 3-11 may have its wire element exposed so that a metal band tightened around the resistor makes contact with the element. This metal band forms the tap connection. (See Fig. 3-13.)

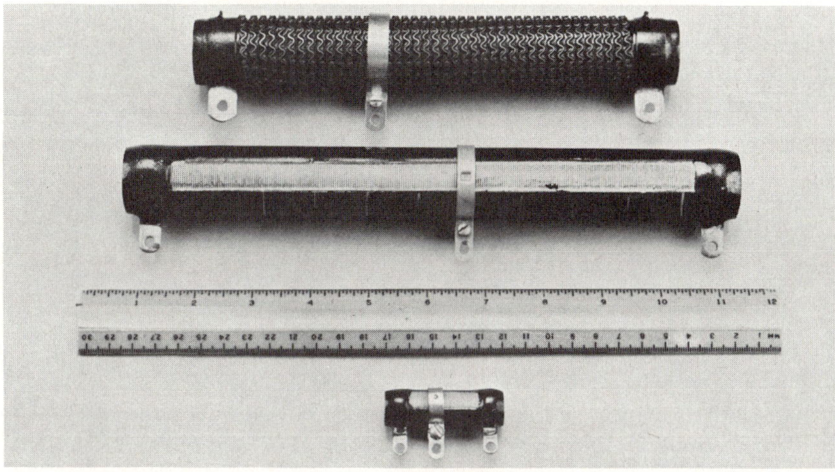

Fig. 3-13. Tapped resistors.

A resistance element, carbon or wirewound, with a metal contact that slides over its surface is a *variable resistor*. When connections are made to one end of the element and to the sliding contact or *wiper,* the variable resistor is used for current control and is called a *rheostat*. When connections are made to each element end and to the wiper to get voltage control, the variable resistor is called a *potentiometer*. There is no basic difference between a rheostat and a potentiometer. They are both variable resistors; only their applications are different. (See Fig. 3-14.)

Fig. 3-14. Variable resistors.

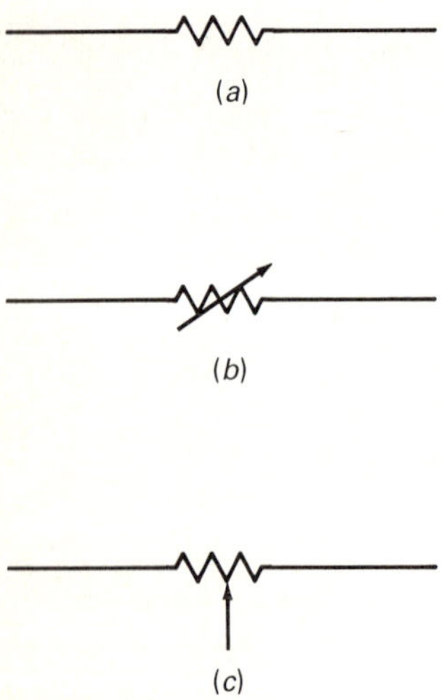

Fig. 3-15. Resistor symbols: *(a)* General-purpose; *(b)* variable; *(c)* potentiometer, a resistor with adjustable contact.

Resistor Symbols

The schematic symbols for resistors are shown in Fig. 3-15.

For a resistor of fixed value, or for any resistor where the type is not important, the symbol of Fig. 3-15*a* is used. Where it is desired to show that resistance value is variable, Fig. 3-15*b* applies. The symbol in Fig. 3-15*c* shows that, in addition to the ordinary two terminals, the resistor has a wiper connection that can be changed in position to give a variable resistance value between the wiper and each end of the resistor.

REVIEW QUESTIONS FOR SECTION 3-5

1. What is a resistor?
2. From what material are most resistors made?
3. What is meant by resistivity?
4. Why are carbon resistors popular?
5. What two purposes does the sheath of a carbon resistor serve?
6. What type of resistor is used for high power dissipation?
7. Resistors wound with manganin wire are used for what purpose? Why?
8. What would be the range of values of a 100-Ω resistor with a 10 percent tolerance?
9. How are the nominal values and tolerances of carbon resistors shown?
10. List the resistor band colors for the numbers 0 to 9.
11. Give the resistor band colors for 10 percent tolerance and 5 percent tolerance.
12. What is the common power rating range of carbon resistors?
13. To what does resistor power rating refer?
14. How can resistors be made to have more than one value?
15. What is the name often given to a variable resistor used for **current** control?
16. What is the name often given to a variable resistor used for **voltage** control?

EXERCISE PROBLEMS FOR SECTION 3-5

1. What are the color bands on the following resistors:
   a. 12 Ω, 10 percent (*ans.* Brown, red, black, silver)
   b. 150 Ω, 5 percent (*ans.* Brown, green, brown, gold)
   c. 56 Ω, 5 percent (*ans.* Green, blue, black, gold)
   d. 820 Ω, 10 percent (*ans.* Gray, red, brown, silver)
   e. 47 kΩ, 10 percent (*ans.* Yellow, violet, orange, silver)
   f. 390 kΩ, 5 percent (*ans.* Orange, white, yellow, gold)
2. What are the maximum and minimum permissible values for resistors with the following nominal values:
   a. 47 Ω, 5 percent
   b. 68 kΩ, 10 percent
   c. 3.3 kΩ, 10 percent
   d. 270 Ω, 5 percent

3. What are the minimum listed power ratings for resistors with the following conditions:
   a. $V = 25$ V, $I = 25$ mA (*ans.* 1 W)
   b. $V = 25$ V, $I = 5.3$ mA (*ans.* $\frac{1}{4}$ W)
   c. $V = 25$ V, $I = 13.9$ mA (*ans.* $\frac{1}{2}$ W)
   d. $V = 12$ V, $I = 17.6$ mA (*ans.* $\frac{1}{4}$ W)
   e. $V = 500$ V, $I = 2.27$ mA (*ans.* 2 W)
   f. $V = 5$ V, $I = 333.3$ mA (*ans.* 2 W)
4. What were the values of the resistors in Prob. 3?
5. What are the minimum listed power ratings for resistors with the following conditions:
   a. $V = 15$ V, $R = 220\ \Omega$ (*ans.* 2 W)
   b. $V = 12$ V, $R = 2.2$ kΩ (*ans.* $\frac{1}{10}$ W)
   c. $V = 20$ V, $R = 1.5$ kΩ (*ans.* $\frac{1}{2}$ W)
6. What are the minimum listed power ratings for resistors with the following conditions:
   a. $I = 50$ mA, $R = 680\ \Omega$
   b. $I = 50$ mA, $R = 330\ \Omega$
   c. $I = 50$ mA, $R = 47\ \Omega$
7. A 120-V soldering iron is rated at 30 W. What is its operating resistance? (*ans.* 480 Ω)
8. Find the conductance of a resistor which dissipates 1.8 W of power and has a current of 6 A.
9. Determine the voltage across a device which has a conductance of 0.166 S and dissipates 29 W of power. (*ans.* 13.22 V)
10. A radio supplied from a 120-V source consumes 75 W of power. What is its equivalent resistance?
11. A voltmeter has a resistance of 5 MΩ on its 250-V range. When the meter reads 150 V, what is
    a. The current through the meter? (*ans.* 30 μA)
    b. The power dissipated by the meter? (*ans.* 4.5 mW)
    c. The conductance of the meter? (*ans.* 0.2 μS)
12. An ammeter on its 100-μA range has a resistance of 2500 Ω. When the meter reads 65 μA, what is
    a. The voltage across the meter?
    b. The power dissipated by the meter?
    c. The conductance of the meter?

SUMMARY OF EQUATIONS

$G = 1/R$ or $R = 1/G$ (3-1)

$I = V/R$ or $V = IR$ or $R = V/I$ (3-2)

$I = VG$ or $V = I/G$ or $G = I/V$ (3-3)

$\Omega = \mathrm{V/A}$ (3-4)

$\mathrm{S} = \mathrm{A/V}$ (3-5)

$P = I^2R$ (3-6)

$P = V^2/R$ (3-7)

$R = \rho l/A$ (3-8)

CHAPTER QUESTIONS

3-1. What is the difference between
a. Resistance? *b*. Conductance? *c*. Resistor?

3-2. How would you *explain* Ohm's law *without* using an equation, or formula?

3-3. What is the ratio
a. Voltage divided by current?
b. Current divided by voltage?

3-4. What is given by
a. Square of current value multiplied by resistance?
b. Square of voltage value divided by resistance?

3-5. In a graph of *voltage* (vertical axis) and *current* (horizontal axis), in what way is *resistance* displayed?

3-6. How would a *nonlinear* resistance show on a voltage-current graph?

3-7. What affects the *resistance* of a *resistor?*

3-8. Two resistors differ only in power rating (same type, same resistance). In what way do the resistors look different?

3-9. What is the difference between *resistance, specific resistance,* and *resistivity?*

3-10. A resistor in a working circuit is too hot. Give three possible reasons for this excessive heat.

3-11. The voltage across a resistor in a working circuit is too high. Give three possible reasons for this excessive voltage.

3-12. The current in a resistor is doubled but its resistance does not change. What happens to the power it dissipates?

3-13. The voltage across a resistor is doubled but its resistance does not change. What happens to the power it dissipates?

3-14. The resistance of a resistor doubles but the voltage across it does not change. What happens to the power it dissipates?

3-15. For use in a measuring instrument, would you expect a resistor to have *high stability of resistance* value and *high temperature coefficient of resistance?*

3-16. What is the point of the resistor color-code system?

3-17. What is the operating difference between a *tapped* resistor and a *variable* resistor?

CHAPTER PROBLEMS

3-1. A variable resistor has a resistance range of 100 to 1000 Ω. What is its conductance range? (*ans*. 1 to 10 mS)

3-2. If the variable resistor of Prob. 3-1 has a fixed 10 V across it, what is its
a. Current range?
b. Power dissipation range?

3-3. If the variable resistor of Prob. 3-1 carries a fixed current of 100 mA, what is the range of
a. The voltage across it (*ans*. 10 to 100 V)
b. Its power dissipation (*ans*. 1 to 10 W)

3-4. For both reliability and economy, which power rating would probably be chosen from a choice of 1, 2, 5, 10, and 20 W for the situations of
a. Prob. 3-2? *b*. Prob. 3-3?

3-5. A number of resistors, each with the color code green, blue, yellow, silver carry the same current of 5 mA. What are the highest and lowest voltages that might be measured across any of the resistors if they are all within tolerance? (*ans*. 2.52 kV, 3.08 kV)

3-6. A 5.6-kΩ resistor carries a 25-mA current when used in a circuit. If the voltage is not to vary outside of the limits 130 V and 150 V, what would the resistor's color code need to be?

4

CAPACITANCE AND CAPACITORS

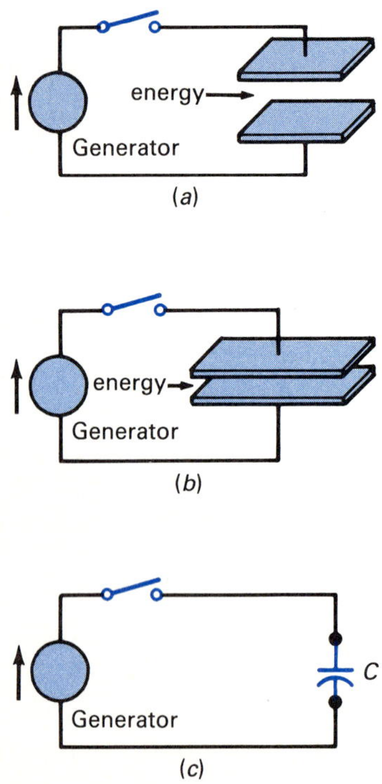

Fig. 4-1. Capacitance. *(a)* Basic capacitor; *(b)* increased capacitance—plates larger and closer together; *(c)* capacitor schematic symbol.

Current flows all the time voltage is applied to a *resistor*. Current flows for a very short time if a voltage is applied to a *capacitor*. This difference arises from the way energy is handled by *resistance* and *capacitance*.

Current causes a *resistor* to get *hot;* electrical energy is converted to heat energy as long as the current is maintained. The heat energy radiates away from the resistor. *Energy leaves a circuit* by being dissipated in resistance. This characteristic of resistance gives rise to power dissipation, often referred to as "I^2R loss."

Current causes an *electric field* to be set up within a *capacitor*. The electric field is a form of *energy* that normally *does not leave the circuit*. This ability to handle energy in the form of an electric field is *capacitance*. Capacitive energy is stored until conditions cause the electric field to collapse, at which time the energy is returned to its original source.

4-1 CAPACITANCE

There is capacitance between any two conducting surfaces separated by insulation. Figure 4-1 shows such a setup connected to a generator through a switch. When the switch is closed, energy is transferred from the generator to the capacitor. More energy is transferred in Fig. 4-1*b* than in Fig. 4-1*a*, as Fig. 4-1*b* has the greater capacitance.

The schematic symbol for the capacitor, shown in Fig. 4-1*c*, is based on the notion of capacitance between two conducting plates separated by air, an insulator. As with most schematic symbols, the one for a capacitor gives no indication of its value, or how much capacitance is present.

The symbol for the quantity capacitance is C. Its unit is the *farad,* symbol F. The farad is a large unit, so that small subdivisions are commonly used: microfarads (μF), nanofarads (nF), and picofarads (pF) are often met in practical equipment.

REVIEW QUESTIONS FOR SECTION 4-1

1. What is the term for the effect between two conducting surfaces separated by insulation?
2. What is the symbol for a capacitor?
3. What is the unit of capacitance and its symbol?
4. What are the three subunits commonly used for the farad?

4-2 CAPACITIVE "OHM'S LAW"

Figure 4-2 shows a generator connected to a capacitor through a switch.

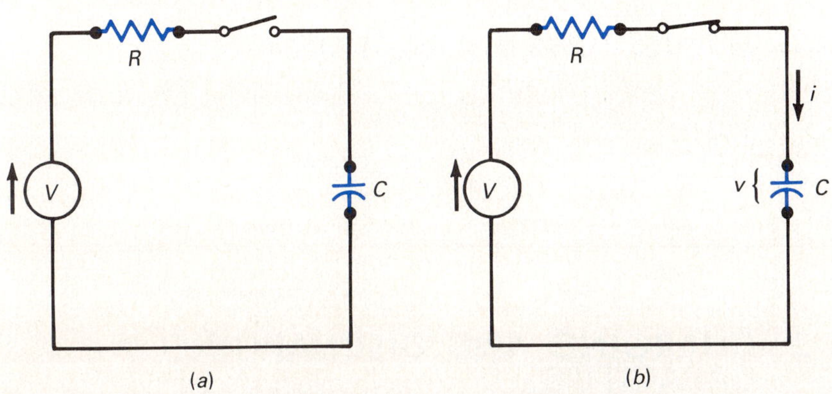

Fig. 4-2. A burst of current as capacitor charges. *(a)* Switch open—capacitor uncharged; *(b)* switch closed—burst of current.

If the capacitor in Fig. 4-2 is initially uncharged, closing the switch will result in a burst of current as electrical energy from the generator enters the capacitor. The current is shown as a lowercase i to indicate that it is not a steady current. It is a short burst of current that lasts just long enough to set up the electric field to charge the capacitor with energy.

Capacitance is usually of importance only when the circuit conditions are changing. The relationship between voltage and current, the "Ohm's law" of capacitance, is concerned with change.

The value of a capacitor's *current* is given by its *capacitance multiplied by the rate of change of its voltage*.

For ease of use in calculating, the symbol dv/dt (change of voltage divided by change of time) is used for the term "rate of change of voltage."

From the above, the value of a capacitive current is

$$i = C\frac{dv}{dt} \tag{4-1}$$

where i = instantaneous current, A
C = capacitance, F
dv/dt = rate of change of voltage, V/s

This concept forms the basis of a definition of capacitance: *Capacitance is the constant of proportionality between the rate of change of voltage across a circuit and the current in the circuit.*

The capacitive Ohm's law emphasizes the importance of voltage change in this type of circuit. Without voltage change there is no capacitive current. Compare capacitive current with the current in a resistive circuit, as given by the ordinary Ohm's law equations (3-2) and (3-3). As can be seen, the current in a resistive circuit is always dependent on the *voltage* magnitude. The current is there whether or not the voltage changes. If the voltage does change, the current follows the change.

This distinction between capacitive and resistive currents determines the behavior of circuits that contain capacitors, resistors, or both.

REVIEW QUESTIONS FOR SECTION 4-2

1. Generally speaking, under what circuit conditions does capacitance have importance?
2. State the "Ohm's law" for capacitive circuits.
3. Why is the *change of voltage* important to the operation of a capacitor?
4. What is required in a capacitive circuit for current to exist?

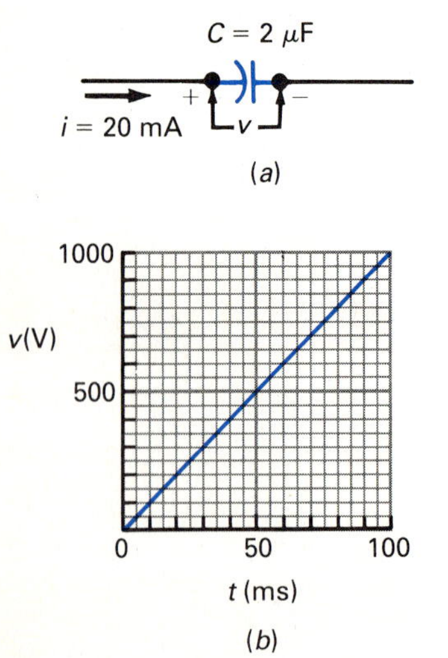

Fig. 4-3. Constant current charge of a capacitor. *(a)* Circuit conditions; *(b)* graph relating voltage and time.

4-3 CHARGING AND DISCHARGING CAPACITORS

Charging

It may be seen from Eq. (4-1) that when a capacitor becomes charged as the result of a constant current, its rate of change in voltage will also be constant. In other words, the capacitor voltage will be directly proportional to time with constant charging current. Figure 4-3 shows this relationship between voltage and time as a linear graph.

Equation 4-1 refers to a *changing* voltage, as that is the situation for capacitive current. However, both the charge held and energy stored in a capacitor are of interest in *static* (nonchanging) and *dynamic* (changing) conditions.

We can rewrite Eq. 4-1 for the situation shown in Fig. 4-3, where the current and the rate of change of voltage are both constant, or fixed in value. The magnitude of the voltage depends on the time the current flows.

In this situation

$$i = C\frac{dv}{dt} \tag{4-1}$$

becomes

$$I = C\frac{V}{t} \qquad \text{or} \qquad It = CV \tag{4-2}$$

Combining Eq. (4-2) with Eq. (2-9), $Q = It$, we get

$$Q = CV \tag{4-3}$$

Equation (4-3) gives the important relationship for the charge held by a capacitor.

In Eqs. (4-2) and (4-3),

C = capacitance, F
I = constant charging current, A
Q = charge, C
t = time of charge, s
V = voltage at end of time t, V

EXAMPLE 4-1 If an uncharged 2-μF capacitor is connected to a 20-mA current source, what voltage is developed across the capacitor after *(a)* 1 μs, *(b)* 10 ms, and *(c)* 100 ms?

Solution (See Fig. 4-4.)

An uncharged capacitor has no energy stored ($W = 0$, $Q = 0$, $V = 0$). Charging increases the amount of energy stored.

$$It = CV \tag{4-2}$$

$$V = \frac{It}{C} = \frac{I}{C} \times t$$

$$V = \frac{20\text{ mA}}{2\ \mu\text{F}} \times t = 10^4\, t$$

or the voltage rises steadily at the rate of 10 kV per second, 10 V per millisecond, 10 mV per microsecond.

a. After time $t = 1\ \mu$s,

$$\mathbf{V = 10\ mV}$$

b. After time $t = 10$ ms,

$$\mathbf{V = 100\ V}$$

c. After time $t = 100$ ms,

$$\mathbf{V = 1\ kV}$$

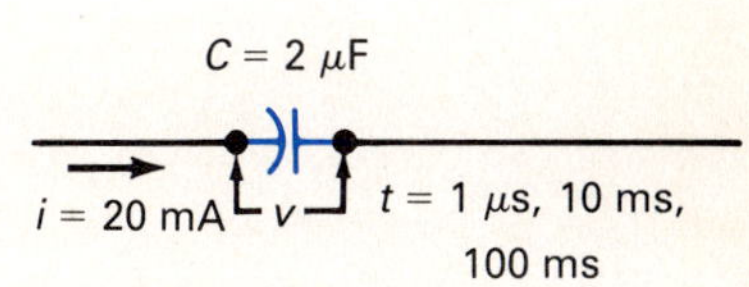

Fig. 4-4. Example 4-1: Data.

EXAMPLE 4-2 The voltage across a 2-μF capacitor rises at a uniform rate from 10 to 60 V in 5 ms. What is *(a)* the change in charge at the capacitor plates and *(b)* the circuit current?

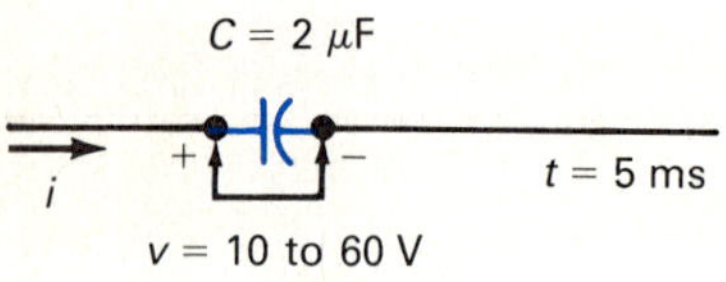

Fig. 4-5. Example 4-2: Data.

Solution (See Fig. 4-5.)

In this example, the voltage rises in a *linear manner* (at a *uniform rate*), giving a steady current. Under such conditions, Eq. (4-3) may be directly applied, the voltage being the 50-V difference between the initial 10 V and the final 60 V. This 50 V is also the *change* of *voltage* in the expression dv/dt in Eq. (4-1).

a.

$$Q = CV \tag{4-3}$$

$$Q = 2\ \mu\text{F} \times 50\ \text{V} = \mathbf{100\ \mu C}$$

b.

$$i = C\frac{dv}{dt} \tag{4-1}$$

$$i = 2\ \mu\text{F} \times \frac{50\ \text{V}}{5\ \text{ms}} = 0.02\ \text{A or } \mathbf{20\ mA}$$

Discharging

When a capacitor is discharging because it is supplying a constant current to a network, there will also be a linear relationship between the capacitor voltage and time. Now, however, the voltage will decrease because energy is being extracted from the capacitor.

Figure 4-6 shows the linear relationship between decreasing capacitor voltage and time when a 2-μF capacitor is discharged at a constant current of 20 mA, the capacitor having been previously charged to 1 kV.

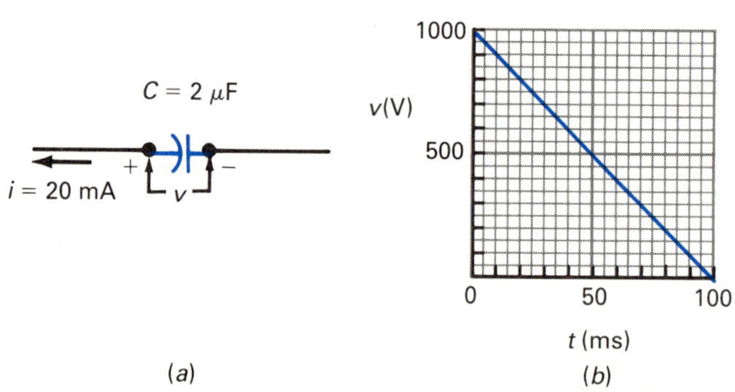

Fig. 4-6. Constant current discharge of a capacitor. *(a)* Circuit conditions; *(b)* graph relating voltage and time.

Comparison of Figs. 4-3*a* and 4-6*a* shows that the voltage polarities are the same but the current directions are opposite. This observation implies that

1. While charging, a capacitor's voltage polarity will be the same as for a load carrying the same direction of current (current from positive to negative within the load).
2. If the charged capacitor is then discharged, its voltage polarity will not change, but its current direction will; it will be the same as for a voltage source (negative to positive within the source).

The voltage polarities and current directions are consistent with the notion that when the capacitor is being charged, it is drawing energy from some source, so it is a load; when it is being discharged, it is supplying energy to a network, so it is a source. Remembering this will help you to avoid confusion in dealing with capacitive current directions and voltage polarities. Figure 4-7 shows these charge and discharge conditions in circuit form.

It is important to note that Fig. 4-7 is only intended to show current directions and voltage polarities. Charging and discharging a capacitor through a resistor will not normally result in a constant current value such as that referred to in Figs. 4-3 and 4-6.

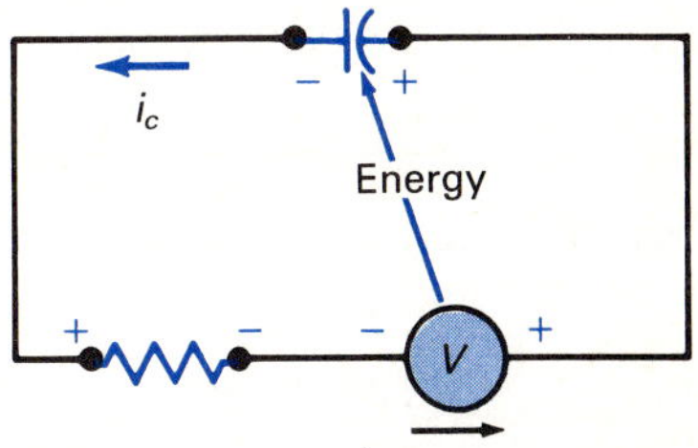

(*a*)

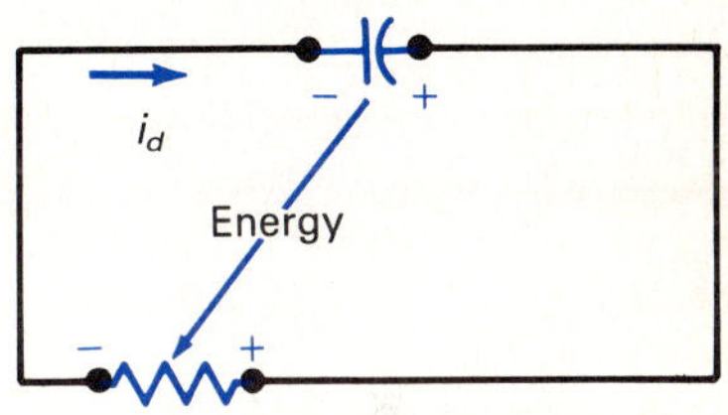

(*b*)

Fig. 4-7. (*a*) Energy direction to the capacitor during charge; the capacitor is a load. (*b*) Energy direction from the capacitor during discharge; the capacitor is an energy source.

REVIEW QUESTIONS FOR SECTION 4-3

1. What rate of change of voltage results from charging a capacitor with a constant current?
2. What is the equation relating capacitance, voltage, and charge?
3. When a capacitor is discharging, what is the capacitive voltage polarity compared with the polarity when it is being charged?
4. When a capacitor is discharging, what is the current direction compared with that when it is being charged?

EXERCISE PROBLEMS FOR SECTION 4-3

1. A 5-mA constant current charges a capacitor for 1 ms. What is the voltage across the capacitor if the capacitor was initially fully discharged and the value of the capacitor is:
   a. 0.05 μF (*ans*. 100 V)
   b. 0.1 μF (*ans*. 50 V)
   c. 0.5 μF (*ans*. 10 V)
2. Find the average current when the voltage across a 1-μF capacitor changes from zero to 15 V in:
   a. 10 ms
   b. 1 ms
   c. 0.1 ms
3. Find the time required for a 1-nF capacitor to charge to 10 V if the charging current is:
   a. 10 μA (*ans*. 1 ms)
   b. 1 mA (*ans*. 10 μs)
4. Find the quantity of charge required to produce the given voltages across the following capacitors:
   a. $C = 1\ \mu F$, $V = 10$ V
   b. $C = 0.05\ \mu F$, $V = 50$ V
   c. $C = 16\ \mu F$, $V = 300$ V
   d. $C = 0.001\ \mu F$, $V = 10$ V
   e. $C = 500\ \mu F$, $V = 3$ V
   f. $C = 68$ pF, $V = 6$ V
   g. $C = 470$ pF, $V = 12$ V
   h. $C = 220$ pF, $V = 50\ \mu V$

5. What would the voltage across the following capacitors be when the charge indicated is present?
   a. $C = 0.1\ \mu F$, $Q = 3 \times 10^{-6}$ C (*ans*. 30 V)
   b. $C = 220$ pF, $Q = 0.001\ \mu C$ (*ans*. 4.545 V)
   c. $C = 15$ pF, $Q = 0.001\ \mu C$ (*ans*. 66.67 V)
   d. $C = 2000\ \mu F$, $Q = 0.01$ C (*ans*. 5 V)
   e. $C = 10$ nF, $Q = 1\ \mu C$ (*ans*. 100 V)
   f. $C = 0.0001\ \mu F$, $Q = 1$ nC (*ans*. 10 V)
6. A 0.2-μF capacitor being charged from a constant-current source has a voltage of 15 V across it after 100 μs. Find the charging current. The capacitor was initially fully discharged.
7. Find the value of current from a constant-current source which will charge a 1500-pF capacitor of a linear ramp generator to 0.75 V in 20 μs. (*ans*. 56.25 μA)
8. Find the time required to charge a 0.05-μF capacitor from 0 to 20 V from a 15-mA constant-current source.
9. Find the time required to increase the voltage across a 20-μF capacitor from 10 to 12 V using a 25-mA constant-current source. (*ans*. 1.6 ms)
10. If the voltage across a 0.02-μF capacitor is changing at a rate of 50 kV/s, find the instantaneous current.
11. If the voltage across a 100-pF capacitor is changing at a rate of 10 MV/s, find the instantaneous current. (*ans*. 1 mA)
12. If the voltage across a 100-pF capacitor is changing at a rate of 10 MV/s, how long will it take for the voltage to rise to 20 V from zero?
13. A 0.002-μF capacitor discharges at a rate of 100 kV/s for a period of 5 μs. Find the quantity of charge lost by the capacitor during this period. (*ans*. 10^{-9} C)
14. A 50-μF capacitor has a voltage of 25 V. Find the time required for the capacitor to discharge to 0 V, if the discharge current is a constant 5 mA.

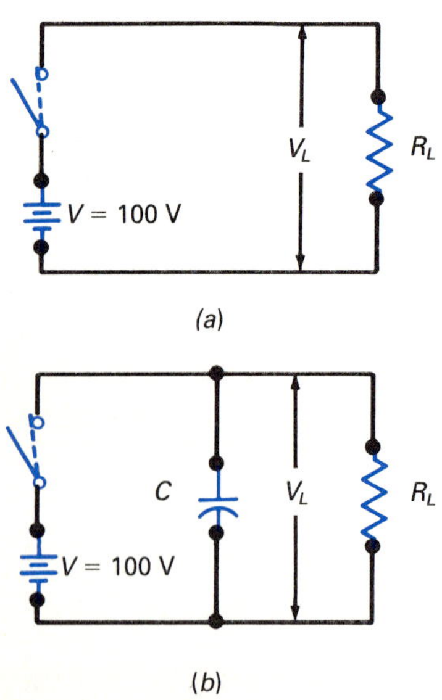

Fig. 4-8. Ripple filter. *(a)* Resistive load; *(b)* capacitor connected across the resistive load.

4-4 OPPOSITION TO VOLTAGE CHANGE

Capacitance is sometimes described as the opposition to voltage change. This notion is based on the following reasoning:

$$i = C\frac{dv}{dt} \quad \text{or} \quad \frac{dv}{dt} = \frac{i}{C} \tag{4-1}$$

For a given value of current the rate of change in voltage across a capacitor is inversely proportional to the capacitance value. In other words, for a stated charge or discharge current, capacitive voltage will change more slowly with a large-value capacitor than with one of small value.

A practical example of the application of this view of capacitance is the ripple filter.

In Fig. 4-8*a* a 100-V emf is applied to a load through a switch. The switch is closed and opened at regular time intervals.

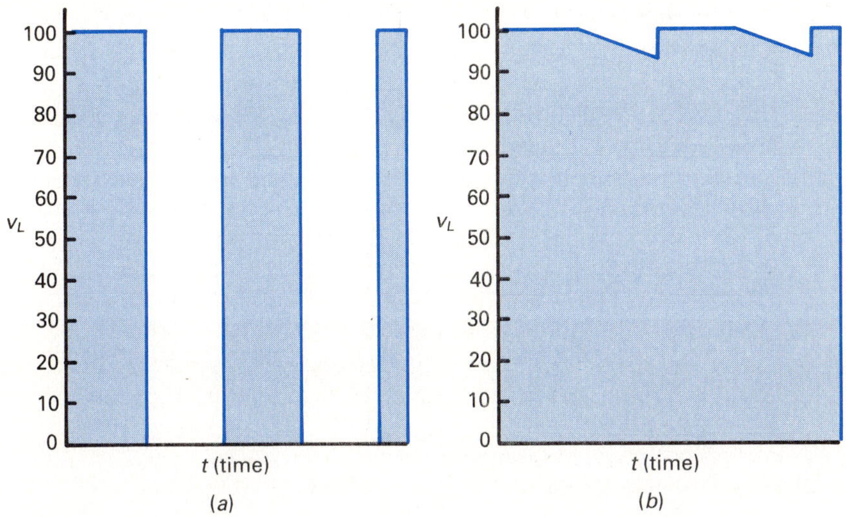

Fig. 4-9. Load voltage waveforms in Fig. 4-8: *(a)* Without the capacitor; *(b)* with the capacitor.

The resulting load voltage is shown in Fig. 4-9*a*. In Fig. 4-8*b* a capacitor of large value is connected across the load. The charge stored in the capacitor maintains load voltage relatively constant, as shown in Fig. 4-9*b*. The details of operation of this circuit are given later.

EXERCISE PROBLEMS FOR SECTION 4-4

1. A 20-V power source with a capacitor ripple filter supplies a constant 20 mA to a load. If the supply voltage is interrupted for 5 ms at 8.33-ms intervals, find the drop in voltage across the filter during the 5 ms if the capacitor value is:

a. 50 μF (*ans*. 2 V)
b. 100 μF (*ans*. 1 V)
c. 500 μF (*ans*. 0.2 V)

2. For the conditions given in Prob. 2, find the value of capacitor that would keep the voltage within 0.2 percent of the desired voltage.

4-5 CAPACITIVE ENERGY STORAGE

Energy is stored in a charged capacitor even if the charging source is disconnected; it is stored in the electric field between the plates. The energy so stored can be extracted from the capacitor by connecting a circuit to its terminals. If the circuit is resistive, the extracted energy will be dissipated into the environment as heat. If the circuit is capacitive or inductive, the energy will be re-stored in an electric or magnetic field.

When a capacitor is being charged, the energy transfer can be made apparent by current and voltage measurements. The circuit current, existing for a certain time, indicates the amount of charge imparted to the capacitor [$Q = It$, Eq. (2-9)]. This charge, coupled with the voltage developed between the plates, is a measure of the energy transferred and stored [$W = VQ$, Eq. (2-6)].

If a capacitor is charged with constant current I, the voltage v between its plates at any time instant will increase at a uniform rate, as shown in Fig. 4-10.

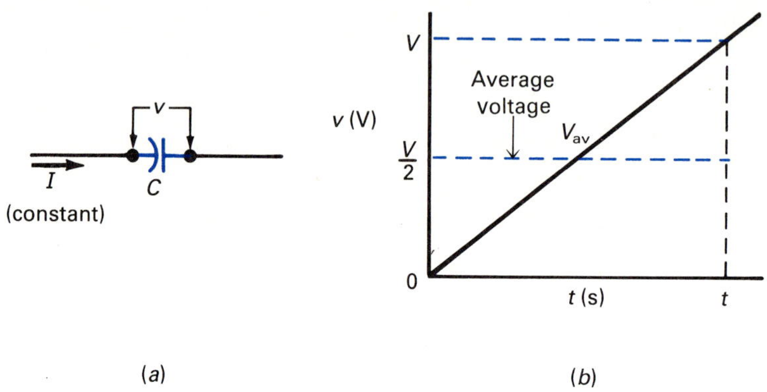

Fig. 4-10. Capacitive energy. *(a)* Circuit conditions; *(b)* graph of voltage with respect to time.

Equation (2-6) ($W = VQ$) gives energy in terms of a constant voltage. When the voltage increases from zero, the energy stored per unit of time also increases from zero. The total energy transferred to the capacitor during a given time will be determined by the average voltage. Since the voltage increases uniformly with constant charge current, the average voltage is one-half the final voltage.

Where the final voltage is V after a time period t seconds, as in Fig. 4-10, the average voltage will be

$$V_{av} = \frac{V}{2}$$

Using Eq. (2-6), the charging energy will be

$$W = V_{av}Q = \frac{V}{2}Q$$

Substituting for Q from Eq. (4-3) gives

$$W = \frac{V}{2}VC$$

$$W = \frac{V^2C}{2} \qquad \textbf{(4-4)}$$

where W = energy stored in the capacitor, J
V = voltage, V
C = capacitance, F

EXAMPLE 4-3 A 5-μF capacitor is charged until its terminal voltage is 100 V. What is the amount of energy stored in it?

Solution (Fig. 4-11)

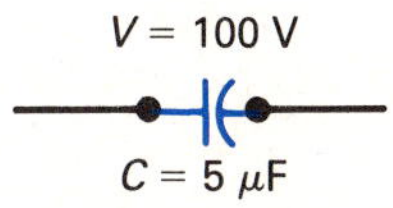

Fig. 4-11. Example 4-3: Data.

$$W = \frac{V^2C}{2} \qquad (4\text{-}4)$$

$$W = \frac{(100\text{ V})^2 \times 5\ \mu\text{F}}{2} = \mathbf{0.025\ J} \quad \text{or} \quad \mathbf{25\ mJ}$$

EXAMPLE 4-4 An electronic power supply has a 30 000-μF capacitor in its filter circuit. If it is charged by a constant current of 100 mA until it stores 50 J of energy, *(a)* what is the capacitor voltage at the end of the charge, and *(b)* how much time does the charging process take?

Solution (Fig. 4-12)

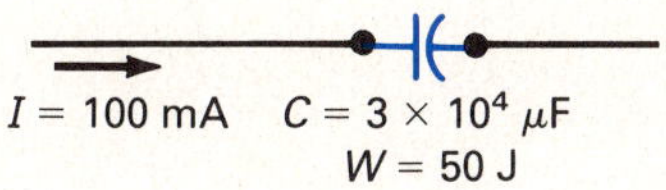

Fig. 4-12. Example 4-4: Data.

a.

$$W = \frac{V^2C}{2} \qquad (4\text{-}4)$$

$$V = \sqrt{\frac{2W}{C}} = \sqrt{\frac{2 \times 50\text{ J}}{3 \times 10^{-2}\text{ F}}} = \mathbf{57.74\ V}$$

b.

$$W = VQ$$

$$Q = \frac{W}{V} = \frac{50\text{ J}}{57.74\text{ V}} = 0.866\text{ C}$$

$$Q = It$$

$$t = \frac{Q}{I} = \frac{0.866\text{ C}}{0.1\text{ A}} = \mathbf{8.66\ s}$$

EXAMPLE 4-5 Figure 4-13 is the circuit of a basic ramp generator (a generator of sawtooth voltages). During the 10-ms periods that the input voltage v_i is negative, the output voltage v_o (which is also the capacitor voltage) rises from approximately zero toward the

25 V of the supply. As v_o rises, the capacitor charges until its voltage at the end of 10 ms is 5 V. If it can be assumed that the increase in v_o is linear from 0 to 5 V, *(a)* what is the charging current, and *(b)* how much energy is stored in the capacitor at the end of the 10-ms charge?

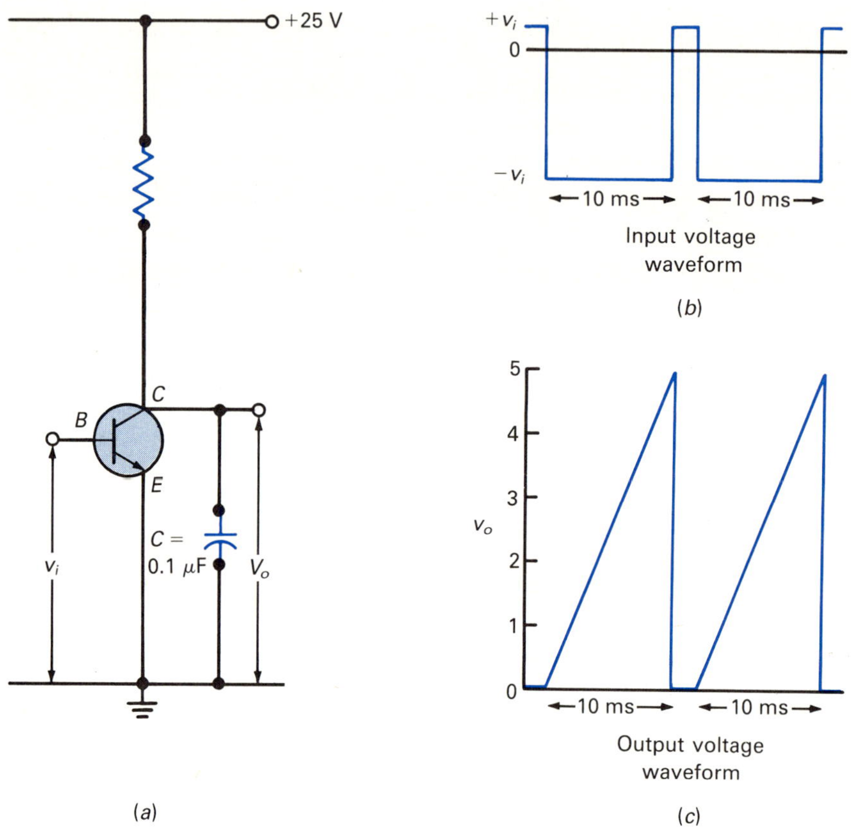

Fig. 4-13. Example 4-5: *(a)* Circuit; *(b)* input voltage waveform; *(c)* output voltage waveform.

Solution *a.* At the end of the 10-ms period, $v_o = 5$ V. Since the voltage increase is assumed to be linear, the rate of change in voltage will be

$$\frac{dv_o}{dt} = \frac{5 \text{ V}}{10 \text{ ms}} = 500 \text{ V/s}$$

For a linear rise in voltage the charging current is constant. From Eq. (4-1),

$$I = C\frac{dv_o}{dt} = 0.1\ \mu\text{F} \times 500 \text{ V/s} = \mathbf{50\ \mu A}$$

b. The energy stored will increase over the 10-ms period, becoming, from Eq. (4-4),

$$W = \frac{V_o{}^2C}{2} = \frac{(5\text{ V})^2 \times 0.1\ \mu\text{F}}{2} = \mathbf{1.25\ \mu J}$$

Ramp generators are widely used to provide constant-velocity electron-beam deflection in cathode-ray tubes. The beam motion causes a spot of light to move steadily across the screen of the tube, usually so fast that it appears to be a line of light. This line of light is called a *trace*.

In an oscilloscope the trace forms the horizontal motion on which is imposed a vertical motion related to the waveform being viewed. The horizontal and vertical motions result in a *display* of the waveform shape.

REVIEW QUESTIONS FOR SECTION 4-5

1. Where in a capacitor is energy stored?
2. What is the equation for the energy stored in a capacitor?

EXERCISE PROBLEMS FOR SECTION 4-5

1. In Prob. 4, Sec. 4-3, find the energy stored in each capacitor. (*ans. a.* 50 μJ; *b.* 62.5 μJ; *c.* 0.72 J; *d.* 0.05 μJ; *e.* 2.25 mJ; *f.* 1224 pJ; *g.* 0.0338 μJ; *h.* 275×10^{-21} J)
2. In Prob. 5, Sec. 4-3, find the energy stored in each capacitor.
3. In Prob. 6, Sec. 4-3, find the energy stored in the capacitor after
   a. 50 μs (*ans.* 5.625 μJ)
   b. 100 μs (*ans.* 22.5 μJ)
4. In Prob. 9, Sec. 4-3, find the energy required to increase the voltage from 10 to 12 V.
5. A disk capacitor with a disk separation of 0.318 cm has a capacitance of 5 pF. Find the energy stored if the capacitor is charged to 2000 V. (*ans.* 10 μJ)

4-6 CAPACITORS

A *capacitor* is a circuit component designed to have a known value of capacitance.

In one form or another, a capacitor consists of the two conductive plates separated by an insulator referred to previously. The plates are usually metal sheet or foil. The insulator, known as the *dielectric,* may be one of many materials, such as plastics, ceramics, oil, or air.

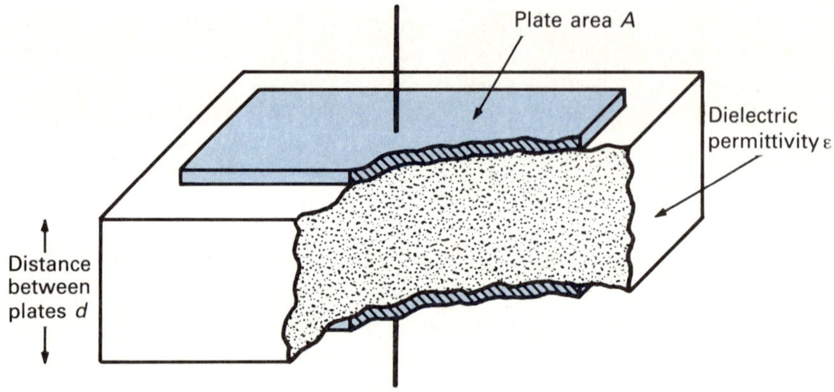

Fig. 4-14. Factors controlling capacitance.

Figure 4-14 illustrates the concept that the dimensions of a capacitor control its capacitance.

The physical size of a capacitor does affect its capacitance. In Fig. 4-14, the main dimensions are shown as the area of each plate and the distance between them. The capacitance is given by

$$C = \frac{\epsilon A}{d} \tag{4-5}$$

where C = capacitance, F
ϵ = dielectric permittivity, F/m*
A = plate area, m^2
d = distance between plates, m

Permittivity

Capacitors having the same plate area and distance between plates will have different capacitances if the material between the plates is different. A capacitor with the plastic polystyrene between its plates will have about 2.5 times the capacitance of the same size capacitor with air between its plates. This difference is due to the *permittivity* of polystyrene being about 2.5 times that of air. Permittivity might be described as *specific capacitance,* in much the same way that resistivity is specific resistance.

To make the capacitance unit, the farad, fit the SI unit system without complex multipliers, the permittivity of a material is given in terms of two factors. One is the permittivity of a vacuum (called the *permittivity of free space*). It gives the capacitance of two plates with nothing between them. This factor is given the symbol ϵ_v (ϵ is the Greek letter epsilon) and has the fixed value 8.854×10^{-12} $[1/(36\pi \times 10^9)]$ farads per meter. With air as the dielectric, the capacitance would be almost the same as with a vacuum. The other factor is the *relative permittivity* of the dielectric material being used. It gives the amount by which capacitance is increased by using that material between the plates instead of a vacuum. Its symbol is ϵ_r. Being a comparative quantity, relative permittivity has

*The unit of permittivity is the *farad per meter* (F/m).

no unit. For example, in the case quoted above, the relative permittivity of polystyrene is 2.5; it gives 2.5 times the capacitance that would be obtained with a vacuum as the dielectric.

The values of relative permittivity or *dielectric constant* for some dielectric materials are listed in Table 4-1.

TABLE 4-1 Permittivities of Dielectric Materials*

Dielectric Material (Insulator)	Relative Permittivity ϵ_r (or Dielectric Constant)	Total Permittivity ϵ (F/m) $\epsilon = \epsilon_v \epsilon_r$ or $\epsilon = 8.854 \times 10^{-12} \times \epsilon_r$
Vacuum	1	8.854×10^{-12}
Air	1.001	8.855×10^{-12}
Ceramic	6 to 2000	53.12×10^{-12} to 17.71×10^{-9}
Mica	3 to 7	26.56×10^{-12} to 61.98×10^{-12}
Mylar	3	26.56×10^{-12}
Oil	4	35.42×10^{-12}
Paper	4	35.42×10^{-12}
Polystyrene	2.5	22.14×10^{-12}
Tantalum oxide	11	97.39×10^{-12}

*In Eq. (4-5) for calculating the value of capacitance, the value of permittivity ϵ for a dielectric material is that shown in the last column of this table.

The value of permittivity ϵ in Eq. (4-5) is found by

$$\epsilon = \epsilon_v \epsilon_r \tag{4-6}$$

where ϵ = total permittivity, F/m
ϵ_v = permittivity of free space, 8.854×10^{-12} F/m
ϵ_r = relative permittivity

EXAMPLE 4-6 Find the capacitance of a capacitor made from two strips of aluminum foil, each 5 cm wide and 10 m long, separated by a 25-μm-thick film of Mylar. The relative permittivity of Mylar is 3.

Solution Commercial capacitors are often made by forming a sandwich of aluminum foil and insulating material and then rolling this sandwich into a cylindrical shape before enclosing it in plastic or a metal can (Fig. 4-15).

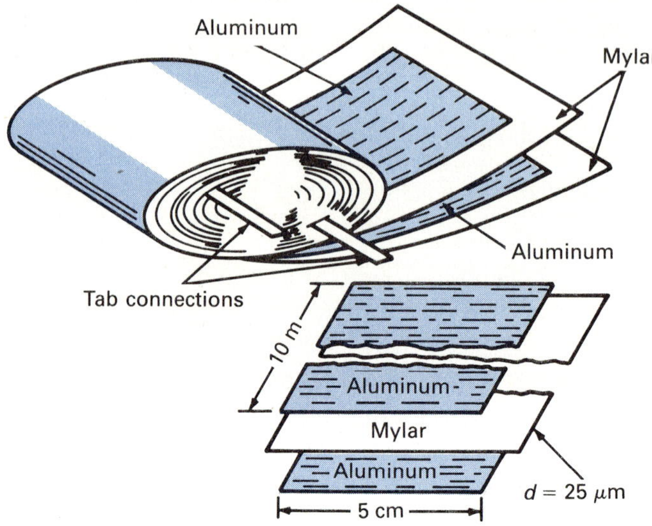

Fig. 4-15. Example 4-6: Capacitor dimensions.

Plate area $\qquad A = 5\text{ cm} \times 10\text{ m} = 0.5\text{ m}^2$

From Eqs. (4-5) and (4-6),

$$C = \frac{\epsilon A}{d} = \frac{\epsilon_v \epsilon_r A}{d}$$

$$C = \frac{3 \times 8.854 \times 10^{-12}\text{ F/m} \times 0.5\text{ m}^2}{2.5 \times 10^{-5}\text{ m}} = \mathbf{0.5312\ \mu F}$$

Thus this capacitor made with foil plates about 2 in wide and 10 yd long, with a dielectric of Mylar film about 1 mil ($\frac{1}{1000}$ in) thick, has a capacitance of approximately 0.5 μF, a common value.

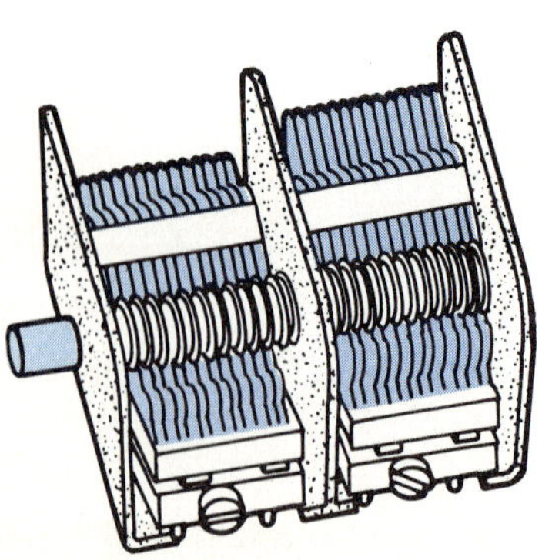

Fig. 4-16. Example 4-7: Variable capacitor.

EXAMPLE 4-7

The two-gang (two-section) variable capacitor shown in Fig. 4-16 is typical of those used to tune radio receivers. The larger of the two sections consists of 14 moving (rotating) plates and 13 fixed plates. As the moving plates rotate, they mesh with the fixed plates, leaving a 1-mm gap between adjacent plates. The dielectric between the plates is air ($\epsilon_r = 1.0001$). What must be the area of the field between adjacent plates if the maximum capacitance is to be 300 pF?

Solution

When the moving plates of a variable capacitor are rotated through 180°, the meshing plate area, and hence the electric field area and the capacitance, varies between maximum and minimum values.

Maximum capacitance occurs when the plates are fully meshed, that is, when the field area is greatest. The following calculation is based on the maximum-capacitance position of the plates:

$$C = \frac{\epsilon A}{d} \qquad (4\text{-}5)$$

$$A = \frac{Cd}{\epsilon} = \frac{Cd}{\epsilon_v \epsilon_r} = \frac{300 \text{ pF} \times 1 \text{ mm}}{8.854 \times 10^{-12} \text{ F/m} \times 1.0001}$$

$$A = 0.0339 \text{ m}^2$$

A is the total area between all adjacent plates (assuming that plate and field area are the same). The area between any adjacent pair of plates is the total area A divided by the number of adjacent pairs. The number of adjacent pairs of plates of a variable capacitor is the total number of plates minus one. The two outside plates each contribute only one surface to the electric field, whereas all other plates have two effective surfaces.

In this example there are 26 adjacent pairs of plates.

$$A_p = \frac{A}{26} = \frac{0.0339 \text{ m}^2}{26} = \mathbf{0.0013\ m^2} \quad \text{or} \quad \mathbf{13\ cm^2}$$

Usually the plates of a variable capacitor are a special shape designed to give a desired relationship between the angular position of the moving plates (relative to the fixed plates) and the capacitance. However, the shape is near enough to a semicircle to enable us to visualize the size of the variable capacitor in Example 4-7.

The area of a semicircle is

$$A = \frac{\pi r^2}{2}$$

The radius will be

$$r = \sqrt{\frac{2A}{\pi}} = \sqrt{\frac{2 \times 13 \text{ cm}^2}{\pi}} = 2.877 \text{ cm}$$

Therefore this ordinary radio tuning capacitor will have plates of the order of a 1-in radius.

REVIEW QUESTIONS FOR SECTION 4-6

1. Describe the basic construction of a capacitor.
2. What is the insulating separator in a capacitor called?
3. State four materials that are used for dielectrics.
4. Do all capacitors with the same plate area and distance between plates have the same capacitance?

5. How does the permittivity of the dielectric affect capacitance?
6. How might permittivity be described?
7. What are the two components determining the permittivity of a material?
8. What is the symbol for the permittivity of a vacuum or free space?
9. How can relative permittivity be described, and what is its symbol?
10. What is another name for relative permittivity?
11. What is the equation for the permittivity of a given dielectric?

EXERCISE PROBLEMS FOR SECTION 4-6

1. Find the capacitance of a capacitor made from two strips of foil 2 cm wide and 5 m long using a dielectric of polyethylene 35 μm thick ($\epsilon_r = 3.3$). (*ans.* 0.0835 μF)
2. Find the capacitance of a capacitor made of two disks with diameters of 5.08 cm separated by an air gap of 0.318 cm ($\epsilon_r = 1.0001$).
3. Find the capacitance of a capacitor made of 29 sheets of foil, 1×2 cm, separated by 15-μm sheets of mica, with a relative permittivity of 6. Construction is such that both sides of all foil sheets except the outside layers are used. (*ans.* 0.0198 μF)
4. A multilayer capacitor to be made from aluminum foil 30 μm thick and sheet Mylar 25 μm thick is to have a capacitance of 0.02 μF $\pm$ 5% and a dimensional area of 2×5 cm. Find the minimum number of plates required.
5. Two sheets of foil 5×5 cm are separated by a Mylar dielectric. If the capacitance is 0.005 μF, what is the thickness of the dielectric? (*ans.* 13.28 μm)
6. If after the capacitor in Prob. 5, Sec. 4-5, has been charged, the separation of the disks is increased to 0.636 cm without changing the charge Q, find
   a. The new value of capacitance.
   b. The new voltage across the capacitance.
   c. The new quantity of energy stored in the capacitance.

4-7 CAPACITOR TYPES

There are many different types of capacitors of which *fixed* and *variable* form two main groups.

Fixed Capacitors

If its capacitance cannot be deliberately controlled, a capacitor is called *fixed*. Like the value of any other circuit component, the capacitance of a fixed capacitor may change with variations of the conditions in which it operates, such as temperature. (See Fig. 4-17.)

Although appearance and methods of manufacture account for some of their differences, the various types of fixed capacitors are usually distinguished by their dielectric material: air, paper, oil, tantalum oxide, mica, etc. Air-dielectric capacitors are very stable in capacitance value and operate well at high frequencies. Electrolytic capacitors have an extremely thin oxide film (e.g., tantalum

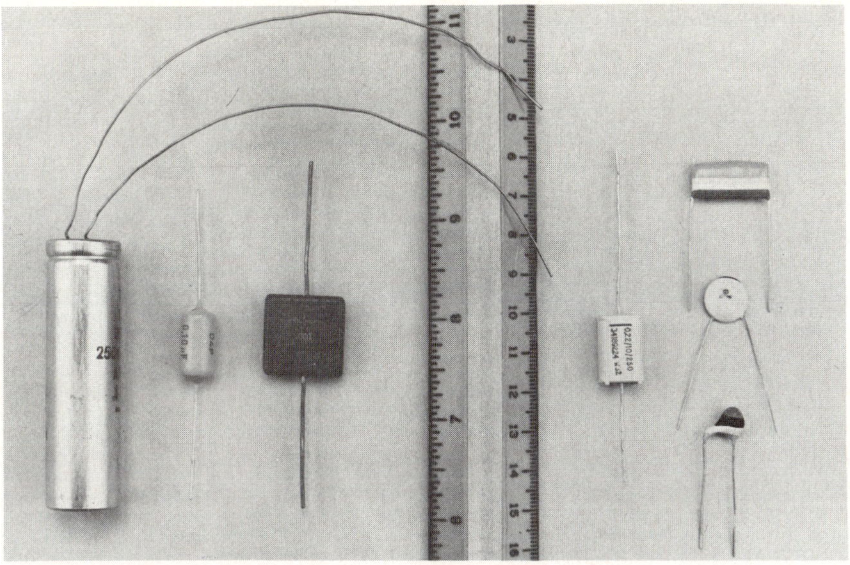

Fig. 4-17. *(Left to right)* Fixed capacitors: Electrolytic, Mylar, mica, polystyrene, and ceramic.

oxide) as a dielectric and thereby get high capacitance in a small space. But electrolytic capacitors have poor high-frequency performance. Capacitors made with very high relative permittivity ceramic dielectric may be used for fairly high capacitance in high-frequency circuits. Certain oils have good characteristics as dielectrics in capacitors for high-voltage circuits. The oil may flow to seal a puncture by an over-voltage breakdown of the dielectric. Even a vacuum, literally nothing, is a useful high-voltage capacitor dielectric for some applications.

Capacitors may be made in the form of the one in Example 4-6, where the plates and the dielectric are like a rolled-up sandwich. This construction, with variations, such as depositing the plates on the dielectric as a film, is popular for use with various dielectric materials. Where the dielectric cannot be rolled up, as with mica sheet, a multilayer flat sandwich can be made, with the plates as separate sheets of metal or deposited as films. Where the dielectric material lends itself to the method, it may be molded into a convenient shape and metal films formed on its surface. Ceramic capacitors are often made this way. (See Fig. 4-18.)

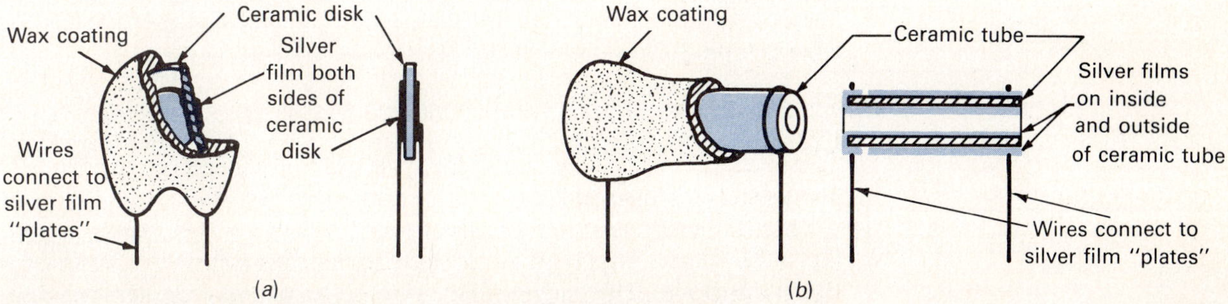

Fig. 4-18. Ceramic dielectric capacitors: *(a)* Ceramic disk capacitor; *(b)* ceramic tube capacitor.

Variable Capacitors

Just as there are different types of fixed capacitors, those whose capacitance value is intended to be constant, so there are different types of *variable* capacitors, those intended to have the capacitance value controllable over a known range. (See Fig. 4-19.)

Fig. 4-19. *(Left to right)* Variable capacitors: Air dielectric, trimmers—mica, Mylar, and ceramic dielectric.

Air-dielectric variable capacitors, such as the type in Example 4-7, have been used for tuning transmitters and receivers since the beginnings of radio. They are still popular, as they are reliable and stable and can be made with almost any desired relationship between the angle of shaft rotation and the capacitance. Variable capacitors, screwdriver adjusted, are used as *trimmers,* giving small, preset values of capacitance. These trimmers may have air, mica, or ceramic materials as the dielectric.

For an increasing number of purposes, *special semiconductor diodes,* electronic components, are used as *voltage-variable* capacitors or *varactors.* The capacitance of a varactor changes as a variable voltage changes the effective thickness of the dielectric, an insulating layer within the diode. Many modern television sets have varactor tuning for channel selection.

(*a*)

(*b*)

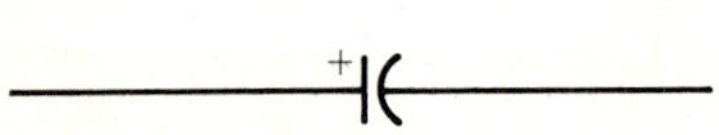

(*c*)

Fig. 4-20. Capacitor symbols: *(a)* General-purpose; *(b)* variable; *(c)* polarized.

Capacitor Symbols

The schematic symbols for capacitors are shown in Fig. 4-20.

For a capacitor where it is not necessary to show the type, the symbol of Fig. 4-20*a* is used. Where it is desired to show that capacitance value is variable, Fig. 4-20*b* applies. The symbol of Fig. 4-20*c* shows that the capacitor must have a voltage across it that causes one plate to always be positive with respect to the other; this usually applies to electrolytic capacitors.

Capacitor Values

Capacitances ranging from a fraction of a picofarad to at least one farad are available in commercial capacitors. This is an enormous range, comparable to a resistance range of one ohm to over one teraohm, or one million million ohms. The capacitances available may follow the same preferred-value scale given for resistors in Table 3-2. The values may be shown by the color code of Fig. 3-12 or they may be printed directly on the capacitor. If color coding is used, the capacitance will be given in picofarads. The colors in Example 3-12 would refer to capacitors of 4.7 pF, 68 pF, 56 nF (56 000 pF), and 3.3 μF (3 300 000 pF).

Due to the many different shapes of capacitors, the colors representing capacitance value may or may not appear as bands. Examples of capacitor color coding are shown in Fig. 4-21.

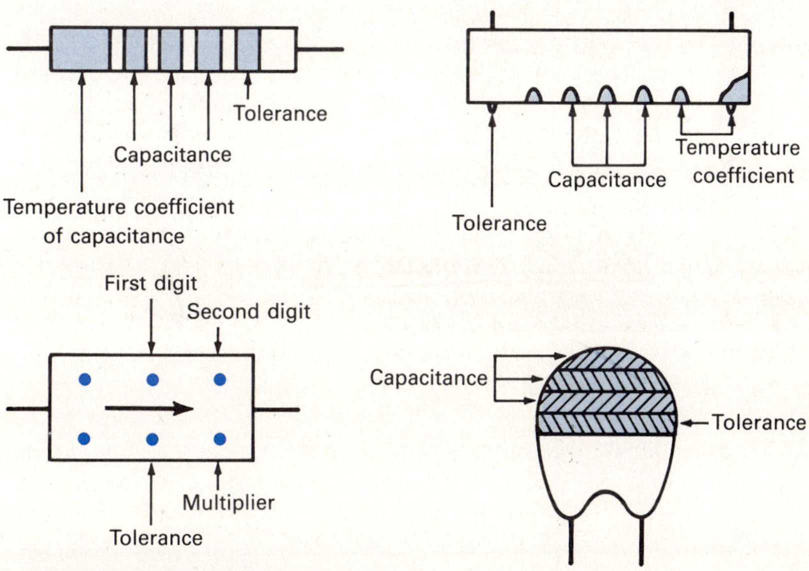

Fig. 4-21. Capacitor color coding.

Characteristics other than capacitance may be shown by color coding. Depending on the capacitor type and application, colors may be assigned for capacitance tolerance, temperature coefficient of capacitance, maximum operating voltage, or other important ratings.

REVIEW QUESTIONS FOR SECTION 4-7

1. List some characteristics of capacitors with the following dielectrics:
   a. Air *b.* Oxide film (electrolytic)
   c. Ceramic *d.* Oil
2. What subunit values are used for capacitors which are color-coded?
3. What other capacitor characteristics may be shown by coded bands or dots on capacitors?
4. What is the name of the semiconductor voltage-variable capacitor?

4-8 STRAY CAPACITANCE

The impression may have been created that capacitance occurs *only* between parallel plates. In fact, it occurs whenever there is a voltage between points or surfaces.

Commercial capacitors are usually of the parallel-plate form, but capacitance is not restricted to capacitors and always exists in circuits. Unintentional capacitance is called *stray capacitance*. Stray capacitances can often be ignored as insignificant in the circuit operation. Equally often such stray capacitances are significant. Whether or not stray capacitance is important depends on the circuit and its operating conditions.

Stray Capacitance in Components

Resistors, inductors, transformers, vacuum tubes, transistors, and other circuit devices have capacitances associated with them. These capacitances, together with varying voltages, cause undesired currents. Two examples, resistors and transistors, will be briefly described in terms of their stray capacitances.

When there is a current I in the resistor in Fig. 4-22*a*, there is a voltage V across it. Thus there is an electric field along the resistor's length, and so there is capacitance between its ends. For most commercial resistors the capacitance value is very small, fractions of a picofarad. The effect of so small a capacitance is not great, except at very high ac frequencies.

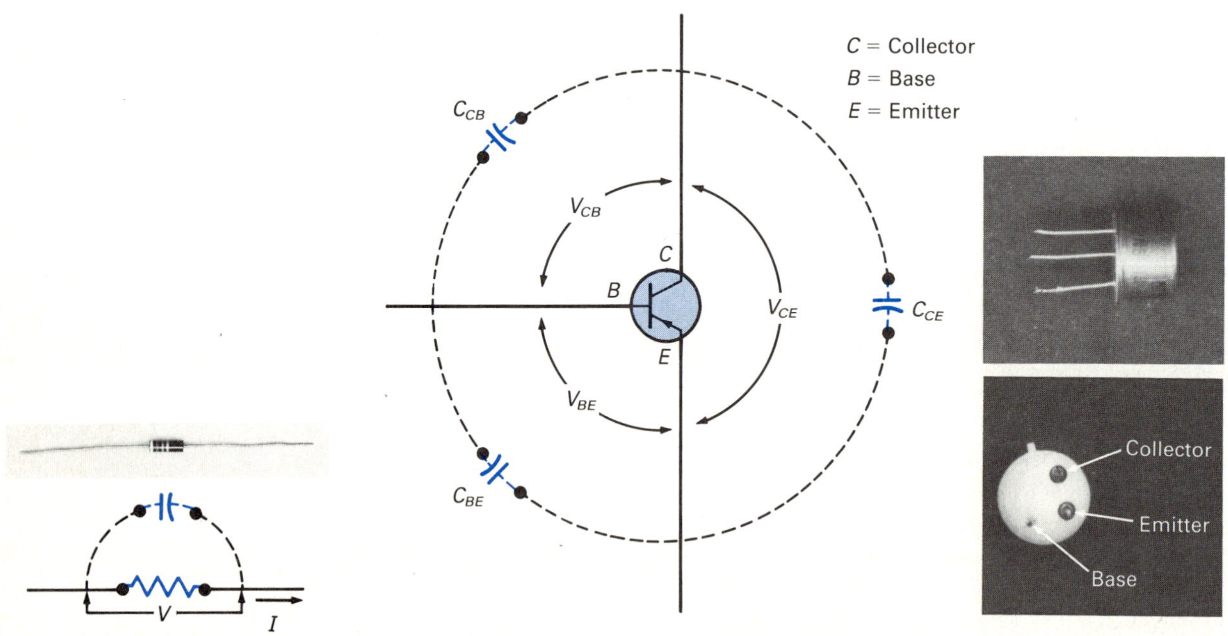

(*a*) (*b*)

Fig. 4-22. Stray capacitance: *(a)* Resistor; *(b)* transistor.

The operating internal voltages of a transistor are shown in Fig. 4-22*b*. Electric fields exist between the collector, base, and emitter, so capacitances exist there too. They are the collector-to-emitter capacitance C_{CE}, the collector-to-base capacitance C_{CB}, and the base-to-emitter capacitance C_{BE}. Transistor capacitances, although only a few picofarads, can significantly affect circuit performance. These capacitances can be so important that transistor manufacturers specify them in their published data, and circuit designers must determine their effects on equipment performance.

Transmission-Line and Cable Capacitance

Electrical energy is usually transmitted by metallic conductors insulated from each other or from ground. There will always be capacitance in such transmission systems, its effect being small or great depending on the circumstances.

Figure 4-23*a* shows a power line, or a telephone line, with two conductors suspended above the ground. There will be capacitance between the conductors and between each conductor and ground. As it is common for such lines to be very long, there can be considerable capacitance, which will influence the line characteristics.

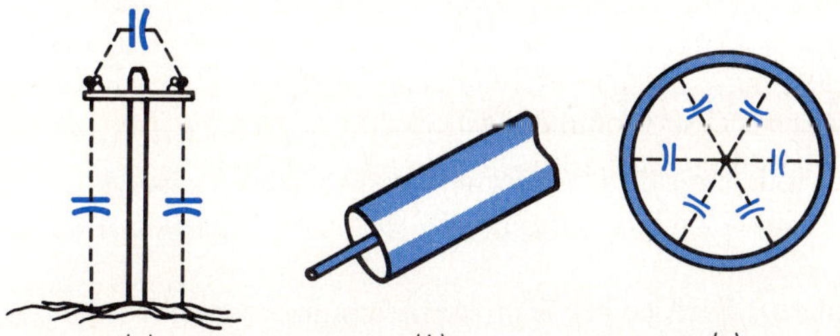

(*a*) (*b*) (*c*)

Fig. 4-23. Stray capacitance in lines and cable.

Figure 4-23*b* shows, in pictorial form, a coaxial cable consisting of an inner conductor suspended in the center of an outer tubular conductor. As in Fig. 4-23*c*, there will be capacitance between the two conductors. Coaxial cables are used extensively for high-frequency applications in which capacitance is an important parameter involved in the transmission of the energy.

Stray Capacitance in Equipment

There is always capacitance between conducting surfaces operating at different voltage levels. Therefore, in electrical equipment there is capacitance between all wires, between all conductors on printed circuit boards, between all terminals, between all components, and between all such conducting surfaces and ground. The last example is significant where, as is usual, the equipment is housed in grounded metal enclosures. There are a great many capacitances present in even a simple item of equipment. The circuit designer must be aware of all these capacitances and allow for them in the design.

The significance of stray capacitance will become more apparent during the study of ac circuits in succeeding chapters.

REVIEW QUESTIONS FOR SECTION 4-8

1. What is the term given to capacitance that is not intentionally put into a circuit in the form of a capacitor?
2. Where in electronic equipment might you expect to find the type of capacitance referred to in the previous question?

SUMMARY OF EQUATIONS

$$i = C\,dv/dt \tag{4-1}$$
$$I = C\,V/t \quad \text{or} \quad It = CV \tag{4-2}$$
$$Q = CV \tag{4-3}$$
$$W = V^2C/2 \tag{4-4}$$
$$C = \epsilon A/d \tag{4-5}$$
$$\epsilon = \epsilon_v \epsilon_r \tag{4-6}$$

CHAPTER QUESTIONS

4-1. Compare the production of a current in a resistor and a capacitor.

4-2. Compare the behavior of a resistor with that of a capacitor from an energy viewpoint.

4-3. What is the main difference between the current in a capacitor when its voltage changes and the current when the voltage stays fixed in value?

4-4. When a capacitor is charged, it stores energy. From where does the energy come?

4-5. What happens to a capacitor's voltage during charge?

4-6. When a capacitor is discharged, it loses energy. Where does the energy go?

4-7. What happens to a capacitor's voltage during discharge?

4-8. What is meant when it is said that a capacitor "opposes voltage change"?

4-9. If the voltage across a capacitor is *doubled,* what happens to its stored *energy?*

4-10. Compare the *energy* in two capacitors charged to the *same voltage* if one has *twice* the *capacitance* of the other.

4-11. Using Table 4-1, compare the capacitance values of similar-dimension capacitors using as dielectrics

a. Air

b. Mylar

c. Tantalum oxide

4-12. If the shaft of a variable capacitor is rotated until its effective plate area is doubled, what happens to its capacitance value?

4-13. Why is the capacitance color code specified in very small unit values?

CHAPTER PROBLEMS

4-1. What is the current in a 2-μF capacitor if the voltage across it rises from zero to 250 V in 50 ms in a linear manner (uniform)? (*ans.* 10 mA)

4-2. What is the current in a 4-μF capacitor if the voltage across it rises from zero to 250 V in 50 ms in a linear manner?

4-3. What is the current in a 2-μF capacitor if the voltage across it rises from zero to 250 V in 100 ms in a linear manner? (*ans.* 5 mA)

4-4. What is the current in a 2-μF capacitor if the voltage across it drops from 250 V to zero in 100 ms in a linear manner?

4-5. How fast will the voltage across a capacitor change if its current is fixed at 10 mA and its capacitance is

a. 2 μF (*ans.* 5 kV/s)

b. 20 μF (*ans.* 500 V/s)

4-6. A 1.25-μF capacitor is to be made by winding into a roll a length of 4-cm-wide Mylar strip, aluminum-coated on each surface. If the Mylar thickness is 25.4 μm (1/1000 in), how much Mylar strip will be needed?

4-7. How much energy is stored in a filter capacitor,

a. For a computer power supply if the capacitance is 60 mF (60 000 μF), operating at 25 V? (*ans.* 18.75 J)

b. For a television picture tube if the capacitance is 0.005 μF, operating at 30 kV? (*ans.* 22.5 J)

4-8. If two small ceramic capacitors, each of exactly the same dimensions, were made from dielectrics at the extremes of the range of permittivity values given in Table 4-1, how would their capacitances compare?

4-9. The lead-in cable from a rooftop antenna to a TV set is 100 ft long (30.48 m). It is made of two stranded copper wires molded into polystyrene insulation that keeps them 1 cm apart. If the wires can be regarded as metal surfaces each 1 mm wide, what is the approximate capacitance of this length of cable? (*ans.* 67.5 pF)

4-10. The manufacture of a semiconductor integrated circuit includes the formation of two metallic silicon surfaces separated by a silicon oxide insulating layer. If the effective area of the silicon surfaces is a square of 1-mm sides and the thickness of the silicon oxide dielectric is 1 μm with a relative permittivity (dielectric constant) of 5, what is the capacitance between the silicon surfaces within the integrated circuit?

5

INDUCTANCE AND INDUCTORS

There is a similarity between inductance and capacitance. Both are involved with energy storage, and both leave power dissipation to the resistance of a circuit. Both are concerned with energy in the form of force fields.

Capacitance is a quantity associated with energy in the form of an electric field. In a somewhat similar way, *inductance* is a quantity related to energy in the form of a *magnetic field*. Just as with a capacitor, energy can be stored in an inductor. Both inductive and capacitive circuits rely on change for the relationship between current and voltage. From a circuit viewpoint, the main difference between inductors and capacitors is that with *inductors, voltage depends on the rate at which current changes,* whereas with capacitors, current depends on the rate at which voltage changes, as was seen in Chap. 4.

People rarely make or repair the resistors and capacitors they use; they usually buy them ready made. Inductors, however, are sometimes made or repaired by hand. The types and values of inductors found in commercial equipment are so varied that a component needed at some time might not be readily available. For this reason, this chapter contains a more detailed introduction to the construction of inductors and the factors controlling their characteristics than was given for resistors and capacitors in Chaps. 3 and 4, respectively.

5-1 INDUCTANCE

Whenever there is an electric current, there is a magnetic field. The magnetic field entirely surrounds a current-carrying conductor with a region in which forces may act. These forces can cause magnetized materials, such as iron, to move into alignment with the field direction. They can cause moving charges to change direction of motion, and they can cause charge movement (current).

These electromagnetic effects are most noticeable when the current-carrying conductor is wound into the form of a coil. In these circumstances, all the magnetic field associated with a long conductor is concentrated into a small space. Such a coil of wire is shown schematically in Fig. 5-1. The switch is open in Fig. 5-1*a*, preventing current in the circuit. When the switch is closed, as in Fig. 5-1*b*, there will be a voltage across the coil and a conducting path through it, but initially there will be no current. The current builds up from zero to a value equal to that which would exist if the resistor R were the only circuit element across the generator. With a current present there is a magnetic field around the coil; it too increases from zero magnitude to a maximum value. The magnetic field thus established is a form of energy which has been transferred from the generator to the coil. This action can be compared with the system in Fig. 4-2, in which energy was transferred from a generator to an electric field between the two plates of a capacitor.

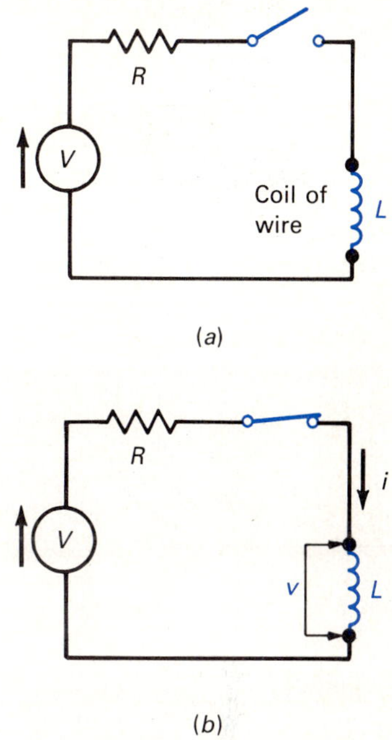

Fig. 5-1. *(a)* Coil of wire without current; *(b)* coil of wire with current.

	Fig. 4-2	Fig. 5-1
Switch open	No current No voltage between the plates No electric field	No current No voltage across the coil No magnetic field
Instant that switch is closed	Current maximum, equals V/R No voltage between the plates No electric field	No current Voltage across the coil maximum, equals V No magnetic field
Switch closed but conditions not stabilized	Current decreasing Voltage between the plates increasing Electric field increasing	Current increasing Voltage across the coil decreasing Magnetic field increasing
Switch closed and conditions stabilized	No current Voltage between the plates maximum, equals V Electric field maximum	Current maximum, equals V/R Voltage across the coil zero Magnetic field maximum

The events in the circuits of Figs. 4-2 and 5-1 are similar but opposite. When conditions have become stabilized, both circuits have maximum energy stored in their fields. The electric field between the capacitor plates is due to its voltage. The magnetic field around the coil is due to its current.

The ability to store energy in a magnetic field is *inductance*. The coil of wire is an *inductor*.

The schematic symbol for an inductor, shown in Fig. 5-1, suggests a coil of wire.

The symbol for the quantity inductance is L. Its unit is the henry, symbol H.

REVIEW QUESTIONS FOR SECTION 5-1

1. What are the two similarities between inductance and capacitance with regard to energy?
2. In what form of field is energy stored in an inductor?
3. Under what conditions does a magnetic field exist?
4. Where is a magnetic field located with respect to a current-carrying conductor?
5. What is the production of a magnetic field around a current-carrying conductor called?
6. When dealing with electric circuits, when are magnetic fields of interest?
7. What configuration should the current-carrying conductor be given to make the electromagnetic effects more noticeable?
8. When the switch is first closed, applying power to a resistive-inductive circuit, what is the value of the current?
9. As time progresses after the switch is closed in a resistive-inductive circuit, what happens to the current and what final value does it assume?
10. Refer to Fig. 5-1*a*. State the conditions as to current, voltage across the resistor, voltage across the inductor, and magnetic field.
11. Refer to Fig. 5-1*b*. State the conditions as to current, voltage across the resistor, voltage across the inductor, and magnetic field at the instant the switch is closed.
12. Repeat Question 11 when the switch is closed and the conditions are stabilized.
13. What is the term applied to the ability of a magnetic field to store energy?
14. What is the coil of wire called?
15. What is the symbol for an inductor?
16. What is the unit of inductance?

5-2 ELECTROMAGNETIC INDUCTION

The interaction between a magnetic field and one or more conductors is called *flux linkage*. For instance, in the circuit of Fig. 5-1, the magnetic field surrounds the coil of wire, so there is interaction and flux linkage. If the magnitude of the flux linkage changes, as it does while the field is building up or dying away, a voltage is induced in the wire. This action is called *electromagnetic induction* and has many important applications, including the operation of electric power generators.

Faraday's law of electromagnetic induction gives the relationship between an *induced voltage* and changing flux linkage:

> **The magnitude of an induced voltage is proportional to the rate of change of flux linkage.**

Faraday's law focuses on the importance of change in the generation of an induced voltage. If there is a steady interaction between a magnetic field and conductors, there will be no induced voltage. A rapid change will result in a relatively high voltage.

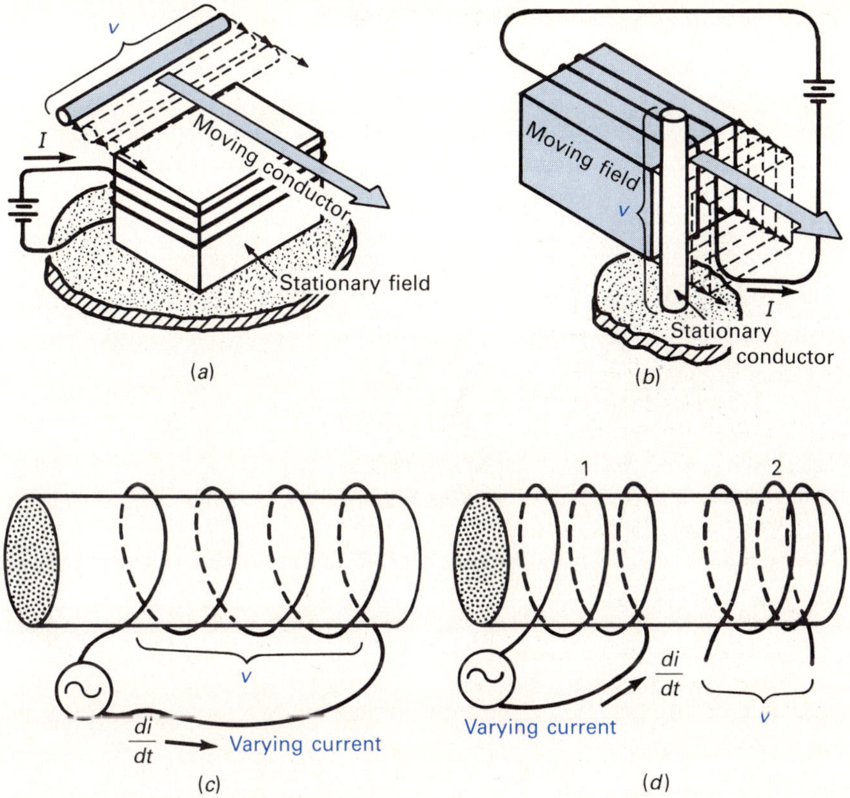

Fig. 5-2. Electromagnetic induction.

There are four basic ways of changing the interaction between a magnetic field and a conductor. These are shown in Fig. 5-2.

In Fig. 5-2*a* a stationary magnetic field is produced by a steady current I in a coil that does not move. A conductor is moved across the field, and a voltage V is induced between the ends of the conductor.

Figure 5-2*b* shows a situation similar to that in Fig. 5-2*a*, except that here the conductor is held stationary and the coil, with its magnetic field, is moved so that the field cuts across the conductor. Again a voltage is induced between the ends of the conductor.

In Fig. 5-2*c* a coil carries a changing current, so its magnetic field changes in strength. The coil is a conductor in its own changing magnetic field. A voltage is induced in the coil because it is in a field of varying intensity.

In Fig. 5-2*d* two coils are wound on a common core. The varying-intensity field produced by the changing current in coil 1 links with the turns of coil 2, inducing a voltage between its ends. Coil 1 also has an induced voltage, as in Fig. 5-2*c*.

The methods of producing induced voltages shown in Fig. 5-2*a* and *b* form the basis of operation of many electric generators.

In Fig. 5-2*c* and *d* we have the principles of *self-* and *mutual-induction*, the basis of inductor and transformer action.

REVIEW QUESTIONS FOR SECTION 5-2

1. What is the interaction between a magnetic field and one or more conductors called?
2. If the magnitude of flux linkage changes around a coil of wire, what is developed?
3. To what is the magnitude of an induced voltage proportional?
4. Describe four methods by which voltage can be induced by changing the interaction between a magnetic field and a conductor.
5. What are the principles of Fig. 5-2*c* and *d* called?

5-3 INDUCTIVE "OHM'S LAW"

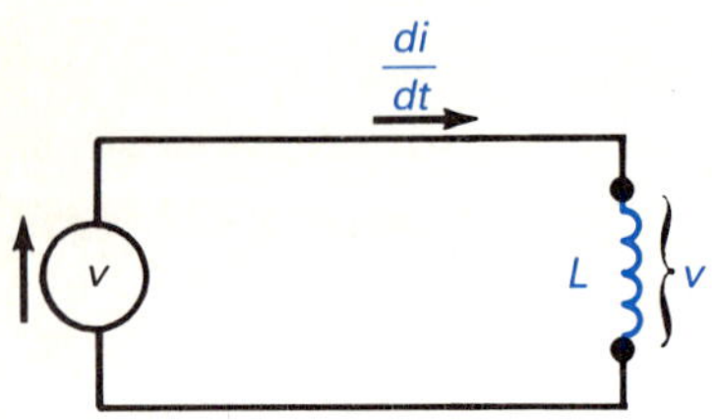

Fig. 5-3. Inductive circuit.

Figure 5-2*c* can be shown in schematic form as Fig. 5-3.

The magnitude of an inductive voltage is given by its *inductance multiplied* by the *rate of change of its current.*

The symbol *di*/*dt* (change of current divided by change of time) abbreviates the term "rate of change of current." This symbol is used in Figs. 5-2 and 5-3. It is also applied in equations for calculating, as in the equation form of the "Ohm's law" of inductance,

$$v = L\frac{di}{dt} \tag{5-1}$$

where v = induced voltage, V
L = inductance, H
di/dt = rate of change of current, A/s

Another way of saying this gives a definition of inductance: *Inductance is the constant of proportionality between the rate of change of current in a circuit and the voltage across the circuit.*

You may find it helpful to your understanding of resistance, capacitance, and inductance to now reread Secs. 3-2 and 4-2 (pages 41 and 61) and compare them with this Sec. 5-3.

EXAMPLE 5-1 A 10-mH inductor is supplied with a current which rises at a uniform rate from 20 to 100 mA in 25 μs. What is the magnitude of the voltage induced between the ends of the inductor winding?

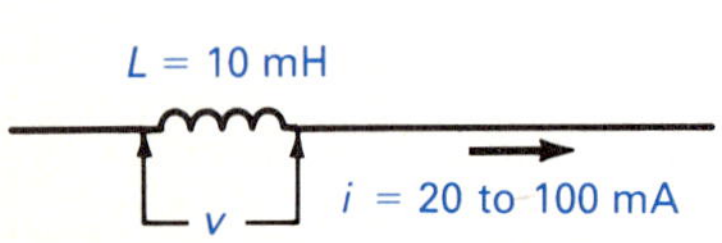

Fig. 5-4. Example 5-1: Data.

Solution (Fig. 5-4)

Since the current rises linearly (uniform rate), we can apply the 80-mA difference between the initial 20 mA and the final 100 mA in the expression *di*/*dt*:

$$v = L\frac{di}{dt} \tag{5-1}$$

$$v = 10\text{ mH} \times \frac{80\text{ mA}}{25\ \mu\text{s}} = \mathbf{32\ V}$$

EXAMPLE 5-2 A constant 32 V is applied across a 10-mH inductor which initially has no magnetic field associated with it. What will be its current (*a*) initially, (*b*) after 1 μs, and (*c*) after 10 ms?

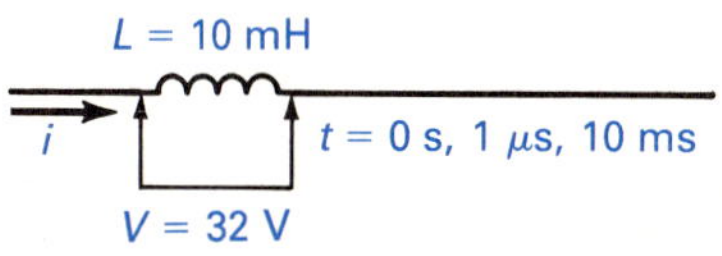

Fig. 5-5. Example 5-2: Data.

Solution (Fig. 5-5)

In Example 5-1, a change of current of 80 mA in 25 μs resulted in a rate of change of current of 3200 A/s. When this was associated with an inductance of 10 mH, it gave an inductance of 32 V. These values can be used in this solution.

a. Initially there is no magnetic field, so the current is at its zero value.

$$i = \mathbf{0\ A}$$

b. After 1 μs, the current has increased for this time period at the rate of 3200 A/s.

$$i = 3200 \text{ A/s} \times 1\ \mu\text{s} = \mathbf{3.2\ mA}$$

c. After 10 ms,

$$i = 3200 \text{ A/s} \times 10 \text{ ms} = \mathbf{32\ A}$$

REVIEW QUESTIONS FOR SECTION 5-3

1. What is the equation for the ''Ohm's law'' of inductance?
2. Of what is inductance the constant of proportionality?
3. What is the relationship between voltage and current in an inductive circuit?

EXERCISE PROBLEMS FOR SECTION 5-3

1. Find the voltage applied to an inductor of 100 mH which causes current to rise at the rate of 500 A/s. (*ans*. 50 V)
2. Find the voltage across an inductor of 6 H if the current is falling at a rate of 40 A/s.
3. Find the rate of change in current in an inductor of 0.2 H if a voltage of 100 V is applied. (*ans*. 500 A/s)
4. The primary winding of a blocking oscillator transformer has an inductance of 400 μH. Find the rate of change in current if the full supply voltage of 6 V is across the winding.
5. Find the ON time of the blocking oscillator in Prob. 4 if the circuit is on during the period while current is rising from 1 mA to 65 mA at a uniform rate. (*ans*. 4.27 μs)
6. Current rises from 0 to 150 mA linearly in 25 μs when 15 V is applied across an inductor. Find the value of the inductor.
7. Find the time for current to rise uniformly from 100 mA to 250 mA in an 8-H conductor while an emf of 50 V is applied. (*ans*. 24 ms)

5-4 CHARGING AND DISCHARGING INDUCTANCE

One way of describing "charged" is "loaded or filled to capacity." When a capacitor is charged, it is loaded or filled with energy. Similarly, an inductor is charged when its magnetic field is loaded with energy.

While there is this energy similarity between a charged capacitor and a charged inductor, the results of these charges are quite different. When the source of energy, the generator, power supply, or battery, is disconnected, the energy remains within the capacitor; it stays charged. Disconnecting the energy source from a charged inductor causes the energy to leave the inductor immediately; it discharges. This difference between the way energy is held or is not held is due to the fact that inductive charge depends on current, and current stops when a circuit is opened.

Charging

An inductor which has no magnetic field associated with it is *uncharged;* it has no stored energy, and it carries no current. *Charging* is the act of increasing the amount of energy stored. Equation (5-1) and Example 5-2 indicate that, when an inductor becomes charged as the result of a constant voltage, its rate of change in current will also be constant. Therefore the inductor's current will be directly proportional to time with constant charging voltage.

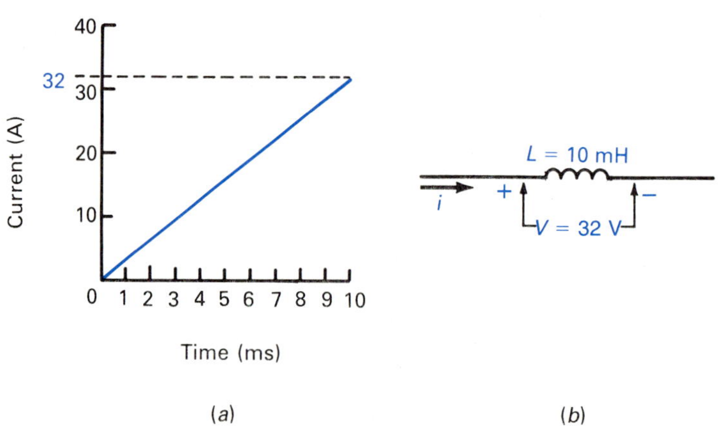

Fig. 5-6. Inductor charging: *(a)* Graph of current with respect to time for a constant voltage; *(b)* schematic diagram.

Figure 5-6*a* shows the linear increase in current in a 10-mH inductor with a constant 32 V across it. The data were obtained from the solutions to Examples 5-1 and 5-2.

Discharging

When an inductor is discharging and is supplying a constant voltage to a network, there will also be a linear relationship between the current and time. The current will decrease because energy is being supplied by the inductor's collapsing magnetic field. The conditions are shown graphically in Fig. 5-7*a*.

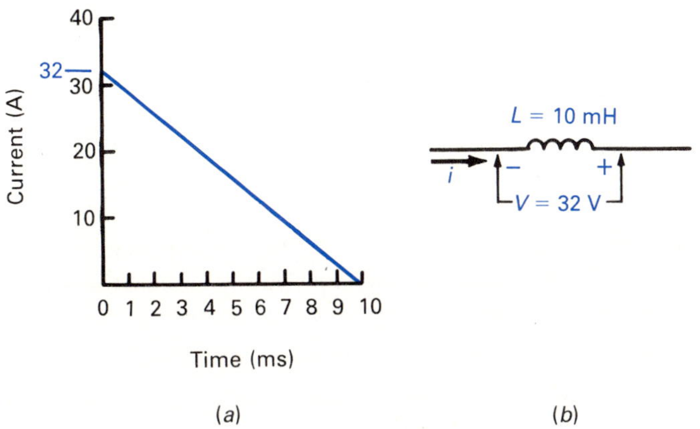

Fig. 5-7. Inductor discharging: *(a)* Graph of current with respect to time for a constant voltage; *(b)* schematic diagram.

Holding Inductive Charge

An inductor can store energy only while it carries current. Thus if the charging source emf is disconnected, the current ceases, the magnetic field collapses, and the energy is released. A comparison of a capacitor and an inductor is given in Fig. 5-8.

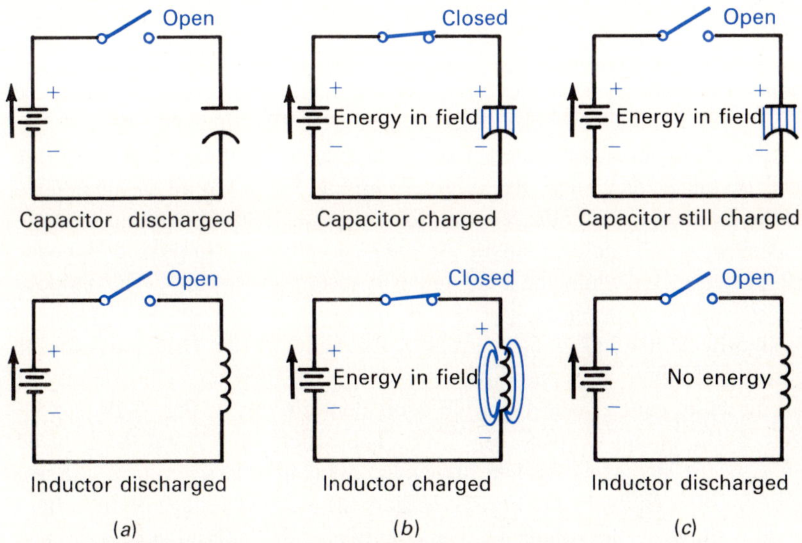

Fig. 5-8. Comparison of a capacitor and an inductor.

Figure 5-8*a* shows initially, with the switch open, both the capacitor and the inductor are discharged (no field energy).

Figure 5-8b shows when the switch is closed, both the capacitor and the inductor become charged (energy is stored in their fields).

Figure 5-8c shows when the switch is then reopened, the capacitor retains its energy but the inductor does not.

Collapsing Magnetic Field

Current is necessary to maintain a magnetic field, so, when current in an inductor ceases, the magnetic field collapses.

A magnetic field which increases in intensity causes an induced voltage. A magnetic field which decreases in intensity (a collapsing field) also causes an induced voltage. However, the effective directions of change in increasing and decreasing fields are opposite and so produce opposite-polarity induced voltages.

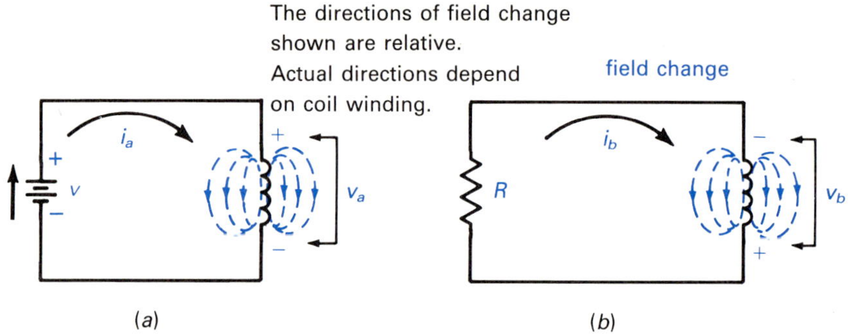

Fig. 5-9. Induced voltage polarities: *(a)* Energy from a source (a battery) to an inductor; *(b)* energy from an inductor to a load.

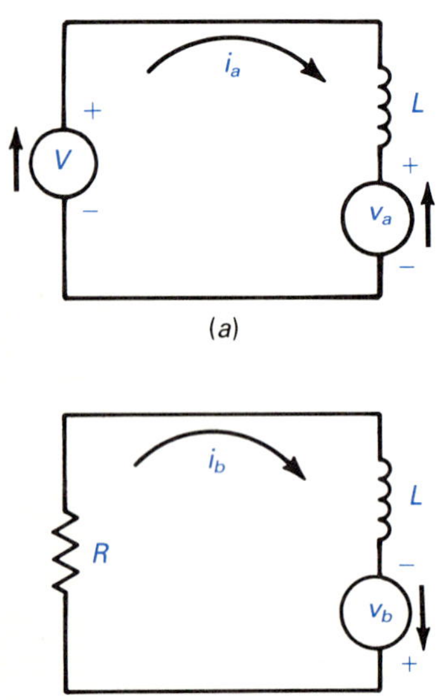

Fig. 5-10. Figure 5-9 shown in schematic diagram form.

Figure 5-9a shows a battery connected across an inductor. Current in the direction shown causes a magnetic field that increases in intensity. An induced voltage v_a of the indicated polarity exists while the field intensity rises.

If, without first becoming open, the circuit changes to that in Fig. 5-9b, the removal of the battery stops the original current, but a circuit is maintained across the inductor through the resistor. The magnetic field cannot hold up without the battery, so the field collapses. An induced voltage v_b, of polarity opposite to that of v_a, exists while the field intensity decreases to zero. The induced voltage v_b causes a current i_b.

It is significant that in Fig. 5-9 currents i_a for the increasing field and i_b for the decreasing field have the *same* direction through the inductor. This identity of current direction can be seen from the equivalent circuits in Fig. 5-10.

Figure 5-10a is equivalent to Fig. 5-9a. The battery emf V and the induced voltage v_a of the increasing field are shown as opposing-polarity voltage sources.

Figure 5-10b is equivalent to Fig. 5-9b. Here the induced voltage of the collapsing field is shown as the voltage v_b. The polarities of the induced voltages v_a and v_b are opposite.

The polarity of the induced voltage v_b is similar to that of the original voltage source V in that they both produce clockwise currents, or currents with a downward direction through the inductor.

Inductance uses no net energy. The energy supplied by the battery to the increasing-intensity field (via the current i_a) will be fully recovered from the collapsing magnetic field (via the current i_b) in the form of heat in the resistor.

To maintain the circuit continuity across the inductance while changing from the battery to the resistance, an arrangement such as that in Fig. 5-11*a* is needed. This diagram shows a *make-before-break* switch in which the connection between contacts 1 and 3 is completed before that between contacts 2 and 3 breaks. If, instead, a *transfer* or *break-before-make* switch were used as in Fig. 5-11*b*, so that the battery is disconnected before the resistance is connected, the induced voltage polarities and current directions would be basically the same, but the values would be unpredictable.

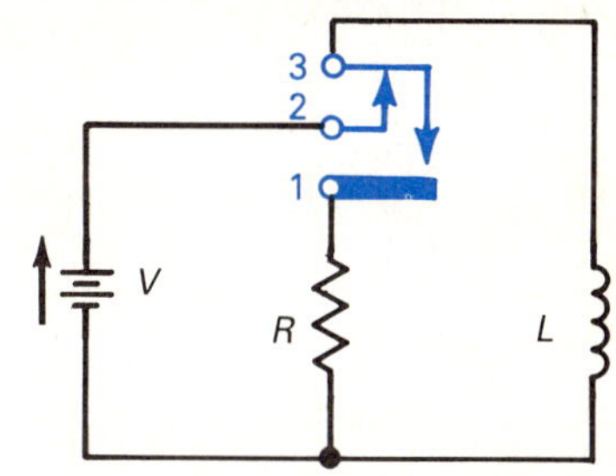

(*a*)

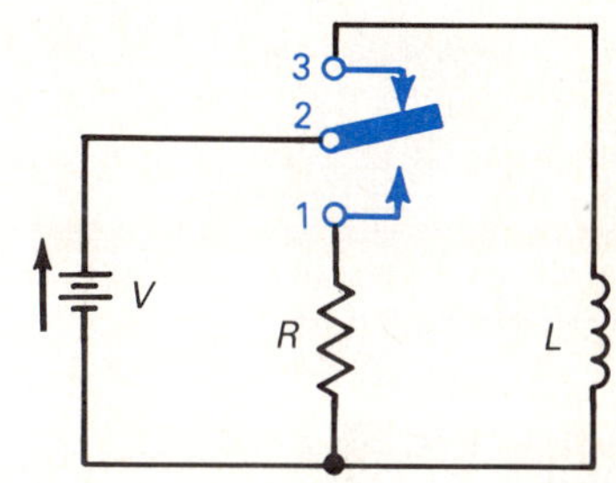

(*b*)

Fig. 5-11. Inductor charge/discharge circuits: *(a)* With make-before-break contacts; *(b)* with break-before-make contacts.

Arcing

If the circuit across a charged inductance is opened, as in Fig. 5-11*b*, a very high induced voltage results because the field collapses rapidly. In practice the field does not collapse in zero time, and so the induced voltage does not become infinite. The induced voltage becomes large enough to develop an arc across the circuit opening. This arc permits current to exist around the circuit, through the open switch contacts.

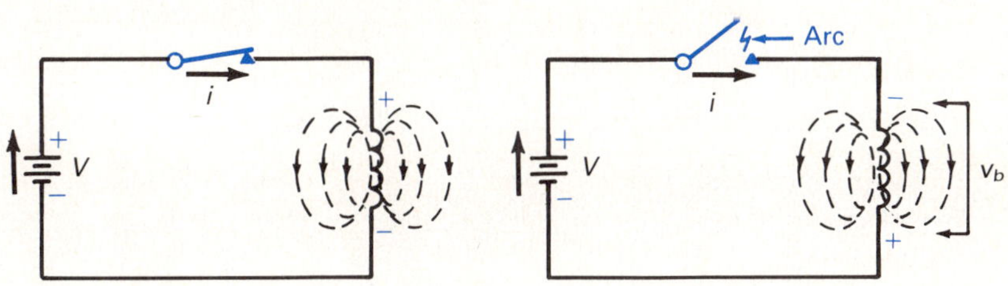

(*a*) (*b*)

Fig. 5-12. *(a)* Inductor charging through switch contacts. *(b)* When the contacts open, an arc may develop.

In Fig. 5-12*a* the battery emf has transferred energy to the inductance via the closed switch, building up a magnetic field.

Figure 5-12*b* shows the conditions existing at the instant the switch opens. The magnetic field collapses rapidly, inducing a high value of voltage v_b. This high voltage appears across the gap between the switch contacts. The air between the contacts *ionizes* (becomes conductive due to the high voltage breaking down the air's insulation resistance). Circuit current is maintained through the conducting arc until the energy of the field has dissipated.

The tendency for a collapsing magnetic field to produce a high-voltage arc is, like most phenomena in electric circuits, a blessing or a curse according to the circumstances. In the automobile an induced voltage of 20 000 V or more is induced by the collapse of a magnetic field produced by the 12-V electric system. This high induced voltage causes an arc across the gap between the spark plug electrodes. The arc ignites the gas-air mixture to make the engine run. Therefore, in a car an arc is useful. In electrical equipment used for switching inductive

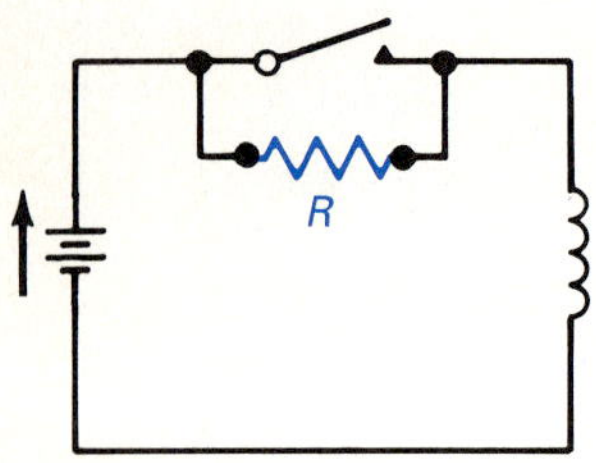

(a)

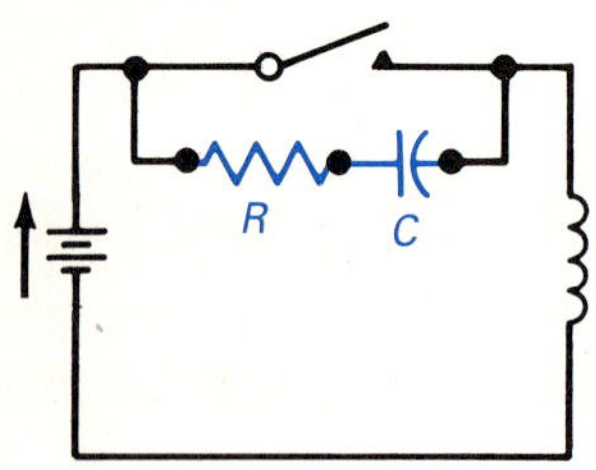

(b)

Fig. 5-13. Arc suppression: *(a)* By resistor; *(b)* by resistor and capacitor.

circuits, the arc between the opening switch contacts causes the contact surfaces to become pitted and carbonized. This erosion results in a poor electric connection in the switch and eventual destruction of the contact material. In such cases, an arc is undesirable.

The formation of an arc between opening contacts can be reduced or suppressed by providing a means for the field to collapse more slowly and thus prevent the development of a high value of induced voltage. Deliberately providing a path for current across the gap can be used to control the rate of field collapse.

In Fig. 5-13*a* a resistor is connected across the contacts so that, when they open, the resistor will conduct a current as a result of the induced voltage of the collapsing field. By suitable choice of resistor, the rate of field collapse, the induced voltage, and the voltage across the resistor can be limited to a value below that required to produce contact arcing. There is a potential problem here because the circuit is never completely open and the current is never reduced to zero. In some applications this permanent small current is unimportant, provided the resistor value is as high as is possible consistent with arc prevention.

The circuit in Fig. 5-13*b* has a series *RC* circuit across the contacts. This arc-suppression circuit ensures that, when the field energy has completely dissipated in the resistor, the circuit really is open, a capacitor being a dc open circuit. This arrangement has its own problems which may prohibit its use for some applications. These problems will be discussed later in the text.

REVIEW QUESTIONS FOR SECTION 5-4

1. When an inductor is charged, where is its energy stored?
2. What happens to the energy in a charged inductor when the energy source is disconnected?
3. What is necessary for an inductor to hold its energy charge?
4. With a constant charging voltage across an inductor, what is the relationship between current and time?
5. If it is desired to remove the charging source and discharge the inductor through a resistor, what form of switch is required?
6. If the circuit across a charged inductor is opened, what is the theoretical maximum voltage produced across the open circuit?
7. In practice, what may happen when the inductive circuit is opened, and why does the voltage not rise to its theoretical maximum?
8. To what practical use can the rapid opening of an inductive circuit be put?
9. What disadvantage is there to the rapid opening of inductive circuits?
10. What may be done to a collapsing magnetic field in order to prevent arcing?

5-5 OPPOSITION TO CURRENT CHANGE

Inductance may be referred to as opposition to current change (capacitance is opposition to voltage change), based on the following:

$$V = L\frac{di}{dt} \qquad \text{or} \qquad \frac{di}{dt} = \frac{v}{L} \tag{5-1}$$

For a given voltage, the magnitude of the rate of change in inductive current is inversely proportional to the inductance value. This proportionality means that a high value of inductance is related to a low rate of change in current.

This opposition to current change can be viewed in two ways. If the magnitude of the current change is fixed by the circuit conditions, the time required to change is greater with higher inductance. If the time in which the current changes is fixed, the magnitude of the change is smaller with high inductance.

The latter view of inductive opposition to current change is the basis for using an inductor as a *choke*. As an example of this application of inductance, consider the situation shown in Fig. 5-14. A generator producing a steady dc voltage V is connected in series with a generator producing an ac voltage v. The *frequency* of the voltage v (the number of times per second its polarity changes) is fixed. The resulting waveform of current in the load is shown in Fig. 5-14*b*.

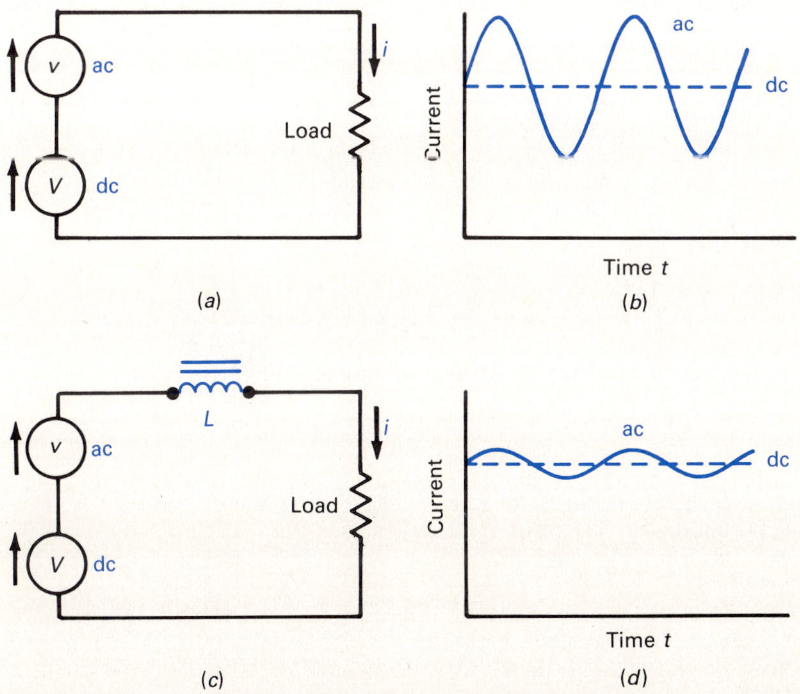

Fig. 5-14. Choke action. *(a)* and *(b)* Large ac component of current without inductance; *(c)* and *(d)* small ac component of current with inductance.

In Fig. 5-14*c* an iron-core inductor of high inductance L has been added to the circuit in Fig. 5-14*a*. The generator voltages are the same as in Fig. 5-14*a*, so the speed of the voltage change, as determined by the voltage v, is the same. The added inductance reduces the rate of change in current. As the speed of the change is fixed, the magnitude of the current change is reduced. Figure 5-14*d* shows the effect of the inductor. This choking action is used in electronic power supplies that produce dc outputs from ac power line inputs. The presence of the inductor (choke) helps to ensure a steady dc output relatively free of fluctuations.

REVIEW QUESTIONS FOR SECTION 5-5

1. To what does the term *choke* refer?
2. What is another name for an inductor?
3. In what type of circuit is an inductor used to help supply a steady dc output?

5-6 INDUCTIVE ENERGY STORAGE

Energy transferred to the magnetic field of an inductance is stored while the current is maintained. When the current decreases, energy is extracted from the magnetic field. It might be dissipated in a resistor, transferred to a capacitor for storage, or passed back to the supply source.

During the process of charging or discharging an inductor (transferring energy to or from it), a constant induced voltage implies a linear relationship between current and time (refer to Figs. 5-6 and 5-7).

If, over a time period t in Fig. 5-15, the current rises from zero to a value I amperes, the energy transferred to the inductance will be related to the average value of that current. Since the current rises in a linear manner, the average value is one-half I.

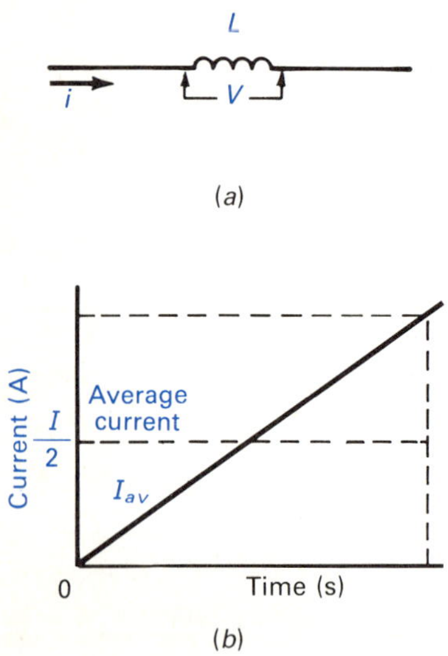

Fig. 5-15. Inductive energy: *(a)* Circuit conditions; *(b)* current-time relationships.

$$I_{av} = \frac{I}{2}$$

Also the voltage at any time is the induced voltage. From Eq. (5-1),

$$V = L\frac{di}{dt}$$

Since the current rise is linear, the rate of change in current di/dt is the current I divided by the time t. Therefore,

$$V = L\frac{I}{t}$$

Substituting for I and V in Eq. (2-14) gives the energy stored in the inductance, $W = VI_{av}t$

$$W = L\frac{I}{t}\frac{I}{2}t$$

$$W = \frac{I^2L}{2} \qquad \textbf{(5-2)}$$

where W = inductive energy, J
I = current, A
L = inductance, H

EXAMPLE 5-3 A 10-H iron-core inductor is used in the filter circuit of an electronic power supply. What is the amount of energy stored in its magnetic field if the current is constant at 100 mA?

Solution (Fig. 5-16)

$$W = \frac{I^2L}{2} \tag{5-2}$$

$$= \frac{(100\text{ mA})^2 \times 10\text{ H}}{2} = \mathbf{0.05\ J}$$

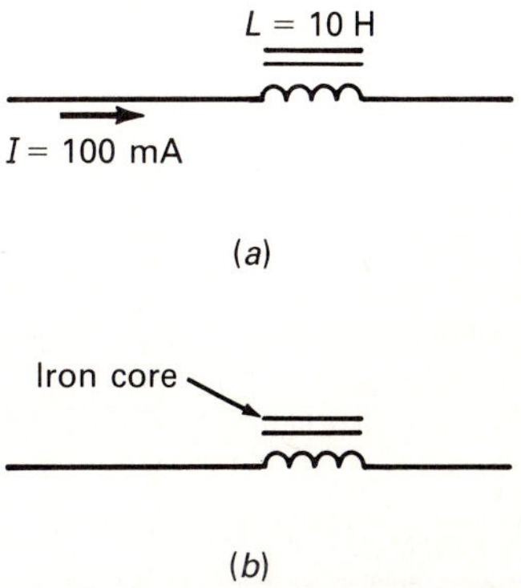

Fig. 5-16. Example 5-3: *(a)* Data; *(b)* standard symbol for iron-core inductance.

EXAMPLE 5-4 When the negative pulse is applied to the base of the transistor in the circuit in Fig. 5-17, the collector current I_c is cut off (reduced to a very low value), the transistor becoming similar to an open switch. If the inductor's field collapses to zero at a uniform rate in 10 μs, *(a)* what amount of energy is dissipated and *(b)* what voltage V_{ce} is developed between the transistor collector and emitter?

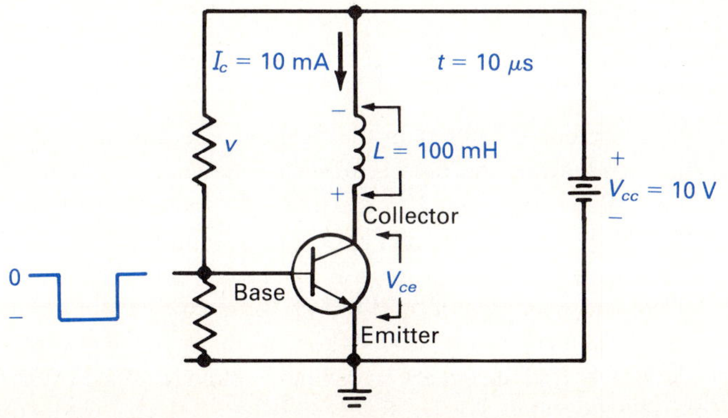

Fig. 5-17. Example 5-4: Circuit.

Solution *a.* From Eq. (5-2),

$$W = \frac{I_c^2L}{2} = \frac{(10\text{ mA})^2 \times 100\text{ mH}}{2} = \mathbf{5\ \mu J}$$

b. The voltage v induced across the inductor by the collapsing field is,

$$v = L\frac{di}{dt} \tag{5-1}$$

$$v = 100\text{ mH}\frac{10\text{ mA}}{10\ \mu\text{s}} = 100\text{ V}$$

The polarities of the induced emf v and the collector supply voltage V_{cc} will be aiding each other. These voltages will add to the value of the transistor voltage V_{ce}.

$$V_{ce} = v + V_{cc} = 100\text{ V} + 10\text{ V} = \mathbf{110\ V}$$

The transistor selected for use in the circuit in Fig. 5-17 must be able to tolerate 110 V between its collector and its emitter when the inductance discharges. This voltage is quite a severe requirement which many transistors cannot handle even though well able to withstand the 10 mA at 10 V when in the ON condition. The discharge time is important, of course. Design precautions would need to be taken to ensure that the time would not be even less than the 10 μs specified, as this would result in an induced voltage greater than 100 V.

REVIEW QUESTIONS FOR SECTION 5-6

1. If a constant voltage is applied across an inductor, current rises at a constant rate. In what way will the energy transferred to the inductor be related to current?
2. State the equation for energy stored in an inductor.

EXERCISE PROBLEMS FOR SECTION 5-6

1. Find the energy stored in the magnetic field of the inductor in Prob. 6, Sec. 5-3, when the current reaches 150 mA. (*ans*. 28.13 μJ)
2. Find the energy added to the magnetic field of the inductor in Prob. 7, Sec. 5-3, when the current rose from 100 mA to 250 mA under the conditions stated.
3. The field windings of a 230-MVA generator have a resistance of 0.15 Ω and an inductance of 1.2 H. When an emf of 145 V produces a field current of 950 A, find the energy stored in the magnetic field of the field windings. (*ans*. 541.5 kJ)

5-7 MUTUAL INDUCTANCE

In Fig. 5-18 a varying current in coil 1 produces a magnetic field of varying intensity. This magnetic field links with coil 2, inducing a voltage between its ends. This type of action in which a varying quantity in one circuit causes the development of a quantity in a different circuit is called *mutual* or *transfer* action. When, as in Fig. 5-18, a varying current in one circuit causes a voltage to be induced in another circuit, the action is called *mutual induction*. The relationship between the changing current and the induced voltage is given by

$$v_2 = M\frac{di_1}{dt} \tag{5-3}$$

where v_2 = voltage induced in coil 2, V
M = mutual inductance between the windings, H
di_1/dt = rate of change in current in coil 1, A/s

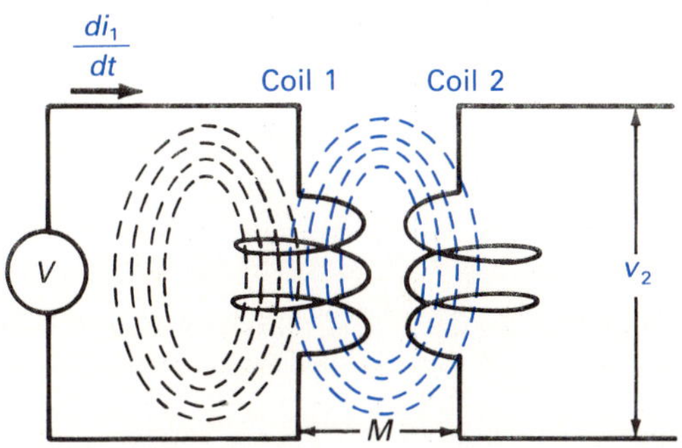

Fig. 5-18. Magnetic field linking two coils.

Equations (5-3), $v_2 = M\,di_1/dt$, and (5-1), $v = L\,di/dt$, should be compared. This comparison will show why M is called *mutual inductance*.

Mutual inductance is the constant of proportionality between the rate of change in current in one circuit and the resulting voltage induced in another circuit.

The Henry as the Mutual-Inductance Unit

The henry, the unit of self-inductance, is also used for mutual inductance as, in both cases, the inductance (self- or mutual-) is 1 H when a rate of change in current of 1 A/s induces a voltage of 1 V.

EXAMPLE 5-5 Two coils of a current transformer are wound on the same iron core. The mutual inductance between them is 0.5 mH. If the current in the line feeding the input coil changes at the rate of 50 A/s, what emf is measured across the other coil?

Solution (Fig. 5-19)

$$v_2 = M\frac{di_1}{dt} \tag{5-3}$$

$$= 0.5\text{ mH} \times 50\text{ A/s} = \mathbf{25\ mV}$$

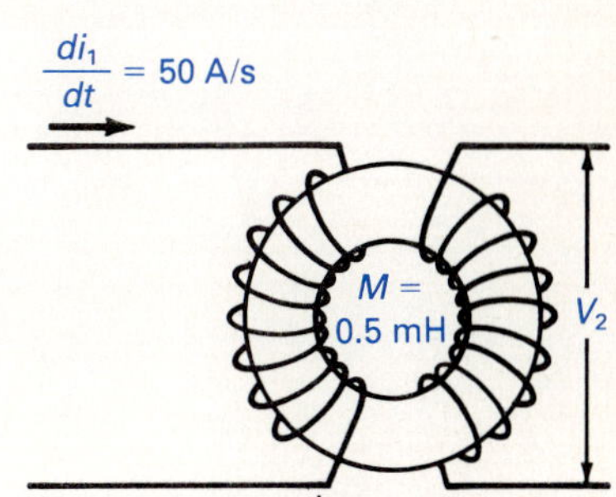

Fig. 5-19. Example 5-5: Data.

A current transformer is used to measure currents in power circuits where, owing to the presence of high voltages, it is not convenient to measure the currents directly.

EXAMPLE 5-6 The collector current i_c in the transistor in Fig. 5-20 changes at a uniform rate from 5 to 10 mA in 100 μs. The mutual inductance between the two windings is 100 mH. What is the output voltage v_o?

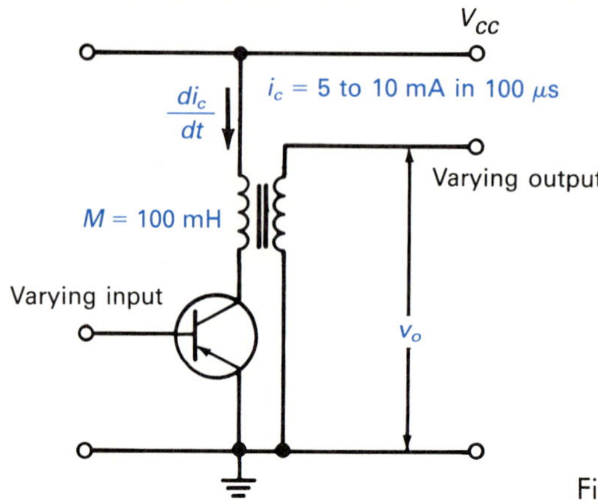

Fig. 5-20. Example 5-6: Circuit.

Solution In this example it is the transistor's collector current that produces the magnetic field in the transformer, so that di_c/dt is the initiating rate of change of current and takes the place of di_1/dt in Eq. (5-3):

$$\frac{di_c}{dt} = \frac{10\text{ mA} - 5\text{ mA}}{100\ \mu\text{s}} = 50\text{ A/s}$$

Similarly, the output voltage v_o is the equivalent of v_2 in Eq. (5-3):

$$v_o = M\frac{di_c}{dt} \qquad \text{from (5-3)}$$

$$v_o = 100\text{ mH} \times 50\text{ A/s} = \mathbf{5\ V}$$

Voltage Polarities

The polarity of a voltage self-induced in a coil due to its own current is the same as that of a voltage across a resistive or capacitive circuit element. But the polarity of a voltage mutually induced in a coil due to current in another coil is not always so obvious. The principle for determining inductive voltage polarities is given by *Lenz' law:*

> **The direction of current resulting from an induced voltage is such that the magnetic field it produces acts in opposition to the magnetic field that induced the voltage.**

Lenz' law may be viewed as a form of the law of physics, "to every action there is an equal and opposite reaction": whatever you try to do, something happens to try to stop you doing it. When a change in current value causes a change in magnetic field strength, an induced voltage appears. This voltage tries to cause a current in opposition to the original current.

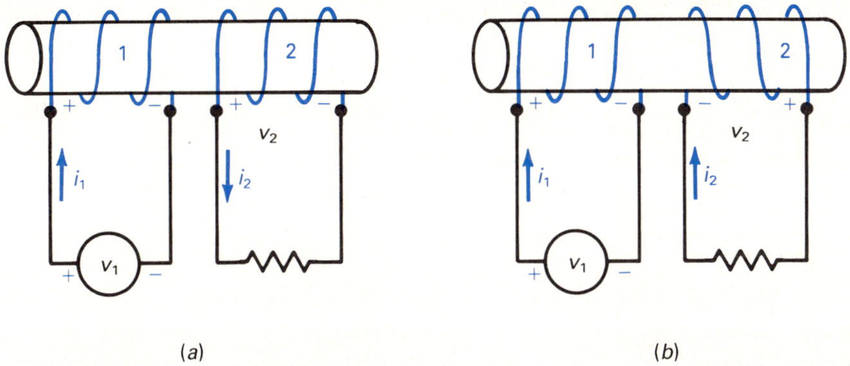

Fig. 5-21. Induced voltage polarities: *(a)* Coils wound in same direction; *(b)* coils wound in opposite directions.

In Fig. 5-21*a*, coil 1 is a load across the generator. For its current i_1 to have the direction shown, the generator's voltage has to have the polarity shown. The voltage v_2 in coil 2 is generated by induction. As both coils are wound on the same core in the same direction and the same magnetic field is acting on them, the induced voltages in both coils will have the same polarity. With this polarity of generated voltage v_2, the current i_2 in the coil 2 circuit must have the direction shown, opposite to the current i_1 in coil 1. The current i_2 will produce a magnetic field opposing that produced by current i_1.

In Fig. 5-21*b* the two coils are wound in opposite directions. The polarity of the induced voltage v_2 and the direction of the current i_2 will be opposite those in Fig. 5-21*a*.

In practice it is not easy to figure the polarities of mutually induced voltages. The way the coils are wound and mounted determine these polarities, and often this information is not known. Schematic symbols do not show these physical characteristics. For practical situations dots may be shown on schematic diagrams to identify terminals of equivalent polarity for mutually inductive coils.

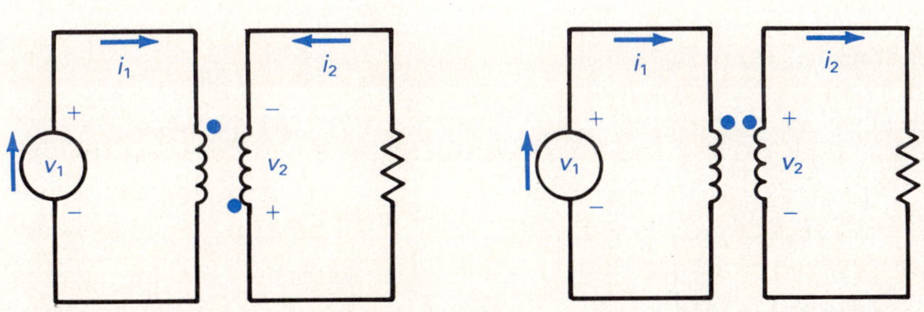

Fig. 5-22. Dots show equivalent polarities.

The dots in Fig. 5-22 show the terminals which have the same instantaneous polarity. For either of the pairs of coils, the terminals with dots are both positive at the same time. They will also both be negative at the same time. They will both have voltages that change in the same direction; if one becomes more positive, so will the other.

Control of Mutual Inductance

Different degrees of mutual inductance, from maximum to zero, are needed for various applications. Often, as in the case of transformers, maximum mutual inductance is necessary for the efficient transfer of energy. Coils in the input and output circuits of amplifiers are normally required to have negligible mutual inductance between them for optimum stability of performance. Variable mutual inductance can be used to control energy transfer between a radio transmitter and its antenna.

There are several methods for controlling mutual inductance.

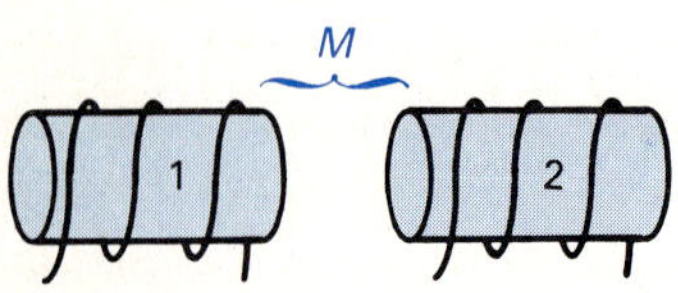

(a)

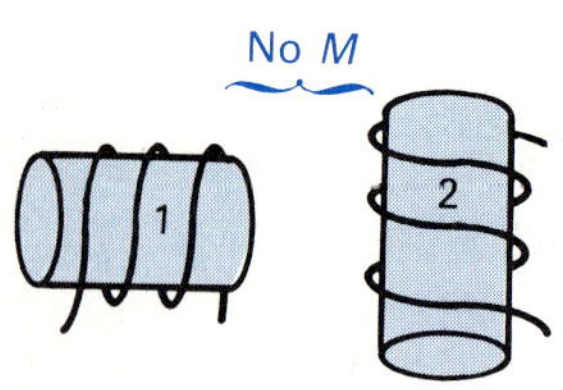

(b)

Fig. 5-23. *(a)* Coils in line result in maximum mutual inductance; *(b)* perpendicular coils result in zero mutual inductance.

Core Efficiency An efficient iron core, such as the core of a toroid-wound (doughnut-shaped) coil, will concentrate the magnetic field into a region within, and close to, the core. Thus any windings on the core will experience high mutual inductance. Nearby coils will not.

Distance between Coils As the distance between coils increases, the mutual inductance between them decreases.

Angle between Coils In Fig. 5-23*a* two *solenoid-wound* coils (cylindrical) are shown with their axes in line. Their mutual inductance will be relatively high. Figure 5-23*b* shows two similar coils with their axes at right angles. Their mutual inductance will be low (theoretically zero). In any other relative positions, the mutual inductance will be a function of the angles between the coils.

Shielding Completely enclosing a coil in an iron-alloy material can substantially reduce any mutual inductance between it and neighboring circuits.

REVIEW QUESTIONS FOR SECTION 5-7

1. What is the term used when a varying quantity in one circuit causes the development of a quantity in a different circuit?
2. What is the action called when a varying current in one circuit causes a voltage to be induced in another circuit?
3. State the equation relating the current in one circuit to the induced voltage in another circuit.
4. Why do both self- and mutual-inductance have the same unit?
5. What is the name of the law determining induced voltage polarities?
6. In practical situations, how are equivalent polarities shown for mutually inductive coils?
7. State three factors which will affect mutual inductance.

EXERCISE PROBLEMS FOR SECTION 5-7

1. The current in the primary of a blocking oscillator transformer rises from 1 mA to 65 mA in 4.27 μs. If the mutual inductance of the transformer is 800 μH, find the secondary voltage. (*ans.* 11.99 V)

2. The transformer in a transistor collector circuit has a mutual inductance of 2.5 mH. Collector current I_c is changing at a rate of 6000 A/s. If a 10-Ω load resistance is connected across the transformer secondary, find I_L and V_L.
3. A linear current waveform which rises from zero to 100 mA in 1 ms, then falls to zero in the next 1 ms, is applied to the primary of a transformer with a mutual inductance of 1 H. Find the secondary voltage for each of the two 1-ms periods. (*ans.* 100 V, −100 V)

5-8 INDUCTORS

An *inductor* is a circuit component consisting of one or more windings of wire. Usually, however, the term *inductor* is applied to single-winding components and *transformers* to those with more than one winding. On this basis an inductor is concerned with self-inductance and a transformer with mutual inductance.

There is no simple way of expressing the inductance of an inductor so that one can be designed for a known inductance value. This is due to the difficulty of concentrating an inductor's magnetic field entirely within the component. Depending on its application, the concentration of an inductor's field may range from very high in the case of an *iron-core* power inductor to very low in an *air-core* radio inductor. This range of field concentration makes it difficult to determine inductance from an inductor's dimensions. However, some of the factors that are important to inductance value can be seen from the formula for calculating the inductance of a "perfect" inductor, one with a perfectly concentrated field. These factors are given in Fig. 5-24.

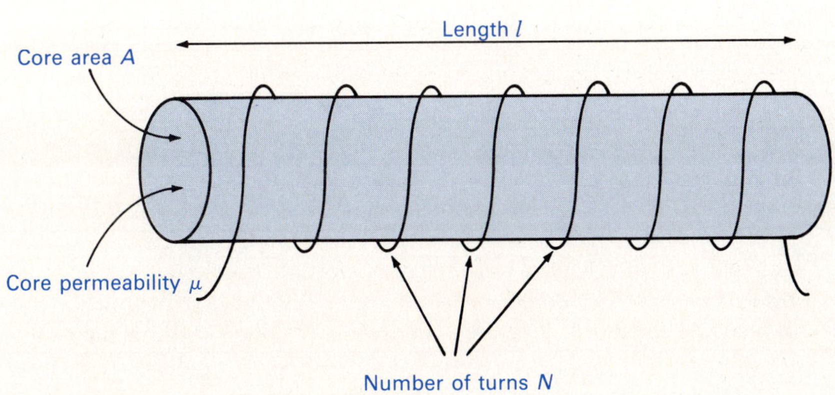

Fig. 5-24. Factors controlling inductance.

Using the factors shown in Fig. 5-24, inductance value is given by

$$L = \frac{N^2 \mu A}{l} \tag{5-4}$$

where L = inductance, H
N = number of turns
μ = permeability, H/m
A = core area, m^2
l = core length, m

Permeability

Two coils with the same dimensions may have different inductance values. One wound on an iron core may have several thousand times the inductance of another wound on a plastic tube. The difference in the inductances is due to the *permeability* of iron being very much higher than that of a plastic tube. Permeability is to inductance what permittivity and resistivity are to capacitance and resistance. The permeability of a material is its *specific inductance;* it determines the inductance of a coil of given dimensions wound on a core of that material.

The value of the permeability of a material is given by two factors. One is the permeability of a vacuum, called the *permeability of free space*. It gives the inductance of a coil wound on a core that has no significant magnetic properties. The plastic tube mentioned above is an example of such a core. The permeability of free space μ_v (μ is the Greek letter mu) has the constant value 1.257×10^{-6} (or $4\pi \times 10^{-7}$) henrys per meter. The other factor is the *relative permeability* μ_r of the core material. It gives the amount by which inductance is increased by using that material instead of a vacuum or other nonmagnetic material. Relative permeability has no unit. If we say that an inductor wound on an iron core has 4000 times the inductance of a similar inductor wound on a plastic tube, the relative permeability of that grade of iron is 4000.

The permeabilities of some magnetic materials are given in Table 5-1.

TABLE 5-1 Typical Permeabilities of Magnetic Materials*

Magnetic Material	Relative Permeability μ_r	Total Permeability μ (H/m) $\mu = \mu_v\mu_r$ or $\mu = 1.257 \times 10^{-6} \times \mu_r$
Nickel	50	62.85×10^{-6}
Cobalt	60	75.42×10^{-6}
Cast iron	50 to 100	62.85×10^{-6} to 125.7×10^{-6}
Cast steel	500 to 1000	628.5×10^{-6} to 1.257×10^{-3}
Nickel-iron	2500 to 5000	3.143×10^{-3} to 6.285×10^{-3}

*The values of relative permeability vary considerably depending on several factors, including the manufacture of a material and its degree of magnetization. In Eq. (5-4) for calculating the value of inductance, the value of permeability μ for a given material is that shown in the last column of this table.

The value of permeability μ in Eq. (5-4) is given by

$$\mu = \mu_v \mu_r \qquad \textbf{(5-5)}$$

where μ = total permeability, H/m*
μ_v = permeability of free space, 1.257×10^{-6} H/m
μ_r = relative permeability

The values of relative permeability for each of the iron-based materials listed in Table 5-1 cover a wide range. This variability may be due to either of two main factors:

1. Variations in the nature of the material. Whether the material is pure iron or an alloy of iron and other metals will affect its permeability. So will the way the material is processed and fabricated.

   Of the materials listed in Table 5-1, only the alloy nickel-iron is normally used for inductor cores. Cast iron and cast steel, both of which are iron processed in different ways, are mentioned for comparison. The metals nickel and cobalt have much lower permeabilities than iron. These and a number of other metals are used to form alloys with iron to modify its characteristics as a core material.
2. Variations in the degree of magnetization. The relative permeability of a magnetic material changes as its magnetization level changes. Applied to an inductor, this means that the permeability of its core, and therefore the magnetization, varies as its coil current changes. If the coil current rises to a high enough value, *saturation* is reached and further increases of magnetization cannot be obtained.

Iron-core permeability can be influenced by the level of steady coil current, or its "dc level." This may seem unimportant, as inductive effects are only of interest when current value changes. However, it often happens that coil current is a combination of steady and varying components. When this is so, the inductive effects may be controlled by the coil current's dc level. In such cases, the dynamic or *incremental permeability* may be of more concern than the static value. Incremental permeability is the effective value for small variations of magnetization close to the dc level.

The differences in the values of permeability for a magnetic material may be seen from Fig. 5-25.

The static permeability taken in the center of the "straight" portion of the magnetization curve is represented by the angle *A* in Fig. 5-25*a*. This is the value most likely to be quoted in a table of permeabilities.

Figure 5-25*b* shows the angle *B* as the incremental, or dynamic, permeability at the same dc level as in Fig. 5-25*a*. Angle *B* is greater than angle *A*, indicating that at this level of magnetization the incremental value of the permeability is greater than the static value.

Two other values of incremental permeability are indicated in Fig. 5-25*c*. At low levels of magnetization the incremental permeability is low, as shown by angle *C*. At high magnetization levels the incremental permeability is also low, as shown by angle *D*. At *very* high levels of magnetization, saturation is reached,

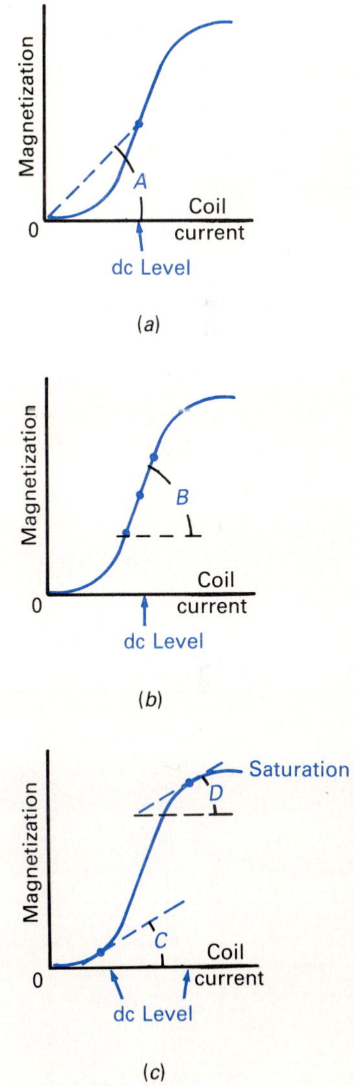

Fig. 5-25. Static and incremental permeability. *(a)* Static permeability; *(b)* incremental permeability; *(c)* effect of dc level on incremental permeability.

*The unit of permeability is the *henry per meter* (H/m).

the magnetization curve becomes almost horizontal, and the incremental permeability approaches zero.

Inductance depends on incremental permeability, and so it depends on the level of magnetization. It is important to realize that the inductance at saturation levels is almost the same as if there were no core. This explains why an inductor may have high measured inductance but be ineffective in a circuit. An inductance "bridge" usually makes measurements at levels of magnetization well below saturation, but the inductor's circuit may give it a high dc level of current, thus running its core into saturation. This situation is sometimes met in electronic power supplies where inductors are used as filter elements. A power supply's dc output current may cause the inductor's iron core to saturate unless the core contains an air gap to reduce the magnetization level.

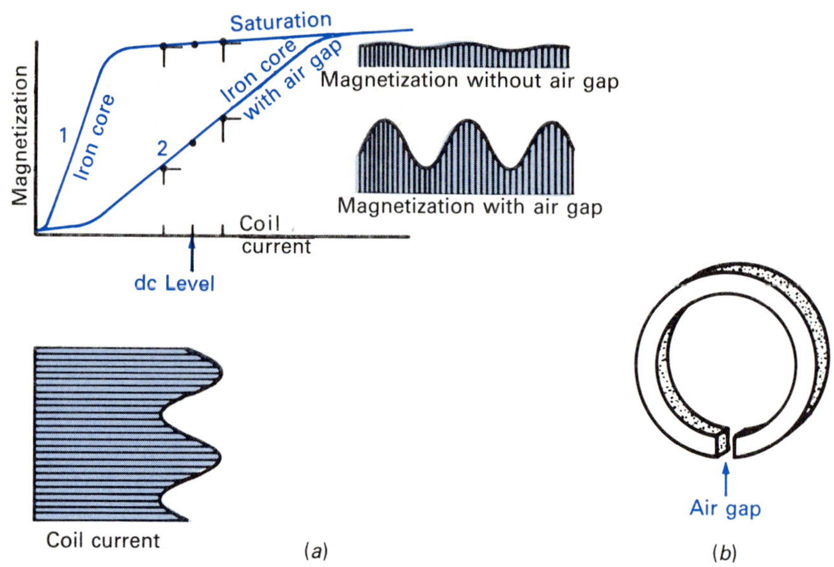

Fig. 5-26. Effect of an air gap in an iron core. *(a)* Magnetization with and without air gap; *(b)* toroidal iron core with air gap.

In Fig. 5-26*a*, curve 1 represents the magnetization of a high-permeability iron core. The dc level of the coil current causes saturation of the iron, so that there is little variation of magnetization and therefore almost no inductance. Curve 2 shows how the effective magnetization may be changed by means of an air gap to avoid saturation. Variation of magnetization is obtained, giving effective inductance. An air-gapped core for a toroidal inductor is shown in Fig. 5-26*b*.

EXAMPLE 5-7 What is the inductance of a 1000-turn filter inductor wound on the nickel-iron alloy core shown in Fig. 5-27? The relative permeability of the core material is 3000.

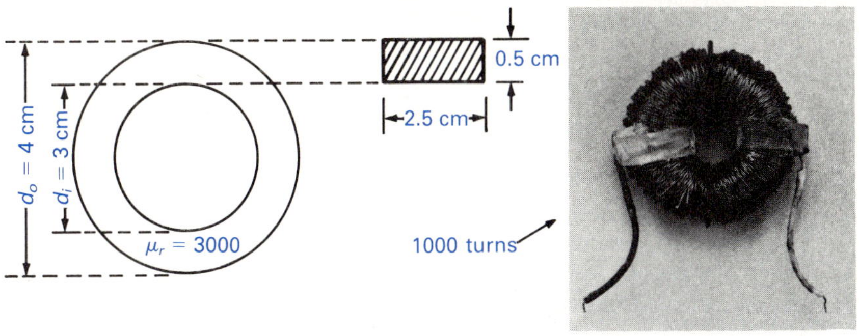

Fig. 5-27. Example 5-7: *(a)* Core dimensions; *(b)* toroidal coil.

Solution The core area A will equal the shaded area in Fig. 5-27*a*:

$$A = 2.5 \text{ cm} \times 0.5 \text{ cm} = 1.25 \text{ cm}^2 = 1.25 \times 10^{-4} \text{ m}^2$$

The effective length of the field l will be the average circumference of the core. The average core diameter is

$$d_{\text{av}} = \frac{d_o + d_i}{2} = \frac{4 \text{ cm} + 3 \text{ cm}}{2} = 3.5 \text{ cm}$$

$$l = \pi d_{\text{av}} = \pi \times 3.5 \text{ cm} = 0.11 \text{ m}$$

Applying Eq. (5-5) to find the value of the core incremental permeability,

$$\mu = \mu_v \mu_r \qquad (5\text{-}5)$$

Substituting the values of μ_v (1.257×10^{-6} H/m) and μ_r,

$$\mu = \mu_v \mu_r = 1.257 \times 10^{-6} \text{ H/m} \times 3000 = 3.77 \times 10^{-3} \text{ H/m}$$

The inductance can be found by substituting values into Eq. (5-4):

$$L = \frac{N^2 \mu A}{l} = \frac{1000^2 \times 3.77 \times 10^{-3} \text{ H/m} \times 1.25 \times 10^{-4} \text{ m}^2}{0.11 \text{ m}}$$

$$L = \mathbf{4.284\ H}$$

A coil wound on a doughnut-shaped core, as in Example 5-7, is called a *toroid*. A toroidally wound coil has efficient magnetic characteristics. Its core, being endless and entirely within the coil winding, concentrates the magnetic field to the maximum degree. A coil of this shape has its measured inductance close to the value calculated using Eq. (5-4). Coils of other shapes may have inductances which differ widely from the calculated values as a result of field inefficiencies.

An air gap in the core of the toroidal inductor of Example 5-7 would help to reduce the effects of saturation at high coil-current levels at the expense of the inductance at lower current levels. To give you some idea of the amount by which an air gap will influence inductance, a gap in the core 1 mm wide (about 1 percent of the core's length) would reduce the calculated inductance from 4.284 H, or 4284 mH, to 152 mH.

EXAMPLE 5-8 An inductor for a radio transmitter consists of a coil of wire wound on a plastic tube. Its dimensions are shown in Fig. 5-28. How many turns of wire are needed to give an inductance of 10 μH?

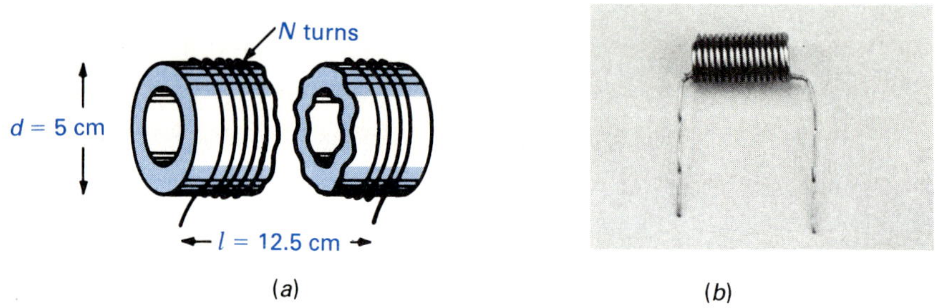

Fig. 5-28. Example 5-8: *(a)* Coil dimensions; *(b)* appearance of inductor.

Solution The core area A is given by

$$A = \pi r^2 = \pi \times (2.5 \times 10^{-2})^2 = 1.963 \times 10^{-3}\ \text{m}^2$$

The plastic tube on which the inductor is wound is nonmagnetic. Magnetically it is similar to a vacuum, with a relative permeability of 1, giving a total permeability $\mu = \mu_v$, 1.257×10^{-6} H/m.

Applying Eq. (5-4),

$$L = \frac{N^2 \mu A}{l} \tag{5-4}$$

$$N = \sqrt{\frac{Ll}{\mu A}} = \sqrt{\frac{10 \times 10^{-6}\ \text{H} \times 12.5 \times 10^{-2}\ \text{m}}{1.257 \times 10^{-6}\ \text{H/m} \times 1.963 \times 10^{-3}\ \text{m}^2}}$$

$$N = 22.5\ \text{turns}$$

In Example 5-8, the application of Eq. (5-4) results in a determination that 142 turns of wire are needed to give 10 μH of inductance on the plastic tube about 5 in long with a 2-in diameter. The equation applies with reasonable accuracy only to inductors that either are very long or are wound on closed iron cores. Therefore the result for the toroidal inductor of Example 5-7 is about right, while that for Example 5-8 is only a rough guide. In practice more than the calculated 22.5 turns would be wound on the plastic tube. Its turns would then be unwound until a measured value of 10 μH was obtained.

EXAMPLE 5-9 Calculate the relative permeability of the ferrite material in the core of the inductor shown in Fig. 5-29. The 1000-turn coil gives an inductance of 0.75 H.

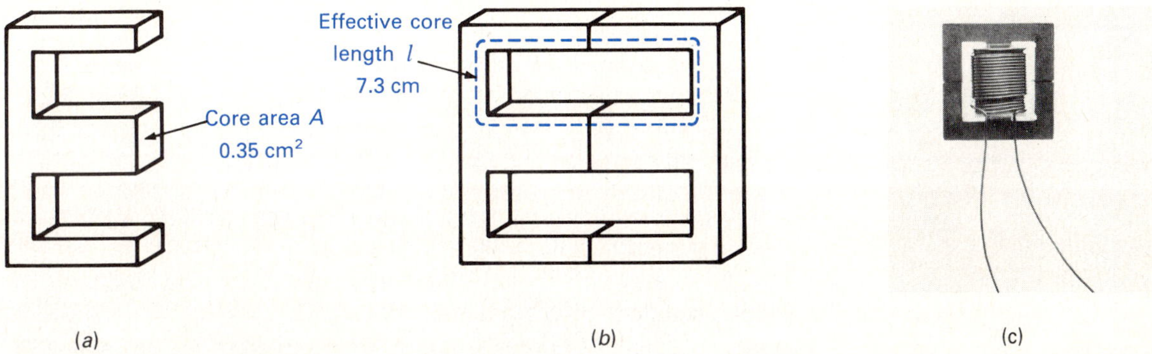

Fig. 5-29. Example 5-9: *(a)* Ferrite core molding. *(b)* Core made up of two moldings; the broken line shows the path of the magnetic field. *(c)* Inductor wound on the ferrite core.

Solution The core area is

$$A = 0.35\ \text{cm}^2 = 3.5 \times 10^{-5}\ \text{m}^2$$

The effective core length is the length of the magnetic field path,

$$l = 7.3\ \text{cm} = 7.3 \times 10^{-2}\ \text{m}$$

From Eq. (5-4),

$$L = \frac{N^2 \mu A}{l} \qquad (5\text{-}4)$$

$$\mu = \frac{Ll}{N^2 A} = \frac{0.75\ \text{H} \times 7.3 \times 10^{-2}\ \text{m}}{10^6 \times 3.5 \times 10^{-5}\ \text{m}^2} = 1.564 \times 10^{-3}\ \text{H/m}$$

Substituting the value of μ in Eq. (5-5),

$$\mu = \mu_v \mu_r \qquad (5\text{-}5)$$

$$\mu_r = \frac{\mu}{\mu_v} = \frac{1.564 \times 10^{-3}\ \text{H/m}}{1.257 \times 10^{-6}\ \text{H/m}}$$

$$\mu_r = 1244$$

REVIEW QUESTIONS FOR SECTION 5-8

1. What is the usual term applied to an inductor with more than one winding?
2. Give the equation for calculating the inductance of a perfect inductor.
3. What is the permeability of a material?
4. What are the two factors that give the value of the permeability of a material?
5. What is the symbol for permeability?

6. What is the unit of permeability?
7. How can changing the magnetization level of a core change its permeability?
8. Does inductance depend on static or incremental permeability?
9. What is the name given to a coil wound on a doughnut-shaped core?
10. An air gap in a core helps prevent what undesirable condition?
11. What effect does the air gap have on the inductance?

EXERCISE PROBLEMS FOR SECTION 5-8

1. Find the permeability of an inductor core made from Sendust, which has a relative permeability of 90 000. (*ans.* 0.1131 H/m)
2. Find the permeability of an inductor core made from Isoperm 36, which has a relative permeability of 62.
3. Assuming no losses, what is the inductance of a coil of 50 turns wound on a Sendust toroid with a cross-sectional area of 0.5 cm^2 and a mean coil length of 6 cm (μ_r of Sendust = 90 000)? (*ans.* 235.6 mH)
4. Assuming no losses, what is the inductance of an air-core coil, 1.5 cm in diameter and 5 cm long, consisting of 100 turns of No. 32 wire which has a resistance of 538 Ω/km?
5. For the coil in Prob. 4, find its inductance if all dimensions remain the same but:
   a. 25 turns are removed. (*ans.* 24.99 μH)
   b. It was made with 25 additional turns. (*ans.* 69.42 μH)
6. Find the equivalent-circuit values (resistance and inductance) for the coil in Prob. 4.

5-9 INDUCTOR TYPES

Commercial inductors range in inductance value from a few microhenrys to many henrys. They also range widely in size, shape, and weight. As the required characteristics of inductors are usually individual for each application, they are often specially designed and made rather than being bought from a line of mass-produced types.

The two main categories of inductors are air-core and iron-core. Air-core inductors are basically simple, but iron-core inductors appear in a large variety of forms.

Air-Core Inductors

Air-core inductors contain no magnetic iron. They are usually simple wire coils wound on plastic forms. These forms are likely to be tubular but may be solid. The inductor of Example 5-8, shown in Fig. 5-28, is air-cored.

The straight spiral coils of air-core inductors are said to be *solenoid wound.*

The absence of an iron core limits the inductance value for a given number of turns. This naturally results in air-core inductors being used in applications where low inductance values are desirable. The radio transmitter inductor of Example 5-8 is such an application. Air-core inductors appear in many types of equipment, such as radio and television sets, radar, and industrial electronic systems.

Photographs of some air-core inductors are shown in Fig. 5-30.

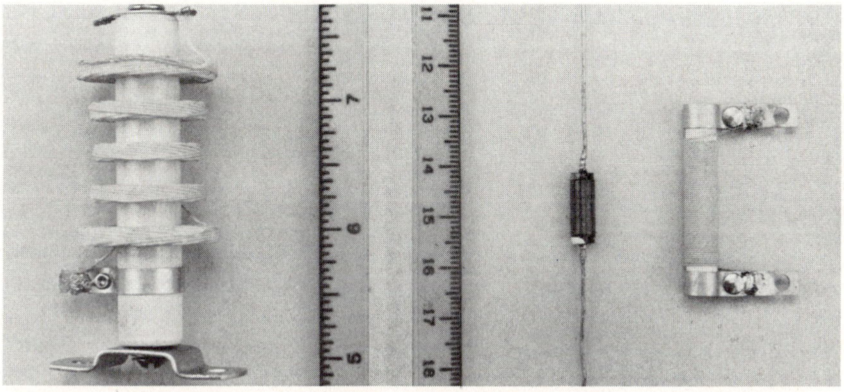

Fig. 5-30. Photo of air-core inductors. *(Left)* Honeycomb-wound. *(Center)* Solenoid-wound on plastic form. *(Right)* Solenoid-wound on ceramic form.

A *perfect* inductor is one that uses no overall energy if its magnetic field is built up and then reduced to zero. The total amount of energy such an inductor takes from a source while its field builds up is given back to the source during the field collapse. Practical inductors are not perfect; they waste some energy, mostly in the form of heat.

If an inductor's energy loss is significant to the operation of a circuit, its presence is recognized as resistance. The resistance is usually shown in a schematic diagram as an element separate from the inductance.

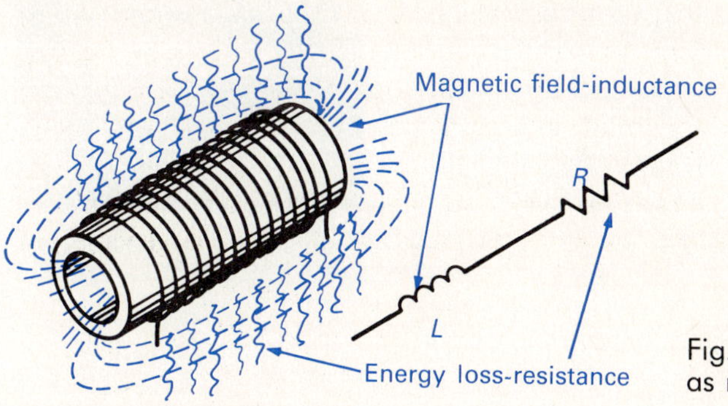

Fig. 5-31. Inductor loss as resistance.

Figure 5-31 shows an inductor's effective circuit made up of two components. One component, the inductance L, relates to the energy of the magnetic field and therefore is the component that is wanted in an inductor. The other, the resistance, represents energy loss and is not wanted.

In an air-core inductor the loss is produced by such factors as:

1. The resistance of the coil wire. This resistance is sometimes called ''ohmic resistance'' to differentiate it from the other factors that make up the inductor's resistance component.
2. Skin effect. This is the tendency for current to flow on the surface of a conductor under some conditions. When this happens in an inductor's winding, the wire diameter is, in effect, made smaller, so its resistance becomes higher.
3. Radiation. Energy radiated by inductors may be lost from the circuit.
4. Dielectric loss. Energy is dissipated by the insulating form on which the inductor's coil is wound. This loss is similar to that which occurs in the dielectric of a capacitor.

Iron-Core Inductors

(a)

Iron-core inductors come in many forms. The cores are made of various alloys of iron and other substances to give the characteristics needed for particular applications. The shapes of the cores also differ according to the application.

Iron-core inductors have all the losses of air-core inductors. In addition, they have loss factors introduced by the iron itself:

1. Eddy currents. Voltages are induced in *any* conductors within magnetic fields. The induced voltages will apply only to the wire coil of an inductor, but the iron core is a conductor too, so voltages are induced in it as well. A large mass of iron has low resistance, and therefore heavy currents may flow within a core, introducing substantial energy wastage. This *eddy-current loss* may be minimized by increasing the iron's resistance with suitable alloying and by *lamination*.

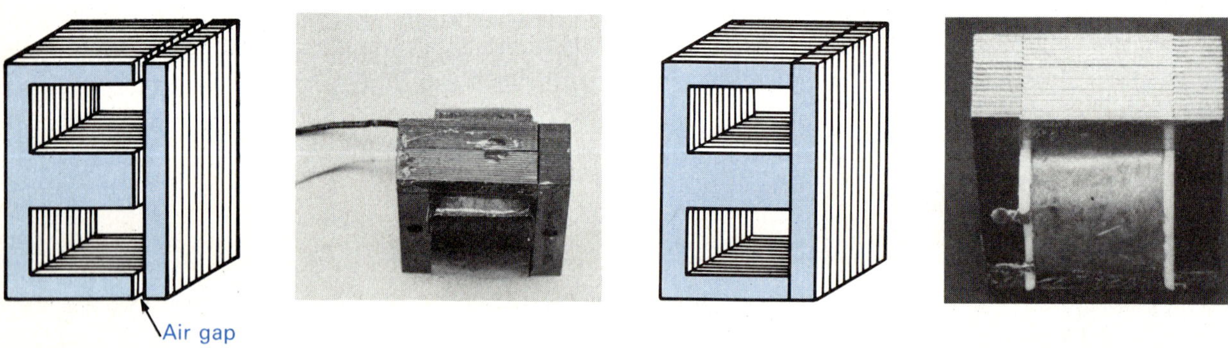

(b) (c)

Fig. 5-32. Laminated cores. *(a)* *E* and *I* laminations of magnetic iron; *(b)* laminations stacked in line give an air gap; *(c)* laminations interleaved (stacked alternately) leave no air gap.

A laminated core is built up from thin sheets of magnetic iron. In Fig. 5-32*a* is shown a pair of *E* and *I* laminations, popular for inductor cores. These laminations may be stacked as shown in Fig. 5-32*b* with all the *E*'s in line with each other and all the I's in line. This method of stacking laminations allows an air gap to be provided if this is desired for the inductor's characteristics. For maximum inductance, but with saturation limitations, the laminations may be *interleaved,* as in Fig. 5-32*c*, so that no air gap is left. The laminations are each coated with a varnish to prevent electrical contact between them and thus stop the circulation of eddy currents through the core stack.

For some purposes, preventing currents between core layers is not sufficient. The iron alloy may then be reduced to powder and mixed with an insulating material for molding into the required core shape. Such *powder, dust,* or *ferrite* cores have carried lamination to the point where eddy currents cannot flow in any direction. The Sendust cores are examples.

Some of the many available ferrite core shapes are shown in Fig. 5-33.

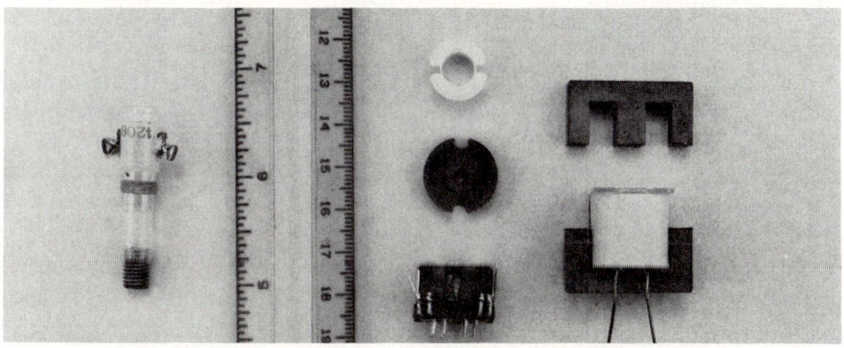

Fig. 5-33. Photo of ferrite core shapes. *(Left)* Threaded to give variable inductance. *(Center)* Pot type to fully enclose the winding. One of the two ferrite moldings that make up the core is shown along with a plastic form on which the coil is wound before enclosure by the core. *(Right)* E-type, molded in two halves.

2. Hysteresis. When the current in a coil increases, a completely demagnetized iron core becomes magnetized in the nonlinear manner shown by the curve in Fig. 5-34*a*.

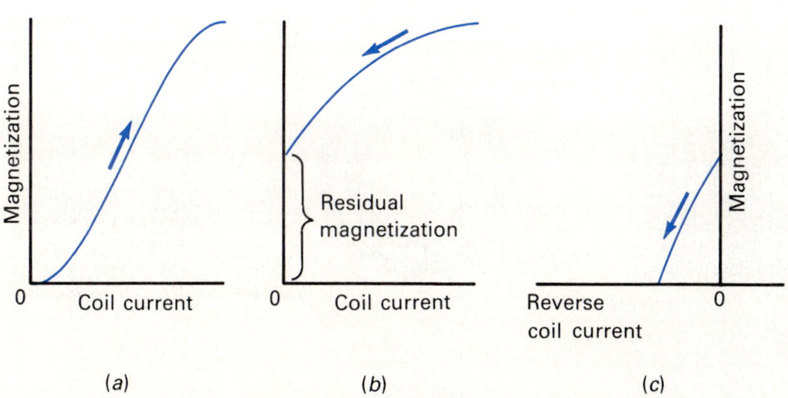

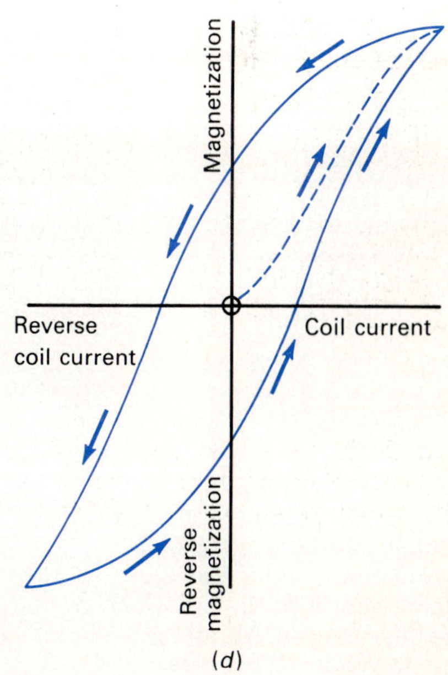

Fig. 5-34. Hysteresis. *(a)* Current rises from zero; magnetization rises from zero. *(b)* Current drops to zero; magnetization does not drop to zero. *(c)* Reversal of current direction is needed to reduce magnetization to zero. *(d)* Hysteresic curve. Change of magnetization does not follow current change. The broken line is the current-increase curve *(a)*. The current-decrease and current-reversal curves *(b)* and *(c)* are shown by the heavy line.

If the coil current is decreased to zero, the magnetization does not return to zero. Some residual magnetism is left in the core, as Fig. 5-34*b* shows.

The coil current must be reversed in direction to get the core magnetization back to zero; see Fig. 5-34*c*.

Alternately varying the magnitude and direction of the coil current, as often happens in inductive circuits, produces a magnetization curve similar to that of Fig. 5-34*d*.

The inability of magnetization to follow current changes is called *hysteresis* and is a characteristic of all magnetic iron alloys. Hysteresis constitutes another form of energy loss within a core.

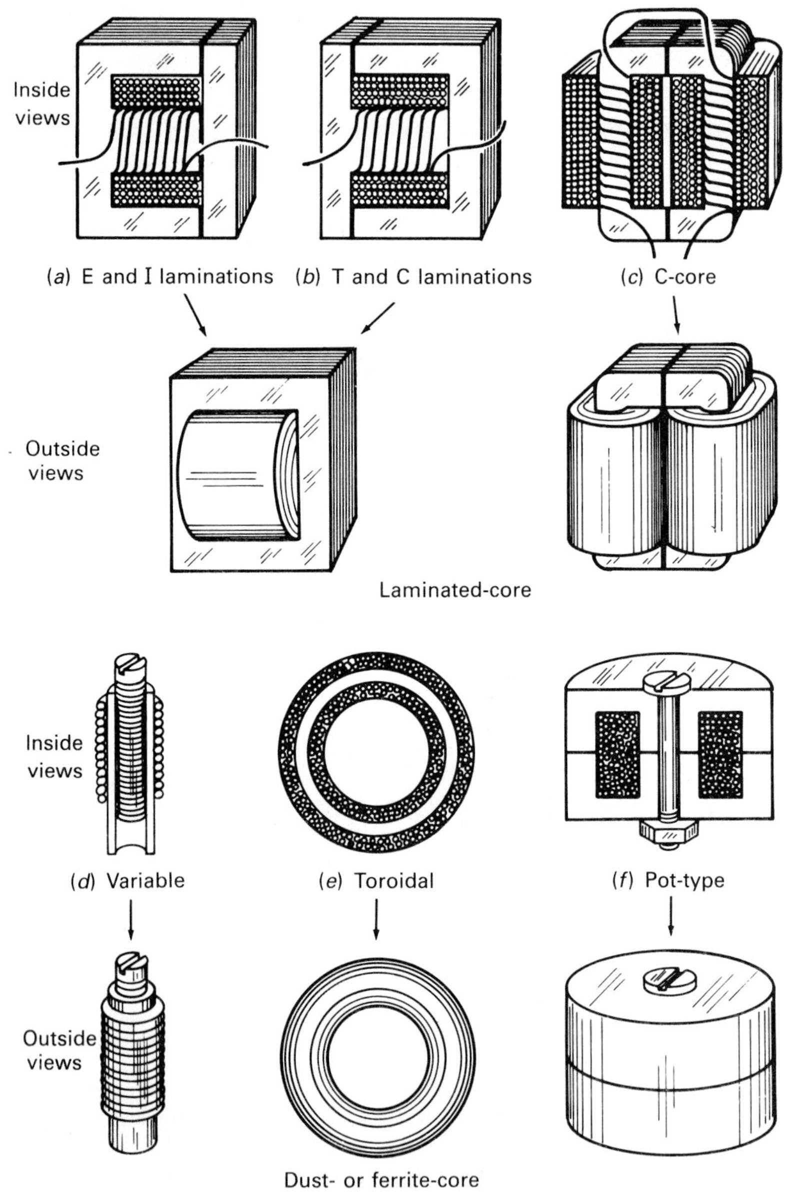

Fig. 5-35. Iron-core inductors. Laminated-core: *(a) E* and *I* laminations; *(b) T* and *C* laminations; *(c)* C core. Dust- or ferrite-core: *(d)* Variable; *(e)* toroidal; *(f)* pot type.

The coil of an iron-core inductor is usually wound in the *windows* of its core. For the laminated cores of Fig. 5-32, the coils will usually be wound around the central limb of the *E* laminations and fill the rectangular windows between it and the outer limbs, as shown in Fig. 5-35*a* and *b*. The *C-core* laminations of Fig. 5-35*c*, made by winding magnetic-iron strip into a roll which is then cut in two, will typically have two coils connected together. Figure 5-35*d* shows one version of a variable-inductance powder-core inductor in which the core is threaded so that its length within the core can be varied. The toroidal inductor of Fig. 5-35*e* has its coil wound around its doughnut-shaped powder-core. In Fig. 5-35*f* the coil is completely enclosed within the pot-type powder-core, resulting in an efficient magnetic design.

Inductor Symbols

The schematic symbols for inductors are shown in Fig. 5-36.

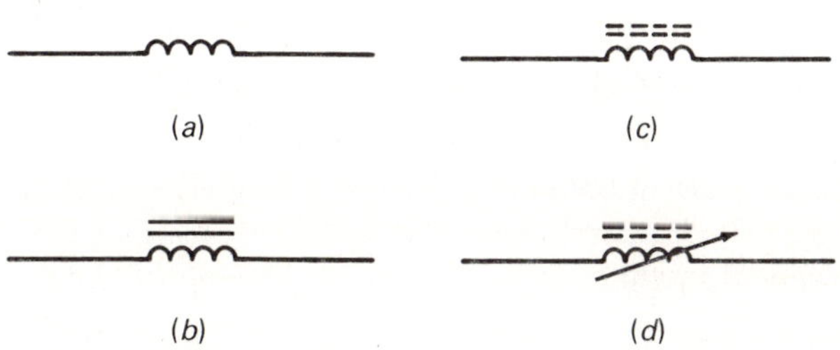

Fig. 5-36. Inductor symbols: *(a)* Air core; *(b)* iron core; *(c)* dust core; *(d)* variable.

For an air-core inductor, or for any inductor where the type is not important, the symbol of Fig. 5-36*a* is used. Where it is desired to show that an inductor has an iron core, Fig. 5-36*b* applies. The symbol of Fig. 5-36*c* shows that the inductor has a dust core, indicated by broken lines. In Fig. 5-36*d* the arrow symbol shows that the inductor has variable inductance.

REVIEW QUESTIONS FOR SECTION 5-9

1. When an inductor has significant losses, how is this shown schematically?
2. What are four factors contributing to inductor losses?
3. In an iron-core inductor, what are eddy currents?
4. How are eddy-current losses minimized?
5. What is the name given to the inability of magnetization to follow current changes in an iron-core inductor?

5-10 STRAY INDUCTANCE

Any current produces a magnetic field and, if the current changes value, inductive effects take place. Commercial inductors and transformers consist of conductors wound in coils to enhance inductive effects. But straight conductors, even the conductive surfaces of printed circuit boards, have inductance, and hence have induced voltages which are usually undesirable. Such undesired inductance is called *stray inductance*. Some examples now follow.

Stray Inductance in Components

All circuit components draw current. When the currents change, inductive effects, which can significantly affect performance, become apparent. The stray inductances of resistors, capacitors, and transistors are now briefly described. They are typical of unwanted inductance in circuit devices.

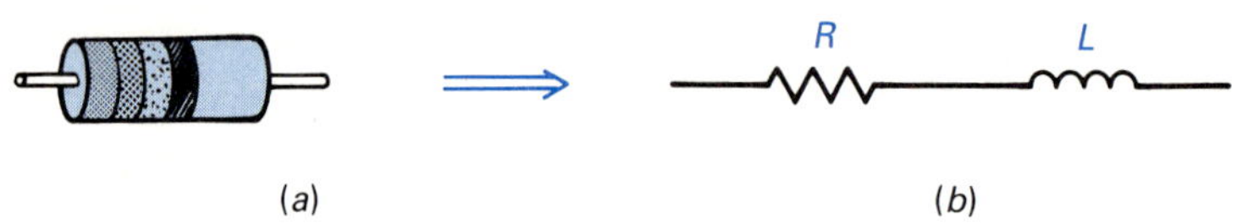

Fig. 5-37. Inductance of a resistor.

A resistor is intended to exhibit only resistance. But a resistor is a conductor so, when its current changes value, it becomes inductive, giving an equivalent circuit similar to that in Fig. 5-37*b*. Except in certain situations, resistors correctly selected for their application have insufficient inductance to affect performance seriously.

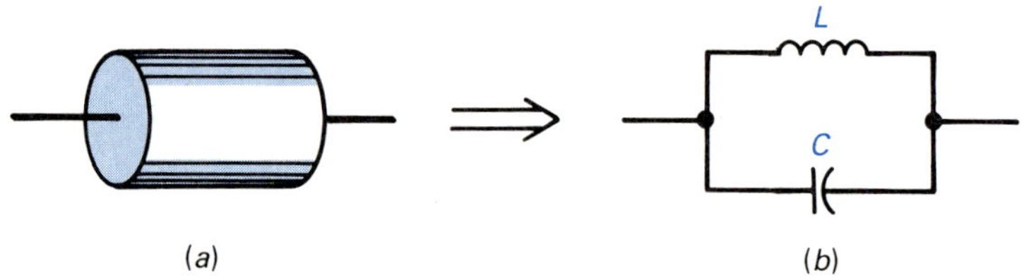

Fig. 5-38. Inductance of a capacitor.

A capacitor too has inductance. Some types of capacitors, particularly those that are of rolled-sandwich construction, are more inductive than others. Figure 5-38*b* shows how the presence of a capacitor's inductance may modify its schematic symbol if this inductance is important in the circuit.

All operating circuit components have some resistance, as they all dissipate power. They all have some capacitance, as they all have electric fields. They all have some inductance, as they all have magnetic fields.

Figure 5-39 is a generalized symbol for a resistor, a capacitor, or an inductor, taking account of power dissipation R, electric field C, and magnetic field L. The amount of resistance, capacitance, and inductance present in any practical component depends on its design. A well-designed capacitor, for example, has little resistance and inductance. The effect of undesirable component characteristics depends on the application. For use in the channel-selection circuits of a TV set,

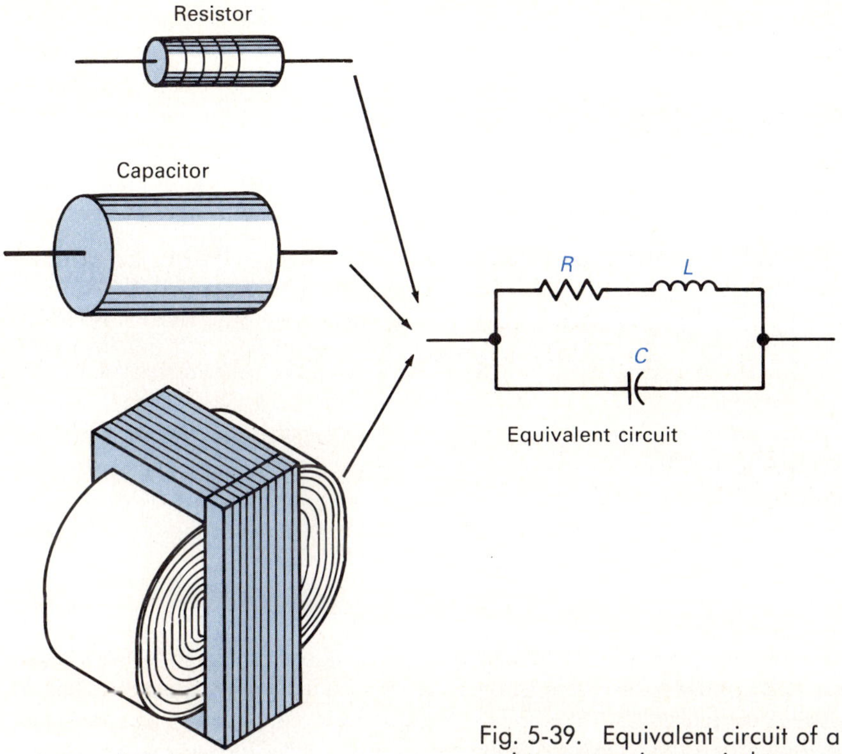

Fig. 5-39. Equivalent circuit of a resistor, capacitor, or inductor.

a capacitor must have very low inductance. A capacitor used in its power-supply circuit need not have such low inductance.

A transistor has three separate connections, one each to its base, emitter, and collector. Each of the small leads from the connecting pins to the internal elements has inductance, as do the elements themselves. Figure 5-40*c* shows the total effective inductance associated with each connection. These inductances, although small in value, have important effects which can render a particular transistor useless for some purposes.

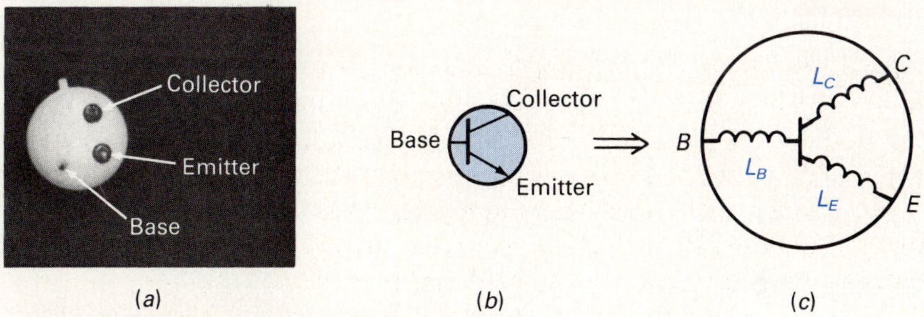

Fig. 5-40. Stray inductances of a transistor.

Integrated circuits are components consisting of many transistors on one small silicon chip. The internal stray inductance of these ICs imposes a limit on the speed at which their circuits can operate. This speed limitation may have a direct bearing on the performance of equipment, for instance, the data-handling capacity of a computer.

Transmission-Line and Cable Inductance

The inductances of the conductors in transmission lines and cables are so important that they must be considered when transmission performance is being determined. Together with its capacitance, the inductance of a line controls the main operating characteristics at high frequencies. The inductance is even involved in the time it takes for a signal to be transferred along a line. This leads to the deliberate design of delay lines in order to delay the time of arrival at the output of a signal applied at the input. Such delay lines are used in radar and other applications.

If the lines are long enough, it is not necessary for the frequency to be high for the inductive effects to be important. Therefore in long-distance power lines, such as those connecting hydroelectric generating stations with cities, the conductor inductance is great enough to be a factor in the planning of the system.

Stray Inductance in Equipment

Inductance is inherent in all conductors, including conducting surfaces. All wiring and all conductors on printed circuit boards have self-inductance. There is also mutual inductance between all of them wherever they appear in equipment. These inductances are usually of low value, and so their effects are usually small except at high operating frequencies or speeds. For example, modern computers operating at high speeds often carry out operations in times less than 1 μs. Such speeds necessitate reducing stray inductance to a minimum.

The importance of stray inductance will become clear when the relationships between inductance, time, and frequency are dealt with in later chapters.

REVIEW QUESTIONS FOR SECTION 5-10

1. What is the term given to inductance that is not intentionally put into a circuit in the form of an inductor?
2. Where might you expect to find the type of inductance referred to in question 1?
3. What are some of the effects of the type of inductance referred to in question 1?

SUMMARY OF EQUATIONS

$$v = L\,di/dt \qquad (5\text{-}1)$$

$$W = I^2L/2 \qquad (5\text{-}2)$$

$$v_2 = M\,di_1/dt \qquad (5\text{-}3)$$

$$L = N^2\mu A/l \qquad (5\text{-}4)$$

$$\mu = \mu_v\mu_r \qquad (5\text{-}5)$$

CHAPTER QUESTIONS

5-1. Compare the production of current in a resistor, a capacitor, and an inductor.

5-2. Compare the production of current in a single-winding inductor and a multiwinding transformer.

5-3. From an energy viewpoint, compare the behavior of a resistor, a capacitor, and an inductor.

5-4. When an inductor receives energy, where is it stored?

5-5. What happens to an inductor's current during charge?

5-6. What is the main difference between an inductor's voltage when the current is changing and when it is fixed in value?

5-7. What happens to an inductor's voltage during discharge?

5-8. What is meant when it is said that an inductor "opposes current change"?

5-9. If the current in an inductor is *doubled,* what happens to its stored *energy?*

5-10. Compare the *energy* in two inductors, each carrying the *same current* but one with *twice* the *inductance* of the other.

5-11. Compare the inductance of similar-dimension inductors with air and iron cores.

5-12. Why is it possible for a very high voltage to appear across the contacts of a switch that is opened in an inductive circuit?

5-13. How may contact arcing in an inductive circuit be minimized?

5-14. What is the essential difference between opening capacitive and inductive circuits?

5-15. Compare *self-* and *mutual-*inductance.

5-16. Why is the *same* unit used for both *self-* and *mutual-*inductance?

5-17. Compare *static* and *incremental* permeability.

5-18. In what circumstances may incremental permeability be important?

5-19. What is the significance of *saturating* an iron core?

5-20. Why is the *effective* resistance of an inductor higher than the *resistance of the wire* of which the winding is made?

5-21. What is the point of *laminating* an iron core?

5-22. What practical significance has the *hysteresis* of an iron core?

5-23. In what major way does the core of an inductor or transformer differ if used for low frequencies and high frequencies?

5-24. What is the effect of introducing a small air gap into the iron core of an inductor or transformer?

5-25. In what circumstances might an air-gapped iron core be used?

CHAPTER PROBLEMS

5-1. What is the voltage across a 2-H inductor if its current rises from zero to 250 mA in 50 ms in a linear manner? (*ans.* 10 V)

5-2. What is the voltage across a 4-H inductor if its current rises from zero to 250 mA in 50 ms in a linear manner?

5-3. What is the voltage across a 2-H inductor if its current rises from zero to 250 mA in 100 ms in a linear manner? (*ans*. 5 V)

5-4. What is the voltage across a 2-H inductor if its current drops from 250 mA to zero in 100 ms?

5-5. What is the voltage across the secondary (output) winding of a transformer if its primary (input) winding current rises from zero to 250 mA in 100 ms? The mutual inductance between the two windings is 2 H. (*ans*. 5 V)

5-6. How much energy is stored in a 2-H filter inductor carrying a current of 250 mA?

5-7. A coil of wire is to be wound on a nickel-iron core with a relative permeability of 5000. The inductance of the coil is to be 2 H. If the effective length of the core is 10 cm and its diameter is 1 in (area of 507×10^{-6} m^2), approximately how many turns of wire will be needed? (*ans*. approximately 250)

5-8. If the inductor in Prob. 5-7 had its core made of nickel-iron of relative permeability 2500, what would be its inductance?

5-9. If an inductor were made with a core of the same dimensions as that of Probs. 5-7 and 5-8 but with a relative permeability of 2500, how many turns of wire would be needed to give the original inductance of 2 H? (*ans*. approximately 350)

5-10. If two identical coils of wire were wound on cores having permeabilities at the extremes of the values given in Table 5-1, compare their inductances.

6

SERIES CIRCUITS

Most of the quantities involved in electric circuits, together with their units and symbols, were introduced in Chaps. 1 to 5. The next few chapters will use this background to help you become familiar with electrical fundamentals.

You have now met generators with fixed polarities and currents with fixed directions: *direct current (dc)*. We have to deal with dc circuits in which voltages and currents change value at times: *varying dc*. Often we meet circuits in which generator polarity changes, giving regular reversals of current direction: *alternating current (ac)*.

Circuits with constant-value dc, *steady dc,* seem to be easiest to understand. Fortunately all the circuit principles that apply to steady dc also apply to varying dc and ac. The calculations are more complicated if the currents change value or direction, but much can be learned about all circuits by learning what happens with steady direct current (usually simply called "direct current").

You will be *analyzing* circuits, studying their behavior by finding out the values of voltage, current, resistance, etc. This is helped by problem solving. Certain steps are necessary to solve problems effectively:

1. Identify the given data; what is known?
2. Identify the required data; what needs to be known?
3. Decide whether any of the given data are redundant, not needed to solve the problem.
4. Where possible, draw a schematic diagram showing the given and required data.
5. Make a data list if a schematic diagram is not needed for the problem solution.
6. Solve the problem in a logical manner, going from the given data to the required data.

When solving problems, get into the habit of using accepted terms, SI units, and correct symbols. At each stage check that your solution is reasonable. Remember the rule: If it is *reasonable,* it may be *right or* it may be *wrong;* if it is *not reasonable, it must be wrong.*

Circuits are formed by interconnecting elements into various patterns or configurations, the most common of these being called *series* and *parallel.* Series circuits are regarded as easier to understand, so they will be dealt with first.

The characteristics of series circuits may seem obvious and not worth your time studying. This would be a mistake, as it is impossible to understand complex circuits without a fine-detail knowledge of simple ones. You must be able to identify a series grouping of elements in a complex network. You must be able to work out series-circuit resistance, current, voltage, and power values and know the effect of changing these values. You need to be able to see how the characteristics of series circuits relate to practical situations. When working with this text, check every example and solve every problem.

6-1 WHAT IS A SERIES CIRCUIT?

Circuit elements are said to be connected in *series* when they are joined together end to end, as in a chain. Figure 6-1 shows a simple series circuit consisting of two elements connected to a battery.

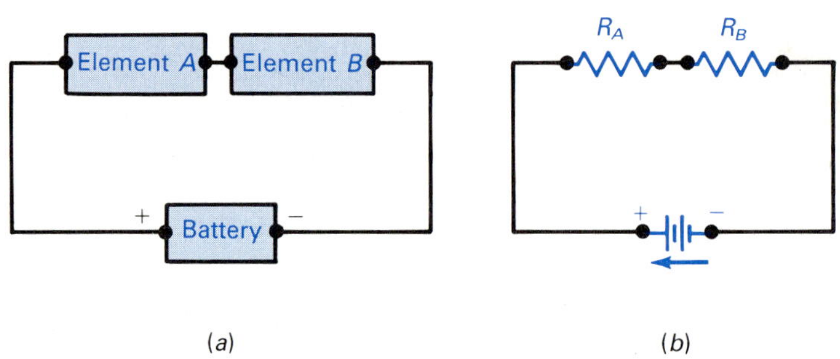

Fig. 6-1. Series circuit: *(a)* Block diagram; *(b)* schematic diagram.

A battery is the most common type of dc generator, giving a steady voltage to the circuit of Fig. 6-1. Elements *A* and *B* in the circuit represent resistors.

In a series circuit any one end, or *terminal,* of an element can be connected to only one other element.

The term "cascade" is an alternative for "series."

6-2 SERIES-CIRCUIT CHARACTERISTICS

The major characteristics of a series circuit are:

1. There is one current at all parts of the circuit.
2. Elements have individual voltages.
3. Voltages are additive.
4. The emf equals the sum of the element voltages.

5. Resistances are additive.
6. Powers and energies are additive.

Each of these characteristics will now be examined in detail. The principles involved will be discussed, and examples given to demonstrate their application.

One Current

In a series circuit there is only one possible current.

In Fig. 6-2 the only possible path for the current I is from the battery V, through resistors R_A and R_B in turn, and then back to the battery. The current is common to all elements.

The current at any point in the circuit equals the amount of charge which moves past that point in one second. The charge movement is a continuous sequence, with electron charges moving progressively from atom to atom. The number of charges moving is constant throughout the circuit.

An obvious result of this common-current characteristic is that in a series circuit it is possible to measure only one current value. If the battery delivers 1 A of current to R_A, then the same 1 A will exist between R_A and R_B and from R_B back to the battery. What might not be so obvious is that the same 1 A of current will exist *within* resistors R_A and R_B and within the battery. It is easy to measure the 1 A of current between the elements, but it is not easy to measure it within them.

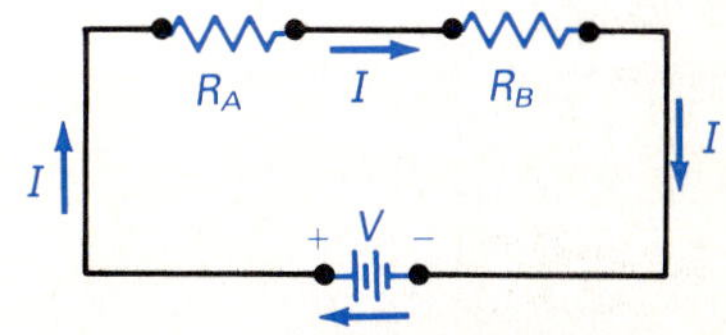

Fig. 6-2. There is only one current in a series circuit.

Individual Voltages

The current I in Fig. 6-2 *develops individual voltages across each* of the *resistance* elements. These are shown in Fig. 6-3 as the voltages V_A and V_B.

By applying Ohm's law, the values of these voltages can be determined:

$$V = IR \qquad (3\text{-}2)$$
$$V_A = IR_A$$
$$V_B = IR_B$$

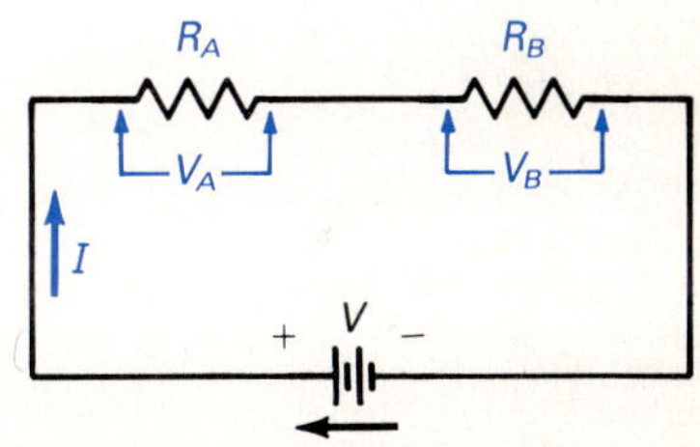

Fig. 6-3. Individual element voltages in a series circuit.

EXAMPLE 6-1 A series circuit consisting of three resistors, 2, 8, and 20 Ω, connected to a battery has a current of 2 A. What voltage exists across each resistor?

Solution (Fig. 6-4)

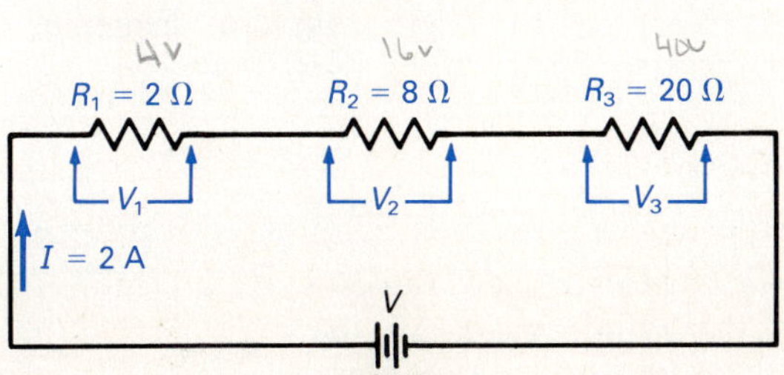

Fig. 6-4. Example 6-1: Circuit.

Applying Eq. (3-2),

$$V = IR \qquad (3\text{-}2)$$
$$V_1 = 2\text{ A} \times 2\ \Omega = \mathbf{4\ V}$$
$$V_2 = 2\text{ A} \times 8\ \Omega = \mathbf{16\ V}$$
$$V_3 = 2\text{ A} \times 20\ \Omega = \mathbf{40\ V}$$

It should be noted from Example 6-1 that *a voltage in a series circuit is proportional to resistance*. $R_2 = 4 \times R_1$; so $V_2 = 4 \times V_1$. $R_3 = 10 \times R_1$; so $V_3 = 10 \times V_1$.

Voltages Additive

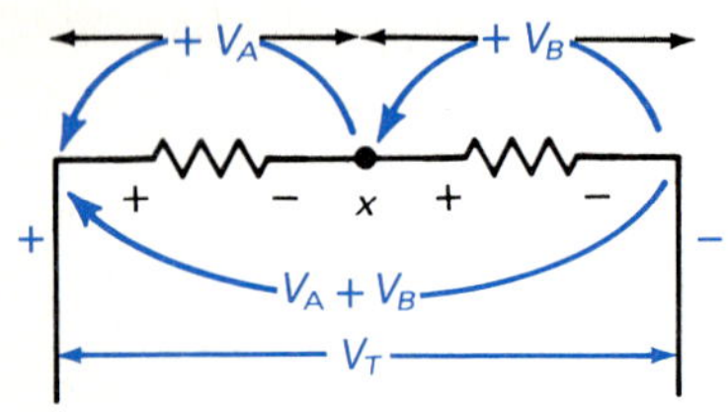

Fig. 6-5. Series-circuit voltages are additive.

In Fig. 6-5 the voltages V_A and V_B are shown with their polarities. With reference to the negative end of the circuit, point x is V_B volts positive. With reference to point x, the positive end of the circuit is V_A volts positive. Therefore the positive end is $V_A + V_B$ volts more positive than the negative end. These statements are summed up by the equation

$$V_T = V_A + V_B$$

This relationship is not limited to circuits with just two elements. The *total voltage across any number of resistors in series is the sum of the voltages across each of the resistors*. This can be written as

$$V_T = V_1 + V_2 + V_3 + \cdots + V_n \qquad \textbf{(6-1)}$$

EXAMPLE 6-2 A series network has a current of 5 A. The network consists of three elements of 10, 20, and 50 Ω. What is the total voltage across the circuit?

Solution (Fig. 6-6)

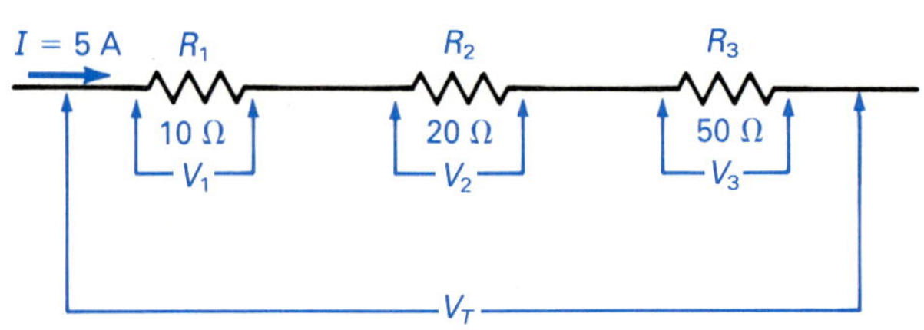

Fig. 6-6. Example 6-2: Data.

Applying Eq. (3-2),

$$V = IR \qquad (3\text{-}2)$$
$$V_1 = IR_1 = 5\text{ A} \times 10\ \Omega = 50\text{ V}$$
$$V_2 = IR_2 = 5\text{ A} \times 20\ \Omega = 100\text{ V}$$
$$V_3 = IR_3 = 5\text{ A} \times 50\ \Omega = 250\text{ V}$$

Applying Eq. (6-1),

$$V_T = V_1 + V_2 + V_3 \qquad (6\text{-}1)$$

$$V_T = 50 \text{ V} + 100 \text{ V} + 250 \text{ V} = \mathbf{400 \text{ V}}$$

EMF Equals the Sum of the Element Voltages (Kirchhoff's Voltage Law)

When the series group of elements makes up the total network[1] across the generator, Eq. (6-1) can be extended to include its emf.

Referring to Fig. 6-7, we can see that the battery emf and the total series voltage must be equal because they are connected across the same points. Therefore,

$$V = V_T \qquad \mathbf{(6\text{-}2)}$$

$$V = V_1 + V_2 + V_3 + \cdots + V_n$$

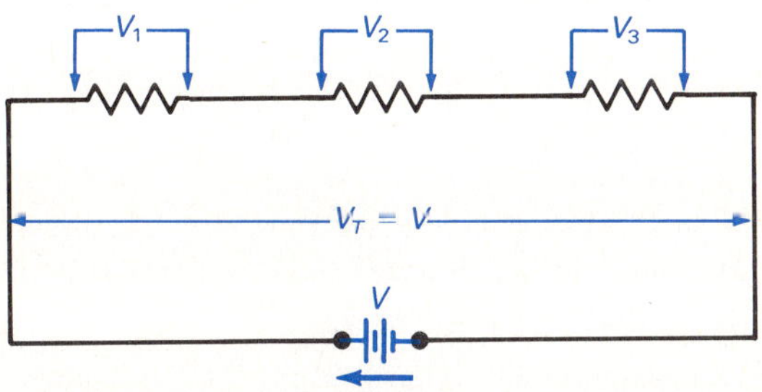

Fig. 6-7. Series-circuit emf equals the sum of the element voltages.

Kirchhoff's voltage law is usually stated:

In any closed loop the algebraic sum of the emfs equals the algebraic sum of the element voltages.

Any continuous conducting path is a *closed loop*. The circuit in Fig. 6-7 is a closed loop, and Eq. (6-2) is the related Kirchhoff's law expression.

The *algebraic sum* is the sum taking into account the polarities.

EXAMPLE 6-3 Four resistors are connected in series. Their resistance values are 1 kΩ, 5 kΩ, 2.5 kΩ, and 500 Ω. The voltage across the 5-kΩ resistor is 10 V. If the battery supplying energy to the network is connected directly across the resistor chain, what is its emf?

Solution As with most problems, there are several ways of obtaining a solution. Two methods will be shown.

[1]It is usual to consider only those elements in a schematic diagram which are separately shown. When the characteristics of some part of a circuit are important for circuit operation, these characteristics must be shown as distinct circuit elements. For example, if the conductors connecting elements have resistances which significantly influence the circuit currents, then resistance elements must be shown to account for these conductors, as will be discussed later.

Method 1 Apply Ohm's law, first to obtain the value of the circuit current, and then to obtain the values of the element voltages.

$$
\begin{aligned}
I &= \frac{V_2}{R_2} = \frac{10\ \text{V}}{5\ \text{k}\Omega} = 2\ \text{mA} \\
V_1 &= IR_1 = 2\ \text{mA} \times 1\ \text{k}\Omega = 2\ \text{V} \\
V_2 &= 10\ \text{V} \qquad \text{(given data)} \\
V_3 &= IR_3 = 2\ \text{mA} \times 2.5\ \text{k}\Omega = 5\ \text{V} \\
V_4 &= IR_4 = 2\ \text{mA} \times 500\ \Omega = 1\ \text{V} \\
V_T &= V_1 + V_2 + V_3 + V_4 \qquad (6\text{-}1) \\
&= 2\ \text{V} + 10\ \text{V} + 5\ \text{V} + 1\ \text{V} = 18\ \text{V} \\
V &= V_T \qquad (6\text{-}2) \\
&= \mathbf{18\ V}
\end{aligned}
$$

Method 2 Resistance ratios can be used to obtain voltage values. As an example, consider V_1 and V_2 in Fig. 6-8:

$$\frac{V_1}{V_2} = \frac{IR_1}{IR_2} = \frac{R_1}{R_2}$$

Therefore $$V_1 = V_2 \frac{R_1}{R_2} \qquad \text{or} \qquad V_2 = V_1 \frac{R_2}{R_1}$$

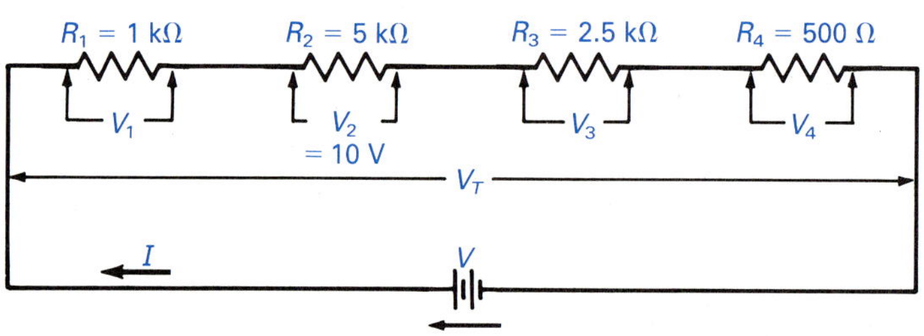

Fig. 6-8. Example 6-3: Circuit.

Applying this principle leads to the solution without calculating the value of the common current.

$$
\begin{aligned}
V_1 &= V_2 \frac{R_1}{R_2} = 10\ \text{V}\,\frac{1\ \text{k}\Omega}{5\ \text{k}\Omega} = 2\ \text{V} \\
V_2 &= 10\ \text{V} \qquad \text{(given data)} \\
V_3 &= V_2 \frac{R_3}{R_2} = 10\ \text{V}\,\frac{2.5\ \text{k}\Omega}{5\ \text{k}\Omega} = 5\ \text{V} \\
V_4 &= V_2 \frac{R_4}{R_2} = 10\ \text{V}\,\frac{0.5\ \text{k}\Omega}{5\ \text{k}\Omega} = 1\ \text{V}
\end{aligned}
$$

$$V = V_T = V_1 + V_2 + V_3 + V_4 = 2\ \text{V} + 10\ \text{V} + 5\ \text{V} + 1\ \text{V} = \mathbf{18\ V}$$

Resistances Additive

If, from an electrical measurement viewpoint, two networks are the same, they are said to be *equivalent*.

If the ammeters and voltmeters (A and V) in Fig. 6-9a and b indicate the same currents and voltages,[1] then network A is equivalent to network B.

Figure 6-10 shows network A as consisting of three resistors in series, while network B consists of just one resistor. If the currents and voltages measured at the terminals are the same in each case, the single resistance R_T is equivalent to the total of the three resistances R_1, R_2, and R_3 in series.

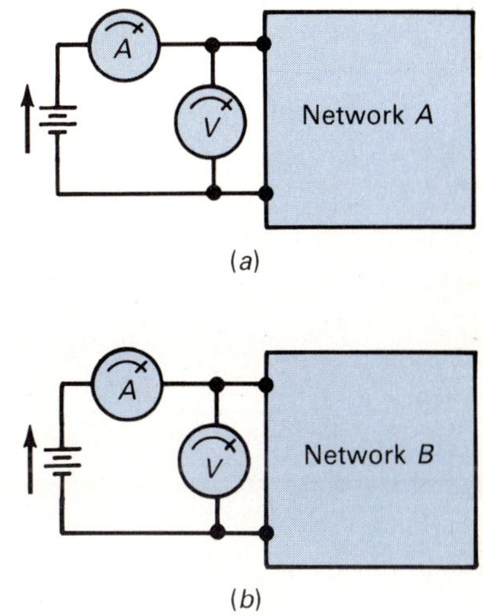

Fig. 6-9. Equivalent networks.

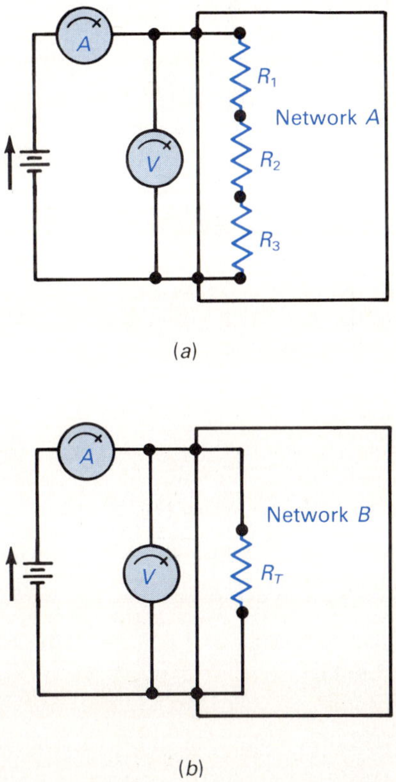

Fig. 6-10. One resistance equivalent to three in series.

The equivalent resistance of a network is often called the *total resistance*, hence the frequently used symbol R_T.

"Equivalent" does not literally mean "the same as." If we take a number of resistors which are physically the same size and use them to construct two net-

[1] "Ammeter" is an abbreviation of "ampere meter"; an *ammeter* is an instrument for measuring the magnitudes of currents. Similarly, a *voltmeter* is an instrument for measuring voltage magnitudes.

works, the resulting network A, with three resistors, will be capable of dissipating more power than network B, with only one resistor. Thus the two networks may be equivalent from the total resistance viewpoint, but not from that of power-handling capability.

In any specific case, it is essential that the conditions of equivalence be clearly understood. The terms of the equivalence must be stated or implied by the situation.

When we say that R_T in Fig. 6-10*b* is the equivalent of R_1, R_2, and R_3 in Fig. 6-10*a*, we only mean that the total resistance is the same in each case. Any relationship which naturally follows from this (for example, the same emf results in the same current) will also be true.

Figure 6-11*a* and *b* represents equivalent networks. With the same current I the total voltage V_T will be the same.

Applying Ohm's law to Fig. 6-11*a*,

$$V_T = IR_T$$

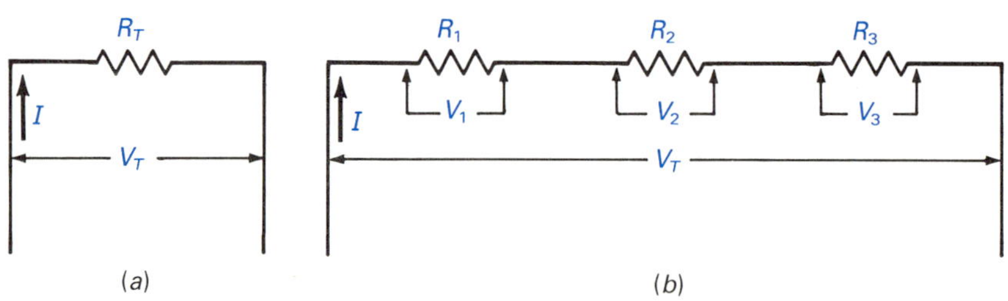

Fig. 6-11. Single-element network *(a)* to be equivalent to three-element network *(b)*.

Applying Eq. (6-1) and Ohm's law to Fig. 6-11*b*,

$$V_T = V_1 + V_2 + V_3 \qquad (6\text{-}1)$$
$$V_T = IR_1 + IR_2 + IR_3 = I(R_1 + R_2 + R_3)$$

We have said that the total voltage V_T is the same for each circuit if the currents are the same, so that

$$IR_T = I(R_I + R_2 + R_3)$$
$$R_T = R_1 + R_2 + R_3$$

This relationship applies to any number of elements connected in series:

$$R_T = R_1 + R_2 + R_3 + \cdots + R_n \qquad \textbf{(6-3)}$$

The *total resistance of a number of resistors* connected *in series* is equivalent to the *sum of the individual resistance values.*

EXAMPLE 6-4 What is the total resistance of four resistors connected in series if their individual resistance values are 1 MΩ, 1.5 MΩ, 150 kΩ, and 50 000 Ω?

Solution (Fig. 6-12)

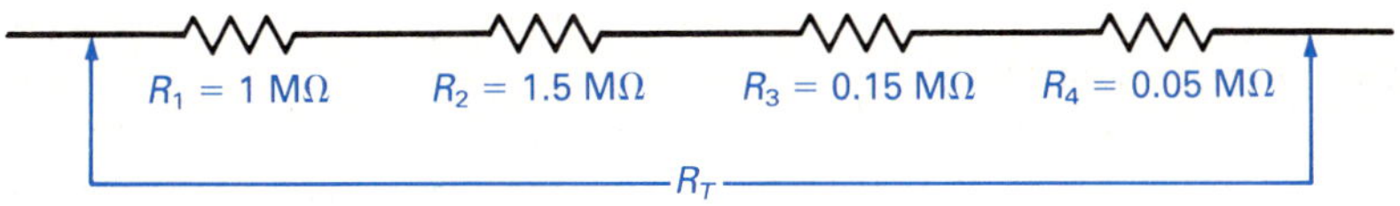

Fig. 6-12. Example 6-4: Data.

$$R_T = R_1 + R_2 + R_3 + R_4 \qquad (6\text{-}3)$$
$$R_T = 1\ \text{M}\Omega + 1.5\ \text{M}\Omega + 0.15\ \text{M}\Omega + 0.05\ \text{M}\Omega = \mathbf{2.7\ M\Omega}$$

EXAMPLE 6-5 What value resistor, connected in series with a 1.8-kΩ resistor, will result in a total resistance of 2 kΩ?

Solution (Fig. 6-13)

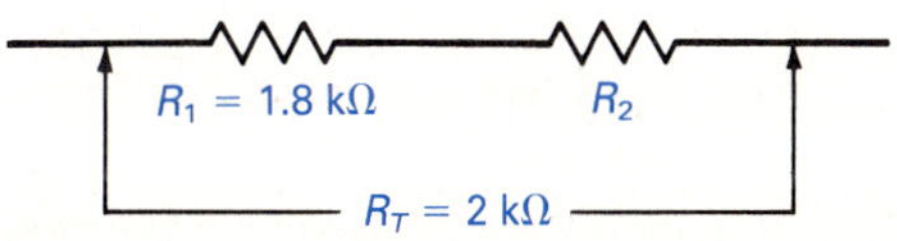

Fig. 6-13. Example 6-5: Data.

$$R_T = R_1 + R_2 \qquad (6\text{-}3)$$
$$R_2 = R_T - R_1 = 2\ \text{k}\Omega - 1.8\ \text{k}\Omega = \mathbf{0.2\ k\Omega} \quad \text{or} \quad \mathbf{200\ \Omega}$$

Powers and Energies Additive

Referring back to Eq. (3-6) ($P = I^2R$), we see that, when the current in a number of elements is the same, the power dissipated by each element is directly proportional to its resistance value. Furthermore, combining Eq. (3-6) with Eq. (2-12) ($W = Pt$) gives $W = I^2Rt$, showing that, over a stated time period, the energy converted by each of the elements will be directly proportional to its resistance value. Hence, since the total equivalent resistance of a number of resistors in series is equal to the sum of the individual resistances, the total power dissipated and the total energy converted will be equal to the sums of the individual powers and energies.

$$P_T = P_1 + P_2 + P_3 + \cdots + P_n \qquad \mathbf{(6\text{-}4)}$$
$$W_T = W_1 + W_2 + W_3 + \cdots + W_n \qquad \mathbf{(6\text{-}5)}$$

EXAMPLE 6-6 Resistors of 10, 20, and 100 Ω are connected in series across an emf of 26 V. What power is dissipated by each resistor, and what total power does the battery supply to the network?

Solution (Fig. 6-14)

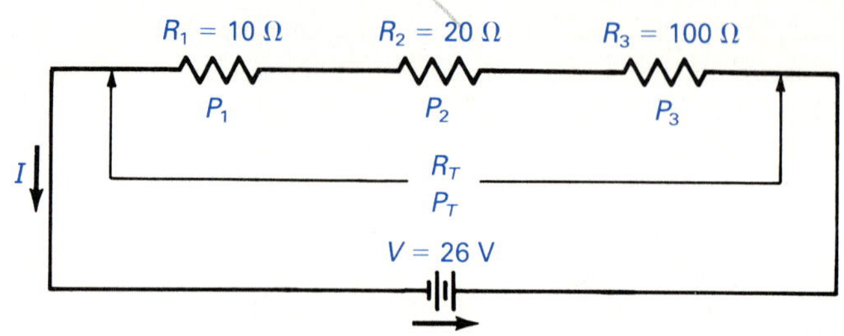

Fig. 6-14. Example 6-6: Circuit.

$$R_T = R_1 + R_2 + R_3 \quad (6\text{-}3)$$
$$R_T = 10\ \Omega + 20\ \Omega + 100\ \Omega = 130\ \Omega$$
$$I = \frac{V}{R_T} = \frac{26\ \text{V}}{130\ \Omega} = 0.2\ \text{A} \quad \text{or} \quad 200\ \text{mA}$$
$$P_1 = I^2R_1 = (0.2\ \text{A})^2 \times 10\ \Omega = \mathbf{0.4\ W}$$
$$P_2 = I^2R_2 = (0.2\ \text{A})^2 \times 20\ \Omega = \mathbf{0.8\ W}$$
$$P_3 = I^2R_3 = (0.2\ \text{A})^2 \times 100\ \Omega = \mathbf{4\ W}$$
$$P_T = P_1 + P_2 + P_3 \quad (6\text{-}4)$$
$$P_T = 0.4\ \text{W} + 0.8\ \text{W} + 4\ \text{W} = \mathbf{5.2\ W}$$

The total power could have been calculated using the emf and the current, as in Eq. (2-14):

$$P_T = VI = 26\ \text{V} \times 0.2\ \text{A} = \mathbf{5.2\ W}$$

EXAMPLE 6-7 Two 120-V 60-W light bulbs are connected in series across a 240-V supply. What total energy is converted per hour?

Solution (Fig. 6-15)

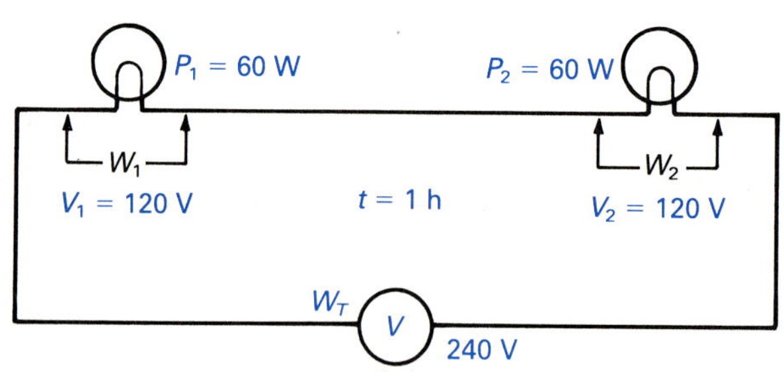

Fig. 6-15. Example 6-7: Circuit.

For each bulb Eq. (2-12) can be applied:

$$W_1 = W_2 = P_1t = P_2t = 60\ \text{W} \times 1\ \text{h} = 60\ \text{Wh} \quad \text{or} \quad 216\ \text{kJ}$$
$$W_T = W_1 + W_2 \quad (6\text{-}5)$$
$$W_T = 2 \times 60\ \text{Wh} = \mathbf{120\ Wh} \quad \text{or} \quad \mathbf{432\ kJ}$$

REVIEW QUESTIONS FOR SECTION 6-2

1. What are six characteristics of series circuits?
2. How many current paths are there in a series circuit?
3. What current exists within a load in a series circuit as compared with that in the generator?
4. To what is the total voltage across a number of series elements equal?
5. What relationship is there between the emfs and the pds in a series circuit?
6. To what law does Question 5 refer?
7. What relationship is there between the total resistance of a series circuit and any individual resistance?
8. With the current constant, what relationship is there between power dissipated by individual resistors and the values of the resistors?

6-3 CALCULATIONS FOR SERIES CIRCUITS

The various quantities found in series circuits are shown in Fig. 6-16. The method used for calculating the value of a quantity depends on the given and required data. Examples show how a method may be chosen and used.

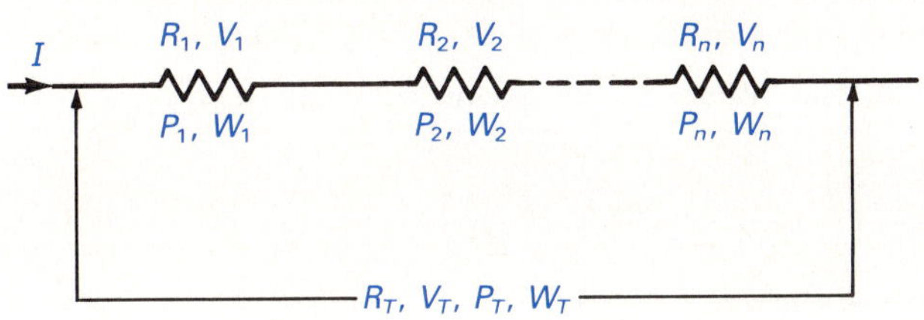

Fig. 6-16. Generalized symbols for a series circuit.

Current I

The circuit current can be calculated by applying Ohm's law to the complete network or to any part of it:

$$I = \frac{V_T}{R_T} \quad \text{or} \quad \frac{V_1}{R_1} \quad \text{or} \quad \frac{V_2}{R_2} \quad \text{or} \quad \cdots \quad \text{or} \quad \frac{V_n}{R_n}$$

EXAMPLE 6-8 Find the current in the circuit in Fig. 6-17.

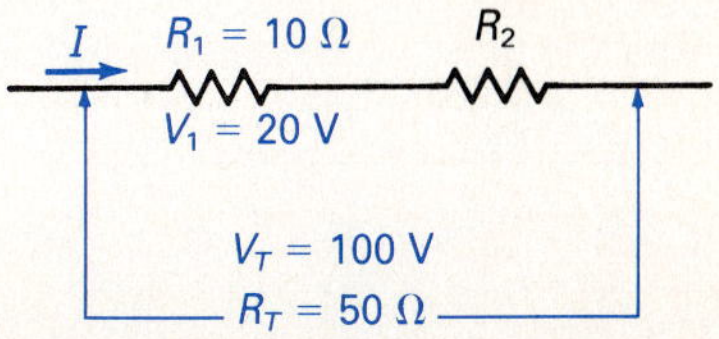

Fig. 6-17. Example 6-8: Data.

Solution Method 1 Applying Ohm's law to the total circuit,

$$I = \frac{V_T}{R_T} = \frac{100\text{ V}}{50\ \Omega} = \mathbf{2\ A}$$

Method 2 Applying Ohm's law to one element (any element may be selected),

$$I = \frac{V_1}{R_1} = \frac{20\ \text{V}}{10\ \Omega} = \mathbf{2\ A}$$

Total Voltage V_T

The total voltage across the network can be calculated by adding the individual element voltages or by the application of Ohm's law:

$$V_T = V_1 + V_2 + \cdots + V_n \qquad (6\text{-}1)$$

or

$$V_T = IR_T$$

The Ohm's law method can be extended to a resistance ratio form:

$$V_T = IR_T = \frac{V_1}{R_1} R_T = V_1 \frac{R_T}{R_1}$$

The resistance ratio equation leads to a general equation:

$$V_T = V_1 \frac{R_T}{R_1} \quad \text{or} \quad V_2 \frac{R_T}{R_2} \quad \text{or} \quad \cdots \quad \text{or} \quad V_n \frac{R_T}{R_n} \qquad \mathbf{(6\text{-}6)}$$

EXAMPLE 6-9 Find the total voltage in the circuit in Fig. 6-18.

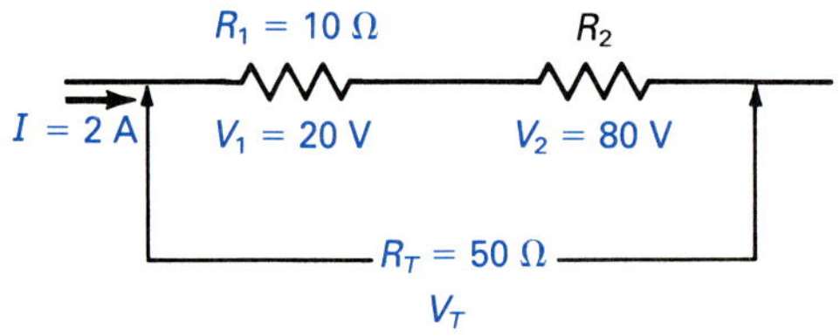

Fig. 6-18. Example 6-9: Data.

Solution Method 1 Sum of the pds:

$$V_T = V_1 + V_2 \qquad (6\text{-}1)$$

$$V_T = 20\ \text{V} + 80\ \text{V} = \mathbf{100\ V}$$

Method 2 Applying Ohm's law to the complete circuit,

$$V_T = IR_T = 2\ \text{A} \times 50\ \Omega = \mathbf{100\ V}$$

Method 3 Resistance ratio:

$$V_T = V_1 \frac{R_T}{R_1} \qquad (6\text{-}6)$$

$$V_T = 20\ \text{V}\ \frac{50\ \Omega}{10\ \Omega} = \mathbf{100\ V}$$

Individual Voltages $V_1, V_2, \ldots, V_n$

The voltage across an individual element can be calculated by the direct application of Ohm's law:

$$V_1 = IR_1$$

or by extending this equation to a resistance ratio form by substituting a suitable ratio V/R for I.

$$V_1 = \frac{V_T}{R_T} R_1$$

or $$V_1 = V_T \frac{R_1}{R_T} \quad \text{or} \quad V_2 \frac{R_1}{R_2} \quad \text{or} \quad \cdots \quad \text{or} \quad V_n \frac{R_1}{R_n} \tag{6-7}$$

The individual voltages V_2 to V_n, if unknown, can be calculated using similar methods.

EXAMPLE 6-10 What is the value of voltage V_1 in the circuit in Fig. 6-19?

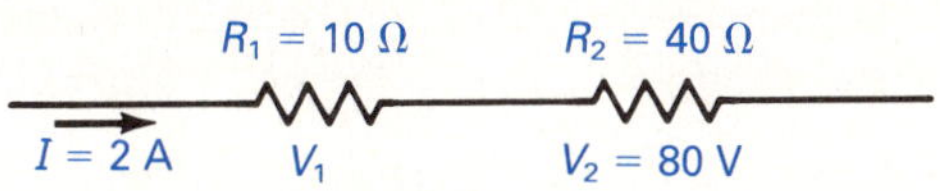

Fig. 6-19. Example 6-10: Data.

Solution Method 1 Ohm's law:

$$V_1 = IR_1 = 2\text{ A} \times 10\ \Omega = \mathbf{20\ V}$$

Method 2 Resistance ratio:

$$V_1 = V_2 \frac{R_1}{R_2} \tag{6-7}$$

$$V_1 = 80\text{ V} \frac{10\ \Omega}{40\ \Omega} = \mathbf{20\ V}$$

Total Resistance R_T

The total network resistance can be calculated by adding the individual resistances or by the application of Ohm's law:

$$R_T = R_1 + R_2 + \cdots + R_n \tag{6-3}$$

or $$R_T = \frac{V_T}{I}$$

Expressions for R_T in terms of voltage ratios can be obtained by substituting (for $1/I$) appropriate ratios R/V in the Ohm's law equation:

$$R_T = \frac{V_T}{I} = V_T \div I = V_T \div \frac{V_1}{R_1} = V_T \frac{R_1}{V_1}$$

$$R_T = R_1 \frac{V_T}{V_1} \quad \text{or} \quad R_2 \frac{V_T}{V_2} \quad \text{or} \quad \cdots \quad \text{or} \quad R_n \frac{V_T}{V_n} \qquad \textbf{(6-8)}$$

EXAMPLE 6-11 What is the total resistance of the circuit in Fig. 6-20?

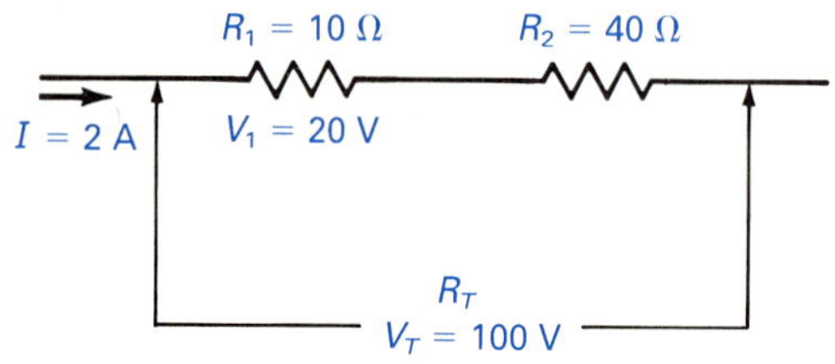

Fig. 6-20. Example 6-11: Data.

Solution Method 1 Sum of the resistances:

$$R_T = R_1 + R_2 \qquad (6\text{-}3)$$

$$R_T = 10\ \Omega + 40\ \Omega = \mathbf{50\ \Omega}$$

Method 2 Ohm's law:

$$R_T = \frac{V_T}{I} = \frac{100\ \text{V}}{2\ \text{A}} = \mathbf{50\ \Omega}$$

Method 3 Voltage ratio:

$$R_T = R_1 \frac{V_T}{V_1} \qquad (6\text{-}8)$$

$$R_T = 10\ \Omega \frac{100\ \text{V}}{20\ \text{V}} = \mathbf{50\ \Omega}$$

Individual Resistances $R_1, R_2, \ldots, R_n$

The individual resistances of the elements can be calculated by Ohm's law:

$$R_1 = \frac{V_1}{I}$$

or by extending this equation to a voltage ratio form by substituting a convenient ratio R/V for $1/I$:

$$R_1 = V_1 \frac{R_T}{V_T} = R_T \frac{V_1}{V_T} \quad \text{or} \quad R_2 \frac{V_1}{V_2} \quad \text{or} \quad \cdots \quad \text{or} \quad R_n \frac{V_1}{V_n} \qquad \textbf{(6-9)}$$

Similar methods can be used for calculating the individual resistances R_2 to R_n.

EXAMPLE 6-12 What is the value of resistor R_1 in the circuit in Fig. 6-21?

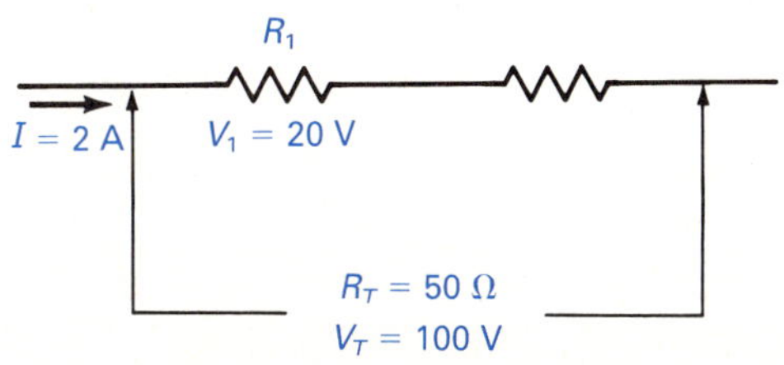

Fig. 6-21. Example 6-12: Data.

Solution Method 1 Ohm's law:

$$R_1 = \frac{V_1}{I} = \frac{20\text{ V}}{2\text{ A}} = \mathbf{10\ \Omega}$$

Method 2 Voltage ratio:

$$R_1 = R_T \frac{V_1}{V_T} \tag{6-9}$$

$$R_1 = 50\ \Omega \frac{20\text{ V}}{100\text{V}} = \mathbf{10\ \Omega}$$

Total Power P_T

The total power can be calculated by summing the individual powers or by applying Ohm's law to the complete network:

$$P_T = P_1 + P_2 + \cdots + P_n \tag{6-4}$$

or, from Eq. (2-14),

$$P_T = V_T I$$

Substituting IR_T for V_T, and V_T/R_T for I, gives two other equations for P_T. From Eq. (3-6),

$$P_T = I^2 R_T$$

or, from Eq. (3-7),

$$P_T = \frac{V_T^2}{R_T}$$

The total power can be expressed in terms of resistance ratios by noting that power is directly proportional to resistance when the circuit current is constant, as in a series circuit.

$$\frac{P_T}{P_1} = \frac{I^2 R_T}{I^2 R_1} = \frac{R_T}{R_1}$$

$$P_T = P_1 \frac{R_T}{R_1} \quad \text{or} \quad P_2 \frac{R_T}{R_2} \quad \text{or} \quad \cdots \quad \text{or} \quad P_n \frac{R_T}{R_n} \tag{6-10}$$

EXAMPLE 6-13 What is the total power dissipated in the series network in Fig. 6-22?

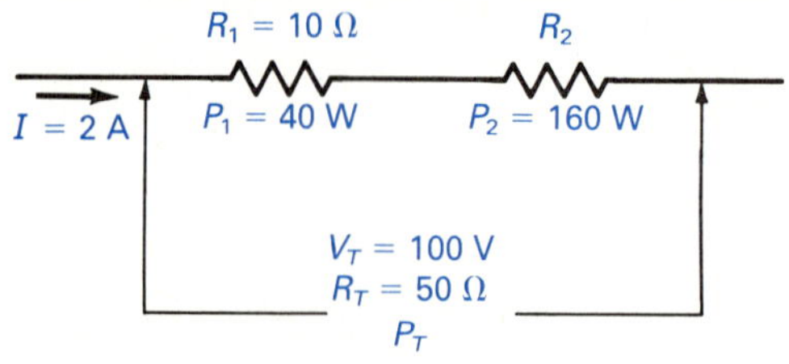

Fig. 6-22. Example 6-13: Data.

Solution Method 1 Sum of the powers:

$$P_T = P_1 + P_2 \tag{6-4}$$
$$P_T = 40\text{ W} + 160\text{ W} = \mathbf{200\ W}$$

Method 2 Product of voltage and current:

$$P_T = V_T I = 100\text{ V} \times 2\text{ A} = \mathbf{200\ W}$$

Method 3 Current and resistance:
From Eq. (3-6),

$$P_T = I^2 R_T = (2\text{ A})^2 \times 50\ \Omega = \mathbf{200\ W}$$

Method 4 Voltage and resistance:
From Eq. (3-7),

$$P_T = \frac{V_T^2}{R_T} = \frac{(100\text{ V})^2}{50\ \Omega} = \mathbf{200\ W}$$

Method 5 Resistance ratio:

$$P_T = P_1 \frac{R_T}{R_1} \tag{6-10}$$
$$P_T = 40\text{ W}\,\frac{50\ \Omega}{10\ \Omega} = \mathbf{200\ W}$$

Individual Powers $P_1, P_2, \ldots, P_n$

The power dissipated by an individual element can be calculated by applying techniques similar to those used for the total power in the last section. Methods are given for calculating the power in R_1, but they can be applied to any of the other elements R_2 to R_n.

$$P_1 = V_1 I \quad \text{or} \quad I^2 R_1 \quad \text{or} \quad \frac{V_1^2}{R_1}$$

$$P_1 = P_T \frac{R_1}{R_T} \quad \text{or} \quad P_2 \frac{R_1}{R_2} \quad \text{or} \quad \cdots \quad \text{or} \quad P_n \frac{R_1}{R_n} \tag{6-11}$$

EXAMPLE 6-14 What is the power dissipated by resistor R_1 in the circuit in Fig. 6-23?

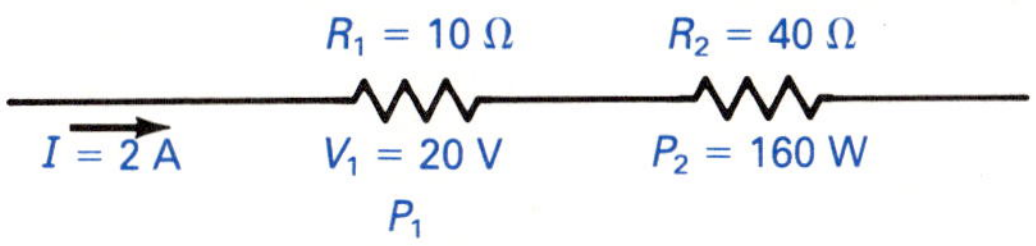

Fig. 6-23. Example 6-14: Data.

Solution Method 1 Product of voltage and current:

$$P_1 = V_1 I = 20\text{ V} \times 2\text{ A} = \mathbf{40\ W}$$

Method 2 Current and resistance:

$$P_1 = I^2 R_1 = (2\text{ A})^2 \times 10\ \Omega = \mathbf{40\ W}$$

Method 3 Voltage and resistance:

$$P_1 = \frac{V_1^2}{R_1} = \frac{(20\text{ V})^2}{10\ \Omega} = \mathbf{40\ W}$$

Method 4 Resistance ratio:

$$P_1 = P_2 \frac{R_1}{R_2} \tag{6-11}$$

$$P_1 = 160\text{ W}\,\frac{10\ \Omega}{40\ \Omega} = \mathbf{40\ W}$$

Energies

The methods for calculating power, given in the last two sections and demonstrated in Examples 6-13 and 6-14, can all be applied to energy. It is only necessary to multiply any power by the time period to obtain the energy for that period.

Series-Circuit Ratios

Ratios are used frequently in series-circuit calculations because the common value of current in all parts of the circuit makes resistance, voltage, power, and energy ratios the same for any two parts of the circuit.

Comparing the quantities relevant to the whole circuit and to R_1 gives relationships which can be applied generally:

$$\frac{V_T}{V_1} = \frac{IR_T}{IR_1} = \frac{R_T}{R_1}$$

$$\frac{P_T}{P_1} = \frac{I^2R_T}{I^2R_1} = \frac{R_T}{R_1}$$

$$\frac{W_T}{W_1} = \frac{I^2R_Tt}{I^2R_1t} = \frac{R_T}{R_1}$$

Therefore
$$\frac{R_T}{R_1} = \frac{V_T}{V_1} = \frac{P_T}{P_1} = \frac{W_T}{W_1} \tag{6-12}$$

This ratio concept can be applied to pairs of elements; for example, $R_2/R_1 = P_2/P_1$.

EXAMPLE 6-15 Using the ratio of R_T to R_1 in Fig. 6-24, determine *(a)* the total voltage, *(b)* the total power, and *(c)* the total energy used in 2 h.

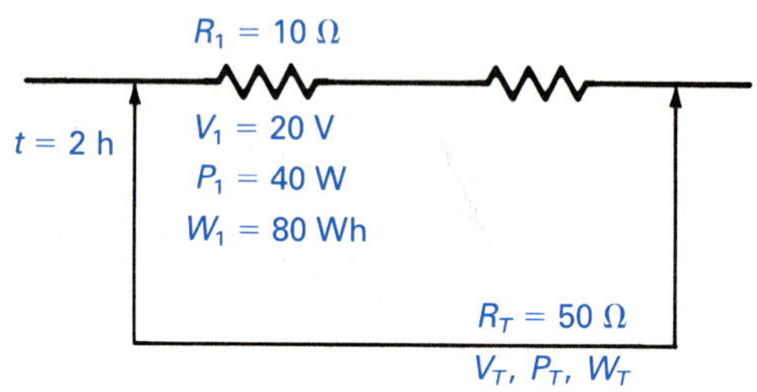

Fig. 6-24. Example 6-15: Data.

Solution The ratio of R_T to R_1 is

$$\frac{R_T}{R_1} = \frac{50\ \Omega}{10\ \Omega} = 5$$

This ratio is shown in parentheses in the following solutions.

a. $V_T = (5)V_1 = 5 \times 20\ \text{V} = \mathbf{100\ V}$

b. $P_T = (5)P_1 = 5 \times 40\ \text{W} = \mathbf{200\ W}$

c. $W_T = (5)W_1 = 5 \times 80\ \text{Wh} = \mathbf{400\ Wh}$

REVIEW QUESTIONS FOR SECTION 6-3

1. State, in equation form, two methods of determining the current in a series circuit.
2. State, in equation form, three methods of determining the total voltage in a series circuit.
3. State, in equation form, two methods of determining the individual voltages across series resistors.
4. State, in equation form, two methods of determining the total resistance of a series circuit.
5. State, in equation form, two methods of determining the individual resistances of a series circuit.
6. State, in equation form, five methods of determining the total power dissipated by a series circuit.
7. State, in equation form, four methods of determining the power dissipated by an individual resistor in a series circuit.

EXERCISE PROBLEMS FOR SECTION 6-3

Resistors R_1 and R_2 are part of a transistor biasing circuit. The current to the transistor base is very small and should be ignored.

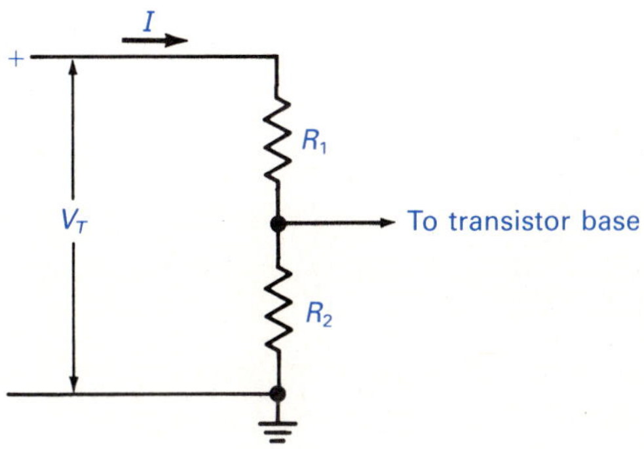

Fig. 6-25. Problems 1 to 6: Network.

1. Find the voltage across each resistor in Fig. 6-25 when
   a. $R_1 = 12\ \text{k}\Omega$, $R_2 = 5.6\ \text{k}\Omega$, $I = 680\ \mu\text{A}$ (*ans.* 8.16 V, 3.808 V)
   b. $R_1 = 8.2\ \text{k}\Omega$, $R_2 = 4.7\ \text{k}\Omega$, $I = 775\ \mu\text{A}$ (*ans.* 6.355 V, 3.643 V)
   c. $R_1 = 56\ \text{k}\Omega$, $R_2 = 18\ \text{k}\Omega$, $I = 0.19\ \text{mA}$ (*ans.* 10.64 V, 3.42 V)
2. In Prob. 1, find the total series voltage and the applied emf for each case.
3. In Prob. 1, find the total resistance for each case.
   (*ans. a.* 17.6 kΩ; *b.* 12.9 kΩ; *c.* 74 kΩ)
4. In Prob. 1, find the power dissipated by each resistor and the total power for each case.
5. In Prob. 1, when $V_T = 15$ V, $R_1 = 3.9\ \text{k}\Omega$, and $R_2 = 1.2\ \text{k}\Omega$, find I.
   (*ans.* 2.941 mA)
6. In Prob. 1, when $I = 1$ mA, $V_1 = 2.2$ V, and $V_2 = 6.8$ V, find R_T.
7. A voltmeter consists of a 99-kΩ resistor in series with a 1000-Ω meter movement. If the current in the resistor is 99 μA, what is the current in the meter movement? (*ans.* 99 μA)
8. A 2000-Ω resistor and a 3000-Ω resistor form a series circuit with a 50-V supply. If the current in the 2000-Ω resistor is 10 mA, what is the current in the 3000-Ω resistor?
9. A 1000-Ω resistor and a 2000-Ω resistor are in series. If the voltage across the 1000-Ω resistor is 50 V, what is the current in the 2000-Ω resistor?
   (*ans.* 50 mA)
10. The 0.625-Ω internal resistance of a generator is in series with a 100-Ω load. If the voltage drop caused by the internal resistance is 7.5 V, what is the voltage across the load?
11. An amplifier producing a signal voltage of 25 V and having an output resistance of 8 Ω is connected in series with an 8-Ω speaker. Find
    a. The current in the amplifier output resistance. (*ans.* 1.563 A)
    b. The current in the speaker. (*ans.* 1.563 A)
    c. The power dissipated by the output resistance. (*ans.* 19.53 W)
    d. The total power dissipated by the circuit. (*ans.* 39.06 W)

12. Eight Christmas-tree lamps, each with a resistance of 37.5 Ω (when lit), are connected in series across a 120-V supply. Find
   a. The total resistance.
   b. The current through the circuit.
   c. The voltage across each lamp.
   d. The total power dissipated.
   e. The power dissipated by each lamp.
   f. The energy used by each lamp per hour.
   g. The total energy used per hour.

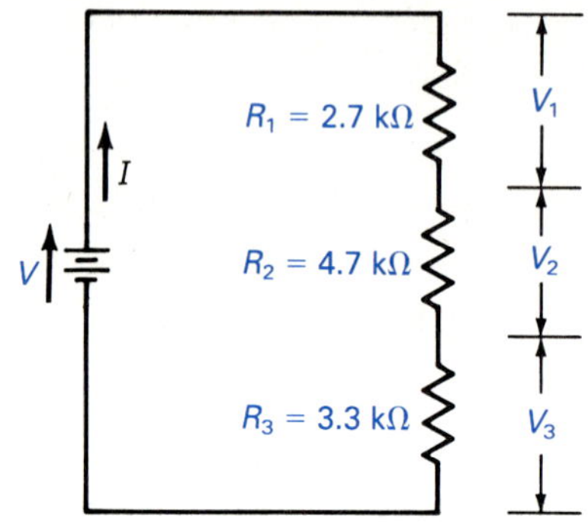

Fig. 6-26. Problem 13: Circuit.

13. Find the current in the circuit in Fig. 6-26 for each of the following cases.
   a. $V_1 = 11$ V (*ans*. 4.074 mA)
   b. $V_2 = 78$ V (*ans*. 16.6 mA)
   c. $V_3 = 66$ V (*ans*. 20 mA)
   d. $V = 220$ V (*ans*. 20.56 mA)
14. Using the circuit for Prob. 13, find the voltage across R_1 and R_3 when the voltage across R_2 is 22 V.
15. Using the circuit for Prob. 13, find the individual voltages, using ratios, when the emf is 24 V. (*ans*. 6.056 V, 10.54 V, 7.402 V)
16. In Fig. 6-27, when $R_a = 680\ \Omega$, $V_a = 3.6$ V, and $V = 18$ V, find R_T.
17. In Fig. 6-27, when $R_c = 4.7\ \text{k}\Omega$, $V_c = 4.2$ V, $V_b = 7.5$ V, and $V = 18$ V, find R_T. (*ans*. 20.14 kΩ)
18. From the data given in Prob. 17, find R_b.
19. An electric saw has an effective resistance, under full load, of 11 Ω. It is fed from a 120-V supply through a 30-m extension cord. If the terminal voltage of the supply is 120 V and the voltage at the saw is 111.2 V, what is the resistance of the extension cord? (*ans*. 0.87 Ω)
20. A 6-V radio is operated from a 12-V supply through a series dropping resistor. If the radio requires 30 W of power for normal operation, what is
   a. The current in the circuit?
   b. The total power drawn from the supply?
   c. The amount of power dissipated by the dropping resistor?

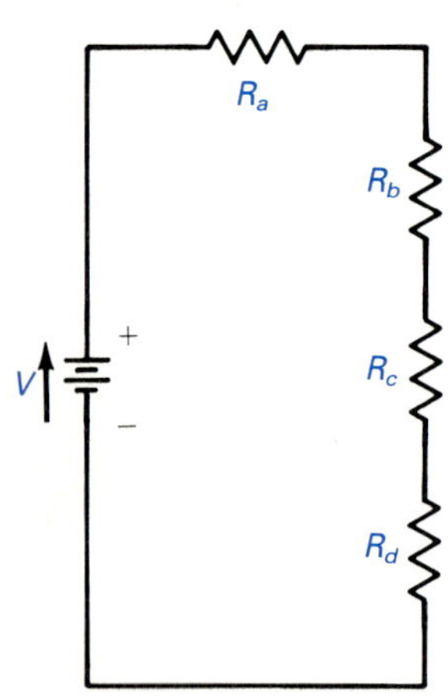

Fig. 6-27. Problems 16 to 18: Circuit.

6-4 RELATIVE POTENTIAL

The discussion on series circuits so far has not emphasized the importance of voltage polarities. They are of particular significance in determining relative potentials.

A *relative potential is the voltage at one point in a circuit with respect to that at another point* (usually called the reference).

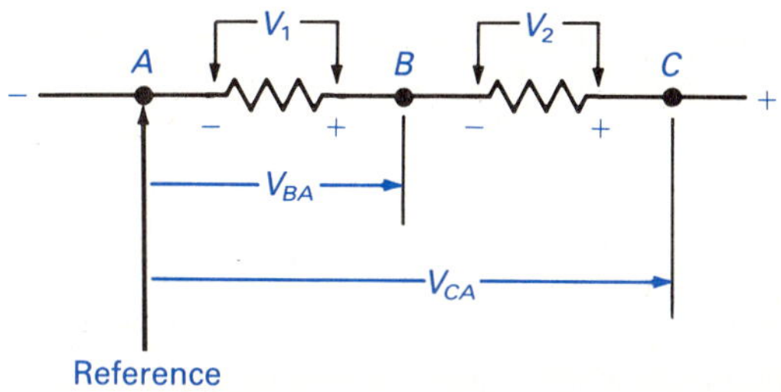

Fig. 6-28. Relative potential, reference at point *A*.

Figure 6-28 shows a two-element series network with the most negative end, point *A*, shown as the reference. With respect to point *A*, both points *B* and *C* are positive.

$$V_{BA}^{*} = +V_1$$
$$V_{CA} = +(V_1 + V_2)$$

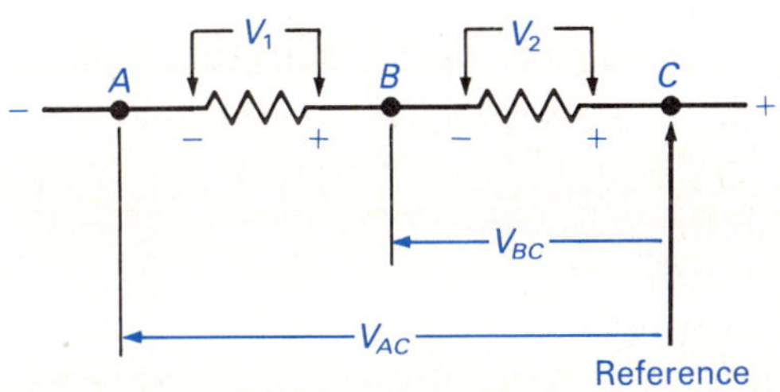

Fig. 6-29. Relative potential, reference at point *C*.

Figure 6-29 shows the same network as Fig. 6-28, but the reference chosen is at point *C*.

With respect to point *C*, both points *A* and *B* are negative.

$$V_{BC} = -V_2$$
$$V_{AC} = -(V_1 + V_2)$$

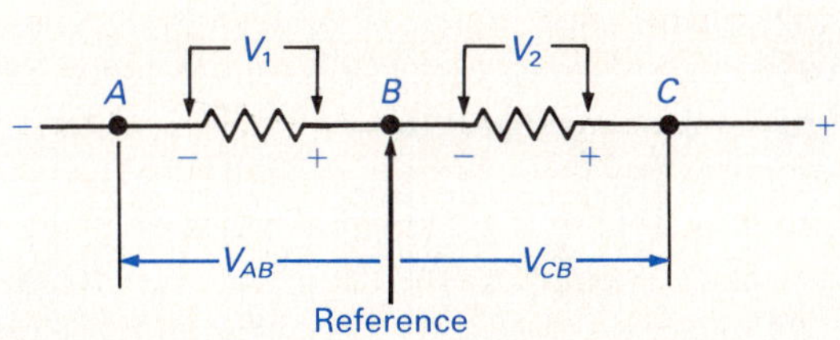

Fig. 6-30. Relative potential, reference at point *B*.

The reference point is shown at point *B* in Fig. 6-30. Point *C* is positive with respect to point *B*, but point *A* is negative with respect to point *B*.

$$V_{CB} = +V_2$$
$$V_{AB} = -V_1$$

*It will be seen that V_{BA} refers to the voltage at *B* with respect to that at *A*. This is the convention used in electronics and which is adopted in this book.

The reference can be chosen at will to be at any point in a circuit. It is selected to suit the situation. Frequently the reference point is called *ground*. Ground is a rather difficult notion to comprehend (despite the fact that many persons use the term quite freely). Originally, it meant a point in a circuit directly connected to the ground (the earth), either for safety in power systems or for efficient communication in radio transmission and reception. This meaning still exists, but it is quite usual today for "ground" to mean a circuit point connected to a relatively large metal object, such as the chassis, frame, or cabinet supporting the equipment. "Common" is another term used as an alternative to "ground."

Where the reference point is known to be grounded, or commoned, we may refer to relative potentials as voltages which are simply positive or negative. It is then understood that they are with respect to ground, or common.

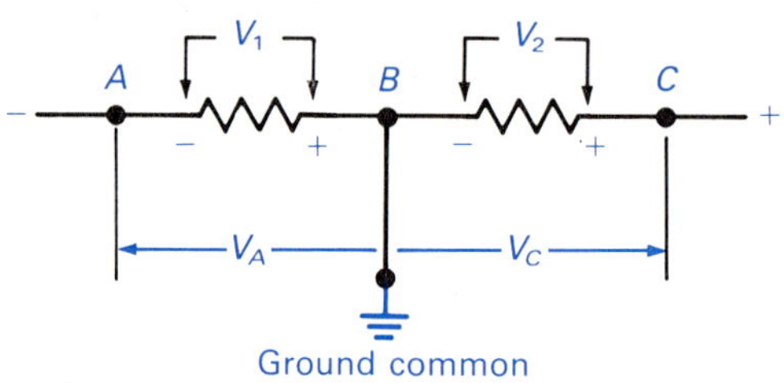

Fig. 6-31. Relative potential with respect to ground.

Figure 6-31 is the same as Fig. 6-30 except for the ground symbol at the reference connection, point *B*. The relative potentials at points *A* and *C* can be written as

$$V_A = -V_1$$
$$V_C = +V_2$$

EXAMPLE 6-16 What are the relative potentials at points *B*, *C*, and *D* with respect to point *A* in Fig. 6-32?

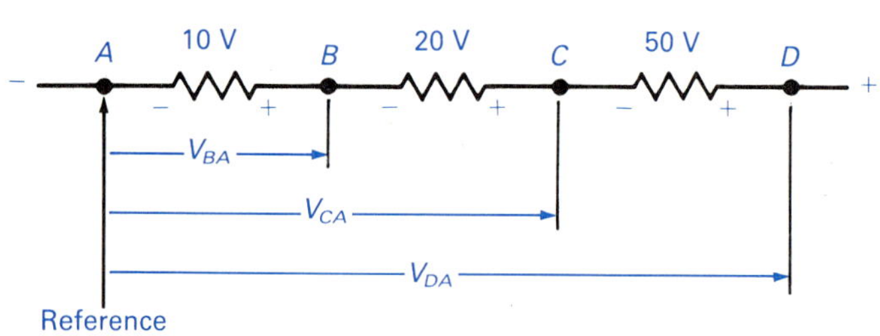

Fig. 6-32. Example 6-16: Data.

Solution

$$V_{BA} = \mathbf{+10\ V}$$
$$V_{CA} = +(20\ \text{V} + 10\ \text{V}) = \mathbf{+30\ V}$$
$$V_{DA} = +(50\ \text{V} + 20\ \text{V} + 10\ \text{V}) = \mathbf{+80\ V}$$

EXAMPLE 6-17 What are the relative potentials at points A, B, and C with respect to point D in Fig. 6-33?

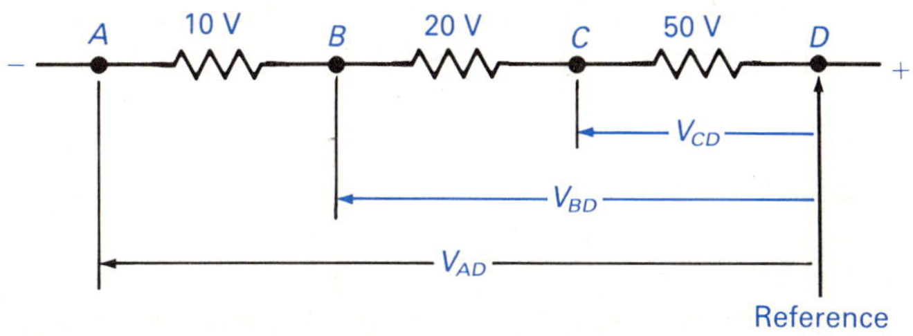

Fig. 6-33. Example 6-17: Data.

Solution

$$V_{AD} = -(50\text{ V} + 20\text{ V} + 10\text{ V}) = \mathbf{-80\ V}$$
$$V_{BD} = -(50\text{ V} + 20\text{ V}) = \mathbf{-70\ V}$$
$$V_{CD} = \mathbf{-50\ V}$$

EXAMPLE 6-18 What are the relative potentials at points A, B, and D with respect to point C in Fig. 6-34?

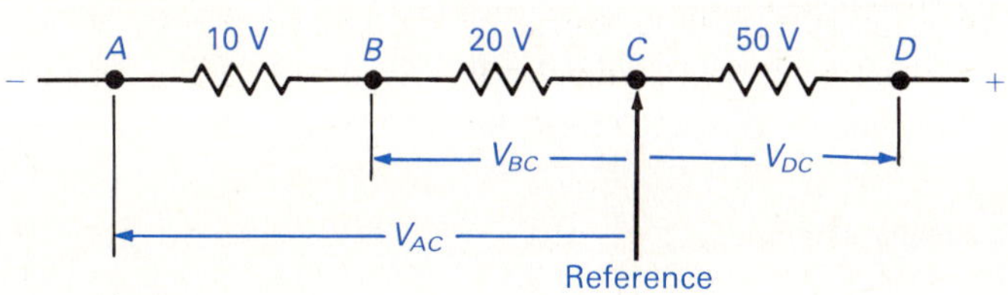

Fig. 6-34. Example 6-18: Data.

Solution

$$V_{AC} = -(20\text{ V} + 10\text{ V}) = \mathbf{-30\ V}$$
$$V_{BC} = \mathbf{-20\ V}$$
$$V_{DC} = \mathbf{+50\ V}$$

EXAMPLE 6-19 What are the potentials at points A, C, and D in Fig. 6-35?

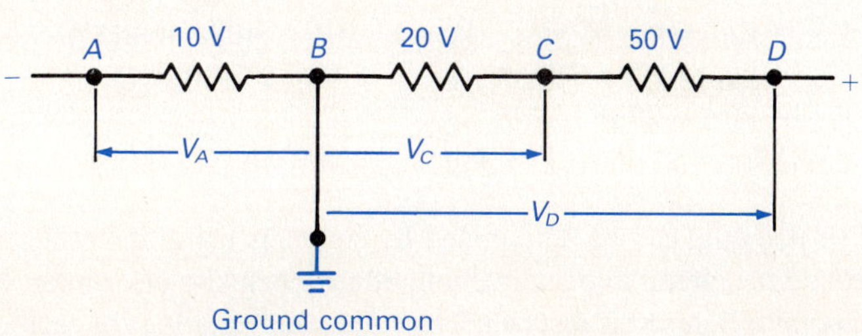

Fig. 6-35. Example 6-19: Data.

Solution

$$V_A = \mathbf{-10\ V}$$
$$V_C = \mathbf{+20\ C}$$
$$V_D = \mathbf{+70\ V}$$

EXAMPLE 6-20 What are the relative potentials in the circuit in Fig. 6-36?

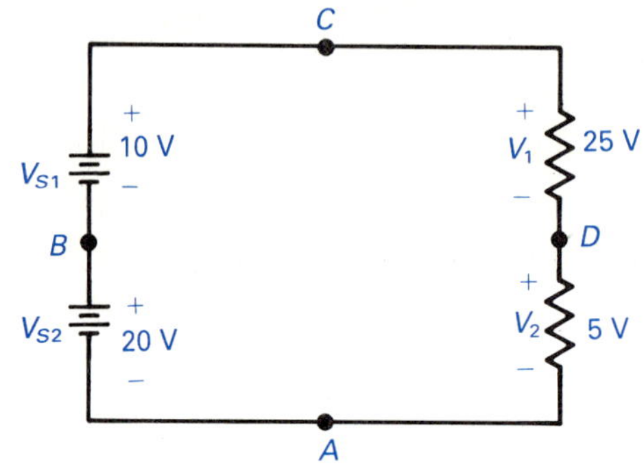

Fig. 6-36. Example 6-20: Circuit.

Solution Reference at point A:

$$V_{BA} = +V_{S2} = \mathbf{+20\ V}$$
$$V_{CA} = +(V_{S1} + V_{S2}) = \mathbf{+30\ V}$$
$$V_{DA} = +V_2 = \mathbf{+5\ V}$$

Reference at point C:

$$V_{AC} = -(V_{S1} + V_{S2}) = \mathbf{-30\ V}$$
$$V_{BC} = -V_{S1} = \mathbf{-10\ V}$$
$$V_{DC} = -V_1 = \mathbf{-25\ V}$$

Reference at point B:

$$V_{AB} = -V_{S2} = \mathbf{-20\ V}$$
$$V_{CB} = +V_{S1} = \mathbf{+10\ V}$$
$$V_{DB} = V_{AB} + V_{DA} = -V_{S2} + V_2 = -20\ \text{V} + 5\ \text{V} = \mathbf{-15\ V}$$
$$\text{or } V_{DB} = V_{CB} + V_{DC} = +V_{S1} - V_1 = +10\ \text{V} - 25\ \text{V} = \mathbf{-15\ V}$$

Reference at point D:

$$V_{AD} = -V_2 = \mathbf{-5\ V}$$
$$V_{BD} = V_{CD} + V_{BC} = +V_1 - V_{S1} = +25\ \text{V} - 10\ \text{V} = \mathbf{+15\ V}$$
$$\text{or } V_{BD} = V_{AD} + V_{BA} = -V_2 + V_{S2} = -5\ \text{V} + 20\ \text{V} = \mathbf{+15\ V}$$
$$V_{CD} = V_1 = \mathbf{+25\ V}$$

Note that a relative potential can be determined by commencing at the reference and moving around the circuit in one direction, either clockwise or counterclockwise, until the point of interest is reached. The values and polarities of each of the voltages are noted in turn and then added algebraically to obtain the result.

EXAMPLE 6-21 Figure 6-37 shows the conditions which determine the base bias of a transitor in an amplifier circuit. What are the potentials at points B, C, and E with respect to ground? What is the value of the base-to-emitter bias V_{BE}?

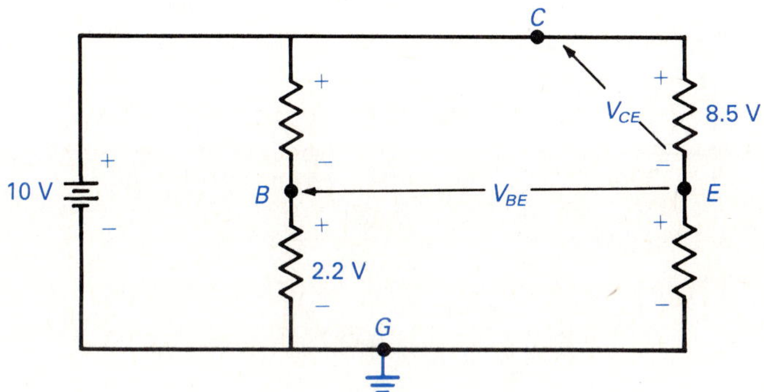

Fig. 6-37. Example 6-21: Circuit.

Solution The potential at point B is

$$V_B = \mathbf{+2.2\ V}$$

The potential at point C is

$$V_C = \mathbf{+10\ V}$$

The potential at point E is

$$V_E = V_C - V_{CE} = +10\text{ V} - 8.5\text{ V} = \mathbf{+1.5\ V}$$

The base-to-emitter voltage V_{BE} is

$$V_{BE} = V_{BG} + V_{GE} = 2.2\text{ V} - 1.5\text{ V} = \mathbf{+0.7\ V}$$

REVIEW QUESTIONS FOR SECTION 6-4

1. What is meant by ''relative potential''?
2. In an electric circuit, what is another name for ''ground''?
3. Compare the relative potentials between two points in a circuit and between a point and ground. In what way does *polarity* influence the value of a relative potential?

EXERCISE PROBLEMS FOR SECTION 6-4

1. Refer to Fig. 6-38 and find

a. V_{ab}	(*ans.* +10 V)	*e.* V_{da}	(*ans.* −40 V)
b. V_{ac}	(*ans.* +25 V)	*f.* V_{bd}	(*ans.* +30 V)
c. V_{cb}	(*ans.* −15 V)	*g.* V_{bc}	(*ans.* +15 V)
d. V_{ca}	(*ans.* −25 V)	*h.* V_{dc}	(*ans.* −15 V)

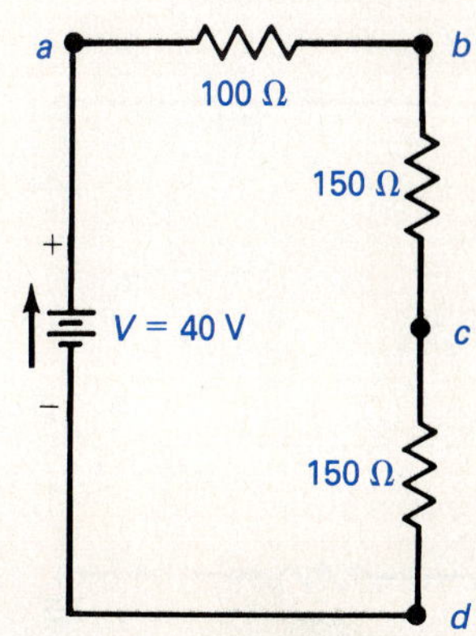

Fig. 6-38. Problem 1: Circuit.

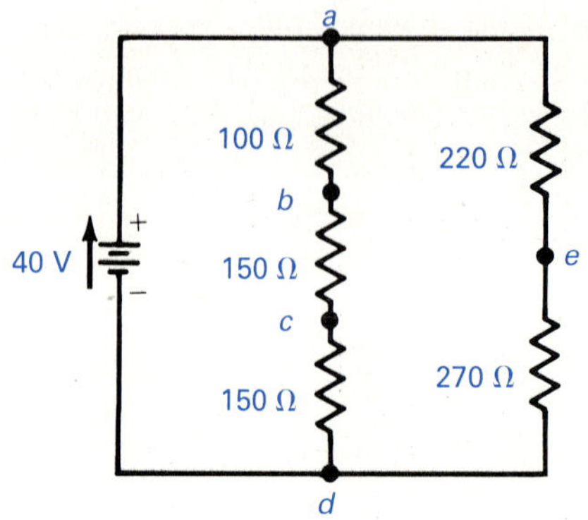

Fig. 6-39. Problem 2: Circuit.

2. Refer to Fig. 6-39 and find
a. V_{be} *b.* V_{ec}

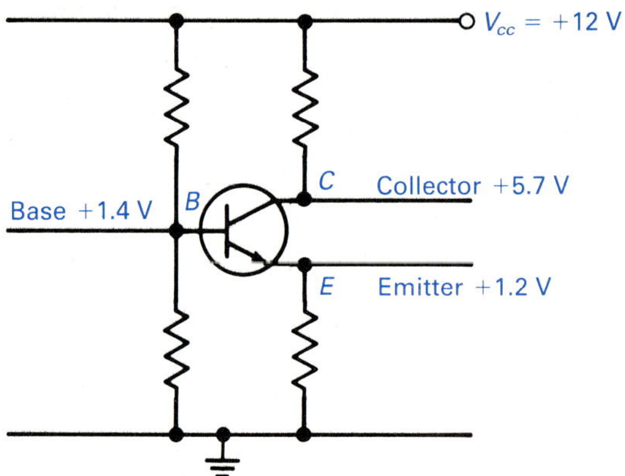

Fig. 6-40. Problem 3: Circuit.

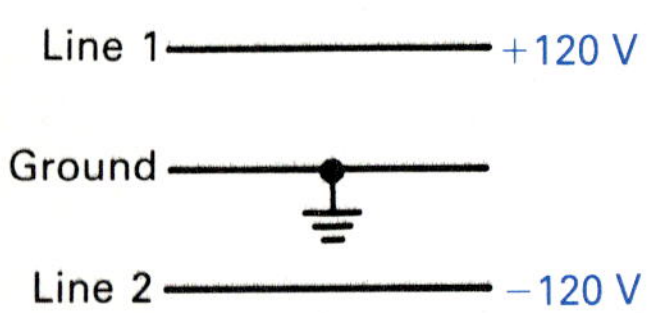

Fig. 6-41. Problem 4: Three-wire distribution voltages.

3. From the transistor amplifier circuit in Fig. 6-40, determine
a. The voltage at the collector with respect to the emitter. (*ans.* +4.5 V)
b. The voltage at the base with respect to the emitter. (*ans.* +0.2 V)
c. The voltage at the base with respect to the collector. (*ans.* −4.3 V)

4. Refer to Fig. 6-41 and determine
a. The voltage at ground with respect to line 1.
b. The voltage at ground with respect to line 2.
c. The voltage at line 2 with respect to line 1.

5. Give the magnitude and polarity of the following voltages in the circuit in Fig. 6-42.

a. V_1	(*ans.* −16 V)	*d.* V_{32}	(*ans.* +29.4 V)
b. V_2	(*ans.* −9.4 V)	*e.* V_{12}	(*ans.* −6.6 V)
c. V_3	(*ans.* +20 V)	*f.* V_{13}	(*ans.* [illegible]36 V)

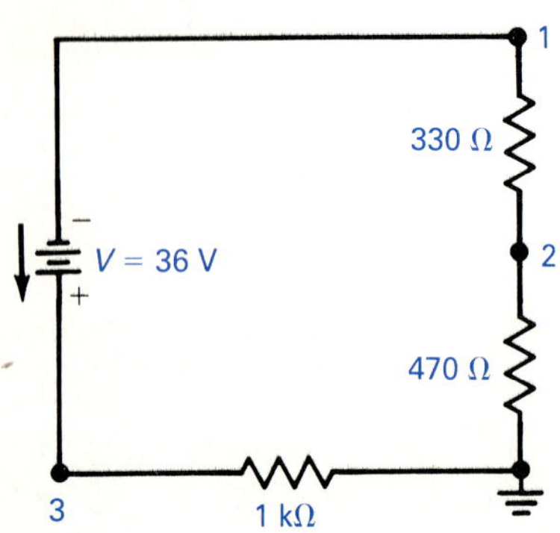

Fig. 6-42. Problem 5: Circuit.

6-5 APPLICATIONS OF SERIES CIRCUITS

The following examples show a few of the vast number of series-circuit applications which exist in practice.

Series Dropping Resistor

It is often necessary to operate a device, or a part of a circuit, at a voltage below that of the supply emf. This lower voltage can be obtained by the use of a series dropping resistor, so that the voltage across this resistor accounts for the unwanted voltage. Examples will best illustrate typical techniques.

EXAMPLE 6-22 A telephone relay, designed to operate from a 24-V dc supply, normally draws a current of 10 mA. What series resistor is necessary to operate this relay, with its correct current, from a supply voltage which may range between 24 and 48 V?

Solution (Fig. 6-43)

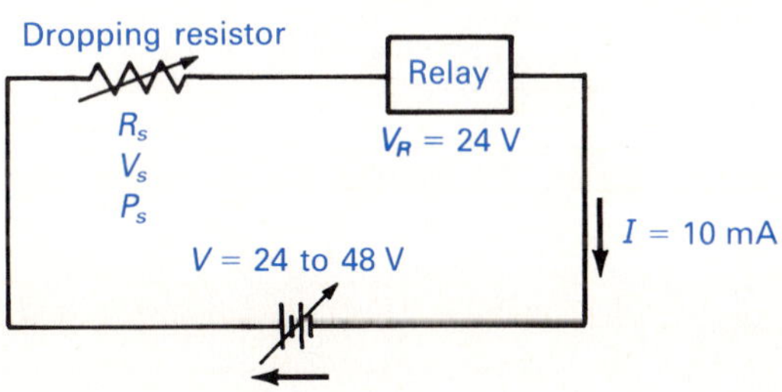

Fig. 6-43. Example 6-22: Circuit.

From Eq. (6-2),

$$V_S = V - V_R$$

When $V = 24$ V,

$$V_S = 24\text{ V} - 24\text{ V} = 0\text{ V}$$

$$R_S = \frac{V_S}{I} = \frac{0\text{ V}}{10\text{ mA}} = \mathbf{0\ \Omega}$$

$$P_S = V_S I = 0\text{ V} \times 10\text{ mA} = \mathbf{0\ W}$$

When $V = 48$ V,

$$V_S = 48\text{ V} - 24\text{ V} = 24\text{ V}$$

$$R_S = \frac{V_S}{I} = \frac{24\text{ V}}{10\text{ mA}} = \mathbf{2.4\ k\Omega}$$

$$P_S = V_S I = 24\text{ V} \times 10\text{ mA} = \mathbf{0.24\ W}$$

In this situation, where the emf can have any value between 24 and 48 V, a variable resistor, or rheostat, will be needed for the series dropping resistor. The rheostat will have a resistance range of 0 to 2.4 kΩ, and a power-handling capability of at least $\frac{1}{4}$ W.

EXAMPLE 6-23 The speed of an electric model train can be varied by changing the voltage applied to its motor. A 12-V supply is available, and it is desired to vary the train voltage between 4 and 10 V. When running with 10 V, the motor draws 0.5 A. The resistance of the motor remains essentially constant over a wide current range. What kind of series resistor is necessary to control the train speed?

Solution (Fig. 6-44)

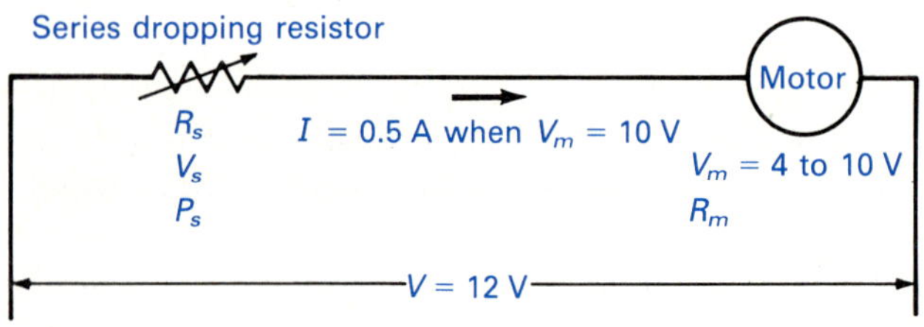

Fig. 6-44. Example 6-23: Data.

When the motor voltage V_m is 10 V,

$$R_m = \frac{V_m}{I} = \frac{10\ \text{V}}{0.5\ \text{A}} = 20\ \Omega$$

From Eq. (6-2),

$$V_S = V_T - V_m$$
$$V_S = 12\ \text{V} - 10\ \text{V} = 2\ \text{V}$$

From Eq. (6-9),

$$R_S = R_m \frac{V_S}{V_m}$$
$$R_S = 20\ \Omega \frac{2\ \text{V}}{10\ \text{V}} = \mathbf{4\ \Omega}$$
$$P_S = V_S I = 2\ \text{V} \times 0.5\ \text{A} = \mathbf{1\ W}$$

When the motor voltage is 4 V, from Eq. (6-2),

$$V_S = V_T - V_m$$
$$V_S = 12\ \text{V} - 4\ \text{V} = 8\ \text{V}$$

From Eq. (6-9),

$$R_S = R_m \frac{V_S}{V_m}$$
$$R_S = 20\ \Omega \frac{8\ \text{V}}{4\ \text{V}} = \mathbf{40\ \Omega}$$
$$P_S = \frac{V_S^2}{R_S} = \frac{(8\ \text{V})^2}{40\ \Omega} = \mathbf{1.6\ W}$$

A *rheostat,* or variable resistor, could control the speed of the train in Example 6-23. Its resistance range would need to be from 4 to 40 Ω. It would have to be capable of handling 1.8 W of power. This power rating is not obvious from the example solution. Maximum power is dissipated by the resistor when its resistance is the same as that of the motor, 20 Ω. The subject of *maximum power transfer* is dealt with in a later chapter.

Generator Internal Resistance

When current is supplied to a load, energy is dissipated within the generator, heating it up, and so this energy is not passed on to the load. The lost energy is accounted for in terms of an equivalent resistance within the generator. It is easy to visualize the resistance of the wire within an automobile generator, for instance, but there are components of generator equivalent resistance which are not so obvious, the friction of bearings being an example. The *internal resistance* of a generator refers to the total energy loss expressed as an equivalent resistance. Internal resistance can be considered as either a series or a parallel element. In this chapter we will deal with it as a series element. In Chap. 7 the parallel view of internal resistance will be investigated.

According to the series internal resistance viewpoint, the generator is made up of two components. One of the components is an emf source (it is not unusual for the emf source to be called a *voltage source, voltage generator,* or just *generator,* and labeled accordingly in diagrams). The other is the internal resistance, which is regarded as a completely separate element connected in series with the emf source. Figure 6-45 shows such a generator connected to a load, the emf source is identified as V.

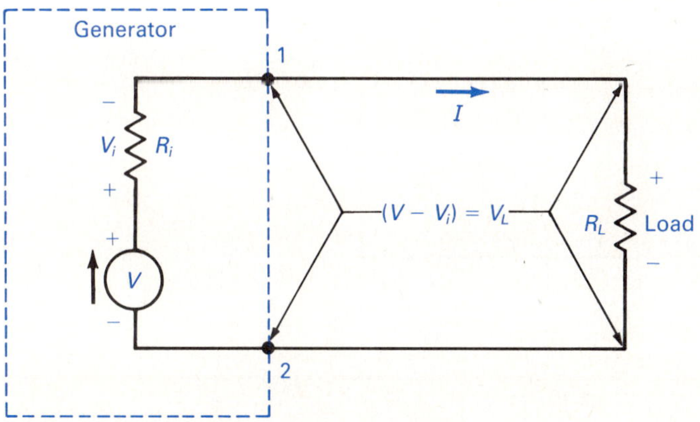

Fig. 6-45. Generator internal resistance.

The voltage across the load V_L must be the same as the voltage appearing across the generator terminals 1 and 2.

When a load is present, as shown in Fig. 6-45, the current I will cause a voltage V_i to be developed across the generator internal resistance R_i.

$$V_i = IR_i$$

The current direction through R_i is the same as that through the load R_L, in this case, clockwise. Therefore the voltages V_i and V_L will have ordinary series-circuit polarities, as in Fig. 6-45. Thus Eq. (6-2) applies, as in any series circuit:

$$V = V_i + V_L \tag{6-13}$$

or

$$V_L = V - V_i$$

It will be seen that $V - V_i$ is the voltage at the generator terminals 1 and 2. This voltage is called the terminal voltage of the generator. The *terminal voltage* of a generator is the voltage across its terminals when it is supplying current to a load.

If the load is disconnected, there will be no current, and so the voltage across the internal resistance $V_i = IR_i$ will be zero. If this is so, then the voltage at the generator terminals will equal the emf V. The voltage at the terminals of a generator when it is supplying no current to a load is called the *open-circuit emf*. Equation (6-13) can be written as

$$\text{Terminal voltage} = \text{open-circuit emf} - \text{internal resistance voltage}$$

EXAMPLE 6-24 The open-circuit emf of an automobile battery is 13.2 V. If its internal resistance is 0.01 Ω, what will be *(a)* the starter motor voltage if its current is 300 A and *(b)* the voltage across the headlights if the total electric load at night is 120 V at a nominal[1] 12 V?

Solution *a.* Starter motor voltage (Fig. 6-46):

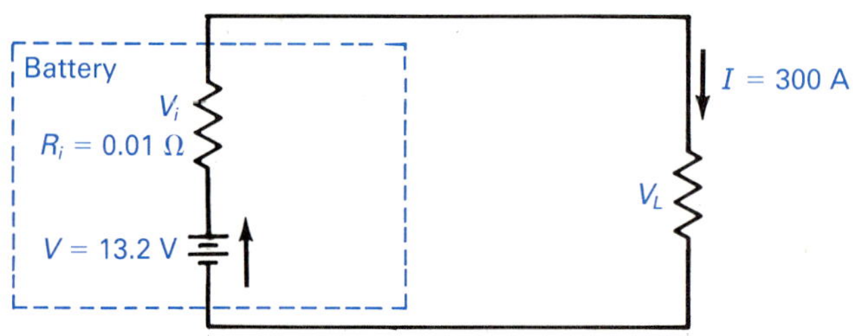

Fig. 6-46. Example 6-24: Starter circuit conditions.

$$V_i = IR_i = 300\ \text{A} \times 0.01\ \Omega = 3\ \text{V}$$
$$V_L = V - V_i$$
$$V_L = 13.2\ \text{V} - 3\ \text{V} = \mathbf{10.2\ V}$$

b. Voltage across headlights (Fig. 6-47):

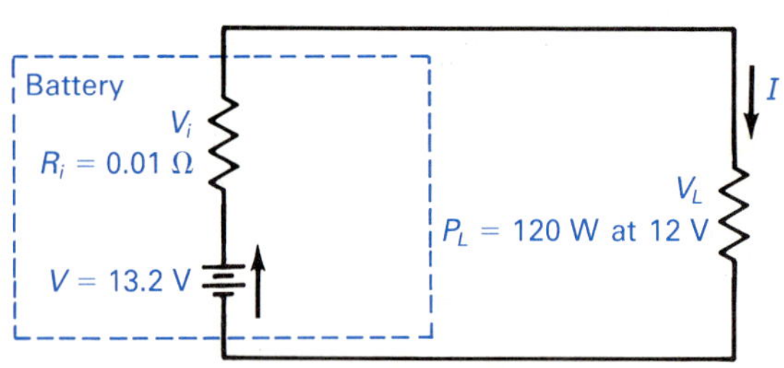

Fig. 6-47. Example 6-24: Lighting circuit conditions.

$$P_L = V_L I$$

If V_L is nominally 12 V,

[1]The word "nominal" is used frequently in electrical work. It means "in name only." A *nominal* value is an assigned value, the actual existing value perhaps being slightly different. The amount of tolerable variation from the nominal value is the *tolerance*.

$$I = \frac{P_L}{V_L} = \frac{120\ \text{W}}{12\ \text{V}} = 10\ \text{A}$$

$$V_i = IR_i = 10\ \text{A} \times 0.01\ \Omega = 0.1\ \text{V}$$

$$V_L = V - V_i \qquad (6\text{-}13)$$

$$V_L = 13.2\ \text{V} - 0.1\ \text{V} = \mathbf{13.1\ V}$$

The values in Example 6-24 are practical for an automobile electric system. The automobile battery, nominally 12 V, is likely to have an emf of about 13.2 V. An internal resistance of 0.01 Ω is reasonable. The load imposed by the starter motor is heavy, about 300 A being common. If the battery resistance is high, because the battery is in poor shape or because of low winter temperatures, the voltage across the starter motor might be as low as 6 V, making it difficult to start the car. The ordinary operating load, even at night with the headlights on high beam, should not reduce the load voltage below 12 V if the battery and the electric system in general are in good condition.

You will see that, although the headlight load was specified at 12 V, it actually receives 13.1 V. This 13.1 V means that the power taken by the lighting system is higher than 120 W. Such variations of voltage from the nominal are common in automotive systems; therefore these systems must be designed to operate over a wide voltage range.

Conductor Resistance

Up to this point it has been assumed that the various circuit elements are connected together by perfect conductors, conductors which have no resistance and therefore do not influence the values of current and voltage in the circuit.

For many applications good physical circuit design aims at ensuring that conductor resistances are so low, compared with the resistances of the elements they interconnect, that they can be ignored. The amount of resistance considered significant will depend on the circuit application, but a fairly general rule is: Conductor resistance is significant if it exceeds 1 percent of the lowest of the element resistances.

When the resistance of a conductor is significant to circuit operation, it must be treated as a separate element so that its characteristics can be considered.

In Fig. 6-48 a generator, and its internal resistance, is connected to a load by two conductors which have significant resistance.

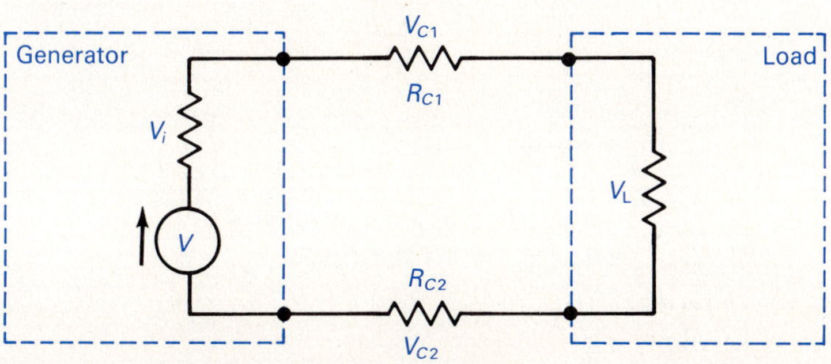

Fig. 6-48. Resistance of connecting leads.

If the conductor resistances R_{C1} and R_{C2} were not present, from Eq. (6-13),

$$V_L = V - V_i$$

With the conductor resistances,

$$V_L = V - V_i - V_{C1} - V_{C2} \qquad \textbf{(6-14)}$$

or

$$V_L = V - (V_i + V_{C1} + V_{C2})$$

Since the objective of the circuit is to provide load current and voltage, V_i, V_{C1}, and V_{C2} are wasted voltages; they contribute nothing useful to the circuit operation. Therefore the conductor resistances (and the generator internal resistance) are often called *circuit losses.*

It is usual in setting up a circuit, to aim at producing a load voltage of maximum and constant value. The presence of significant conductor resistance can stop this happening.

The examples that follow will show some of the effects of conductor resistance.

EXAMPLE 6-25 A battery of 10-V emf and 1-Ω internal resistance is connected to a 100-Ω load. What is the load voltage *(a)* if the conductor resistance can be ignored and *(b)* if each conductor connecting generator and load has a resistance of 9.5 Ω?

Solution *a.* (Fig. 6-49)

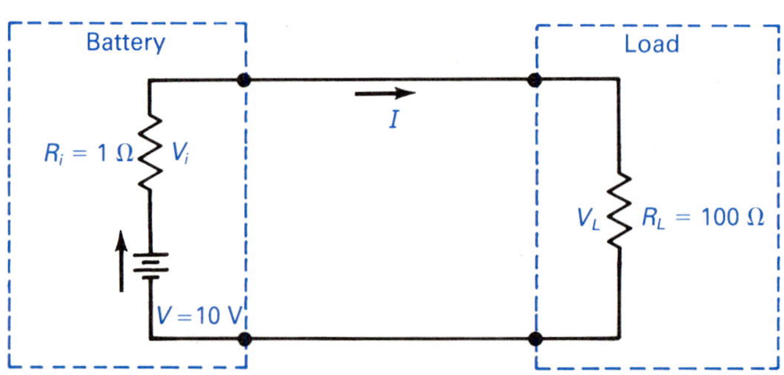

Fig. 6-49. Example 6-25: Circuit without conductor resistance.

From Eq. (6-3),

$$R_T = R_i + R_L$$

$$I = \frac{V}{R_T} = \frac{V}{R_i + R_L} = \frac{10\text{ V}}{1\ \Omega + 100\ \Omega} \simeq 0.1\text{ A*}$$

$$V_i = IR_i \simeq 0.1\text{ A} \times 1\ \Omega \simeq 0.1\text{ V}$$

$$V_L = V - V_i \simeq 10\text{ V} - 0.1\text{ V} \simeq \mathbf{9.9\ V}$$

*The symbol ≃ means "approximately equal to." In electric circuit work, if two quantities are within about 1 percent, they are usually considered to be approximately equal to each other.

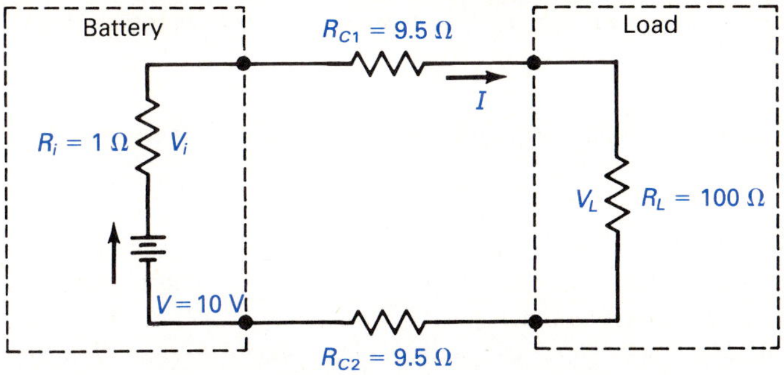

Fig. 6-50. Example 6-25: Circuit with conductor resistance.

b. (Fig. 6-50)
From Eq. (6-3),

$$R_T = R_i + R_L + R_{C1} + R_{C2}$$

$$I = \frac{V}{R_T} = \frac{V}{R_i + R_L + R_{C1} + R_{C2}}$$

$$I = \frac{10\text{ V}}{1\ \Omega + 100\ \Omega + 9.5\ \Omega + 9.5\ \Omega} = \frac{10\text{ V}}{120\ \Omega} = 0.08333\text{ A}$$

$$V_L = V - (V_i + V_{C1} + V_{C2}) = V - I(R_i + R_{C1} + R_{C2}) \qquad (6\text{-}14)$$

$$V_L = 10\text{ V} - 0.08333\text{ A}\,(1\ \Omega + 9.5\ \Omega + 9.5\ \Omega) = 10\text{ V} - 1.666\text{ V}$$

$$V_L = \mathbf{8.334\ V}$$

If the generator and load in Example 6-25 had been connected by resistanceless conductors, the load voltage would have been 9.01 V rather than the 8.334 V obtained with the 9.5-Ω wires. Therefore the presence of these conductor resistances reduced the available load voltage by 15.83 percent. It is evident, then, that conductor resistances can cause large-percentage load voltage losses. The relationship between the load resistance and the total conductor resistance determines the magnitude of this effect. For example, had the load resistance been 1000 Ω instead of 100 Ω, the voltage reduction due to the 9.5-Ω conductors would only have been 1.86 percent.

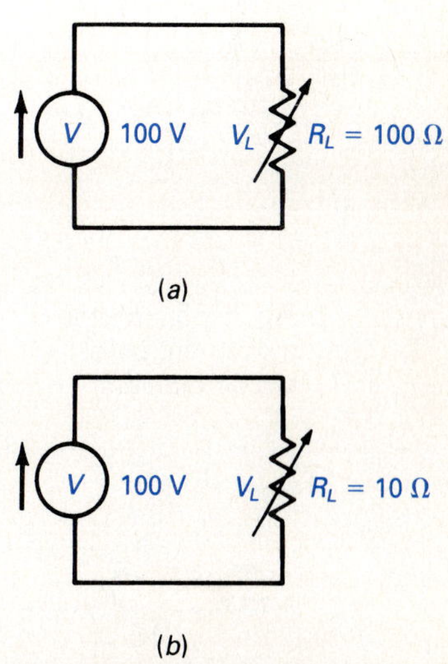

Fig. 6-51. Example 6-26: Circuit without conductor resistance.

EXAMPLE 6-26 A 100-V generator of negligible internal resistance is connected to a load which varies in resistance from 10 to 100 Ω. What is the load voltage if the total conductor resistance is *(a)* insignificant and *(b)* 10 Ω?

Solution *a.* Figures 6-51*a* and *b* show the extremes of the load resistance with zero conductor resistance. Clearly, in both cases, as there are no losses, the load voltage will equal the generator emf.

$$V = V_L$$

$$V_L = \mathbf{100\ V}$$

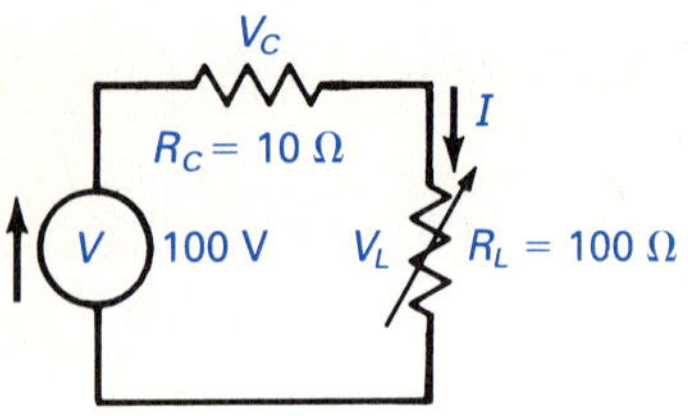

(a)

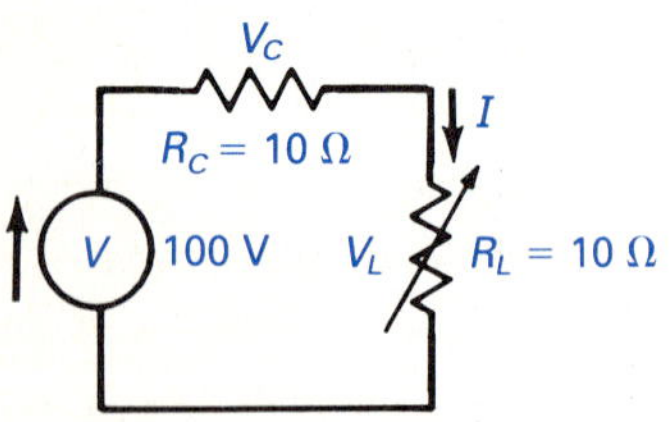

(b)

Fig. 6-52. Example 6-26: Circuit with conductor resistance.

b. In Fig. 6-52*a*, the load resistance is at its maximum value of 100 Ω.

$$I = \frac{V}{R_C + R_L} = \frac{100\ \text{V}}{10\ \Omega + 100\ \Omega} = 0.9091\ \text{A}$$

$$V_L = IR_L = 0.9091\ \text{A} \times 100\ \Omega = \mathbf{90.91\ V}$$

In Fig. 6-52*b*, the load resistance is at its minimum value of 10 Ω.

$$I = \frac{V}{R_C + R_L} = \frac{100\ \text{V}}{10\ \Omega + 10\ \Omega} = 5\ \text{A}$$

$$V_L = IR_L = 5\ \text{A} \times 10\ \Omega = \mathbf{50\ V}$$

The significance of Example 6-26 is that, if the conductor resistance is negligible, a load fed from a constant-voltage source will have a constant voltage across it, even though the load resistance varies. If, on the other hand, the conductor resistance is significant, the load voltage will vary as its resistance varies. The circuit with the significant conductor resistance is said to have poor voltage regulation because the load voltage, which is meant to be constant despite load-resistance variations, varies considerably.

Voltmeters

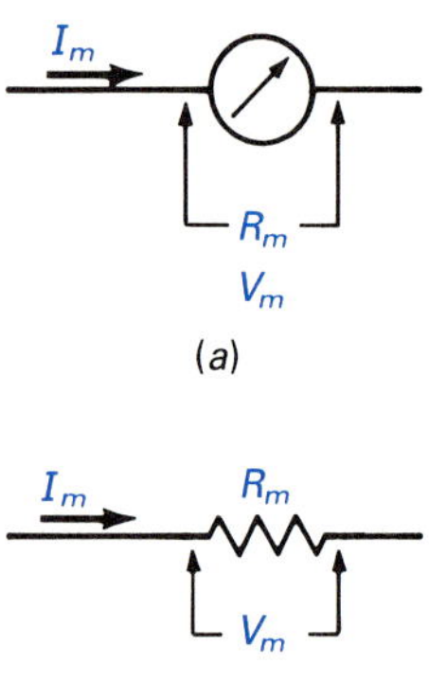

Fig. 6-53. Meter movement as a circuit element.

Usually a voltmeter consists of a basic meter movement with a resistor connected in series with it.

The meter movement measures the current flowing through it, and is thus an ammeter. The position of its pointer on the calibrated scale is directly proportional to the current. A common type of movement used for voltmeters has a full-scale deflection of 50 μA, meaning that its pointer will be deflected to the end of the calibrated scale with a current of 50 μA. As an electric circuit element, the movement has a resistance R_m, a full-scale current I_m, and a full-scale voltage V_m. These components are shown in Fig. 6-53.

Figure 6-53*a* shows a frequently used symbol for a meter movement. The resistance of the movement may be shown as in Fig. 6-53*b*.

EXAMPLE 6-27 A meter has a full-scale current of 50 μA and a resistance of 2 kΩ. What voltage will produce a full-scale deflection?

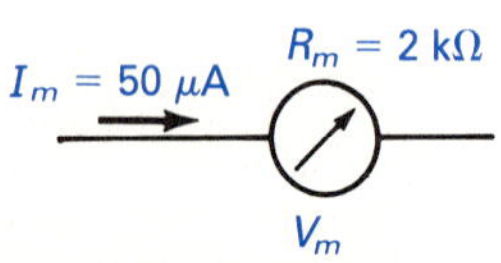

Fig. 6-54. Example 6-27: Data.

Solution (Fig. 6-54)

$$V_m = I_m R_m = 50\ \mu\text{A} \times 2\ \text{k}\Omega = \mathbf{0.1\ V}$$

The pointer of this 50-μA movement will therefore be deflected to the end of the calibrated scale if 0.1 V is applied across it.

Because the full-scale condition represents the safe limit to which the movement can be operated, and because the range of a meter is specified in terms of its full-scale deflection, only full-scale values are considered in calculations. However, it should be noted that, since voltage is directly proportional to current, deflections other than full-scale will be produced by voltages proportionally less than the full-scale voltage. In the case of the basic voltmeter in Example 6-27, 0.05 V will produce a half-scale reading.

As you can see, the basic movement is of limited value as a voltmeter. If it is limited to a maximum voltage of 0.1 V, as in Example 6-27, it cannot be of much use. A resistor connected in series with the movement will result in a useful voltmeter, and this arrangement is typical of most commercial voltmeters. In this application the series resistor is called a *multiplier*.

Figure 6-55 shows the basic equivalent network of a voltmeter.

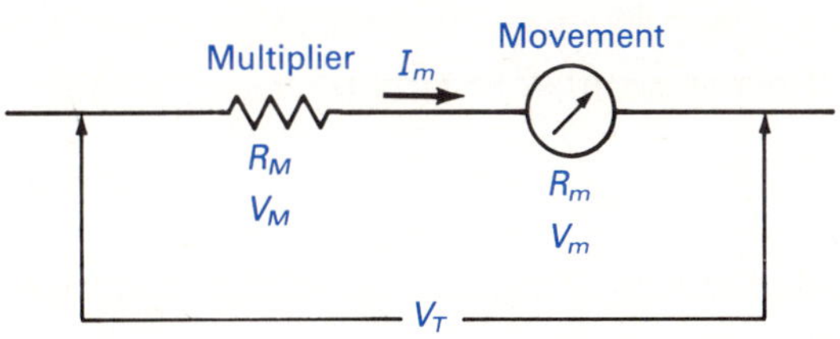

Fig. 6-55. Basic voltmeter.

$$V_T = V_M + V_m \tag{6-15}$$

By applying Ohm's law to this relationship, meters can be designed to measure virtually any range of voltage.

EXAMPLE 6-28 It is desired to make a voltmeter with a range of 0 to 1 V (1 V full-scale), using a 50-μA 2-kΩ movement. What value of multiplier resistor will be needed?

Solution (Fig. 6-56)

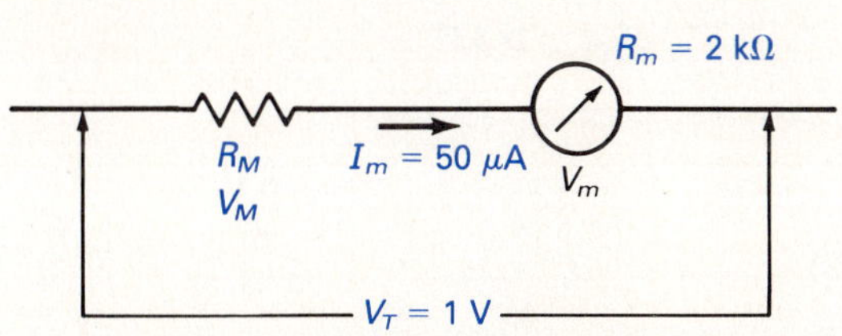

Fig. 6-56. Example 6-28: Data.

Only the full-scale conditions need to be considered, as all other conditions are directly proportional to them. From Example 6-27,

$$V_m = 0.1\ \text{V}$$

From Eq. (6-15),

$$V_M = V_T - V_m = 1\ \text{V} - 0.1\ \text{V} = 0.9\ \text{V}$$

From Eq. (6-9),

$$R_M = R_m \frac{V_M}{V_m} = 2\ \text{k}\Omega \times \frac{0.9\ \text{V}}{0.1\ \text{V}} = \mathbf{18\ k\Omega}$$

EXAMPLE 6-29 Using the same 50-μA 2-kΩ movement as in Examples 6-27 and 6-28, design a voltmeter with three ranges, 1, 10, and 100 V, full-scale.

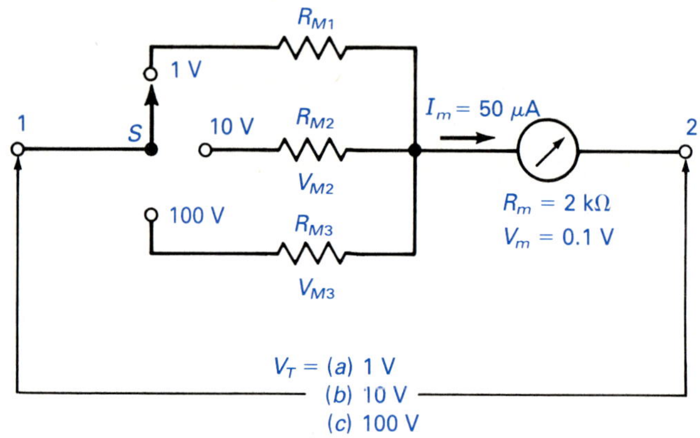

Fig. 6-57. Example 6-29: Data.

Solution Figure 6-57 shows the movement with three multipliers capable of being connected, one at a time, in series with the movement by means of the switch S. For any of the three switch positions, one series circuit will exist between the voltmeter terminals 1 and 2.

a. For 1-V full-scale, the multiplier R_{M1} will be in series with the movement. This range is the same as that in Example 6-28.

$$R_{M1} = \mathbf{18\ k\Omega}$$

b. For 10-V full-scale, from Eq. (6-15),

$$V_{M2} = V_T - V_m = 10\ \text{V} - 0.1\ \text{V} = 9.9\ \text{V}$$

From Eq. (6-9),

$$R_{M2} = R_m \frac{V_{M2}}{V_m} = 2\ \text{k}\Omega \times \frac{9.9\ \text{V}}{0.1\ \text{V}} = \mathbf{198\ k\Omega}$$

c. For 100-V full-scale, from Eq. (6-15),

$$V_{M3} = V_T - V_m = 100\ \text{V} - 0.1\ \text{V} = 99.9\ \text{V}$$

From Eq. (6-9),

$$R_{M3} = R_m \frac{V_{M3}}{V_m} = 2\ \text{k}\Omega \times \frac{99.9\ \text{V}}{0.1\ \text{V}} = \mathbf{1998\ k\Omega} \text{ or } \mathbf{1.998\ M\Omega}$$

Voltmeter Sensitivity

In Example 6-29, the total resistance $R_M + R_m$ divided by the full-scale voltage V_T is the same for each range. This relationship is not coincidental, but is bound to be true because in any series circuit the current is the same throughout. Therefore the ratio voltage/resistance (or resistance/voltage) will be the same for the entire voltmeter as for the movement alone. The quantity ohms per volt, in voltmeter specifications, is known as the *sensitivity* of the instrument. A 50-μA movement will always result in a voltmeter sensitivity of 20 kΩ/V, as in our examples. This is a higher sensitivity than the 1 kΩ/V which would be obtained by using a movement with a full-scale current of 1 mA (1000 μA).

Ammeter Loading

An ammeter always has to be connected as a series-circuit element so that it will be directly in the path of the current being measured.

The network in Fig. 6-58*a* draws a current I_1 from the battery. To measure this current, an ammeter must be connected in series with the network, as shown in Fig. 6-58*b*. The ammeter can be connected on either side of the network, as the current will be the same. The schematic diagram in Fig. 6-59 is the equivalent of Fig. 6-58*b*.

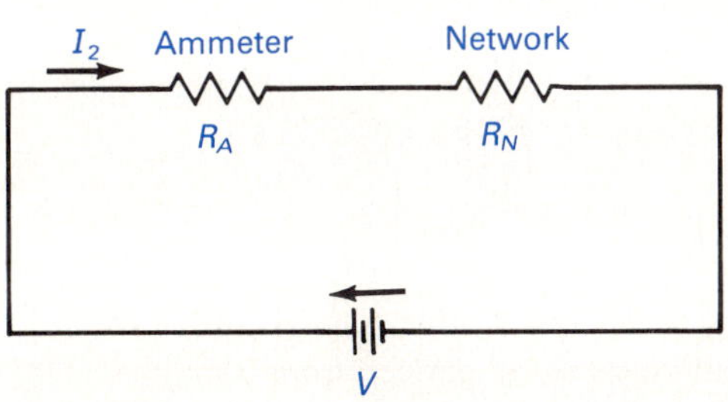

Fig. 6-59. Schematic of ammeter connection.

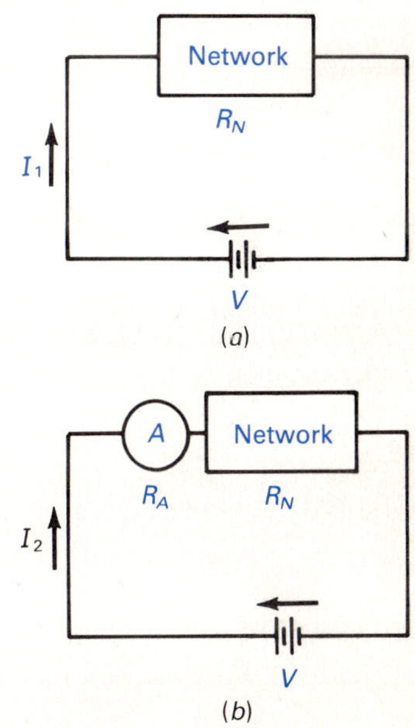

Fig. 6-58. *(a)* Network connected to a battery; *(b)* ammeter connected to read the network current.

The intention in using the ammeter is to measure the current I_1 in the network, as in Fig. 6-58*a*.

$$I_1 = \frac{V}{R_N}$$

But the existence of the ammeter modifies this current, reducing it to the value I_2, as in Fig. 6-58*b* or 6-59:

$$I_2 = \frac{V}{R_A + R_N}$$

This interference with the quantity being measured is a problem with all measuring instruments and is unavoidable. What has to be done in practice is to reduce the *loading* effect by the choice of an instrument of acceptable characteristics. An ammeter should be chosen so that its resistance introduces only a small measurement error. If the ammeter resistance R_A is 1 percent of the network resistance R_N, it will introduce an error of about 1 percent.

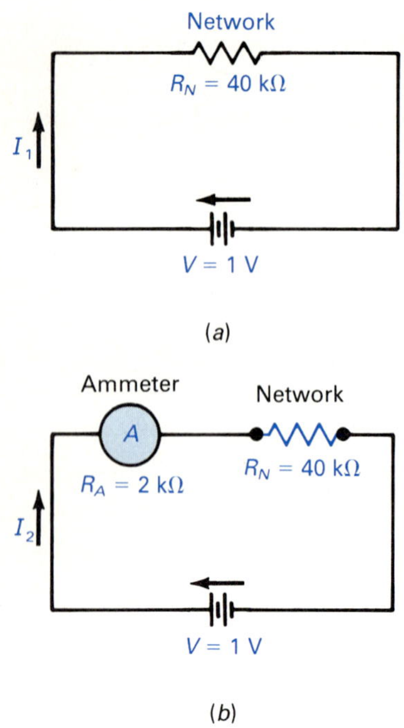

Fig. 6-60. Example 6-30: Circuits. *(a)* Without ammeter; *(b)* with ammeter.

EXAMPLE 6-30 The resistance of a network connected across a 1-V supply is 40 kΩ. A 50-μA 2-kΩ meter movement is to be used as an ammeter to measure the current in the network. What percentage of error is caused by the presence of the ammeter?

Solution Figure 6-60*a* shows the conditions which exist without the meter. The current to be measured will be

$$I_1 = \frac{V}{R_N} = \frac{1\text{ V}}{40\text{ k}\Omega} = 2.5 \times 10^{-5}\text{ A or } 25\ \mu\text{A}$$

This current of 25 μA should result in half-scale deflection of the meter pointer.

In Fig. 6-60*b*, the meter movement is shown as an ammeter, introducing 2 kΩ of series resistance into the circuit.

$$I_2 = \frac{V}{R_A + R_N} = \frac{1\text{ V}}{2\text{ k}\Omega + 40\text{ k}\Omega} = 2.38 \times 10^{-5}\text{ A or } 23.8\ \mu\text{A}$$

The percentage error due to the movement resistance will be

$$\text{Percentage error} = \frac{\text{current error}}{\text{correct current}} \times 100 = \frac{I_1 - I_2}{I_1} \times 100$$

$$= \frac{25\ \mu\text{A} - 23.8\ \mu\text{A}}{25\ \mu\text{A}} \times 100 = \frac{1.2\ \mu\text{A}}{25\ \mu\text{A}} \times 100$$

Percentage error = **4.8 percent**

If this basic error of 4.8 percent is not acceptable, then an ammeter of lower resistance will have to be used.

Unloaded Voltage Divider

The *unloaded* voltage divider is a series network which can be used to feed to other networks a number of different voltages, all derived from a single voltage source. There is a constraint on the use of the unloaded voltage divider which, if rigidly imposed, would render it useless for practical application; namely, that the loads (networks) fed by the voltage divider can draw no current from it. Fortunately, the principle of the unloaded voltage divider holds within useful limits if the loads fed by it draw very small currents.

First, the principle will be developed assuming no current loading.

Figure 6-61 shows a simple voltage divider with two output voltages V_1 and

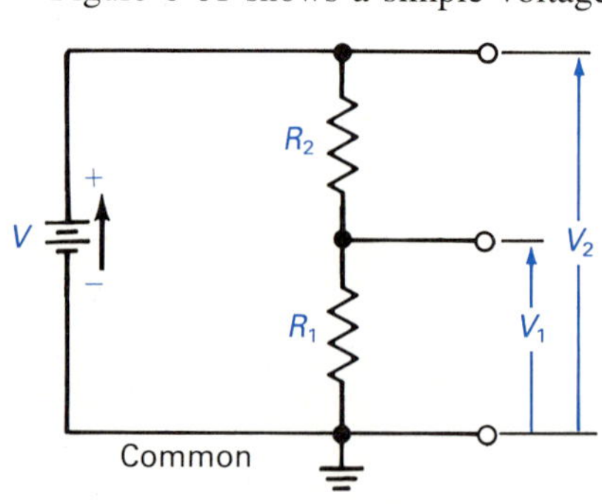

Fig. 6-61. Unloaded voltage divider.

V_2, both derived from the emf V. No loads are connected to the output terminals, the output voltages V_1 and V_2 being directly derived from the resistance network R_1 and R_2.

$$V_2 = V$$

If no current is drawn by the V_1 output lead, from Eq. (6-7),

$$V_1 = V\frac{R_1}{R_T} = V\frac{R_1}{R_1 + R_2}$$

This concept can be applied to voltage dividers with any number of outputs and is called the "voltage-divider principle."

$$V_n = V\frac{R_n}{R_T} \tag{6-16}$$

where V_n = voltage between the output terminal and ground
V = network supply voltage
R_n = resistance between the output terminal and ground
R_T = total resistance of the voltage divider

EXAMPLE 6-31 What are the output voltages of the unloaded voltage divider in Fig. 6-62?

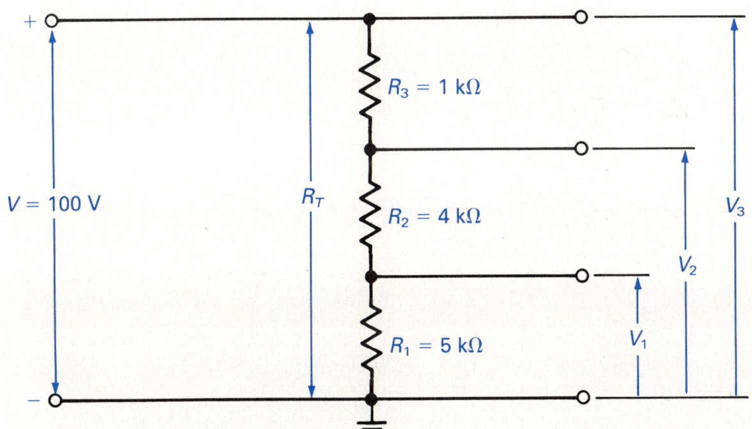

Fig. 6-62. Example 6-31: Network.

Solution

$$R_T = R_1 + R_2 + R_3 \tag{6-3}$$
$$R_T = 5\text{ k}\Omega + 4\text{ k}\Omega + 1\text{ k}\Omega = 10\text{ k}\Omega$$

From Eq. (6-16),

$$V_1 = V\frac{R_1}{R_T} = 100\text{ V} \times \frac{5\text{ k}\Omega}{10\text{ k}\Omega} = \mathbf{50\ V}$$

$$V_2 = V\frac{R_1 + R_2}{R_T} = 100\ \text{V} \times \frac{9\ \text{k}\Omega}{10\ \text{k}\Omega} = \mathbf{90\ V}$$

$$V_3 = V = \mathbf{100\ V}$$

EXAMPLE 6-32 What are the resistances of the elements of the unloaded voltage divider in Fig. 6-63 if the divider chain (the resistances of the divider) draws a current of 0.1 A from the supply?

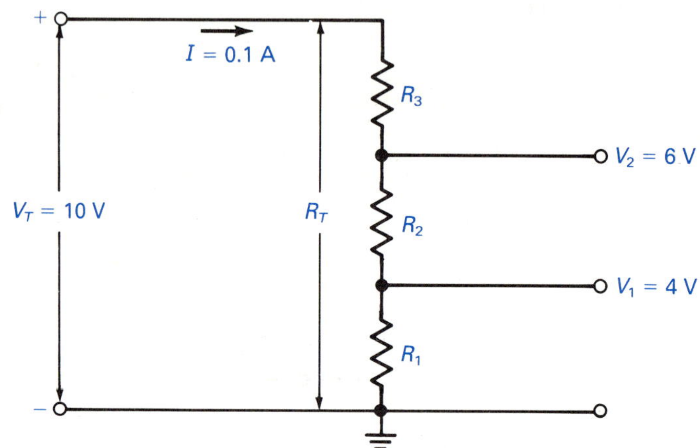

Fig. 6-63. Example 6-32: Network.

Solution

$$R_T = \frac{V_T}{I} = \frac{10\ \text{V}}{0.1\ \text{A}} = 100\ \Omega$$

From Eq. (6-9),

$$R_1 = R_T\frac{V_1}{V_T} = 100\ \Omega \times \frac{4\ \text{V}}{10\ \text{V}} = \mathbf{40\ \Omega}$$

$$R_1 + R_2 = R_T\frac{V_2}{V_T} = 100\ \Omega \times \frac{6\ \text{V}}{10\ \text{V}} = 60\ \Omega$$

$$R_2 = (R_1 + R_2) - R_1 = 60\ \Omega - 40\ \Omega = \mathbf{20\ \Omega}$$

$$R_3 = R_T - (R_1 + R_2) = 100\ \Omega - 60\ \Omega = \mathbf{40\ \Omega}$$

Voltage Regulation

It is common for the voltage at a load to be different from the value desired. It is also common for a load voltage to vary when it is desired that it be constant. There are three basic causes for these problems.

1. The presence of resistances other than the load resistance. These resistances include the internal resistance of the energy source and the resistance of the connecting conductors (often called *line resistance*).

2. The source emf being different from its nominal value.
3. The load resistance being different from its nominal value, or the load demand being different from its nominal value. The load *demand* is the current or power taken by the load, both of which are dependent on the load resistance and the load voltage.

The degree to which the actual load voltage differs from the desired value is called *voltage regulation*. A low value of regulation, or good regulation, means a small difference.

In general, voltage regulation is given by the ratio

$$\frac{\text{Voltage deviation from that desired}}{\text{Voltage desired}}$$

Voltage regulation can be expressed in different ways according to the circumstances. One situation is shown in Fig. 6-64, where the values of the generator emf and the load resistance may vary. In this case,

$$\text{Voltage regulation (percent)} = \frac{\text{maximum variation in } V_L}{\text{nominal } V_L} \times 100$$

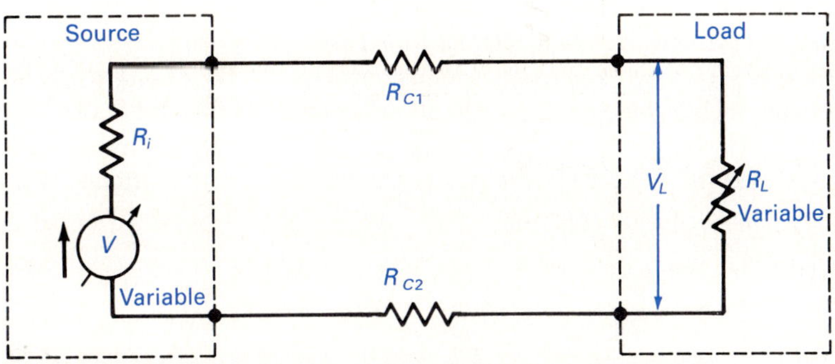

Fig. 6-64. Voltage regulation, load voltage varying.

EXAMPLE 6-33 The maximum output rating of an electronic power supply is 12 V at 5 A. Its voltage regulation is 1 percent for load currents between 1 and 5 A. What are the effective internal resistance and the emf of the power supply?

Solution

$$\text{Voltage regulation} = \frac{\text{maximum variation in } V_L}{\text{nominal } V_L}$$

$$\text{Maximum variation in } V_L = \text{voltage regulation} \times \text{nominal } V_L$$
$$= 1 \text{ percent of } 12 \text{ V} = 0.12 \text{ V}$$

It is specified that $V_L = 12$ V when the output current is 5 A. V_L will rise to 12 V + 0.12 V = 12.12 V as the current drops to 1 A.

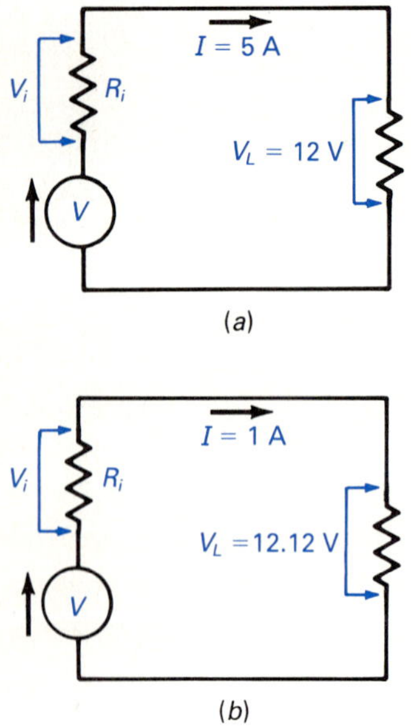

Fig. 6-65. Example 6-33: Circuit. *(a)* When the current is 5 A; *(b)* when the current is 1 A.

When $I = 5$ A (Fig. 6-65*a*),

$$V_i = IR_i = 5R_i \qquad (5 \text{ A} \times R_i)$$

Applying Eq. (6-13) ($V = V_i + V_L$),

$$V = 5R_i + 12 \tag{A}$$

When $I = 1$ A (Fig. 6-65*b*),

$$V_i = IR_i = R_i \qquad (1 \text{ A} \times R_i)$$

$$V = R_i + 12.12 \tag{B}$$

The emf V is a constant for any energy source, so the right sides of Eqs. (A) and (B) are equal.

$$5R_i + 12 = R_i + 12.12$$

$$R_i = \mathbf{0.03\ \Omega}$$

Substituting for R_i in Eq. (A),

$$V = 5R_i + 12 = 5 \times 0.03 + 12 = \mathbf{12.15\ V}$$

Voltage Sources in Series

When voltage sources (generators) are connected in series, their emfs are added algebraically (polarities) and their internal resistances are added.

Figure 6-66*a* shows two voltage generators connected in *series-aiding*, their polarities being such that their emfs can be added together. In Fig. 6-66*b* the two sources are shown with their emfs grouped together and their internal resistances similarly grouped. A single source which is the equivalent of the original two generators is shown in Fig. 6-66*c*.

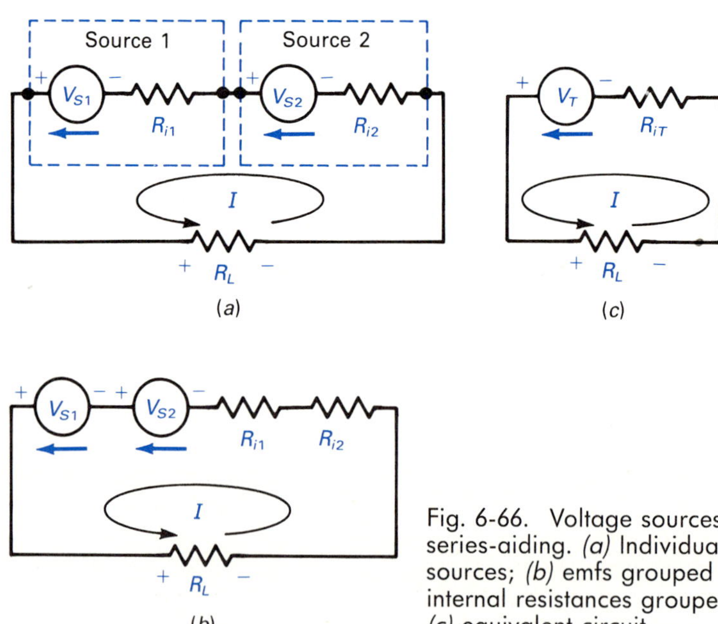

Fig. 6-66. Voltage sources in series-aiding. *(a)* Individual sources; *(b)* emfs grouped and internal resistances grouped; *(c)* equivalent circuit.

$$R_{iT} = R_{i1} + R_{i2}$$
$$V_T = V_{S1} + V_{S2}$$

When the polarities of generators are such that each tends to produce opposite-direction currents, the connection is called *series-opposing*. This type of connection is shown in Fig. 6-67, the diagrams in the sequence being the series-opposing equivalents of those in Fig. 6-66.

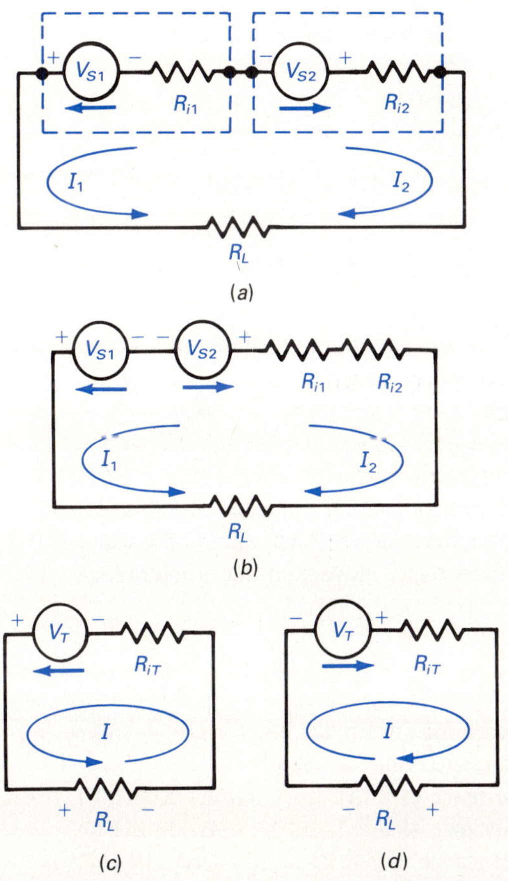

Fig. 6-67. Voltage sources in series-opposing. *(a)* Individual sources; *(b)* emfs grouped and internal resistances grouped; *(c)* equivalent circuit when $V_{S1} > V_{S2}$; *(d)* equivalent circuit when $V_{S1} < V_{S2}$.

$$R_{iT} = R_{i1} + R_{i2}.$$

If $V_{S1} > V_{S2}$ (Fig. 6-67*c*) $\quad V_T = V_{S1} - V_{S2}$

If $V_{S1} < V_{S2}$ (Fig. 6-67*d*) $\quad V_T = V_{S2} - V_{S1}$

REVIEW QUESTIONS FOR SECTION 6-5

1. What is the name given to a device used to reduce the load voltage in a series circuit?

2. When current is drawn from a generator, why is the generator's terminal voltage lower than its open circuit emf?

3. What is the approximate emf of an automobile battery that has a nominal rating of 12 V?
4. If the lead resistance is relatively high compared to the load resistance, what effect does it have on the load voltage?
5. What is required in addition to a meter movement to make a useful voltmeter? What is this addition called?
6. In what way does an ammeter affect the circuit in which it is connected?
7. An unloaded voltage divider fed by a supply V has two resistors R_1 and R_2. State an equation for the voltage across R_2.
8. What is the main limitation on the usefulness of an unloaded voltage divider?
9. What is meant by "voltage regulation"?
10. Which is better, a voltage regulation of 5 percent or 10 percent?

EXERCISE PROBLEMS FOR SECTION 6-5

1. A 50-μA meter movement with an internal resistance of 5 kΩ is to be used to make a multirange voltmeter. Find the required multiplier resistances for ranges of
   a. 5 V (*ans.* 95 kΩ)
   b. 15 V (*ans.* 295 kΩ)
   c. 50 V (*ans.* 995 kΩ)
2. A 100-μA meter movement with an internal resistance of 850 Ω is equalized to 1000 Ω by adding 150 Ω in series. What values of resistance are required to make this equalized meter movement into a multirange voltmeter with ranges of
   a. 1.5 V
   b. 5 V
   c. 15 V
3. When the following meter movements are adapted for use as voltmeters, what is their sensitivity (ohms per volt)?
   a. 50 μA with internal resistance of 2 kΩ (*ans.* 20 kΩ/V)
   b. 50 μA with internal resistance of 5 kΩ (*ans.* 20 kΩ/V)
   c. 100 μA with internal resistance of 850 Ω (*ans.* 10 kΩ/V)
   d. 1 mA with internal resistance of 250 Ω (*ans.* 1 kΩ/V)
   e. 10 μA with internal resistance of 20 kΩ (*ans.* 100 kΩ/V)
4. One volt is applied to a 1000-Ω resistance. If the current is to be measured by a 0-to-1-mA meter movement with an internal resistance of 250 Ω, what current will be indicated by the meter? What percent error does this current reading represent?
5. A 100-μA meter movement with an internal resistance of 850 Ω is used to plot the low-current characteristics of a forward-biased semiconductor diode. If, when the applied voltage is 0.18 V, the microammeter reads 100 μA, what is
   a. The resistance of the diode? (*ans.* 950 Ω)
   b. The voltage across the diode? (*ans.* 95 mV)
   c. The actual current through the diode? (*ans.* 100 μA)

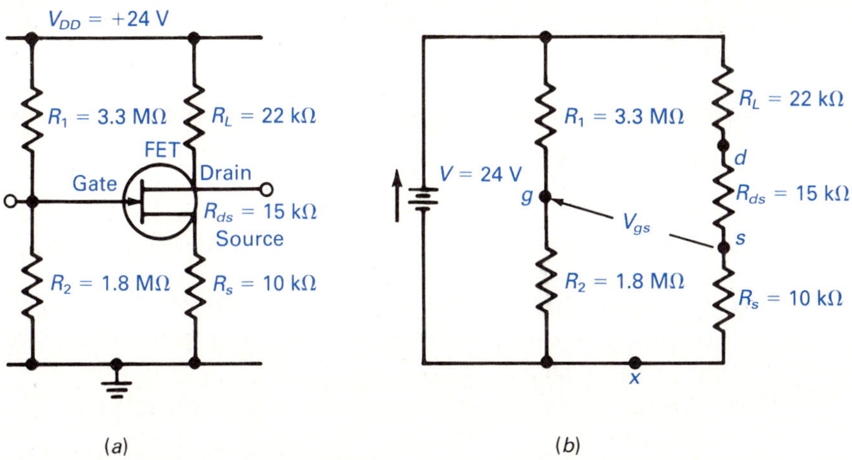

Fig. 6-68. Problem 6: *(a)* Original circuit; *(b)* equivalent circuit.

6. The current drawn by the gate of a field-effect transistor (FET) is essentially zero. What is the relative potential, gate to source, V_{gs}, for the FET in Fig. 6-68*a*? Figure 6-68*b* is an equivalent circuit.

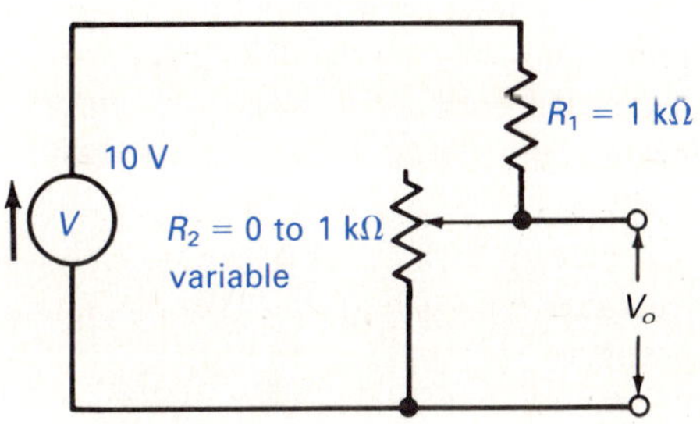

Fig. 6-69. Problem 7: Circuit.

7. Figure 6-69 is the circuit of an unloaded voltage divider using a fixed resistor and a variable resistor. Plot a graph of V_0 as R_2 is varied in 100-Ω steps from 0 to 1000 Ω. (*ans*. 0 V, 0.9191 V, 1.667 V, 2.308 V, 2.857 V, 3.333 V, 3.75 V, 4.118 V, 4.444 V, 4.737 V, 5 V)

8. The open-circuit voltage of a generator is 120 V, and its internal resistance is 0.3 Ω. It is connected to a 14.4-Ω heating element through an extension cord. Find the current in the circuit if the terminal voltage is 117.6 V when the heater is connected.

9. For the conditions stated in Prob. 8, find

a. The voltage across the heating element. (*ans*. 115.2 V)
b. The resistance of the extension cord. (*ans*. 0.3 Ω)
c. The power dissipated by the heating element. (*ans*. 921.6 W)
d. The power lost by the internal resistance and the extension cord. (*ans*. 38.4 W)

10. The terminal voltage of a generator is 117 V when a load is connected. The total resistance of the connecting wires is 0.506 Ω, and the voltage across the load is 113.4 V. What is

a. The circuit current?
b. The resistance of the load?

11. A 100-μA movement with an internal resistance of 2500 Ω is connected in series with a 10 000-Ω load. When the meter reads 60 μA, what is
a. The applied voltage? (*ans*. 0.75 V)
b. The voltage drop across the meter? (*ans*. 0.15 V)
c. The voltage drop across the load? (*ans*. 0.6 V)

12. In Prob. 11 the meter is removed and the circuit reconnected with the same applied voltage. Find
a. The voltage across the load.
b. The current through the load.

13. A 100-μA meter movement with an internal resistance of 2500 Ω is connected in series with a 270-kΩ load. The voltage drop across the load is found to be 24 V. What is
a. The voltage across the meter? (*ans*. 0.2222 V)
b. The percentage error introduced by the meter? (*ans*. 0.9174 percent)

14. A meter movement with a full-scale deflection of 1 mA and an internal resistance of 115 Ω is used in series with a resistor to form a voltmeter. What must the total series resistance be for the meter to read full-scale when the voltmeter is placed across 5 V?

15. A 120-V generator with an internal resistance of 0.2 Ω feeds a 1-kW heater through a 15-m extension cord which has a total resistance of 0.4 Ω. Find
a. The power dissipated by the generator. (*ans*. 13.89 W)
b. The power dissipated by the extension cord. (*ans*. 27.78 W)

16. For the circuit in Prob. 15, if a short circuit occurs at the end of the extension cord, find
a. The short-circuit current.
b. The power dissipated in the generator.
c. The power dissipated in the extension cord.

6-6 SERIES CIRCUITS IN ELECTRONICS

Series circuits are often used in electronics. They may be made up of ordinary resistors, as, for instance, a voltage divider is. They are more likely to include semiconductor devices, such as diodes and transistors.

One of the objectives of this discussion of series circuits in electronics is to show that the presence of semiconductors does not upset ordinary dc circuit principles. Ohm's and Kirchhoff's laws, the characteristics of series circuits, and other principles still to be introduced all apply to circuits containing semiconductors. Furthermore, it is not necessary to understand the details of operation of the semiconductor devices. Just as with ordinary resistors, it is only necessary to know the voltage, current, and resistance relationships of the devices to follow the behavior of dc circuits containing them.

Unlike an ordinary resistor, a semiconductor device does not have about the same resistance under all conditions. Its resistance may be changed by several means, such as

Changing its operating voltages or currents
Changing the intensity of light that falls on it
Changing its temperature

The voltage-divider principle is often applied to series circuits containing semiconductors, but the presence of their varying resistances produces effects not found in circuits with ordinary resistors only. You will see how useful these effects may be in the examples that follow.

Voltage Regulator

A voltage regulator maintains a voltage nearly constant in spite of variations in other circuit characteristics. The simplest electronic voltage regulator consists of a resistor and a *zener diode* connected in series. The voltage-regulator circuit shown in Fig. 6-70 uses typical values for a 5-V 5 percent zener diode.

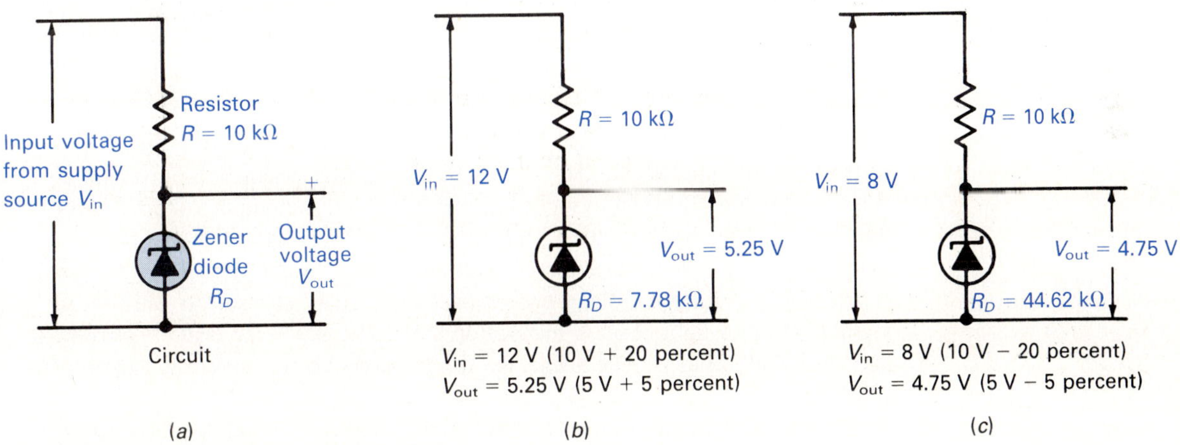

Fig. 6-70. Zener-diode voltage regulator.

The resistance of the zener diode in Fig. 6-70 changes as the input voltage V_{in} changes from its nominal 10 V.

If the input voltage happens to be 12 V (Fig. 6-70*b*), the diode resistance R_D is 7.78 kΩ. Applying the voltage-divider principle [$V_{out} = V_{in}R_D/(R_D + R)$], the output voltage V_{out} is 5.25 V.

If the input voltage becomes as low as 8 V (Fig. 6-70*c*), R_D is 14.62 kΩ and V_{out} is 4.75 V.

The zener diode, by automatically changing resistance, keeps the circuit's output voltage within 5 percent of 5 V (4.75 V to 5.25 V), even though the voltage supplied to the circuit varies by 20 percent from its nominal 10 V (8 V to 12 V).

Zener diodes are available for regulated voltages from about 3 V to over 100 V. They may be used to give a *reference* voltage to which other circuit voltages can be compared, or they may supply equipment with relatively constant operating voltages.

Light Detectors

When light falls on a *photodiode,* the device's resistance drops. This change of resistance can be used to detect light intensity levels or to measure them.

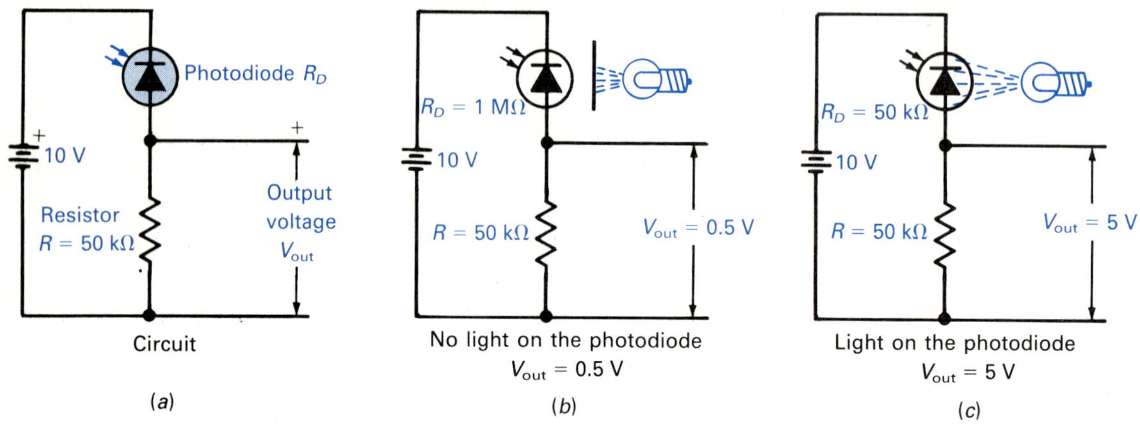

Fig. 6-71. Photodiode light detector.

The circuit of Fig. 6-71 contains a photodiode connected in series with a resistor. If this photodiode has a resistance of 1 MΩ when it is in darkness (Fig. 6-71*b*) and 50 kΩ in a bright light (Fig. 6-71*c*), the output voltage across the 50-kΩ resistor will, by the voltage-divider principle, vary between approximately 0.5 V and 5 V under these conditions.

The varying voltage from a photodiode circuit may be used to activate an alarm system. A light beam is focused on the diode. When the beam is broken, perhaps by an intruder in the house, the rise in output voltage causes the alarm to sound.

A similar system may be used to count items on a production line. A counter is arranged to record the number of pulses of output voltage as the light beam is broken by each item.

A light meter for photography may be made by replacing the series resistor by a meter movement. The meter reading at any time indicates the light intensity.

Transistor Bias

A transistor is a three-terminal semiconductor device. The relationship between the voltages at the three terminals is important to transistor operation. Figure 6-72 shows a circuit used to control a transistor's dc voltages.

Figure 6-72*a* is not just a series circuit. However, as the *base* current I_B may be less than 10 μA, very much lower than the other circuit currents, it is often in order to ignore it. When this is done, the effective result is to separate the circuit into two series groups of elements, as shown in Fig. 6-72*b*.

The two-resistor series group, R_1 and R_2, divides the 10 V supplied to the circuit so that there is 0.64 V across R_2. When connected to the transistor base, this 0.64 V becomes the *base-bias* voltage, the voltage between the base and the *emitter* V_{BE}.

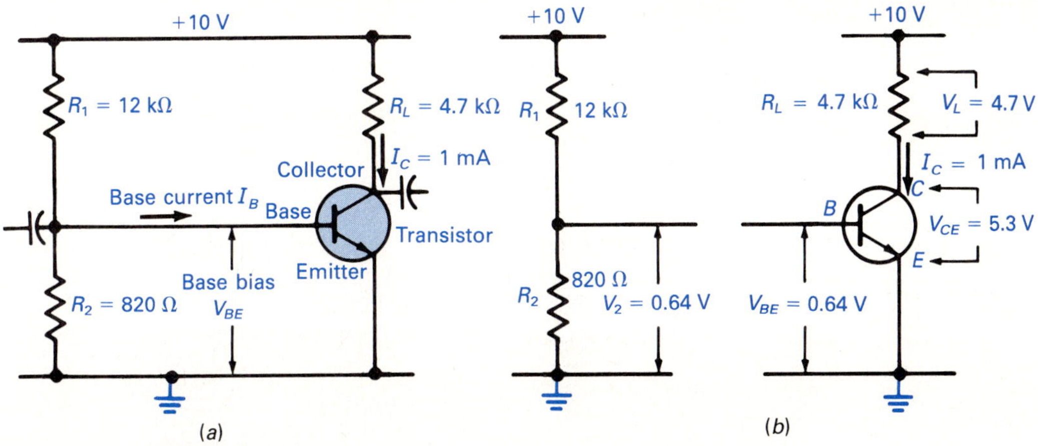

Fig. 6-72. Transistor bias. *(a)* Transistor bias circuit; *(b)* equivalent: two series circuits.

The *collector load* resistor R_L forms a series group with the collector-to-emitter resistance of the transistor. If R_L is 4.7 kΩ and the collector current I_C is 1 mA, the dc load voltage will be 4.7 V. The transistor's collector-to-emitter dc voltage V_{CE} will then be 10 − 4.7 = 5.3 V.

Load Line

The resistance of a semiconductor device depends on its circuit conditions, but these conditions may be difficult to ascertain by ordinary methods. Graphical methods sometimes help in such situations.

In Fig. 6-73*a* a diode and a 1-kΩ resistor are shown connected in series. The total voltage V_T across them is 2 V.

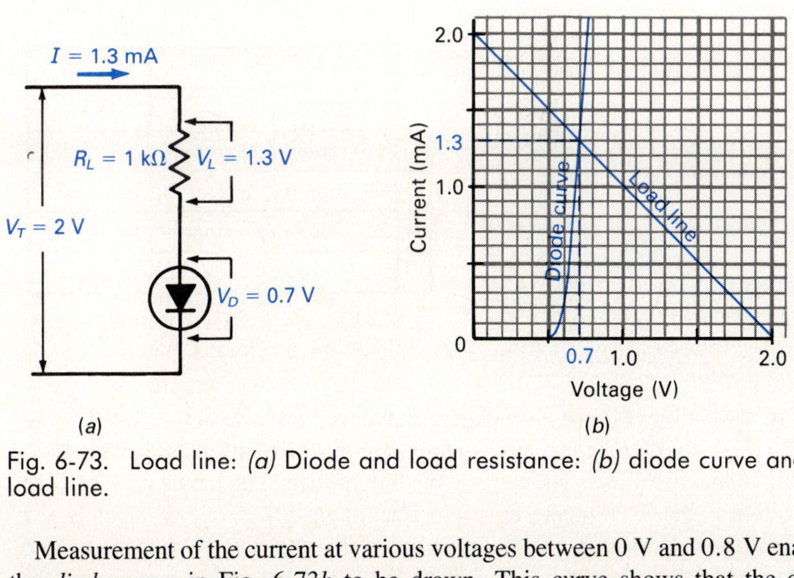

Fig. 6-73. Load line: *(a)* Diode and load resistance: *(b)* diode curve and load line.

Measurement of the current at various voltages between 0 V and 0.8 V enables the *diode curve* in Fig. 6-73*b* to be drawn. This curve shows that the diode resistance varies between an infinitely high value with a voltage up to 0.5 V

(0.5 V/0 mA) to the low value of 380 Ω when the voltage is 0.76 V (0.76 V/2 mA).

The *load line* in Fig. 6-73*b* represents the 1-kΩ resistance of the load resistor R_L. It is drawn between the 2-mA and 2-V axis points (2 V/2mA = 1 kΩ).

The intersection of the diode curve and the load line gives the conditions existing in the circuit: the current $I = 1.3$ mA, the diode voltage $V_D = 0.7$ V, and the resistor voltage $V_L = 1.3$ V (this is 2 V − 0.7 V).

When the conditions become known from the diode curve and the load line, we find that ordinary circuit principles apply to Fig. 6-73*a*,

$$I = \frac{V_L}{R_L} = \frac{1.3\text{ V}}{1\text{ k}\Omega} = 1.3\text{ mA}$$

$$R_D = \frac{V_D}{I} = \frac{0.7\text{ V}}{1.3\text{ mA}} = 538.5\ \Omega$$

$$I = \frac{V_T}{R_T} = \frac{2\text{ V}}{(1 + 0.5385)\text{ k}\Omega} = 1.3\text{ mA}$$

AND Gate

Whether in a computer, a watch, a television set, an industrial control system, or some other application, digital electronics always depends on a few types of *logic gates*. A gate is an electronic switch that performs a "logical" function in that it responds to particular sets of input conditions by giving an output signal. If the conditions are right for a certain gate type, a path will open up for signals to go through it from input to output, just as a wooden gate opens up to allow cattle to go through it from one field to another.

In digital electronics, gates use semiconductor devices, diodes and transistors, to decide whether conditions are right to pass signals in the form of voltage-level input signals. However, basic gate actions may be seen by using ordinary switches to set up the input conditions.

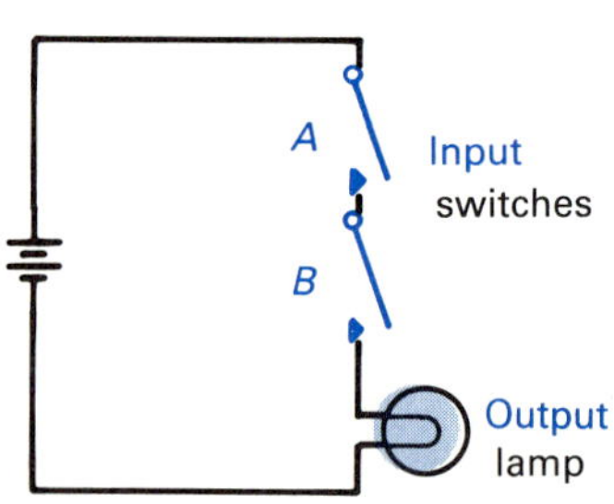

(*a*)

	Input switches A	Input switches B	Output lamp
1	OFF (open)	OFF (open)	OFF (unlit)
2	OFF (open)	ON (closed)	OFF (unlit)
3	ON (closed)	OFF (open)	OFF (unlit)
4	ON (closed)	ON (closed)	ON (lit)

(*b*)

Fig. 6-74. Two-input switch AND gate. *(a)* Circuit; *(b)* truth table.

In Fig. 6-74*a* a two-input AND gate is shown as two switches, *A* and *B*, connected in series with a battery and a lamp. The positions of the switches determine the input conditions. The lamp is the output load. The inputs must *both* be ON (switches closed) to make an ON output (lamp lit). If either input is OFF (switches open), the output is OFF (lamp unlit). In other words, inputs *A* AND *B* must be ON for the output to be ON. The *truth table* of Fig. 6-74*b* summarizes the action of the two-input AND gate for all four possible input conditions.

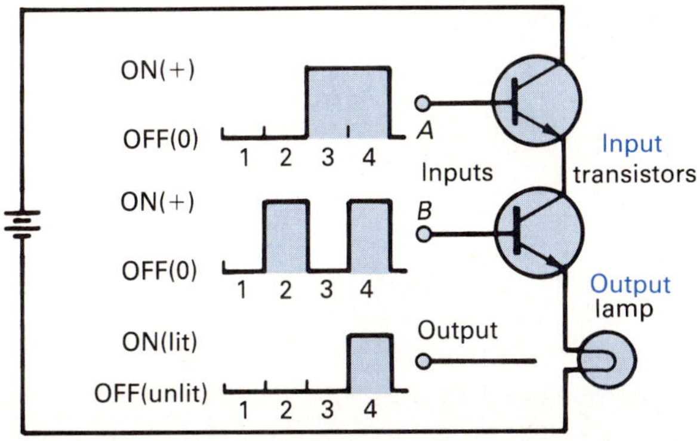

(a)

	Input voltages A	Input voltages B	Output lamp
1	OFF (0)	OFF (0)	OFF (unlit)
2	OFF (0)	ON (+)	OFF (unlit)
3	ON (+)	OFF (0)	OFF (unlit)
4	ON (+)	ON (+)	ON (lit)

(b)

Fig. 6-75. Two-input electronic AND gate. *(a)* Circuit; *(b)* truth table.

A simple electronic two-input AND gate is shown in Fig. 6-75. It has two voltage levels for its inputs, 0 for OFF and + for ON. Four time intervals are shown, each representing an input condition similar to that of Fig. 6-74. You will see that the output is ON (lamp lit) only at interval 4, where both inputs are in the ON (+) state. The truth table of the electronic AND gate is the same as that of the series switch AND gate.

An AND gate may have more than two inputs. If it does, *all* its inputs must be in the ON condition for its output to be ON.

6-7 CAPACITORS IN SERIES

The characteristics of series circuits apply with capacitors. Of course, the conditions important to capacitance must be observed; for instance, the importance of change. A steady current is of little interest in a capacitive circuit, and so the *common current* characteristic of series resistive circuits becomes *common instantaneous current* in capacitive circuits.

The series capacitors in Fig. 6-76 have the common instantaneous current i. At any time instant the total voltage across both capacitors V_T is the sum of the voltages across each of them $v_1 + v_2$.

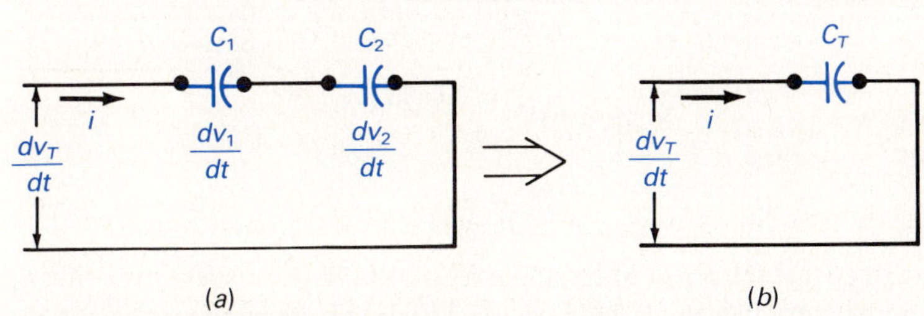

Fig. 6-76. *(a)* Capacitors in series; *(b)* single capacitor equivalent.

Over a given time period the total *voltage change* is the sum of the individual *voltage changes,*

$$\frac{dv_T}{dt} = \frac{dv_1}{dt} + \frac{dv_2}{dt}$$

From Eq. (4-1), $i = C\,dv/dt$, we can get

$$\frac{i}{C} = \frac{dv}{dt}$$

Substituting this expression for i/C in the voltage-change equation,

$$\frac{i}{C_T} = \frac{i}{C_1} + \frac{i}{C_2}$$

or

$$\frac{1}{C_T} = \frac{1}{C_1} + \frac{1}{C_2}$$

This equation can be applied to any number of capacitors connected in series:

$$\frac{1}{C_T} = \frac{1}{C_1} + \frac{1}{C_2} + \frac{1}{C_3} + \cdots + \frac{1}{C_n} \qquad \textbf{(6-17)}$$

You should particularly note that Eq. (6-17), for the *total capacitance* of a number of capacitors in *series,* is quite different in *form* from Eq. (6-3), which gives the *total resistance* of a number of resistors in *series*. In the next chapter you will see that the *form* of the equations is the same for *parallel capacitors* and *series resistors* [Eqs. (7-26) and (6-3)]. Similarly, the equation *form* is the same for *parallel resistors* and *series capacitors* [Eqs. (7-3) and (6-17)].

EXAMPLE 6-34 What is the total capacitance of two capacitors, 0.1 μF and 0.22 μF, connected in series?

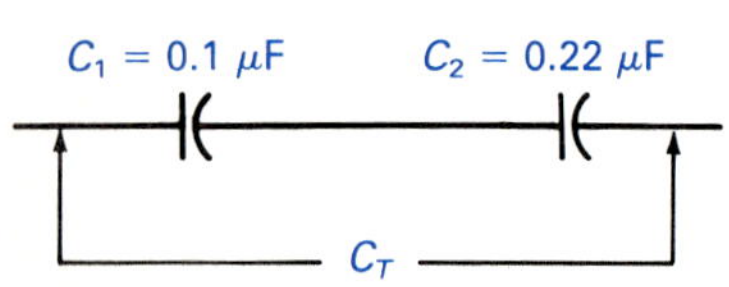

Fig. 6-77. Example 6-34: Data.

Solution (Fig. 6-77)

$$\frac{1}{C_T} = \frac{1}{C_1} + \frac{1}{C_2} \qquad (6\text{-}17)$$

$$\frac{1}{C_T} = \frac{1}{0.1\ \mu\text{F}} + \frac{1}{0.22\ \mu\text{F}} = 10^{-5} + 4.545 \times 10^{-6}*$$

$$= 14.55 \times 10^{-6}*$$

$$C_T = \frac{1}{14.55 \times 10^{-6}} = \mathbf{0.0688\ \mu F}$$

You may notice that the result of Example 6-34, 0.0688 μF, is less than either of the individual capacitances, 0.1 μF and 0.22 μF. This applies to any series capacitive circuit.

*The reciprocal of μF is MF^{-1}, an awkward unit of little practical value. It is omitted, as it is not helpful.

The total capacitance of a number of series-connected capacitors is less than any of the individual capacitances.

EXAMPLE 6-35 The voltage applied to two series-connected capacitors, 0.5 μF and 0.75 μF, is 100 V. What is the voltage across each capacitor?

Solution (Fig. 6-78)

From Eq. (6-17), the total capacitance of the circuit is

$$\frac{1}{C_T} = \frac{1}{C_1} + \frac{1}{C_2} \tag{6-17}$$

$$\frac{1}{C_T} = \frac{1}{0.5\ \mu\text{F}} + \frac{1}{0.75\ \mu\text{F}} = 3.333 \times 10^{-6}$$

$$C_T = 0.3\ \mu\text{F}$$

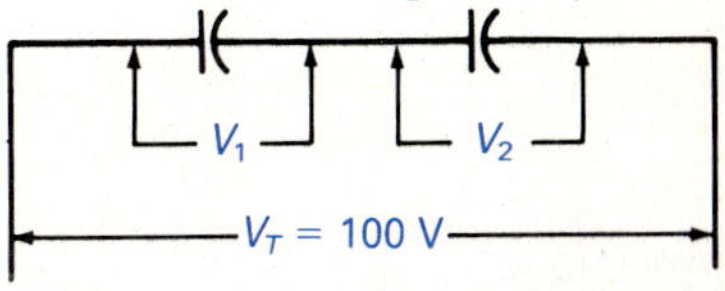

Fig. 6-78. Example 6-35: Data.

From Eq. (4-3), the circuit charge is

$$Q_T = C_T V \tag{4-3}$$

$$Q_T = 0.3\ \mu\text{F} \times 100\ \text{V} = 30\ \mu\text{C}$$

Since during charge the instantaneous current is the same throughout the series circuit, the charge on each capacitor is the same as the total circuit charge [from Eq. (2-9), $Q = it$]:

$$Q_1 = Q_2 = Q_T = 30\ \mu\text{C}$$

The voltage across capacitor C_1 is

$$V_1 = \frac{Q_1}{C_1} = \frac{30\ \mu\text{C}}{0.5\ \mu\text{F}} = \mathbf{60\ V}$$

The voltage across capacitor C_2 is

$$V_2 = \frac{Q_2}{C_2} = \frac{30\ \mu\text{C}}{0.75\ \mu\text{F}} = \mathbf{40\ V}$$

The results of Example 6-35 show that in a series capacitive circuit, voltages are inversely proportional to capacitance; C_2 has 1.5 times the capacitance of C_1, but C_1 has 1.5 times the voltage of C_2.

The sum of the capacitor voltages in the series circuit of Example 6-35 is 100 V, the total circuit voltage, as would be expected from Kirchhoff's voltage law.

If, as often happens with practical capacitive circuits, the total voltage changes, the individual capacitor voltages will follow the changes. In Example 6-35, if the total voltage V_T were to increase from 100 V to 300 V in 1 ms, V_1 would increase from 60 V to 180 V and V_2 from 40 V to 120 V in the same 1 ms. This means that the *rates* of voltage change would be

$$\text{Rate of change of total voltage} = \frac{(300 - 100)\text{ V}}{1\text{ ms}} = 200\text{ V/ms or }200\text{ kV/s}$$

$$\text{Rate of change of } C_1 \text{ voltage} = \frac{(180 - 60)\text{ V}}{1\text{ ms}} = 120\text{ V/ms or }120\text{ kV/s}$$

$$\text{Rate of change of } C_2 \text{ voltage} = \frac{(120 - 40)\text{ V}}{1\text{ ms}} = 80\text{ V/ms or }80\text{ kV/s}$$

As you might expect, Kirchhoff's voltage law applies to *rates* of voltage change too.

EXAMPLE 6-36 What value of capacitance should be connected in series with a 16-μF filter capacitor to produce a total capacitance of 4 μF?

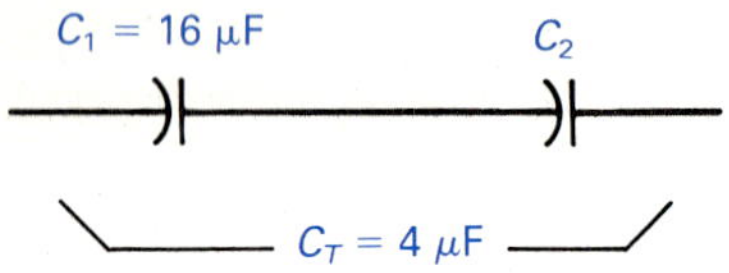

Fig. 6-79. Example 6-36: Data.

Solution (Fig. 6-79)

$$\frac{1}{C_T} = \frac{1}{C_1} + \frac{1}{C_2} \qquad (6\text{-}17)$$

$$\frac{1}{C_2} = \frac{1}{C_T} - \frac{1}{C_1} = \frac{C_1 - C_T}{C_1 C_T}$$

$$C_2 = \frac{C_1 C_T}{C_1 - C_T} = \frac{16\ \mu\text{F} \times 4\ \mu\text{F}}{16\ \mu\text{F} - 4\ \mu\text{F}} = \mathbf{5.333\ \mu F}$$

EXAMPLE 6-37 What is the total capacitance of three capacitors, each 0.015 μF, connected in series?

Solution (Fig. 6-80)

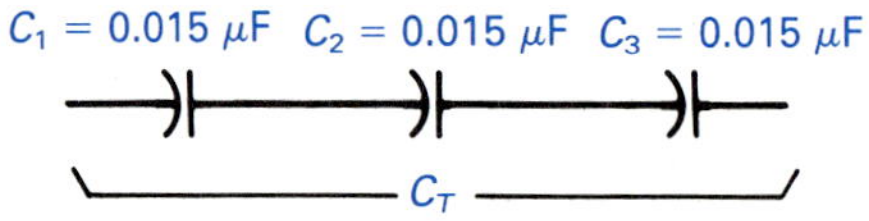

Fig. 6-80. Example 6-37: Given data. Equal capacitors connected in series.

$$\frac{1}{C_T} = \frac{1}{C_1} + \frac{1}{C_2} + \frac{1}{C_3} \qquad (6\text{-}17)$$

$$\frac{1}{C_T} = \frac{1}{0.015\ \mu\text{F}} + \frac{1}{0.015\ \mu\text{F}} + \frac{1}{0.015\ \mu\text{F}} = \frac{3}{0.015\ \mu\text{F}}$$

$$C_T = \frac{0.015\ \mu\text{F}}{3} = \mathbf{0.005\ \mu F}$$

Example 6-37 demonstrates that *when several equal-value capacitors are connected in series, the total capacitance is the value of any one of them divided by the number of capacitors.*

REVIEW QUESTIONS FOR SECTION 6-7

1. What is the equation for the total capacitance of two capacitors in series?
2. What is the equation for the total voltage of two capacitors in series?
3. Compare the total capacitance of a number of capacitors in series to the capacitance of any of the capacitors.

SUMMARY OF EQUATIONS

$$V_T = V_1 + V_2 + V_3 + \cdots + V_n \qquad (6\text{-}1)$$
$$V = V_1 + V_2 + V_3 + \cdots + V_n \qquad (6\text{-}2)$$
$$R_T = R_1 + R_2 + R_3 + \cdots + R_n \qquad (6\text{-}3)$$
$$P_T = P_1 + P_2 + P_3 + \cdots + P_n \qquad (6\text{-}4)$$
$$W_T = W_1 + W_2 + W_3 + \cdots + W_n \qquad (6\text{-}5)$$
$$V_T = V_1R_T/R_1 \quad \text{or} \quad V_2R_T/R_2 \quad \text{or} \quad \cdots \quad \text{or} \quad V_nR_T/R_n \qquad (6\text{-}6)$$
$$V_1 = V_TR_1/R_T \quad \text{or} \quad V_2R_1/R_2 \quad \text{or} \quad \cdots \quad \text{or} \quad V_nR_1/R_n \qquad (6\text{-}7)$$
$$R_T = R_1V_T/V_1 \quad \text{or} \quad R_2V_T/V_2 \quad \text{or} \quad \cdots \quad \text{or} \quad R_nV_T/V_n \qquad (6\text{-}8)$$
$$R_1 = R_TV_1/V_T \quad \text{or} \quad R_2V_1/V_2 \quad \text{or} \quad \cdots \quad \text{or} \quad R_nV_1/V_n \qquad (6\text{-}9)$$
$$P_T = P_1R_T/R_1 \quad \text{or} \quad P_2R_T/R_2 \quad \text{or} \quad \cdots \quad \text{or} \quad P_nR_T/R_n \qquad (6\text{-}10)$$
$$P_1 = P_TR_1/R_T \quad \text{or} \quad P_2R_1/R_2 \quad \text{or} \quad \cdots \quad \text{or} \quad P_nR_1/R_n \qquad (6\text{-}11)$$
$$R_T/R_1 = V_T/V_1 = P_T/P_1 = W_T/W_1 \qquad (6\text{-}12)$$
$$V = V_i + V_L \quad \text{or} \quad V_L = V - V_i \qquad (6\text{-}13)$$
$$V_L = V - (V_i + V_{C1} + V_{C2}) \qquad (6\text{-}14)$$
$$V_T = V_M + V_m \qquad (6\text{-}15)$$
$$V_n = VR_n/R_T \qquad (6\text{-}16)$$
$$\frac{1}{C_T} = \frac{1}{C_1} + \frac{1}{C_2} + \frac{1}{C_3} + \cdots + \frac{1}{C_n} \qquad (6\text{-}17)$$

CHAPTER QUESTIONS

6-1. What are the two most common circuit forms?

6-2. Describe a series circuit.

6-3. Why might it be said that the *common current* in each element best defines what is meant by a series circuit?

6-4. Compare the currents in a series circuit from the viewpoints of its generator (source), load, and other elements.

6-5. What is the characteristic common to voltages, powers, and energies in a series circuit?

6-6. To what does Kirchhoff's voltage law refer?

6-7. Distinguish between the terms "reference," "ground," and "common" when applied to relative potential.

6-8. Compare relative potentials obtained by adding voltages counted in one direction around a circuit with those obtained using the other direction.

6-9. How can a series resistor "drop" voltage?

6-10. Distinguish between "open-circuit emf" and "terminal voltage." To what types of energy source do these terms refer?

6-11. What is meant when it is said that the nominal voltage of a car battery is 12 V?

6-12. Why is it satisfactory to use a relatively fine wire cable to supply electricity to a table lamp, but a very heavy cable is needed to connect a car battery?

6-13. A fairly small-diameter wire might carry the current required by a power saw. Why, then, might a much heavier wire be used if the power saw were in a garage several hundred feet from the house?

6-14. If a small-diameter wire were used to feed electricity to the saw in the garage (Question 6-13), what might we say about the resulting voltage regulation?

6-15. How many meter movements are needed to make a voltmeter that covers three voltage ranges?

6-16. In what circumstances may an unloaded voltage divider be used to advantage?

6-17. In what way does the relationship for total series capacitance differ from that for total series resistance?

CHAPTER PROBLEMS

6-1. What are the total voltage and internal resistance of two series-connected 1.5-V 100-Ω flashlight batteries? (*ans.* 3 V, 200 mΩ)

6-2. An ordinary 12-V car battery is actually made up of six cells, each with a voltage that ranges between about 1.8 V and 2.3 V, depending on the state of charge (low charge—low voltage). What is the real battery voltage?

6-3. Which voltmeter draws the greater current,
a. One with 20-kΩ/V sensitivity?
b. One with 50-kΩ/V sensitivity? (*ans.* 20 kΩ/V)

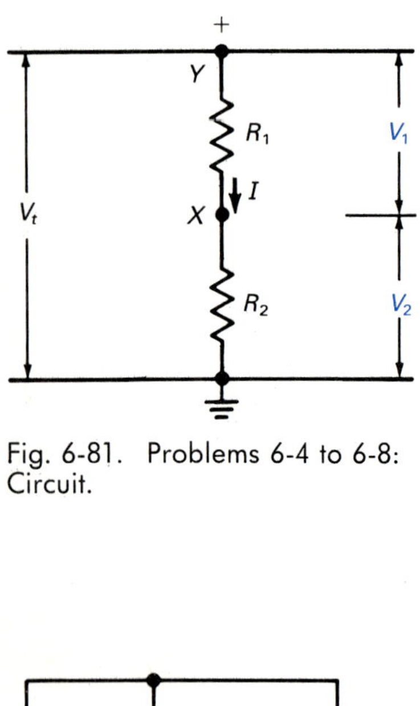

Fig. 6-81. Problems 6-4 to 6-8: Circuit.

6-4. What are the voltages across each resistor in Fig. 6-81 if $R_1 = 560$ kΩ, $R_2 = 1.8$ MΩ, and $I = 20\ \mu$A?

6-5. What is the voltage across each resistor in Fig. 6-81 if $R_1 = 560$ kΩ, $R_2 = 1.8$ MΩ, and $V_t = 50$ V? (*ans.* 11.9 V, 38.1 V)

6-6. In Fig. 6-81, what is the relative potential at point X with respect to point Y?

6-7. In Fig. 6-81, what is the potential of point X? (*ans.* +38.1 V)

6-8. In Fig. 6-81, what power is dissipated by the circuit if the total voltage $V_t = 50$ V, $R_1 = 560$ kΩ, and $V_2 = 10$ V?

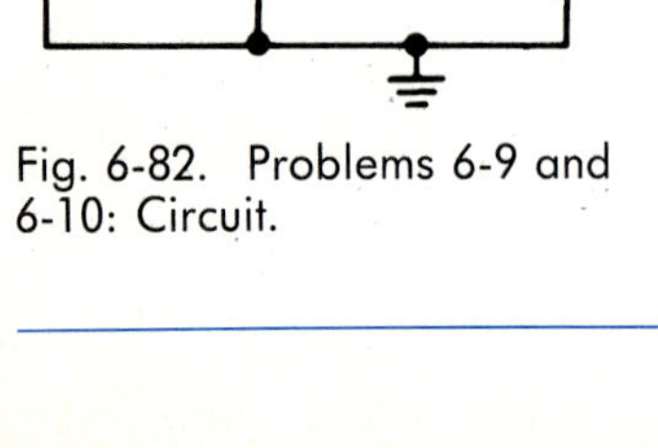

Fig. 6-82. Problems 6-9 and 6-10: Circuit.

6-9. What is the relationship between the four resistors in Fig. 6-82 if the relative potential between points X and Y is to be zero? (*ans.* $R_1 \times R_4 = R_2 \times R_3$)

6-10. One set of values that would satisfy Prob. 6-9 and give 0 V between points X and Y is $R_1 = 1$ kΩ, $R_2 = 10$ kΩ, $R_3 = 2.2$ kΩ, and $R_4 = 22$ kΩ. What would be the polarity of the relative potential between points X and Y if the values of R_1 and R_2 were reversed ($R_1 = 10$ kΩ and $R_2 = 1$ kΩ)?

6-11. A man has a power saw for use in his home. It takes 10 A when operating at its normal 120 V (equivalent resistance 12 Ω). When he uses the saw in his house, the total resistance of the connecting conductors is 0.1 Ω. When the saw is used in his garage at the end of his yard, the total resistance of the conductors is 1 Ω. What would be the saw voltage at the house and at the garage? (Take the saw resistance to be constant.)
(*ans.* 119 V, 102.9 V)

6-12. Two voltmeters, each with a full-scale range of 100 V, have different movements. One has a sensitivity of 20 kΩ/V, the other 50 kΩ/V. If the movement resistance is 2000 Ω in each case, what multiplier resistances are needed?

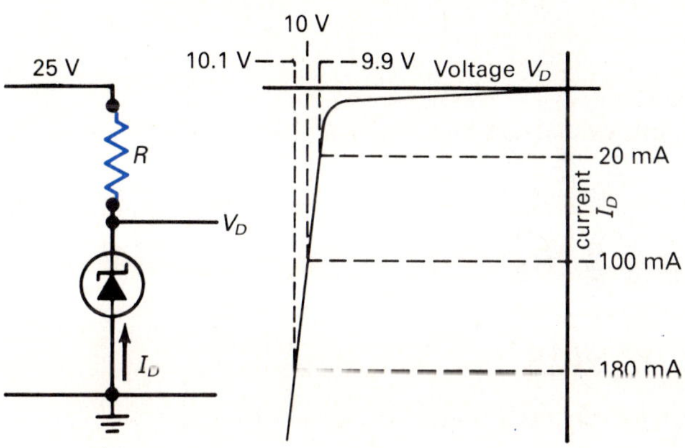

V_D diode voltage	I_D diode current
9.9 V	20 mA
10 V	100 mA
10.1 V	180 mA

(*a*) (*b*) (*c*)

Fig. 6-83. Problem 6-13: Circuit.

6-13. A certain 10-V zener diode has the characteristic shown in Fig. 6-83*b*. What value of series dropping resistor is needed so that the voltage across the diode is 10 V when operating from a 25-V supply and there is no load across the diode? (*ans.* 150 Ω)

7

PARALLEL CIRCUITS

Parallel and series circuits have opposite characteristics. For instance, what applies to voltage in a series circuit, applies to current in a parallel circuit. It follows that what applies to resistance in a series circuit, applies to conductance in a parallel circuit.

For many people, series circuits seem to be easier to understand than parallel circuits. This is somehow related to a preference for thinking in terms of resistance rather than conductance. Perhaps this preference develops because the resistors used to make up both series and parallel practical circuits are sold only in resistance values. To help you understand the characteristics of parallel circuits, we will deal with parallel-circuit principles from a conductance viewpoint. We will use resistance values for most of the practical application examples.

7-1 WHAT IS A PARALLEL CIRCUIT?

When all the elements of a circuit are connected between two common points, they are said to be connected in *parallel*.

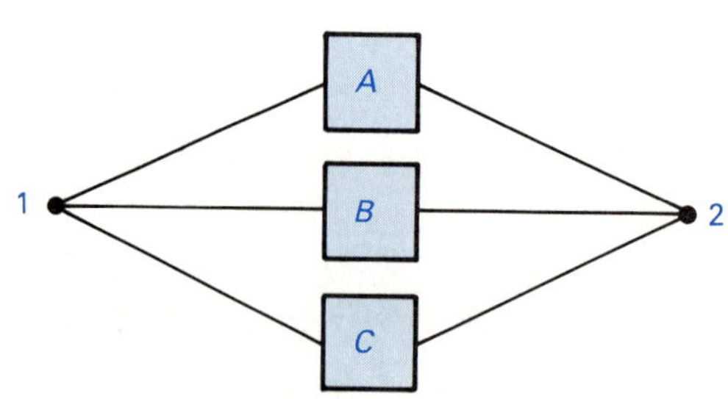

Fig. 7-1. Elements between common points.

In Fig. 7-1 the elements A, B, and C are connected in parallel between points 1 and 2. Figure 7-2 shows this parallel circuit as it is more likely to appear in electrical or electronic diagrams. The Fig. 7-2 presentation illustrates the meaning of the word "parallel"; all the elements are depicted as being geometrically parallel, as are the conductors between which they are connected.

An alternative term for "parallel" is "shunt."

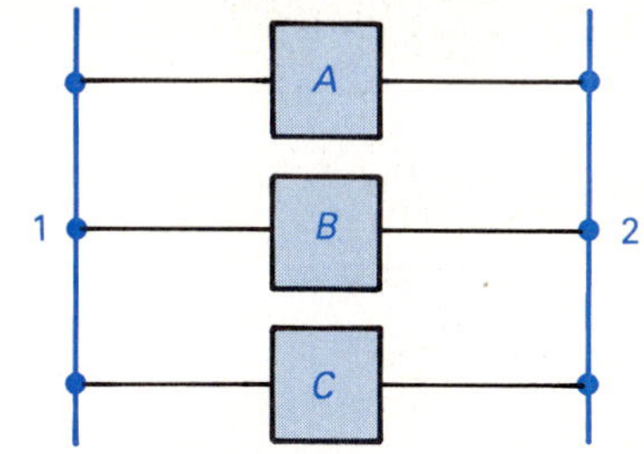

Fig. 7-2. Elements between common conductors.

7-2 CHARACTERISTICS OF PARALLEL CIRCUITS

The most important characteristics of parallel circuits are:

1. There is only one voltage across all parts of the circuit.
2. Elements have individual currents.
3. Currents are additive.
4. Conductances are additive.
5. Resistance reciprocals are additive.
6. Powers and energies are additive.

Each of these characteristics is now dealt with in detail.

One Voltage

In a parallel circuit, the voltages across each of the elements are the same.

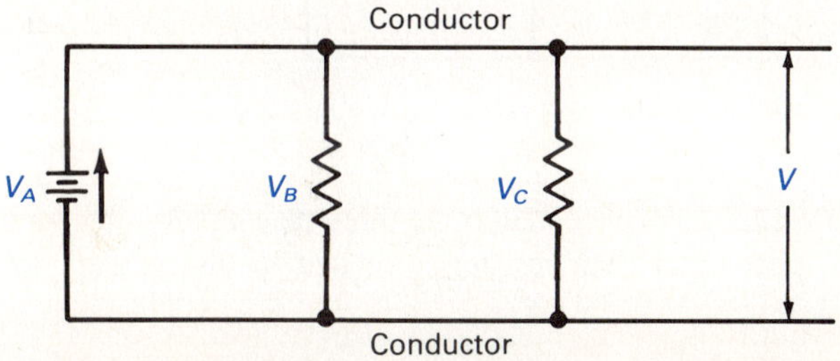

Fig. 7-3. Common voltage across a parallel circuit.

Figure 7-3 is a schematic-diagram equivalent of Figs. 7-1 and 7-2. Connected in parallel are a battery and two resistors. The voltage V between the conductors is the only possible circuit voltage. Therefore the battery emf V_A and the voltages across the resistances V_B and V_C are all equal to one another and equal to V. Any voltage measurement in this circuit must be made between the conductors. It does not matter where this measurement is made along the length of the conductors, since the result will always be the same.

Currents Individual

Individual currents exist in each of the elements of a parallel circuit. These are the currents I_A and I_B in Fig. 7-4.

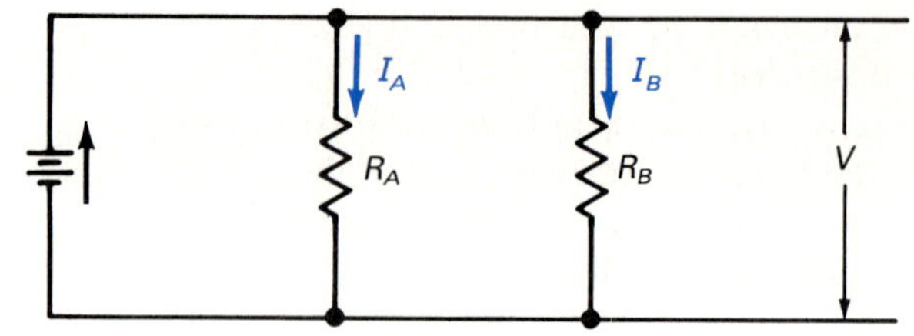

Fig. 7-4. Individual currents in a parallel circuit.

Applying Ohm's law to each of the resistance elements,

$$I_A = \frac{V}{R_A} \quad \text{and} \quad I_B = \frac{V}{R_B}$$

EXAMPLE 7-1 A battery with a 12-V emf and negligible internal resistance forms a parallel circuit with a network of three resistors, 2, 4, and 6 Ω. What is the current in each resistor?

Solution (Fig. 7-5)

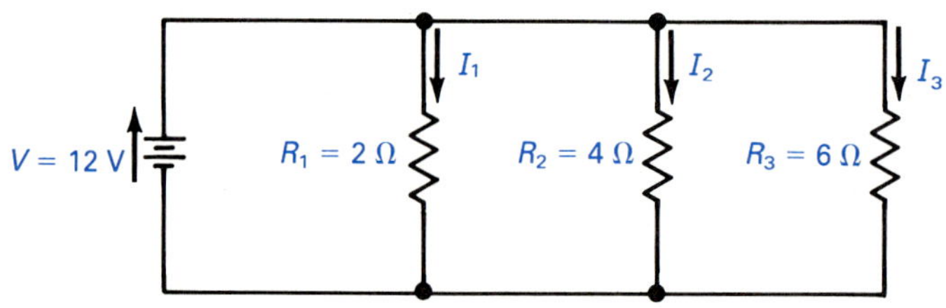

Fig. 7-5. Example 7-1: Circuit.

The 12-V emf V is connected to the three resistors R_1, R_2, and R_3. The current in each resistor is related to that voltage by Ohm's law:

$$I_1 = \frac{V}{R_1} = \frac{12\ \text{V}}{2\ \Omega} = \mathbf{6\ A}$$

$$I_2 = \frac{V}{R_2} = \frac{12\ \text{V}}{4\ \Omega} = \mathbf{3\ A}$$

$$I_3 = \frac{V}{R_3} = \frac{12\ \text{V}}{6\ \Omega} = \mathbf{2\ A}$$

Currents Additive (Kirchhoff's Current Law)

The current leaving the network at B in Fig. 7-6*a* must be the same as that entering it at A. There is no way in which the total current can change; it cannot collect additional current and so leave greater than it entered; no current can be lost, causing the amount leaving to be less than the amount entering.

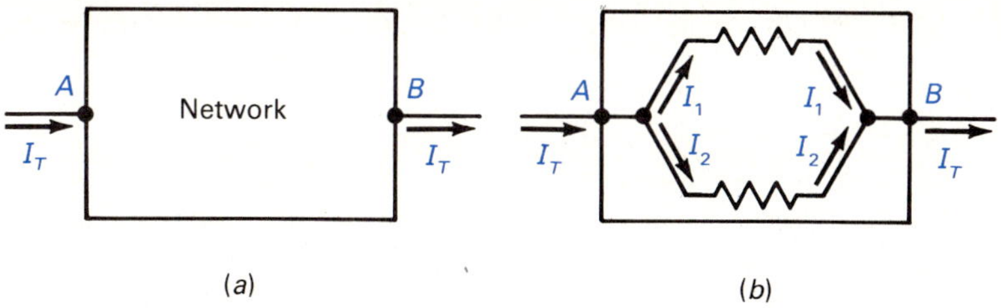

Fig. 7-6. *(a)* Current enters and leaves a network; *(b)* current divides at a branching point, or node, and then recombines.

When the network consists of elements in parallel, as in Fig. 7-6*b*, the current which enters splits up among the elements, in this case into two components I_1 and I_2. The two components recombine and leave as the original current. At both points *A* and *B*,

$$I_T = I_1 + I_2$$

This addition of current is a general characteristic of parallel resistive circuits, no matter how many elements are involved. In general form it can be written as

$$I_T = I_1 + I_2 + I_3 + \cdots + I_n \quad \textbf{(7-1)}$$

Kirchhoff's current law is usually stated:

The algebraic sum of the currents entering a node equals the algebraic sum of the currents leaving it.

Equation (7-1) is an expression of Kirchhoff's current law. It applies to points *A* and *B* in Fig. 7-6, since these points are conductor junctions or *nodes.*

EXAMPLE 7-2 What is the total current entering and leaving the three-element resistive network in Example 7-1?

Solution (Fig. 7-7)

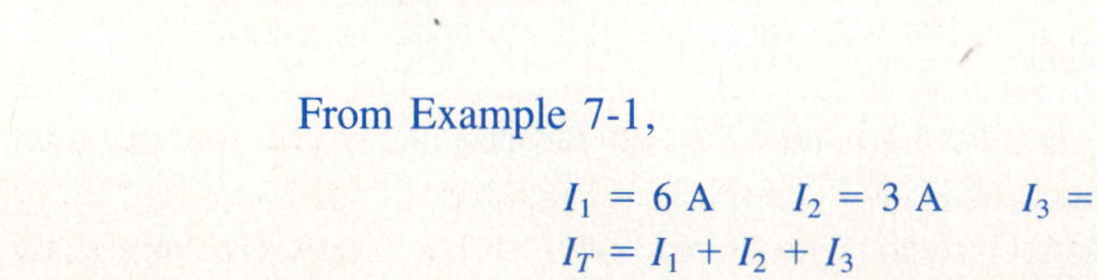

Fig. 7-7. Example 7-2: Circuit.

From Example 7-1,

$$I_1 = 6\text{ A} \qquad I_2 = 3\text{ A} \qquad I_3 = 2\text{ A}$$

$$I_T = I_1 + I_2 + I_3 \quad (7\text{-}1)$$

$$I_T = 6\text{ A} + 3\text{ A} + 2\text{ A} = \mathbf{11\ A}$$

When, as in Examples 7-1 and 7-2, the total circuit consists of a generator and a parallel network, the total current entering and leaving the network will be the total current drawn from the generator.

EXAMPLE 7-3 A generator is applied to a parallel network consisting of four resistors, 10, 20, 25, and 50 Ω. The current in the 25-Ω resistor is 4 A. What are the other circuit currents?

Solution (Fig. 7-8)

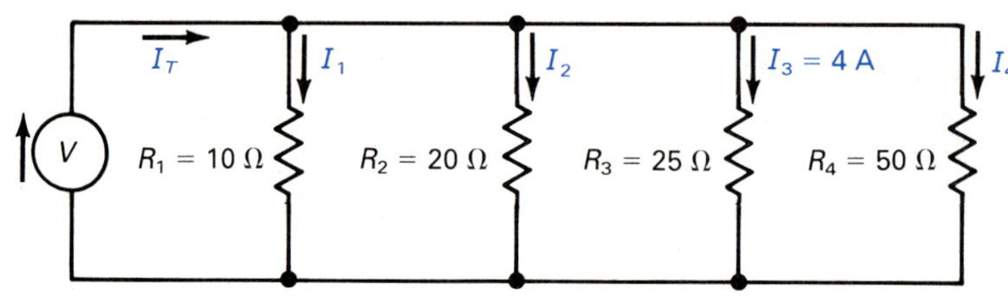

Fig. 7-8. Example 7-3: Circuit.

Since this is a parallel network, there is only one circuit voltage, and so the voltage across R_3 must be the same as the generator emf V.

$$V = I_3R_3 = 4\ \text{A} \times 25\ \Omega = 100\ \text{V}$$

This emf of 100 V will also be the voltage across the other three resistors R_1, R_2, and R_4.

$$I_1 = \frac{V}{R_1} = \frac{100\ \text{V}}{10\ \Omega} = \mathbf{10\ A}$$

$$I_2 = \frac{V}{R_2} = \frac{100\ \text{V}}{20\ \Omega} = \mathbf{5\ A}$$

$$I_4 = \frac{V}{R_4} = \frac{100\ \text{V}}{50\ \Omega} = \mathbf{2\ A}$$

$$I_T = I_1 + I_2 + I_3 + I_4 \tag{7-1}$$

$$I_T = 10\ \text{A} + 5\ \text{A} + 4\ \text{A} + 2\ \text{A} = \mathbf{21\ A}$$

Conductances Additive

Figure 7-9*a* shows a network with three parallel elements of conductance G_1, G_2, and G_3. Figure 7-9*b* is its single-element equivalent. The voltages V and the currents I_T are the same in each network, so the conductance G_T is the total equivalent conductance of G_1, G_2, and G_3.

Conductance is the reciprocal of resistance, or $G = 1/R$. Therefore Ohm's law can be written in the form $I = VG$ instead of $I = V/R$. Applying this to Fig. 7-9,

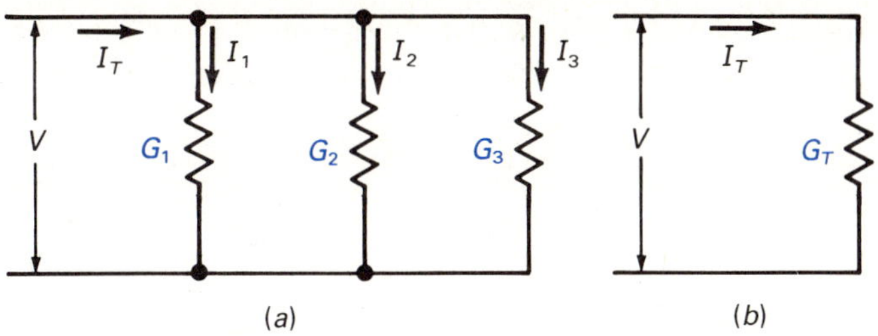

Fig. 7-9. *(a)* Three-element parallel network; *(b)* single-element equivalent.

$$I_T = VG_T \qquad I_1 = VG_1 \qquad I_2 = VG_2 \qquad I_3 = VG_3$$

$$I_T = I_1 + I_2 + I_3 \tag{7-1}$$

$$VG_T = VG_1 + VG_2 + VG_3 = V(G_1 + G_2 + G_3)$$

$$G_T = G_1 + G_2 + G_3$$

This formula applies to any number of parallel elements. Thus,

$$G_T = G_1 + G_2 + G_3 + \cdots + G_n \tag{\textbf{7-2}}$$

The total equivalent conductance of a number of elements connected in parallel is the sum of their individual conductances.

Resistance Reciprocals Additive

Because it is usual to express element values in resistance rather than conductance, the resistance equivalent of Eq. (7-2) can be useful.

$$G_T = G_1 + G_2 + G_3 + \cdots + G_n \tag{7-2}$$

$$\frac{1}{R_T} = \frac{1}{R_1} + \frac{1}{R_2} + \frac{1}{R_3} + \cdots + \frac{1}{R_n} \tag{\textbf{7-3}}$$

Equation (7-3) may look rather complicated, but it is only Eq. (7-2) put in resistance form. Since resistors come in resistance values, it is likely that you will carry out your parallel-circuit calculations using Eq. (7-3).

Working with reciprocals ($1/R$ rather than G), as in Eq. (7-3), used to be an arithmetic difficulty, but calculators now make it easy.

You will see that Eq. (7-3) does not directly give the resistance of a parallel circuit; to obtain this, Eq. (7-3) must be inverted.

$$R_T = \frac{1}{1/R_1 + 1/R_2 + 1/R_3 + \cdots + 1/R_n}$$

This equation is given only to illustrate how awkward matters can become if the resistance viewpoint is used throughout in dealing with parallel circuits. The statement which relates to this equation makes this clear: The total equivalent resistance of a number of elements connected in parallel is the reciprocal of the sum of the reciprocals of the resistances of the individual elements.

It is important to realize that, as the total conductance of a parallel circuit is the sum of the conductances of its individual elements, this total is greater than any individual conductance. Therefore the *total* circuit *resistance is smaller than the resistance of any individual element.*

EXAMPLE 7-4 A parallel network consists of three elements with conductances of 0.5, 0.25, and 0.2 S. What is the total equivalent *(a)* conductance and *(b)* resistance?

Solution *(a)* Fig. 7-10

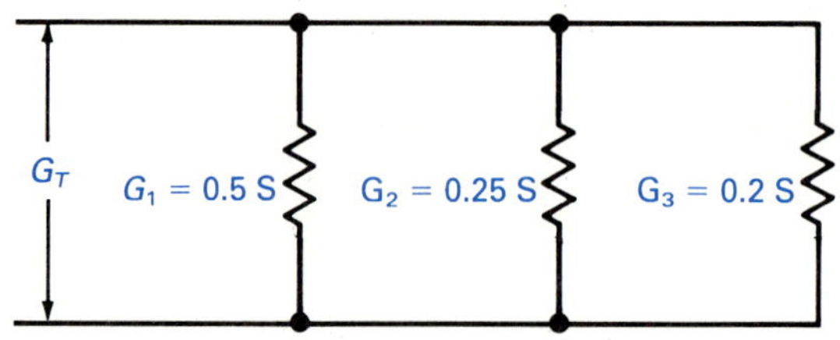

Fig. 7-10. Example 7-4: Data in conductance form.

$$G_T = G_1 + G_2 + G_3 \tag{7-2}$$
$$G_T = 0.5\text{ S} + 0.25\text{ S} + 0.2\text{ S} = \mathbf{0.95\ S}$$

(b) Fig. 7-11

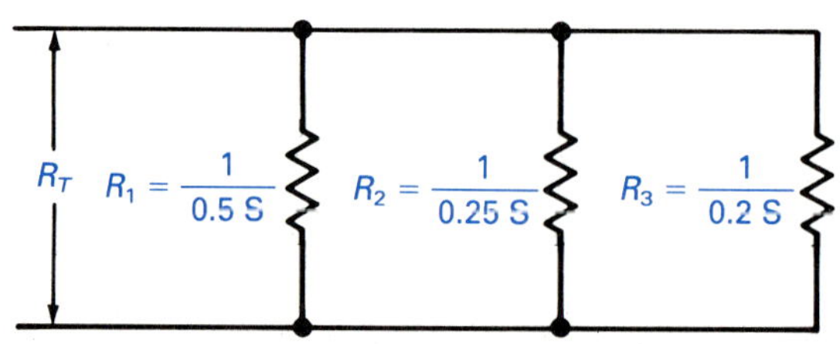

Fig. 7-11. Example 7-4: Data in resistance form.

$$R_1 = \frac{1}{G_1} = \frac{1}{0.5\text{ S}} = 2\ \Omega$$
$$R_2 = \frac{1}{G_2} = \frac{1}{0.25\text{ S}} = 4\ \Omega$$
$$R_3 = \frac{1}{G_3} = \frac{1}{0.2\text{ S}} = 5\ \Omega$$
$$\frac{1}{R_T} = \frac{1}{R_1} + \frac{1}{R_2} + \frac{1}{R_3} \tag{7-3}$$
$$\frac{1}{R_T} = \frac{1}{2\ \Omega} + \frac{1}{4\ \Omega} + \frac{1}{5\ \Omega} = 0.5 + 0.25 + 0.2 = 0.95$$
$$R_T = \frac{1}{0.95}\ \Omega = \mathbf{1.053\ \Omega}$$

By comparing the solutions for Example 7-4*a* and *b*, you can see how working with conductances can be simpler than working with resistances in solving parallel-circuit problems.

EXAMPLE 7-5 A parallel network has elements with resistances of 30, 40, 60, and 120 Ω. What is the network's total equivalent *(a)* resistance and *(b)* conductance?

Solution (Fig. 7-12)

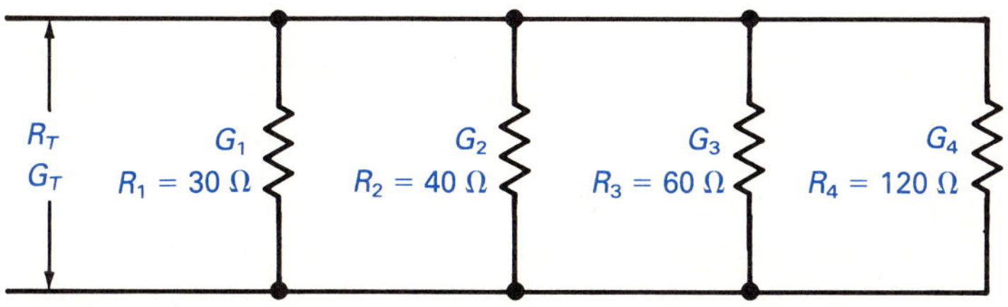

Fig. 7-12. Example 7-5: Data.

a.

$$\frac{1}{R_T} = \frac{1}{R_1} + \frac{1}{R_2} + \frac{1}{R_3} + \frac{1}{R_4} \qquad (7\text{-}3)$$

$$= \frac{1}{30\ \Omega} + \frac{1}{40\ \Omega} + \frac{1}{60\ \Omega} + \frac{1}{120\ \Omega}$$

$$= \frac{4 + 3 + 2 + 1}{120\ \Omega} = \frac{10}{120\ \Omega} = \frac{1}{12\ \Omega}$$

$$R_T = \mathbf{12\ \Omega}$$

b.

$$G_T = G_1 + G_2 + G_3 + G_4 \qquad (7\text{-}2)$$

$$= \frac{1}{30}\ \text{S} + \frac{1}{40}\ \text{S} + \frac{1}{60}\ \text{S} + \frac{1}{120}\ \text{S}$$

$$G_T = \frac{4 + 3 + 2 + 1}{120}\ \text{S} = \mathbf{\frac{1}{12}\ S}$$

The solution to Example 7-5 is given in arithmetic form, adding fractions by the ''least common denominator'' method. This way you can see the basic difference between series- and parallel-circuit calculations. With an electronic calculator, parallel circuits are made easy to handle. In Example 7-5*a*, the calculator is used to add the reciprocals of 30, 40, 60, and 120 by using the 1/x button, giving a total conductance of 0.08333 S. Then pressing the 1/x button again gives the total resistance, 12 Ω.

EXAMPLE 7-6 Four 100-kΩ resistors are connected in parallel. What is the total equivalent *(a)* conductance and *(b)* resistance?

Solution (Fig. 7-13)

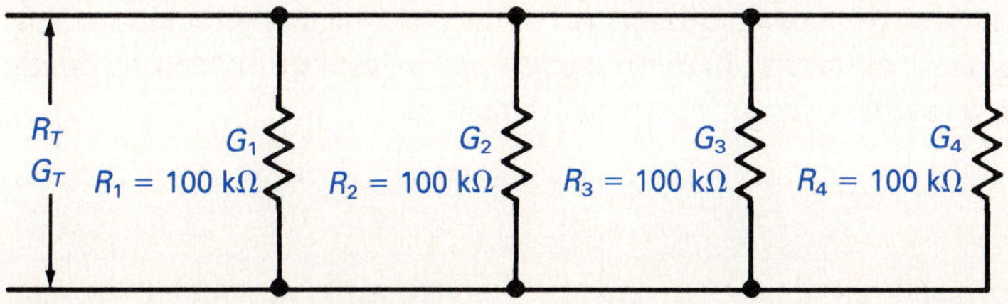

Fig. 7-13. Example 7-6: Data.

a. Each resistor has a conductance of

$$G = \frac{1}{R} = \frac{1}{100\ \text{k}\Omega} = 10\ \mu\text{S}$$

$$G_T = G_1 + G_2 + G_3 + G_4 \qquad (7\text{-}2)$$

$$= 10\ \mu\text{S} + 10\ \mu\text{S} + 10\ \mu\text{S} + 10\ \mu\text{S}$$

$$G_T = 4 \times 10\ \mu\text{S} = \mathbf{40\ \mu S}$$

b.

$$R_T = \frac{1}{G_T} = \frac{1}{40\ \mu\text{S}} = 2.5 \times 10^4\ \Omega \text{ or } \mathbf{25\ k\Omega}$$

or

$$R_T = \frac{R}{4} = \frac{100\ \text{k}\Omega}{4} = \mathbf{25\ k\Omega}$$

You will see from Example 7-6 that the *total conductance* of a number of equal elements in parallel is the conductance of one element multiplied by the number of elements. Also, the *total resistance* is the resistance of one element divided by the number of elements.

$$G_T = nG, \text{ S} \qquad \mathbf{(7\text{-}4)}$$

$$R_T = \frac{R}{n} \qquad \mathbf{(7\text{-}5)}$$

where G_T = total equivalent conductance, S
n = number of elements connected in parallel
G = conductance of each equal individual element, S
R_T = total equivalent resistance, Ω
R = resistance of each equal individual element, Ω

Powers and Energies Additive

The power dissipated in a resistor is

$$P = \frac{V^2}{R} \qquad (3\text{-}7)$$

Equation (3-7) can be written in conductance form as

$$P = V^2G$$

From this equation it can be seen that when elements have the same voltage, the power dissipated by each element will be directly proportional to its conductance value. Therefore, since the total conductance of a parallel network is equal to the sum of the individual conductances, the total power dissipated will be equal to the sum of the powers dissipated by each element. Furthermore, this summation applies to energy, because energy and power are related by time, which will be a constant for any particular situation.

$$P_T = P_1 + P_2 + P_3 + \cdots + P_n \qquad \mathbf{(7\text{-}6)}$$

$$W_T = W_1 + W_2 + W_3 + \cdots + W_n \qquad \mathbf{(7\text{-}7)}$$

It should be noted that Eqs. (7-6) and (7-7), which apply to parallel circuits, are the same as Eqs. (6-4) and (6-5), which give the power and energy relationships for series circuits. The concept can be carried further. In any resistive network, no matter what its configuration, the total power or energy is the sum of the individual powers or energies.

EXAMPLE 7-7 Resistors of 12, 3, and 24 kΩ are connected in parallel across a 400-V power supply. What power is supplied to the parallel network?

Solution (Fig. 7-14)

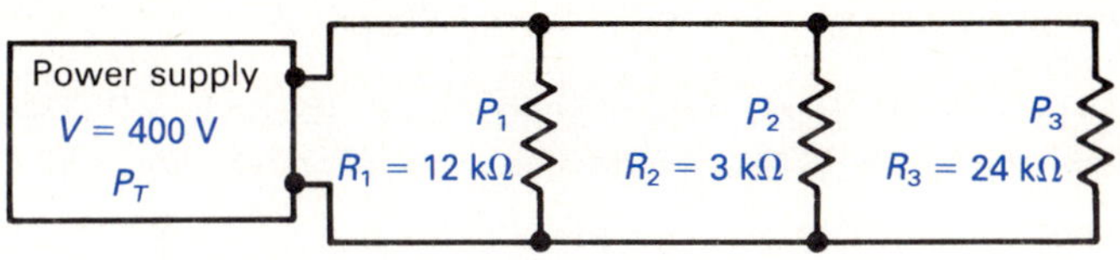

Fig. 7-14. Example 7-7: Circuit.

Method 1 Sum the powers directly:

$$\begin{aligned} P_T &= P_1 + P_2 + P_3 \qquad (7\text{-}6)\\ &= V^2G_1 + V^2G_2 + V^2G_3 \\ &= V^2(G_1 + G_2 + G_3) \\ &= (400\ \text{V})^2\left(\frac{1}{12\ \text{k}\Omega} + \frac{1}{3\ \text{k}\Omega} + \frac{1}{24\ \text{k}\Omega}\right) \\ &= 1.6 \times 10^5\left(\frac{2+8+1}{2.4 \times 10^4}\right) = 1.6 \times 10^5 \times \frac{11}{2.4 \times 10^4} \end{aligned}$$

$$P_T = \mathbf{73.33\ W}$$

Method 2 Calculate the individual powers and then sum them:

$$P_1 = \frac{V^2}{R_1} = \frac{1.6 \times 10^5\ \text{V}^2}{1.2 \times 10^4\ \Omega} = 13.33\ \text{W}$$

$$P_2 = \frac{V^2}{R_2} = \frac{1.6 \times 10^5\ \text{V}^2}{3 \times 10^3\ \Omega} = 53.33\ \text{W}$$

$$P_3 = \frac{V^2}{R_3} = \frac{1.6 \times 10^5\ \text{V}^2}{2.4 \times 10^4\ \Omega} = 6.667\ \text{W}$$

$$\begin{aligned} P_T &= P_1 + P_2 + P_3 \qquad (7\text{-}6)\\ &= 13.33\ \text{W} + 53.33\ \text{W} + 6.667\ \text{W} \end{aligned}$$

$$P_T = \mathbf{73.33\ W}$$

EXAMPLE 7-8 Two 120-V 60-W light bulbs are connected to a 120-V supply. What total energy is converted per hour? If the bulbs are to operate normally on the 120-V supply, they will have to be connected in parallel across it.

Solution (Fig. 7-15)

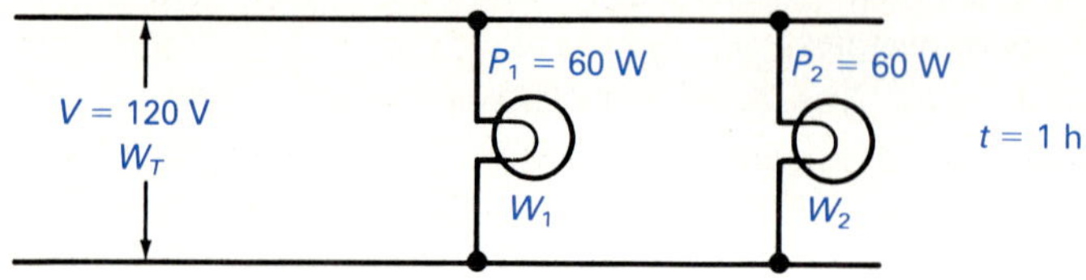

Fig. 7-15. Example 7-8: Data.

For each bulb,

$$W_1 = W_2 = P_1t = P_2t = 60 \text{ W} \times 1 \text{ h} = 60 \text{ Wh} \quad \text{or} \quad 216 \text{ kJ}$$

$$W_T = W_1 + W_2 \qquad (7\text{-}7)$$

$$W_T = 2 \times 60 \text{ Wh} = \mathbf{120\ Wh} \quad \text{or} \quad \mathbf{432\ kJ}$$

This example is the parallel equivalent of Example 6-7. The same total energy is converted because, although the supply voltage is different in each case, the bulbs are the same, and each is operating normally, dissipating the rated 60 W. The total power is 120 W in each case. (In this example it is 120 V × 1 A; in Example 6-7 it was 240 V × 0.5 A.)

EXAMPLE 7-9 Two electric baseboard heaters are used to heat a living room. Each operates at 240 V. One is rated at 1.25 kW, and the other at 0.5 kW. What is the cost of operating the heaters for 1 month (30 days) if both are on for an average of 10 h/day and the electricity rate is 5 cents/kWh?

Solution (Fig. 7-16)

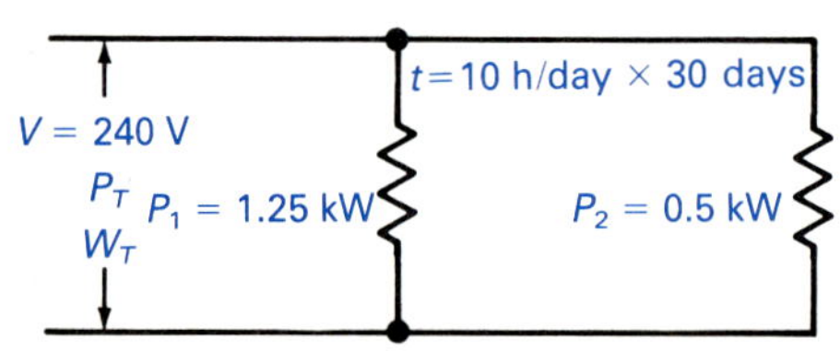

Fig. 7-16. Example 7-9: Data.

$$P_T = P_1 + P_2 \qquad (7\text{-}6)$$

$$P_T = 1.25 \text{ kW} + 0.5 \text{ kW} = 1.75 \text{ kW}$$

For 30 days at 10 h/day,

$$t = 10 \frac{\text{h}}{\text{day}} \times 30 \text{ days} = 300 \text{ h}$$

$$W_T = P_Tt = 1.75 \text{ kW} \times 300 \text{ h} = 525 \text{ kWh}$$

$$\text{Cost of electricity} = \text{energy} \times \text{rate} = W_T \times 5 \text{ cents/kWh}$$

$$= 525 \text{ kWh} \times \frac{5 \text{ cents}}{\text{kWh}}$$

$$\text{Cost} = 2625 \text{ cents} \quad \text{or} \quad \mathbf{\$26.25}$$

Electrical utilities sell energy, even though they are often called power companies. The cost of the electricity is, as implied in the solution to Example 7-9, obtained by taking the basic rate for an energy unit and multiplying it by the sum of the energies taken by the individual elements (heaters, lamps, etc.).

It is usual to quote electricity for domestic loads in kilowatthours, rather than in joules, the SI unit of energy. The reason for using kilowatthours is made clear by comparing the energy in Example 7-9 given in kilowatthours with the same amount of energy expressed in joules. 525 kWh is easier to write or say than 1.89×10^9 J.

The figures given in Example 7-9 are typical of what may be involved in heating a living room of about 250-ft^2 area in winter.

REVIEW QUESTIONS FOR SECTION 7-2

1. How would you describe the characteristics of parallel circuits?
2. What configuration must a network have to be considered a parallel circuit?
3. List six characteristics of a parallel circuit.
4. How many current paths are there in a parallel circuit that consists of three resistors and a generator?
5. In a parallel circuit, what is the relation between the:
   a. Total conductance and the individual conductances?
   b. Total conductance and the individual resistances?
   c. Total resistance and the individual conductances?
   d. Total resistance and the individual resistances?
6. Three equal resistors are connected in parallel. What is the total resistance?
7. Give an equation, in terms of voltage and conductance, for the total power dissipated in a parallel circuit.

7-3 CALCULATIONS FOR PARALLEL CIRCUITS

Calculations for parallel circuits involve the relationships that apply to single elements, which were dealt with in Chap. 2, and their association with the characteristics of the parallel configuration.

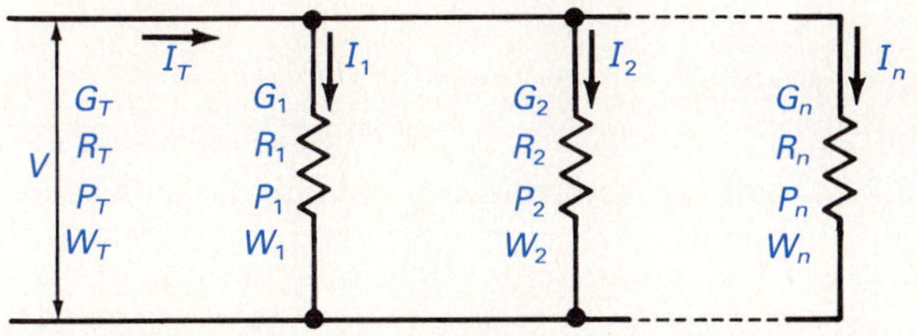

Fig. 7-17. Generalized parallel circuit.

A generalized parallel circuit is shown in Fig. 7-17, together with the quantities that can be calculated. There are many different ways in which calculations can be performed in such a circuit, depending on what data are known and what information is required. The techniques described, and demonstrated by examples, are typical of those frequently used in practice; by observing these methods, you can learn to devise others to fit any parallel-circuit needs.

The examples intentionally demonstrate several different ways of achieving the same result, each using different given data. There is no need for you to memorize each of these methods, but you should check them all thoroughly so that you can see how they are derived from basic circuit principles. In this way you will get practice in the use of fast, direct methods of calculating.

Voltage V

The only voltage in the circuit is V, which appears across each element. It can be calculated by applying Ohm's law to the complete circuit, or to any part of it:

$$V = I_T R_T \quad \text{or} \quad I_1 R_1 \quad \text{or} \quad I_2 R_2 \quad \text{or} \quad \cdots \quad \text{or} \quad I_n R_n$$

EXAMPLE 7-10 What is the common voltage across the circuit in Fig. 7-18?

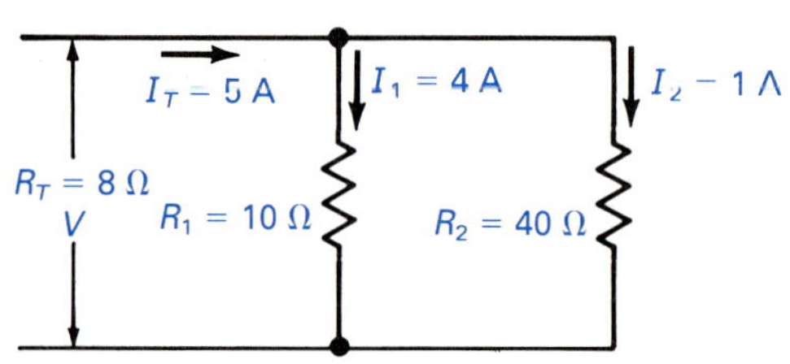

Fig. 7-18. Example 7-10: Data.

Solution Method 1 Applying Ohm's law to the entire circuit:

$$V = I_T R_T = 5\ \text{A} \times 8\ \Omega = \mathbf{40\ V}$$

Method 2 Applying Ohm's law to one element (any element can be selected):

$$V = I_1 R_1 = 4\ \text{A} \times 10\ \Omega = \mathbf{40\ V}$$

or $$V = I_2 R_2 = 1\ \text{A} \times 40\ \Omega = \mathbf{40\ V}$$

Total Current I_T

The total current in a parallel circuit can be obtained by adding the individual element currents:

$$I_T = I_1 + I_2 + \cdots + I_n \tag{7-1}$$

or by applying Ohm's law to the complete circuit:

$$I_T = VG_T$$

The latter equation can be converted into conductance ratio or resistance ratio form by using the fact that the common voltage is the ratio of an element's current and conductance (for example, $V = I_1/G_1$ or $V = I_2/G_2$):

$$I_T = VG_T = \frac{I_1}{G_1} G_T = I_1 \frac{G_T}{G_1}$$

This equation can be written as one of a set of similar equations:

$$I_T = I_1 \frac{G_T}{G_1} \quad \text{or} \quad I_2 \frac{G_T}{G_2} \quad \text{or} \quad \cdots \quad \text{or} \quad I_n \frac{G_T}{G_n} \tag{7-8}$$

Because resistance and conductance are reciprocal quantities ($R = 1/G$), the Ohm's law equation for the total current can be written in resistance form as

$$I_T = \frac{V}{R_T}$$

Another set of equations for the total current can be obtained by using the $V = IR$ form of Ohm's law:

$$I_T = I_1 \frac{R_1}{R_T} \quad \text{or} \quad I_2 \frac{R_2}{R_T} \quad \text{or} \quad \cdots \quad \text{or} \quad I_n \frac{R_n}{R_T} \tag{7-9}$$

EXAMPLE 7-11 Determine the total current in the circuit in Fig. 7-19.

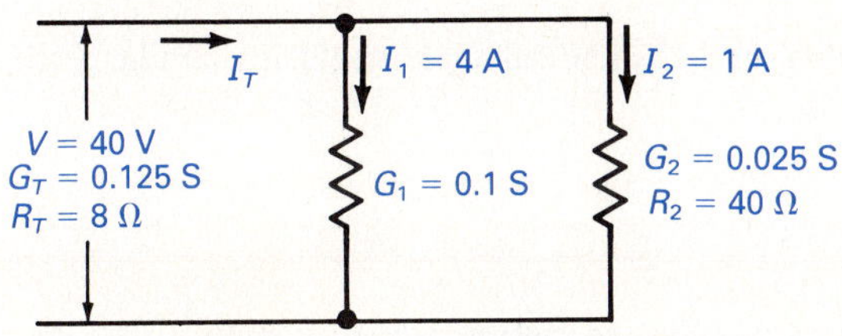

Fig. 7-19. Example 7-11: Data.

Solution Method 1 Kirchhoff's current law:

$$I_T = I_1 + I_2 \tag{7-1}$$

$$I_T = 4\text{ A} + 1\text{ A} = \mathbf{5\text{ A}}$$

Method 2 Ohm's law (conductance form):

$$I_T = VG_T = 40\text{ V} \times 0.125\text{ S} = \mathbf{5\text{ A}}$$

Method 3 Conductance ratio:

$$I_T = I_2 \frac{G_T}{G_2} \tag{7-8}$$

$$I_T = 1\text{ A} \frac{0.125\text{ S}}{0.025\text{ S}} = \mathbf{5\text{ A}}$$

Method 4 Ohm's law (resistance form):

$$I_T = \frac{V}{R_T} = \frac{40\ \text{V}}{8\ \Omega} = \mathbf{5\ A}$$

Method 5 Resistance ratio:

$$I_T = I_2 \frac{R_2}{R_T} \tag{7-9}$$

$$I_T = 1\ \text{A}\ \frac{40\ \Omega}{8\ \Omega} = \mathbf{5\ A}$$

Individual Currents $I_1, I_2, \ldots, I_n$

The currents in each of the elements can be calculated by the application of Ohm's law:

$$I_1 = VG_1 \quad \text{or} \quad I_1 = \frac{V}{R_1}$$

Substituting I/G for V will give a conductance ratio form:

$$I_1 = I_T \frac{G_1}{G_T} \quad \text{or} \quad I_2 \frac{G_1}{G_2} \quad \text{or} \quad \cdots \quad \text{or} \quad I_n \frac{G_1}{G_n} \tag{7-10}$$

Alternatively, resistance ratio forms can be obtained:

$$I_1 = I_T \frac{R_T}{R_1} \quad \text{or} \quad I_2 \frac{R_2}{R_1} \quad \text{or} \quad \cdots \quad \text{or} \quad I_n \frac{R_n}{R_1} \tag{7-11}$$

The other individual currents I_2 to I_n can be calculated by using similar techniques.

EXAMPLE 7-12 What is the current I_1 in the circuit in Fig. 7-20?

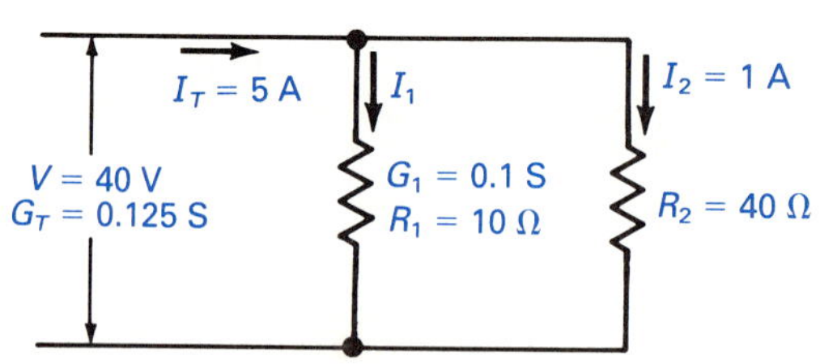

Fig. 7-20. Example 7-12: Data.

Solution Method 1 Ohm's law (conductance form):

$$I_1 = VG_1 = 40\ \text{V} \times 0.1\ \text{S} = \mathbf{4\ A}$$

Method 2 Conductance ratio:

$$I_1 = I_T \frac{G_1}{G_T} \tag{7-10}$$

$$I_1 = 5 \text{ A} \frac{0.1 \text{ S}}{0.125 \text{ S}} = \mathbf{4\ A}$$

Method 3 Ohm's law (resistance form):

$$I_1 = \frac{V}{R_1} = \frac{40 \text{ V}}{10\ \Omega} = \mathbf{4\ A}$$

Method 4 Resistance ratio:

$$I_1 = I_2 \frac{R_2}{R_1} \tag{7-11}$$

$$I_1 = 1 \text{ A} \frac{40\ \Omega}{10\ \Omega} = \mathbf{4\ A}$$

Total Conductance G_T

The equivalent conductance of the complete network can be obtained by summing the conductances of the individual elements

$$G_T = G_1 + G_2 + \cdots + G_n \tag{7-2}$$

or by applying Ohm's law to the entire network. From $I_T = VG_T$, we get

$$G_T = \frac{I_T}{V}$$

G_T can be expressed in current ratio form by substituting appropriate ratios of G and I. For example, as $V = I_1/G_1$, $1/V = G_1/I_1$ so that

$$G_T = \frac{I_T}{V} = I_T \frac{1}{V} = I_T \frac{G_1}{I_1} = G_1 \frac{I_T}{I_1}$$

This technique results in a set of equations for the total conductance G_T:

$$G_T = G_1 \frac{I_T}{I_1} \quad \text{or} \quad G_2 \frac{I_T}{I_2} \quad \text{or} \quad \cdots \quad \text{or} \quad G_n \frac{I_T}{I_n} \tag{7-12}$$

EXAMPLE 7-13 Find the total conductance of the circuit shown in Fig. 7-21.

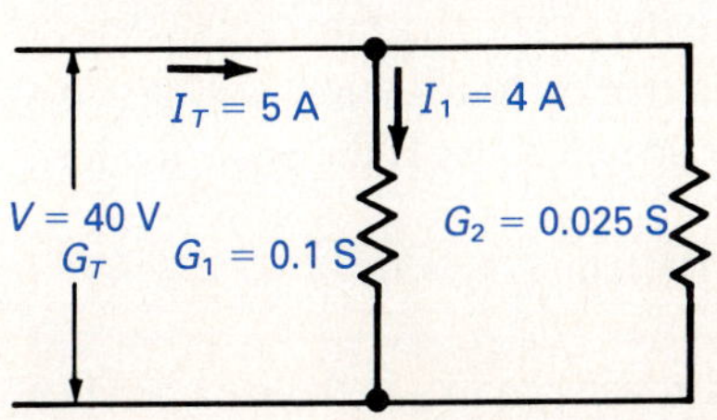

Fig. 7-21. Example 7-13: Data.

Solution Method 1 Sum of the conductances:

$$G_T = G_1 + G_2 \quad (7\text{-}2)$$
$$G_T = 0.1\ \text{S} + 0.025\ \text{S} = \mathbf{0.125\ S}$$

Method 2 Ohm's law:

$$G_T = \frac{I_T}{V} = \frac{5\ \text{A}}{40\ \text{V}} = \mathbf{0.125\ S}$$

Method 3 Current ratio:

$$G_T = G_1 \frac{I_T}{I_1} \quad (7\text{-}12)$$
$$G_T = 0.1\ \text{S}\ \frac{5\ \text{A}}{4\ \text{A}} = \mathbf{0.125\ S}$$

Total Resistance R_T

The total equivalent resistance of a parallel circuit may be calculated by first computing the total conductance and then taking its reciprocal.

Thus the total resistance of the two-element, parallel network in Examples 7-10 to 7-13 will be

$$R_T = \frac{1}{G_T} = \frac{1}{0.125\ \text{S}} = 8\ \Omega$$

Practical circuits are made up of resistors, which are sold in resistance values (ohms) more often than in conductances (siemens). Calculating parallel-circuit conditions using resistance values without first finding conductances speeds up the work. The direct method of doing this involves the use of reciprocals ($1/R$) in Eq. (7-3). As mentioned in Sec. 7-2 when referring to this equation, a calculator gives the reciprocal of a number by simply pressing the 1/x button. The steps in using a calculator with Eq. 7-3 are:

1. Key in the first resistance value R_1 by pressing the numeral buttons, and then press the 1/x button to get its reciprocal ($1/R_1$).
2. *a.* If your calculator is of *RPN* type (Hewlett-Packard and a few other makes), obtain the reciprocal of the second resistance value ($1/R_2$) as above, and then press the + button.
   b. If your calculator operates on the *algebraic* system (most makes), press the + button *and then* obtain the reciprocal of the second resistance value ($1/R_2$).
   When you have done *(a)* or *(b)*, you will have the sum of the reciprocals of the first two resistance values ($1/R_1 + 1/R_2$). You should note that the difference between the RPN and algebraic operations is the point at which the + button is pressed.
3. Repeat step 2 for each additional resistance, giving $1/R_t = 1/R_1 + 1/R_2 + 1/R_3 + \cdots + 1/R_n$.
4. Press the 1/x button to obtain the total resistance (R_T is the reciprocal of $1/R_T$).

Here is an example of the use of reciprocals to find the total resistance of a parallel circuit made up of preferred-value resistors.

EXAMPLE 7-14 Calculate the total equivalent resistance of the parallel network shown in Fig. 7-22.

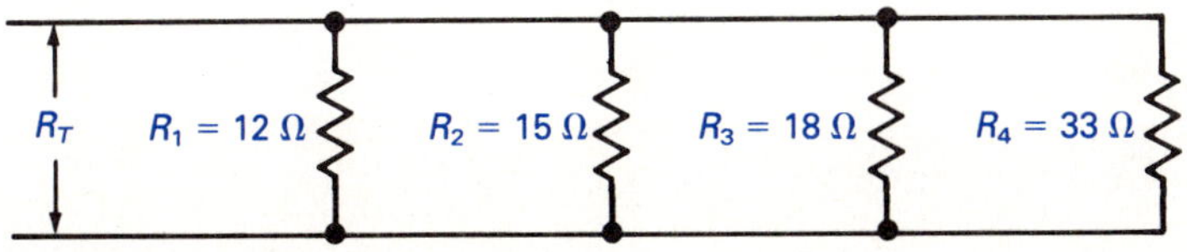

Fig. 7-22. Example 7-14: Data.

Solution Using the calculator to find reciprocals,

$$\text{Reciprocal of } R_1 = \frac{1}{R_1} = 0.0833$$

$$\text{Reciprocal of } R_2 = \frac{1}{R_2} = 0.0667$$

$$\text{Reciprocal of } R_3 = \frac{1}{R_3} = 0.0556$$

$$\text{Reciprocal of } R_4 = \frac{1}{R_4} = 0.0303$$

$$\frac{1}{R_T} = \frac{1}{R_1} + \frac{1}{R_2} + \frac{1}{R_3} + \frac{1}{R_4} \quad (7\text{-}3)$$

$$\frac{1}{R_T} = 0.0833 + 0.0667 + 0.0556 + 0.0303 = 0.2359$$

$$R_T = \text{reciprocal of } 0.2359 = \mathbf{4.24\ \Omega}$$

In electrical or electronic circuits it is more usual to find just two elements in parallel rather than a larger number. When this is so, a simplification of Eq. (7-3) can be applied to further speed up calculation. (See Fig. 7-23.)

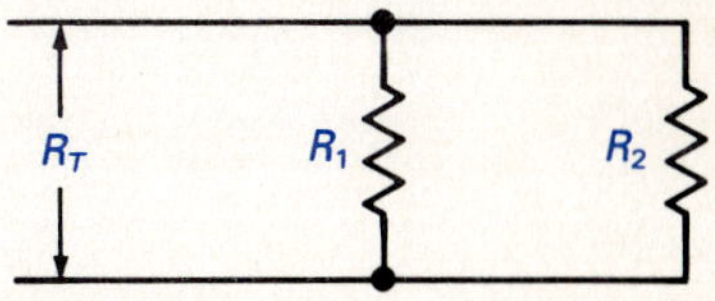

Fig. 7-23. Two parallel elements.

$$\frac{1}{R_T} = \frac{1}{R_1} + \frac{1}{R_2} \quad (7\text{-}3)$$

$$\frac{1}{R_T} = \frac{R_2 + R_1}{R_1 \times R_2}$$

$$R_T = \frac{R_1 \times R_2}{R_1 + R_2} \quad \mathbf{(7\text{-}13)}$$

From Eq. (7-13), it can be seen that the *total equivalent resistance of two resistances connected in parallel* is their *product divided by their sum*. To use Eq. (7-13) with a calculator:

1. Add the values of the two resistances ($R_1 + R_2$) and then take the reciprocal [$1/(R_1 + R_2)$].
2. Multiply by R_1 and then by R_2 to get the total resistance R_T.

EXAMPLE 7-15 Two resistors, 22 and 47 Ω, are connected in parallel. What is the total equivalent resistance of the network?

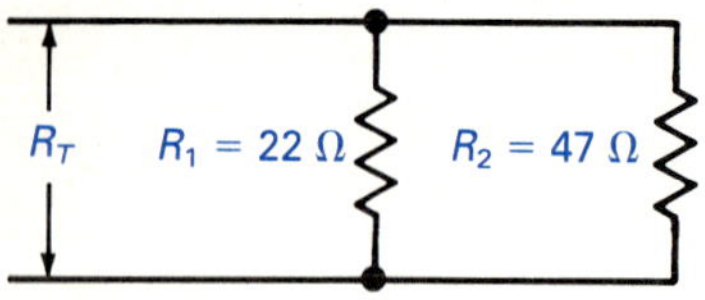

Fig. 7-24. Example 7-15: Data.

Solution (Fig. 7-24)

$$R_T = \frac{R_1 \times R_2}{R_1 + R_2} \qquad (7\text{-}13)$$

$$R_T = \frac{22\ \Omega \times 47\ \Omega}{22\ \Omega + 47\ \Omega} = \frac{1}{69} \times 22 \times 47 \simeq \mathbf{15\ \Omega}$$

Individual Conductances $G_1, G_2, \ldots, G_n$

The conductance of an individual element can be calculated by taking the reciprocal of its resistance

$$G_1 = \frac{1}{R_1}$$

or by the application of Ohm's law to the element

$$G_1 = \frac{I_1}{V}$$

or by extending the Ohm's law equation into a series of current ratio equations by substituting a suitable ratio G/I for $1/V$

$$G_1 = G_T \frac{I_1}{I_T} \quad \text{or} \quad G_2 \frac{I_1}{I_2} \quad \text{or} \quad \cdots \quad \text{or} \quad G_n \frac{I_1}{I_n} \qquad \mathbf{(7\text{-}14)}$$

Similar methods may be used to calculate the conductances G_2 to G_n.

EXAMPLE 7-16 What is the value of conductance G_1 in the circuit in Fig. 7-25?

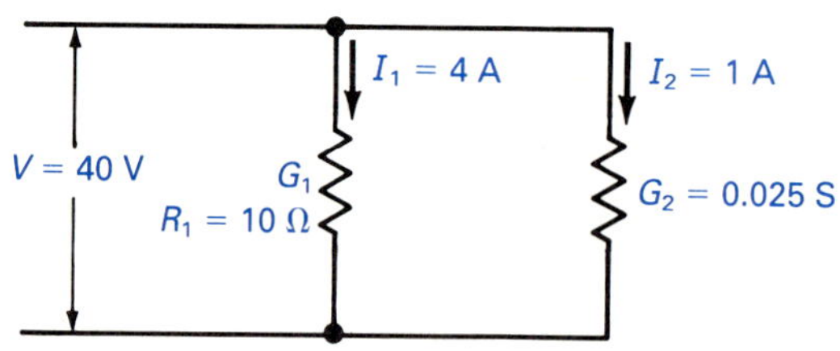

Fig. 7-25. Example 7-16: Data.

Solution Method 1 Resistance reciprocal:

$$G_1 = \frac{1}{R_1} = \frac{1}{10\ \Omega} = \mathbf{0.1\ S}$$

Method 2 Ohm's law:

$$G_1 = \frac{I_1}{V} = \frac{4\ \text{A}}{40\ \text{V}} = \mathbf{0.1\ S}$$

Method 3 Current ratio:

$$G_1 = G_2 \frac{I_1}{I_2} \tag{7-14}$$

$$G_1 = 0.025 \text{ S} \frac{4 \text{ A}}{1 \text{ A}} = \mathbf{0.1 \text{ S}}$$

Individual Resistances $R_1, R_2, \ldots, R_n$

The resistance of an individual element can be calculated by taking the reciprocal of its conductance

$$R_1 = \frac{1}{G_1}$$

or by the use of Ohm's law

$$R_1 = \frac{V}{I_1}$$

or by the formation of a set of equations by substituting for V ($V = R_T I_T$ or $R_2 I_2$, etc.)

$$R_1 = R_T \frac{I_T}{I_1} \quad \text{or} \quad R_2 \frac{I_2}{I_1} \quad \text{or} \quad \cdots \quad \text{or} \quad R_n \frac{I_n}{I_1} \tag{7-15}$$

The individual resistances R_2 to R_n can be similarly treated.

EXAMPLE 7-17 What is the value of resistance R_1 in the circuit in Fig. 7-26?

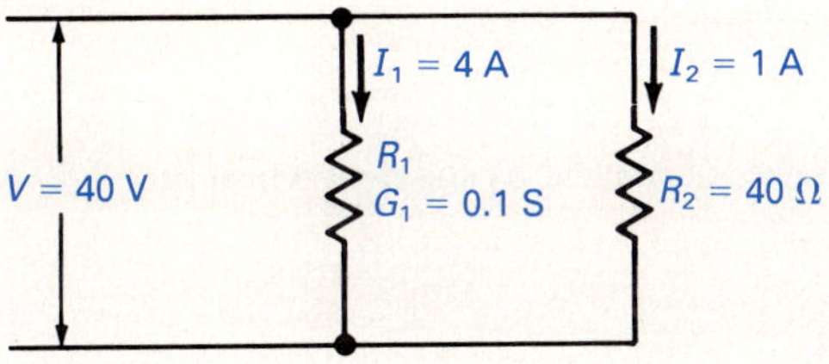

Fig. 7-26. Example 7-17: Data.

Solution Method 1 Conductance reciprocal:

$$R_1 = \frac{1}{G_1} = \frac{1}{0.1 \text{ S}} = \mathbf{10\ \Omega}$$

Method 2 Ohm's law:

$$R_1 = \frac{V}{I_1} = \frac{40 \text{ V}}{4 \text{ A}} = \mathbf{10\ \Omega}$$

Method 3 Current ratio:

$$R_1 = R_2 \frac{I_2}{I_1} \qquad (7\text{-}15)$$

$$R_1 = 40\ \Omega \frac{1\ \text{A}}{4\ \text{A}} = \mathbf{10\ \Omega}$$

Total Power P_T

The value of the total power in the circuit can be obtained by adding the individual powers

$$P_T = P_1 + P_2 + \cdots + P_n \qquad (7\text{-}6)$$

or by applying the basic power relationships given in Eqs. (2-14), (3-6), and (3-7)

$$P_T \quad \text{or} \quad VI_T \quad \text{or} \quad I_T{}^2R_T \quad \text{or} \quad \frac{V^2}{R_T}$$

The voltage in a parallel circuit is common. Therefore, if an expression for the voltage is derived in any convenient manner and applied to the last equation, power can be expressed in resistance ratio form. For instance, the power dissipated by resistor R_1 is given by

$$P_1 = \frac{V^2}{R_1} \quad \text{or} \quad V^2 = P_1R_1$$

Substituting $V^2 = P_1R_1$ in the equation $P_T = V^2/R_T$,

$$P_T = \frac{P_1R_1}{R_T} = P_1 \frac{R_1}{R_T}$$

The same procedure can be applied to any of the other elements, so that

$$P_T = P_1 \frac{R_1}{R_T} \quad \text{or} \quad P_2 \frac{R_2}{R_T} \quad \text{or} \quad \cdots \quad \text{or} \quad P_n \frac{R_n}{R_T} \qquad \mathbf{(7\text{-}16)}$$

EXAMPLE 7-18 What is the total power dissipated by the network in Fig. 7-27?

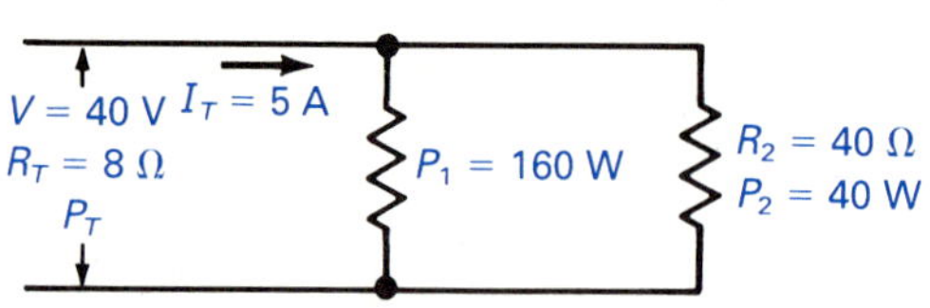

Fig. 7-27. Example 7-18: Data.

Solution Method 1 Sum of the powers:

$$P_T = P_1 + P_2 \qquad (7\text{-}6)$$
$$P_T = 160\ \text{W} + 40\ \text{W} = \mathbf{200\ W}$$

Method 2 Current and voltage:

$$P_T = VI_T = 40\ \text{V} \times 5\ \text{A} = \mathbf{200\ W}$$

Method 3 Current and resistance:

$$P_T = I_T^2 R_T = (5\ \text{A})^2 \times 8\ \Omega = \mathbf{200\ W}$$

Method 4 Voltage and resistance:

$$P_T = \frac{V^2}{R_T} = \frac{(40\ \text{V})^2}{8\ \Omega} = \mathbf{200\ W}$$

Method 5 Resistance ratio:

$$P_T = P_2 \frac{R_2}{R_T} \tag{7-16}$$

$$P_T = 40\ \text{W}\ \frac{40\ \Omega}{8\ \Omega} = \mathbf{200\ W}$$

Individual Powers $P_1, P_2, \ldots, P_n$

The methods demonstrated in Example 7-18 for the total power can be used to obtain the power dissipated by individual elements. The following relationships are for P_1, the power in element R_1, but apply equally well to the other elements R_2 to R_n.

The basic relationships for P_1 are

$$P_1 = VI_1 \quad \text{or} \quad I_1^2 R_1 \quad \text{or} \quad \frac{V^2}{R_1}$$

As we have seen, $V^2 = PR$ throughout the circuit ($V^2 = P_T R_T = P_1 R_1$, etc.). Substituting for V^2 in the equation $P_1 = V^2/R_1$, we can obtain a set of resistance ratio equations for P_1:

$$P_1 = P_T \frac{R_T}{R_1} \quad \text{or} \quad P_2 \frac{R_2}{R_1} \quad \text{or} \quad \cdots \quad \text{or} \quad P_n \frac{R_n}{R_1} \tag{7-17}$$

EXAMPLE 7-19 What power is dissipated in element R_1 in Fig. 7-28?

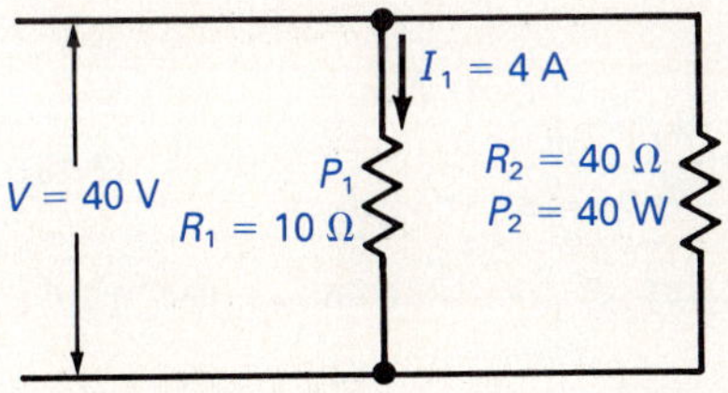

Fig. 7-28. Example 7-19: Data.

Solution Method 1 Voltage and current:

$$P_1 = VI_1 = 40\ \text{V} \times 4\ \text{A} = \mathbf{160\ W}$$

Method 2 Current and resistance:

$$P_1 = I_1^2 R_1 = (4\ \text{A})^2 \times 10\ \Omega = \mathbf{160\ W}$$

Method 3 Voltage and resistance:

$$P_1 = \frac{V^2}{R_1} = \frac{(40\ \text{V})^2}{10\ \Omega} = \mathbf{160\ W}$$

Method 4 Resistance ratio:

$$P_1 = P_2 \frac{R_2}{R_1} \tag{7-17}$$

$$P_1 = 40\ \text{W}\ \frac{40\ \Omega}{10\ \Omega} = \mathbf{160\ W}$$

Energy

Energy values can be calculated by using any of the methods in Examples 7-18 and 7-19 to obtain the appropriate power, and then multiplying by the time, which for any situation will be common to the whole circuit.

Ratios in Parallel Circuits

Ratios developed from simple formulas, such as Ohm's law, can sometimes help in understanding circuits. This ratio concept may be applied to parallel circuits as follows:

$$\frac{I_T}{I_1} = \frac{VG_T}{VG_1} = \frac{G_T}{G_1}$$

$$\frac{P_T}{P_1} = \frac{VI_T}{VI_1} = \frac{I_T}{I_1}$$

$$\frac{W_T}{W_1} = \frac{P_T t}{P_1 t} = \frac{P_T}{P_1}$$

These ratios can be coupled with the inverse ratio

$$\frac{G_T}{G_1} = \frac{R_1}{R_T}$$

to give the result

$$\frac{G_T}{G_1} = \frac{I_T}{I_1} = \frac{P_T}{P_1} = \frac{W_T}{W_1} = \frac{R_1}{R_T} \tag{7-18}$$

Instead of G_1, G_T, etc., the values for any of the elements or the whole network can be used with equal validity.

EXAMPLE 7-20 Using the ratio G_T/G_1, determine *(a)* the total current, *(b)* the total power, *(c)* the total energy, and *(d)* the total resistance associated with the circuit in Fig. 7-29.

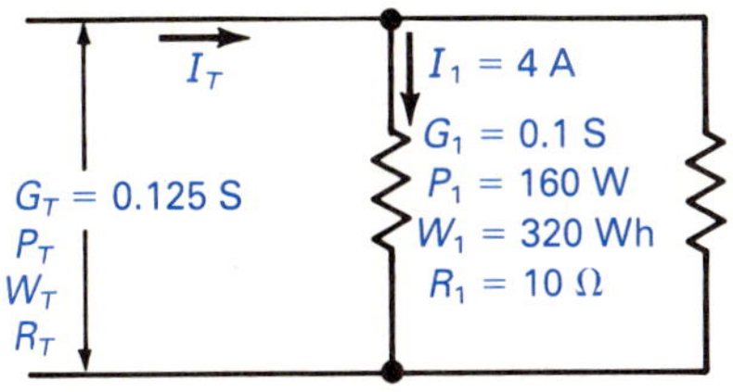

Fig. 7-29. Example 7-20: Data.

Solution The ratio G_T/G_1 is given by

$$\frac{G_T}{G_1} = \frac{0.125 \text{ S}}{0.1 \text{ S}} = 1.25$$

This value is shown in parentheses in the following solutions.

a. $I_T = (1.25)I_1 = (1.25)4 \text{ A} = \mathbf{5 \text{ A}}$

b. $P_T = (1.25)P_1 = (1.25)160 \text{ W} = \mathbf{200 \text{ W}}$

c. $W_T = (1.25)W_1 = (1.25)320 \text{ Wh} = \mathbf{400 \text{ Wh}}$

d. $R_T = \frac{1}{(1.25)} R_1 = \frac{1}{(1.25)} 10 \ \Omega = \mathbf{8 \ \Omega}$

REVIEW QUESTIONS FOR SECTION 7-3

1. State in equation form:
   a. Two methods for determining the voltage across a parallel circuit.
   b. Five methods for determining the total current in a parallel circuit.
   c. Four methods for determining the current in an individual resistor in a parallel circuit.
   d. Three methods for determining the total conductance of a parallel circuit.
   e. One method for determining the total resistance of a parallel circuit.
   f. Four methods for determining the conductance of an individual element in a parallel circuit.
   g. Four methods for determining the resistance of an individual element in a parallel circuit.
   h. Four methods for determining the total power dissipated by a parallel circuit.
   i. Five methods for determining the power dissipated by an individual element of a parallel circuit.
2. Give a simple formula for the total resistance of two resistors connected in parallel.
3. State the formula of Question 2 in words.

EXERCISE PROBLEMS FOR SECTION 7-3

1. For the circuit in Fig. 7-30, find the total conductance when
   a. $G_1 = 0.1$ S, $G_2 = 0.5$ S, $G_3 = 2$ S (*ans.* 2.6 S)
   b. $G_1 = 2$ mS, $G_2 = 6$ mS, $G_3 = 500$ μS (*ans.* 8.5 mS)
   c. $G_1 = 2500$ μS, $G_2 = 1000$ μS, $G_3 = 0.25$ mS (*ans.* 3.75 mS)
   d. $G_1 = G_2 = G_3 = 0.005\ 56$ S (*ans.* 16.68 mS)

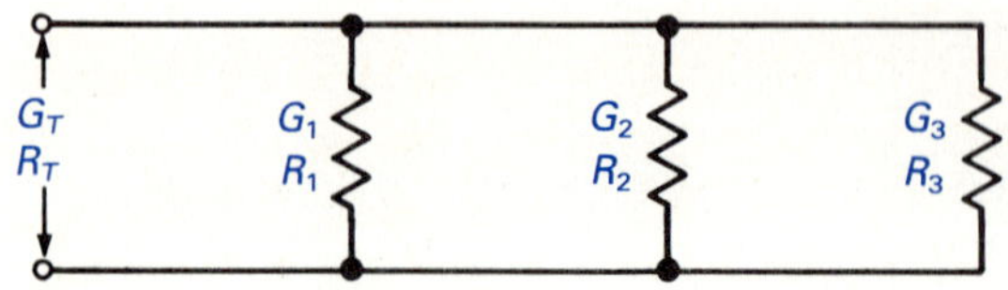

Fig. 7-30. Problems 1 to 3: Network.

2. For the circuit in Fig. 7-30, find the total resistance when
a. $G_1 = 0.1$ S, $G_2 = 0.5$ S, $G_3 = 2$ S
b. $G_1 = 0.883$ mS, $G_2 = 370\ \mu$S, $G_3 = 666\ \mu$S
c. $G_1 = G_2 = G_3 = 30.3$ mS

3. For the circuit in Fig. 7-30, find the total resistance when
a. $R_1 = 100\ \Omega$, $R_2 = 150\ \Omega$, $R_3 = 180\ \Omega$ (*ans.* 45 Ω)
b. $R_1 = 1200\ \Omega$, $R_2 = 2700\ \Omega$, $R_3 = 1800\ \Omega$ (*ans.* 568.4 Ω)
c. $R_1 = 560$ kΩ, $R_2 = 1.5$ MΩ, $R_3 = 120$ kΩ (*ans.* 92.72 kΩ)
d. $R_1 = R_2 = R_3 = 18$ kΩ (*ans.* 6 kΩ)

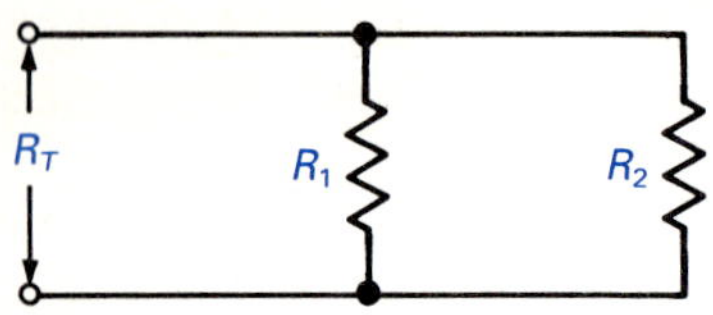

Fig. 7-31. Problems 4 and 5: Network.

4. For the circuit in Fig. 7-31, find the total resistance, using the conductance method, when
a. $R_1 = 10\ \Omega$, $R_2 = 22\ \Omega$
b. $R_1 = 2.2$ MΩ, $R_2 = 680$ kΩ
c. $R_1 = 82$ kΩ, $R_2 = 39$ kΩ

5. For the circuit in Fig. 7-31, find the total resistance, using the product-over-sum method, when
a. $R_1 - 33\ \Omega$, $R_2 = 82\ \Omega$ (*ans.* 23.53 MΩ)
b. $R_1 = 150$ kΩ, $R_2 = 68$ kΩ (*ans.* 46.79 kΩ)
c. $R_1 = 15$ kΩ, $R_2 = 150$ kΩ (*ans.* 13.64 kΩ)
d. $R_1 = 1.8$ MΩ, $R_2 = 2.7$ MΩ (*ans.* 1.08 MΩ)

6. For the circuit in Fig. 7-32, find the total current and the total power drawn from the generator when
a. $V = 30$ V, $I_1 = 3$ A, $I_2 = 6$ A
b. $V = 5$ V, $G_1 = 0.1$ S, $I_2 = 0.1$ A
c. $I_1 = 8$ mA, $G_T = 0.004$ S, $G_1 = 0.001$ S
d. $V_1 = 25$ V, $R_2 = 1000\ \Omega$, $I_1 = 50$ mA
e. $V = 12$ V, $R_1 = 22$ kΩ, $R_2 = 15$ kΩ
f. $V_1 = 10$ V, $G_T = 0.0129$ S, $R_2 = 220\ \Omega$

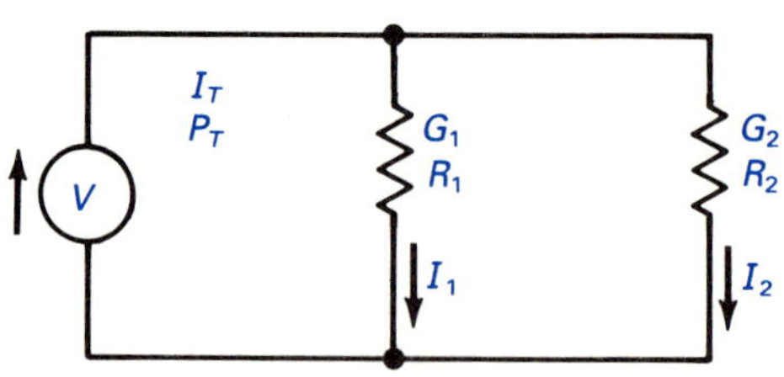

Fig. 7-32. Problems 6 and 7: Circuit.

7. For the circuit in Fig. 7-32, find I_1 when
a. $V_1 = 12$ V, $G_1 = 0.005\ 56$ S, $G_2 = 0.066$ S (*ans.* 66.72 mA)
b. $V = 48$ V, $I_T = 48$ mA, $I_2 = 4$ mA (*ans.* 44 mA)
c. $I_T = 18$ mA, $R_T = 11$ kΩ, $R_1 = 33$ kΩ (*ans.* 6 mA)
d. $I_T = 150$ mA, $R_2 = 15$ kΩ, $I_2 = 45$ mA (*ans.* 105 mA)
e. $V_1 = 100$ V, $R_2 = 1$ kΩ, $I_T = 100$ mA (*ans.* 0 A)

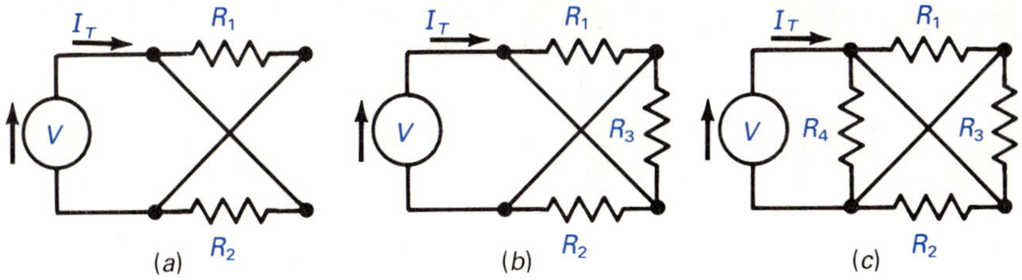

Fig. 7-33. Problems 8 to 10: Circuits.

8. Find I_T in Fig. 7-33*a* when $V = 20$ V, $R_1 = 680\ \Omega$, and $R_2 = 560\ \Omega$.
9. Find I_T in Fig. 7-33*b* when $V = 20$ V, $R_1 = 680\ \Omega$, $R_2 = 560\ \Omega$, and $R_3 = 390\ \Omega$. (*ans*. 116.4 mA)
10. Find I_T in Fig. 7-33*c* when $V = 20$ V, $R_1 = 680\ \Omega$, $R_2 = 560\ \Omega$, $R_3 = 390\ \Omega$, and $R_4 = 820\ \Omega$.
11. Four 6-V power sources, each capable of delivering 150 mA, are connected in parallel.
a. What is the total voltage? (*ans*. 6 V)
b. What is the maximum current available? (*ans*. 600 mA)

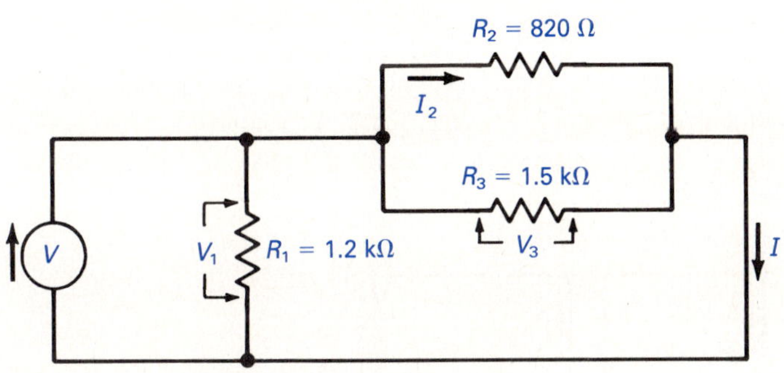

Fig. 7-34. Problems 12 to 14: Circuit.

12. Find I in Fig. 7-34 when $V_3 = 12$ V.
13. Find I in Fig. 7-34 when $V_1 = 10$ V. (*ans*. 18.86 mA)
14. Find I in Fig. 7-34 when $I_2 = 10$ mA.
15. Three constant-current sources of 10, 15, and 20 mA are in parallel across a 1000-Ω load. Find
a. The total current supplied to the load. (*ans*. 45 mA)
b. The voltage across the load. (*ans*. 45 V)

7-4 APPLICATIONS OF PARALLEL CIRCUITS

Current Divider

The current entering a parallel circuit divides so that part of the current flows through each of the elements (Fig. 7-35). High-conductance (low-resistance) elements will have a greater share of the total current than will those of low conductance.

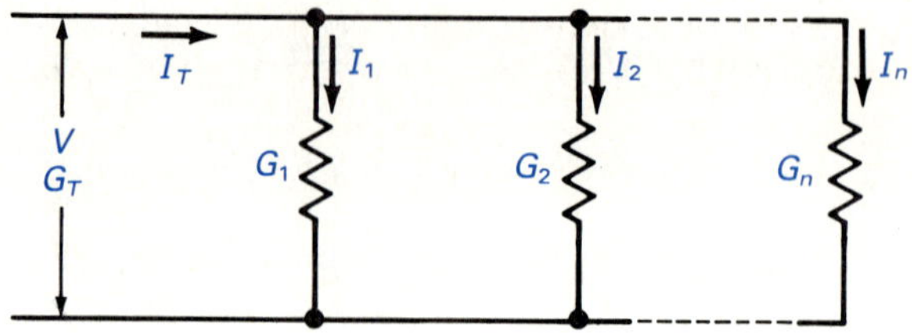

Fig. 7-35. Current division in a parallel network.

$$V = \frac{I_T}{G_T}$$

$$I_1 = VG_1 = \frac{I_T}{G_T} G_1$$

$$I_1 = I_T \frac{G_1}{G_T} \quad \text{or} \quad I_T \frac{G_1}{G_1 + G_2 + \cdots + G_n} \tag{7-19}$$

Equation (7-19) is basically the same as Eq. (7-10). It can be used to find the value of any other current I_2 to I_n.

EXAMPLE 7-21 Three resistors, 10, 25, and 50 Ω, are connected in parallel. If the total network current is 4 A, what is the current in each element?

Solution (Fig. 7-36)

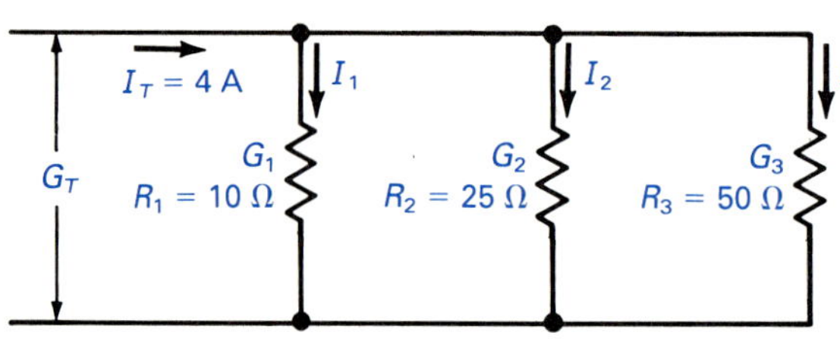

Fig. 7-36. Example 7-21: Data.

$$G_1 = \frac{1}{R_1} = \frac{1}{10\ \Omega} = 0.1\ \text{S}$$

$$G_2 = \frac{1}{R_2} = \frac{1}{25\ \Omega} = 0.04\ \text{S}$$

$$G_3 = \frac{1}{R_3} = \frac{1}{50\ \Omega} = 0.02\ \text{S}$$

$$G_T = G_1 + G_2 + G_3 \tag{7-2}$$

$$G_T = 0.1\ \text{S} + 0.04\ \text{S} + 0.02\ \text{S} = 0.16\ \text{S}$$

$$I_1 = I_T \frac{G_1}{G_T} = 4\ \text{A} \frac{0.1\ \text{S}}{0.16\ \text{S}} = \mathbf{2.5\ A} \tag{7-19}$$

$$I_2 = I_T \frac{G_2}{G_T} = 4\ \text{A} \frac{0.04\ \text{S}}{0.16\ \text{S}} = \mathbf{1\ A}$$

$$I_3 = I_T \frac{G_3}{G_T} = 4\ \text{A} \frac{0.02\ \text{S}}{0.16\ \text{S}} = \mathbf{0.5\ A}$$

A simple formula can be derived for calculating the division of current between *two* parallel elements. (See Fig. 7-37.)

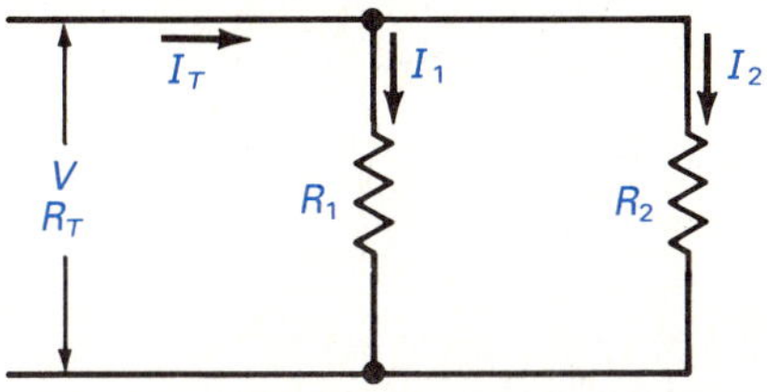

Fig. 7-37. Current division in a two-element parallel network.

From Eq. (7-13) $R_T = R_1 R_2/(R_1 + R_2)$,

$$V = I_T R_T = I_T \frac{R_1 R_2}{R_1 + R_2}$$

$$I_1 = \frac{V}{R_1} = I_T \frac{R_1 R_2}{R_1 + R_2} \times \frac{1}{R_1}$$

$$I_1 = I_T \frac{R_2}{R_1 + R_2} \qquad \textbf{(7-20)}$$

The formula for I_2 will be similar $[I_T \times R_1/(R_1 + R_2)]$.

In a *two-element parallel network, the current in one element is equal to the total current multiplied by the resistance of the* **other** *element divided by the sum of the two resistances. With more than two elements this rule does not apply.*

EXAMPLE 7-22 A two-element parallel network has element values of 15 and 47 Ω. If the total network current is 5 A, what is the current in each of the elements?

Solution (Fig. 7-38)

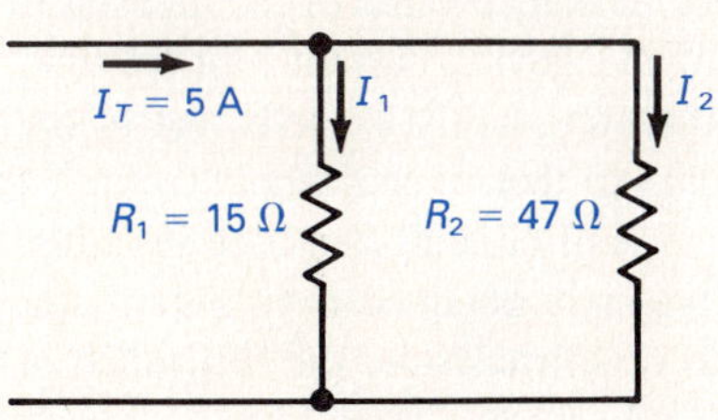

Fig. 7-38. Example 7-22: Data.

$$I_1 = I_T \frac{R_2}{R_1 + R_2} \qquad (7\text{-}20)$$

$$I_1 = 5\text{ A}\frac{47\ \Omega}{15\ \Omega + 47\ \Omega} = 5\text{ A}\frac{47\ \Omega}{62\ \Omega} = \mathbf{3.79\ A}$$

$$I_2 = I_T \frac{R_1}{R_1 + R_2} = 5\text{ A}\frac{15\ \Omega}{15\ \Omega + 47\ \Omega} = 5\text{ A}\frac{15\ \Omega}{62\ \Omega} = \mathbf{1.21\ A}$$

EXAMPLE 7-23 A small electric motor has a variable shunt of 5 to 100 Ω connected across it. If the motor has a constant resistance of 10 Ω, and the network is fed from a 1-A constant-current source, what is the range of current in the motor?

Solution (Fig. 7-39)

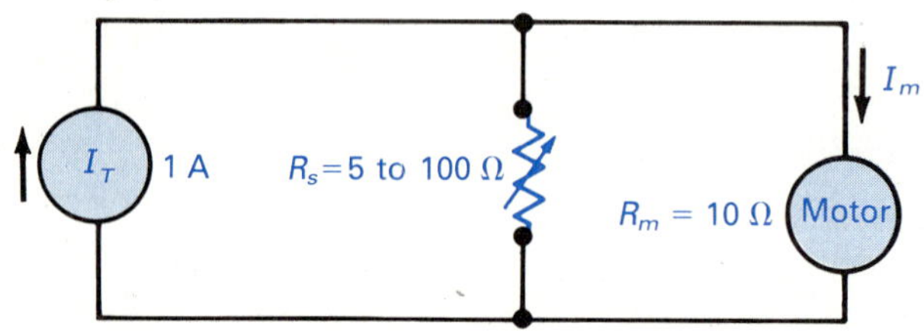

Fig. 7-39. Example 7-23: Circuit.

When R_s is at its minimum value of 5 Ω,

$$I_m = I_T \frac{R_s}{R_s + R_m} \qquad (7\text{-}20)$$

$$I_m = 1\text{ A}\frac{5\ \Omega}{5\ \Omega + 10\ \Omega} = 1\text{ A}\frac{5\ \Omega}{15\ \Omega} = \mathbf{0.3333\ A}$$

When R_s is at its maximum value of 100 Ω,

$$I_m = I_T \frac{R_s}{R_s + R_m} = 1\text{ A}\frac{100\ \Omega}{100\ \Omega + 10\ \Omega}$$

$$I_m = 1\text{ A}\frac{100\ \Omega}{110\ \Omega} = \mathbf{0.9091\ A}$$

The motor current range in Example 7-23 is 0.3333 to 0.9091 A. This means that, as the shunt resistance is varied from one-half the motor resistance to 10 times its resistance, the motor current varies from one-third of the supply current to almost the whole of it. Therefore, when a motor (or any other device) is fed from a constant-current source, its current can be varied by shunting it with a variable resistor, or rheostat. When the lowest resistance value of the rheostat is very much lower than the device resistance, the device current will be low. Conversely, when the highest value of the rheostat greatly exceeds that of the device, the device current will be almost equal to that of the supply.

Example 7-23 gives, in simplified form, a type of circuit often used in industrial control. The constant-current source is typically obtained from a transistor circuit. The shunt resistance is also likely to be a transistor, the resistance of which can be varied by a control voltage.

Current Generator

Current generators are frequently used in electronic circuits. In its ideal state a voltage generator is typified as supplying a constant voltage regardless of the nature of the load. An ideal current generator always supplies a constant current to a load. We saw in Sec. 6-5 that a practical voltage generator does not in fact produce a constant voltage, because its internal resistance reduces its output voltage (terminal voltage) by an amount which depends on the load current. Similarly, because of its internal conductance, a practical current generator does not actually supply a constant current to its load.

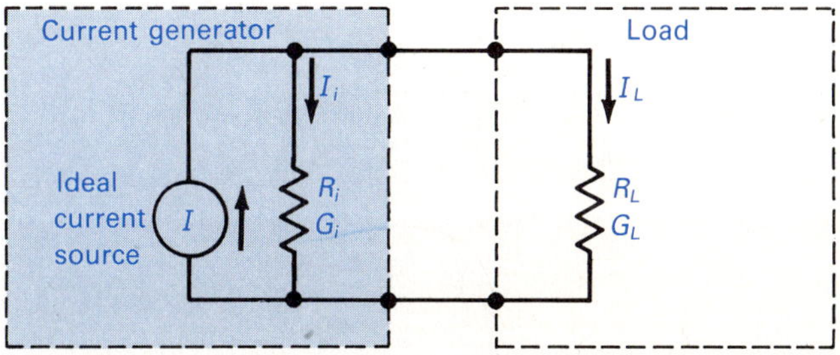

Fig. 7-40. Current generator and load.

Figure 7-40 shows a current generator connected to its load. The ideal current source supplies a constant current I to a two-element parallel network consisting of the generator internal conductance G_i and the load G_L. The source current divides between G_i and G_L. The load current can be determined by using the current-divider principle in:

conductance form,
$$I_L = I\frac{G_L}{G_L + G_i} \qquad \text{from (7-19)}$$

or, in resistance form,
$$I_L = I\frac{R_i}{R_L + R_i} \qquad (7\text{-}20)$$

Kirchhoff's current law can be applied to give another statement of the conditions in a loaded practical current generator:

$$I = I_i + I_L \qquad \text{or} \qquad I_L = I - I_i \qquad \textbf{(7-21)}$$

You will see that the current supplied by a practical current generator is affected by its internal conductance (or resistance).

EXAMPLE 7-24 A current generator has an ideal output of 5 A. If its internal conductance is 0.1 S, what current will it supply to a load of 0.15 S?

Solution (Fig. 7-41)

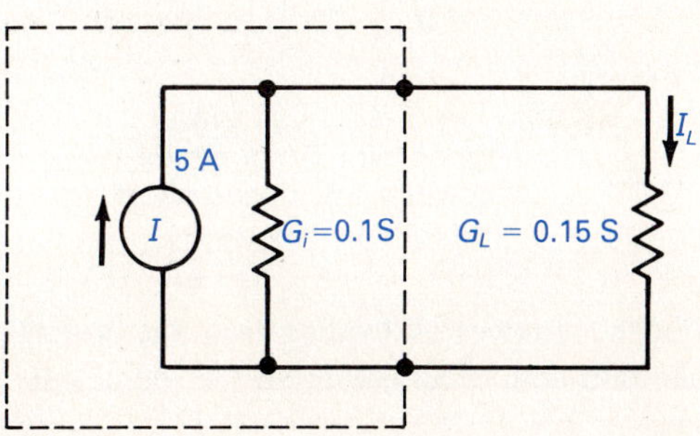

Fig. 7-41. Example 7-24: Circuit.

$$I_L = I\frac{G_L}{G_L + G_i} \tag{7-19}$$

$$I_L = 5\text{ A}\frac{0.15\text{ S}}{0.15\text{ S} + 0.1\text{ S}} = 5\text{ A}\frac{0.15\text{ S}}{0.25\text{ S}} = \mathbf{3\text{ A}}$$

EXAMPLE 7-25 Referring to Example 7-24, what would be the current in the load if its conductance were 10 S?

Solution (Fig. 7-42)

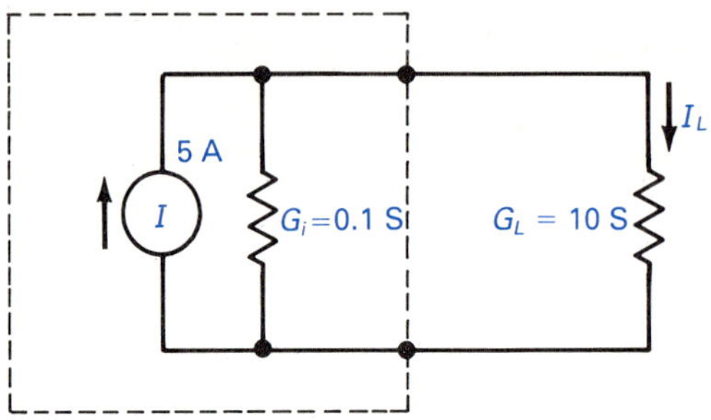

Fig. 7-42. Example 7-25: Circuit.

$$I_L = I\frac{G_L}{G_L + G_i} \tag{7-19}$$

$$I_L = 5\text{ A}\frac{10\text{ S}}{10\text{ S} + 0.1\text{ S}} = 5\text{ A}\frac{10\text{ S}}{10.1\text{ S}} = \mathbf{4.95\text{ A}}$$

Example 7-25 shows that, if the load conductance is high compared with the internal conductance of the generator, the load current will almost equal the current supplied by an ideal current generator. If the load conductance were infinitely high (load resistance = 0 Ω), the load current would literally equal the source current.

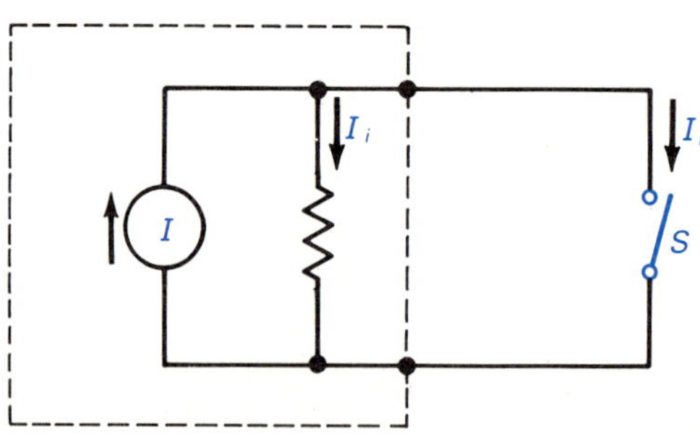

Fig. 7-43. Current generator open circuit (switch open) and short circuit (switch closed).

In Fig. 7-43 the switch S, when open, removes all loading from the current generator. The open-circuit load condition now occurs, with all the source current existing in the internal conductance:

$$I_L = 0$$
$$I_i = I$$

When the switch is closed, a solid connection is formed between the generator terminals. No load element is shown, so it is implied that the load resistance is zero, the load conductance being infinite. This zero load resistance is the short-circuit[1] load condition, all the source current now being in the load.

$$I_L = I$$
$$I_i = 0$$

EXAMPLE 7-26 A current generator has a variable resistor connected across its terminals, its resistance value ranging from 0 to 10 Ω. If the generator source current is 10 A and its internal resistance is 10 mΩ, what is the load current when the load resistance is *(a)* 0 Ω, *(b)* 1 Ω, and *(c)* 10 Ω?

Solution (Fig. 7-44)

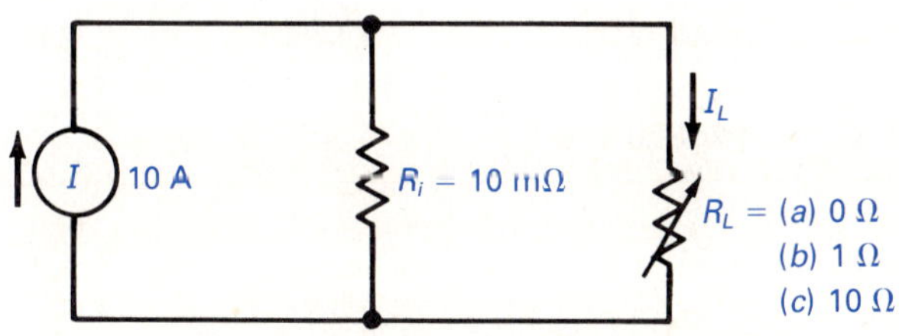

Fig. 7-44. Example 7-26: Circuit.

a. When $R_L = 0\ \Omega$,

$$I_L = I\frac{R_i}{R_i + R_L} \tag{7-20}$$

$$I_L = 10\text{ A}\frac{10\text{ m}\Omega}{10\text{ m}\Omega + 0\ \Omega} = \mathbf{10\ A}$$

b. When $R_L = 1\ \Omega$,

$$I_L = I\frac{R_i}{R_i + R_L} = 10\text{ A}\frac{10\text{ m}\Omega}{10\text{ m}\Omega + 1\ \Omega} = 10\text{ A}\frac{10\text{ m}\Omega}{1.01\ \Omega} \simeq \mathbf{0.1\ A}$$

c. When $R_L = 10\ \Omega$,

$$I_L = I\frac{R_i}{R_i + R_L} = 10\text{ A}\frac{10\text{ m}\Omega}{10\text{ m}\Omega + 10\ \Omega} = 10\text{ A}\frac{10\text{ m}\Omega}{10.01\ \Omega} \simeq \mathbf{0.01\ A}$$

As the load resistance varies from a value of 0 Ω until it becomes high relative to the generator internal resistance (in Example 7-26 it becomes 10 Ω, which is

[1]The term "short circuit" is often misused to imply a faulty condition. Actually it is quite specific and does not necessarily mean that there is something wrong. When a short circuit exists between two points in a circuit, the resistance between them is 0 Ω. A short circuit occurs in practice because of a direct connection by a conductor, such as a wire or a metal plate.

1000 times the internal resistance), the load current decreases from an initial value equal to the basic source current down to a value which is by comparison very low.

EXAMPLE 7-27 A certain transistor behaves as a current generator with a source current of 10 mA and an internal resistance of 1 kΩ. What load current will exist if the load resistance is *(a)* 100 Ω, *(b)* 1 kΩ, and *(c)* 4 kΩ?

Solution (Fig. 7-45)

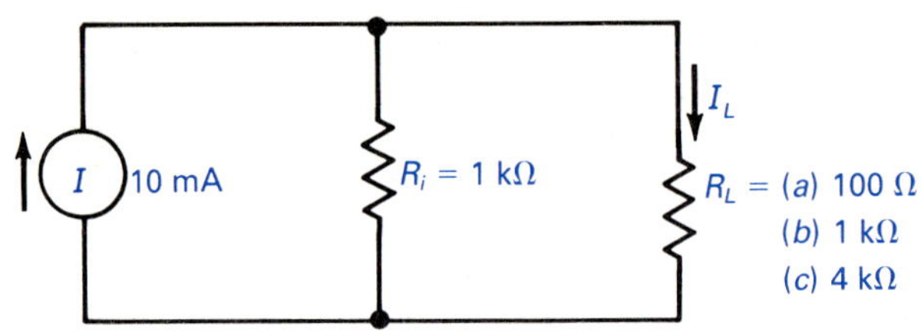

Fig. 7-45. Example 7-27: Circuit.

a. When $R_L = 100\ \Omega\ (0.1\ \text{k}\Omega)$,

$$I_L = I \frac{R_i}{R_i + R_L} \tag{7-20}$$

$$I_L = 10\ \text{mA} \frac{1\ \text{k}\Omega}{1\ \text{k}\Omega + 0.1\ \text{k}\Omega} + 10\ \text{mA} \frac{1\ \text{k}\Omega}{1.1\ \text{k}\Omega} \simeq \mathbf{9\ mA}$$

b. When $R_L = 1\ \text{k}\Omega$,

$$I_L = I \frac{R_i}{R_i + R_L} = 10\ \text{mA} \frac{1\ \text{k}\Omega}{1\ \text{k}\Omega + 1\ \text{k}\Omega} = 10\ \text{mA} \frac{1\ \text{k}\Omega}{2\ \text{k}\Omega} = \mathbf{5\ mA}$$

c. When $R_L = 4\ \text{k}\Omega$,

$$I_L = I \frac{R_i}{R_i + R_L} = 10\ \text{mA} \frac{1\ \text{k}\Omega}{1\ \text{k}\Omega + 4\ \text{k}\Omega} = 10\ \text{mA} \frac{1\ \text{k}\Omega}{5\ \text{k}\Omega} = \mathbf{2\ mA}$$

Comparison of Voltage and Current Generators

Much has been said about voltage generators and current generators, but what are they in practice? Is a battery a voltage generator or a current generator? What about an automobile generator; what type is it, voltage or current? All real generators are both voltage and current generators. The distinction is a matter of convenience; either view can be chosen to suit a situation.

A discussion of the interchangeability of voltage and current generators was postponed until now, because the principles that have been developed for these generators are needed to show how to perform the conversion from one to the other. The need to do this will arise when you work with different circuit types and analysis methods.

Schematic diagrams for a voltage and current generator are shown in Fig. 7-46*a* and *b*, respectively. Resistances are shown in both circuits to simplify comparison between them.

If the internal resistance is made the same for each generator:

$$\text{Voltage generator (Fig. 7-46}a\text{):}\quad I_L = \frac{V}{R_i + R_L} \tag{7-22}$$

$$\text{Current generator (Fig. 7-46}b\text{):}\quad I_L = I\frac{R_i}{R_i + R_L} \tag{7-23}^*$$

For the generators to be interchangeable, they must be equivalent from the load-current viewpoint. (If their internal resistances are the same and they give the same current to the same load resistance, they are equivalent.)

The expressions for the load currents from Eqs. (7-22) and (7-23) are equated:

$$\frac{V}{R_i + R_L} = I\frac{R_i}{R_i + R_L}$$

$$V = IR_i \qquad \text{or} \qquad I = \frac{V}{R_i}$$

Thus, if the emf of the voltage generator equals the product of the source current and the internal resistance of the current generator, and their internal resistances are equal, the two generators will be equivalent.

Alternatively, if the source current of the current generator equals the emf divided by the internal resistance of the voltage generator, and their internal resistances are equal, the two generators will be equivalent.

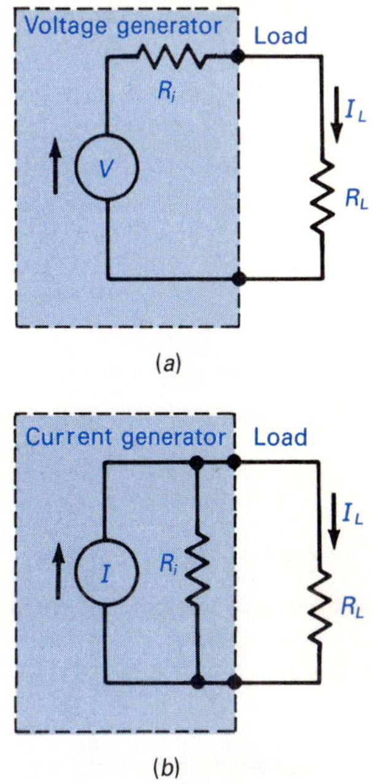

Fig. 7-46. *(a)* Voltage generator; *(b)* current generator.

Generator Loading

The *loading* on a generator *is* said to be *increased when its* load *current is increased*. Therefore a *high-conductance* load results in a high, or *heavy, loading*.

Put in resistance terms, a *low-resistance* load means a *heavy loading*. This last viewpoint is important, since there seems to be a tendency to regard an increase in load resistance as an increase in loading, the opposite of what is actually true.

When *loading is increased,* the *load voltage goes down* for either voltage or current generators, as now shown.

Voltage generator (Fig. 7-47*a*); using the voltage-divider principle:

$$V_L = V \times \frac{R_L}{R_i + R_L} = \frac{V}{R_i/R_L + 1}$$

$$V_L = \frac{V}{R_iG_L + 1} \tag{7-24}$$

As the load conductance G_L increases (or the load resistance R_L decreases), the term R_iG_L in Eq. (7-24) becomes larger. Therefore a loading increase causes the voltage across the load of a voltage generator to become smaller.

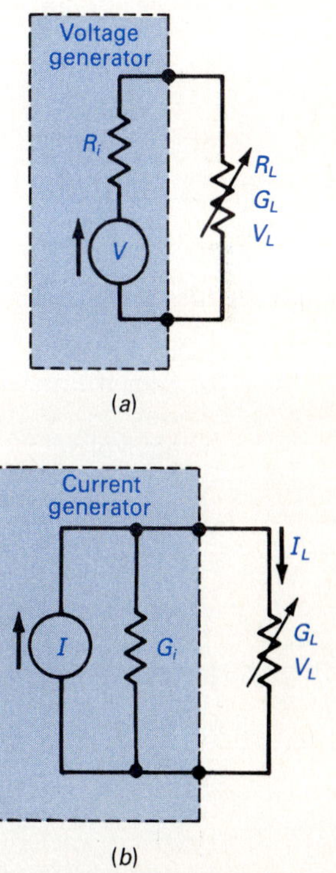

Fig. 7-47. Generator loading: *(a)* Voltage generator; *(b)* current generator.

*Equation (7-23) is specific to a current-generator-driven load. The same equation appears previously as an application of the two-element current divider [Eq. (7-20)].

Current generator (Fig. 7-47*b*):

$$V_L = \frac{I_L}{G_L} = I\frac{G_L}{G_i + G_L} \times \frac{1}{G_L}$$

$$V_L = \frac{I}{G_i + G_L} \qquad \textbf{(7-25)}$$

From Eq. (7-25) it can be seen that, if the load conductance G_L increases (or the load resistance R_L decreases), the voltage across the load of a current generator becomes smaller.

Whether an energy source is viewed as a voltage generator or as a current generator, its load voltage will go down if the loading is increased.

Ammeter

A moving-coil meter movement, such as was referred to in Sec. 6-5 when voltmeter multipliers were discussed, is a current-operated device. As such, it is an ammeter. A typical basic movement, however, is of limited use as an ammeter because the range of current it can handle is too small. The movement which forms the basis of many commercial meters is driven to full-scale deflection with a current of 50 μA. To be able to measure a current in excess of this small value, the movement needs to be combined with a parallel resistor, called a *shunt*.

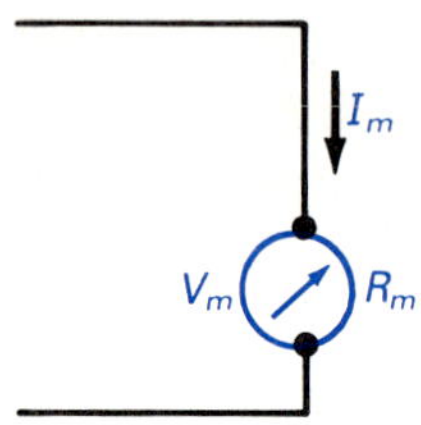

(*a*)

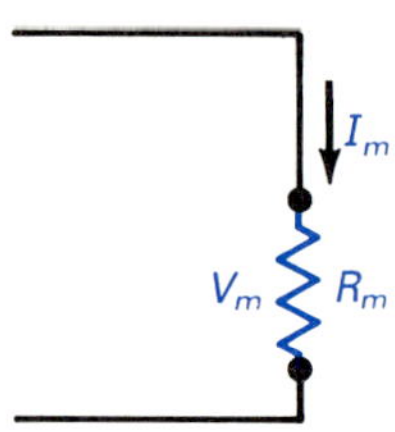

(*b*)

Fig. 7-48. Meter movement as circuit element: (*a*) General meter symbol; (*b*) equivalent resistance.

Figure 7-48 is a schematic diagram of a basic meter movement. The symbol in Fig. 7-48*a* is a representation of a meter face, showing the pointer; this symbol will be used in the ammeter diagrams. The electric circuit equivalent of the movement is the resistance in Fig. 7-48*b*. It is common to specify the full-scale deflection current I_m and the resistance of the movement R_m; its full-scale voltage V_m can be calculated by using Ohm's law ($V_m = I_mR_m$).

Figure 7-49 is a schematic diagram of an ammeter. It consists of a meter movement with a shunt R_s connected in parallel with it.

There are several different methods of computing the resistance value of the shunt, some involving formulas. As it is unlikely that you will remember special formulas when you need to use them, it is better to use circuit principles. However, the following will show you how one ammeter formula can be developed.

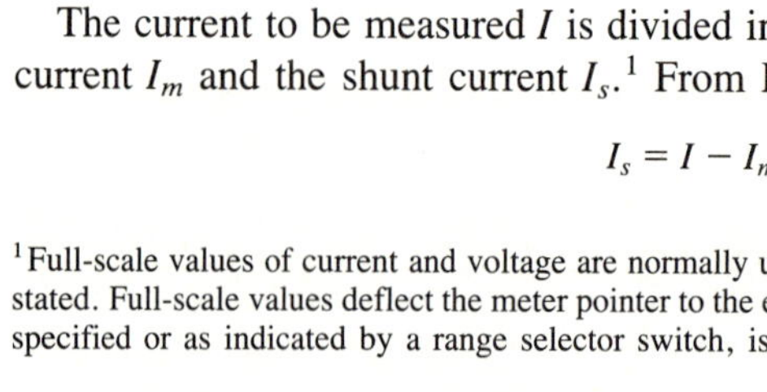

Fig. 7-49. Basic ammeter circuit.

The current to be measured I is divided into two components, the movement current I_m and the shunt current I_s.[1] From Eq. (7-1),

$$I_s = I - I_m$$

[1] Full-scale values of current and voltage are normally used in meter computations unless otherwise stated. Full-scale values deflect the meter pointer to the end of the calibrated scale. A meter range, as specified or as indicated by a range selector switch, is also a full-scale value.

From Eq. (7-15),

$$R_s = R_m \frac{I_m}{I_s}$$

EXAMPLE 7-28 An ammeter with a full-scale deflection of 1 mA (range of measurement, 0 to 1 mA) is required. What value shunt resistance should be used with a 50-μA 2-kΩ movement?

Solution (Fig. 7-50)

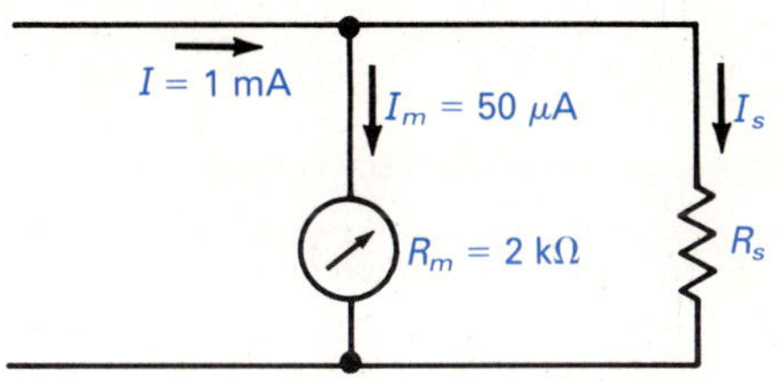

Fig. 7-50. Example 7-28: Data.

$$I_s = I - I_m \qquad (7\text{-}1)$$

$$I_s = 1000\ \mu A - 50\ \mu A = 950\ \mu A$$

$$R_s - R_m \frac{I_m}{I_s} \qquad (7\text{-}15)$$

$$R_s = 2\ k\Omega \frac{50\ \mu A}{950\ \mu A} = \mathbf{105.3\ \Omega}$$

EXAMPLE 7-29 Using the same movement as in Example 7-28 design an ammeter with three ranges, *(a)* 1 mA, *(b)* 10 mA, and *(c)* 100 mA.

Solution

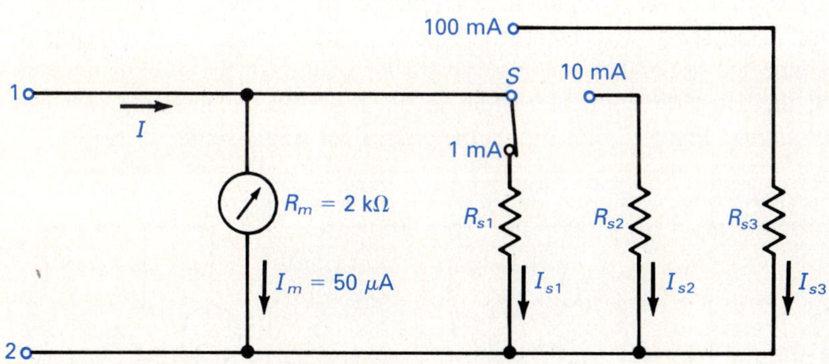

Fig. 7-51. Example 7-29: Network.

Figure 7-51 shows the movement with three shunts which can be individually connected in parallel with the movement by the switch *S*. For any of the three switch positions, a two-element parallel circuit will exist between the ammeter terminals 1 and 2.

a. For 1 mA full scale, the shunt R_{s1} will be in parallel with the movement. This range is the same as that in Example 7-28:

$$R_{s1} = \mathbf{105.3\ \Omega}$$

b. For 10 mA (10^4 μA):

$$I_{s2} = I - I_m \qquad (7\text{-}1)$$
$$I_{s2} = 10^4\ \mu\text{A} - 50\ \mu\text{A} = 9950\ \mu\text{A}$$
$$R_{s2} = R_m \frac{I_m}{I_{s2}} \qquad (7\text{-}15)$$
$$R_{s2} = 2\ \text{k}\Omega \frac{50\ \mu\text{A}}{9950\ \mu\text{A}} \simeq \mathbf{10\ \Omega}$$

c. For 100 mA (10^5 μA):

$$I_{s3} = I - I_m = 10^5\ \mu\text{A} - 50\ \mu\text{A} = 99\,950\ \mu\text{A}$$
$$R_{s3} = R_m \frac{I_m}{I_{s3}} = 2\ \text{k}\Omega \frac{50\ \mu\text{A}}{99\,950\ \mu\text{A}} \simeq \mathbf{1\ \Omega}$$

Comparison of Voltmeter and Ammeter

Referring to Sec. 6-5, it can be seen that a *voltmeter* conforms to *series-circuit principles*. As the *range increases* (higher full-scale voltage), the *mutiplier resistance increases*. The overall circuit resistance is directly proportional to the range value. In this section, we see that an *ammeter* conforms to *parallel-circuit principles*. A *range increase* is associated with a *decrease in shunt resistance*. The total circuit conductance is directly proportional to the range value.

Current Sources in Parallel

When current sources are connected in parallel, the total source current is the algebraic sum of the individual currents, and the total equivalent internal conductance is the sum of the individual internal conductances.

Figure 7-52*a* shows two current generators, each with its internal conductance, connected in parallel and supplying a load. Each of the sources tends to produce the same direction of current in the load. This connection is called *parallel-aiding*. Figure 7-52*b* shows the equivalent single-source circuit.

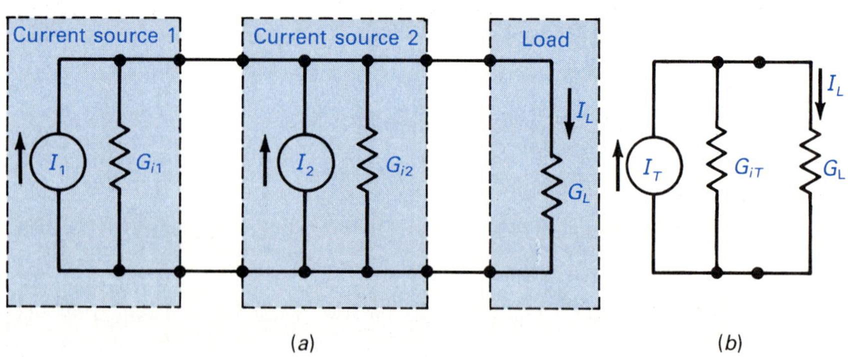

Fig. 7-52. Current sources in parallel-aiding.

$$G_{iT} = G_{i1} + G_{i2}$$
$$I_T = I_1 + I_2$$

A *parallel-opposing* connection, each current source tending to produce load currents of opposite directions, is shown in Fig. 7-53*a*.

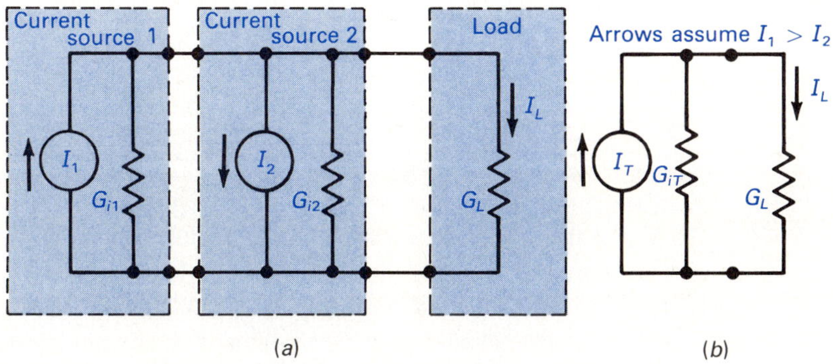

Fig. 7-53. Current sources in parallel-opposing.

$$G_{iT} = G_{i1} + G_{i2}$$

If $I_1 > I_2$ (as implied by the arrows in Fig. 7-53*b*),

$$I_T = I_1 - I_2$$

If $I_1 < I_2$ (the arrows in Fig. 7-53*b* will have directions opposite those shown)

$$I_T = I_2 - I_1$$

EXAMPLE 7-30 An electronic power supply will deliver 10 A if its output terminals are short-circuited. When two of these power supplies are connected in parallel, together they cause a current of 15 A in a 2-Ω load. What is the internal resistance of each power supply?

Solution As each power supply delivers 10 A to a short circuit, it can be regarded as a 10-A current generator. Figure 7-54*a* shows the circuit consisting of two of these generators connected in parallel-aiding; Fig. 7-54*b* is the equivalent single-source circuit.

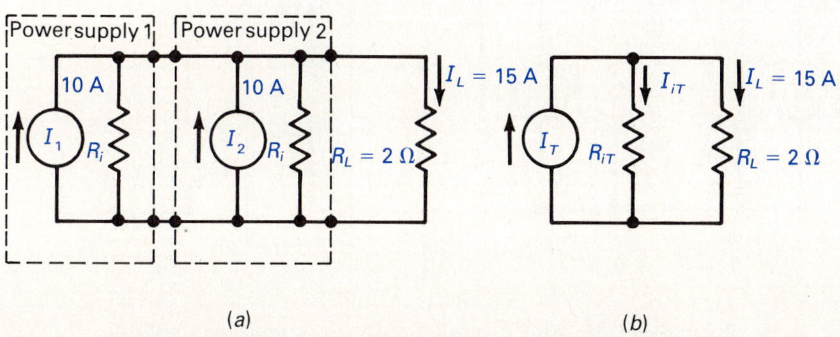

Fig. 7-54. Example 7-30: *(a)* Original circuit; *(b)* equivalent circuit.

$$I_T = I_1 + I_2 = 20 \text{ A}$$
$$I_{iT} = I_T - I_L = 20 \text{ A} - 15 \text{ A} = 5 \text{ A}$$

$$\text{Since } I_{iT} = 5 \text{ A} = \frac{I_L}{3}, \text{ then } R_{iT} = 3R_L$$

$$R_{iT} = 3 \times 2\ \Omega = 6\ \Omega$$

R_{iT} is the two individual internal resistances in parallel, so

$$R_i = 2 \times R_{iT} = 2 \times 6\ \Omega = \mathbf{12\ \Omega}$$

EXAMPLE 7-31 A car will not start because its battery is discharged. The emf is 9.5 V, and the internal resistance is 0.2 Ω. A fully charged battery is jumped across the discharged battery to start the car. The charged battery has an emf of 13.2 V, and its internal resistance is 20 mΩ. If the starter motor draws 300 A at 12 V, *(a)* what starter current is supplied by both batteries together, and *(b)* what is the starter voltage?

Solution The discharged battery with the charged battery boosting it is shown in Fig. 7-55*a*. The current-generator equivalent circuit is given in Fig. 7-55*b*.

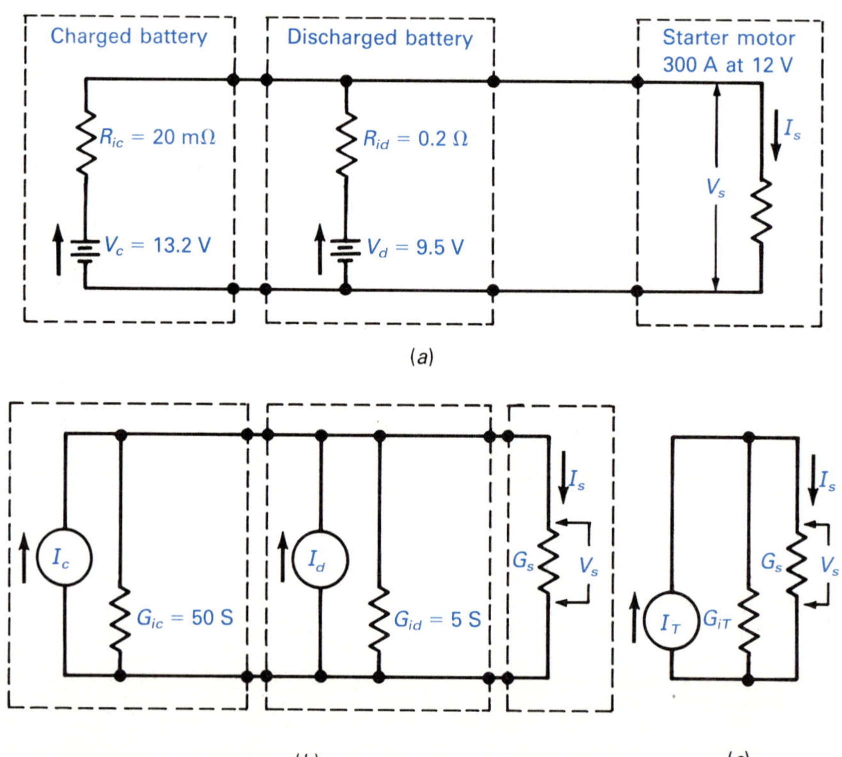

Fig. 7-55. Example 7-31: *(a)* Original circuit; *(b)* circuit with voltage sources converted to current sources; *(c)* equivalent single-source circuit.

Current when charged

$$I_c = \frac{V_c}{R_{ic}} = \frac{13.2\ \text{V}}{20\ \text{m}\Omega} = 660\ \text{A}$$

Current when discharged

$$I_d = \frac{V_d}{R_{id}} = \frac{9.5\ \text{V}}{0.2\ \Omega} = 47.5\ \text{A}$$

Starter conductance

$$G_s = \frac{300\ \text{A}}{12\ \text{V}} = 25\ \text{S}$$

Figure 7-55*c* shows a single-source equivalent circuit of Fig. 7-55*b*.

$$I_T = I_c + I_d = 660\ \text{A} + 47.5\ \text{A} = 707.5\ \text{A}$$

$$G_{iT} = G_{ic} + G_{id} = 50\ \text{S} + 5\ \text{S} = 55\ \text{S}$$

a.

$$I_s = I_T \frac{G_s}{G_{iT} + G_s} \tag{7-19}$$

$$I_s = 707.5\ \text{A} \frac{25\ \text{S}}{55\ \text{S} + 25\ \text{S}} = \mathbf{221.1\ A}$$

b.

$$V_s = \frac{I_s}{G_s} = \frac{221.1\ \text{A}}{25\ \text{S}} = \mathbf{8.844\ V}$$

The way in which jump-starting a car works can be seen by referring to the solution of Example 7-31. The discharged battery, with its 9.5-V emf and 0.2-Ω internal resistance, could only give a starter voltage and current of about 1.6 V and 40 A, not nearly enough to start the car. The connection of the charged battery in parallel with the discharged one raises the starter voltage and current to about 9 V and 200 A, sufficient for the car to start. Once the car has started, its alternator will keep it going with the discharged battery alone.

REVIEW QUESTIONS FOR SECTION 7-4

1. Resistors R_1 and R_2 are connected in parallel. Their total current is I_T. Give a simple formula for computing the current in R_1.
2. State the formula of Question 1 in words.
3. Why does a practical current generator not supply a constant current to a varying resistance load?
4. What is the difference between a current generator and a voltage generator?
5. State the requirement for equivalence between a current generator and a voltage generator.
6. When the load on a generator is increased, is the conductance of the load increased?
7. What is an element called when its function is to permit measuring a current in excess of the full-scale current of a meter movement?

8. For an ammeter to read 10 times the current rating of the basic movement, should the parallel resistor have a resistance greater or smaller than that of the movement?

EXERCISE PROBLEMS FOR SECTION 7-4

1. Two current generators are connected in parallel. $I_{g1} = 50$ mA and $I_{g2} = 25$ mA. Each generator has an internal conductance of 100 μS in parallel with its current source. Find the open-circuit voltage. (*ans*. 375 V)
2. Find the current and power that would be supplied to the following loads by the parallel-connected generators in Prob. 1.
   a. 1000 Ω
   b. 5000 Ω
   c. 10 kΩ
3. The basic movement of a microammeter has a full-scale deflection of 100 μA and a resistance of 850 Ω. Find the value for a shunt resistance which will give the instrument a full-scale deflection of
   a. 10 mA (*ans*. 8.586 Ω)
   b. 50 mA (*ans*. 1.703 Ω)

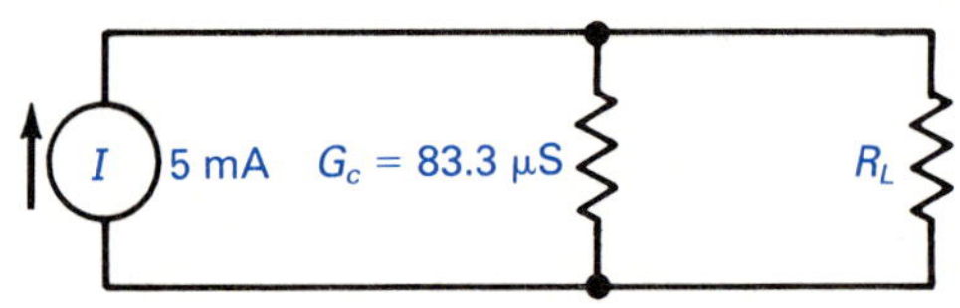

Fig. 7-56. Problems 4 and 5: Circuit.

4. The output equivalent circuit of an amplifier employing a type 2N3905 transistor is as shown in Fig. 7-56. Find the voltage across R_L when R_L is
   a. 600 Ω
   b. 1500 Ω
5. From the equivalent circuit in Fig. 7-56, what must the value of R_L be to obtain an output voltage across R_L of 15 V? (*ans*. 4 kΩ)

7-5 PARALLEL CIRCUITS IN ELECTRONICS

If circuits are resistive (with no inductance or capacitance), dc circuit methods can usually be adapted to follow their operation even if they have varying voltages and currents. This is fortunate, as, in electronics, parallel element groups are often found in situations where the voltages and currents are ac or a combination of ac and dc. In the examples that follow, resistive parallel networks are treated in the dc manner, but the techniques can be used for any form of voltage or current.

Before we look at specific electronic circuits, a principle that often arises in electronics should be introduced.

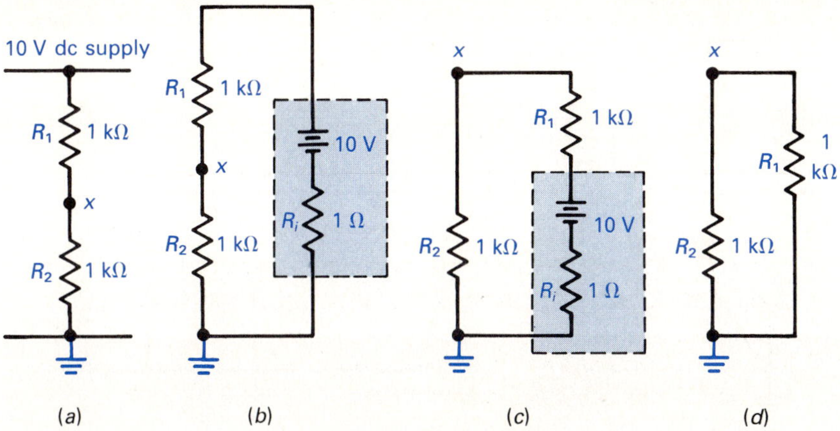

Fig. 7-57. Series element sometimes regarded as parallel.

In Fig. 7-57*a*, two 1-kΩ resistors are connected in series across a dc source. Although it may seem unlikely, the two resistors can be regarded as connected in *parallel, from a resistance viewpoint,* if so desired. This viewpoint does not change the fact that the two resistors have a total series resistance of 2 kΩ. The current drawn from the 10-V supply is the 5 mA you would expect. The reasoning behind this apparent contradiction is as follows:

If this were in a practical electronic circuit, the dc supply would have very low internal resistance. In Fig. 7-57*b* this source is shown as a 10-V battery with 1 Ω internal resistance.

The circuit connections in Fig. 7-57*c* are the same as those in Fig. 7-57*a* and *b*; point X is the connection between R_1 and R_2 in each diagram.

If it happens that, in a certain situation, we are interested only in the effective resistance between point X and ground, the source voltage and the current it produces may be ignored. Also, when the 1-Ω internal resistance R_i is added to the 1-kΩ resistance of R_2, the difference is negligible (1.001 kΩ $\simeq$ 1 kΩ). As they are unimportant to this resistance viewpoint, both the battery voltage and the internal resistance have been omitted in Fig. 7-57*d*. This final diagram shows that the effective resistance between point X and ground in each diagram of Fig. 7-57 is equivalent to the resistance of R_1 and R_2 connected in parallel, 500 Ω.

Another point of explanation before we get to electronic examples: To make this introduction to parallel circuits in electronics easier to follow, some of the network terms and symbols are simplified. For instance, the effective resistance between a transistor's base and emitter is shown as R_{BE}, whereas in a later chapter you will meet it as h_{ie} (the "hybrid common-emitter input parameter"). This again is not contradictory, as either term is acceptable, but at this stage R_{BE} may be more comfortable for you.

Amplifier Input Resistance

The effective input resistance of a transistor amplifier is usually complicated by the presence of the base-emitter bias resistors. Part of the circuit of Fig. 6-72 is given again as Fig. 7-58*a*.

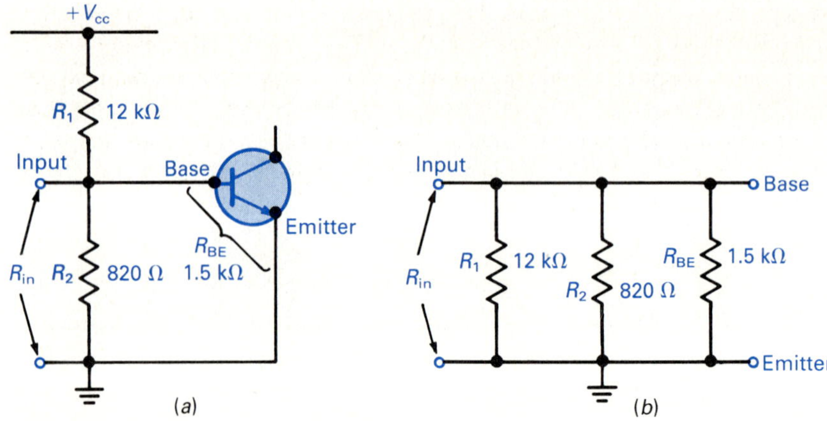

Fig. 7-58. Amplifier input resistance: *(a)* Actual circuit; *(b)* equivalent network.

Using the principle that, in certain circumstances, series elements may be regarded as being in parallel, bias resistors R_1 and R_2 can be considered to be in parallel between the circuit input terminals and ground. As the transistor's base-emitter resistance R_{BE} is also connected between these points, the input resistance consists of R_1, R_2, and R_{BE}, all connected in parallel, as shown in Fig. 7-58*b*.

The input resistance of this amplifier circuit is found as follows:

$$\frac{1}{R_{\text{in}}} = \frac{1}{R_1} + \frac{1}{R_2} + \frac{1}{R_{BE}} = \frac{1}{12\ \text{k}\Omega} + \frac{1}{820\ \Omega} + \frac{1}{1.5\ \text{k}\Omega}$$

$$R_{\text{in}} = 507.7\ \Omega$$

Amplifier Output Resistance

Two transistors are connected in Fig. 7-59*a* in such a way that the collector of transistor Q_1 feeds ac signals to the base of transistor Q_2 through a capacitor.

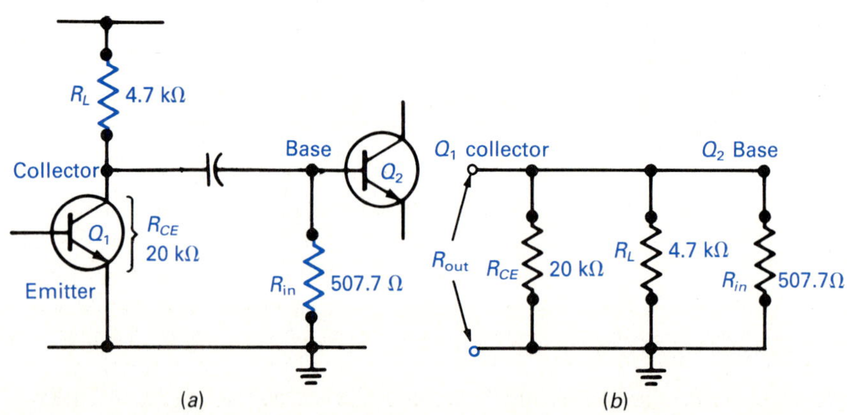

Fig. 7-59. Amplifier output resistance: *(a)* Actual circuit; *(b)* equivalent network.

A capacitor in a circuit such as that in Fig. 7-59 will have very low effective resistance to the signals and so can be ignored. This concept will become clear in later chapters when ac circuits are discussed. With the capacitor omitted, the equivalent circuit of Fig. 7-59*b* is a resistive parallel network that can be treated in the dc manner. This network is connected between the collector and the emitter of transistor Q_2 and so constitutes its effective output load. The value of this load resistance can be found:

$$\frac{1}{R_{\text{out}}} = \frac{1}{R_{CE}} + \frac{1}{R_L} + \frac{1}{R_{\text{in}}} = \frac{1}{20\ \text{k}\Omega} + \frac{1}{4.7\ \text{k}\Omega} + \frac{1}{507.7\ \Omega}$$

$$R_{\text{out}} = 447.9\ \Omega$$

Meter Protection

A meter movement can be given some protection by the characteristics of a diode. A popular meter movement has a full-scale current rating of 50 μA and typical resistance of 2 kΩ. Such a movement has a full-scale voltage of 100 mV and would be damaged by a much higher voltage. An inexpensive way to give the movement some protection if it meets too high a voltage is to connect two germanium diodes across it back to back, as shown in Fig. 7-60.

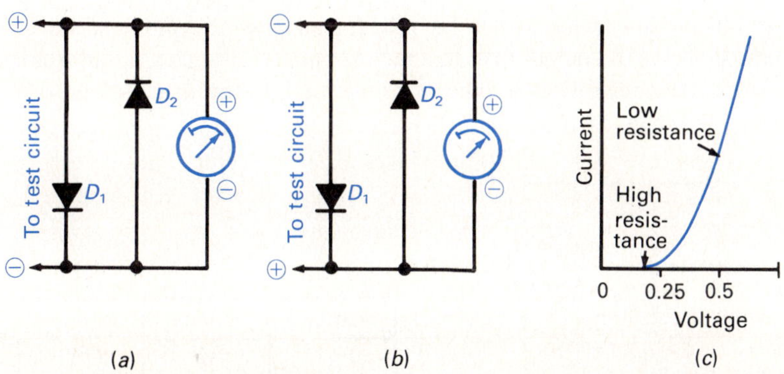

Fig. 7-60. Diode meter movement protection. *(a)* Meter polarity correct: diode D_1 forward biased; diode D_2 reverse biased. *(b)* Meter polarity wrong: diode D_1 reverse biased; diode D_2 forward biased. *(c)* Germanium diode forward biased; high resistance at low voltage and low resistance at higher voltage.

Up to the 100-mV limit for meter-movement safety, the resistance of a forward-connected germanium diode is much higher than the 2-kΩ movement resistance and so will not interfere with a measurement. If the voltage becomes higher, the diode resistance drops, diverting current away from the movement.

If the circuit voltage exceeds about 250 mV, the diode resistance will be very low, so that it will take most of the current. The movement is protected provided that the voltage is never high enough to damage the diode—perhaps around 1 V.

Two diodes are shown in Fig. 7-60. Diode D_1 is connected so that, if the meter polarity is correct for the test being carried out, the diode is forward-biased and protects the movement during normal use. If, however, the movement is incorrectly connected, diode D_2 will give the same protection.

OR Gate

In digital electronics, a gate is a circuit that makes the *logical* decision whether or not its input states are correct to give an output. An OR gate gives an ON output if *any* of its inputs is in the ON state. An OR gate with two inputs *A* and *B* will give an ON output if *A* OR *B* is in the ON state.

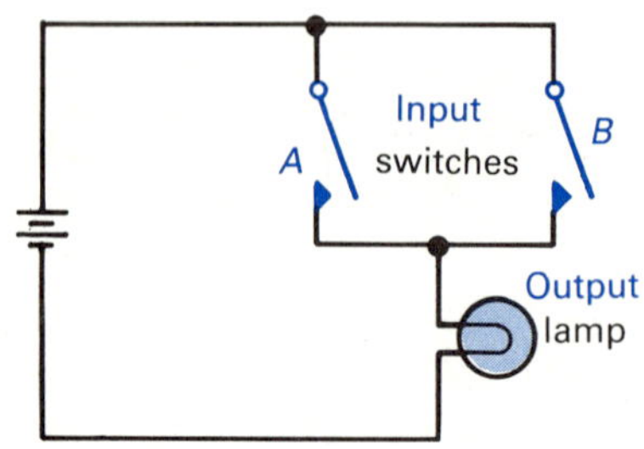

(*a*)

	Input switch *A*	Input switch *B*	Output lamp
1	OFF(open)	OFF(open)	OFF(unlit)
2	OFF(open)	ON(closed)	ON(lit)
3	ON(closed)	OFF(open)	ON(lit)
4	ON(closed)	ON(closed)	ON(lit)

(*b*)

Fig. 7-61. Switch OR gate: *(a)* Circuit; *(b)* truth table.

The simplest form of two-input OR gate is made from switches, as shown in Fig. 7-61*a*. If *either* switch *A* OR switch *B* is closed, the lamp will light. The lamp will be unlit only if *both* switches are open. The truth table of Fig. 7-61*b* summarizes the action of the switch OR gate for all possible combinations of input states. It shows the output OFF (lamp unlit) only when inputs *A* and *B* are both OFF (switches open). For the other possible input states, one ON, one OFF and both ON, the output is ON.

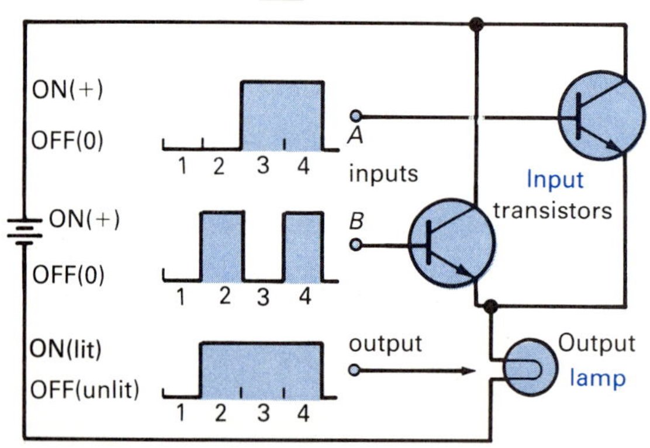

(*a*)

	Input voltage *A*	Input voltage *B*	Output lamp
1	OFF(0)	OFF(0)	OFF(unlit)
2	OFF(0)	ON (+)	ON (lit)
3	ON (+)	OFF (0)	ON (lit)
4	ON (+)	ON (+)	ON (lit)

(*b*)

Fig. 7-62. Electronic OR gate: *(a)* Circuit; *(b)* truth table.

One way of forming an electronic two-input OR gate is the circuit of Fig. 7-62*a*. Here there are two transistors, connected in parallel. The OFF state of either input voltage is 0, in which case the related transistor does not conduct. In the ON state, an input is a positive voltage and its transistor conducts. If either transistor is ON, the output is ON, or the lamp is lit. The truth table of Fig. 7-62*b* shows all possible input combinations and the resulting output states.

Compare Figs. 7-61 and 7-62 with Figs. 6-74 and 6-75 to see the essential differences between OR and AND gates.

7-6 CAPACITORS IN PARALLEL

Parallel capacitive circuits can be handled in much the same manner as parallel resistive circuits. But it must be remembered that capacitance is of primary importance when current or voltage *change* is taking place. For instance, Kirchhoff's current law applies for *instantaneous* current values.

The parallel capacitors in Fig. 7-63 have the common rate of change of voltage dv/dt. At any time instant, Kirchhoff's current law applies,

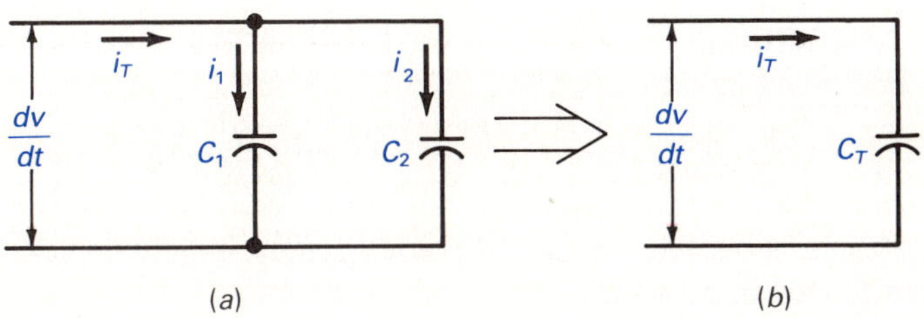

Fig. 7-63. *(a)* Capacitors in parallel; *(b)* single capacitor equivalent.

$$i_T = i_1 + i_2$$

From Eq. (4-1), $i = C\,dv/dt$,

$$C_T \frac{dv}{dt} = C_1 \frac{dv}{dt} + C_2 \frac{dv}{dt}$$

$$C_T = C_1 + C_2$$

For any number of capacitors connected in parallel,

$$C_T = C_1 + C_2 + C_3 + \cdots + C_n \qquad \textbf{(7-26)}$$

The total capacitance of a number of capacitors connected in parallel is the sum of the individual capacitances.

EXAMPLE 7-32 What is the total capacitance of two capacitors of 0.1 μF and 0.22 μF connected in parallel?

Solution (Fig. 7-64)

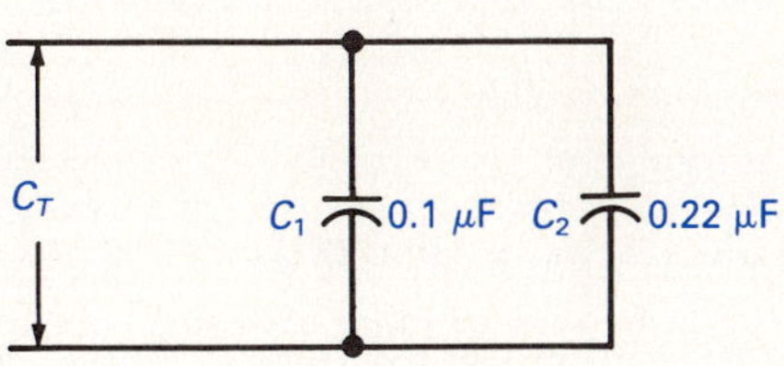

Fig. 7-64. Examples 7-32 and 7-33: Data.

$$C_T = C_1 + C_2 \qquad (7\text{-}26)$$

$$= 0.1\ \mu\text{F} + 0.22\ \mu\text{F}$$

$$C_T = \mathbf{0.32\ \mu F}$$

EXAMPLE 7-33 If 100 V is applied to the circuit of Fig. 7-64, what is the energy stored in each capacitor and in the parallel network?

Solution Applying Eq. (4-4), $W = V^2C/2$,

$$W_1 = \frac{V^2C}{2} = \frac{(100\text{ V})^2 \times 0.1\ \mu\text{F}}{2} = \mathbf{500\ \mu J}$$

$$W_2 = \frac{V^2C}{2} = \frac{(100\text{ V})^2 \times 0.22\ \mu\text{F}}{2} = \mathbf{1.1\ mJ}$$

The total energy stored in the network is the sum of the energies stored in each capacitor.

$$W_T = W_1 + W_2 = 500\ \mu\text{J} + 1.1\text{ mJ} = \mathbf{1.6\ mJ}$$

EXAMPLE 7-34 What is the total capacitance of three capacitors, each 0.015 μF, connected in parallel?

Solution *a.* Refer to Fig. 7-65,

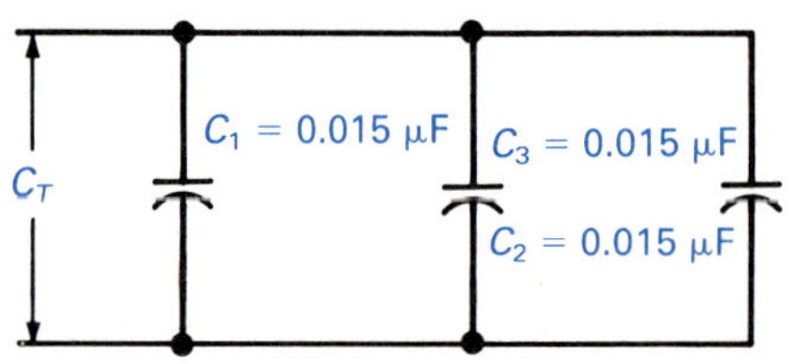

Fig. 7-65. Example 7-34: Data.

$$\begin{aligned} C_T &= C_1 + C_2 + C_3 \qquad (7\text{-}26)\\ &= 0.015\ \mu\text{F} + 0.015\ \mu\text{F} + 0.015\ \mu\text{F}\\ C_T &= 3 \times 0.015\ \mu\text{F} = \mathbf{0.045\ \mu F}\end{aligned}$$

Example 7-34 demonstrates that when several equal-value capacitors are connected in parallel, the total capacitance is the value of any one of them multiplied by the number of capacitors.

EXAMPLE 7-35 What value capacitor should be connected in parallel with an 8-μF filter capacitor to increase the total capacitance to 24 μF?

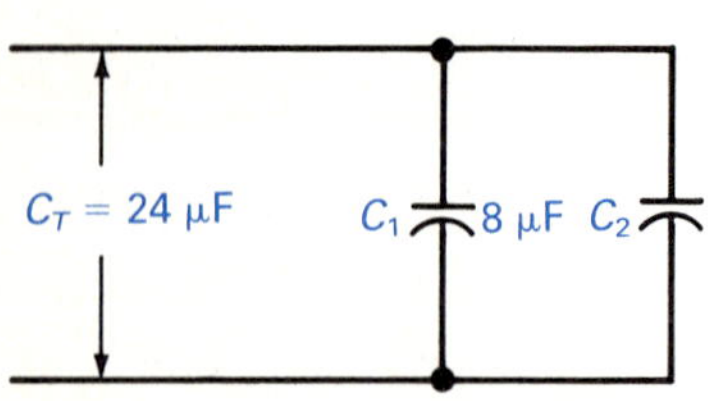

Fig. 7-66. Example 7-35: Data.

Solution (Fig. 7-66)

$$\begin{aligned} C_T &= C_1 + C_2 \qquad (7\text{-}26)\\ C_2 &= C_T - C_1 = 24\ \mu\text{F} - 8\ \mu\text{F} = \mathbf{16\ \mu F}\end{aligned}$$

REVIEW QUESTIONS FOR SECTION 7-6

1. What is the main point to remember when applying resistive-circuit principles to a parallel capacitive circuit?

2. What is the equation for the total capacitance of four parallel-connected capacitors?
3. Express the equation of Question 2 in words.

EXERCISE PROBLEM FOR SECTION 7-6

1. For the network in Fig. 7-67, find
   a. The total capacitance,
   b. The total charge stored,
   c. The total energy stored,
   when
   i. $C_1 = 0.5\ \mu F$, $C_2 = 0.5\ \mu F$, $C_3 = 0.5\ \mu F$
   (*ans*. 1.5 μF, 150 μC, 7.5 mJ)
   ii. $C_1 = 470$ pF, $C_2 = 820$ pF, $C_3 = 220$ pF
   (*ans*. 1510 pF, 0.151 μC, 7.55 μJ)
   iii. $C_1 = 15$ pF, $C_2 = 68$ pF, $C_3 = 150$ pF
   (*ans*. 233 pF, 0.0233 μC, 1.165 μJ)

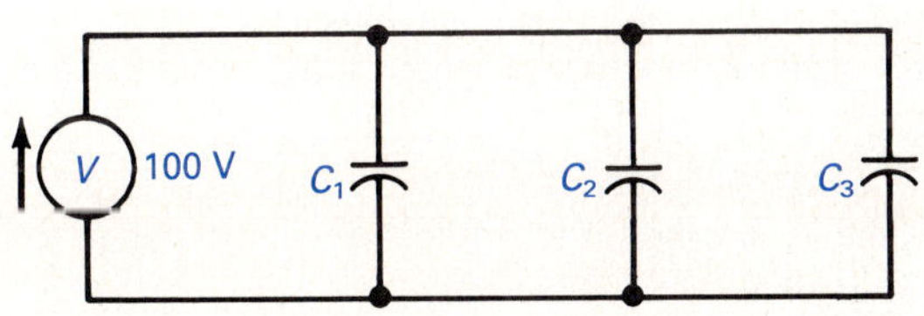

Fig. 7-67. Problem 1: Circuit.

SUMMARY OF EQUATIONS

$I_T = I_1 + I_2 + I_3 + \cdots + I_n$ (7-1)
$G_T = G_1 + G_2 + G_3 + \cdots + G_n$ (7-2)
$1/R_T = 1/R_1 + 1/R_2 + 1/R_3 + \cdots + 1/R_n$ (7-3)
$G_T = nG$ (7-4)
$R_T = R/n$ (7-5)
$P_T = P_1 + P_2 + P_3 + \cdots + P_n$ (7-6)
$W_T = W_1 + W_2 + W_3 + \cdots + W_n$ (7-7)
$I_T = I_1G_T/G_1$ or I_2G_T/G_2 or I_nG_T/G_n (7-8)
$I_T = I_1R_1/R_T$ or I_2R_2/R_T or I_nR_n/R_T (7-9)
$I_1 = I_TG_1/G_T$ or I_2G_1/G_2 or I_nG_1/G_n (7-10)
$I_1 = I_TR_T/R_1$ or I_2R_2/R_1 or I_nR_n/R_1 (7-11)
$G_T = G_1I_T/I_1$ or G_2I_T/I_2 or G_nI_T/I_n (7-12)
$R_T = (R_1 \times R_2)/(R_1 + R_2)$ (7-13)
$G_1 = G_TI_1/I_T$ or G_2I_1/I_2 or G_nI_1/I_n (7-14)
$R_1 = R_TI_T/I_1$ or R_2I_2/I_1 or R_nI_n/I_1 (7-15)
$P_T = P_1R_1/R_T$ or P_2R_2/R_T or P_nR_n/R_T (7-16)
$P_1 = P_TR_T/R_1$ or P_2R_2/R_1 or P_nR_n/R_1 (7-17)
$G_T/G_1 = I_T/I_1 = P_T/P_1 = W_T/W_1 = R_1/R_T$ (7-18)
$I_1 = I_TG_1/G_T$ or $I_TG_1/(G_1 + G_2 + \cdots + G_n)$ (7-19)
$I_1 = I_TR_2/(R_1 + R_2)$ (7-20)

$I = I_i + I_L$ or $I_L = I - I_i$ (7-21)

$I_L = V/(R_i + R_L)$ (7-22)

$I_L = IR_i/(R_i + R_L)$ (7-23)

$V_L = V/(R_iG_L + 1)$ (7-24)

$V_L = I/(G_i + G_L)$ (7-25)

$C_T = C_1 + C_2 + C_3 + \cdots + C_n$ (7-26)

CHAPTER QUESTIONS

7-1. When it is said that parallel and series circuits have opposite characteristics, what is meant?

7-2. Describe a parallel circuit.

7-3. Why might it be said that the common voltage across each element best defines what is meant by a parallel circuit?

7-4. Compare the voltages across a parallel circuit from the viewpoints of its source, load, and other elements.

7-5. If a parallel network is part of a more complex circuit, in what manner would it make sense to refer to its source and its load?

7-6. What is the characteristic common to voltages, powers, and energies in a parallel circuit?

7-7. To what does Kirchhoff's current law refer?

7-8. If it is not correct to add the resistances of a parallel network to obtain its total resistance, what equivalent would be correct?

7-9. What is the relationship between the total resistance of a parallel network and the resistance of any of its elements?

7-10. Why is it often believed that parallel circuits are more difficult to deal with than series circuits?

7-11. Why is the belief expressed in Question 7-10 less reasonable these days?

7-12. How does a *current* divider compare with a *voltage* divider?

7-13. Which is the better way to depict an energy source (e.g., a battery), as a voltage generator or as a current generator?

7-14. How can a parallel resistor bypass current?

7-15. If an *open circuit* is described as meaning *zero current,* in what way would a *short circuit* be similarly described?

7-16. Compare a current generator with a voltage generator from the viewpoints of open-circuit emf and terminal voltage.

7-17. In what way does the relationship for total parallel capacitance differ from that for total parallel resistance?

7-18. If it were desired to increase the capacitance of a circuit, why would it be better to connect a capacitor in parallel with the circuit rather than in series?

CHAPTER PROBLEMS

7-1. What is the total voltage and internal resistance of two parallel-connected 1.5-v, 100 mΩ, flashlight batteries? (*ans.* 1.5 V, 50 mΩ)

7-2. Which ammeter has the greater voltage across it,
a. One based on a 50-μA, 2-kΩ movement?
b. One based on a 20-μA, 2-kΩ movement?

7-3. In Fig. 7-68, what are the currents in each resistor if $R_1 = 560$ kΩ, $R_2 = 1.8$ MΩ, and $I_t = 20$ μA? (*ans*. 15.25 μA, 4.746 μA)

7-4. In Fig. 7-68, what is the voltage across each resistor if the total current is 20 μA? The resistor values are the same as in Prob. 7-3.

7-5. What power is dissipated by the network in Fig. 7-68 if $R_1 = 560$ kΩ, $R_2 = 1.8$ MΩ, and $I_1 = 15.25$ μA? (*ans*. 171.8 μW)

7-6. In Fig. 7-68, what is the resistance of R_1 if $R_2 = 1.8$ MΩ and $R_t = 720$ kΩ?

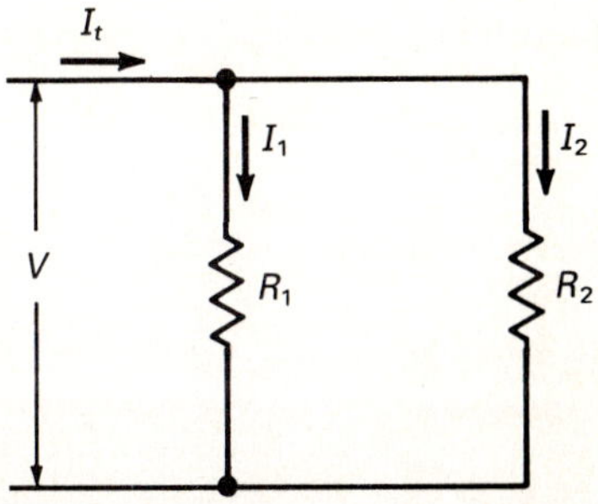

Fig. 7-68. Problems 7-3 to 7-6: Circuit.

7-7. Two ammeters, each with a full-scale range of 1 mA, have different movements. One has a 50-μA movement, the other a 20-μA movement. Both are of 2-kΩ resistance. What shunt resistance values are needed for each ammeter? (*ans*. 105.3 Ω, 40.82 Ω)

7-8. The filter capacitor in a power supply is of 500-μF capacitance. What value capacitor is needed to raise the total capacitance to 1500 μF?

7-9. The circuit of Fig. 7-69 develops a fault when in use. A check shows that the total resistance R_t measures 427 kΩ. What seems to be wrong with the circuit? (*ans*. the 680-kΩ resistor is open-circuit)

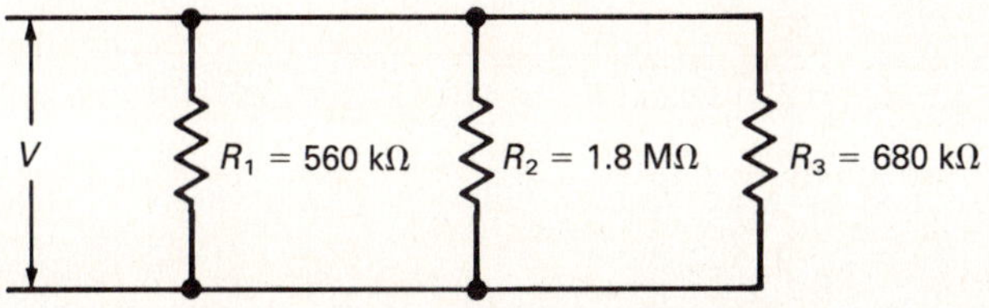

Fig. 7-69. Problems 7-9 and 7-10: Circuit.

7-10. With a circuit such as that in Fig. 7-69, what might be the point of tracing the fault by resistance measurement?

7-11. Figure 7-70 shows the equivalent circuit of a dc electric motor. If the combined voltage across both armature and series field is 440 V and the series and shunt field resistances are 1 mΩ and 200 Ω, respectively, what is the shunt field current? (*ans*. 2.2 A)

7-12. If the motor in Fig. 7-70 has its effective armature resistance very much greater than that of the series field, and the total current taken by the motor is 50 A, what is the approximate armature resistance? The operating conditions are similar to those of Prob. 7-11.

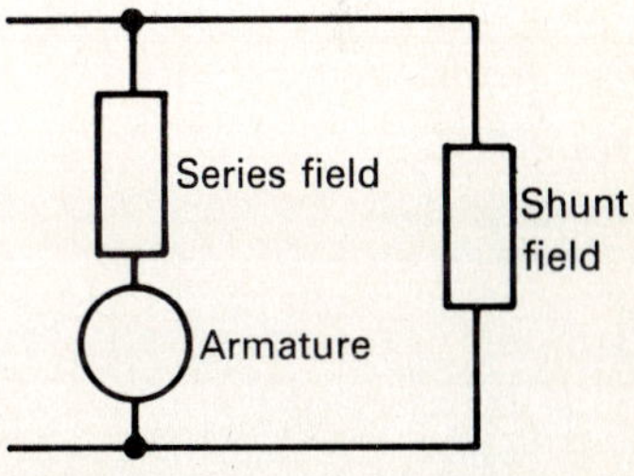

Fig. 7-70. Problems 7-11 and 7-12: Circuit.

7-13. A fault has occurred in the circuit in Fig. 7-71. One resistor is known to be burned out and open-circuit. The current I is 564 mA. Which resistor is burned out? (*ans*. 39 Ω)

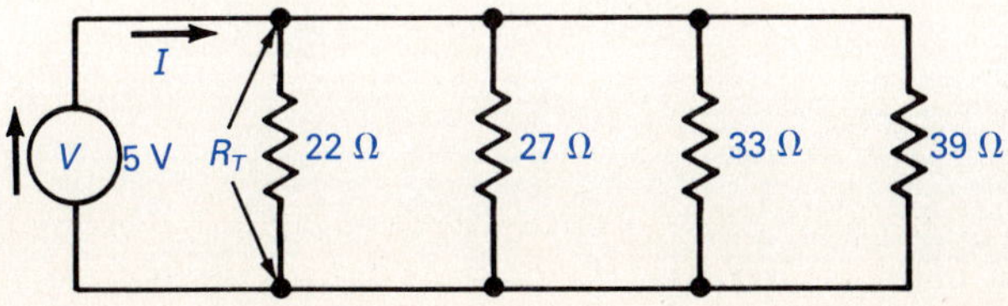

Fig. 7-71. Problem 7-13: Circuit.

7-14. In the circuit in Fig. 7-72, if the resistance between A and B measures 59 Ω, which resistor is open-circuit?

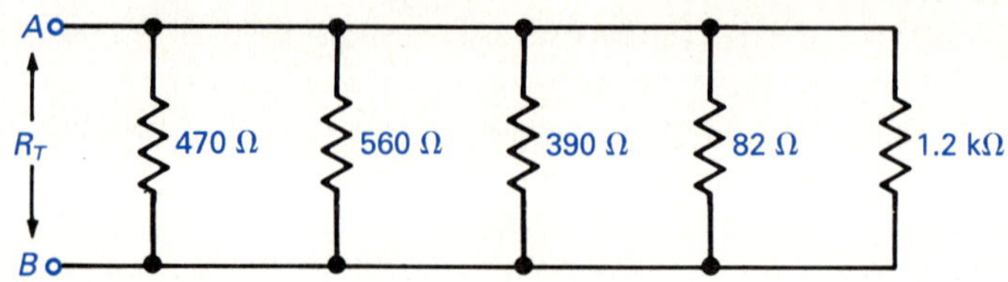

Fig. 7-72. Problem 7-14: Network.

7-15. For the circuit in Fig. 7-73, what is the fault if the voltage across R_1 is 50 V? (*ans.* R_1 open)

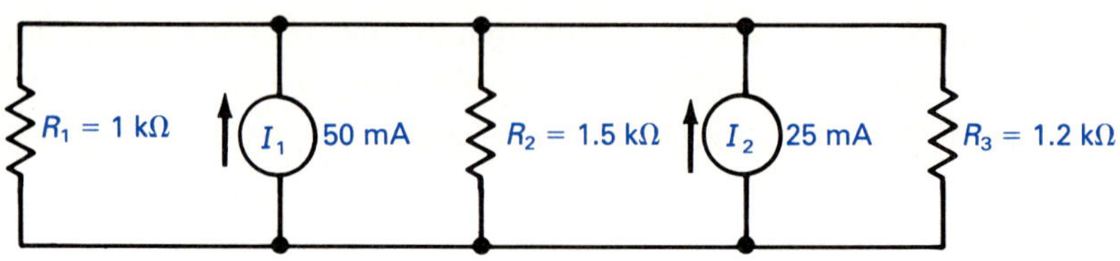

Fig. 7-73. Problems 7-15, 7-16, and 7-18: Circuit.

7-16. For the circuit in Fig. 7-73, what is the fault if the current through R_2 is 13.3 mA?

7-17. In Fig. 7-32, $R_1 = 270\ \Omega$ and $R_2 = 470\ \Omega$. Both are $\frac{1}{2}$-W resistors. If V is increased from zero, which resistor will reach its maximum power rating first and at what voltage will this power rating be reached? (*ans.* R_1, 11.6 V)

7-18. In Fig. 7-73 each resistor has a rating of 1 W. If I_1 is held constant and I_2 is increased, which resistor will be the first to dissipate 1 W and at what value of I_2 will this dissipation occur?

8

Series-Parallel Circuits

Simple series and parallel groups of elements do not usually constitute the whole of a practical circuit. They are often connected together to form *series-parallel* or *combination* circuits.

There are no characteristics of series-parallel circuits which differ basically from those of the series and parallel circuits described in Chaps. 6 and 7. This chapter is therefore primarily concerned with some practical applications.

8-1 WHAT IS A SERIES-PARALLEL CIRCUIT?

There are two basic methods by which series and parallel networks can be combined. They may be handled as follows:

Parallel Groups Connected in Series

In Fig. 8-1, two separate parallel circuits are connected in series with each other.

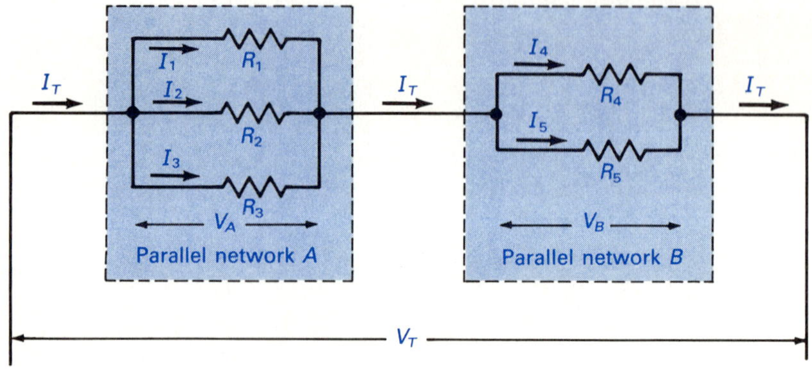

Fig. 8-1. Two parallel circuits connected in series.

Applying the principles developed in Chaps. 6 and 7 to the circuit of Fig. 8-1,

1. The total current I_T in each of the parallel networks is the same ($I_T = I_1 + I_2 + I_3$ and $I_T = I_4 + I_5$).
2. The voltage across each element in either one of the parallel groups is the same (V_A is the voltage across R_1, R_2, and R_3; V_B is the voltage across R_4 and R_5).
3. The total voltage across the series-parallel network is the sum of the voltages across the individual parallel groups ($V_T = V_A + V_B$).

The methods for calculating the quantities involved in this type of circuit depend on the known and the unknown data. They will be essentially the same as those for simple series and parallel circuits.

EXAMPLE 8-1

What is the current in R_3 in Fig. 8-2?

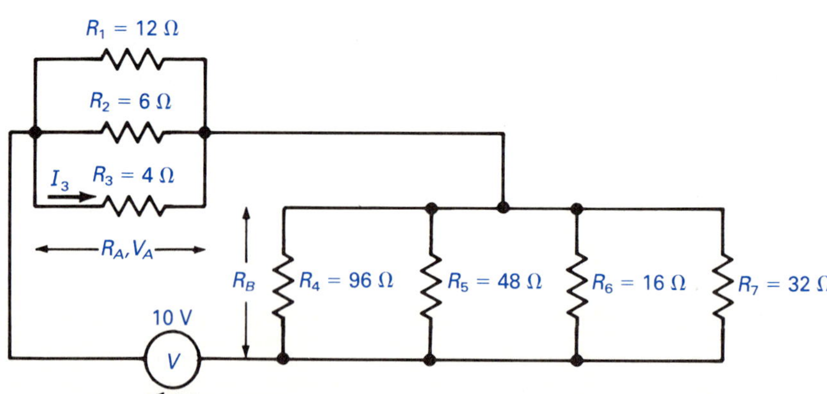

Fig. 8-2. Example 8-1: Circuit.

Solution

$$\frac{1}{R_A} = \frac{1}{R_1} + \frac{1}{R_2} + \frac{1}{R_3} = \frac{1}{12\ \Omega} + \frac{1}{6\ \Omega} + \frac{1}{4\ \Omega} = \frac{1}{2\ \Omega}$$

$$R_A = 2\ \Omega$$

$$\frac{1}{R_B} = \frac{1}{R_4} + \frac{1}{R_5} + \frac{1}{R_6} + \frac{1}{R_7}$$

$$= \frac{1}{96\ \Omega} + \frac{1}{48\ \Omega} + \frac{1}{16\ \Omega} + \frac{1}{32\ \Omega} = \frac{1}{8\ \Omega}$$

$$R_B = 8\ \Omega$$

$$V_A = V\frac{R_A}{R_A + R_B} = 10\ \text{V}\frac{2\ \Omega}{2\ \Omega + 8\ \Omega} = 2\ \text{V}$$

$$I_3 = \frac{V_A}{R_3} = \frac{2\ \text{V}}{4\ \Omega} = \mathbf{0.5\ A}$$

EXAMPLE 8-2 What is the voltage at point A in the circuit in Fig. 8-3?

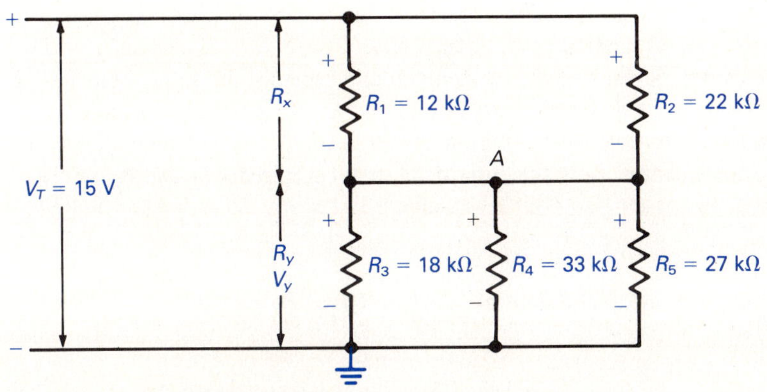

Fig. 8-3. Example 8-2: Network.

Solution R_x is the parallel combination R_1 and R_2, and R_y is the parallel combination R_3, R_4, and R_5.

$$R_x = \frac{R_1R_2}{R_1 + R_2} = \frac{12\ \text{k}\Omega \times 22\ \text{k}\Omega}{12\ \text{k}\Omega + 22\ \text{k}\Omega} = 7.765\ \text{k}\Omega$$

$$\frac{1}{R_y} = \frac{1}{\text{R}_3} + \frac{1}{R_4} + \frac{1}{R_5} = \frac{1}{18\ \text{k}\Omega} + \frac{1}{33\ \text{k}\Omega} + \frac{1}{27\ \text{k}\Omega} = 123\ \mu\text{S}$$

$$R_y = 8.137\ \text{k}\Omega$$

Applying the voltage-divider principle,

$$V_y = V_T\frac{R_y}{R_x + R_y} = 15\ \text{V}\frac{8.137\ \text{k}\Omega}{7.765\ \text{k}\Omega + 8.137\ \text{k}\Omega} = 7.675\ \text{V}$$

The potential at point A (with respect to ground) = **+7.675 V**

Series Groups Connected in Parallel

Figure 8-4 shows two separate series networks (R_1, R_2, and R_3, R_4, R_5) connected in parallel.

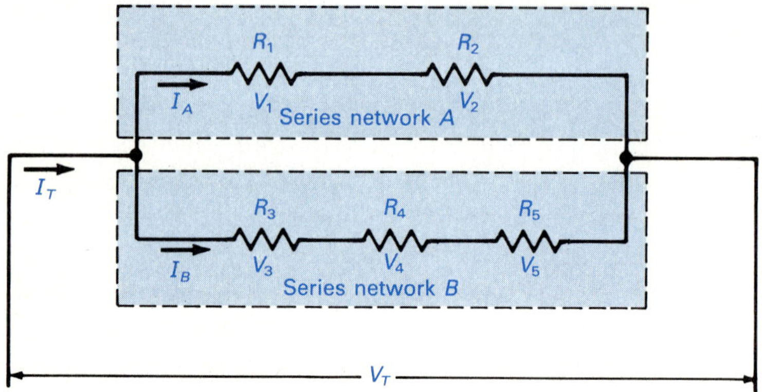

Fig. 8-4. Two series circuits connected in parallel.

The main characteristics of this type of series-parallel circuit are:

1. The total voltage V_T across each of the series networks is the same ($V_T = V_1 + V_2$ and $V_T = V_3 + V_4 + V_5$).
2. The current in each of the elements in either one of the series branches is the same (I_A is the current in both R_1 and R_2; I_B is the current in R_3, R_4, and R_5).
3. The total current I_T is the sum of the branch currents I_A and I_B.

EXAMPLE 8-3 What is the current I_A in the circuit in Fig. 8-5?

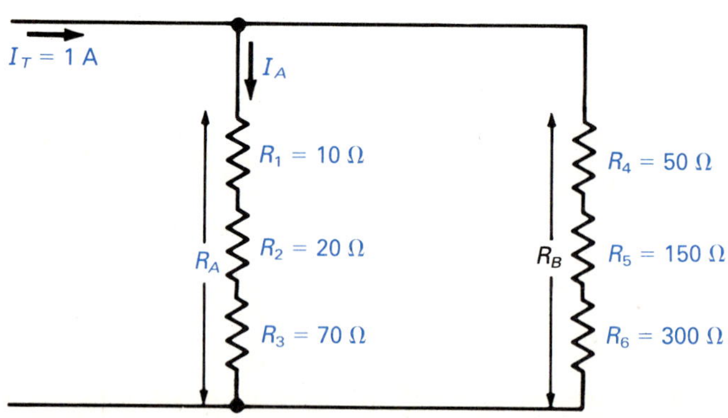

Fig. 8-5. Example 8-3: Network.

Solution

$$R_A = R_1 + R_2 + R_3 = 10\ \Omega + 20\ \Omega + 70\ \Omega = 100\ \Omega$$
$$R_B = R_4 + R_5 + R_6 = 50\ \Omega + 150\ \Omega + 300\ \Omega = 500\ \Omega$$

Applying the current-divider principle,

$$I_A = I_T \frac{R_B}{R_A + R_B} = 1\ \text{A} \frac{500\ \Omega}{100\ \Omega + 500\ \Omega} = \mathbf{0.833\ A}$$

EXAMPLE 8-4 What is the emf of the battery in the circuit in Fig. 8-6? The voltage across the 6.8-MΩ resistance is 5 V.

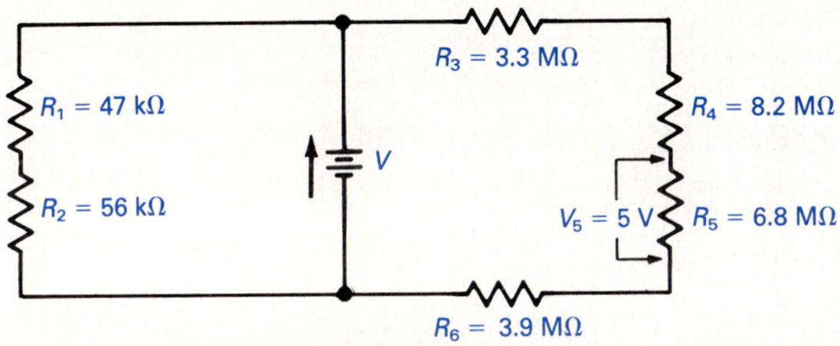

Fig. 8-6. Example 8-4: Given circuit.

Solution The emf of the battery V is the voltage across each of the two branch networks. As there are sufficient data in the branch containing resistances R_3 to R_6, only this branch need be considered, as shown in Fig. 8-7.

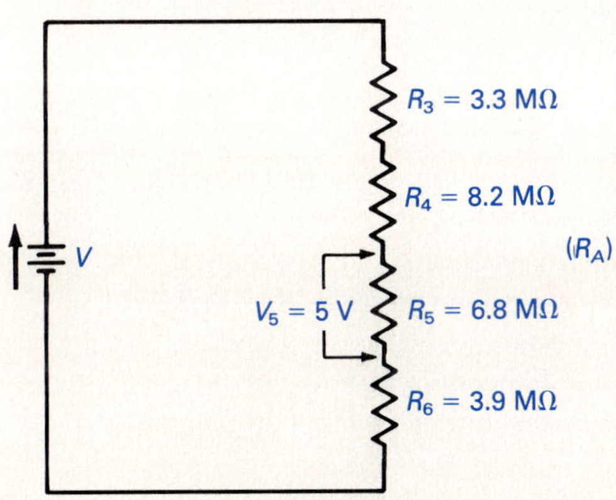

Fig. 8-7. Example 8-4: Significant circuit.

$$R_A = R_3 + R_4 + R_5 + R_6 = 3.3\ \text{M}\Omega + 8.2\ \text{M}\Omega + 6.8\ \text{M}\Omega + 3.9\ \text{M}\Omega$$
$$R_A = 22.2\ \text{M}\Omega$$

Applying ratios,

$$\frac{V}{V_5} = \frac{R_A}{R_5}$$

$$V = V_5 \frac{R_A}{R_5} = 5\ \text{V} \frac{22.2\ \text{M}\Omega}{6.8\ \text{M}\Omega} = \mathbf{16.32\ V}$$

REVIEW QUESTIONS FOR SECTION 8-1

1. What are the basic differences in the characteristics of series-parallel circuits compared with series and parallel circuits?
2. What are the two basic series-parallel-circuit configurations?
3. With parallel element groups connected in series, what is the relationship between the total currents in each of the groups?
4. With parallel element groups connected in series, what is the relationship between the total circuit voltage and the voltages across each element group?
5. With series element groups connected in parallel, what are the relationships between the element groups?

EXERCISE PROBLEMS FOR SECTION 8-1

1. For the circuit in Fig. 8-8, find the total resistance when
   a. $R_1 = 180\ \text{k}\Omega$, $R_2 = 1.2\ \text{M}\Omega$, $R_3 = 330\ \text{k}\Omega$ (*ans.* 438.8 kΩ)
   b. $R_1 = 120\ \Omega$, $R_2 = 390\ \Omega$, $R_3 = 680\ \Omega$ (*ans.* 367.9 Ω)
   c. $G_1 = 2\ \text{mS}$, $G_2 = 0.002\ 13\ \text{S}$, $G_3 = 1.47\ \text{mS}$ (*ans.* 777.8 Ω)
   d. $G_1 = 2.56\ \mu\text{S}$, $G_2 = 4.55\ \mu\text{S}$, $R_3 = 390\ \text{k}\Omega$ (*ans.* 531.2 kΩ)

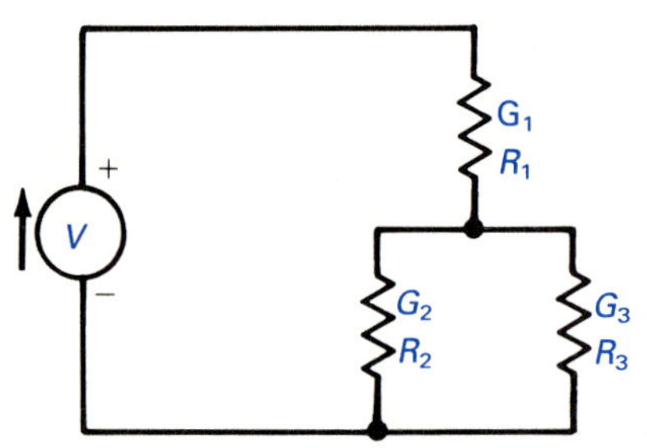

Fig. 8-8. Problems 1 and 2: Circuit.

2. For the circuit in Fig. 8-8, let $R_1 = 47\ \Omega$, $R_2 = 68\ \Omega$, and $R_3 = 27\ \Omega$.
   a. Find the current in R_1 when $V = 6$ V.
   b. Find the current in R_3 when $V = 10$ V.
   c. Find the voltage across R_1 when the current in $R_3 = 0.7$ A.
   d. Find V when the voltage across $R_2 = 21$ V.
   e. Find the current in R_2 when the current in $R_1 = 30$ mA.
   f. Find the current in R_1 when the current in $R_2 = 30$ mA.

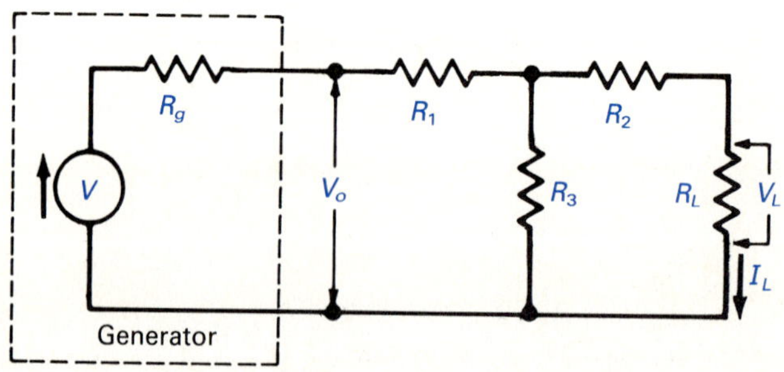

Fig. 8-9. Problem 3: Circuit.

3. For the circuit in Fig. 8-9, find I_L and V_L when

a. $V = 10$ V, $R_g = 600\ \Omega$, $R_1 = 400\ \Omega$, $R_2 = 400\ \Omega$, $R_3 = 250\ \Omega$, $R_L = 600\ \Omega$ (*ans.* 1.667 mA, 1 V)

b. $V_o = 3$ V, $R_g = 50\ \Omega$, $R_1 = 0$, $R_2 = 100\ \Omega$, $R_3 = 30\ \Omega$, $R_L = 50\ \Omega$ (*ans.* 20 mA, 1 V)

c. $V = 16$ V, $R_g = 600\ \Omega$, $R_1 = 200\ \Omega$, $R_2 = 400\ \Omega$, $R_3 = 800\ \Omega$, $R_L = 1200\ \Omega$ (*ans.* 4 mA, 4.8 V)

4. For the circuit in Fig. 8-10, when

a. $V = 10$ V, $R_g = 600\ \Omega$, $R_1 = 900\ \Omega$, $R_2 = 1440\ \Omega$, $R_3 = 900\ \Omega$, $R_L = 600\ \Omega$, find I_L and V_L.

b. $V_o = 6$ V, $R_g = 50\ \Omega$, $R_1 = 230\ \Omega$, $R_2 = 69\ \Omega$, $R_3 = 230\ \Omega$, $R_L = 50\ \Omega$, find I_L and V_L.

c. $V = 20$ V, $R_g = 600\ \Omega$, $R_1 = 360\ \Omega$, $R_2 = 360\ \Omega$, $R_3 = 320\ \Omega$, $R_L = 600\ \Omega$, find V_o.

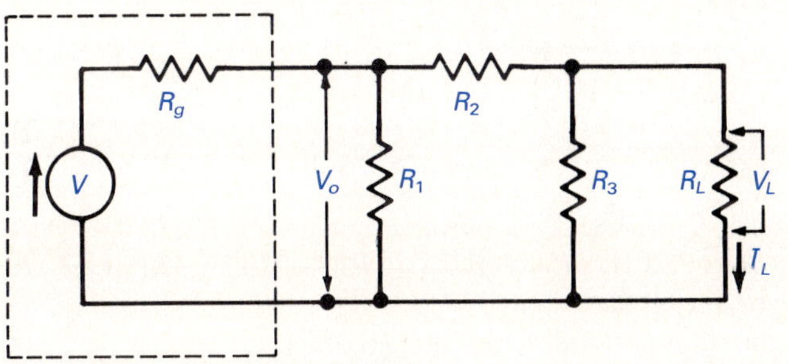

Fig. 8-10. Problem 4: Circuit.

5. For the circuit in Fig. 8-11*a*, when $R_1 = 150\ \Omega$, $R_2 = 220\ \Omega$, $R_3 = 120\ \Omega$, $R_4 = 270\ \Omega$, and $V_{R4} = 16.9$ V, find

a. I_{R1} (*ans.* 121 mA)

b. I_T (*ans.* 203.4 mA)

c. V (*ans.* 35.04 V)

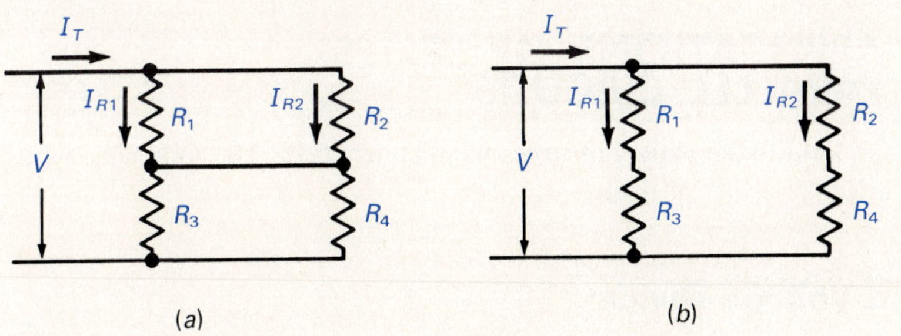

Fig. 8-11. *(a)* Problems 5 and 7; *(b)* Probs. 6 and 8.

6. Repeat Prob. 5 using Fig. 8-11*b*.

7. For the circuit in Fig. 8-11*a*, when $R_1 = 220\ \Omega$, $R_2 = 470\ \Omega$, $R_3 = 560\ \Omega$, $V = 50$ V, and $V_{R4} = 29$ V, find

a. I_{R1} (*ans*. 95.46 mA)
b. I_{R2} (*ans*. 44.69 mA)
c. I_T (*ans*. 140.1 mA)
d. R_4 (*ans*. 328 Ω)

8. Repeat Prob. 7 using Fig. 8-11*b*.

9. For the circuit in Fig. 8-12, find R_T when
a. $R_1 = 12\ \Omega$, $R_2 = 18\ \Omega$, $R_3 = 27\ \Omega$, $R_4 = 18\ \Omega$, $R_5 = 22\ \Omega$, $R_6 = 15\ \Omega$ (*ans*. 27.67 Ω)
b. $R_1 = 680\ \Omega$, $R_2 = 120\ \Omega$, $R_3 = 560\ \Omega$, $R_4 = 470\ \Omega$, $R_5 = 820\ \Omega$, $R_6 = 220\ \Omega$ (*ans*. 935.1 Ω)

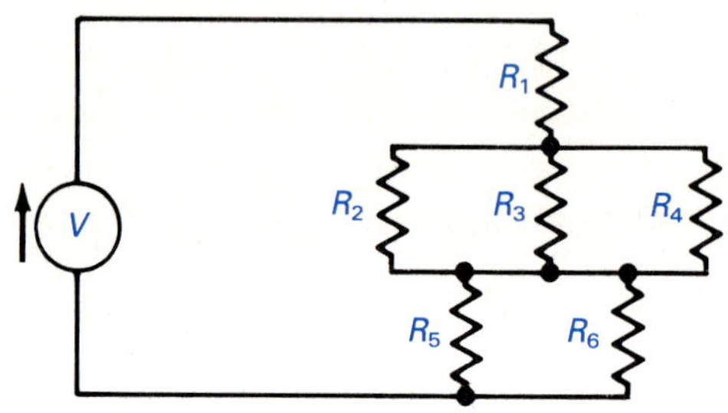

Fig. 8-12. Problems 9 to 12.

10. Find the current in R_3 in Fig. 8-12 when
a. $V = 25$ V, $R_1 - 82\ \Omega$, $R_2 = 180\ \Omega$, $R_3 - 220\ \Omega$, $R_4 = 68\ \Omega$, $R_5 = 82\ \Omega$, $R_6 = 68\ \Omega$
b. $V = 50$ V and all resistance values remain as in *a*

11. Find the voltage across R_5 in Fig. 8-12 when $V = 220$ V, $R_1 = 120\ \Omega$, $R_2 = 150\ \Omega$, $R_3 = 330\ \Omega$, $R_4 = 180\ \Omega$, $R_5 = 82\ \Omega$, $R_6 = 120\ \Omega$. (*ans*. 45.74 V)

12. In Prob. 11, one resistor is open-circuit. If the voltage measured across R_5 is 86.5 V, which resistor is open?

8-2 APPLICATIONS OF SERIES-PARALLEL CIRCUITS

There is no limit to the variety of series-parallel networks. The following applications illustrate a few of them.

Loaded Voltage Divider

A voltage divider can be used to supply one or more loads with voltages that are less than the output voltage of an energy source. The series-circuit voltage divider, described in Chap. 6, is of limited use because the currents the loads can draw must be very small. Where substantial load currents are required, a voltage divider must be a series-parallel circuit.

A loaded voltage divider with one load is shown in Fig. 8-13.

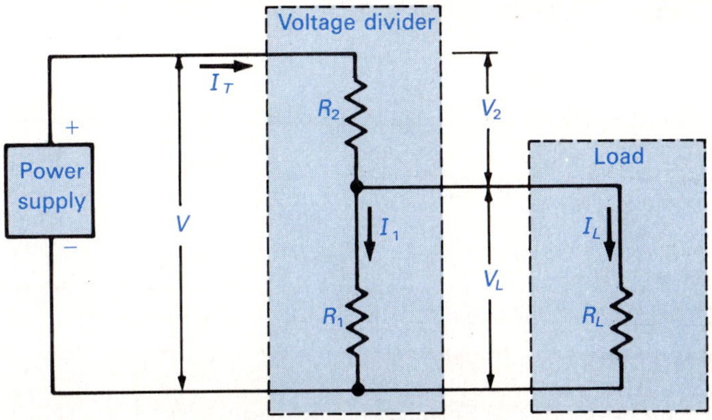

Fig. 8-13. Loaded voltage divider.

The *power supply* in Fig. 8-13 is the source of energy for supplying the load via the *voltage-divider chain* (R_1 and R_2). The aim in designing the voltage divider is to reduce the supply voltage V to the value required by the load V_L, while maintaining an acceptable level of voltage regulation.

The voltage divider originated as a *bleeder* resistor, connected across a capacitor to dissipate its energy and thus make it safe to handle when not in use. It was said to bleed current out of the capacitor. The bleeder resistor was often tapped to obtain reduced voltages for various circuit needs. The term *bleeder current* now refers to the lowest current in any of the voltage-divider elements. The choice of bleeder current is a compromise; if it is too low, small variations in load current will seriously affect load voltage (poor regulation); if it is too high, much power will be dissipated by the divider (an expense and heating problem). The bleeder current is often chosen arbitrarily as 10 to 20 percent of the total load current. After the bleeder current I_1 has been determined, the design can proceed.

$$I_T = I_1 + I_L$$

The voltage across R_1 is the load voltage V_L. So the voltage across R_2 will be the difference between the power-supply voltage and the load voltage:

$$V_2 = V - V_L$$

$$R_1 = \frac{V_L}{I_1}$$

$$R_2 = \frac{V_2}{I_T}$$

If the power ratings for R_1 and R_2 are required, they can be calculated by any of the usual methods (the product of voltage and current, for example).

For voltage dividers supplying more than one load, the technique of design is essentially the same.

EXAMPLE 8-5 A 12-V battery is to be used to supply three loads with 6 V at 0.75 A, 9 V at 0.5 A, and 12 V at 1 A. Design a suitable voltage divider.

Solution In Fig. 8-14 load R_{L3} is connected directly across the power supply. Its current I_{L3} is not supplied by the voltage divider and, therefore, is not included in the bleeder-current calculation.

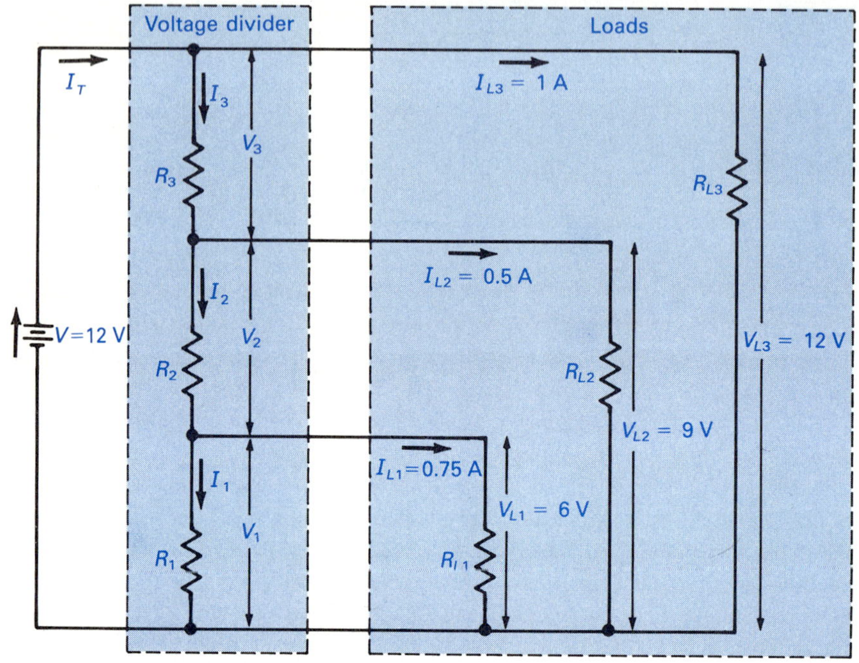

Fig. 8-14. Example 8-5: Circuit.

The total load current supplied by the divider is

$$I_{DL} = I_{L1} + I_{L2} = 0.75 \text{ A} + 0.5 \text{ A} = 1.25 \text{ A}$$

Selecting a bleeder current which is 20 percent of the total divider load current.

$$I_1 = 20\% \; I_{DL} = 0.2 \times 1.25 \text{ A} = 0.25 \text{ A}$$
$$I_2 = I_1 + I_{L1} = 0.25 \text{ A} + 0.75 \text{ A} = 1 \text{ A}$$
$$I_3 = I_2 + I_{L2} = 1 \text{ A} + 0.5 \text{ A} = 1.5 \text{ A}$$

Note particularly in Fig. 8-14 that the voltage across R_1 is the load voltage V_{L1}, while the voltages across R_2 and R_3 are *not* the load voltages V_{L2} and V_{L3}. R_2 and R_3 are *not* connected directly across the loads.

$$V_1 = V_{L1} = 6 \text{ V}$$
$$V_2 = V_{L2} - V_{L1} = 9 \text{ V} - 6 \text{ V} = 3 \text{ V}$$
$$V_3 = V_{L3} - V_{L2} \text{ (or } V - V_{L2}) = 12 \text{ V} - 9 \text{ V} = 3 \text{ V}$$
$$R_1 = \frac{V_1}{I_1} = \frac{6 \text{ V}}{0.25 \text{ A}} = \mathbf{24\ \Omega}$$
$$R_2 = \frac{V_2}{I_2} = \frac{3 \text{ V}}{1 \text{ A}} = \mathbf{3\ \Omega}$$

$$R_3 = \frac{V_3}{I_3} = \frac{3\ \text{V}}{1.5\ \text{A}} = \mathbf{2\ \Omega}$$

The power dissipations for R_1, R_2, and R_3 are

$$P_1 = V_1 I_1 = 6\ \text{V} \times 0.25\ \text{A} = \mathbf{1.5\ W}$$
$$P_2 = V_2 I_2 = 3\ \text{V} \times 1\ \text{A} = \mathbf{3\ W}$$
$$P_3 = V_3 I_3 = 3\ \text{V} \times 1.5\ \text{A} = \mathbf{4.5\ W}$$

EXAMPLE 8-6 For operation, a solid-state amplifier needs +25 V at 1 A and −25 V at 2 A. What resistance values are needed for a voltage divider to obtain these outputs from a 60-V power supply?

Solution When both positive and negative load voltages are to be supplied from a voltage divider, it is important that the principles of relative potential be remembered. (See Chap. 6, Fig. 6-31 and Example 6-19 being relevant.) Note that in this example there is a total of 50 V between the +25-V and the −25-V outputs, and that the voltage across R_3 in Fig. 8-15 is the difference between this total and the power-supply emf (60 V − 50 V = 10 V).

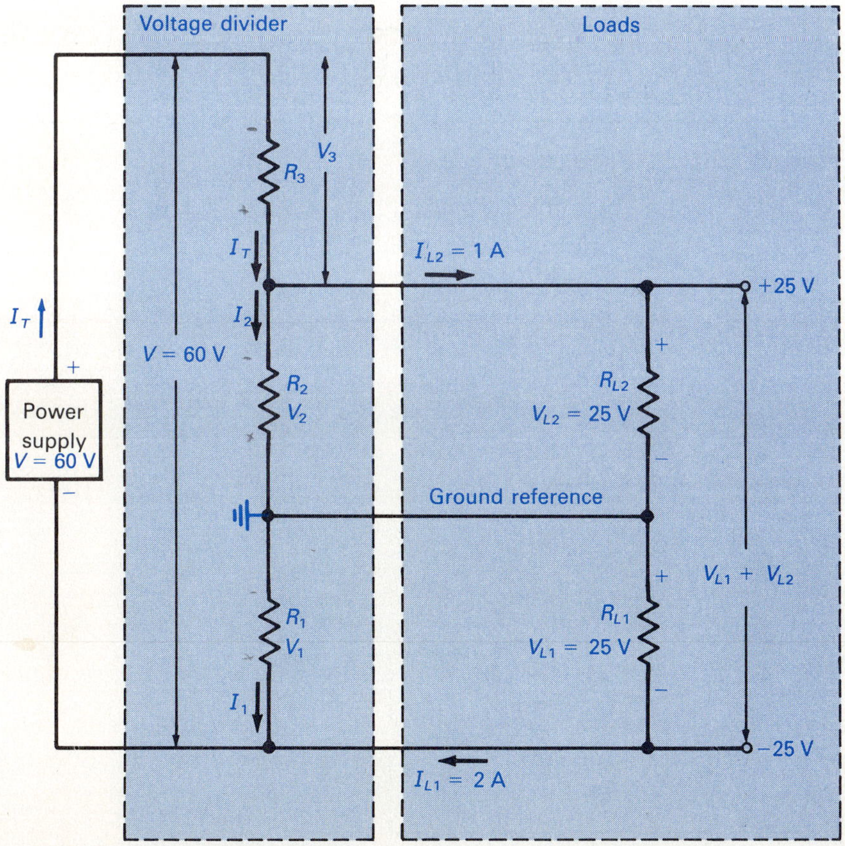

Fig. 8-15. Example 8-6: Circuit.

The current directions should also be noted. For instance, the direction of I_{L2}, on the *positive side, is out of the voltage-divider chain,* while that of I_{L1}, on the *negative side, is into the chain.*

Again, because of the existence of positive and negative load voltages in this example, it is necessary to determine on which side of the ground reference, positive or negative, the greater total load current occurs. This value is then used to calculate the bleeder current. The *bleeder resistor* is *connected to the reference on the side of maximum load current.*

The total load current on the negative side is $I_{L1} = 2$ A.

The total load current on the positive side is $I_{L2} = 1$ A.

The total load current on the negative side is the greater, and so R_1 and I_1 are the bleeder resistor and current.

If we let the bleeder current be 10 percent of the total negative load current I_{L1},

$$I_1 = 10\%\ I_{L1} = 0.1 \times 2\text{ A} = 0.2\text{ A}$$

$$I_T = I_1 + I_{L1} = 0.2\text{ A} + 2\text{ A} = 2.2\text{ A}$$

$$I_2 = I_T - I_{L2} = 2.2\text{ A} - 1\text{ A} = 1.2\text{ A}$$

(note that the current in R_3 is I_T)

$$V_1 = V_{L1} = 25\text{ V}$$

$$V_2 = V_{L2} = 25\text{ V}$$

$$V_3 = V - (V_{L1} + V_{L2}) = 60\text{ V} - (25\text{ V} + 25\text{ V}) = 10\text{ V}$$

$$R_1 = \frac{V_1}{I_1} = \frac{25\text{ V}}{0.2\text{ A}} = \mathbf{125\ \Omega}$$

$$R_2 = \frac{V_2}{I_2} = \frac{25\text{ V}}{1.2\text{ A}} = \mathbf{20.83\ \Omega}$$

$$R_3 = \frac{V_3}{I_T} = \frac{10\text{ V}}{2.2\text{ A}} = \mathbf{4.545\ \Omega}$$

EXAMPLE 8-7 In the circuit in Fig. 8-16, the load represents a transistor switching circuit which draws 1 A at 9 V when it is on and zero current when it is off. What is the load voltage regulation?

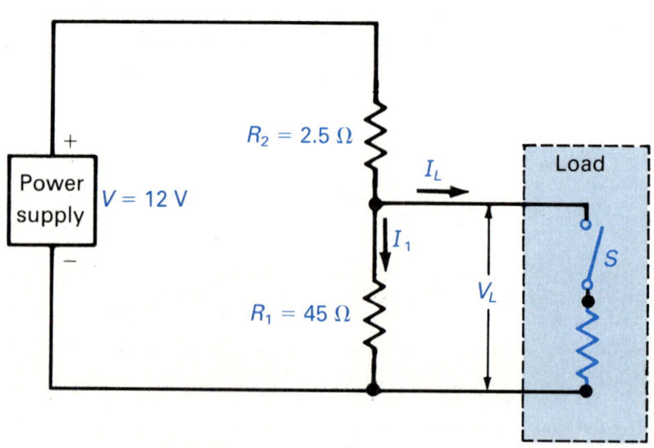

Fig. 8-16. Example 8-7: Circuit.

Solution In Fig. 8-16, the switch S simulates the ON-OFF conditions of the load.

When the switch is closed (load ON),

$$\text{ON} \qquad V_L = 9 \text{ V}$$

When the switch is open (load OFF), the network across the power supply consists only of R_1 and R_2 in series.

$$\text{OFF} \qquad V_L = V\frac{R_1}{R_1 + R_2} = 12 \text{ V}\frac{45\ \Omega}{45\ \Omega + 2.5\ \Omega} = 11.37 \text{ V}$$

The ON load voltage is 9 V. It rises to 11.37 V when the load is OFF. In this application, the ON voltage would be regarded as the desired voltage (required to be constant). A significant voltage rise during the OFF situation might be a cause for concern; it could damage the transistors.

The voltage regulation is

$$\text{Percent regulation} = \frac{\text{OFF voltage} - \text{ON voltage}}{\text{ON voltage}} \times 100$$

$$= \frac{11.37 \text{ V} - 9 \text{ V}}{9 \text{ V}} \times 100$$

Voltage regulation = **26.33 percent**

Three-Wire Power in the Home

The electrical wiring in most homes is carried out in a three-wire configuration. While it is true that these domestic installations use alternating current, the basic principle underlying three-wire distribution can be seen by regarding it as a series-parallel dc circuit.

Two virtually separate 120-V supplies are available to operate the lights and most appliances. Although this depends on the home equipment, typically the water heater, furnace, clothes dryer, and some of the stove elements operate at 240 V; this voltage is derived from the two 120-V supplies in series.

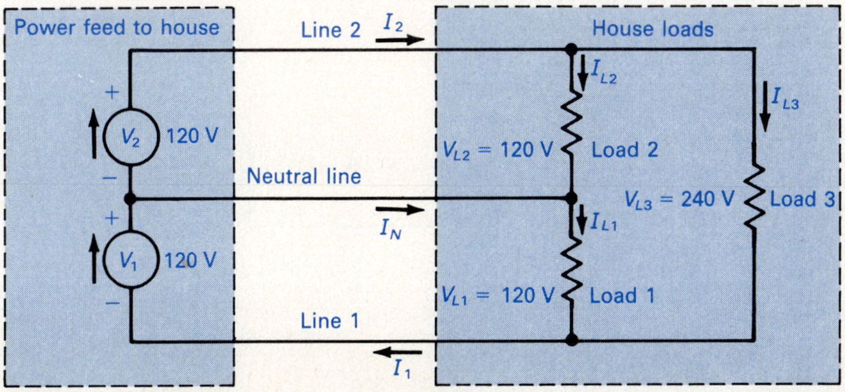

Fig. 8-17. Three-wire power distribution system.

Figure 8-17 shows the equivalent circuit of a three-wire system. The two 120-V generators, connected in series, simulate the power feed to the home. The three wires, or *lines,* provide three separate supplies, two of 120 V and one of 240 V. One of the wires is usually called the *neutral* line because if the two 120-V loads (loads 1 and 2) are the same, the current I_N will be zero (*neutral*). The main relationships are

$$V_{L1} = V_{L2} = 120 \text{ V}$$
$$V_{L3} = 240 \text{ V}$$
$$I_1 = I_{L1} + I_{L3}$$
$$I_2 = I_{L2} + I_{L3}$$

If $I_{L1} > I_{L2}$, $I_N = I_{L1} - I_{L2}$ with the direction shown in Fig. 8-17.
If $I_{L2} > I_{L1}$, $I_N = I_{L2} - I_{L1}$ with its direction opposite that shown.
If $I_{L1} = I_{L2}$, I_N will be zero. This is called a *balanced* condition.

EXAMPLE 8-8 The electrical equipment in a house puts the following load on a three-wire supply.

Line 1 (120 V): Six light bulbs, one refrigerator, one radio, and one stove element; total power $P_{L1} = 2.99$ kW.
Line 2 (120 V): Four light bulbs, one washing machine, one television set, and one stove element; total power $P_{L2} = 2.473$ kW.
Lines 1 and 2 (240 V): One clothes dryer and one stove oven; total power $P_{L3} = 5$ kW.

What is the current in each of the three lines?

Solution

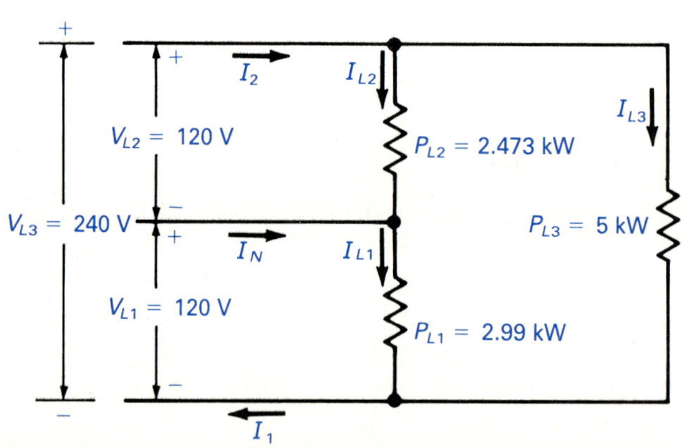

Fig. 8-18. Example 8-8: Network.

$$I_{L1} = \frac{P_{L1}}{V_{L1}} = \frac{2990 \text{ W}}{120 \text{ V}} = 24.92 \text{ A}$$

$$I_{L2} = \frac{P_{L2}}{V_{L2}} = \frac{2473 \text{ W}}{120 \text{ V}} = 20.61 \text{ A}$$

$$I_{L3} = \frac{P_{L3}}{V_{L3}} = \frac{5000 \text{ W}}{240 \text{ V}} = 20.83 \text{ A}$$

$$I_1 = I_{L1} + I_{L3} = 24.92 \text{ A} + 20.83 \text{ A} = \mathbf{45.75\ A}$$

$$I_2 = I_{L2} + I_{L3} = 20.61 \text{ A} + 20.83 \text{ A} = \mathbf{41.44\ A}$$

$$I_N = I_{L1} - I_{L2} = 24.92 \text{ A} - 20.61 \text{ A} = \mathbf{4.31\ A}$$

As I_{L1} is greater than I_{L2}, the direction of the netural line current I_N will be from left to right, as indicated in Fig. 8-18.

Ammeter and Voltmeter Connections

When current and voltage measurements are made, errors may occur if an ammeter and a voltmeter are both used at the same time, even if both are perfectly accurate. The series connection of the ammeter and the parallel voltmeter connection give a series-parallel measurement circuit. Either the ammeter measures both the test current and the current taken by the voltmeter, or the voltmeter measures both the test voltage and the voltage across the ammeter.

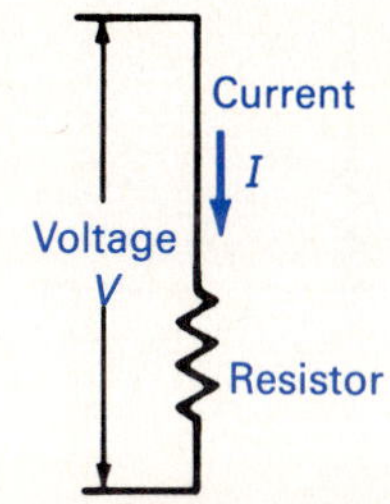

Fig. 8-19. Measuring current and voltage at the same time. Resistor current and voltage to be measured.

In Fig. 8-19, the current I and the voltage V are to be measured at the same time.

One way of connecting the ammeter and the voltmeter is shown in Fig. 8-20*a*. In this series-parallel circuit, the ammeter current includes the current taken by the voltmeter. If the voltmeter current is not small compared with the resistor's current, an important error will result. To reduce this effect, the resistance of the voltmeter must be very much higher than that of the resistor.

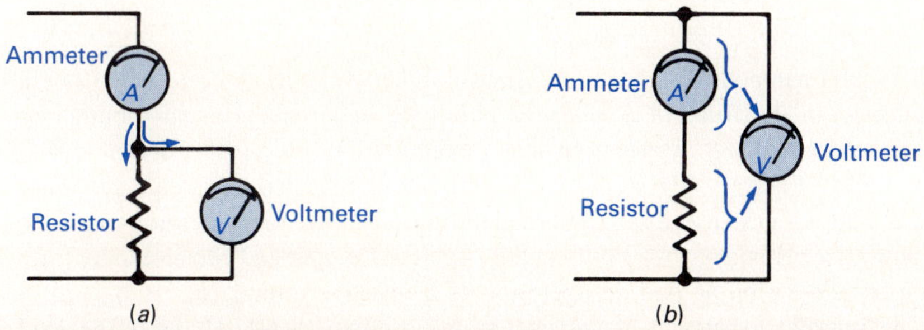

Fig. 8-20. Measuring current and voltage at the same time. *(a)* Ammeter measures resistor current and voltmeter current; *(b)* voltmeter measures resistor voltage and ammeter voltage.

Alternatively, the meters might be connected as in Fig. 8-20*b*. The voltmeter in this series-parallel circuit measures the voltage across the ammeter as well as the resistor's voltage. If the ammeter resistance is not very much lower than that of the resistor, the voltmeter will not show the resistor's voltage correctly.

An *ammeter measures the current through it;* a *voltmeter measures the voltage across it*. The most that can be expected from a meter is that it accurately shows what it measures. The use of an ammeter and a voltmeter at the same time involves one of the two series-parallel circuits shown in Fig. 8-20*a* and *b*. The accuracy of the results depends on the circuit values.

Even connecting one meter may upset the circuit conditions enough to interfere with measurements. This problem was referred to in Sec. 6-5 (Ammeter Loading) and is raised again later in this section (Voltmeter Loading).

T and π Networks

Certain configurations of circuit elements appear often in practical circuits. You can expect to find series, parallel, and various series-parallel groupings in equipment schematic diagrams. Two series-parallel formats, T and π, are used for such purposes as *attenuation,* the reduction of signal level.

An attenuator circuit that may be found in broadcasting equipment for controlling audio (sound) signal level is shown in Fig. 8-21.

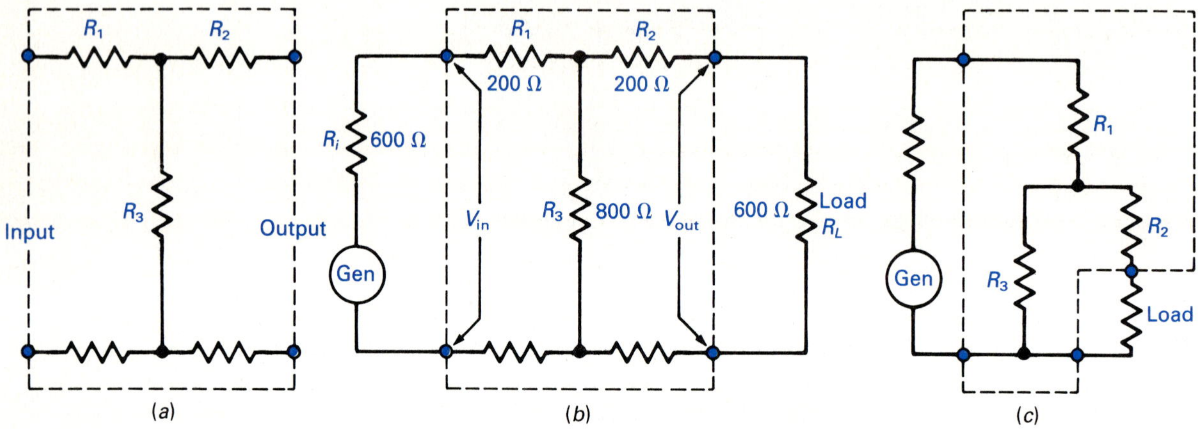

Fig. 8-21. T network as an attenuator. *(a)* T network; *(b)* generator input to the T network; T network output to a load; *(c)* Fig. 8-21*b* redrawn to show its series-parallel form.

The attenuator circuit in Fig. 8-21*a* is a T network, its three resistors being connected in the form of a letter T. This type of circuit is usually built into equipment to connect two sections that are required to operate at different levels, or volumes. Figure 8-21*b* shows the T attenuator connected between a generator and a load. In practice, as in the broadcasting example, the generator would be an ac source. However, we can treat the circuit as dc-operated because there is no difference between ac and dc operation of a resistive circuit.

The internal resistance of the generator is shown as 600 Ω, as is the load resistance. The circuit would normally appear as in Fig. 8-21*b*, but it is more easily seen to be of series-parallel type when it is redrawn as in Fig. 8-21*c*.

Using series-parallel-circuit treatment, you may see that the attenuator output voltage v_{out} is one-half its input voltage v_{in}.

The load and R_2 are connected in series. If their combined resistance is R_s,

$$R_s = R_L + R_2 = (600 + 200)\ \Omega = 800\ \Omega$$

This series combination is connected in parallel with R_3. As each is 800 Ω, their combined resistance R_p is

$$R_p = \frac{R_3}{2} \quad \text{or} \quad \frac{R_s}{2} = \frac{800\ \Omega}{2} = 400\ \Omega$$

The resistor R_1 and the R_p group are series-connected across the input voltage v_{in}. Using the voltage-divider principle, the voltage across R_p (across R_3 or across R_2 and R_L in series) is

$$v_p = v_{in}\frac{R_p}{R_1 + R_p} = v_{in}\frac{400\ \Omega}{(200 + 400)\ \Omega} = \frac{2}{3}v_{in}$$

Similarly, the output voltage is

$$v_{out} = v_p\frac{R_L}{R_s} = \frac{2}{3}v_{in} \times \frac{600\ \Omega}{800\ \Omega} = \frac{1}{2}v_{in}$$

This level reduction is attenuation.

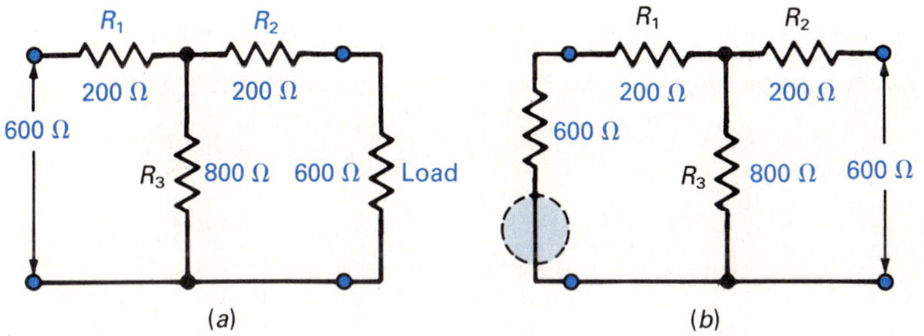

Fig. 8-22. Matching: 600-Ω input and output (same as generator and load). *(a)* Resistance at input terminals = 600 Ω; *(b)* resistance at output terminals = 600 Ω.

Another aspect of Fig. 8-21 and some other often-used circuits is *matching*, or maintaining a certain circuit resistance at different parts of the circuit. Figure 8-22 looks at the attenuator circuit of Fig. 8-21 from this viewpoint. With the load connected and the generator disconnected, as in Fig. 8-22*a*, the resistance between the input terminals is 600 Ω. With the generator internal resistance connected across the input terminals (its voltage being switched off) and the load disconnected, as in Fig. 8-22*b*, the resistance between the output terminals is also 600 Ω. This design feature is deliberate. The 600-Ω generator is connected to the 600-Ω attenuator input. The 600-Ω attenuator output is connected to the 600-Ω load. This 600-Ω matching is common in professional audio equipment. Other values, such as 50 Ω, 75 Ω, and 300 Ω, are used for other applications.

The amount of attenuation may be varied to give a range of output voltages, while still maintaining the matching. If variable resistors are used for each of the three elements and they are operated from the same shaft, they will change value together. To increase the attentuation (decrease the output voltage), the two equal-value resistors, R_1 and R_2 in Figs. 8-21 and 8-22, must increase in value while the resistance of the third, R_3, decreases.

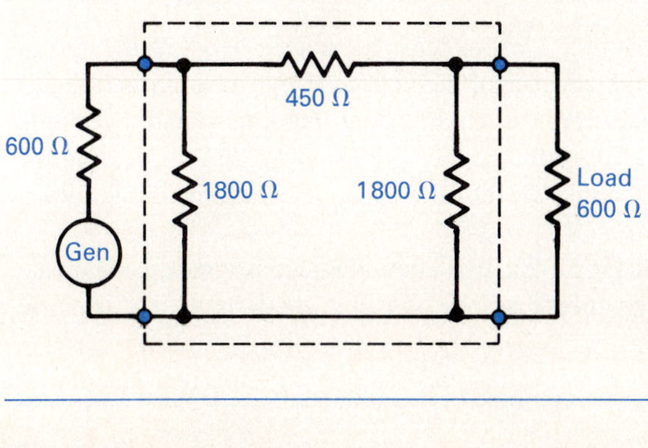

Fig. 8-23. π network as attenuator.

A π network also consists of three elements. It too may be used as an attenuator. Figure 8-23 shows the three elements connected in the form of the Greek letter π. The resistance values are chosen so that the attenuator has the same characteristics as the T attenuator of Figs. 8-21 and 8-22. The output voltage is one-half the input voltage; the input and output resistances are matched at 600 Ω. These values may be checked by methods similar to those described for the T attenuator.

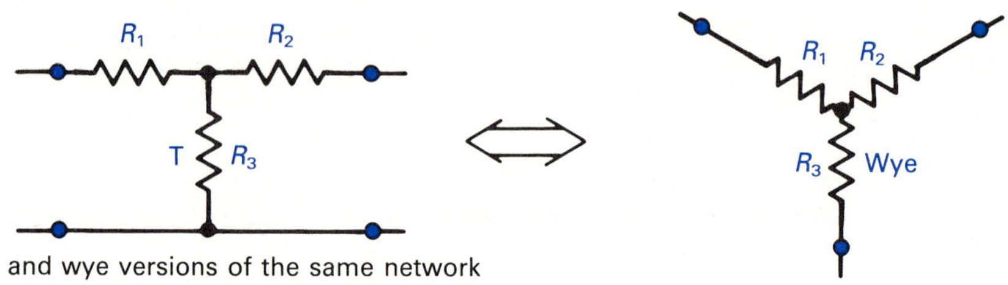

T and wye versions of the same network

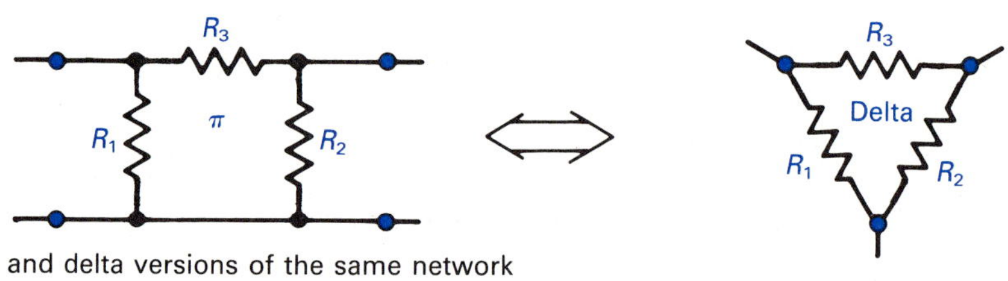

π and delta versions of the same network

Fig. 8-24. T, wye, π, and delta networks.

The networks of Figs. 8-21 to 8-23 are usually shown in the T and π forms, as attenuators are normally associated with electronic equipment. In electrical power equipment diagrams, the same networks are likely to appear in the *wye* and *delta* forms, in which the letter T becomes Y (wye) and the Greek letter π becomes an upside-down version of the Greek letter Δ (delta). Only the method of drawing is changed for convenience in the types of equipment schematic diagrams met in power work. Figure 8-24 compares the two ways of showing T or wye and π or delta networks.

Voltmeter Loading

When a voltage is to be measured, a voltmeter is connected between the appropriate points in the circuit. The presence of the voltmeter may change the magnitude of the voltage being measured to such an extent that a reasonably accurate measurement is not possible. This effect, called *voltmeter loading,* is determined by the relationship between the resistances of the voltmeter and the circuit being measured.

In each of the diagrams of Fig. 8-25, two equal-value resistors are connected across a 100-V supply. The voltage across one of the resistors, R_1, is to be

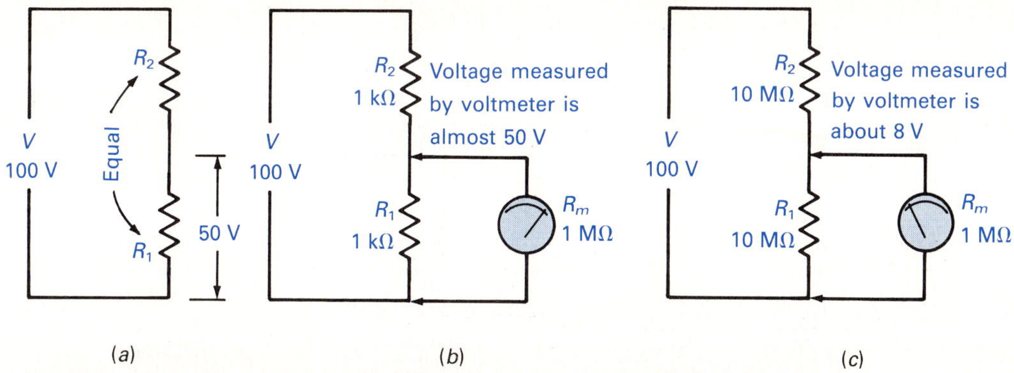

Fig. 8-25. Voltmeter loading. *(a)* Two equal-value resistors; each has half the 100 V of the supply. *(b)* Meter resistance high compared with circuit resistances; voltmeter loading low, measurement accuracy high. *(c)* Meter resistance low compared with circuit resistances; voltmeter loading high, measurement accuracy low.

measured by a voltmeter. Since a reading of 50 V (one-half of the 100 V) is expected, the 20-kΩ/V voltmeter is switched to its 50-V range, giving a meter resistance R_m of 1 MΩ. Each of the circuits is of series-parallel form, so that the actual voltage at the voltmeter terminals V_m can be determined by series-parallel-circuit methods.

The combined resistance of R_1 and R_m in parallel is given by

$$R_p = \frac{R_1 R_m}{R_1 + R_m} = \frac{R_1 \times 1\ \text{M}\Omega}{R_1 + 1\ \text{M}\Omega}$$

The total resistance, with R_2 in series with the parallel group R_p, is

$$R_T = R_2 + R_p$$

The voltage at the voltmeter terminals is found using the voltage-divider principle,

$$V_m = V\frac{R_p}{R_T} = 100\ \text{V}\,\frac{R_p}{R_T}$$

Applying these relationships to the diagrams in Fig. 8-25 gives:
Figure 8-25*b*

$$R_1 = R_2 = 1\ \text{k}\Omega$$

$$R_p = \frac{1\ \text{k}\Omega \times 1\ \text{M}\Omega}{1\ \text{k}\Omega + 1\ \text{M}\Omega} = 0.999\ \text{k}\Omega$$

$$R_T = 1\ \text{k}\Omega + 0.999\ \text{k}\Omega = 1.999\ \text{k}\Omega$$

$$V_m = \frac{100\ \text{V} \times 0.999\ \text{k}\Omega}{1.999\ \text{k}\Omega} = 49.97\ \text{V}$$

49.97 V is 0.06 percent lower than 50 V.
Figure 8-25*c*

$$R_1 = R_2 = 10\ \text{M}\Omega$$

$$R_p = \frac{10\ \mathrm{M\Omega} \times 1\ \mathrm{M\Omega}}{10\ \mathrm{M\Omega} + 1\ \mathrm{M\Omega}} = 0.909\ \mathrm{M\Omega}$$

$$R_T = 10\ \mathrm{M\Omega} + 0.909\ \mathrm{M\Omega} = 10.91\ \mathrm{M\Omega}$$

$$V_m = \frac{100\ \mathrm{V} \times 0.909\ \mathrm{M\Omega}}{10.91\ \mathrm{M\Omega}} = 8.33\ \mathrm{V}$$

8.33 V is 83.33 percent lower than 50 V.

Even if it is perfectly accurate, a voltmeter can do no better than exactly indicate the voltage at its terminals; it cannot decide what the voltage *should be*. The calculations show that if the voltmeter resistance is very much higher than the resistance across which it is measuring, the measurements can be accurate (only 0.06 percent basic error in Fig. 8-25*b* with the 1-kΩ resistors). However, with higher circuit resistances, the voltmeter loading becomes greater and the possible measurement accuracy gets lower. This is shown clearly with the 10-MΩ resistors in Fig. 8-25*c*, where the measurement error due to voltmeter loading is very high, over 80 percent.

D/A Converter

A D/A (digital-to-analog) converter takes a number expressed in binary form and converts it to analog form. In binary form the number is represented by a coded sequence of two different voltage levels. When converted to analog form, the number is represented by a single voltage proportional to the number.

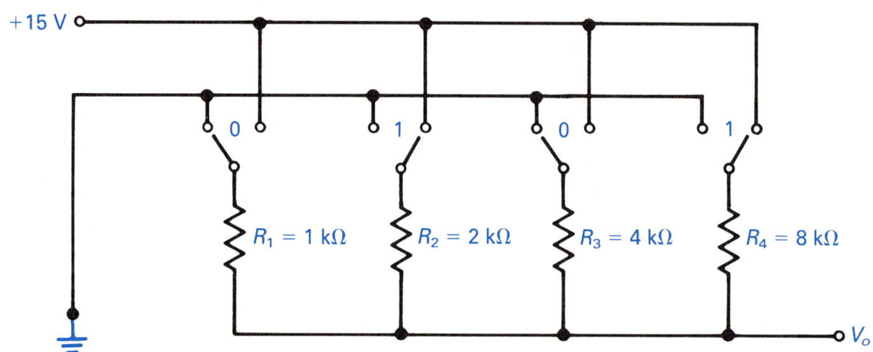

Fig. 8-26. Four-bit D/A converter with the number 5 in its binary form 0101 set by the switches.

Figure 8-26 shows a 4-bit D/A converter in which the number in binary form is coded in a sequence of 4 two-level digits. The switches are shown in the positions corresponding to the number 5, coded 0101 in its binary form. Binary 0 is represented by 0 V (ground) and binary 1 by 15 V.

You may see that the circuit of Fig. 8-26, with the switches in the positions shown, can be drawn as the series-parallel equivalent of Fig. 8-27.

The output voltage V_o (the analog representation of the number 5) can be obtained as follows: R_A is R_2 and R_4 in parallel. Similarly, R_B is R_1 and R_3 in parallel.

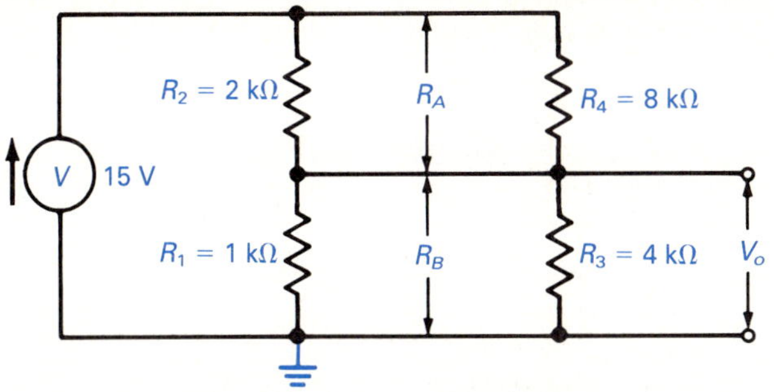

Fig. 8-27. Series-parallel equivalent of the D/A converter circuit, Fig. 8-26.

$$R_A = \frac{R_2 R_4}{R_2 + R_4} = \frac{2\ k\Omega \times 8\ k\Omega}{2\ k\Omega + 8\ k\Omega} = 1.6\ k\Omega$$

$$R_B = \frac{R_1 R_3}{R_1 + R_3} = \frac{1\ k\Omega \times 4\ k\Omega}{1\ k\Omega + 4\ k\Omega} = 0.8\ k\Omega$$

$$V_o = V\frac{R_B}{R_A + R_B} = 15\ V\ \frac{0.8\ k\Omega}{1.6\ k\Omega + 0.8\ k\Omega} = \mathbf{5\ V}$$

It is not a coincidence that a 5-V output represents the number 5. The supply was selected as 15 V because the maximum number that can be stored in four binary digits is 15.

Any number from 0 to 15 can be set in binary form with the switches. The output voltage for any of the numbers will be the same number of volts.

REVIEW QUESTIONS FOR SECTION 8-2

1. What is the limitation on the usefulness of a series voltage divider?
2. What is the result if the bleeder current in a voltage divider is too high?
3. What is the result if the bleeder current in a voltage divider is too low?
4. What is a frequently used value for voltage-divider bleeder current?
5. For the schematic diagram of Fig. 8-28, express the voltages across the following resistors in terms of V_{L1}, V_{L2}, and V:

   a. R_1
   b. R_2
   c. R_3
   d. R_1 *and* R_2
   e. R_1, R_2, *and* R_3

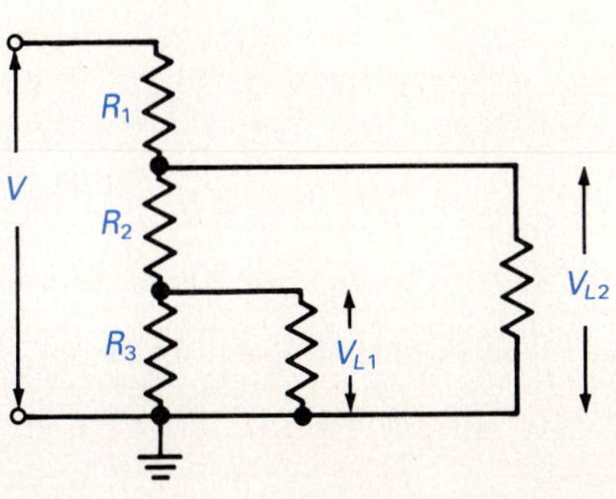

Fig. 8-28. Question 5, Sec. 8-2: Data.

6. How can both negative and positive voltages be obtained at the same time from a power supply?
7. In the case of a voltage divider supplying both negative and positive voltages, on which side of the ground reference is the bleeder resistor?
8. How might a voltmeter be connected to cause a current measurement to be incorrect?
9. How might an ammeter be connected to cause a voltage measurement to be incorrect?
10. What is meant by "matching"?

EXERCISE PROBLEMS FOR SECTION 8-2

1. For the circuit in Fig. 8-29
   a. With line *B* open-circuit, find the resistance between lines *A* and *C*. (*ans*. 54.25 Ω)
   b. With line *C* open-circuit, find the resistance between lines *A* and *B*. (*ans*. 40.68 Ω)
   c. With line *A* open-circuit, find the resistance between lines *B* and *C*. (*ans*. 43.77 Ω)
   d. With 20 V applied between lines *A* and *C*, find the voltage between lines *A* and *B*. (*ans*. 9.432 V)
   e. With 20 V applied between lines *A* and *B*, find the voltage between lines *B* and *C*. (*ans*. 7.425 V)
   f. With 20 V applied between lines *A* and *C*, find the power dissipated by each resistor and the total power dissipated by the circuit. (*ans*. 2.447 W, 0 W, 1.631 W, 0.7319 W, 0.8983 W, 0.5206 W, 1.145 W, 7.374 W)

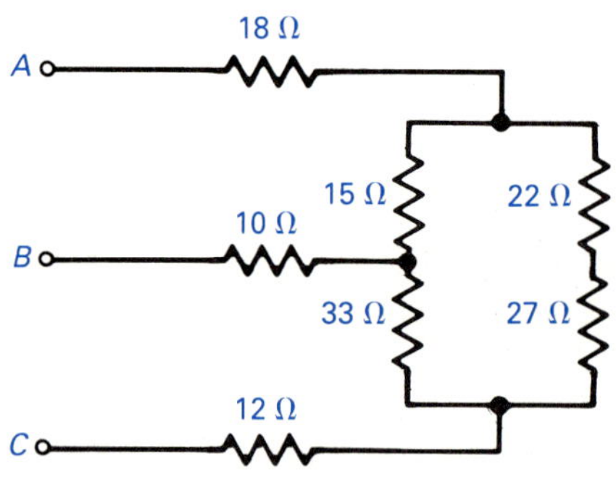

Fig. 8-29. Problem 1.

2. For the circuit in Fig. 8-30*a*, find
   a. The equivalent resistance
   b. The input current I_i
   c. The output current I_L
   d. The voltage across R_L
   e. The power dissipated by R_5
3. Repeat Prob. 2 using Fig. 8-30*b*. (*ans*. *a*. 296.5 Ω; *b*. 337.2 mA; *c*. 13.21 mA; *d*. 7.926 V; *e*. 4.719 W)

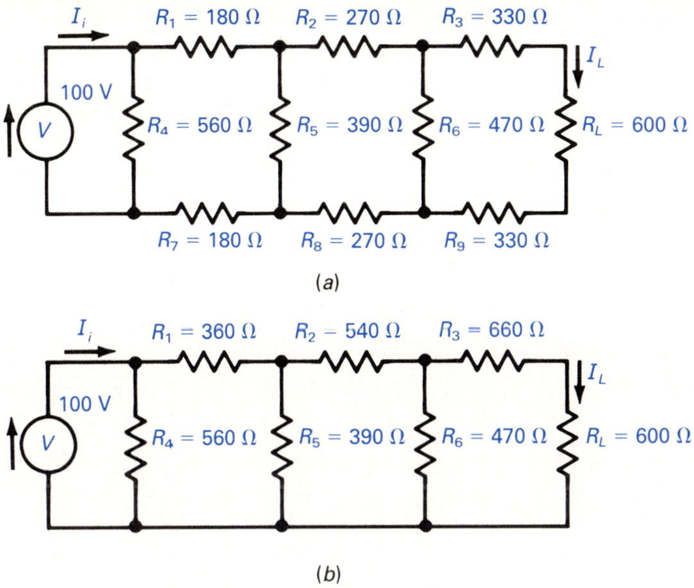

Fig. 8-30. *(a)* Problem 2; *(b)* Prob. 3.

4. A 24-V battery is to be used to supply 15 V at 1.6 A to an amplifier system. Design a suitable voltage divider, with resistance values, using a 20 percent bleeder current.

a. Find the power dissipated by each resistor of the voltage divider.

b. Find the voltage supplied to the amplifier from this circuit if the amplifier current dropped to 1.2 A.

c. Calculate the value of a simple series dropping resistor which would supply 15 V at 1.6 A to the amplifier, and then find the voltage to the amplifier when the current drops to 1.2 A.

d. Repeat *b* and *c* for the condition in which the amplifier current drops to 0.8 A.

5. A 30-V 1-A power source is to supply three loads: +12 V at 300 mA, −12 V at 300 mA, and −8 V at 200 mA. Design a suitable voltage divider using a bleeder current of 20 percent of the greater of the positive and negative total load currents. (*ans.* 10 Ω, 40 Ω, 80 Ω, 13.33 Ω)

6. Find the three output voltages of the voltage divider designed in Prob. 5 when

a. The −8-V load resistance rises to 80 Ω.

b. The −8-V load resistance drops to 25 Ω.

7. Power is fed to a house, from a 120-V 125-A power source, by a pair of No. 4 copper wires, each 400 ft long. (This copper wire has a resistance of 0.2485 Ω/1000 ft.) What voltage would be measured at the house if a 1500-W 120-V water heater and two 1000-V 120-V stove elements were on. Assume that there is no change in the resistance of the heating elements due to minor voltage changes. (*ans.* 114.5 V)

8. Figure 8-31 is the circuit for an ohmmeter. To zero the instrument, terminals *A* and *B* are connected together and R_s is adjusted for full-scale deflection (100 μA).

a. What value of R_s is required to obtain full-scale deflection?
b. What is the current I (short-circuit current) when A and B are connected together?
c. What is the current in the microammeter when A and B are connected together?
d. What current is read on the microammeter when the following resistances are connected between A and B? *(i)* 5 Ω, *(ii)* 10 Ω, *(iii)* 15 Ω, *(iv)* 20 Ω, *(v)* 25 Ω.

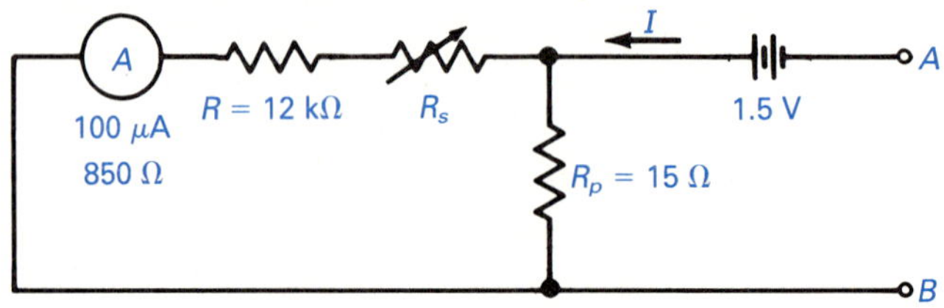

Fig. 8-31. Problem 8.

9. When measuring the characteristics of a semiconductor device, it is required that the voltage across the device and the current in the device be measured as accurately as possible. For the circuit in Fig. 8-32*a*, $V = 40$ V, ammeter A is a 50-μA movement with a resistance of 5 kΩ, voltmeter V is an electronic voltmeter with a resistance of 11 MΩ, and resistance R represents a reverse-biased semiconductor diode and equals 5 MΩ. Calculate

a. The actual voltage across the diode. (*ans*. 39.94 V)
b. The voltage measured by the voltmeter. (*ans*. 39.94 V)
c. The actual current in the diode. (*ans*. 7.988 μA)
d. The current measured by the ammeter. (*ans*. 11.62 μA)

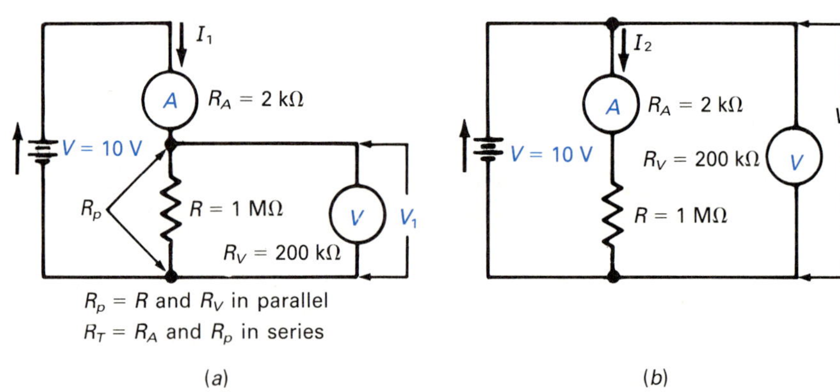

Fig. 8-32. *(a)* Problems 9 and 11: Voltmeter across resistor; *(b)* Probs. 10 and 12: Ammeter in series with resistor.

10. Repeat Prob. 9 using the circuit in Fig. 8-32*b*.

11. For the circuit in Fig. 8-32*a*, $V = 1$ V, ammeter A is a 10-mA movement with a resistance of 25 Ω, voltmeter V is an electronic voltmeter with a resistance of 11 MΩ, and resistance R represents a forward-biased semiconductor diode and equals 100 Ω. Calculate

a. The actual voltage across the diode. (*ans*. 0.8 V)
b. The voltage measured by the voltmeter. (*ans*. 0.8 V)
c. The actual current in the diode. (*ans*. 8 mA)
12. *d*. The current measured by the ammeter. (*ans*. 8 mA)
Repeat Prob. 11 using the circuit in Fig. 8-32*b*.

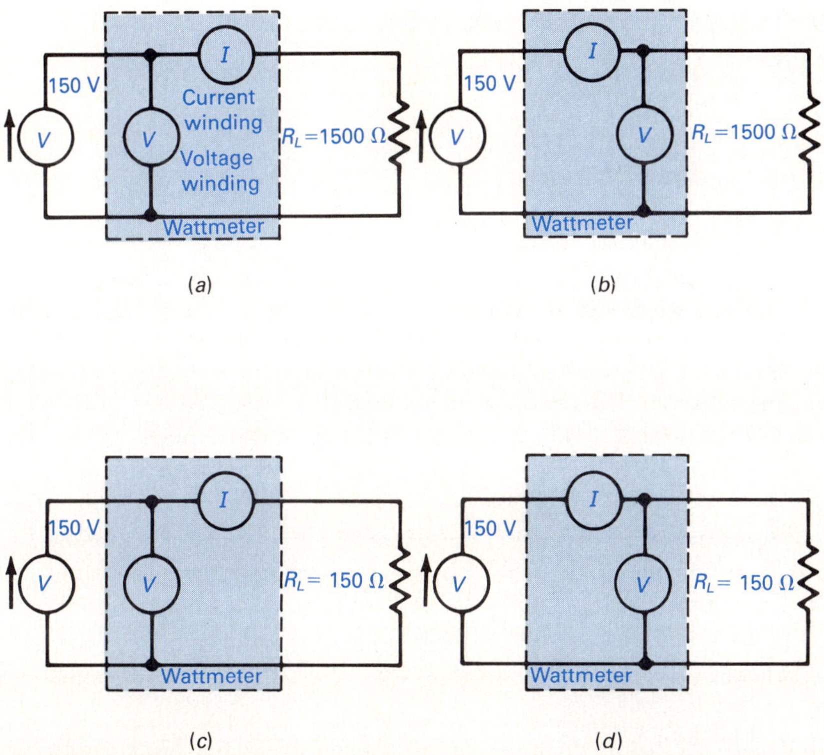

Fig. 8-33. Problem 13.

13. A wattmeter measures current and voltage and displays the product of the two quantities. The wattmeter used in Fig. 8-33 has a voltage-winding resistance of 9 kΩ and a current-winding resistance of 6 Ω.

i. What is the power measured by the wattmeter in each of the four circuits? (*ans*. *a*. 14.94 W; *b*. 17.33 W; *c*. 144.2 W; *d*. 140.8 W)

ii. What is the actual power dissipated by the load resistance in each of the four circuits?
(*ans*. *a*. 14.88 W; *b*. 14.84 W; *c*. 138.6 W; *d*. 138.5 W)

14. The output resistance of a transistor amplifier is 8 Ω. It is required that four equal speakers be connected in series-parallel across the output. What must the resistance of each speaker be so that the series-parallel combination will have the same resistance as the amplifier?

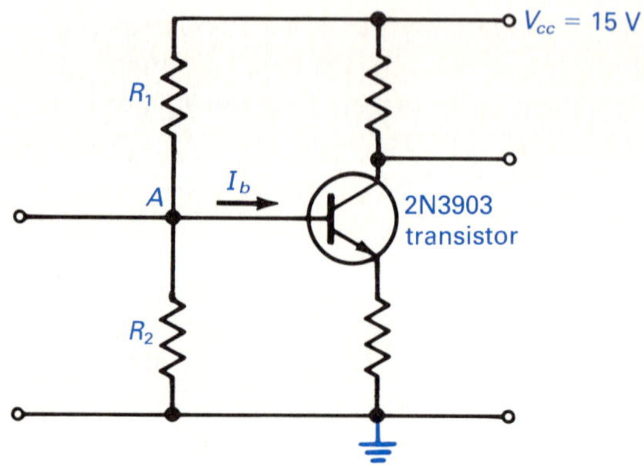

Fig. 8-34. Problem 15.

15. Figure 8-34 is the circuit of a transistor amplifier. I_b is to be 0.1 mA, and the voltage at point A is to be 6 V with respect to ground. If the bleeder current is to be five times I_b, find R_1 and R_2 (R_1 and R_2 form a voltage divider). (*ans.* 15 kΩ, 12 kΩ)

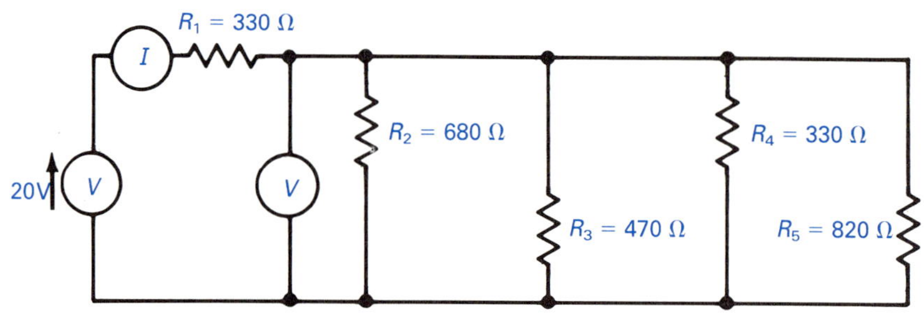

Fig. 8-35. Problem 16.

16. In the circuit in Fig. 8-35, which resistor is open-circuit or disconnected when

a. $V = 7.72$ V?
b. $V = 6.45$ V?
c. $I = 39.6$ mA?
d. $I = 41.1$ mA?

17. One resistor has been removed from the circuit in Fig. 8-2 (page 230). If I_3 measures 0.394 A, which resistor is missing? (*ans.* R_7)

18. Find the total resistance between points A and B in Fig. 8-36.

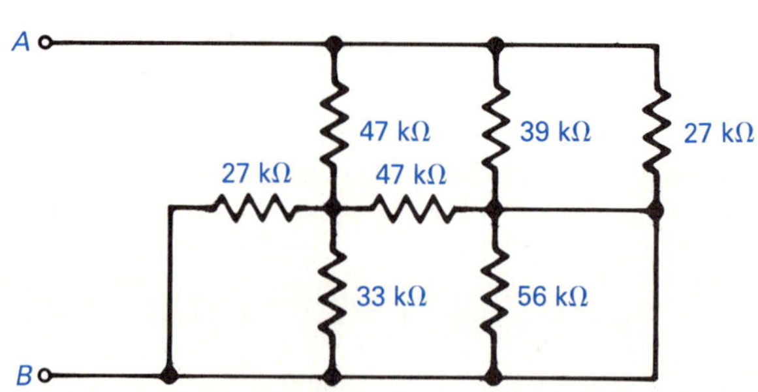

Fig. 8-36. Problem 18.

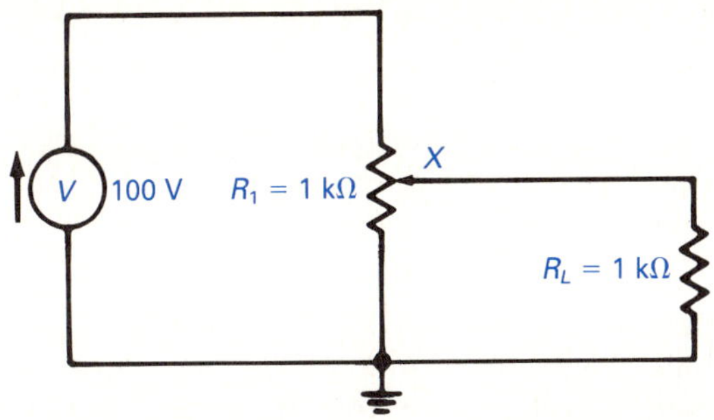

Fig. 8-37. Problem 19.

19. For the circuit in Fig. 8-37, find the resistance of R_1 between point X and ground so that the voltage across R_L is 50 V. (*ans*. 618 Ω)

8-3 SERIES-PARALLEL CIRCUITS IN ELECTRONICS

Electronic circuits may operate with voltages and currents that are direct (dc), alternating (ac), or a combination of direct and alternating. The dc circuit methods dealt with until now can be used with series or parallel groups of resistive elements (resistors) no matter what types of voltages and currents are present. This principle is true for series-parallel resistive circuits too.

The following examples show how dc series-parallel circuit techniques may be applied in the treatment of electronic circuits. You will see that the only difficulty is in taking an electronic circuit and recognizing the series-parallel grouping of resistors.

Transistor Bias

In Chap. 6 a transistor amplifier was introduced as a dc series-circuit application.

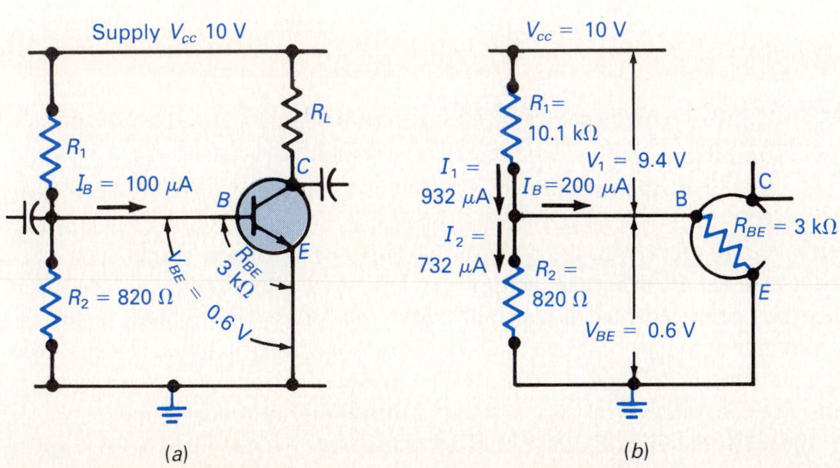

Fig. 8-38. Transistor base bias. *(a)* Transistor bias circuit; *(b)* equivalent series-parallel circuit.

Figure 6-72 is shown again as Fig. 8-38*a*. When this circuit was dealt with in Chap. 6, the base current I_B was ignored because its very low value, less than 10 μA, had little effect on the circuit operation. This technique enabled the circuit to be treated as if it were two separate series-element groups, with R_1 and R_2 as one of them, R_L and the transistor collector-to-emitter resistance as the other. However, the circuit is really more complicated than that, and this fact must be recognized if the base current is too high to be ignored.

In Fig. 8-38*b*, the base current I_B is shown as 200 μA. If the base-to-emitter voltage V_{BE} is to have the desirable value of 0.6 V as before, the equivalent base-to-emitter resistance R_{BE} is 0.6 V/200 μA = 3 kΩ. This resistance, with R_1 and R_2, forms a series-parallel network, as shown in Fig. 8-38*b*. The conditions in this network may be found as follows:

If R_2 is kept at 820 Ω, its current with the 0.6 V of the base-to-emitter voltage V_{BE} across it is

$$I_2 = \frac{V_{BE}}{R_2} = \frac{0.6\text{ V}}{820\ \Omega} = 732\ \mu\text{A}$$

The current in R_1 is the sum of I_2 and the base current I_B,

$$I_1 = I_2 + I_B = (732 + 200)\ \mu\text{A} = 932\ \mu\text{A}$$

The voltage across R_1 is the difference between the supply voltage V_{cc} and the base voltage V_{BE},

$$V_1 = V_{cc} - V_{BE} = (10 - 0.6)\text{ V} = 9.4\text{ V}$$

To give these conditions, the resistance of R_1 would have to be

$$R_1 = \frac{V_1}{I_1} = \frac{9.4\text{ V}}{932\ \mu\text{A}} = 10.1\text{ k}\Omega$$

Therefore, if the base current I_B is high enough, its effect on the current I_1 in resistor R_1 may make it necessary to use a lower-value resistor than would be needed if I_B could be ignored. For the 200-μA value of I_B, a 10-kΩ resistor would be wanted for R_1 rather than the 12 kΩ required when I_B is very small.

Operational Amplifier Bias

An *operational amplifier* (op amp) is a complex electronic network, usually made as an *integrated circuit* (IC) containing a number of transistors and other components. An op amp has three terminals, as shown in Fig. 8-39, known as the *inverting* input (−), the *noninverting* input (+), and the output.

Operational amplifiers are used in so many types of electronic equipment that they are widely regarded as general-purpose amplifiers. Therefore the circuits in which they appear are extremely varied. However, these circuits can be resolved into simple subcircuits that can be treated by ordinary principles. An example is Fig. 8-39, extracted from a complicated measuring instrument circuit.

The network of Fig. 8-39*a* applies a variable "zero-adjust" bias to the + input of the op amp. It is shown as a series-parallel group of elements in Fig. 8-39*b*.

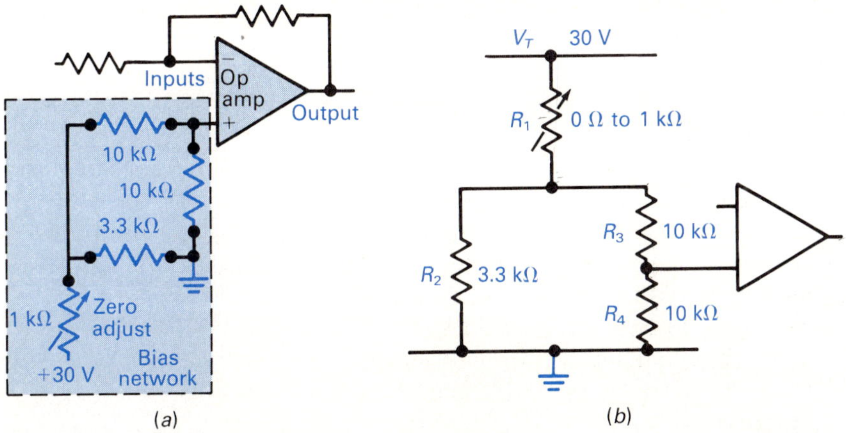

Fig. 8-39. Part of an electronic measuring instrument circuit. *(a)* Circuit as shown in the equipment schematic diagram; *(b)* equivalent series-parallel circuit for applying bias to the operational amplifier + terminal.

When R_1 is zero ohms, R_3 and R_4 are equal-value resistors connected across the 30-V supply. The bias voltage, therefore, is 30 V/2 = 15 V.

When R_1 is 1 kΩ, R_3 and R_4 in series are connected in parallel with R_2. The three-element group (R_2, R_3, and R_4) is connected in series with R_1.

The combined resistance of R_2, R_3, and R_4 is

$$R_A = \frac{R_2(R_3 + R_4)}{R_2 + R_3 + R_4} = \frac{3.3\ \text{k}\Omega(10 + 10)\ \text{k}\Omega}{(3.3 + 10 + 10)\ \text{k}\Omega} = 2.83\ \text{k}\Omega$$

The total network resistance is

$$R_T = R_A + R_1 = (2.83 + 1)\ \text{k}\Omega = 3.83\ \text{k}\Omega$$

The voltage across R_2 and R_3 in series is

$$V_A = V_T\frac{R_A}{R_T} = 30\ \text{V}\ \frac{2.83\ \text{k}\Omega}{3.83\ \text{k}\Omega} = 22.17\ \text{V}$$

The bias voltage is 22.17 V/2 = 11.08 V.

Therefore, as the zero-adjust resistor is varied between zero resistance and 1 kΩ, the bias applied to the noninverting input terminal changes between 15 V and 11.08 V.

Cathode-Ray Tube Supplies

A *cathode-ray tube* (CRT) is the display, or picture, tube used in oscilloscopes, television, radar, computers, and many other types of electronic equipment. The operation of a CRT requires several dc voltages, as the tubes used for different purposes have different voltage needs. An *oscilloscope* is an important electronic measuring instrument, usually containing an *electrostatic* CRT. In this type of tube, an electron beam is controlled by voltages applied to internal electrodes.

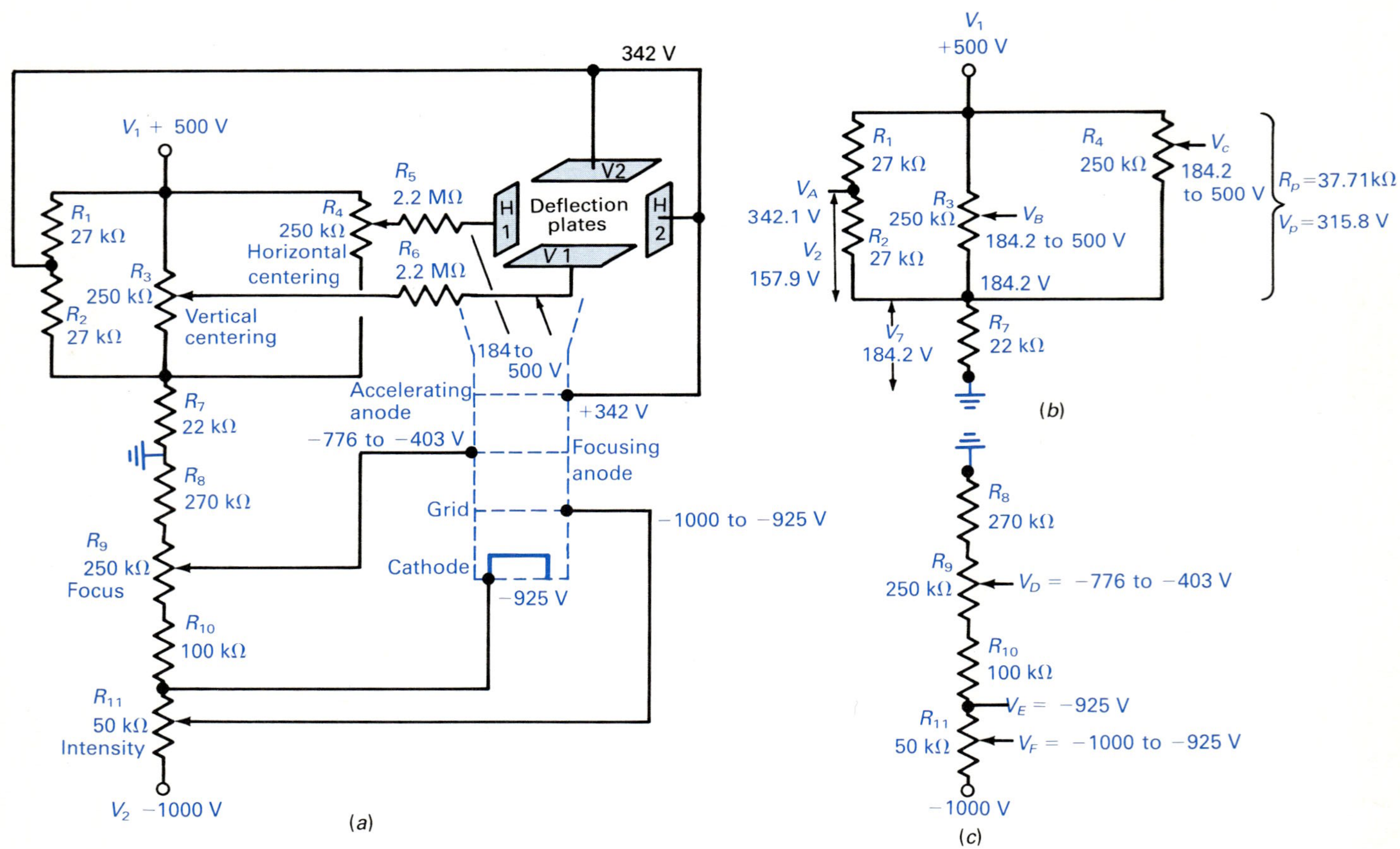

Fig. 8-40. Cathode-ray-tube supplies. *(a)* Distribution of voltage supplies; *(b)* deflection-plate supplies; *(c)* supplies to grid, cathode, and focus anode.

Figure 8-40 gives a circuit for deriving and distributing dc voltages for a type of CRT used in an oscilloscope. Figure 8-40*a* shows the complete circuit for supplying the dc voltages. Without knowledge of how a CRT works, it is possible to follow the operation of this circuit by adopting ordinary dc circuit methods. This is made easier by dividing the circuit into two parts, the series-parallel network R_1 to R_4 supplying beam-centering voltages to the deflecting plates, and the series network R_7 to R_{11} supplying voltages for beam formation and control. The production of the various voltages may be seen as follows:

Referring to Fig. 8-40*b*, the supply to this network is V_1 of positive polarity and 500 V in magnitude. R_1 and R_2 are both 27-kΩ resistors connected in series, giving 54 kΩ. This R_1, R_2 combination is connected in parallel with R_3 and R_4, each of which is a 250-kΩ variable resistor. This parallel group has a combined resistance $R_p = 37.71$ kΩ. By the voltage-divider principle, the voltage V_7 across R_7 is

$$V_7 = V_1 \frac{R_7}{R_7 + R_p} = 500 \text{ V} \frac{22 \text{ k}\Omega}{(22 + 37.71) \text{ k}\Omega} = 184.2 \text{ V}$$

The voltages V_B and V_C applied to the deflecting plates $V1$ and $H1$ are variable between the limits $V_7 = 184.2$ V and $V_1 = 500$ V due to R_3 and R_4 being variable (R_5 and R_6 have no effect).

The voltage V_p across the series-parallel group R_1 to R_4 is 315.8 V, the difference between V_1 and V_7. One-half of this V_p, 157.9 V, will appear across both R_1 and R_2. Therefore the voltage V_A applied to deflection plates $V2$ and $H2$ will be the sum of voltages V_7 and V_2, equal to 342.1 V.

The position of the CRT display is controlled in the vertical and horizontal directions by the relative potential between the appropriate pair of deflection plates. This is the relative potential of V_B or V_C with respect to V_A. As V_B or V_C can vary between 184 V and 500 V while V_A is fixed at 342 V, the relative potential between a pair of plates can vary between (184 − 342) V = −158 V and (500 − 342) V = +158 V.

As Fig. 8-40*c* is a series circuit, the beam-forming and controlling voltages V_D, V_E, and V_F can be found by direct use of the voltage-divider principle. Since the supply voltage V_2 is negative, all those voltages will be negative. The focusing voltage V_D can vary between −766 V and −403 V. The cathode voltage V_E is −925.4 V. The grid voltage V_F can vary between −1000 V and −925 V.

For a bright display, the CRT accelerating anode voltage needs to be high with respect to its cathode voltage. The accelerating anode voltage is +342 V, while that of the cathode is −925 V. Therefore the relative potential of the anode with respect to the cathode is +1267 V.

CHAPTER QUESTIONS

8-1. What is a series-parallel circuit?

8-2. Give another popular name for "series-parallel."

8-3. When a series-parallel network consists of element groups connected in series with each other, what is the important network characteristic?

8-4. When a series-parallel network consists of element groups connected in parallel with each other, what is the important network characteristic?

8-5. Why would many applications refer to a *loaded* voltage divider rather than the *unloaded* type?

8-6. What is meant by the *bleeder current* in a voltage-divider circuit?

8-7. What is the basic difference between the bleeder currents in loaded and unloaded voltage dividers?

8-8. A voltage-divider application requires it to supply both positive and negative voltages from a common power supply. In what way must the power supply voltage relate to the load voltages?

8-9. Refer to Fig. 8-17. Your house probably has a three-wire feed for its electricity. If this is so, why does some of your equipment (e.g., stove elements) operate between lines 1 and 2, while other items (e.g., light bulbs) are connected between the neutral line and either line 1 or line 2?

8-10. What might be the effect of measuring with a voltmeter that has a low sensitivity or an ammeter that has high resistance?

8-11. Figure 8-41 is a series-parallel circuit. Describe a method by which the current in resistor R_2 could be obtained *by measurement without* opening the circuit (without disconnecting either end of any resistor).

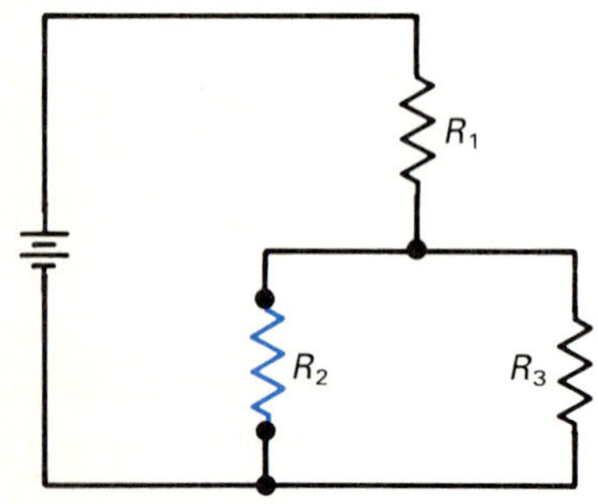

Fig. 8-41. Question 8-11.

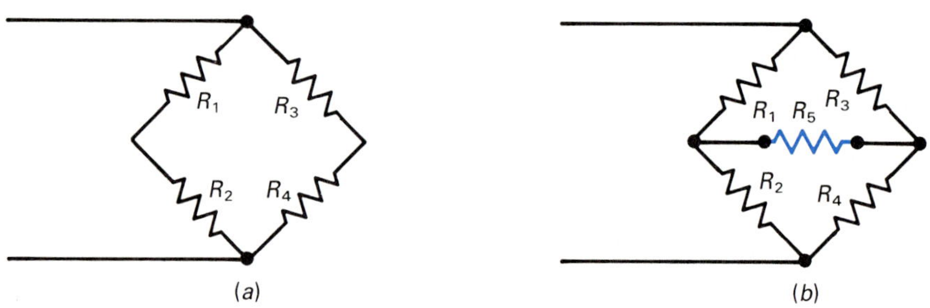

Fig. 8-42. Question 8-12.

8-12. Figure 8-42*a is* a series-parallel circuit. Due to R_5, Fig. 8-42*b is not*. Why?

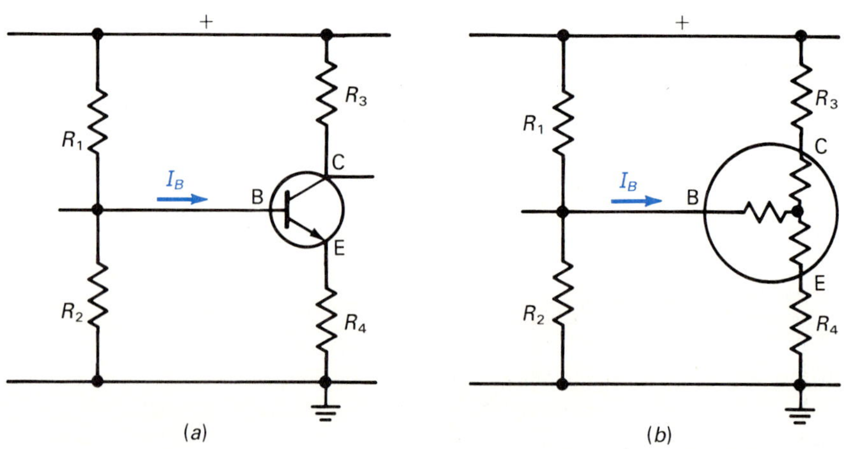

Fig. 8-43. Question 8-13.

8-13. A transistor circuit is shown in Fig. 8-43*a*, and its dc equivalent is given in Fig. 8-43*b*. The circuit *may* or *may not* be of series-parallel form, depending on the value of current I_B. What is this dependency?

CHAPTER PROBLEMS

8-1. In Fig. 8-44, what is the total resistance of the series-parallel network? (*ans*. 4.377 kΩ)

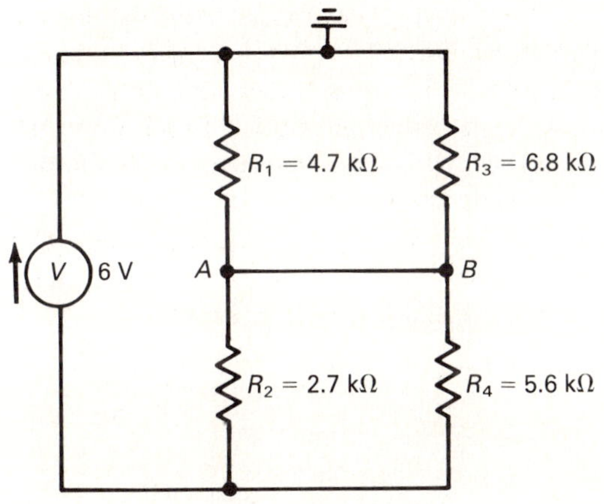

Fig. 8-44. Problems 8-1 to 8-6.

8-2. In Fig. 8-44, how much power is dissipated by resistor R_1?
8-3. In Fig. 8-44, what is the voltage across resistor R_2? (*ans*. 1.822 V)
8-4. In Fig. 8-44, what will be the current in resistor R_4 if resistor R_2 becomes *a*. Open-circuit? *b*. Short-circuit?
8-5. What is the potential of point *B* in Fig. 8-44? (*ans*. −2.555 V)
8-6. What is the current in the wire connecting points *A* and *B* in Fig. 8-44?
8-7. In Fig. 8-45, what is the voltage across resistor R_2? (*ans*. 33.82 V)

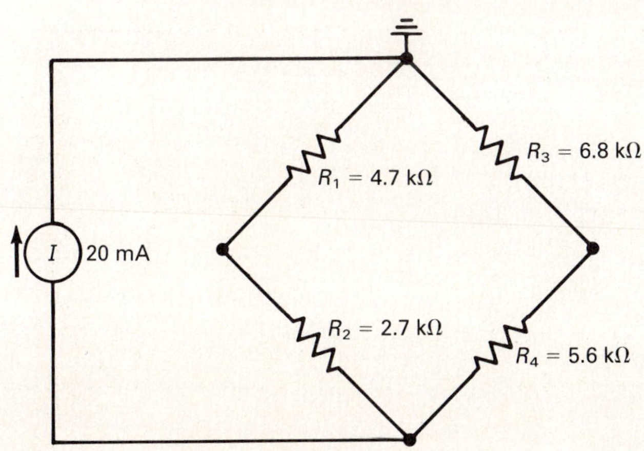

Fig. 8-45. Problems 8-7 to 8-10.

8-8. What will be the voltage across resistor R_2 in Fig. 8-45 if it becomes
a. Open-circuit?
b. Short-circuit?

8-9. How much power is dissipated by resistor R_3 in Fig. 8-45? (*ans*. 379.9 mW)

8-10. In Fig. 8-45, what is the relative potential between points *A* and *B*?

8-11. Design a voltage divider that operates off a 15-V power supply to give outputs of +5 V at 5 A and −5 V at 3 A. Use a bleeder current of 0.5 A (10 percent of the higher of the load currents).
(*ans*. 909.1 mΩ, 10 Ω, 2 Ω)

8-12. Referring to the circuit of Prob. 8-11, what value would the voltage across the +5-V load become if the −5-V load were disconnected? Assume that the resistance of the +5-V load remains the same.

8-13. Referring to Prob. 8-11, what value would the voltage across the −5-V load become if the bleeder resistor became open-circuit? Assume that the resistances of the loads remain unchanged. (*ans*. 3.837 V)

9

NETWORK ANALYSIS BY DIRECT METHODS

Network analysis is concerned with methods for finding the quantity values in a circuit or network (voltages, currents, etc.). There are two general approaches.

1. Direct methods. The network is left in its original form and its values are calculated directly, mainly by Ohm's and Kirchhoff's laws. These methods are best for fairly simple circuits that are essentially series-parallel. The calculations in Chaps. 6 to 8 were carried out by direct methods.
2. Equivalent-circuit methods. The network is converted into an equivalent form that is easier to handle, often for the application of techniques that will be introduced later. The equivalent-circuit approach is specially helpful in analyzing complex circuits containing element groups that are neither series nor parallel.

There are always alternative ways of finding a circuit quantity. Usually one way will be quicker or easier than the others, or you may just prefer one of them. This chapter introduces several different direct methods of circuit analysis to give you some familiarity with various ways of tackling a job and, perhaps, enable you to decide which to use in a certain situation.

9-1 CURRENT ASSUMPTION

Some networks consist of a generator feeding a load via a long string of elements. The ladder network of Fig. 9-1 is a typical example. It is used extensively in communications as an *attenuator,* a circuit for reducing and controlling signal strength.

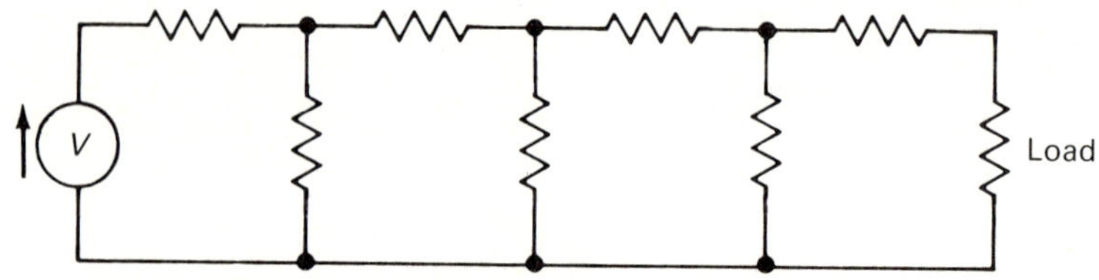

Fig. 9-1. Ladder network.

The currents or voltages in circuits like this can be calculated in many ways. In these long-string circuits, with only one generator, the current-assumption method is fast and involves only simple arithmetic.

It depends on the principle that a current in a linear circuit may be assumed to have any desired value if the following conditions are observed:

1. Other currents or voltages obtained by using the assumed current value are accepted as valid only in their relationship to that current.
2. Derived currents and voltages are converted to correct values by multiplying by a common ratio.
3. The common ratio is obtained from some known value of current or voltage.

The use of this principle for circuit analysis can best be seen by working examples.

EXAMPLE 9-1 What are the element currents in the circuit in Fig. 9-2?

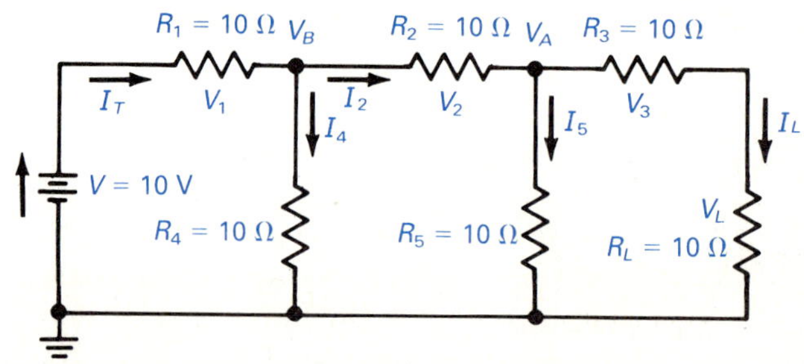

Fig. 9-2. Example 9-1: Circuit.

Solution Simplicity is the aim, so an assumed current is chosen which is easy to use. Assume I_L to be 1 A; then,

$$V_L = I_L R_L = 1\text{ A} \times 10\ \Omega = 10\text{ V}$$

We already know that I_L is *not* 1 A, because V_L cannot be 10 V, the same as the battery emf. This does not matter for now, and so we continue:

$$V_3 = I_L R_3 = 1\text{ A} \times 10\ \Omega = 10\text{ V}$$
$$V_A = V_L + V_3 = 10\text{ V} + 10\text{ V} = 20\text{ V}$$
$$I_5 = \frac{V_A}{R_5} = \frac{20\text{ V}}{10\ \Omega} = 2\text{ A}$$
$$I_2 = I_5 + I_L = 2\text{ A} + 1\text{ A} = 3\text{ A}$$
$$V_2 = I_2 R_2 = 3\text{ A} \times 10\ \Omega = 30\text{ V}$$
$$V_B = V_2 + V_A = 30\text{ V} + 20\text{ V} = 50\text{ V}$$
$$I_4 = \frac{V_B}{R_4} = \frac{50\text{ V}}{10\ \Omega} = 5\text{ A}$$
$$I_T = I_4 + I_2 = 5\text{ A} + 3\text{ A} = 8\text{ A}$$
$$V_1 = I_T R_1 = 8\text{ A} \times 10\ \Omega = 80\text{ V}$$
$$V = V_1 + V_B = 80\text{ V} + 50\text{ V} = 130\text{ V}$$

If the load current I_L were 1 A, the battery emf would have to be 130 V. But the emf is only 10 V. In a linear network, all the currents and voltages due to a 130-V battery will be 13 times higher than for a 10-V battery. So, in this example, all the currents are actually $\frac{1}{13}$ of the values obtained from the 1-A assumption for I_L.

$$I_T = \frac{8\text{ A}}{13} = \mathbf{0.6154\ A}$$

$$I_4 = \frac{5\text{ A}}{13} = \mathbf{0.3846\ A}$$

$$I_2 = \frac{3\text{ A}}{13} = \mathbf{0.2308\ A}$$

$$I_5 = \frac{2\text{ A}}{13} = \mathbf{0.1538\ A}$$

$$I_L = \frac{1\text{ A}}{13} = \mathbf{0.0769\ A}$$

A lot of writing is shown for these assumed-current solutions. When these methods are actually used, only the assumed and real values of current and voltage need be noted, not the details of how they are calculated.

Example 9-1 intentionally used simple numbers to demonstrate the technique. All assumed values could have been worked out mentally. Even when the numbers are not as simple, only a few simple operations are involved. The following examples have commercial, preferred-resistance values which are arithmetically awkward but no trouble with a calculator.

EXAMPLE 9-2 In the circuit in Fig. 9-3, how much of the generator voltage V is delivered to the load?

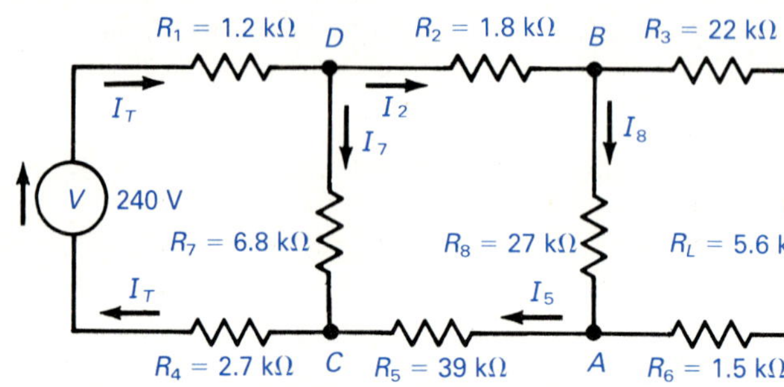

Fig. 9-3. Example 9-2: Circuit.

Solution Assuming I_L to be 1 mA, then

$$V_{BA} = I_L(R_3 + R_L + R_6) = 1\text{ mA}(22\text{ k}\Omega + 5.6\text{ k}\Omega + 1.5\text{ k}\Omega) = 29.1\text{ V}$$

$$I_8 = \frac{V_{BA}}{R_8} = \frac{29.1\text{ V}}{27\text{ k}\Omega} = 1.078\text{ mA}$$

$$I_2 = I_5 = I_8 + I_L = 1.078\text{ mA} + 1\text{ mA} = 2.078\text{ mA}^*$$

$$V_{DC} = V_{BA} + I_2(R_2 + R_5) = 29.1\text{ V} + 2.078\text{ mA}(1.8\text{ k}\Omega + 39\text{ k}\Omega) = 113.9\text{ V}$$

$$I_7 = \frac{V_{DC}}{R_7} = \frac{113.9\text{ V}}{6.8\text{ k}\Omega} = 16.75\text{ mA}$$

$$I_T = I_7 + I_2 = 16.75\text{ mA} + 2.078\text{ mA} = 18.83\text{ mA}$$

$$V = V_{DC} + I_T(R_1 + R_4) = 113.9\text{ V} + 18.83\text{ mA}(1.2\text{ k}\Omega + 2.7\text{ k}\Omega) = 187.3\text{ V}$$

The emf, calculated on an assumed load current of 1 mA, is 187.3 V. But the real emf is 240 V, so the calculated load current (and all the other currents and voltages) is too low by the factor 240 V/187.3 V = 1.281. The real value of the load current is

$$I_L = 1\text{ mA} \times 1.281 = 1.281\text{ mA}$$

$$V_L = I_L R_L = 1.281\text{ mA} \times 5.6\text{ k}\Omega = \mathbf{7.174\ V}$$

or 3 percent of the generator voltage V is delivered to the load.

EXAMPLE 9-3 Figure 9-4 is the schematic diagram of an attentuator in an item of electronic test equipment. Resistor R_4 has become faulty. What power rating is required for the replacement resistor?

*I_2 and I_5 are *exactly* the same currents ($I_8 + I_L$). When elements carry the same current (not just currents of the same value), they are *virtually in series,* even though they are not connected together.

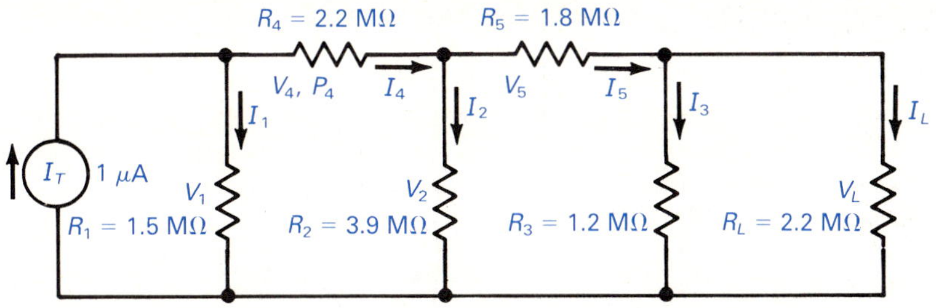

Fig. 9-4. Example 9-3: Circuit.

Solution Assuming that $I_L = 1\ \mu A$, then

$$V_L = I_L R_L = 1\ \mu A \times 2.2\ M\Omega = 2.2\ V$$

$$I_3 = \frac{V_L}{R_3} = \frac{2.2\ V}{1.2\ M\Omega} = 1.833\ \mu A$$

$$I_5 = I_3 + I_L = 1.833 + 1\ \mu A = 2.833\ \mu A$$

$$V_5 = I_5 R_5 = 2.833\ \mu A \times 1.8\ M\Omega = 5.099\ V$$

$$V_2 = V_5 + V_L = 5.099\ V + 2.2\ V = 7.299\ V$$

$$I_2 = \frac{V_2}{R_2} = \frac{7.299\ V}{3.9\ M\Omega} = 1.872\ \mu A$$

$$I_4 = I_2 + I_5 = 1.872\ \mu A + 2.833\ \mu A = 4.705\ \mu A$$

$$V_4 = I_4 R_4 = 4.705\ \mu A \times 2.2\ M\Omega = 10.35\ V$$

$$V_1 = V_4 + V_2 = 10.35\ V + 7.299\ V = 17.65\ V$$

$$I_1 = \frac{V_1}{R_1} = \frac{17.65\ V}{1.5\ M\Omega} = 11.77\ \mu A$$

$$I_T = I_1 + I_4 = 11.77\ \mu A + 4.705\ \mu A = 16.47\ \mu A$$

But I_T is actually 1 μA, so all the calculated currents and voltages are too high by the factor 16.47. The real values of V_4 and I_4 are

$$V_4 = \frac{10.35\ V}{16.47} = 0.6284\ V$$

$$I_4 = \frac{4.705\ \mu A}{16.47} = 0.2857\ \mu A$$

The power dissipated by R_4 is

$$P_4 = V_4 I_4 = 0.6284\ V \times 0.2857\ \mu A = \mathbf{0.1795\ \mu W}$$

This is a very small power. Any commercial resistor will be able to dissipate it without difficulty.

REVIEW QUESTIONS FOR SECTION 9-1

1. What is network analysis concerned with?
2. Name two general approaches to network analysis.
3. What types of circuit are suited for analysis by current assumption?
4. Using current-assumption analysis, if the current is found to be incorrect, what adjustment is made to the calculated values of voltage and current?

EXERCISE PROBLEMS FOR SECTION 9-1

1. For the circuit in Fig. 9-5, by assuming a current, find I_L when

a. $V = 20$ V, $R_1 = 47\ \Omega$, $R_2 = 68\ \Omega$, $R_3 = 100\ \Omega$, $R_L = 100\ \Omega$ (*ans.* 68.03 mA)

b. $V = 24$ V, $R_1 = 3.3$ kΩ, $R_2 = 3.9$ kΩ, $R_3 = 1.2$ kΩ, $R_L = 1$ kΩ (*ans.* 1.108 mA)

c. $V = 6$ V, $R_1 = 120$ kΩ, $R_2 = 270$ kΩ, $R_3 = 82$ kΩ, $R_L = 56$ kΩ (*ans.* 6.5 μA)

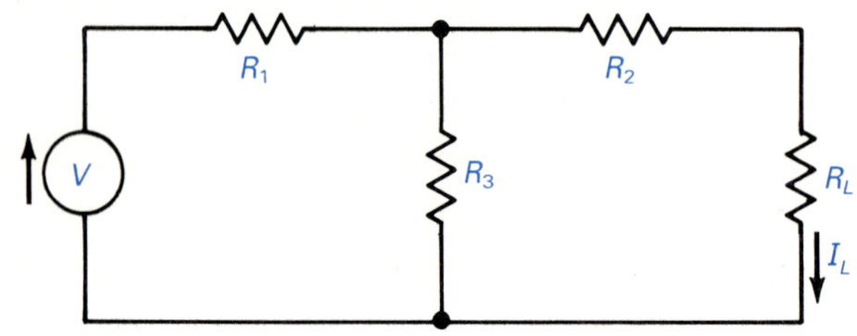

Fig. 9-5. Problem 1: Circuit.

2. For the circuit in Fig. 9-6 by assuming a current, find I_L when

a. $R_L = 100\ \Omega$

b. $R_L = 1000\ \Omega$

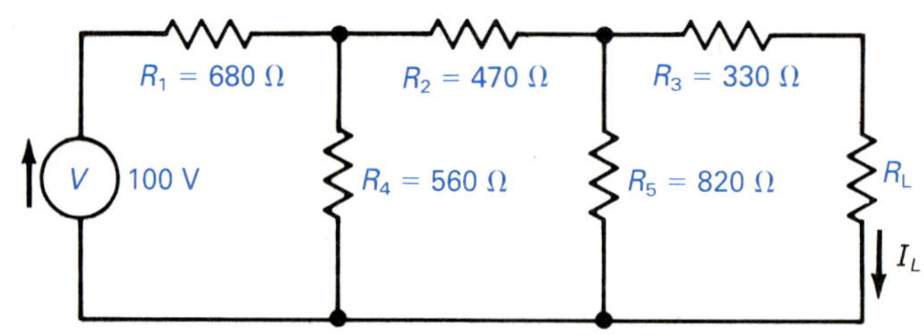

Fig. 9-6. Problems 2 and 3: Circuit.

3. For the circuit in Fig. 9-6, by assuming a current, find the current through the 820-Ω resistor when

a. $R_L = 150\ \Omega$ (*ans.* 15.43 mA)

b. $R_L = 680\ \Omega$ (*ans.* 20.27 mA)

4. For the circuit in Fig. 9-7, by assuming a current, find the power dissipated by G_1 and G_L when

a. $G_L = 0.01$ S

b. $G_L = 0.0455$ S

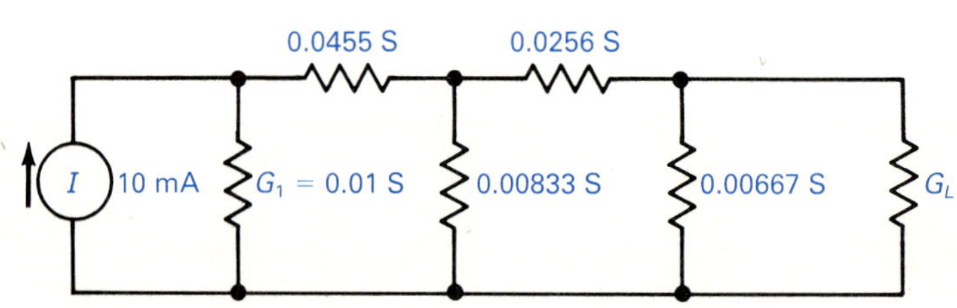

Fig. 9-7. Problem 4: Circuit.

5. By assuming a current, find

a. The power dissipation of R_1 and R_L in Fig. 9-8*a* when $V = 20$ V (*ans*. 166.7 mW, 10.4 mW)

b. The power dissipation of R_1 and R_L in Fig. 9-8*b* when $V = 20$ V (*ans*. 166.7 mW, 10.4 mW)

c. The voltages V_1 and V_2 in Fig. 9-8*a* when $V = 100$ V (*ans*. 50 V, 12.5 V)

d. The voltages V_1 and V_2 in Fig. 9-8*b* when $V = 100$ V (*ans*. 50 V, 12.5 V)

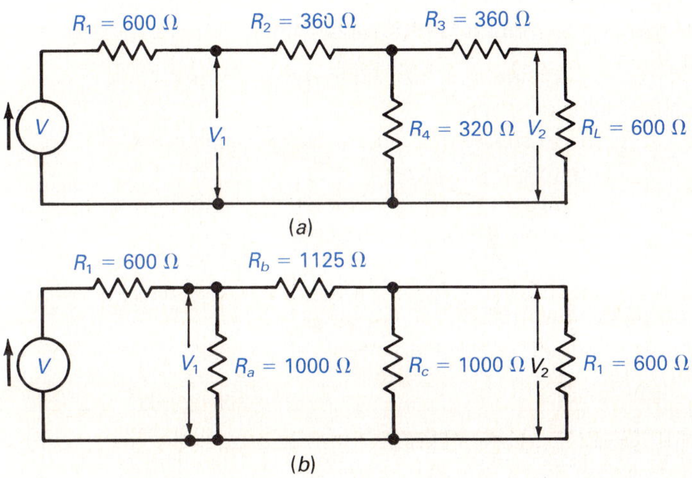

Fig. 9-8. Problem 5: *(a)* Circuit with T network; *(b)* circuit with π network.

6. Solve Prob. 2, Sec. 8-1 (page 234) by the current-assumption method.

7. By assuming a current in R_2, R_3, or R_4, solve Prob. 10, Sec. 8-1 (page 236). (*ans*. 28.72 mA, 57.45 mA)

8. By assuming a current in R_2, R_3, or R_4, solve Prob. 11, Sec. 8-1 (page 236).

9. By assuming a current, find the three load voltages in the voltage-divider circuit in Fig. 9-9. (*ans*. 50.45 V, 23.3 V, −35.32 V)

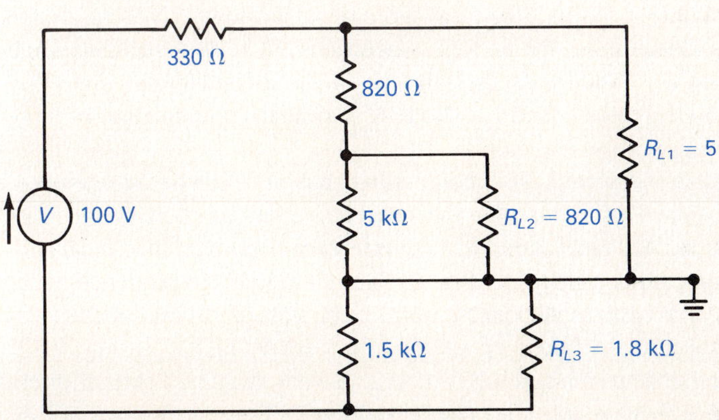

Fig. 9-9. Problem 9: Circuit.

9-2 NODAL ANALYSIS

Nodal analysis and loop analysis (Sec. 9-3) can be regarded as complementary. Nodal analysis relates to current sources and conductances, while loop analysis is concerned with voltage sources and resistances. This duality leads to some general guidelines on when to use nodal analysis.

Note that these are only guidelines, not rigid rules. Such guidelines will aid in choosing the most suitable of several different methods of analyzing the same circuit. A good choice of method can reduce the amount of work and improve the reliability of the results.

Nodal analysis should be considered for a network if

1. The energy sources are predominantly current generators.
2. Voltage values are to be found.
3. There are a number of different quantities to be calculated.
4. There are two or more energy sources.
5. The number of independent nodes is less than the number of independent loops.

The first step in nodal analysis is to identify the nodes. A *node* is a junction, or connection point, of three or more conductors.

Figure 9-10 is a π network connected between two current generators.

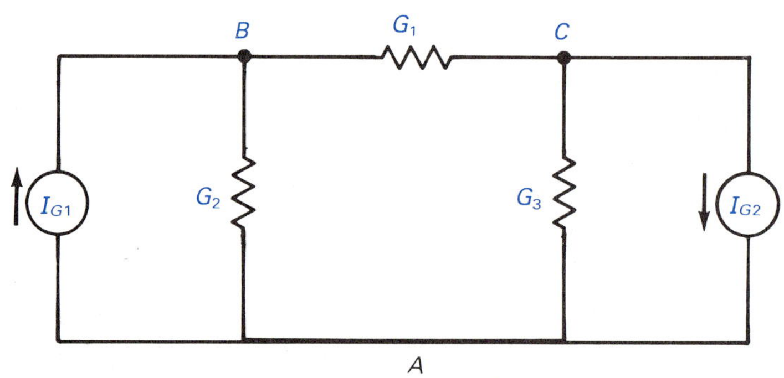

Fig. 9-10. Identifying nodes.

In the circuit in Fig. 9-10, there are three separate nodes. Node A is the junction of four conductors, the bottom connections to the two generators and elements G_2 and G_3. Node B is the junction of three conductors connecting generator I_{G1} to elements G_1 and G_2. Node C is similarly the junction of three conductors.

A reference node is selected. The choice is free; any node can be the reference. It may be convenient to choose for the reference node a circuit common or ground connection. All nodes other than the reference node are called *independent* nodes. If the common line A in Fig. 9-10 is chosen as the reference node, points B and C will be the independent nodes. The voltages at the independent nodes, with respect to the reference node, are labeled V_B and V_C in Fig. 9-11.

To aid in the initial understanding of nodal analysis, assumed current directions and element voltage polarities can be indicated. These are shown in Fig. 9-11. When assumptions are made, specific current directions must be consistent

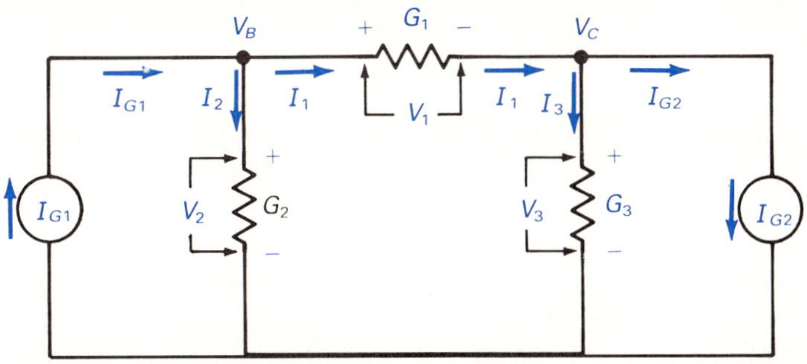

Fig. 9-11. Circuit in Fig. 9-10 with currents and voltages identified.

with the related voltage polarities (+ → − or − ← +). Whether or not the assumptions are correct will be established by the analysis results. A positive result confirms an assumption; a negative result indicates that the direction or polarity is opposite that assumed.

Figure 9-11 is the same as Fig. 9-10, except for the addition of current and voltage symbols, directions, and polarities.

The conditions existing at the independent nodes are written as equations. The development of the equations for nodes B and C in Fig. 9-11 is as follows.

At Node B Generator current I_{G1} is entering the node. Element currents I_1 and I_2 are leaving the node.

The Kirchhoff's current law equation for node B is

$$I_1 + I_2 = I_{G1}$$

All currents, except generator currents, are now expressed in Ohm's law terms, $I = VG$. The voltage across G_1 will be $V_B - V_C$, because the assumed direction of current I_1, left to right, implies that V_B is greater than V_C.

$$I_1 = V_1G_1 = (V_B - V_C)G_1$$
$$I_2 = V_2G_2 = V_BG_2$$

Therefore the equation for node B is

$$(V_B - V_C)G_1 + V_BG_2 = I_{G1}$$

or

$$V_B(G_1 + G_2) - V_CG_1 = I_{G1} \tag{9-1}$$

At Node C Element current I_1 is entering the node. Generator current I_{G2} and element current I_3 are leaving the node.

$$I_3 = V_3G_3 = V_CG_3$$

The equation for node C is

$$I_{G2} + I_3 = I_1$$
$$I_{G2} + V_CG_3 = (V_B - V_C)G_1$$

or

$$V_C(G_1 + G_3) - V_BG_1 = -I_{G2} \tag{9-2}$$

Solving the pair of simultaneous equations, Eqs. (9-1) and (9-2), will give the values of V_B and V_C. All other circuit quantities can then be worked out.

It was assumed that nodes B and C in Fig. 9-11 are both positive with respect to reference node A. The node equations for any circuit can be developed quickly by assuming that all independent nodes are positive with respect to the chosen reference node, and noting what is implied by Eqs. (9-1) and (9-2). With this method there is no need to assign current directions or voltage polarities. The node voltage polarities will be established by the equation solutions. If a solution is positive, the relevant node voltage is positive, and vice versa. Element polarities and current directions can be determined by applying basic principles and conventions.

Writing the Node Equation

The recommended procedure for writing the equation for a node is as follows. The term "local node" is used to apply to the node for which the equation is being written. The term "remote node" implies a node connected to the local node by a conductance.

Element Currents

1. Place on the left side of the equation currents in conductances.
2. Multiply the local node voltage (with respect to the reference) by the sum of the conductances connected to the node. Show the product as a positive quantity. Ignore elements in series with current sources, as they cannot affect circuit currents.
3. Multiply each remote node voltage by the conductance of the element connecting it to the local node. Show the product as a negative quantity.

Source Currents

1. Place source currents on the right side of the equation.
2. Show currents entering the local node as positive quantities.
3. Show currents leaving the local node as negative quantities.
4. If there are no source currents at the local node, show as zero.

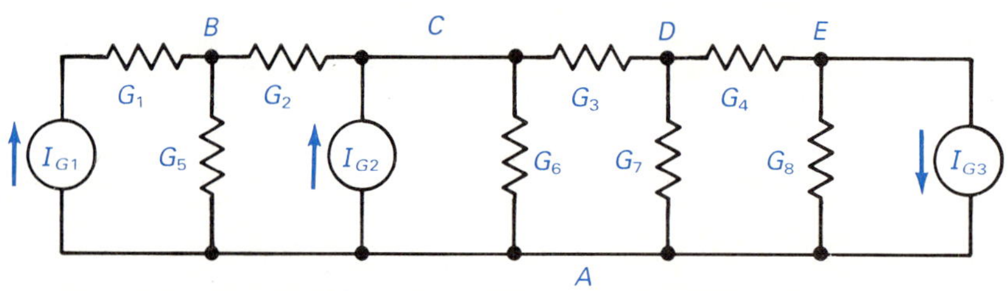

Fig. 9-12. Multisource network with nodes identified.

The circuit in Fig. 9-12 is fairly complex, but its nodal equations can be written rapidly. If point (line) A is the reference node, points B, C, D, and E are independent nodes. Using the steps just described, the node equations are

Node B: $$V_B(G_2 + G_5) - V_C G_2 = I_{G1}$$
(G_1 is omitted as not affecting the currents)

Node C: $$V_C(G_2 + G_3 + G_6) - V_B G_2 - V_D G_3 = I_{G2}$$

Node D: $$V_D(G_3 + G_4 + G_7) - V_C G_3 - V_E G_4 = 0$$

Node E: $$V_E(G_4 + G_8) - V_D G_4 = -I_{G3}$$

EXAMPLE 9-4 What are the values of each voltage and current in the schematic diagram in Fig. 9-13?

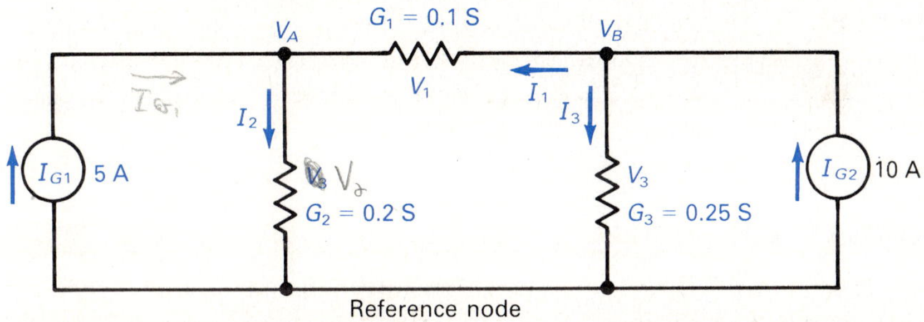

Fig. 9-13. Example 9-4: Circuit.

Solution

At node A: $$V_A(G_1 + G_2) - V_B G_1 = I_{G1}$$
$$V_A(0.1\text{ S} + 0.2\text{ S}) - V_B 0.1\text{ S} = 5\text{ A}$$

At node B: $$V_B(G_1 + G_3) - V_A G_1 = I_{G2}$$
$$V_B(0.1\text{ S} + 0.25\text{ S}) - V_A 0.1\text{ S} = 10\text{ A}$$

If units are omitted, the equations reduce to

$$0.3V_A - 0.1V_B = 5$$
$$-0.1V_A + 0.35V_B = 10$$

Solving the pair of simultaneous equations by any convenient method gives the values of V_A and V_B

$$V_A = \mathbf{28.95\ V} = V_2$$
$$V_B = \mathbf{36.84\ V} = V_3$$
$$V_1 = V_B - V_A = 36.84\text{ V} - 28.95\text{ V} = \mathbf{7.89\ V}$$
$$I_1 = V_2 G_2 = 7.89\text{ V} \times 0.1\text{ S} = \mathbf{0.789\ A}$$
$$I_2 = V_A G_3 = 28.95\text{ V} \times 0.2\text{ S} = \mathbf{5.79\ A}$$
$$I_3 = V_B G_4 = 36.84\text{ V} \times 0.25\text{ S} = \mathbf{9.21\ A}$$

The results of Example 9-4 satisfy the Kirchhoff's law requirements at all three nodes (currents entering equal currents leaving).

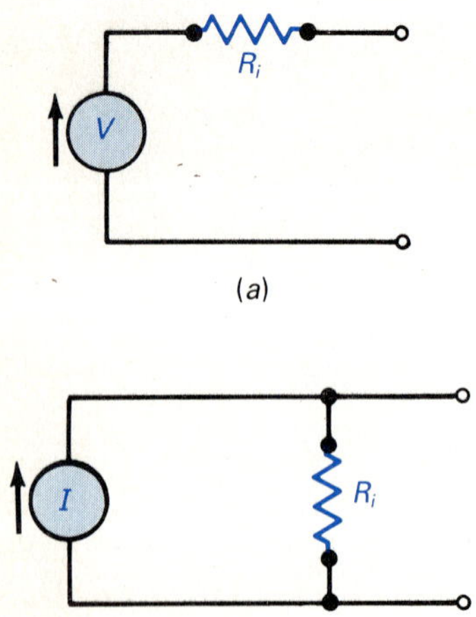

Fig. 9-14. *(a)* Voltage generator; *(b)* current generator.

Node *A*:

$$I_{G1} + I_1 = I_2$$
$$5\ \text{A} + 0.789\ \text{A} \simeq 5.79\ \text{A}$$

Node *B*:

$$I_{G2} = I_1 + I_3$$
$$10\ \text{A} \simeq 0.789\ \text{A} + 9.21\ \text{A}$$

Reference node:

$$I_2 + I_3 = I_{G1} + I_{G2}$$
$$5.79\ \text{A} + 9.21\ \text{A} = 5\ \text{A} + 10\ \text{A}$$

If the circuit includes voltage generators, these should be converted to equivalent current generators, using the principle established in Sec. 7-4. This principle is summarized as follows:

The current generator in Fig. 9-14*b* will be the equivalent of the voltage generator in Fig. 9-14*a* if

1. The internal resistance R_i of each generator is the same.
2. The source current I of the current generator is related to the voltage generator emf V by the equation

$$I = \frac{V}{R_i}$$

EXAMPLE 9-5

What is the voltage V_2 in the circuit in Fig. 9-15?

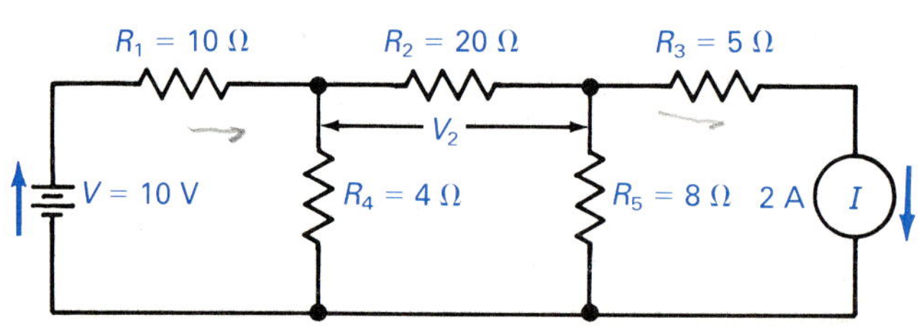

Fig. 9-15. Example 9-5: Given circuit.

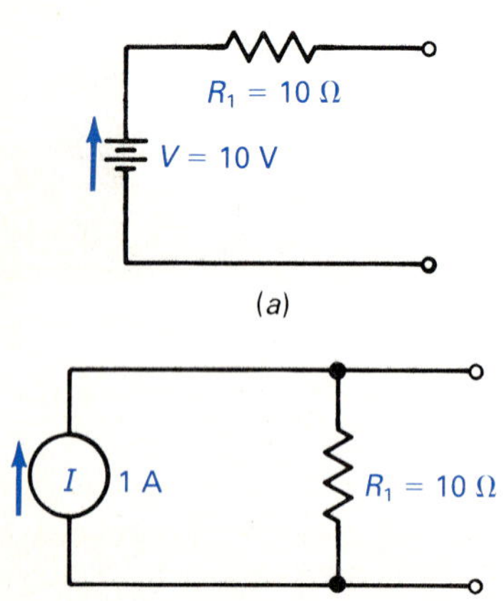

Fig. 9-16. Example 9-5: Voltage- to current-generator conversion.

Solution Convert the battery to an equivalent current generator, with R_1 regarded as the battery's series internal resistance (see Fig. 9-16).

Redraw Fig. 9-15 with all resistance values changed to conductances, omitting R_3, since it is an element in series with a current generator. The results are shown in Fig. 9-17.

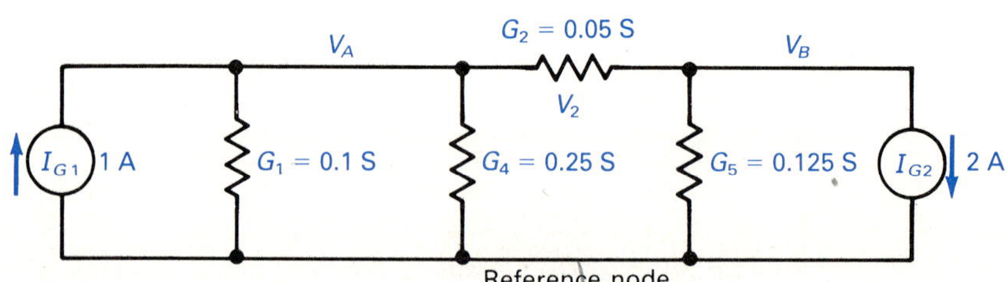

Fig. 9-17. Figure 9-15 with a current generator replacing the battery.

The equations for the two independent nodes are

$$V_A\,(0.1\text{ S} + 0.25\text{ S} + 0.05\text{ S}) - V_B\,(0.05\text{ S}) = 1\text{ A}$$
$$V_B\,(0.125\text{ S} + 0.05\text{ S}) - V_A\,(0.05\text{ S}) = -2\text{ A}$$

Reduce the equations to their simplest forms:

$$0.04V_A - 0.05V_B = 1$$
$$-0.05V_A + 0.175V_B = -2$$

The solutions to these equations are

$$V_A = 1.11\text{ V}$$
$$V_B = -11.1\text{ V}$$

These solutions mean that the voltages at the ends of element R_2, with respect to the reference node, are as shown in Fig. 9-18.

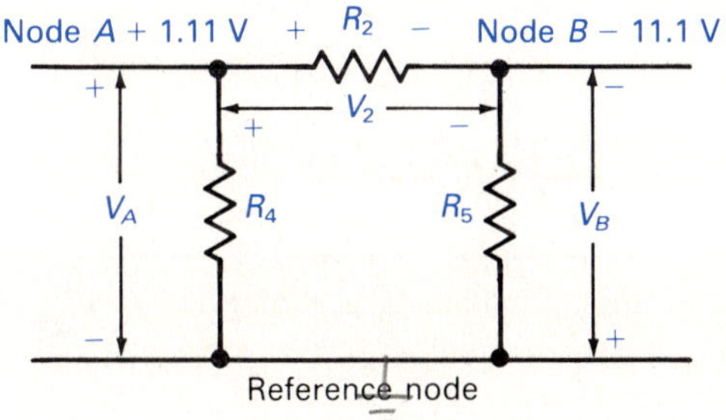

Fig. 9-18. Voltage across R_2 in Fig. 9-15.

$$V_2 = V_A - V_B = 1.11\text{ V} - (-11.1\text{ V}) = \mathbf{12.21\ V}$$

The polarity of V_2 will be as shown.

EXAMPLE 9-6 Figure 9-19 shows the basic elements of an analog adding circuit. Show that the output voltage V_o is approximately proportional to the sum of the input voltages V_1 to V_3.

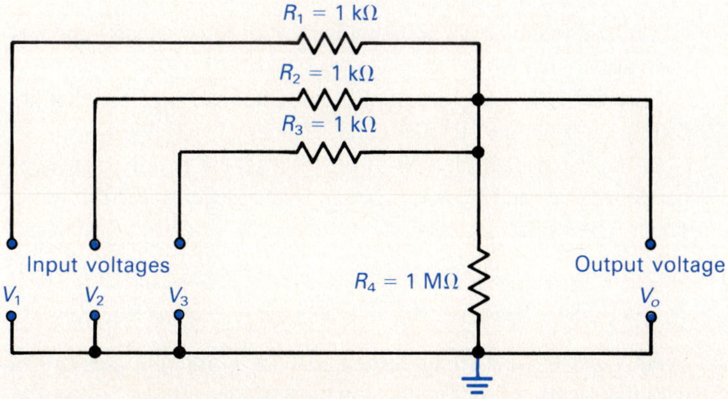

Fig. 9-19. Example 9-6: Circuit.

Solution Redrawing the circuit in Fig. 9-19 with the input voltages and resistances converted to current sources and conductances results in the equivalent circuit in Fig. 9-20*a*.

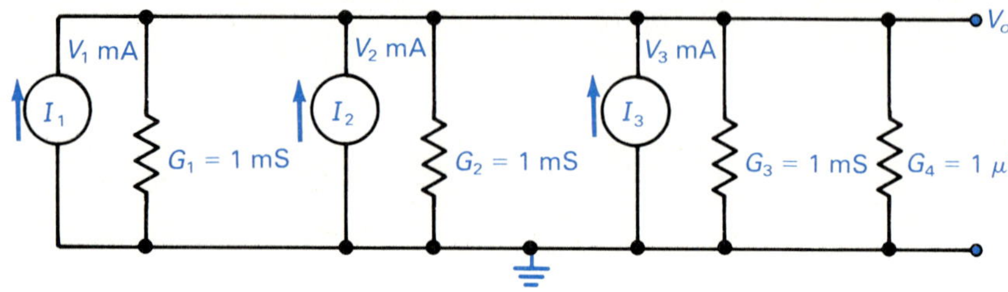

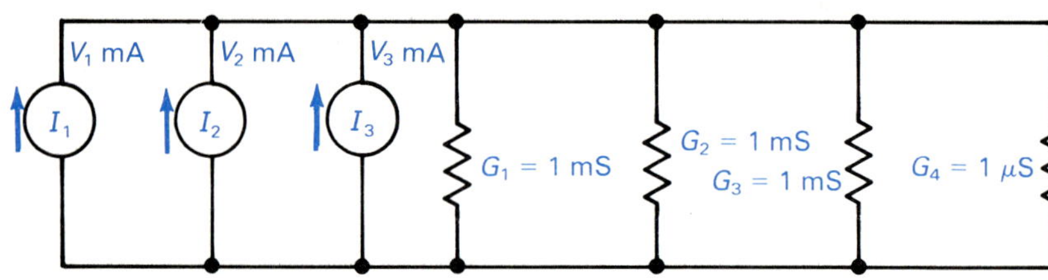

Fig. 9-20. Equivalent circuits of Fig. 9-19.

In Fig. 9-20*a* each of the current sources I_1 to I_3 and the associated internal conductances G_1 to G_3 is obtained from a voltage source consisting of the related voltage V_1 to V_3 and resistance R_1 to R_3.

Each of the resistances R_1 to R_3 has the value 1 kΩ, *so that each source current has a value in milliamperes numerically equal to its related voltage (for example,* $I_1 = V_1/R_1 = V_1$ volts/1 kΩ $= V_1$ milliamperes).

Each of the conductances G_1 to G_3 is the reciprocal of the related resistance R_1 to R_3, and hence equals 1 mS (e.g., $G_1 = 1/R_1 = 1/1\ \text{k}\Omega = 1\ \text{mS}$).

The conductance G_4 is the reciprocal of resistance R_4 and has the value 1 μS.

Figure 9-20*b* is the circuit of Fig. 9-20*a* rearranged so that the source currents and source conductances are grouped.

In the equivalent circuits in Fig. 9-20 there is only one independent node, and V_o is its voltage with respect to the reference node. The nodal equation is

$$V_o(G_1 + G_2 + G_3 + G_4) = I_1 + I_2 + I_3$$
$$V_o(1\ \text{mS} + 1\ \text{mS} + 1\ \text{mS} + 1\ \mu\text{S}) = (V_1 + V_2 + V_3)\quad \text{mA}$$

As 1 μS is very small compared with 1 mS, the quantity in parentheses on the left of the equation is approximately equal to 3 mS.

$$V_o \times 3\ \text{mS} \simeq (V_1 + V_2 + V_3) \quad \text{mA}$$

$$V_o \simeq \frac{V_1 + V_2 + V_3}{3} \quad \text{V}$$

In an adding circuit such as that in Fig. 9-19, with any number of inputs, the output voltage will be approximately equal to the sum of the input voltages divided by the number of inputs. Thus the output voltage will be approximately proportional to the sum of the input voltages.

EXAMPLE 9-7 The circuit in Fig. 9-21 is a transistor switch. It is to be biased by R_1 and R_2 in the ON condition. In this state, the base-to-emitter voltage V_{BE} is to be 0.7 V, and the base current I_B is to be 1 mA. If a value of 1 kΩ is chosen for R_1, what resistance would be suitable for R_2?

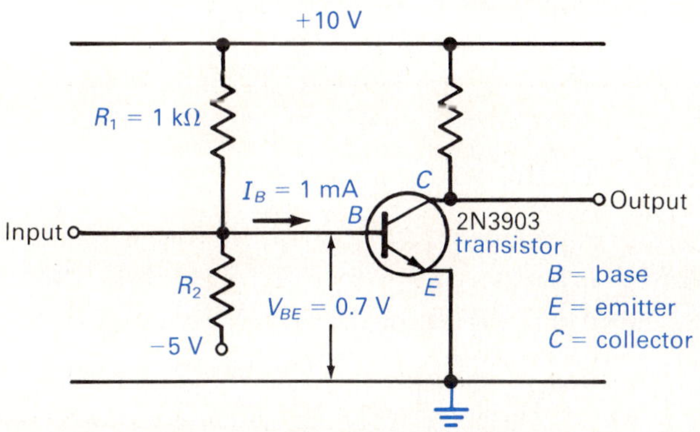

Fig. 9-21. Example 9-7: Circuit.

Solution The base-to-emitter resistance is

$$R_{BE} = \frac{V_{BE}}{I_B} = \frac{0.7\ \text{V}}{1\ \text{mA}} = 700\ \Omega$$

The bias circuit can be shown as in Fig. 9-22.

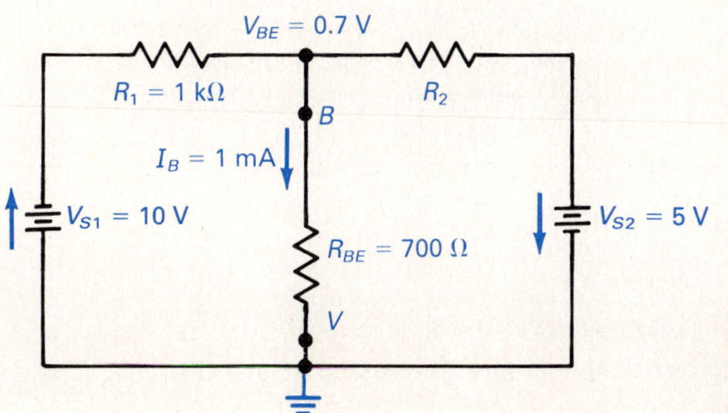

Fig. 9-22. Equivalent circuit of Fig. 9-21.

If V_{S1} with R_1, and V_{S2} with R_2 are considered voltage generators converted to current generators, Fig. 9-23 results.

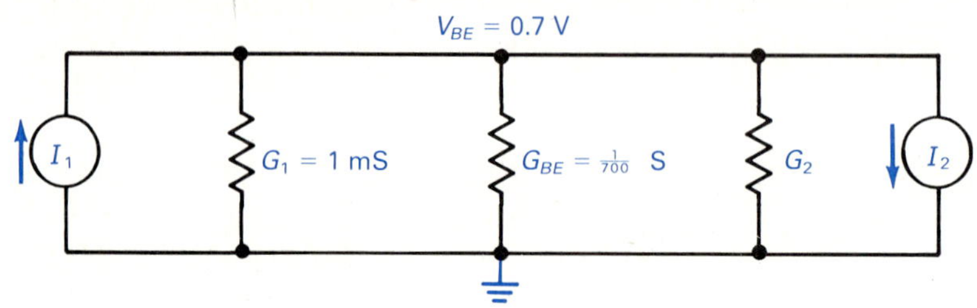

Fig. 9-23. Current-source equivalent of Fig. 9-22.

$$I_1 = V_{S1}G_1 = 10 \text{ V} \times 1 \text{ mS} = 10 \text{ mA}$$
$$I_2 = V_{S2}G_2 = 5 \text{ V} \times G_2 \text{ S} = 5G_2 \text{ A}$$

The equation for the single node is

$$V_{BE}(G_1 + G_{BE} + G_2) = I_1 - I_2$$
$$0.7 \text{ V}(10 \text{ mS} + \tfrac{1}{700} \text{ S} + G_2 \text{ S}) = 10 \text{ mA} - 5G_2 \text{ A}$$
$$7 \times 10^{-4} + 10^{-3} + 0.7\, G_2 = 10^{-2} - 5G_2$$
$$G_2 = 1.456 \text{ mS}$$
$$R_2 = \mathbf{686.8\ \Omega}$$

The preferred value, 680 Ω, would probably be used.

EXAMPLE 9-8 Part of the circuit of an airplane rudder control system is given in Fig. 9-24. What is the required resistance range for R_5 to maintain the bridge in balance with V_{AC} equal to zero?

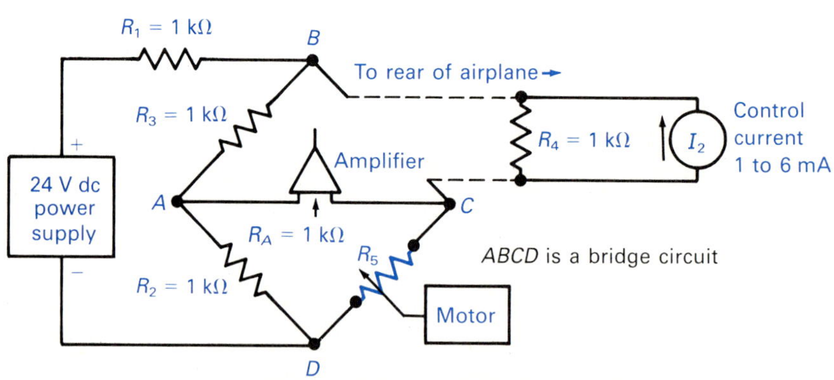

Fig. 9-24. Example 9-8: Circuit.

Solution Making D the reference node, converting the power supply (voltage source) and R_1 to a current source, and changing resistances to conductances gives Fig. 9-25, the equivalent circuit.

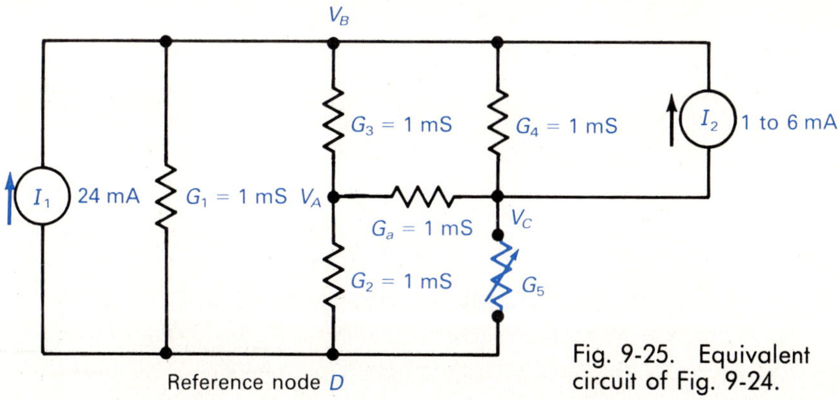

Fig. 9-25. Equivalent circuit of Fig. 9-24.

With the voltages in volts, currents in milliamperes, and conductances in millisiemens, the node equations are

Node A: $\quad 3V_A - V_B - V_C = 0$

Node B: $\quad -V_A + 3V_B - V_C = 24 + I_2$

Node C: $\quad -V_A - V_B + (2 + G_5)V_C = -I_2$

These equations apply whether or not the bridge is balanced, and for any value of the control current I_2. In this example, the bridge is balanced, so

$$V_{AC} = 0$$

or

$$V_A = V_C$$

Substituting V_A for V_C, the equations become

$$2V_A - V_B = 0 \qquad \textbf{(9-3)}$$

$$-2V_A + 3V_B = 24 + I_2 \qquad \textbf{(9-4)}$$

$$(1 + G_5)V_A - V_B = -I_2 \qquad \textbf{(9-5)}$$

From Eq. (9-3),

$$V_B = 2V_A$$

Substituting $2V_A$ for V_B in Eqs. (9-4) and (9-5),

$$4V_A = 24 + I_2 \qquad \textbf{(9-6)}$$

$$(1 - G_5)V_A = I_2 \qquad \textbf{(9-7)}$$

Equations (9-6) and (9-7) are valid for any value of I_2, if the bridge is balanced. It remains to insert the values of I_2.

When $I_2 = 1$ mA, we obtain from Eq. (9-6)

$$V_A = \frac{24 + 1}{4} = 6.25 \text{ V}$$

Substituting for V_A and I_2 in Eq. (9-7),

$$6.25(1 - G_5) = 1$$

$$G_5 = 0.84 \text{ mS}$$

$$R_5 = \mathbf{1.19\ k\Omega}$$

When $I_2 = 6$ mA, Eq. (9-6) gives

$$V_A = \frac{24 + 6}{4} = 7.5 \text{ V}$$

Substituting for V_A and I_2 in Eq. (9-7),

$$7.5(1 - G_5) = 6$$
$$G_5 = 0.2 \text{ mS}$$
$$R_5 = \mathbf{5\ k\Omega}$$

Therefore the motor-driven variable resistor R_5 must have a range of 1.19 to 5 kΩ to maintain the bridge in balance while the control current varies between 1 and 6 mA.

Example 9-8 illustrates the use of subunits to simplify equations. When powers of 10 are used, the equations appear to be quite complicated.

REVIEW QUESTIONS FOR SECTION 9-2

1. To what other form of analysis is nodal analysis considered complementary?
2. What is a node?
3. Name five conditions under which nodal analysis could be considered as a suitable analysis method.
4. When a reference node is selected, what term is used for the other nodes?
5. If an assumed current direction in nodal analysis is not correct, how will this show up at the end of the analysis?
6. When considering node currents in nodal analysis, what electrical law is used?
7. How should R_1 in Fig. 9-26 be treated in a nodal equation?

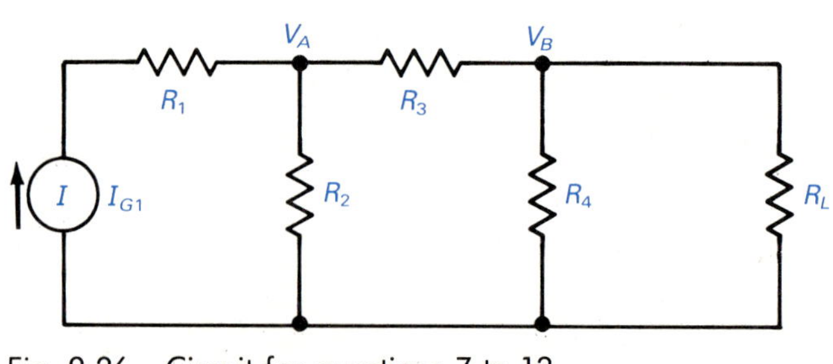

Fig. 9-26. Circuit for questions 7 to 13.

8. For nodal analysis, what should be the first step taken with respect to resistors and sources?
9. In a nodal equation for Fig. 9-26, R_1, R_2, R_3, and V_a are assembled to make one term. What is this term with its sign?

10. Repeat Question 9-9 with respect to R_3, R_4, R_L, and V_b.
11. How many equations are required for the nodal-analysis solution of the circuit of Fig. 9-26?
12. When converting a voltage generator to a current generator, how does the internal conductance of the current generator compare with the internal resistance of the voltage generator?
13. When converting a voltage generator to a current generator, how is the source current of the current generator related to the emf of the voltage generator?

EXERCISE PROBLEMS FOR SECTION 9-2

1. For Prob. 1*a* through *c*, Sec. 9-1, convert the voltage generator *V* and series resistance R_1 to an equivalent current source, convert all resistances to conductances, and solve for I_L using nodal analysis.
(*ans*. *a*. 68.02 mA; *b*. 1.108 mA; *c*. 6.501 μA)
2. For the circuit in Fig. 9-11, using nodal analysis, find the current through G_2 when $I_{G1} = 100$ mA, $I_{G2} = 200$ mA, $G_1 = 2.5$ mS, $G_2 = 2$ mS, and $G_3 = 3.33$ mS.
3. For Prob. 2*a* and *b*, Sec. 9-1, convert the voltage generator *V* and the series 680-Ω resistance to an equivalent current source, convert all resistances to conductances, and solve for I_L using nodal analysis.
(*ans*. *a*. 27.97 mA; *b*. 13.41 mA)
4. Solve Prob. 4*a* and *b*, Sec. 9-1, using nodal analysis.
5. Using nodal analysis and Fig. 9-12 (page 272), find the current in each of the conductances when $I_{G1} = 50$ mA, $I_{G2} = 100$ mA, $I_{G3} = 50$ mA, $G_1 = 10$ mS, $G_2 = 5$ mS, $G_3 = 2$ mS, $G_4 = 5$ mS, and $G_5 = G_6 = G_7 = G_8 = 1$ mS. (*ans*. 50 mA, 5.844 mA, 62.86 mA, 51.3 mA, 44.16 mA, 42.99 mA, 11.56 mA, 1.3 mA)
6. The domestic power system in Example 9-14 and Fig. 9-45 (pp. 293–294) can be redrawn as in Fig. 9-27. Using nodal analysis, find the powers dissipated by each of the three loads R_A, R_B, and R_C.
7. Using nodal analysis, find the current supplied by each generator in Fig. 9-27 if R_C is changed to 7 Ω.

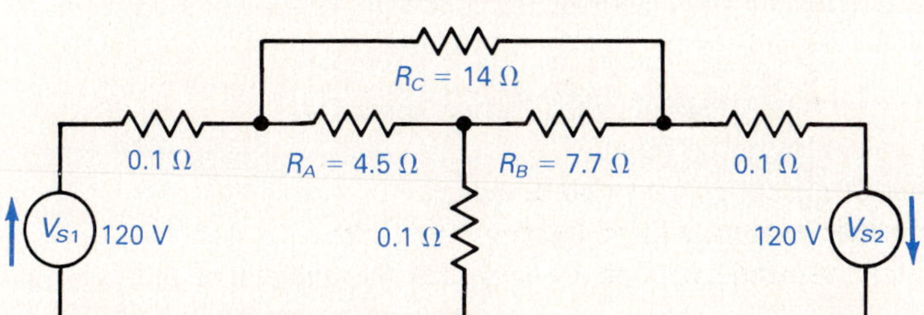

Fig. 9-27. Problems 6 and 7: Circuit.

8. The circuit in Fig. 9-28 is a type of transistor switch. When the switch is ON, $V_{BE} = 0.2$ V and $I_B = 150$ μA. Using nodal analysis, find a value for

R_S which will allow I_B to be 150 μA when a 4-V positive pulse is applied to the input. To ensure that I_B is at least 150 μA, should the preferred-value resistor selected for R_S be higher than the calculated value?

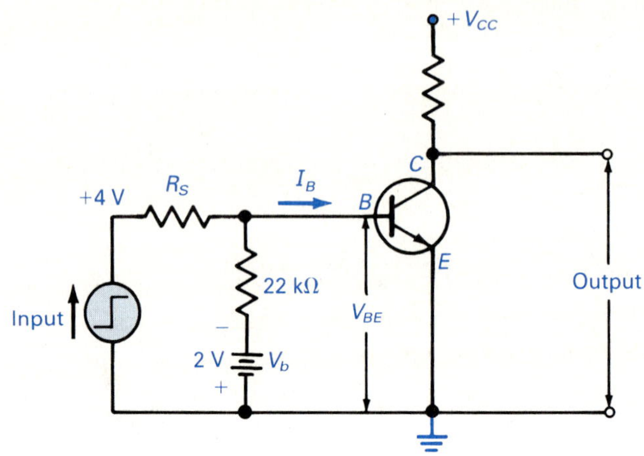

Fig. 9-28. Problems 8 to 10: Circuit.

9. If V_b in Prob. 8 is increased to 3 V, using nodal analysis find the value to which R_S must be changed for I_B to remain at 150 μA.

(*ans.* 12.86 kΩ)

10. If R_S in Prob. 8 is left at the value calculated and V_b is increased to 3 V, to what value must the input pulse be changed for I_B to remain at 150 μA? Solve by nodal analysis.

9-3 LOOP ANALYSIS

Where nodal analysis uses currents at nodes as its basis, loop analysis uses voltages around loops for the same ends, finding current and voltage values. Deciding which to use in a particular situation can be difficult, but you will probably get to prefer one and use it more often. Only by giving both a good try can you form an opinion.

Some guidelines to help you determine the conditions under which loop analysis is most applicable are:

1. Energy sources are predominantly voltage generators
2. Current values are to be found
3. A number of quantities are to be calculated
4. There are two or more energy sources
5. Number of independent loops are less than the number of independent nodes

A *loop* or *mesh* is a continuous path that can be traced in a circuit from one point, through a number of elements, back to the starting point. The path can go in any direction and through elements, including sources, carrying different currents; the only thing that matters is that the path is complete, with no breaks.

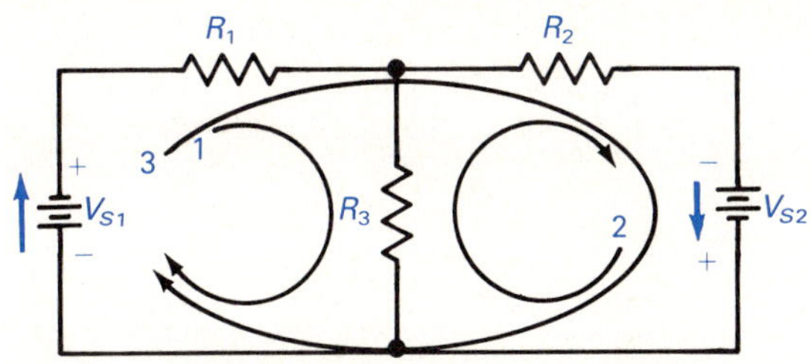

Fig. 9-29. Identifying circuit loops or meshes.

Figure 9-29, a T network connected between two voltage sources, is a simple example for showing loops.

There are three possible loops:

1. From the positive of V_{S1}, through R_1 and R_3, and back to negative V_{S1}
2. From the positive of V_{S2}, through R_3 and R_2, and back to negative V_{S2}
3. From the positive of V_{S1}, through R_1, R_2, and V_{S2}, and back to negative V_{S1}

The number of loops needed for analysis is the minimum which will include each element at least once. Any two of the three loops in Fig. 9-29 embrace all the elements, so only two are needed.

Because of the large variety of circuit configurations, there can be no general rule for deciding which loops to use to give the simplest analysis. Usually, selection of clearly visible loops is the most productive and least likely to result in errors. Therefore loops 1 and 2 are the logical selection for the circuit in Fig. 9-29 (not loops 1 and 3 or 2 and 3).

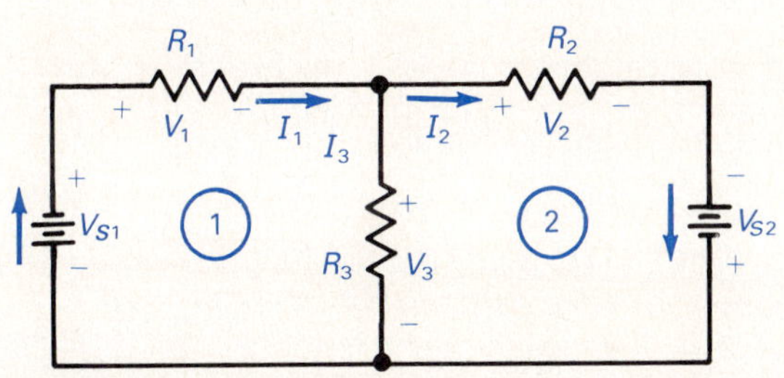

Fig. 9-30. Current directions and voltage polarities within loops.

Figure 9-30 shows the same circuit as Fig. 9-29, but current directions and voltage polarities are included. For elements R_1 and R_2 these directions and polarities are compatible with the emf polarities, and so are probably correct. The emfs V_{S1} and V_{S2} each tend to produce opposite current directions in R_3. The current direction and voltage polarity shown for R_3 are assumed. Provided that the current direction and voltage polarity are consistent for each individual element, the analysis results will show whether these assumptions are correct.

Kirchhoff's voltage law is applied to develop an equation for each loop. The algebraic sum of the source emfs equals the algebraic sum of the element pds.

Voltage Polarity

The polarity of a voltage for determining an algebraic sum is established as follows.

Source Voltages (emfs)

1. If the loop contains one emf, its polarity is considered positive.
2. If there is more than one emf in a loop, one polarity is assumed to be positive. Any emf of opposing polarity is then negative.
3. If there are no emf sources in a loop, the sum is zero.

Element Voltages (pds)

1. An element's voltage is positive if its current direction is compatible with a positive emf.
2. Opposite current directions produce negative pds.

In Fig. 9-30 the polarities are

Loop 1	Loop 2
V_{S1} positive	V_{S2} positive
V_1 positive	V_2 positive
V_3 positive	V_3 negative

In loop 1, I_3 (assumed to be downward) is compatible with the polarity of V_{S1}; V_3 is positive. *In loop 2, I_3 is not compatible with the polarity of V_{S2}; V_3 is negative.*

The loop equations are

Loop 1: $$V_{S1} = V_1 + V_3 = I_1R_1 + I_3R_3$$

But $I_3 = I_1 - I_2$ (Kirchhoff's current law),

$$V_{S1} = I_1R_1 + (I_1 - I_2)R_3$$

or $$(R_1 + R_3)I_1 - R_3I_2 = V_{S1} \tag{9-8}$$

Loop 2: $$V_{S2} = V_2 - V_3 = I_2R_2 - I_3R_3$$
$$= I_2R_2 - (I_1 - I_2)R_3$$

or $$-R_3I_1 + (R_2 + R_3)I_2 = V_{S2} \tag{9-9}$$

The solutions of Eqs. (9-8) and (9-9) will give the values of I_1 and I_2, and hence all the other circuit quantities.

Examination of Eqs. (9-8) and (9-9) leads to a speedy method of writing the loop equations. The technique is known as *Maxwell's cyclic currents,* in honor of the physicist who developed it about 100 years ago. You will see that it bears a similarity to the method of writing nodal equations used in Sec. 9-2. In fact it reinforces the duality of nodal and loop analyses.

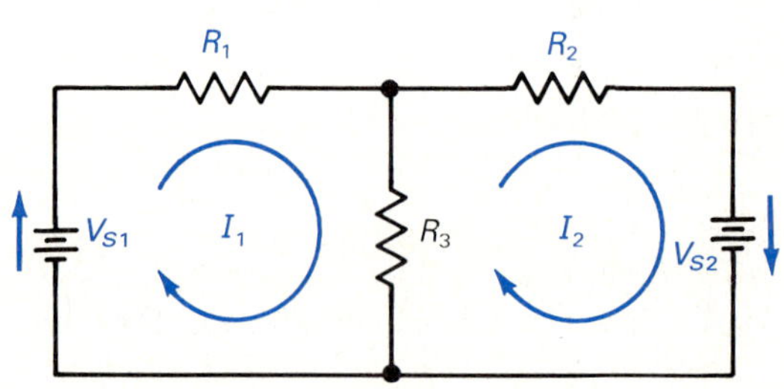

Fig. 9-31. Cyclic currents within loops.

The circuit is redrawn in Fig. 9-31 to show I_1 and I_2 as loop circulating currents rather than actual element currents. The directions selected for the currents are not important; the analysis results will show their validity.

Writing a Loop Equation

In the following procedure for directly writing the equation for a loop, ''local'' refers to the loop being considered, and ''remote'' refers to a loop which shares an element with the local loop.

Element Voltages

1. Place element voltages on the left side of the equation.
2. Multiply the local loop current by the sum of the resistances in that loop. Show the product as a positive quantity. Ignore any elements in parallel with voltage sources, as they cannot affect the circuit voltages.
3. For an element common to the local loop and some other loop, multiply the remote loop current by the resistance of the common element. Show the product as a positive quantity if the local and remote loop currents tend to be in the same direction in the common element. Show as a negative quantity if the current directions are opposing.

Source Voltages

1. Place source voltages on the right side of the equation.
2. Show voltages supporting the direction of the local loop current as positive quantities.
3. Show opposing emfs as negative quantities.
4. If there are no voltage sources in the loop, show as zero.

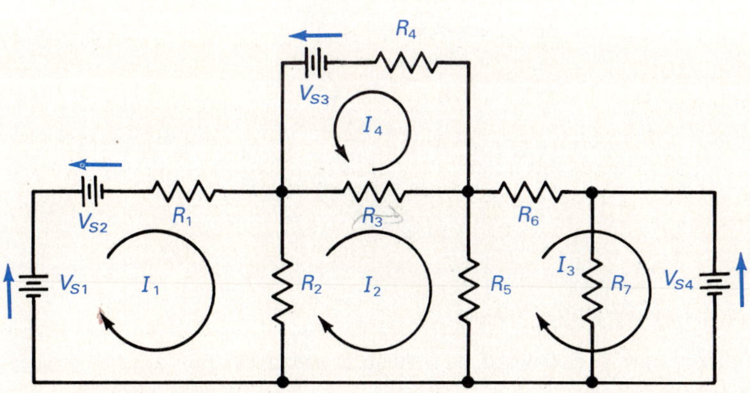

Fig. 9-32. Multiloop circuit.

In Fig. 9-32, element R_7 is ignored because it is in parallel with V_{S4}. The loop equations are

Loop 1: $(R_1 + R_2)I_1 - R_2I_2 = V_{S1} - V_{S2}$

Loop 2: $(R_2 + R_3 + R_5)I_2 - R_2I_1 + R_3I_4 - R_5I_3 = 0$

Loop 3: $(R_5 + R_6)I_3 - R_5I_2 = -V_{S4}$

Loop 4: $(R_3 + R_4)I_4 + R_3I_2 = V_{S3}$

EXAMPLE 9-9 What are the values of each current and voltage in the circuit in Fig. 9-33?

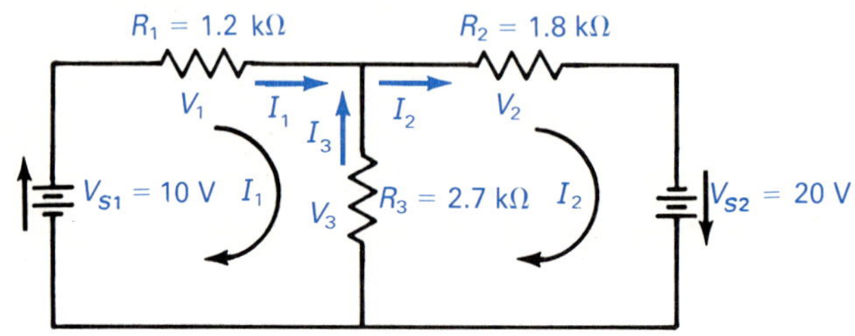

Fig. 9-33. Example 9-9: Circuit.

Solution Assigning loop current directions as shown, the equations are

Loop 1:
$$(R_1 + R_3)I_1 - R_3I_2 = V_{S1}$$
$$(1.2\ \text{k}\Omega + 2.7\ \text{k}\Omega)I_1 - (2.7\ \text{k}\Omega)I_2 = 10\ \text{V}$$

Loop 2:
$$(R_2 + R_3)I_2 - R_3I_1 = V_{S2}$$
$$(1.8\ \text{k}\Omega + 2.7\ \text{k}\Omega)I_2 - (2.7\ \text{k}\Omega)I_1 = 20\ \text{V}$$

Using kilohms, volts, and milliamperes, these equations reduce to

$$3.9I_1 - 2.7I_2 = 10$$
$$-2.7I_1 + 4.5I_2 = 20$$

The solutions of this pair of equations give:

$$I_1 = \mathbf{9.65\ mA}$$
$$I_2 = \mathbf{10.23\ mA}$$
$$I_3 = I_2 - I_1 = \mathbf{0.58\ mA}$$
$$V_1 = I_1R_1 = 9.65\ \text{mA} \times 1.2\ \text{k}\Omega = \mathbf{11.58\ V}$$
$$V_2 = I_2R_2 = 10.23\ \text{mA} \times 1.8\ \text{k}\Omega = \mathbf{18.41\ V}$$
$$V_3 = I_3R_3 = 0.58\ \text{mA} \times 2.7\ \text{k}\Omega = \mathbf{1.566\ V}$$

Points to note in Example 9-9 are:

1. The direction of I_3 is upward, as indicated, because I_2 proves to be greater than I_1.
2. Taking account of the polarity of V_3, Kirchhoff's voltage law is substantiated for each loop within the normal limits of arithmetic accuracy.
3. Omitting the letter k in equations may simplify calculations, but you must be sure that the currents are recognized as milliamperes.

4. I_1 and I_2 are, in addition to being the loop circulating currents, the actual currents in elements R_1 and R_2.

If a circuit includes current generators, they must be changed to equivalent voltage generators for loop analysis. Conductances should be changed to resistances.

The voltage generator in Fig. 9-34*b* will be equivalent to the current generator in Fig. 9-34*a* if

$$V = \frac{I}{G_i} \qquad \text{and} \qquad R_i = \frac{1}{G_i}$$

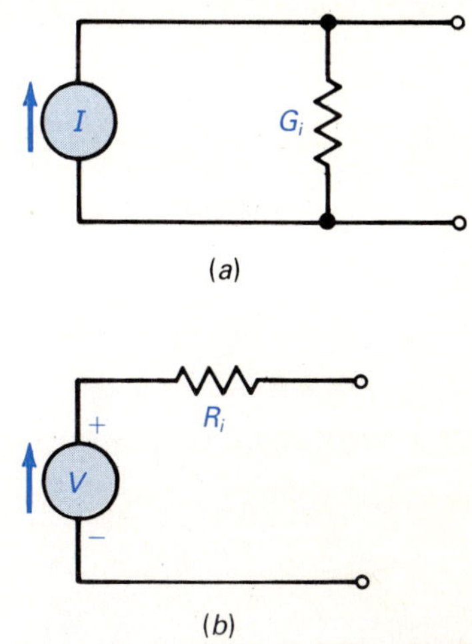

Fig. 9-34. Conversion of a current generator to a voltage generator.

EXAMPLE 9-10 What is the current supplied by the battery in Fig. 9-35?

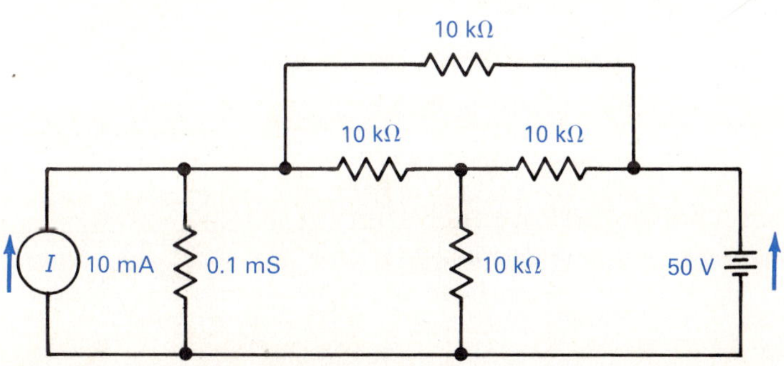

Fig. 9-35. Example 9-10: Circuit.

Solution Redraw Fig. 9-35 with the current generator changed to a voltage source and show the loop currents (see Fig. 9-36).

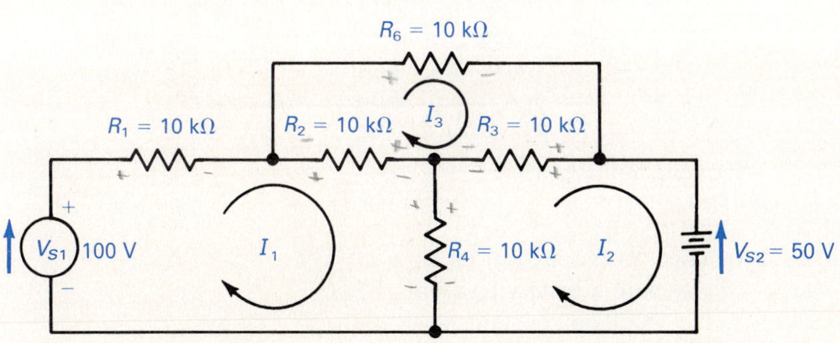

Fig. 9-36. Circuit in Fig. 9-35 with the current generator converted to a voltage generator.

The loop equations are

Loop 1: $(R_1 + R_2 + R_4)I_1 - R_4 I_2 - R_2 I_3 = V_{S1}$

$$30\ \text{k}\Omega \times I_1 - 10\ \text{k}\Omega \times I_2 - 10\ \text{k}\Omega \times I_3 = 100\ \text{V}$$

Loop 2: The battery emf V_{S2} has its polarity such that it opposes the indicated direction of the loop current I_2. Thus V_{S2} is shown as negative in the loop equation

$$(R_3 + R_4)I_2 - R_4 I_1 - R_3 I_3 = -V_{S2}$$
$$20\ \text{k}\Omega \times I_2 - 10\ \text{k}\Omega \times I_1 - 10\ \text{k}\Omega \times I_3 = -50\ \text{V}$$

Loop 3: $$(R_2 + R_3 + R_6)I_3 - R_2 I_1 - R_3 I_2 = 0$$
$$30\ \text{k}\Omega \times I_3 - 10\ \text{k}\Omega \times I_1 - 10\ \text{k}\Omega \times I_2 = 0$$

The solutions of these three simultaneous equations give

$$I_1 = 3.75\ \text{mA} \qquad I_2 = 0 \qquad I_3 = 1.25\ \text{mA}$$

The battery current is the circulating current I_2, which is zero. The 10 mA from the current generator, acting via the resistive bridged-T network, offsets the effect of the 50 V of the battery so that it supplies no current.

This example would have been difficult to solve by nodal analysis. The 50-V battery, without series resistance (nominally perfect), could not be converted into an equivalent current generator; it would generate infinite current (50 V/0 Ω), which is impossible.

EXAMPLE 9-11 What are the currents in the circuit in Fig. 9-37? (The current direction arrows were inserted after the solution was obtained.)

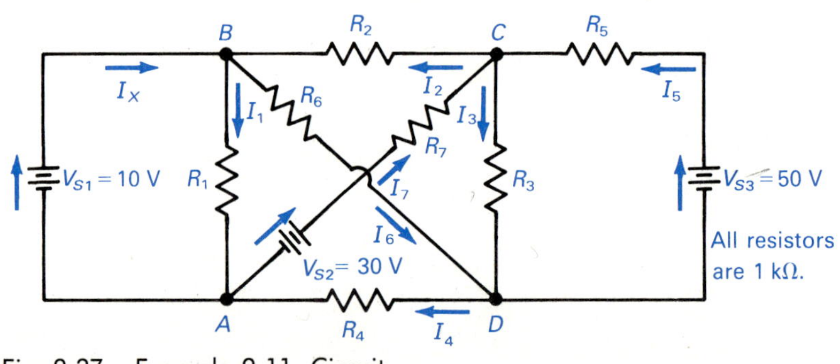

Fig. 9-37. Example 9-11: Circuit.

Solution Redraw the circuit with R_1 omitted because it is directly in parallel with a battery (see Fig. 9-38).

The loop equations are

$$\begin{array}{lllll}
(R_4 + R_6)I_A & - R_6 I_B & & & = V_{S1} \\
-R_6 I_A & + (R_2 + R_3 + R_6)I_B & -R_2 I_C & -R_3 I_D & = 0 \\
 & -R_2 I_B & + (R_2 + R_7)I_C & & = V_{S2} - V_{S1} \\
 & -R_3 I_B & & + (R_3 + R_5)I_D & = -V_{S3}
\end{array}$$

Insert quantities (with units of kilohms, volts, and milliamperes):

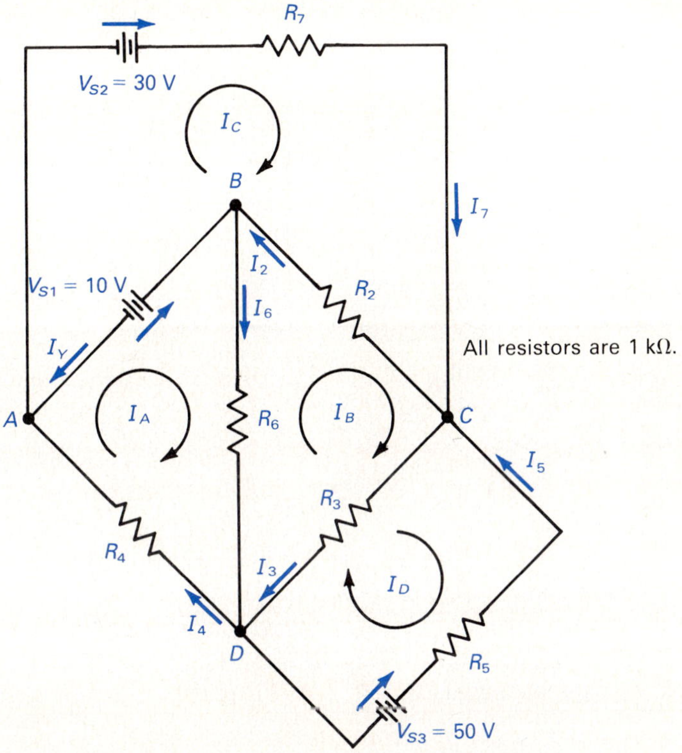

Fig. 9-38. Planar circuit equivalent to Fig. 9-37.

$$
\begin{aligned}
2I_A \quad -I_B \qquad\qquad\qquad &= 10 \\
-I_A + 3I_B \quad -I_C \quad -I_D &= 0 \\
-I_B + 2I_C \qquad\qquad &= 20 \\
-I_B \qquad\quad + 2I_D &= -50
\end{aligned}
$$

The solutions of the equations are

$$
\begin{aligned}
I_A &= 1\tfrac{2}{3}\text{ mA} \\
I_B &= -6\tfrac{2}{3}\text{ mA} \qquad (6\tfrac{2}{3}\text{ mA counterclockwise}) \\
I_C &= 6\tfrac{2}{3}\text{ mA} \\
I_D &= -28\tfrac{1}{3}\text{ mA} \qquad (28\tfrac{1}{3}\text{ mA counterclockwise})
\end{aligned}
$$

The circuit currents are

$$I_1 = \frac{V_{S1}}{R_1} = \frac{10\text{ V}}{1\text{ k}\Omega} = \mathbf{10\text{ mA}} \qquad B \to A$$

$$I_2 = I_C - I_B = 6\tfrac{2}{3}\text{ mA} - (-6\tfrac{2}{3}\text{ mA}) = \mathbf{13\tfrac{1}{3}\text{ mA}} \qquad C \to B$$

$$I_3 = I_B - I_D = -6\tfrac{2}{3}\text{ mA} - (-28\tfrac{1}{3}\text{ mA}) = \mathbf{21\tfrac{2}{3}\text{ mA}} \qquad C \to D$$

$$I_4 = I_A = \mathbf{1\tfrac{2}{3}\text{ mA}} \qquad D \to A$$

$$I_5 = -I_D = -(-28\tfrac{1}{3}\text{ mA}) = \mathbf{28\tfrac{1}{3}\text{ mA}} \qquad D \to C$$

$$I_6 = I_A - I_B = 1\tfrac{2}{3}\text{ mA} - (-6\tfrac{2}{3}\text{ mA}) = \mathbf{8\tfrac{1}{3}\text{ mA}} \qquad B \to D$$

$$I_7 = I_C = \mathbf{6\tfrac{2}{3}\text{ mA}} \qquad A \to C$$

$$I_Y = I_C - I_A = 6\tfrac{2}{3}\text{ mA} - 1\tfrac{2}{3}\text{ mA} = \mathbf{5\text{ mA}} \qquad B \to A$$

$$I_X = I_1 + I_6 - I_2 = 10\text{ mA} + 8\tfrac{1}{3}\text{ mA} - 13\tfrac{1}{3}\text{ mA} = \mathbf{5\text{ mA}} \qquad A \to B$$

I_Y, the current in the 10-V battery in Fig. 9-38, has a direction opposing the battery emf V_{S1} as the effect of the 30- and 50-V batteries overcomes that of I_Y. But V_{S1} is *not* the actual 10-V battery current in the original circuit in Fig. 9-37, because element R_1 and current I_1 are omitted from Fig. 9-38. To determine the value of I_X, the real 10-V battery current, it is desirable to look at node B in Fig. 9-37. Node B and its current components are detailed in Fig. 9-39.

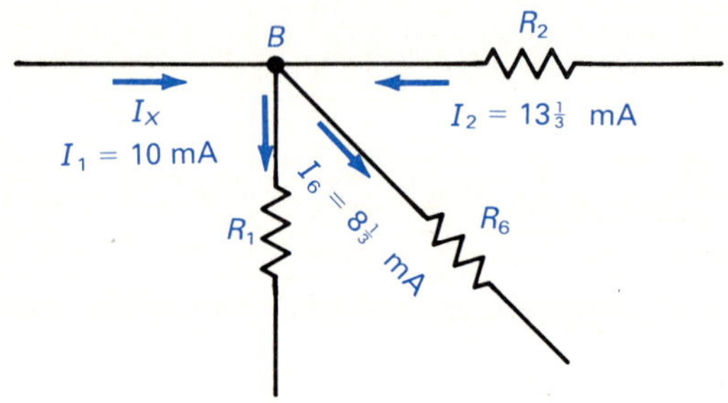

Fig. 9-39. Node B in Fig. 9-37.

The sum of the currents entering node B must equal the sum of the currents leaving it.

$$I_X + I_2 = I_1 + I_6$$
$$I_X = I_1 + I_6 - I_2 = 10\text{ mA} + 8\tfrac{1}{3}\text{ mA} - 13\tfrac{1}{3}\text{ mA}$$
$$I_X = 5\text{ mA}$$

Since I_X is a positive quantity, its assumed direction toward node B is correct.

The result for I_X is valid. Without R_1 the network would supply 5 mA to the battery V_{S1}. R_1 draws 10 mA from the battery. The net result is **5 mA** from the battery.

Significant points which arise from Example 9-11 are

1. It is desirable to redraw a circuit so that it becomes *planar* before attempting an analysis. A planar circuit has no crossing conductors.
2. If the assumed direction of a circulating current is wrong, it will present no problem. The negative values of I_B and I_D expose wrong assumptions about their directions.
3. If an element is omitted because it does not influence the main analysis, great care must be taken to ensure its reinsertion if its presence is important to some of the required data. Element R_1 was not significant in the general analysis, but was important in determining its current I_1 and the current associated with the 10-V battery I_X.
4. It is all right to use sub- and superunits, like milliamperes and kilohms, to simplify the writing and calculations, provided this is remembered when evaluating the circuit quantities.

EXAMPLE 9-12 The bridge circuit in Fig. 9-40 is the basis of commercial instruments for measuring resistance. The *unknown* (the device or network being measured) is connected between terminals X and Y. The element resistances (R_1, R_2, and R_3) are adjusted to give a balance condition. Balance is identified by a zero-current indication on the meter. The value of the unknown resistance is then read off a scale associated with the variable resistor R_3. During a particular measurement, the conditions shown in the diagram existed. What is the meter current, and what is the polarity of the voltage across it?

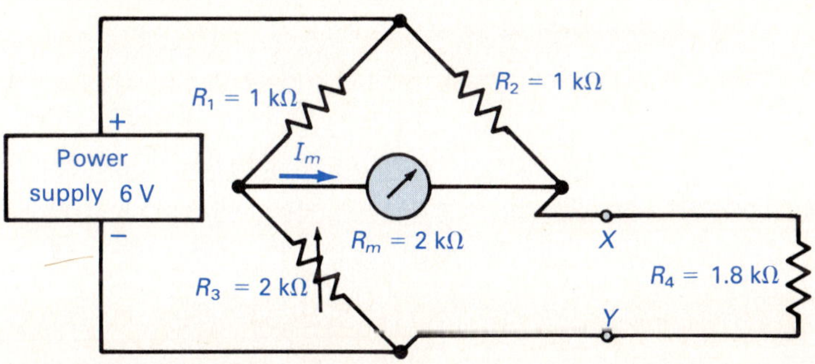

Fig. 9-40. Example 9-12: Circuit.

Solution Redraw the circuit as in Fig. 9-41.

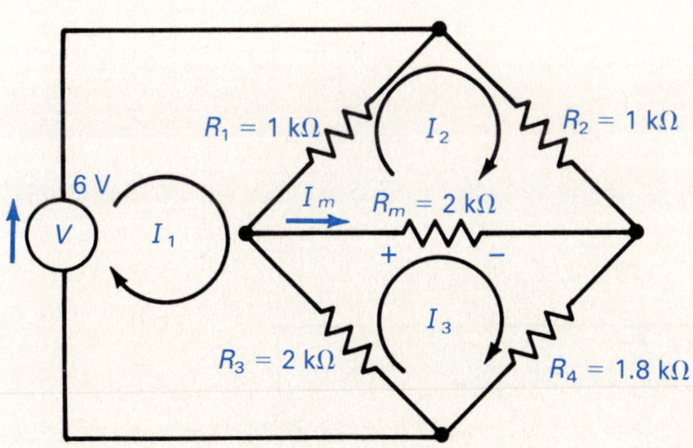

Fig. 9-41. Equivalent circuit of Fig. 9-40.

With the units volts, kilohms, and milliamperes, the loop equations are

$$3I_1 - I_2 - 2I_3 = 6$$
$$-I_1 + 4I_2 - 2I_3 = 0$$
$$-2I_1 - 2I_2 + 5.8I_3 = 0$$

From these equations are obtained

$$I_1 = 4.144 \text{ mA}$$
$$I_2 = 2.115 \text{ mA}$$
$$I_3 = 2.158 \text{ mA}$$

The meter current I_m is given by

$$I_m = I_3 - I_2 = 2.158 \text{ mA} - 2.115 \text{ mA} = \mathbf{0.043\ mA} \quad \text{or} \quad \mathbf{43\ \mu A}$$

The direction of the meter current is from left to right, as in the diagrams. Therefore the meter voltage will have the polarity $+ \rightarrow -$, as shown in Fig. 9-41.

EXAMPLE 9-13 The circuit in Fig. 9-42 is designed to provide base bias for a linear transistor amplifier in a stereo radio. What is the transistor's base-to-ground resistance R_{in} if its base current I_B is 10 μA?

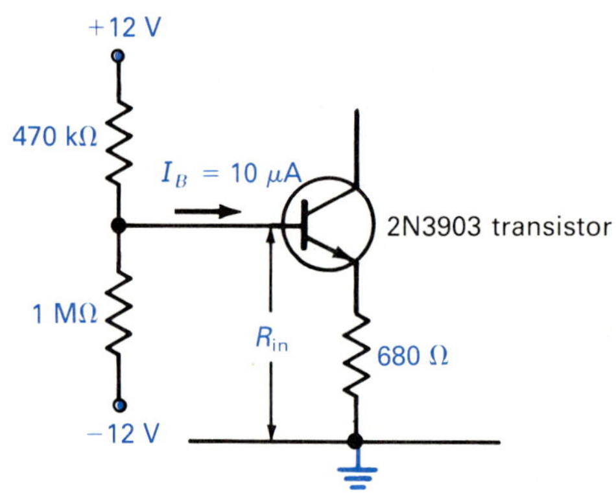

Fig. 9-42. Example 9-13: Circuit.

Solution The equivalent of the base-bias circuit is shown in Fig. 9-43.

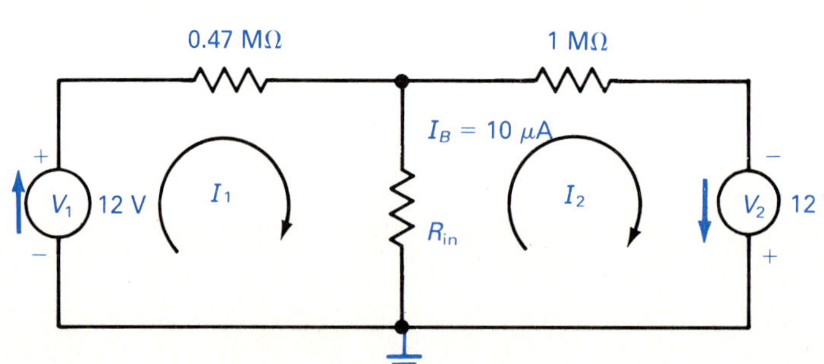

Fig. 9-43. Equivalent circuit of Fig. 9-42.

The loop equations, with the units volts, megohms, and microamperes, are

$$(0.47 + R_{in})I_1 - R_{in}I_2 = 12$$
$$-R_{in}I_1 + (1 + R_{in})I_2 = 12$$

As $I_1 = I_2 + I_B$ or $I_2 + 10\ \mu A$, this expression for I_1 can be substituted in the loop equations, giving

$$0.47I_2 + 10R_{in} = 7.3$$
$$I_2 - 10R_{in} = 12$$

Solving these equations for R_{in} results in

$$R_{in} = \mathbf{0.1129\ M\Omega} \qquad \text{or} \qquad \mathbf{112.9\ k\Omega}$$

EXAMPLE 9-14 A domestic three-wire system with three resistive loads is shown in Fig. 9-44. Each of the lines leading to the house has a resistance of 0.1 Ω. The load powers indicated are based on 120 V and 240 V being present at the loads. What actually is each load power? Assume that the load resistances are not affected by the reduced voltages as a result of line resistance.

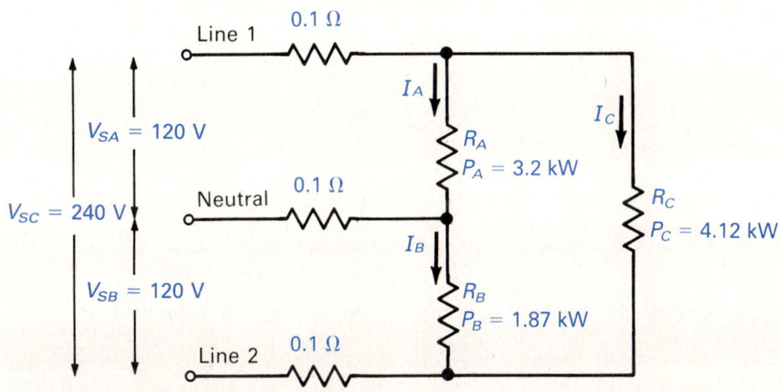

Fig. 9-44. Example 9-14: Network.

Solution

$$P = \frac{V^2}{R} \qquad \text{or} \qquad R = \frac{V^2}{P}$$

$$R_A = \frac{V_{SA}^2}{P_A} = \frac{(120\ \text{V})^2}{3.2\ \text{kW}} = 4.5\ \Omega$$

$$R_B = \frac{V_{SB}^2}{P_B} = \frac{(120\ \text{V})^2}{1.87\ \text{kW}} = 7.7\ \Omega$$

$$R_C = \frac{V_{SC}^2}{P_C} = \frac{(240\ \text{V})^2}{4.12\ \text{kW}} = 14\ \Omega$$

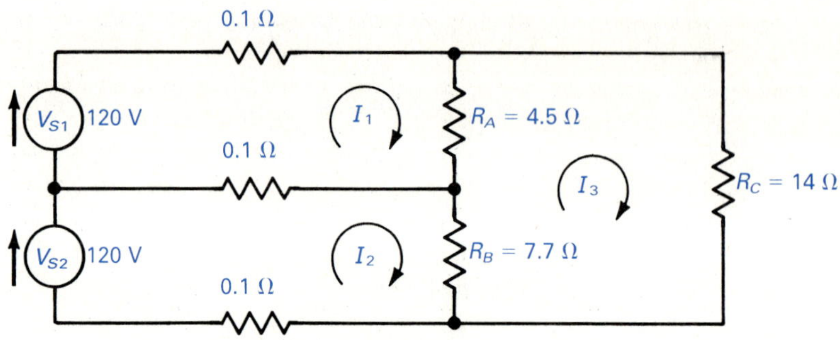

Fig. 9-45. Equivalent circuit of Fig. 9-44.

The circuit equivalent to Fig. 9-44 is shown in Fig. 9-45. The loop equations are

$$4.7I_1 - 0.1I_2 - 4.5I_3 = 120$$
$$-0.1I_1 + 7.9I_2 - 7.7I_3 = 120$$
$$-4.5I_1 - 7.7I_2 + 26.2I_3 = 0$$

The solutions for these equations are

$$I_1 = 42.12 \text{ A}$$
$$I_2 = 31.92 \text{ A}$$
$$I_3 = 16.61 \text{ A}$$

The load currents are

$$I_A = I_1 - I_3 = 25.59 \text{ A}$$
$$I_B = I_2 - I_3 = 15.31 \text{ A}$$
$$I_C = I_3 = 16.61 \text{ A}$$

The load powers therefore are

$$P_A = I_A{}^2R_A = (25.59 \text{ A})^2 \times 4.5\ \Omega = \mathbf{2.947\ kW}$$
$$P_B = I_B{}^2R_B = (15.31 \text{ A})^2 \times 7.7\ \Omega = \mathbf{1.804\ kW}$$
$$P_C = I_C{}^2R_C = (16.61 \text{ A})^2 \times 14\ \Omega = \mathbf{3.862\ kW}$$

These results imply that the effect of the line resistances, in this case, is to reduce the total power delivered to the loads by about 6 percent.

REVIEW QUESTIONS FOR SECTION 9-3

1. Name five conditions under which loop analysis could be considered as suitable for analyzing a network.
2. Give another name for "loop" when applied to an electrical circuit.
3. What is a network loop?
4. In the circuit shown in Fig. 9-22, how many loops are possible?
5. Should an element be included in more than one loop when analyzing a network?
6. What is the minimum number of loops required for analysis?

7. In loop analysis, how will an incorrect assumption of current direction show up?
8. To obtain the equations for the cyclic currents method of loop analysis, what currents are used rather than the actual element currents?
9. By what resistance is a local loop current multiplied in the cyclic current form of loop analysis?
10. If a source emf opposes the direction of local loop current, how is this indicated in a loop-analysis equation?
11. How does a loop equation show that there is no emf in the loop?
12. How should current generators be treated in loop analysis?

EXERCISE PROBLEMS FOR SECTION 9-3

1. Solve Prob. 1*a* through *c*, Sec. 9-1, using loop analysis.
(*ans*. *a*. 68.05 mA; *b*. 1.107 mA; *c*. 6.498 μA)
2. Solve Prob. 3*a* and *b*, Sec. 9-1, using loop analysis.

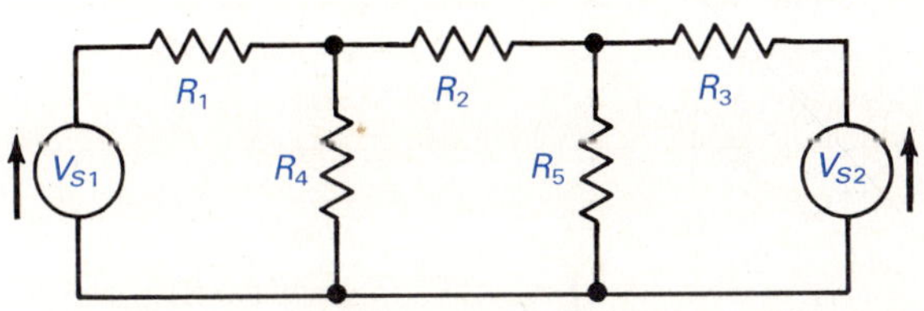

Fig. 9-46.
Problem 3.

3. For the circuit in Fig. 9-46, using loop analysis, find the direction and magnitude of current through R_2 when
a. $V_{S1} = 10$ V, $V_{S2} = 20$ V, $R_1 = 180\ \Omega$, $R_2 = 150\ \Omega$, $R_3 = 120\ \Omega$, $R_4 = 470\ \Omega$, $R_5 = 560\ \Omega$ (*ans*. left, 24.41 mA)
b. $V_{S1} = 40$ V, $V_{S2} = 50$ V, $R_1 = 12$ kΩ, $R_2 = 27$ kΩ, $R_3 = 22$ kΩ, $R_4 = 18$ kΩ, $R_5 = 33$ kΩ (*ans*. left, 126.6 μA)
4. Repeat Example 9-12 (page 291), finding the meter current by loop analysis when R_1 is 100 Ω and all the other values remain the same.
5. Repeat Example 9-12 (page 291), finding the meter current by loop analysis when R_1 is set to 1000 Ω and R_2 is set to 10 Ω.
(*ans*. left, 735 μA)
6. Convert the rudder control current source in Example 9-8 (page 278) to a voltage source, and using loop analysis find the resistance of R_5 for a balanced condition when the rudder is at mid-position. [Control current = (6 + 1)/2 = 3.5 mA.]
7. For Example 9-7 (page 277), find R_2 using loop analysis if the -5-V source is changed to -10 V. (*ans*. 1.289 kΩ)
8. In Prob. 7 if an input signal of +2 V is applied through a 1000-Ω resistor, to what value will I_B change, provided R_{BE} remains constant?
9. In Example 9-14 (page 293), if an additional 5-kW load is connected between lines 1 and 2, find the three line currents using loop analysis.
(*ans*. $I_1 = 61.26$ A, $I_2 = 51.26$ A, $I_N = 10$ A)

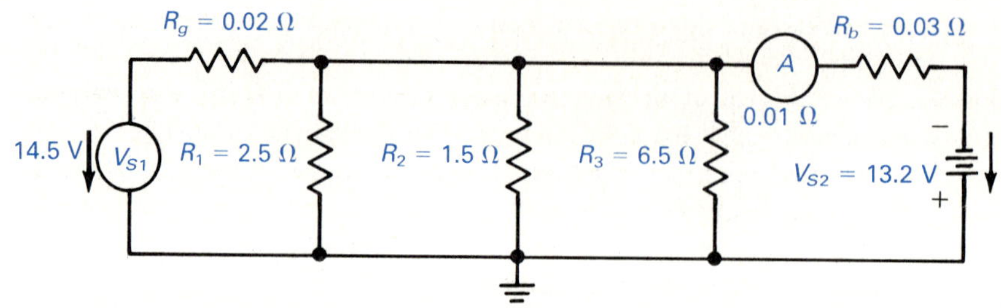

Fig. 9-47. Problem 10: Circuit.

10. The circuit in Fig. 9-47 represents a condition in the electric system of an automobile. V_{S1} is the generator with an internal resistance $R_g = 0.02\ \Omega$; R_1, R_2, and R_3 represent individual loads being drawn; A is the ammeter with an internal resistance of 0.01 Ω; and V_{S2} is the battery with an internal resistance R_b of 0.03 Ω. What is the magnitude of the current through the ammeter? Is the battery charging?

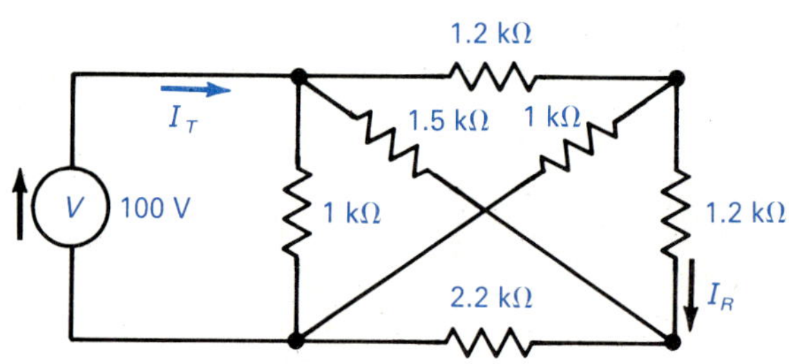

Fig. 9-48. Problem 11: Circuit.

11. Using loop analysis, find I_T and I_R in Fig. 9-48.
(*ans.* I_T = 173.2 mA, I_R = 5.31 mA)

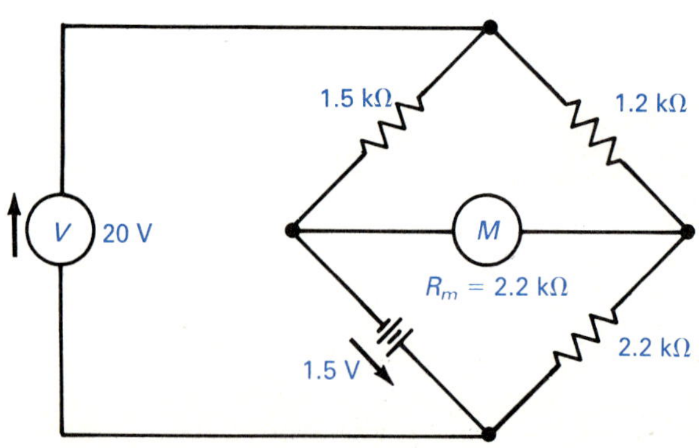

Fig. 9-49. Problem 12: Circuit.

12. Using loop analysis, find the meter current in the bridge circuit in Fig. 9-49.

9-4 SUPERPOSITION

Superposition is an important circuit concept which is of value in understanding dc and ac circuits.

In its effect on circuit quantities, any energy source can be considered separately from other sources. All individual effects can be combined to give the total effect.

This concept can be used to simplify circuit analysis.

Some guidelines on circuits which are most suitable for superposition analysis are

1. More than one energy source
2. Voltage or current sources
3. Simple circuit structure

While any of the nodal or loop examples with more than one source could have been solved by superposition, most of them involved circuits too complex for any advantage to have been gained. When any one of the energy sources is considered separately from the others, as described below, the remaining circuit will be series-parallel. Where a circuit is suitable for superposition analysis, this method is often preferred because it does not involve simultaneous equations.

There are six basic operations in applying superposition to circuit analysis.

1. Select one energy source.
2. Consider all other source voltages and currents to be removed.
   a. Source voltages to be short-circuited.
   b. Source currents to be open-circuited.
3. Leave internal resistances of removed sources in the circuit.
4. Determine the current in, or the voltage across, each element. Note the directions and polarities.
5. Repeat 1 to 4 for each source.
6. Algebraically add the individual results (take into account the direction or polarity).

Numerical examples best illustrate the use of this technique.

EXAMPLE 9-15 What are the currents in the circuit in Fig. 9-50?

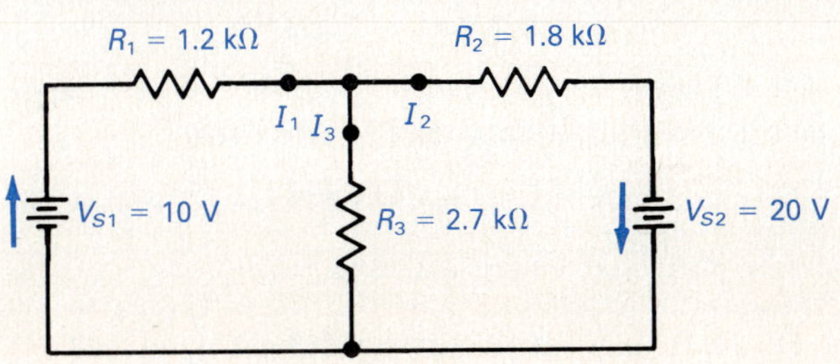

Fig. 9-50. Example 9-15: Circuit.

Solution Select source V_{S1} and redraw the circuit with V_{S2} removed, as in Fig. 9-51.

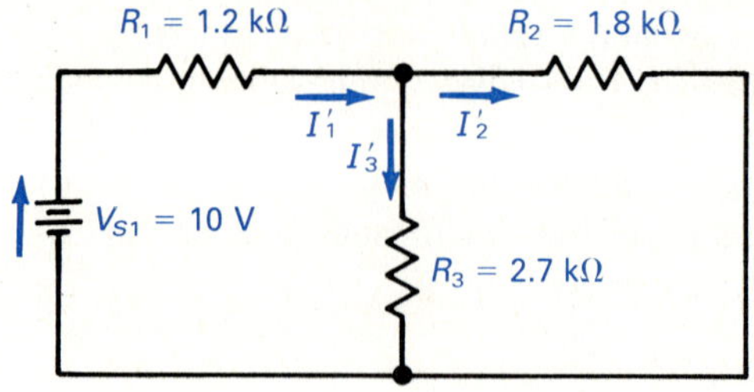

Fig. 9-51. Figure 9-50 with V_{S2} removed.

Calculate the current components due to V_1.

$$I_1' = \frac{V_{S1}}{R_1 + \dfrac{R_2R_3}{R_2 + R_3}} = \frac{10\text{ V}}{1.2\text{ k}\Omega + \dfrac{1.8\text{ k}\Omega \times 2.7\text{ k}\Omega}{1.8\text{ k}\Omega + 2.7\text{ k}\Omega}} = 4.386\text{ mA}$$

$$I_2' = I_1' \frac{R_3}{R_2 + R_3} = 4.386\text{ mA}\,\frac{2.7\text{ k}\Omega}{1.8\text{ k}\Omega + 2.7\text{ k}\Omega} = 2.632\text{ mA}$$

$$I_3' = I_1' - I_2' = 4.386\text{ mA} - 2.632\text{ mA} = 1.754\text{ mA}$$

Select source V_{S2} and redraw the circuit with V_{S1} removed as in Fig. 9-52.

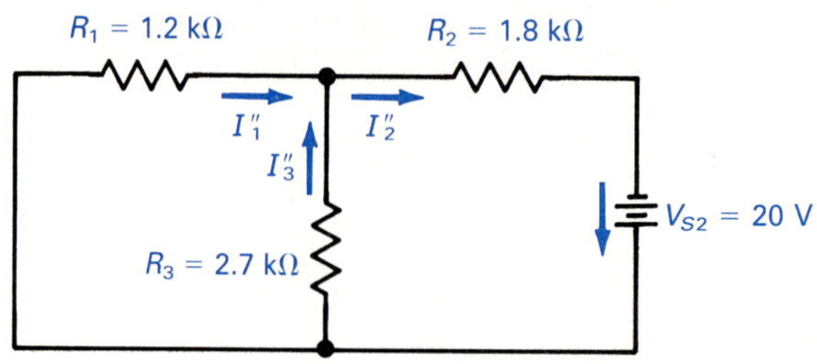

Fig. 9-52. Figure 9-50 with V_{S1} removed.

Calculate the current components due to V_{S2}.

$$I_2'' = \frac{V_{S2}}{R_2 + \dfrac{R_1R_3}{R_1 + R_3}} = \frac{20\text{ V}}{1.8\text{ k}\Omega + \dfrac{1.2\text{ k}\Omega \times 2.7\text{ k}\Omega}{1.2\text{ k}\Omega + 2.7\text{ k}\Omega}} = 7.602\text{ mA}$$

$$I_1'' = I_2'' \frac{R_3}{R_1 + R_3} = 7.602\text{ mA}\,\frac{2.7\text{ k}\Omega}{1.2\text{ k}\Omega + 2.7\text{ k}\Omega} = 5.263\text{ mA}$$

$$I_3'' = I_2'' - I_1'' = 7.602\text{ mA} - 5.263\text{ mA} = 2.339\text{ mA}$$

The actual current in any element is the sum of its components, taking into account their relative directions.

$$I_1 = I_1' + I_1'' = 4.386\text{ mA} + 5.263\text{ mA} = \mathbf{9.649\text{ mA}} \quad \text{to the right}$$

$$I_2 = I_2' + I_2'' = 2.632\text{ mA} + 7.602\text{ mA} = \mathbf{10.23\text{ mA}} \quad \text{to the right}$$

$$I_3 = I_3'' - I_3' = 2.339\text{ mA} - 1.754\text{ mA} = \mathbf{0.585\text{ mA}} \quad \text{upward}$$

The circuit in Example 9-15 is the same as that in Example 9-9. The results are the same within ordinary arithmetic limits. This shows the validity of the superposition technique.

EXAMPLE 9-16 What are the currents in the circuit in Fig. 9-53? (The current direction arrows were added after the solution was obtained.)

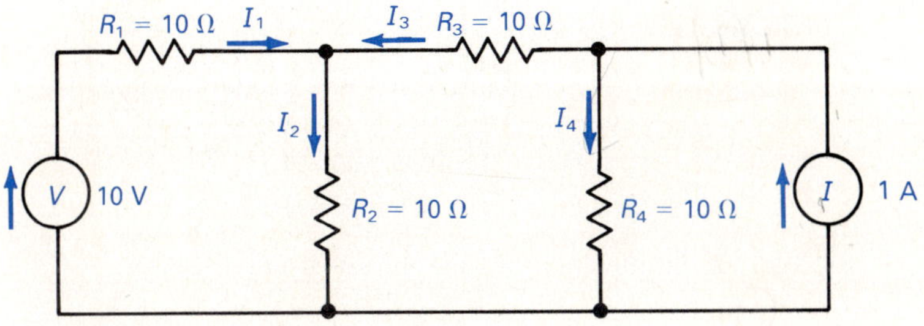

Fig. 9-53. Example 9-16: Circuit.

Solution The voltage generator on its own gives Fig. 9-54.

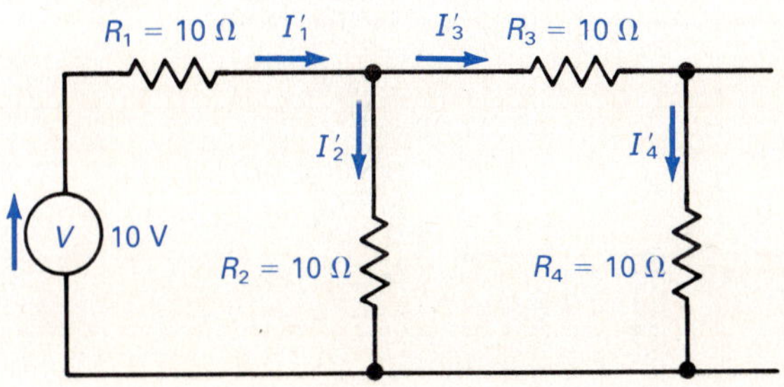

Fig. 9-54. Figure 9-53 with the current generator removed and replaced by its internal conductance (0 *S*)—an open circuit.

$$I'_1 = \frac{V}{R_1 + \dfrac{R_2(R_3 + R_4)}{R_2 + R_3 + R_4}} = \frac{10\ \text{V}}{10\ \Omega + \dfrac{10\ \Omega \times 20\ \Omega}{30\ \Omega}} = 0.6\ \text{A}$$

$$I'_2 = I'_1 \frac{R_3 + R_4}{R_2 + R_3 + R_4} = 0.6\ \text{A}\ \frac{20\ \Omega}{30\ \Omega} = 0.4\ \text{A}$$

$$I'_3 = I'_4 = I'_1 - I'_2 = 0.6\ \text{A} - 0.4\ \text{A} = 0.2\ \text{A}$$

The current generator on its own gives Fig. 9-55.

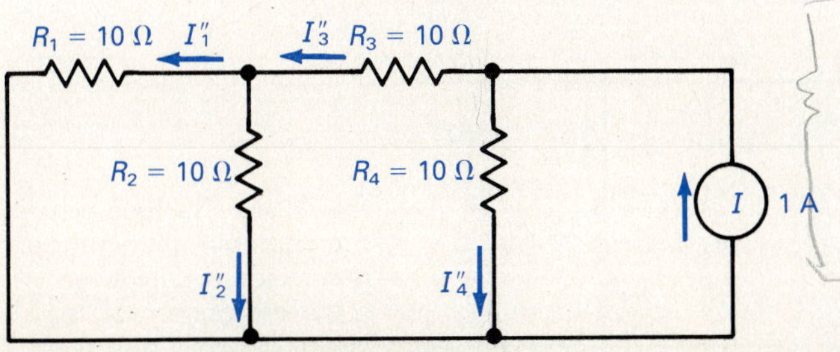

Fig. 9-55. Figure 9-53 with the voltage generator removed and replaced by its internal resistance (0 Ω)—a short circuit.

Apply the current-divider principle

$$I_3'' = I \frac{R_4}{\left(R_3 + \dfrac{R_1R_2}{R_1 + R_2}\right) + R_4} = 1\ \text{A}\ \frac{10\ \Omega}{25\ \Omega} = 0.4\ \text{A}$$

$$I_1'' = I_2'' = \frac{I_3''}{2} = 0.2\ \text{A}$$

$$I_4'' = I - I_3'' = 1\ \text{A} - 0.4\ \text{A} = 0.6\ \text{A}$$

The total currents are

$$I_1 = I_1' - I_1'' = 0.6\ \text{A} - 0.2\ \text{A} = \mathbf{0.4\ A}$$
$$I_2 = I_2' + I_2'' = 0.4\ \text{A} + 0.2\ \text{A} = \mathbf{0.6\ A}$$
$$I_3 = I_3'' - I_3' = 0.4\ \text{A} - 0.2\ \text{A} = \mathbf{0.2\ A}$$
$$I_4 = I_4' + I_4'' = 0.2\ \text{A} + 0.6\ \text{A} = \mathbf{0.8\ A}$$

The current directions are as shown in Fig. 9-53.

The superposition treatment can, in some cases, reduce the chance of error. If Example 9-16 were handled by nodal analysis, the voltage generator would be changed to a current generator. The resulting current generator (Fig. 9-56*b*) would then be incorporated into the original circuit in Fig. 9-53.

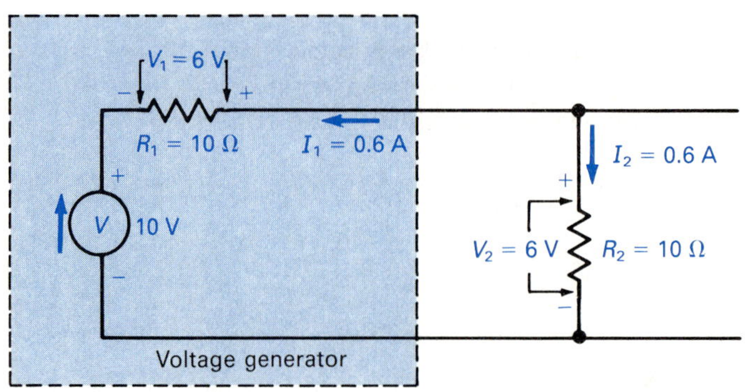

(*a*)

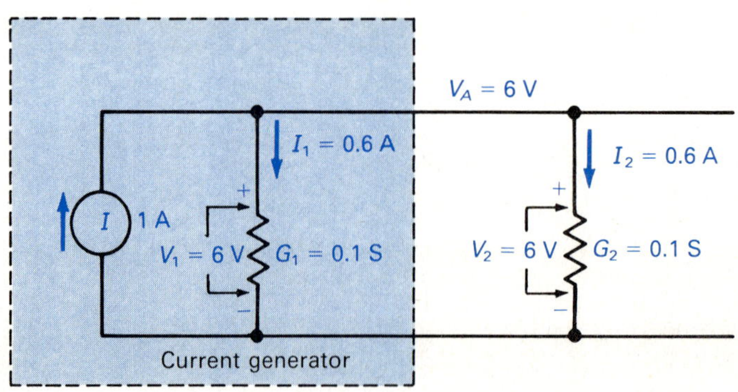

(*b*)

Fig. 9-56. Serious errors are possible if a voltage generator is converted to a current generator for nodal analysis and then converted back to a voltage generator with incorrect interpretation.

Upon solving the nodal equations, node voltage V_A (Fig. 9-56*b*) = 0.6 A/0.1 S = +6 V and $I_1 = I_2 = 0.6$ A. If V_A and I_1 are referred back to the voltage generator in Fig. 9-53 without care, they might appear as in Fig. 9-56*a*.

In checking further, it is noted that, if this is so, there are two odd situations:

1. The generator current direction is opposite that of the emf polarity. This result is possible in some circuits, but it does not appear to be correct in this one, in which the current source equivalent of the voltage generator has the same source current as the real original current generator.
2. The voltage V_1 across R_1 is 6 V (0.6 A × 10 Ω), with the polarity shown, so that the voltage across R_2 is $V + V_1 = 16$ V. But these data were obtained from a nodal voltage of 6 V across R_2.

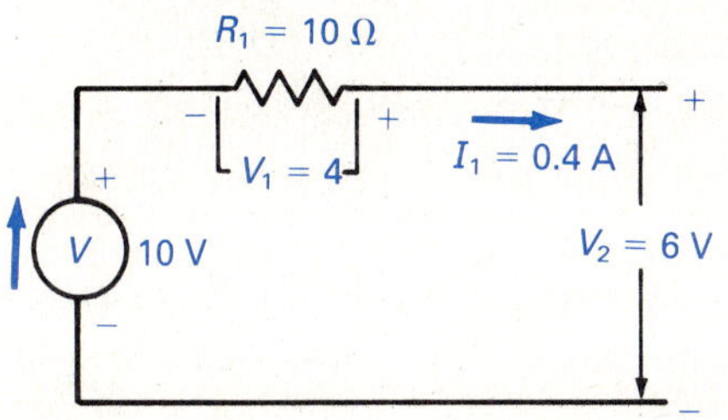

Fig. 9-57. Voltage generator in Fig. 9-53 with correct voltage and current polarities and direction.

The real voltage across R_2 is 6 V (not 16 V), with the polarity shown in Fig. 9-56*a*. This is the voltage across the generator and R_1 in series, as in Fig. 9-57.

The actual voltage across R_1 is 10 V − 6 V = 4 V, with 0.4 A current in same direction as Fig. 9-57. This result is the same as for Example 9-16.

For this circuit, nodal analysis is faster than the superposition technique. The superposition method eliminates the error tendency described above. Thus the decision regarding which method you should use could hinge on whether you want speed (nodal analysis) or caution (superposition).

EXAMPLE 9-17 Figure 9-58 is the circuit of an electronic voltage regulator. Its purpose is to keep the output constant at 20 V, despite changes in the current taken by the load or changes in the input voltage. If the base-to-emitter resistance of transistor Q_1 is 25 kΩ, what are its base current and base-to-emitter voltage?

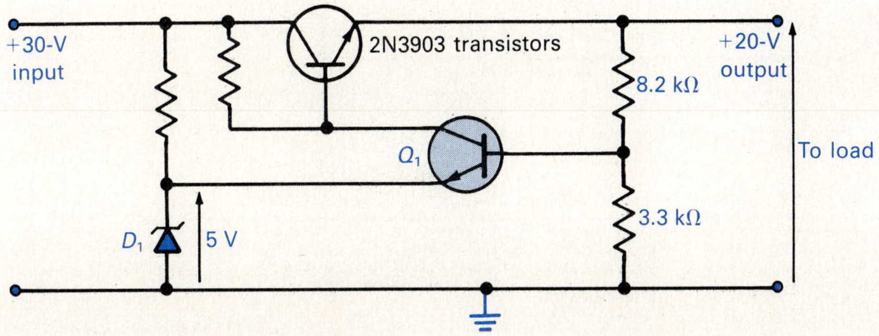

Fig. 9-58. Example 9-17: Circuit.

Solution The network which determines the base current I_B and the base-to-emitter voltage V_{BE} of Q_1 is shown in Fig. 9-59.

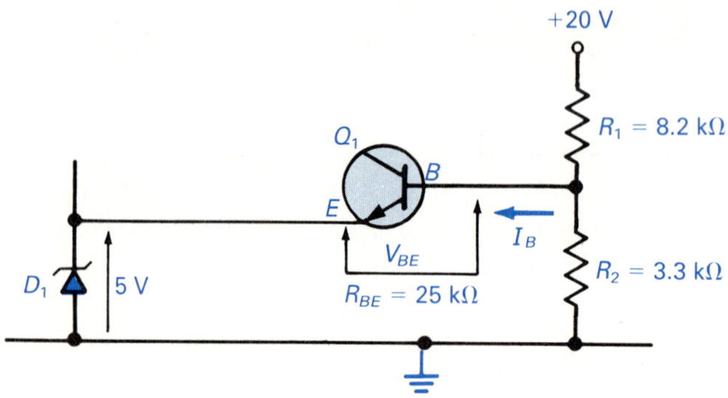

Fig. 9-59. Base-to-emitter circuit of transistor Q_1 in Fig. 9-58.

The zener diode D_1 maintains a steady voltage between the transistor emitter and ground, and is thus virtually a voltage source. Figure 9-59 can be reduced to the equivalent circuit in Fig. 9-60.

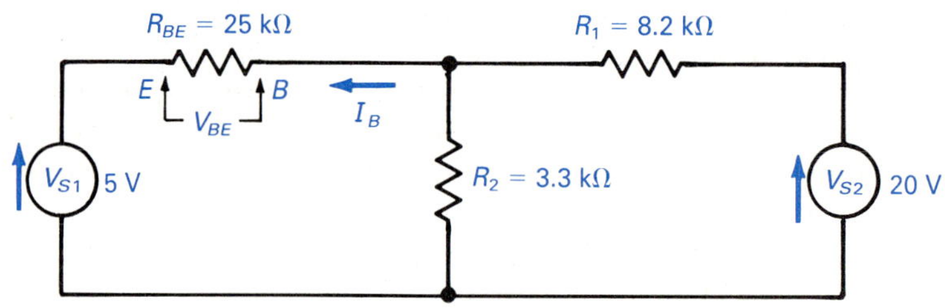

Fig. 9-60. Reduced equivalent circuit of Fig. 9-59.

Taking V_{S1} alone gives Fig. 9-61.

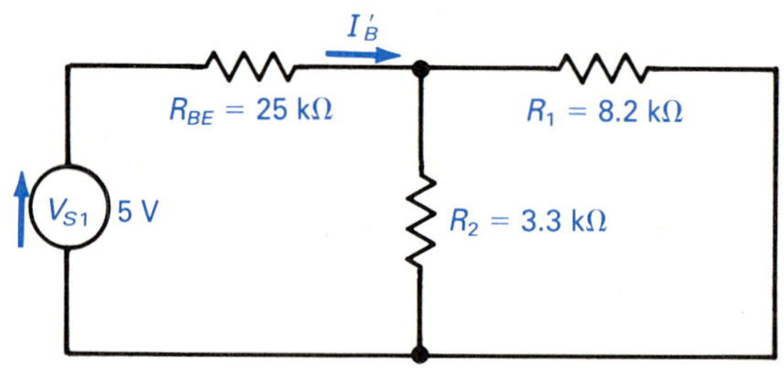

Fig. 9-61. Figure 9-60 with V_{S1} alone.

$$I'_B = \frac{V_{S1}}{R_{BE} + \dfrac{R_1 R_2}{R_1 + R_2}} = \frac{5\ \text{V}}{25\ \text{k}\Omega + \dfrac{8.2\ \text{k}\Omega \times 3.3\ \text{k}\Omega}{8.2\ \text{k}\Omega + 3.3\ \text{k}\Omega}}$$

$$= 0.1828\ \text{mA}$$

Taking V_{S2} alone gives Fig. 9-62.

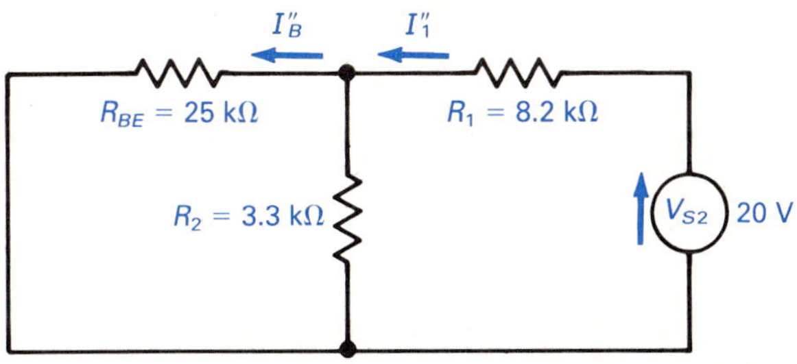

Fig. 9-62. Figure 9-61 with V_{S2} alone.

$$I''_1 = \frac{V_{S2}}{R_1 + \dfrac{R_2R_{BE}}{R_2 + R_{BE}}} = \frac{20\text{ V}}{8.2\text{ k}\Omega + \dfrac{3.3\text{ k}\Omega \times 25\text{ k}\Omega}{3.3\text{ k}\Omega + 25\text{ k}\Omega}} = 1.799\text{ mA}$$

$$I''_B = I''_1 \frac{R_2}{R_2 + R_{BE}} = 1.799\text{ mA} \frac{3.3\text{ k}\Omega}{3.3\text{ k}\Omega + 25\text{ k}\Omega} = 0.2098\text{ mA}$$

$$I_B = I''_B - I'_B = 0.2098\text{ mA} - 0.1828\text{ mA} = \mathbf{0.027\text{ mA}}$$

$$V_{BE} = I_BR_{BE} = 0.027\text{ mA} \times 25\text{ k}\Omega = \mathbf{0.675\text{ V}}$$

These are typical values for a silicon transistor used in this type of circuit.

The final expression for I_B in Example 9-17 may be used to show how important arithmetic accuracy can be. When it is necessary to subtract two quantities of almost the same value, accuracy can be all-important. For instance, if the values of I'_B and I''_B were calculated to be 0.18 mA and 0.21 mA, the final value of I_B would be 0.03 mA. Thus a maximum calculating error of 1.5 percent would cause a final error of 10 percent.

EXAMPLE 9-18 A compound-wound electric motor is connected to a 120-V dc supply. The series and shunt field windings have resistances of 10 and 1000 Ω, respectively. The motor's back emf is 100 V, and its armature resistance is 5 Ω. What are the field and armature currents?

Solution The circuit of the electric motor will be as in Fig. 9-63.

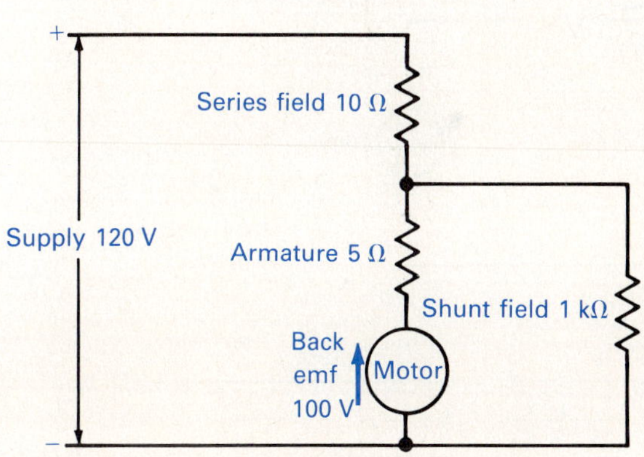

Fig. 9-63. Example 9-18: Network.

This diagram can be redrawn as an equivalent circuit (Fig. 9-64).

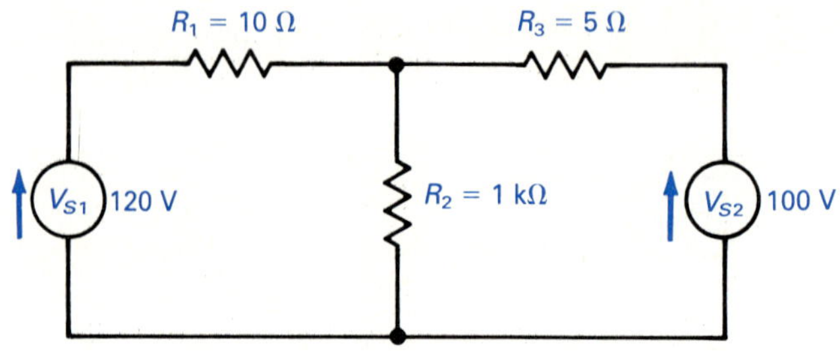

Fig. 9-64. Equivalent circuit of Fig. 9-63.

With V_{S1} alone (Fig. 9-65),

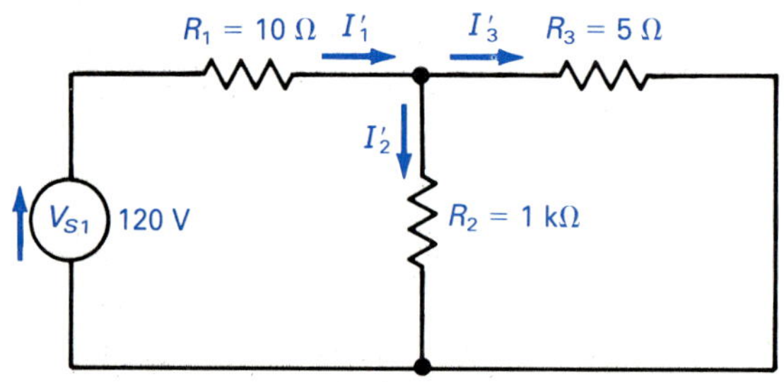

Fig. 9-65. Figure 9-64 with V_{S2} removed.

$$I_1' = \frac{V_{S1}}{R_1 + \dfrac{R_2R_3}{R_2 + R_3}} = \frac{120\text{ V}}{10\ \Omega + \dfrac{5\ \Omega \times 1\text{ k}\Omega}{5\ \Omega + 1\text{ k}\Omega}} = 8.013\text{ A}$$

$$I_2' = I_1' \frac{R_3}{R_2 + R_3} = 8.013\text{ A}\, \frac{5\ \Omega}{5\ \Omega + 1\text{ k}\Omega} = 0.0399\text{ A}$$

$$I_3' = I_1' - I_2' = 8.013\text{ A} - 0.0399\text{ A} = 7.973\text{ A}$$

With V_{S2} alone (Fig. 9-66),

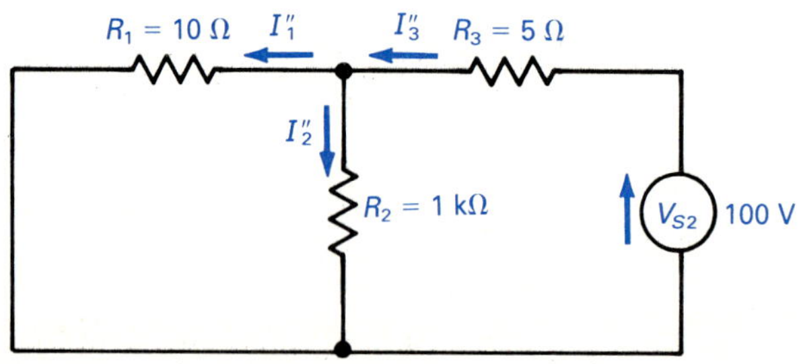

Fig. 9-66. Figure 9-64 with V_{S1} removed.

$$I_3'' = \frac{V_{S2}}{R_3 + \dfrac{R_1R_2}{R_1 + R_2}} = \frac{100\text{ V}}{5\ \Omega + \dfrac{10\ \Omega \times 1\text{ k}\Omega}{10\ \Omega + 1\text{ k}\Omega}} = 6.711\text{ A}$$

$$I_1'' = I_3''\frac{R_2}{R_1 + R_2} = 6.711\text{ A}\frac{1\text{ k}\Omega}{10\ \Omega + 1\text{ k}\Omega} = 6.644\text{ A}$$
$$I_2'' = I_1'' - I_3'' = 6.711\text{ A} - 6.644\text{ A} = 0.067\text{ A}$$
$$I_1 = I_1' - I_1'' = 8.013\text{ A} - 6.644\text{ A} = 1.369\text{ A}$$
$$I_2 = I_2' + I_2'' = 0.0399\text{ A} + 0.067\text{ A} = 0.1069\text{ A}$$
$$I_3 = I_3' - I_3'' = 7.973\text{ A} - 6.711\text{ A} = 1.262\text{ A}$$

The motor currents are

Series field (and supply) current I_1 = **1.369 A**
Shunt field current I_2 = **0.1069 A**
Armature current I_3 = **1.262 A**

REVIEW QUESTIONS FOR SECTION 9-4

1. When using superposition, how are the individual sources applied to obtain the solution?
2. List the six basic operations in the application of superposition analysis.

EXERCISE PROBLEMS FOR SECTION 9-4

1. Find the magnitude and polarity of the voltage at point A in Fig. 9-67, using superposition, when
   a. $V_{S1} = 50$ V, $V_{S2} = 100$ V, $R_1 = 1$ kΩ, $R_2 = 500\ \Omega$, $R_3 = 200\ \Omega$
   (*ans.* −18.75 V)
   b. $V_{S1} = 25$ V, $V_{S2} = 25$ V, $R_1 = 47$ kΩ, $R_2 = 15$ kΩ, $R_3 = 10$ kΩ
   (*ans.* −6.038 V)

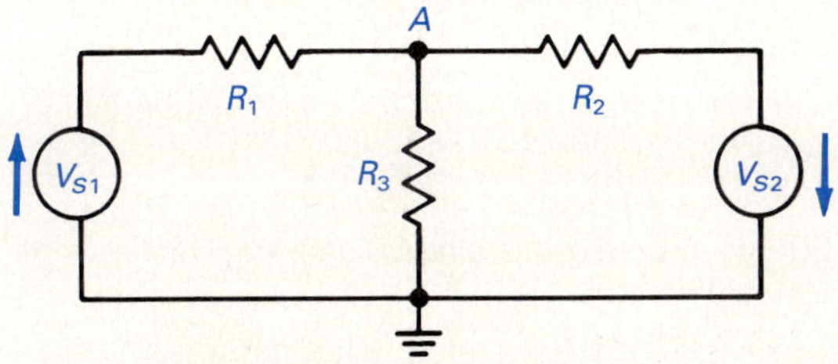

Fig. 9-67. Problem 1: Circuit.

2. Using superpositioning, find the voltage drop across R_5 in Fig. 9-46 (page 295) when $V_{S1} = 15$ V, $V_{S2} = 10$ V, $R_1 = 4.7$ kΩ, $R_2 = 1.8$ kΩ, $R_3 = 2.7$ kΩ, $R_4 = 2.2$ kΩ, $R_5 = 3.3$ kΩ.
3. Solve Prob. 9, Sec. 9-3 (page 295) using superpositioning.
   (*ans.* 61.27 A, 51.24 A, 10.03 A)

   In Fig. 9-68 the input signal V varies between +1 and +2 V. Find the base current I_B, using superposition, when
   a. $V = +1$ V
   b. $V = +1.5$ V
   c. $V = +2$ V

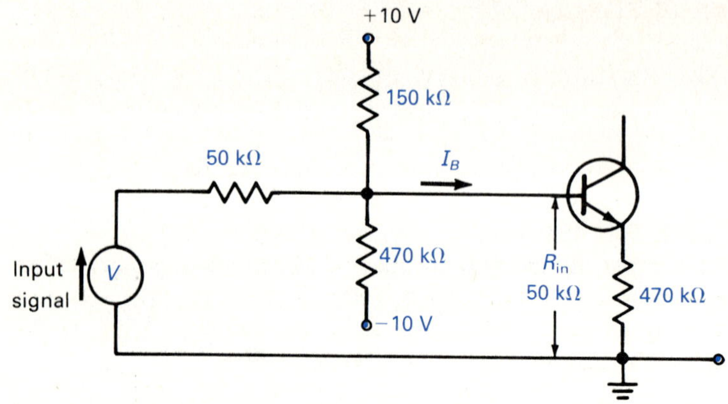

Fig. 9-68. Problem 4: Circuit.

Refer to Example 9-13 (page 292) for hints on drawing an equivalent circuit for Fig. 9-68.

CHAPTER QUESTIONS

9-1. Why is a series-parallel circuit usually better analyzed by current assumption than by nodal or loop analysis?

9-2. How many energy sources (generators or batteries) should there be in a circuit for the current-assumption analysis method to be appropriate?

9-3. If a circuit contains groups of elements that are not connected in series or parallel with each other, which analysis methods may be used
a. Current assumption?
b. Nodal analysis?
c. Loop analysis?

9-4. Which is the best method of circuit analysis?

9-5. What is the basic principle that makes the current-assumption method of circuit analysis satisfactory?

9-6. Are the circuits in Figs. 9-1 to 9-4 series-parallel?

9-7. What is the essential difference between the nodal and loop methods of circuit analysis?

9-8. In what circumstances do negative signs appear when writing
a. Nodal equations?
b. Loop equations?

9-9. In which circuit-analysis method is it the aim to
a. Determine the voltages between circuit junctions and a common?
b. Determine the currents tending to be in the circuit meshes?

9-10. For what circuit-analysis methods would it be desirable to convert
a. Current sources to voltage sources?
b. Voltage sources to current sources?

9-11. Why is it necessary to be extremely careful when reconverting sources to their original forms before finally determining voltage and current values in circuit analysis?

9-12. How could Examples 9-7 and 9-13 be solved by Ohm's law without using one of the methods described in this chapter?

9-13. Why does the text omit the details of solving simultaneous equations?

9-14. Which of the examples given for nodal analysis could have been solved by loop analysis?

9-15. Network problems involving three-wire circuits were solved in Chap. 8 as examples of series-parallel circuits. What is different about the circuit of Example 9-14 that prevents it being similarly treated?

9-16. In what way can the superposition method of circuit analysis be simpler to use than either the nodal or loop methods?

9-17. If the superposition method can be simpler than the nodal or loop methods of circuit analysis, why is it not always used instead?

CHAPTER PROBLEMS

9-1. Refer to Fig. 9-69. Applying the current-assumption method, find the

a. Current in resistor R_3 (*ans*. 6.981 μA)
b. Voltage across resistor R_1 (*ans*. 8.109 V)
c. Power supplied by the generator (*ans*. 81.09 μW)
d. Power in the load R_L (*ans*. 1.529 μW)

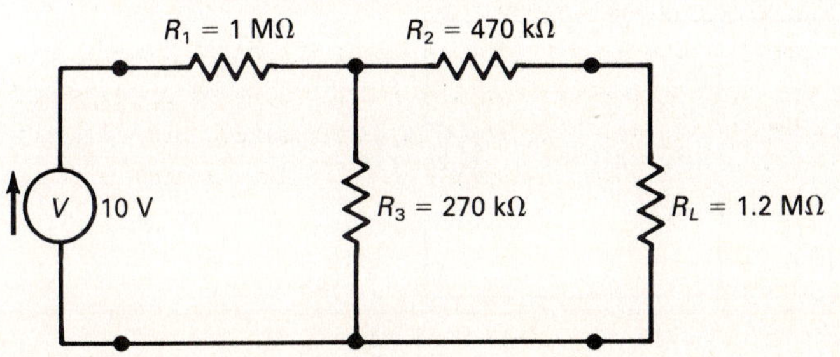

Fig. 9-69.

9-2. Refer to Fig. 9-70. Using the current-assumption method, find the

a. Current in resistor R_3
b. Voltage across resistor R_1
c. Power supplied by the generator
d. Power in the load R_L

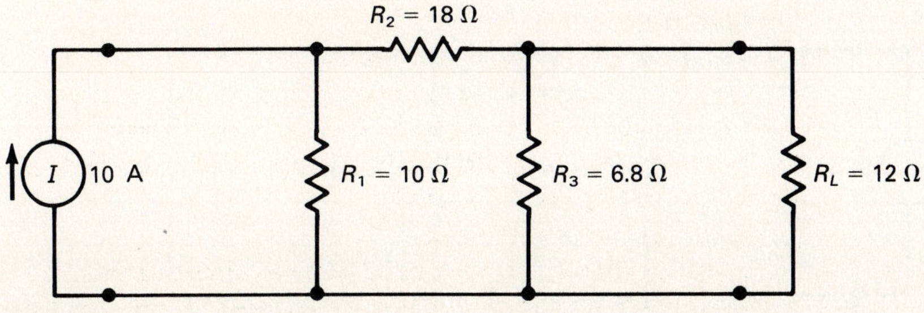

Fig. 9-70.

9-3. Refer to Fig. 9-71. Find the current in resistor R_3, applying
a. Nodal analysis
b. Loop analysis
c. Superposition
} (*ans*. 16.95 μA)

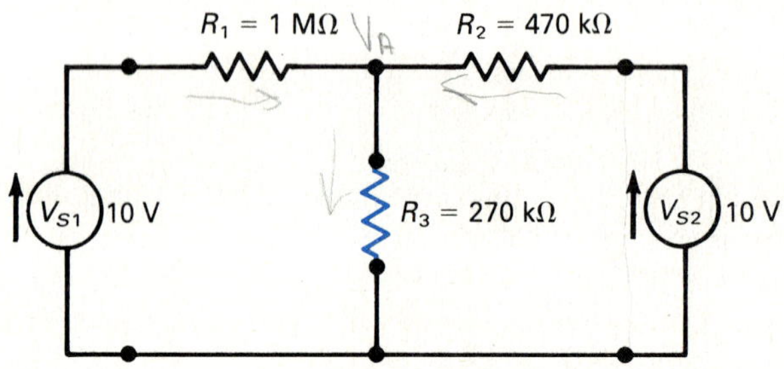

Fig. 9-71.

9-4. Refer to Fig. 9-72. Find the current in resistor R_3, using
a. Nodal analysis
b. Loop analysis
c. Superposition

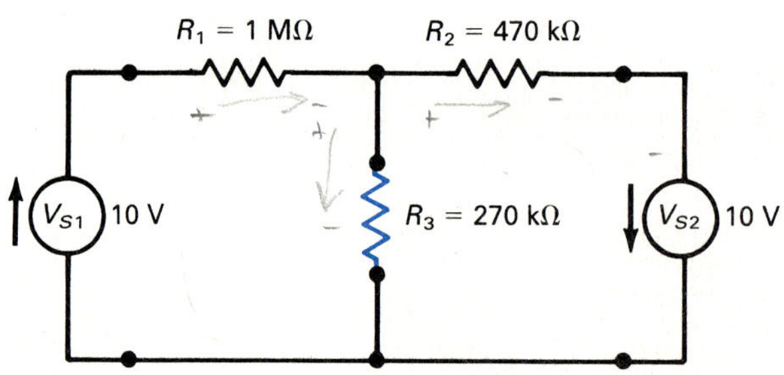

Fig. 9-72.

9-5. Refer to Fig. 9-73. Find the current in resistor R_3 by
a. Nodal analysis
b. Loop analysis
c. Superposition
} (*ans*. 10.92 A)

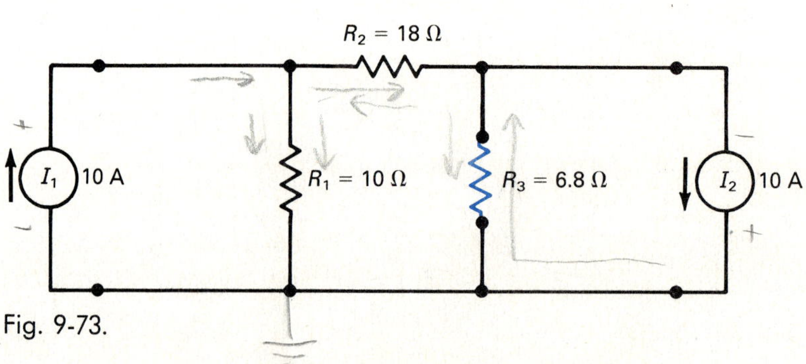

Fig. 9-73.

9-6. Refer to Fig. 9-74. Find the current in resistor R_3, using
a. Nodal analysis *b*. Loop analysis *c*. Superposition

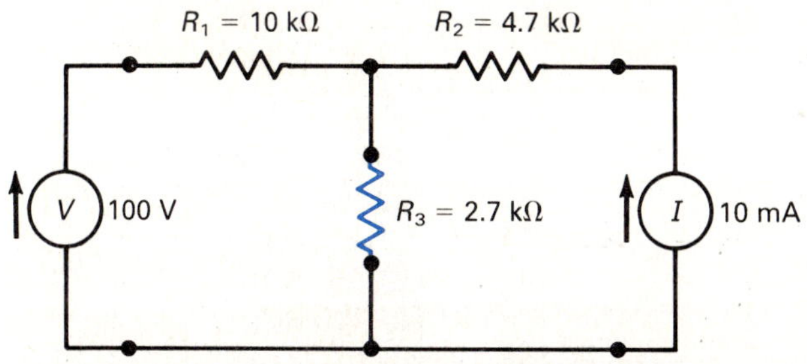

Fig. 9-74.

9-7. Refer to Fig. 9-75. Find the current in resistor R_3, applying
a. Nodal analysis
b. Loop analysis (*ans*. 3.225 mA)
c. Superposition

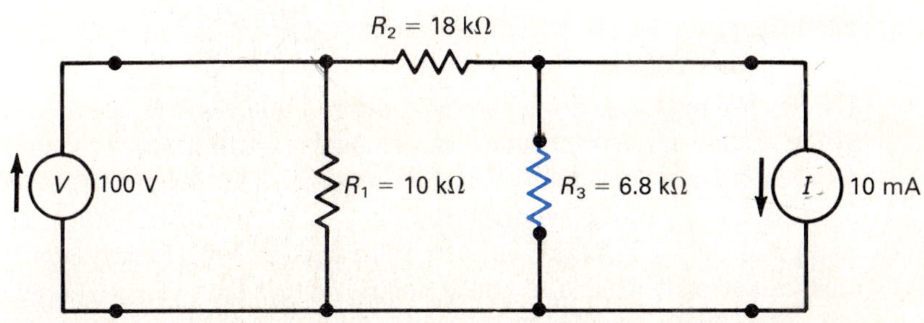

Fig. 9-75.

9-8. An outline diagram of a three-wire electrical system in a house is given in Fig. 9-76. The total load of 18 kW is made up of two 120-V loads of 1500 W each and one 240-V load of 15 kW. If the three wires in the connecting cable are all the same, what would be their maximum allowable resistance for the voltage at any of the loads to be within 5 percent of the rated voltage? Assume the supply voltages to be correct and the load resistances to be as calculated with correct voltages.

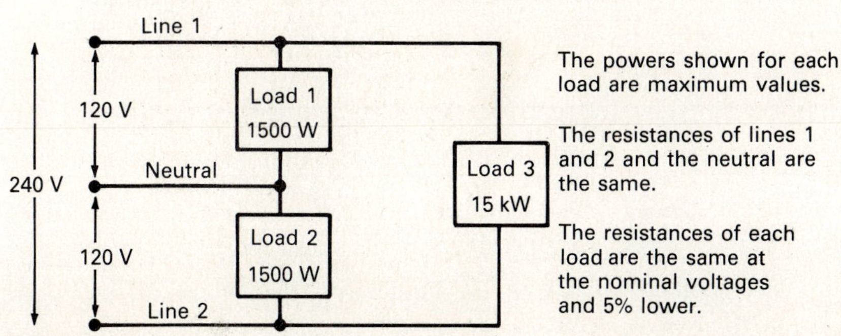

Fig. 9-76.

9-9. Figure 9-77 shows part of the circuit of an electronic control system.

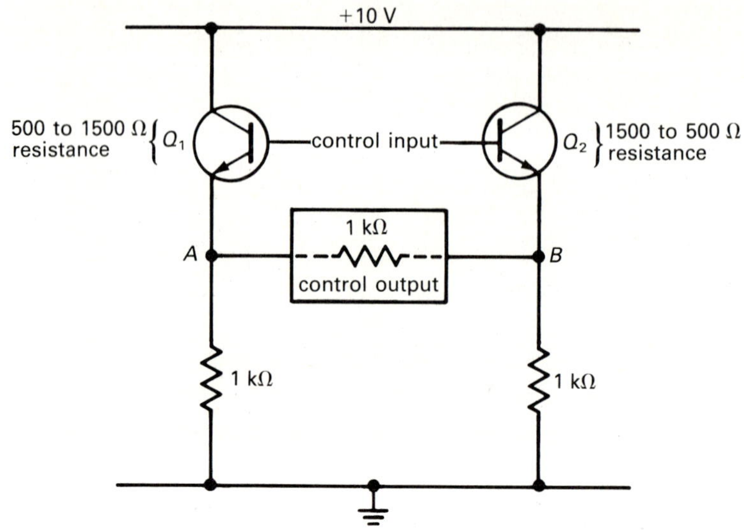

Fig. 9-77.

The transistors Q_1 and Q_2 behave as resistors whose value is determined by the voltage developed by the control input. These resistances can vary between 500 and 1500 Ω in an inverse manner: When the resistance of A_1 is 500 Ω, the resistance of Q_2 is 1500 Ω, and vice versa. The resistances change in a uniform manner over their range. What is the range of relative potential of side B of the control output with respect to side A? (*ans.* −1.38 V to +1.38 V)

10

EQUIVALENT-CIRCUIT NETWORK ANALYSIS

You have now been introduced to four methods of analyzing an electric circuit: current assumption and nodal, loop, and superposition analysis. Which is the best? You have received the message if you realize that there is no *best* method. Each has advantages in certain situations.

It is likely that you will prefer one of the techniques and use it more frequently than the others, even when it is not the most efficient. This is all right, and quite in keeping with the philosophy that it is only through the knowledge of many aspects of a problem that you can make confident, wise decisions. You should be familiar with a number of techniques, each of which can be used for the same purpose. This familiarity gives a degree of circuit consciousness which cannot be obtained through learning how to solve each individual problem with a single preferred method.

The direct network analysis methods of Chap. 9 are convenient for solving many circuit problems. However, if a circuit cannot easily be reduced to a series-parallel form, the direct methods may not work. Again, you may need to investigate the behavior of a network without knowing its circuit details and can only determine its characteristics by measurements. In situations where the direct methods cannot be used, equivalent-circuit methods may be employed. This chapter introduces several of these methods.

You will see that a number of the examples of equivalent-circuit analysis involve finding the conditions existing in a load. Often the sole purpose of an electric or electronic system is to supply controlled energy to a load. Examples are a radio transmitter supplying energy to an antenna, a hydroelectric station supplying energy to a city, and a microwave oven heating food. You will also see that some of the equivalent-circuit methods are useful for rapidly calculating the effects of changing component values, important in assessing equipment performance under varying conditions.

10-1 T $\leftrightarrow$ π CONVERSIONS

The equivalent-circuit methods of network analysis rely on the fact that it is possible to reduce complex networks to simple ones for special purposes. It has already been shown that a two-terminal (single-port)[1] network can be reduced to a single equivalent element. A three-terminal (two-port) network with many elements can be reduced to a three-element network by the use of T $\leftrightarrow$ π conversions. The acceptance of this technique implies the validity of using three-element circuits to represent networks with any number of elements. This permits some general circuit principles to be derived by fairly simple means with the confidence that they even apply to complex circuits.

T, π, Wye, and Delta Networks

Figure 10-1*a* shows a T network and Fig. 10-1*b* a wye network. Both are exactly the same; only the drawing layout is different. Similarly, the π network in Fig. 10-1*c* is exactly the same as the delta network in Fig. 10-1*d*.

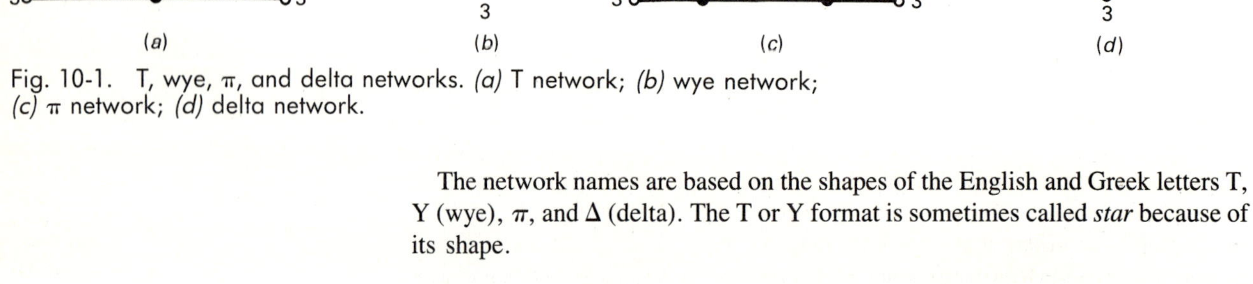

Fig. 10-1. T, wye, π, and delta networks. *(a)* T network; *(b)* wye network; *(c)* π network; *(d)* delta network.

The network names are based on the shapes of the English and Greek letters T, Y (wye), π, and Δ (delta). The T or Y format is sometimes called *star* because of its shape.

[1] A *port* is a network location where a connection can be made externally to the network.

In electronics the T and π formats are adopted because they conveniently fit the input-output type of external connection so common in that branch of the industry. Typically, terminals 1 and 3 constitute the input port, while terminals 2 and 3 form the output port. Often terminal 3, the common between the input and output, is grounded.

In electric power systems the wye and delta views are customary because no distinction is made between specific pairs of terminals. None of the terminals is likely to be grounded.

The terms "T" and "π" will be used here because modern industry is predominantly electronic. However, the wye and delta diagrams will be used for some explanations, as they can aid understanding.

Equivalence

Figure 10-2 shows a T (or wye) and a π (or delta) network which are to be equivalent to each other. For this to be so, it must be impossible to distinguish between them by electrical measurement at the terminals. If the resistances between all related pairs of terminals are the same, this equivalence will be established.

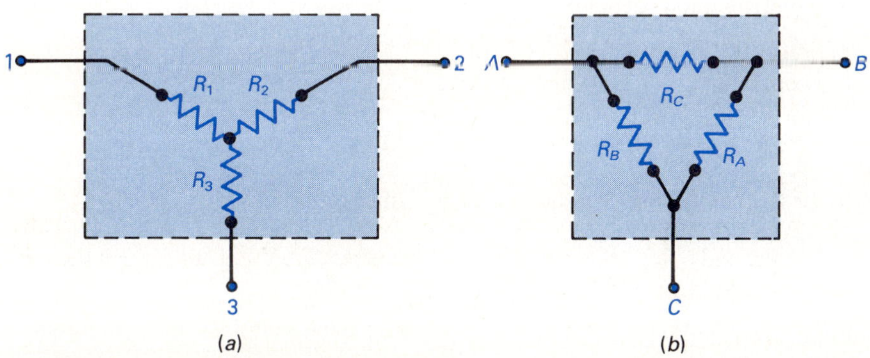

Fig. 10-2. *(a)* T network; *(b)* π network.

The conditions for equivalence will be:

T Network, Fig. 10-2*a*		π Network, Fig. 10-2*b*
Resistance 1 to 2	=	Resistance *A* to *B*
Resistance 2 to 3	=	Resistance *B* to *C*
Resistance 3 to 1	=	Resistance *C* to *A*

When one of the terminals of the T network in Fig. 10-2*a* is left disconnected, a series circuit remains between the other two terminals (see Fig. 10-3*a*, in which terminal 2 is left disconnected and the circuit between terminals 1 and 3 consists of R_1 and R_3 in series).

If one terminal of the π network in Fig. 10-2*b* is left open, the remaining circuit is series-parallel. The circuit between the other two terminals will consist of two resistors in series, this pair being in parallel with the third resistor (see Fig. 10-3*b*, in which terminal B is left disconnected and the circuit between terminals A and C is the series combination of R_A and R_C connected in parallel with R_B).

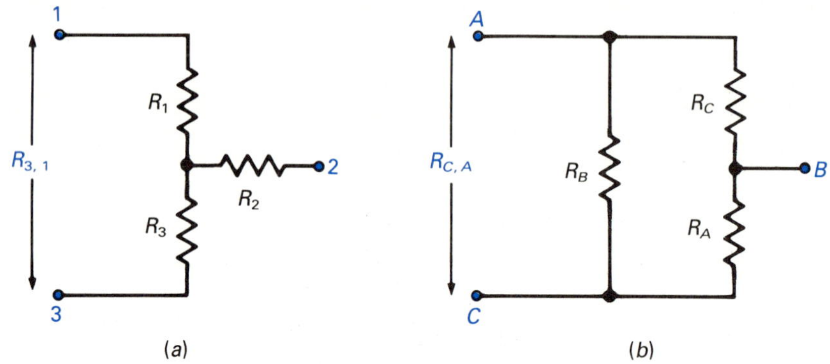

Fig. 10-3. Figure 10-2 redrawn to show equivalent circuits when one terminal is left open.

Figure 10-3 compares the resistance between a pair of terminals in each of the networks when the third terminal is left disconnected.

T network: $$R_{3,1} = R_3 + R_1$$

π network: $$R_{C,A} = \frac{R_B(R_C + R_A)}{R_A + R_B + R_C}$$

For the networks to be equivalent, $R_{3,1}$ must equal $R_{C,A}$.

$$R_3 + R_1 = \frac{R_A R_B + R_B R_C}{R_A + R_B + R_C}$$

By repeating this process for the other two pairs of terminals, the three equations for equivalence are obtained.

$$R_1 + R_2 = \frac{R_B R_C + R_C R_A}{\Sigma R_\nabla} \qquad \textbf{(10-1)}$$

$$R_2 + R_3 = \frac{R_A R_B + R_C R_A}{\Sigma R_\nabla} \qquad \textbf{(10-2)}$$

$$R_3 + R_1 = \frac{R_A R_B + R_B R_C}{\Sigma R_\nabla} \qquad \textbf{(10-3)}$$

The symbol ΣR_∇ is used for the sum of the elements in the π network. It is common practice to use Σ (capital Greek letter sigma) to indicate a constantly repeated sum of several terms.

π-to-T Conversion

To convert the π network in Fig. 10-4*a* to its T equivalent, Fig. 10-4*b*, it is necessary to express each of the elements of the T network (R_1, R_2, and R_3) in terms of the elements of the π network (R_A, R_B, and R_C). This is done by manipulating Eqs. (10-1) to (10-3) to give

$$R_1 = \frac{R_B R_C}{\Sigma R_\nabla} \tag{10-4}$$

$$R_2 = \frac{R_C R_A}{\Sigma R_\nabla} \tag{10-5}$$

$$R_3 = \frac{R_A R_B}{\Sigma R_\nabla} \tag{10-6}$$

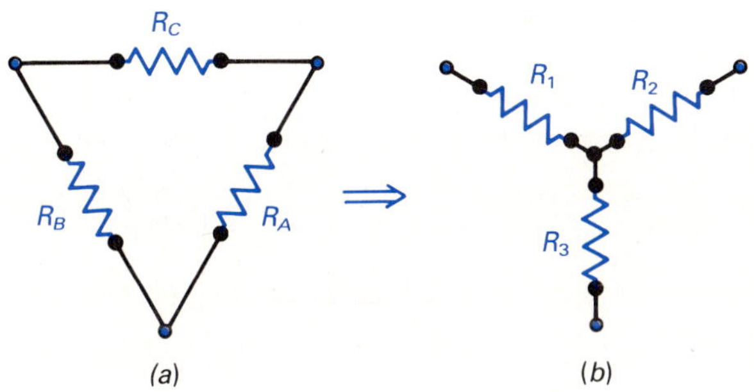

Fig. 10-4. π-to-T conversion.

Note that

1. The denominator ΣR_∇ equals $R_A + R_B + R_C$ and is the same for all three equations.
2. The numerator for any of the equations is the product of the two π-network elements adjacent to the related terminal. For example, R_3 connects to the bottom terminal of the T network, while R_A and R_B connect to the bottom terminal of the π network.

T-to-π Conversion

To convert the T network in Fig. 10-5*a* to its π equivalent, Fig. 10-5*b*, each of the π-network elements (R_A, R_B, and R_C) must be expressed in terms of the T-network elements (R_1, R_2, and R_3). From Eqs. (10-4) to (10-6) the following equations can be derived:

$$R_A = \frac{R_1R_2 + R_2R_3 + R_3R_1}{R_1} = \frac{\Sigma R_Y}{R_1} \tag{10-7}$$

$$R_B = \frac{\Sigma R_Y}{R_2} \tag{10-8}$$

$$R_C = \frac{\Sigma R_Y}{R_3} \tag{10-9}$$

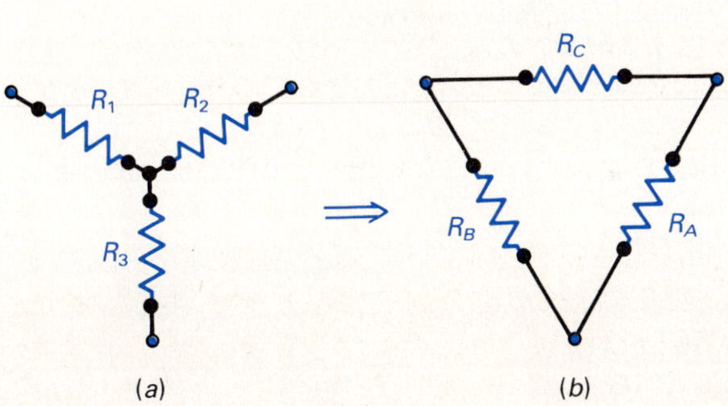

Fig. 10-5. T-to-π conversion.

Note that

1. The numerator is the same for each equation. $\Sigma R_Y = R_1R_2 + R_2R_3 + R_3R_1$ (the sum of the product pairs of the T-network elements).
2. The denominator for any of the equations is the opposite T element. For instance, R_C is at the top of the π network, while R_3 is at the bottom of the T network.

The following examples first show how to perform direct conversions, and then show how to reduce complex networks. Some examples of the use of network conversion for circuit analysis are also given.

EXAMPLE 10-1 Convert the π network in Fig. 10-6*a* to the T format in Fig. 10-6*b*.

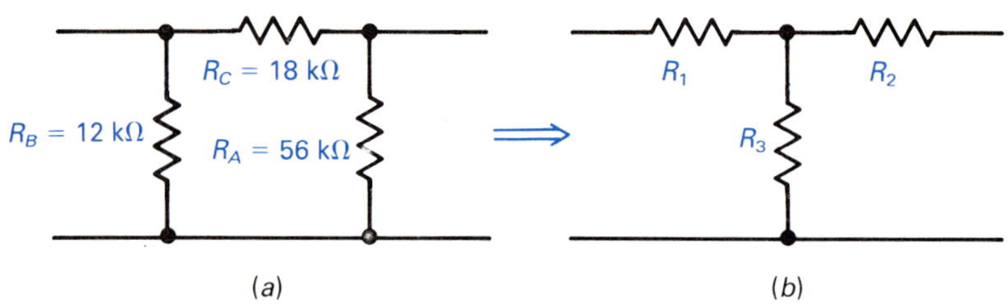

Fig. 10-6. Example 10-1: *(a)* π network; *(b)* equivalent T network.

Solution

$$\Sigma R_\nabla = R_A + R_B + R_C = 56\text{ k}\Omega + 12\text{ k}\Omega + 18\text{ k}\Omega = 86\text{ k}\Omega$$

$$R_1 = \frac{R_BR_C}{\Sigma R_\nabla} \tag{10-4}$$

$$R_1 = \frac{12\text{ k}\Omega \times 18\text{ k}\Omega}{86\text{ k}\Omega} = \mathbf{2.512\text{ k}\Omega}$$

$$R_2 = \frac{R_CR_A}{\Sigma R_\nabla} \tag{10-5}$$

$$R_2 = \frac{18\text{ k}\Omega \times 56\text{ k}\Omega}{86\text{ k}\Omega} = \mathbf{11.72\text{ k}\Omega}$$

$$R_3 = \frac{R_AR_B}{\Sigma R_\nabla} \tag{10-6}$$

$$R_3 = \frac{56\text{ k}\Omega \times 12\text{ k}\Omega}{86\text{ k}\Omega} = \mathbf{7.814\text{ k}\Omega}$$

EXAMPLE 10-2 Convert the T network in Fig. 10-7*a* to its π equivalent, Fig. 10-7*b*.

Solution

$$\begin{aligned}\Sigma R_Y &= R_1R_2 + R_2R_3 + R_3R_1 \\ &= (33\ \Omega \times 47\ \Omega) + (47\ \Omega \times 68\ \Omega) + (68\ \Omega \times 33\ \Omega) = 6991\ \Omega^2\end{aligned}$$

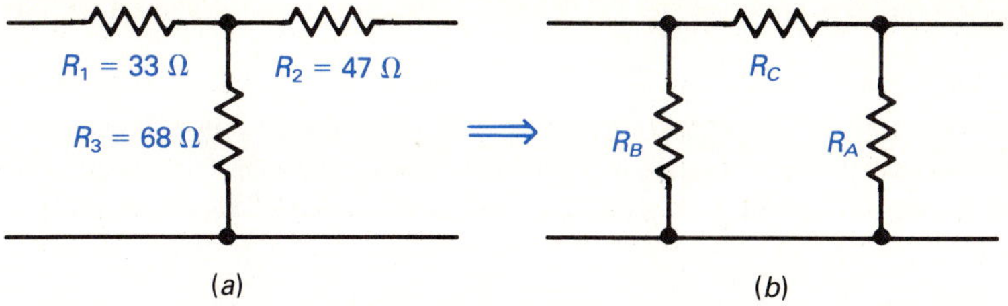

Fig. 10-7. Example 10-2: *(a)* T network; *(b)* equivalent π network.

$$R_A = \frac{\Sigma R_Y}{R_1} \tag{10-7}$$

$$R_A = \frac{6991\ \Omega^2}{33\ \Omega} = \mathbf{211.8\ \Omega}$$

$$R_B = \frac{\Sigma R_Y}{R_2} \tag{10-8}$$

$$R_B = \frac{6991\ \Omega^2}{47\ \Omega} = \mathbf{148.7\ \Omega}$$

$$R_C = \frac{\Sigma R_Y}{R_3} \tag{10-9}$$

$$R_C = \frac{6991\ \Omega^2}{68\ \Omega} = \mathbf{102.8\ \Omega}$$

You will see from the results for Examples 10-1 and 10-2 that the resistances of the T network are lower than those of the equivalent π network. This conclusion is to be expected. Both networks must have the same resistances between related terminal pairs. Each T-network element has another effectively in series with it, giving additional resistance. Each π-network element has elements in parallel with it, making a lower total resistance.

EXAMPLE 10-3 Reduce the six-element network in Fig. 10-8*a* to an equivalent three-element network, maintaining the original terminals.

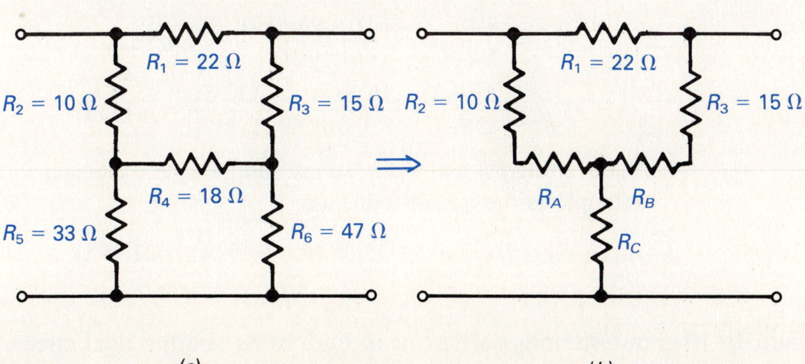

Fig. 10-8. Example 10-3: *(a)* Given network; *(b)* first reduction.

Solution

1. Convert the π group R_4, R_5, R_6 into the T network R_A, R_B, R_C in Fig. 10-8*b*.

$$\Sigma R_\nabla = R_4 + R_5 + R_6 = 18\ \Omega + 33\ \Omega + 47\ \Omega = 98\ \Omega$$

$$R_A = \frac{R_4 R_5}{\Sigma R_\nabla} = \frac{18\ \Omega \times 33\ \Omega}{98\ \Omega} = 6.061\ \Omega$$

$$R_B = \frac{R_4 R_6}{\Sigma R_\nabla} = \frac{18\ \Omega \times 47\ \Omega}{98\ \Omega} = 8.633\ \Omega$$

$$R_C = \frac{R_5 R_6}{\Sigma R_\nabla} = \frac{33\ \Omega \times 47\ \Omega}{98\ \Omega} = 15.83\ \Omega$$

2. Two series groups, R_2, R_A and R_3, R_B, are shown as elements R_D and R_E in Fig. 10-9*a*.

$$R_D = R_2 + R_A = 10\ \Omega + 6.061\ \Omega = 16.06\ \Omega$$
$$R_E = R_3 + R_B = 15\ \Omega + 8.633\ \Omega = 23.63\ \Omega$$

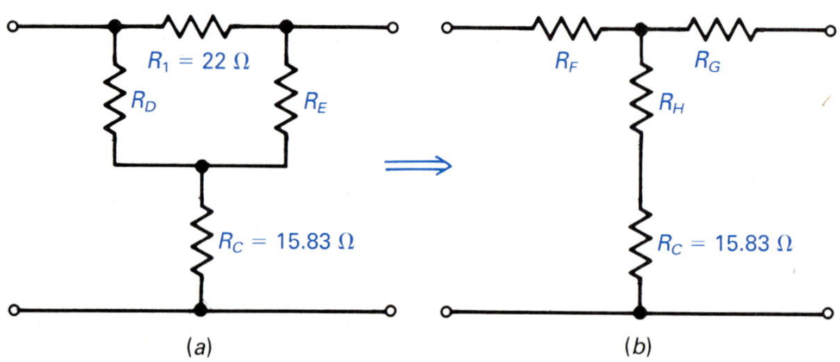

Fig. 10-9. *(a)* Second reduction of Fig. 10-8*a*; *(b)* third reduction of Fig. 10-8*a*.

3. Convert the π group R_1, R_D, R_E in Fig. 10-9*a* into the T network R_F, R_G, R_H in Fig. 10-9*b*.

$$\Sigma R_\nabla = R_1 + R_D + R_E = 22\ \Omega + 16.06\ \Omega + 23.63\ \Omega = 61.69\ \Omega$$

$$R_F = \frac{R_1 R_D}{\Sigma R_\nabla} = \frac{22\ \Omega \times 16.06\ \Omega}{61.69\ \Omega} = 5.727\ \Omega$$

$$R_G = \frac{R_1 R_E}{\Sigma R_\nabla} = \frac{22\ \Omega \times 23.63\ \Omega}{61.69\ \Omega} = 8.427\ \Omega$$

$$R_H = \frac{R_D R_E}{\Sigma R_\nabla} = \frac{16.06\ \Omega \times 23.63\ \Omega}{61.69\ \Omega} = 6.152\ \Omega$$

4. Converting R_C and R_H in series into the single element R_J completes the circuit reduction:

$$R_J = R_C + R_H = 15.83\ \Omega + 6.152\ \Omega = \mathbf{21.98\ \Omega}$$

Figure 10-10 shows the original circuit in Fig. 10-8*a* and the final circuit which is its three-element equivalent. Measurements taken at a pair of terminals of the original network will be the same as those taken at the equivalent pair of terminals of the final network. Also, the relationship between input and output voltages and currents would be the same in each case.

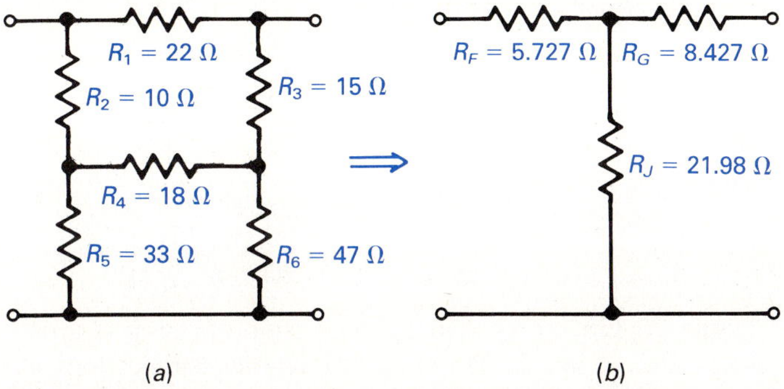

Fig. 10-10. Example 10-3: *(a)* Original network; *(b)* final three-element T equivalent network.

EXAMPLE 10-4 What are the magnitude and direction of the meter current in the resistance bridge circuit in Fig. 10-11?

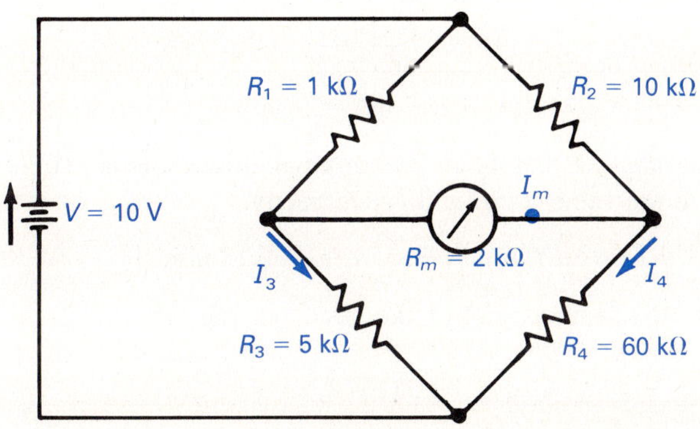

Fig. 10-11. Example 10-4: Circuit.

Solution Convert the π network R_1, R_2, R_m into an equivalent T network, as in Fig. 10-12*a*.

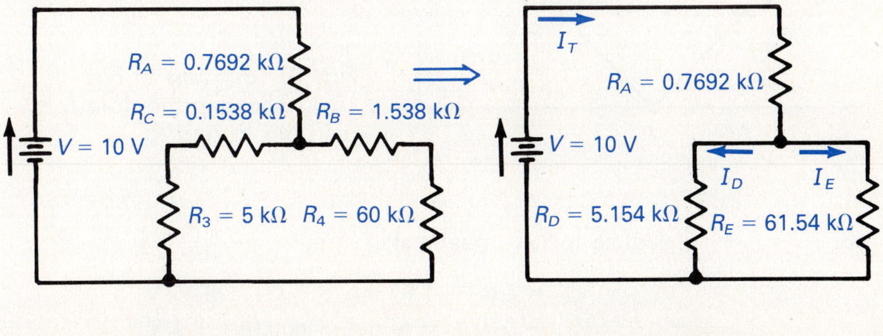

Fig. 10-12. *(a)* First reduction of Fig. 10-11; *(b)* bridge network in Fig. 10-11 reduced to a three-element equivalent network.

$$\Sigma R_\nabla = R_1 + R_2 + R_m = 1\text{ k}\Omega + 10\text{ k}\Omega + 2\text{ k}\Omega = 13\text{ k}\Omega$$

$$R_A = \frac{R_1 R_2}{\Sigma R_\nabla} = \frac{1\text{ k}\Omega \times 10\text{ k}\Omega}{13\text{ k}\Omega} = 0.7692\text{ k}\Omega$$

$$R_B = \frac{R_2 R_m}{\Sigma R_\nabla} = \frac{10\text{ k}\Omega \times 2\text{ k}\Omega}{13\text{ k}\Omega} = 1.538\text{ k}\Omega$$

$$R_C = \frac{R_m R_1}{\Sigma R_\nabla} = \frac{2\text{ k}\Omega \times 1\text{ k}\Omega}{13\text{ k}\Omega} = 0.1538\text{ k}\Omega$$

In Fig. 10-12*b*,

$$R_D = R_3 + R_C = 5\text{ k}\Omega + 0.1538\text{ k}\Omega = 5.154\text{ k}\Omega$$
$$R_E = R_4 + R_B = 60\text{ k}\Omega + 1.538\text{ k}\Omega = 61.54\text{ k}\Omega$$

Figure 10-12*b* shows that the conversion of the π network reduces the original bridge circuit to a series-parallel format. The resistance values of this equivalent series-parallel circuit are now known. The currents in R_D and R_E can be calculated:

$$I_T = \frac{V}{R_A + \dfrac{R_D R_E}{R_D + R_E}} = \frac{10\text{ V}}{0.7692\text{ k}\Omega + \dfrac{5.154\text{ k}\Omega \times 61.54\text{ k}\Omega}{5.154\text{ k}\Omega + 61.54\text{ k}\Omega}}$$
$$= 1.81\text{ mA}$$

$$I_D = I_T \frac{R_E}{R_D + R_E} = 1.81\text{ mA}\,\frac{61.54\text{ k}\Omega}{66.69\text{ k}\Omega} = 1.67\text{ mA}$$
$$I_E = I_T - I_D = 1.81\text{ mA} - 1.67\text{ mA} = 0.14\text{ mA}$$

The currents I_3 and I_4 in the original bridge circuit (Fig. 10-11) are the same as I_D and I_E, respectively.

$$I_3 = 1.67\text{ mA} \qquad \text{and} \qquad I_4 = 0.14\text{ mA}$$

Figure 10-13 shows part of the circuit in Fig. 10-11.

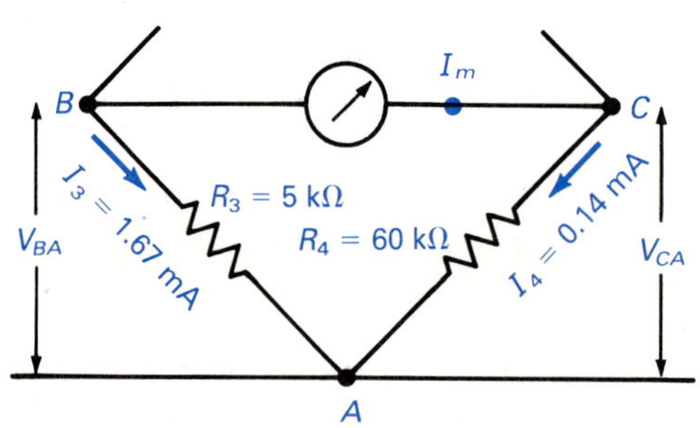

Fig. 10-13. Part of Fig. 10-11 showing the data for resistors R_3 and R_4.

Calculate relative potentials:

$$V_{BA} = I_3 R_3 = 1.67\text{ mA} \times 5\text{ k}\Omega = 8.35\text{ V}$$
$$V_{CA} = I_4 R_4 = 0.14\text{ mA} \times 60\text{ k}\Omega = 8.4\text{ V}$$

Therefore node C is positive with respect to node B, as shown in Fig. 10-14.

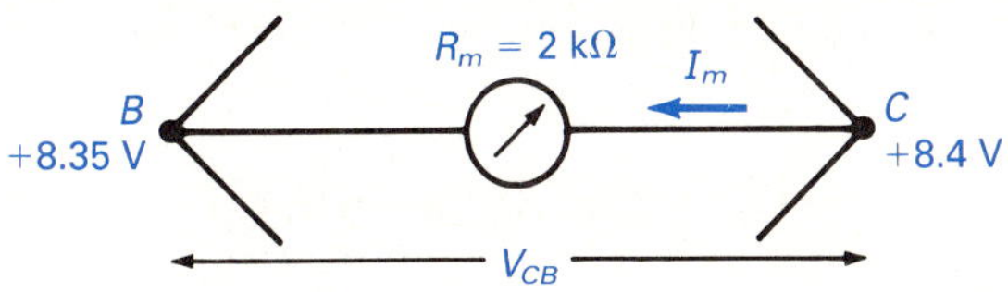

Fig. 10-14. Meter circuit in Fig. 10-11.

The voltage across the meter will be

$$V_{CB} = V_{CA} - V_{BA} = 8.4 \text{ V} - 8.35 \text{ V} = 0.05 \text{ V}$$

The current in the meter will have the direction shown in Fig. 10-14, *right to left*. Its value will be

$$I_m = \frac{V_{CB}}{R_m} = \frac{0.05 \text{ V}}{2 \text{ k}\Omega} = \mathbf{0.025 \text{ mA}} \qquad \text{or} \qquad \mathbf{25\ \mu A}$$

Although long, the solution of Example 10-4 involves only simple calculator operations without simultaneous equations; you may like it for that reason. This use of π-to-T conversions is one way of solving circuit problems.

REVIEW QUESTIONS FOR SECTION 10-1

1. What type of circuit would be classed as single-port?
2. What is required for a circuit to be classed as two-port?
3. By what other names are T and π networks known?
4. Using resistance measurements, what conditions establish the equivalence between two circuits?
5. State the equations for converting a π network to a T network.
6. State the equations for converting a T network to a π network.
7. The resistances of the equivalent T network are higher than those of the π network from which it has been converted. Correct?

EXERCISE PROBLEMS FOR SECTION 10-1

1. Convert the π network in Fig. 10-15*a* to the T network in Fig. 10-15*b* when
   a. $R_A = 68\ \Omega$, $R_B = 68\ \Omega$, $R_C = 68\ \Omega$
   (*ans*. 22.7 Ω, 22.7 Ω, 22.7 Ω)
   b. $R_A = 47$ kΩ, $R_B = 100$ kΩ, $R_C = 47$ kΩ
   (*ans*. 24.2 kΩ, 11.4 kΩ, 24.2 kΩ)
   c. $R_A = 560$ kΩ, $R_B = 220$ kΩ, $R_C = 330$ kΩ
   (*ans*. 111 kΩ, 166.5 kΩ, 65.44 kΩ)
2. Convert the T network in Fig. 10-15*b* to the π network in Fig. 10-15*a* when
   a. $R_1 = 68\ \Omega$, $R_2 = 68\ \Omega$, $R_3 = 68\ \Omega$
   b. $R_1 = 1$ MΩ, $R_2 = 470$ kΩ, $R_3 = 100$ kΩ
   c. $R_1 = 2.51$ kΩ, $R_2 = 11.7$ kΩ, $R_3 = 7.82$ kΩ

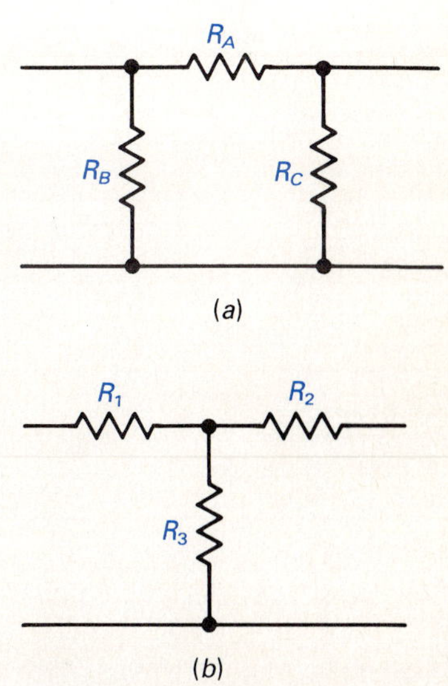

Fig. 10-15. Problems 1 and 2: *(a)* π network; *(b)* T network.

3. Convert the network in Fig. 10-16 to an equivalent π network when
a. $R_1 = 100\ \Omega$, $R_2 = 220\ \Omega$, $R_3 = 100\ \Omega$, $R_4 = 150\ \Omega$, $R_5 = 150\ \Omega$
(*ans. a.* 945 Ω; 250 Ω; 250 Ω)
b. $R_1 = 560\ \text{k}\Omega$, $R_2 = 390\ \text{k}\Omega$, $R_3 = 820\ \text{k}\Omega$, $R_4 = 560\ \text{k}\Omega$, $R_5 = 680\ \text{k}\Omega$ (*ans.* 4.596 MΩ, 1.093 MΩ, 1.547 MΩ)

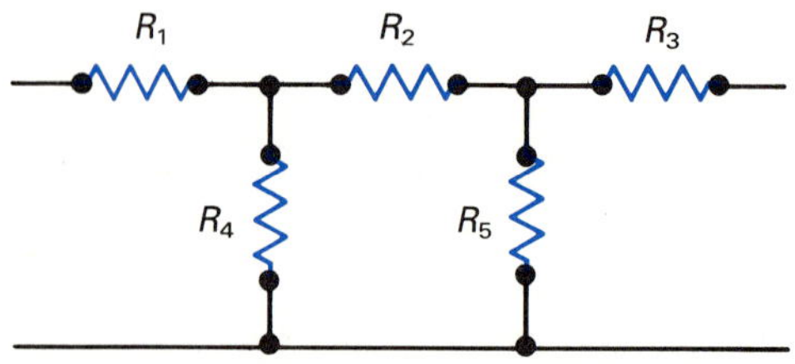

Fig. 10-16. Problems 3 and 4: Network.

4. Convert the network in Fig. 10-16 to an equivalent T network when
a. $R_1 = 47\ \Omega$, $R_2 = 68\ \Omega$, $R_3 = 47\ \Omega$, $R_4 = 33\ \Omega$, $R_5 = 33\ \Omega$
b. $R_1 = 1.8\ \text{k}\Omega$, $R_2 = 220\ \Omega$, $R_3 = 1.5\ \text{k}\Omega$, $R_4 = 2.7\ \text{k}\Omega$, $R_5 = 390\ \Omega$

5. Convert the network (R_1 through R_5) in Fig. 9-6 (page 268) to an equivalent T network and find I_L when
a. $R_L = 120\ \Omega$ (*ans.* 27.3 mA)
b. $R_L = 1200\ \Omega$ (*ans.* 12 mA)
c. $R_L = 1800\ \Omega$ (*ans.* 9.16 mA)

6. Convert the T network (R_2, R_3, and R_4) in Fig. 9-8*a* (page 269) to an equivalent π network and compare with Fig. 9-8*b*.

7. Convert the π network (R_a, R_b, and R_c) in Fig. 9-8*b* (page 269) to an equivalent T network and compare with Fig. 9-8*a*. (*ans.* same)

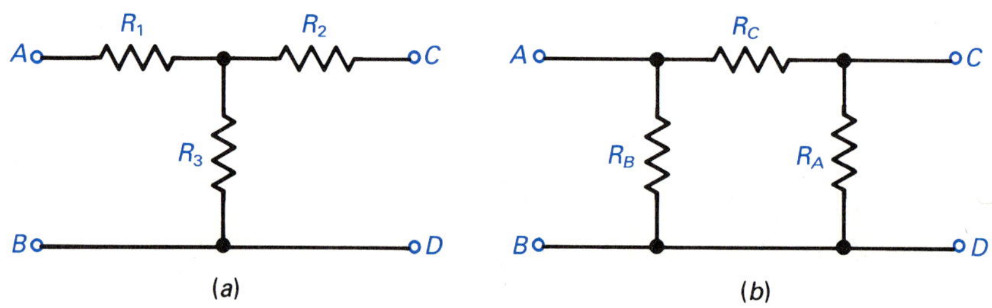

Fig. 10-17. Problem 8: *(a)* T network; *(b)* π network.

8. Choose any T network previously converted to a π network and use those resistance values for the values in Fig. 10-17. Under the stated conditions, for both circuits, calculate the resistance between
a. *A* and *B* with *C* and *D* open-circuit.
b. *A* and *B* with *C* and *D* short-circuit.

9. Reduce the circuit in Fig. 10-18 to its minimum.
(*ans. A-B*, 446.5 Ω; *C-D*, 185.9 Ω)

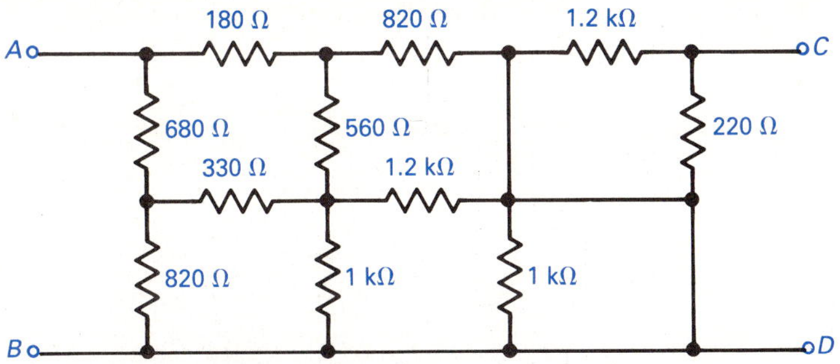

Fig. 10-18. Problem 9: Network.

10-2 THEVENIN'S THEOREM

The ultimate in network reduction occurs when a circuit is reduced to just two items: the element about which information is required, and the remainder of the circuit considered as a unit. An example of this type of circuit reduction is depicted in Fig. 10-19.

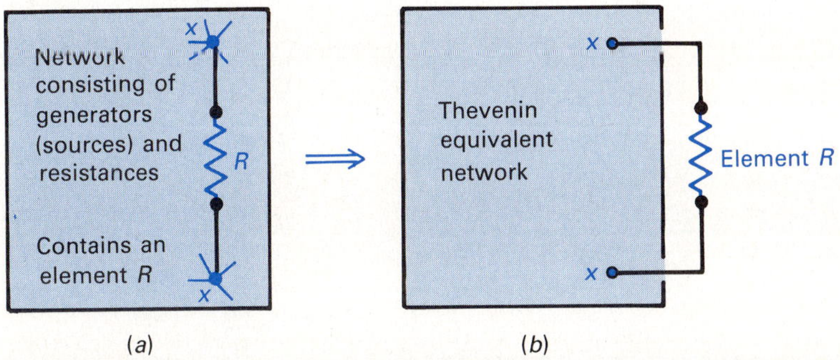

Fig. 10-19. *(a)* Network of unspecified type; *(b)* network with one element isolated.

Thevenin's theorem is concerned with achieving this reduction and then using it to arrive at the current, voltage, or power associated with the element R. Thevenin's theorem states that the current in any resistor R connected to two terminals of a network is the same as if R were connected to a voltage source where

1. The emf is the open-circuit voltage at the terminals of R.
2. The internal resistance is the resistance of the network at the terminals of R, with all other sources replaced by resistances equal to their internal resistances.

Thevenin Equivalent Circuit

The circuit which results from the Thevenin approach consists of a voltage source with series resistance, as in Fig. 10-20.

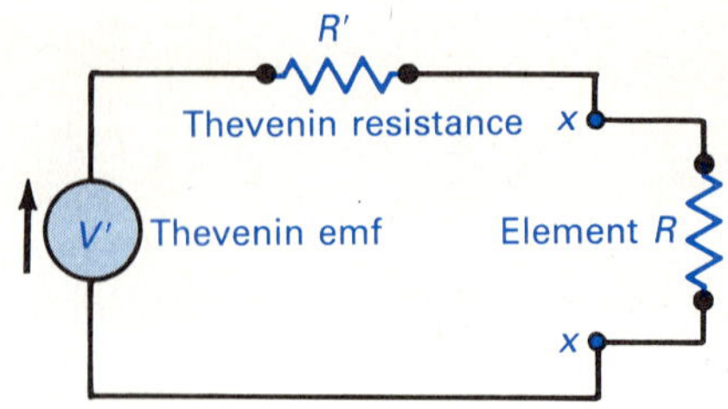

Fig. 10-20. Basic Thevenin circuit.

Referring to Figs. 10-19 and 10-20, the *Thevenin emf* V' is the open-circuit voltage at the x,x terminals (the voltage with R disconnected).

The *Thevenin resistance* R' is the resistance at the x,x terminals with these terminals open-circuit (R disconnected) and all energy sources in the remaining circuit removed and replaced by their internal resistances.

To see how the Thevenin voltage and resistance are determined, let us look at a circuit consisting of a battery, a T network, and a load, as in Fig. 10-21. Since a T network represents all networks which can be reduced to a T format, the circuit in Fig. 10-21 is considered typical of all such networks.

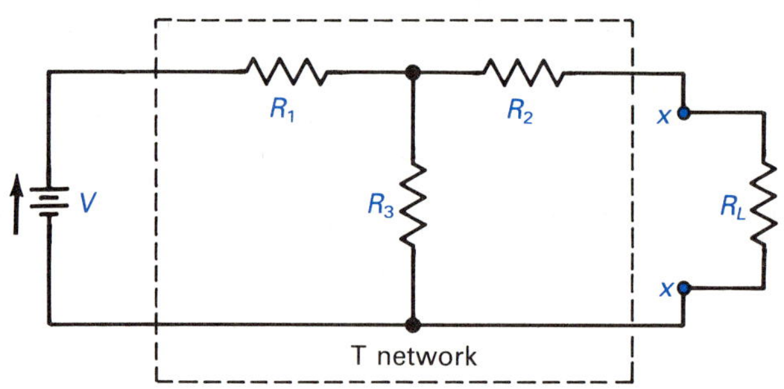

Fig. 10-21. Battery, T network, and load.

To obtain the value of the Thevenin emf V',

1. Disconnect R_L at the x,x terminals.
2. Determine the voltage at the x,x terminals.

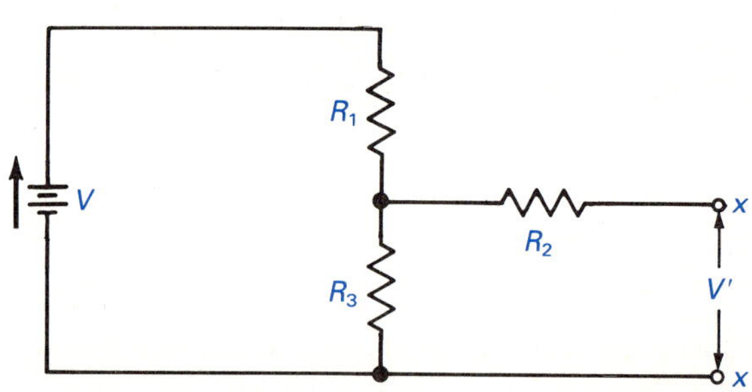

Fig. 10-22. Thevenin voltage in the circuit in Fig. 10-21.

When R_L is disconnected (Fig. 10-22), the voltage at the terminals is the same as the voltage across R_3 (since there can be no current in R_2, there is no voltage across it).

$$V' = V \frac{R_3}{R_1 + R_3} \tag{10-10}$$

To obtain the value of the Thevenin resistance R',

1. Disconnect the load R_L at the x,x terminals.
2. Remove the battery V. As no internal resistance is shown for this battery, it is assumed to be 0 Ω, a short circuit.
3. Determine the resistance between the x,x terminals.

With R_L removed and the battery replaced by a short circuit, the network across the terminals will be series-parallel, as shown in Fig. 10-23.

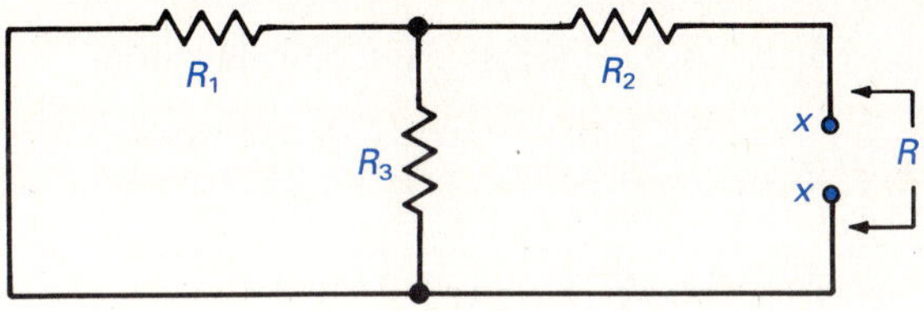

Fig. 10-23. Thevenin resistance in the circuit in Fig. 10-21.

$$R' = R_2 + \frac{R_1 R_3}{R_1 + R_3} \tag{10-11}$$

The Thevenin circuit equivalent to Fig. 10-21 will be as in Fig. 10 24.

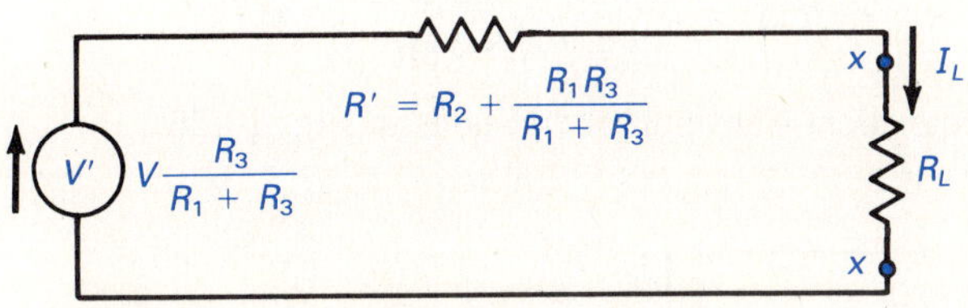

Fig. 10-24. Thevenin equivalent of the circuit in Fig. 10-21.

When the Thevenin circuit has been drawn, the load current, voltage, power, or energy can be computed for any load resistance.

Guidelines

Guidelines for judging the suitability of a circuit for Thevenin analysis include:

1. Data for one element are required.
2. The resistance of that element has various alternative values.
3. There are more voltage sources than current sources.
4. The circuit details are not known.

The last guideline is significant. Provided that the Thevenin voltage and resistance can be measured, it is not necessary to know anything about the network. Thus the current, voltage, and power of an element can be determined, even though the equipment with which it is associated is too sophisticated to be analyzed easily.

Numerical examples will demonstrate the application of Thevenin's theorem.

EXAMPLE 10-5 What is the load current in the circuit in Fig. 10-25?

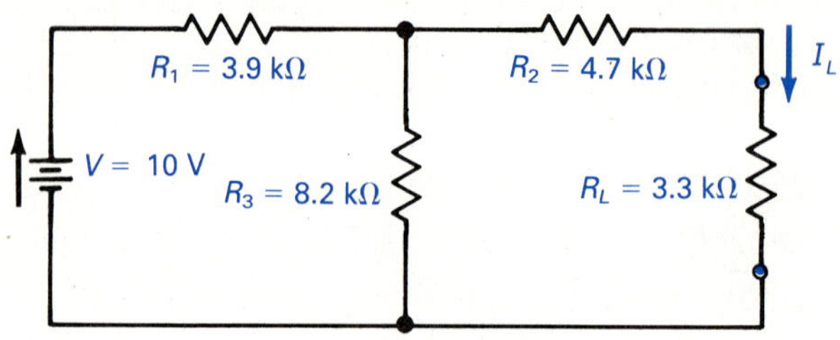

Fig. 10-25. Example 10-5: Circuit.

Solution With R_L removed, the voltage across the terminals will be

$$V' = V \frac{R_3}{R_1 + R_3} \tag{10-10}$$

$$V' = 10\ \text{V} \frac{8.2\ \text{k}\Omega}{3.9\ \text{k}\Omega + 8.2\ \text{k}\Omega} = 6.777\ \text{V}$$

With R_L and the battery removed (its internal resistance is zero), the resistance across the load terminals will be

$$R' = R_2 + \frac{R_1 R_3}{R_1 + R_3} \tag{10-11}$$

$$R' = 4.7\ \text{k}\Omega + \frac{3.9\ \text{k}\Omega \times 8.2\ \text{k}\Omega}{3.9\ \text{k}\Omega + 8.2\ \text{k}\Omega} = 7.343\ \text{k}\Omega$$

The equivalent Thevenin circuit is as in Fig. 10-26.

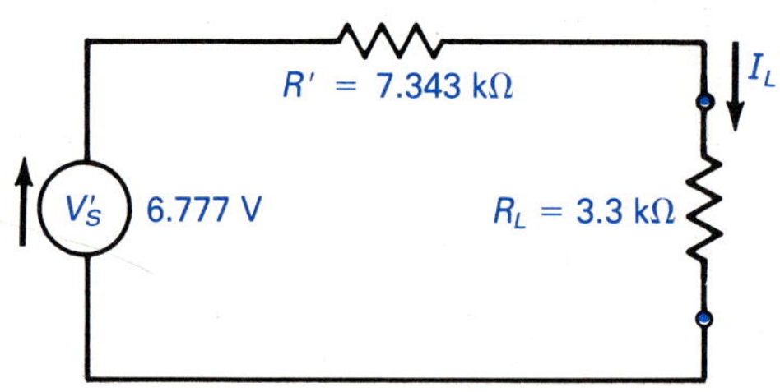

Fig. 10-26. Example 10-5: Thevenin equivalent of Fig. 10-25.

$$I_L = \frac{V'_S}{R' + R_L} = \frac{6.777\ \text{V}}{7.343\ \text{k}\Omega + 3.3\ \text{k}\Omega} = \mathbf{0.636\ mA}$$

EXAMPLE 10-6 The bridged-T network in Fig. 10-27*a* can be reduced to the simple T network in Fig. 10-27*b* using preferred-value 10 percent resistors. If the load resistance can be 100 Ω, 1 kΩ, or 10 kΩ, what are the related load currents?

Solution Applying Eq. (10-10) to Fig. 10-27*b*, the Thevenin voltage will be

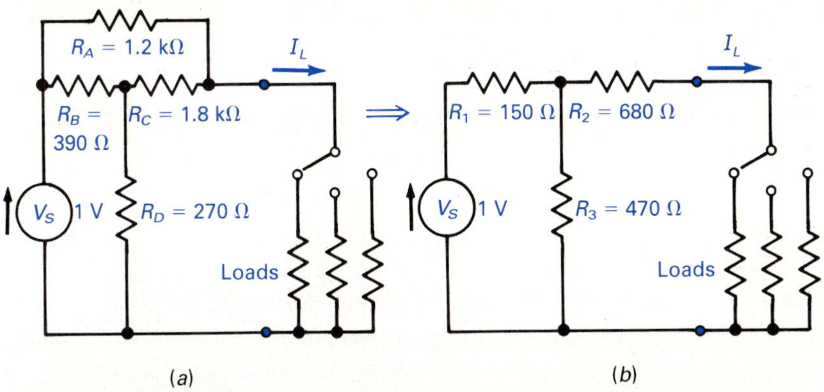

Fig. 10-27. Example 10-6: *(a)* Original circuit with a bridged-T network; *(b)* Fig. 10-27*a* with the bridged-T network reduced to a simple T network by π-to-T conversion.

$$V_S' = V_S \frac{R_3}{R_1 + R_3} = 1\text{ V}\,\frac{470\ \Omega}{150\ \Omega + 470\ \Omega} = 0.7581\text{ V}$$

Applying Eq. (10-11) to Fig. 10-27*b*, the Thevenin resistance will be

$$R' = R_2 + \frac{R_1 R_3}{R_1 + R_3} = 680\ \Omega + \frac{150\ \Omega \times 470\ \Omega}{150\ \Omega + 470\ \Omega} = 793.7\ \Omega$$

The equivalent Thevenin circuit is shown in Fig. 10-28.

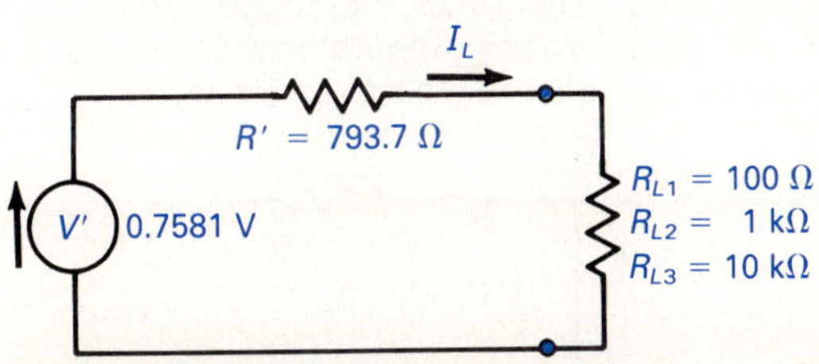

Fig. 10-28. Example 10-6: Thevenin equivalent of Fig. 10-27.

When R_L is 100 Ω,

$$I_L = \frac{V_S'}{R' + R_{L1}} = \frac{0.7581\text{ V}}{793.7\ \Omega + 100\ \Omega} = \mathbf{0.8483\ mA}$$

When R_L is 1 kΩ,

$$I_L = \frac{V_S'}{R' + R_{L2}} = \frac{0.7581\text{ V}}{793.7\ \Omega + 1\text{ k}\Omega} = \mathbf{0.4226\ mA}$$

When R_L is 10 kΩ,

$$I_L = \frac{V_S'}{R' + R_{L3}} = \frac{0.7581\text{ V}}{793.7\ \Omega + 10\text{ k}\Omega} = \mathbf{0.0702\ mA} \quad \text{or} \quad \mathbf{70.2\ \mu A}$$

Example 10-6 illustrates a particular advantage of the Thevenin method of circuit analysis. The Thevenin voltage and resistance have to be calculated just once for any network. When this calculation has been made, a series of values associated with the element of interest can be determined using a simple series circuit. This method is useful in many practical situations.

EXAMPLE 10-7 An electronic voltage regulator has within its circuit a 1-kΩ 1-W resistor. A circuit designer wishes to reduce the resistor's current by 50 percent by connecting another resistor in series with it. The 1 kΩ is first disconnected. With the power on, 125 V is measured across the resistor circuit's terminals. With the power off, 4 kΩ is measured across these terminals. What value series resistor should be used?

Solution A representation of measured conditions (Fig. 10-29) defines the Thevenin voltage as 125 V and the resistance as 4 kΩ. Thevenin's equivalent of the regulator viewed from the 1-kΩ resistor R_1 is shown in Fig. 10-30*a*. This circuit with a series resistor R_2 added (Fig. 10-30*b*) reduces the current in R_1 by 50 percent.

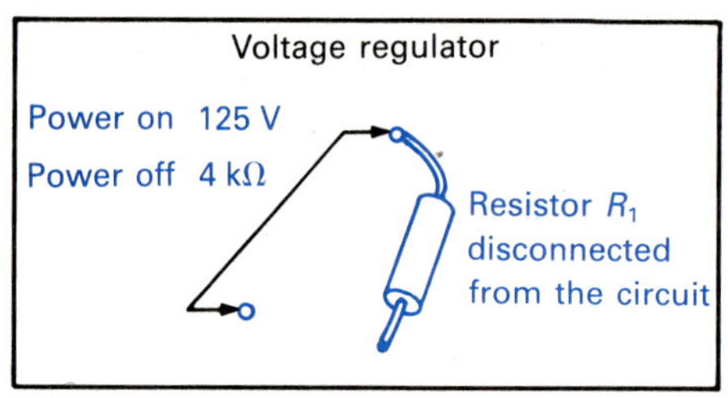

Fig. 10-29. Example 10-7: Voltage regulator with resistor disconnected.

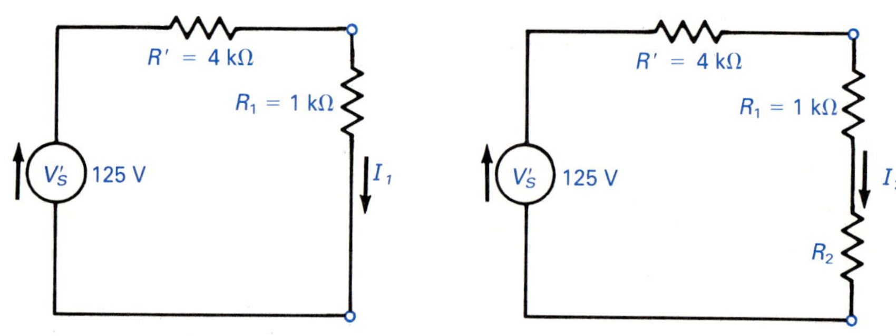

Fig. 10-30. Example 10-7: *(a)* Thevenin circuit of original voltage regulator; *(b)* Thevenin circuit with the current-reducing resistor added.

From the Thevenin equivalent of the original circuit, Fig. 10-30*a*.

$$I_1 = \frac{V_S'}{R' + R_1} = \frac{125\text{ V}}{4\text{ k}\Omega + 1\text{ k}\Omega} = 25\text{ mA}$$

The current I_2 in the circuit in Fig. 10-30*b*, which includes the added resistor R_2, is to be 50 percent of the original current I_1, or 12.5 mA.

The total circuit resistance R_T in Fig. 10-30*b* will be

$$R_T = \frac{V_S'}{I_2} = \frac{125 \text{ V}}{12.5 \text{ mA}} = 10 \text{ k}\Omega$$

$$R_2 = R_T - (R' + R_1) = 10 \text{ k}\Omega - (4 \text{ k}\Omega + 1 \text{ k}\Omega) = \mathbf{5 \text{ k}\Omega}$$

In Example 10-7 it was not necessary to know anything about the details of the voltage regulator circuit. The Thevenin characteristics of the regulator circuit, from the viewpoint of the 1-kΩ resistor, were determined by measurement. Once these Thevenin values were known, it was a simple matter to calculate the value of the resistor to be added to meet the designer's requirements.

EXAMPLE 10-8

What is the meter current in the circuit in Fig. 10-31?

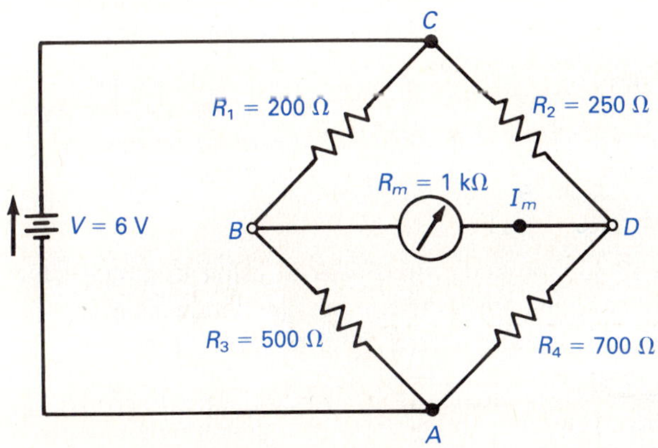

Fig. 10-31. Example 10-8: Given circuit.

Solution Redrawing the circuit with the meter removed, to determine the Thevenin emf, gives Fig. 10-32.

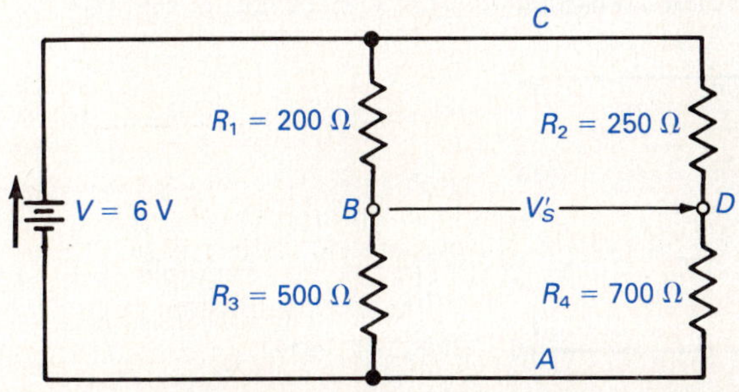

Fig. 10-32. Thevenin voltage in Fig. 10-31.

The Thevenin emf will be the voltage across the open-circuit meter terminals, $V_{DB} = V_{DA} - V_{BA}$:

$$V_{DA} = V\frac{R_4}{R_2 + R_4} = 6\text{ V}\frac{700\ \Omega}{250\ \Omega + 700\ \Omega} = 4.421\text{ V}$$

$$V_{BA} = V\frac{R_3}{R_1 + R_3} = 6\text{ V}\frac{500\ \Omega}{200\ \Omega + 500\ \Omega} = 4.286\text{ V}$$

$$V_S' = V_{DA} - V_{BA} = 4.421\text{ V} - 4.286\text{ V} = 0.135\text{ V}$$

As V_{DA} is greater than V_{BA}, terminal D will be positive with respect to terminal B.

With the meter terminals open-circuit and no battery, the circuit will be as shown in Fig. 10-33*a*. Note the short circuit between points C and A due to the zero battery resistance.

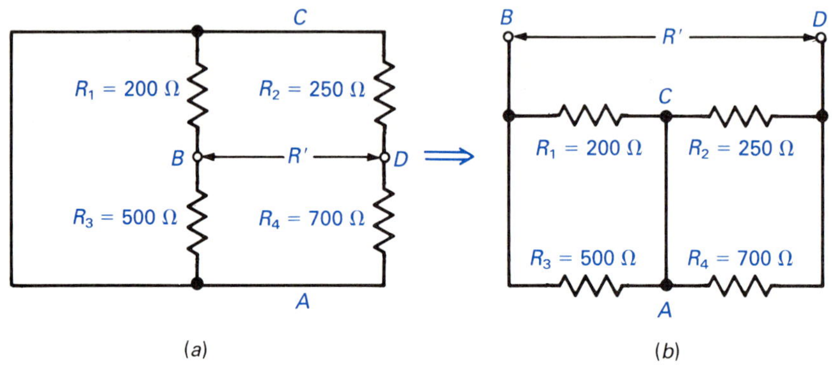

Fig. 10-33. Thevenin resistance in Fig. 10-31. *(a)* Original circuit layout; *(b)* circuit rearranged to show R' as the resistance of a series-parallel network.

Figure 10-33*b* shows the circuit with its elements arranged for ease in determining the resistance between terminals B and D. This is a series-parallel circuit. The resistance B to D is the Thevenin resistance:

$$R' = \frac{R_1R_3}{R_1 + R_3} + \frac{R_2R_4}{R_2 + R_4}$$

$$= \frac{200\ \Omega \times 500\ \Omega}{200\ \Omega + 500\ \Omega} + \frac{250\ \Omega \times 700\ \Omega}{250\ \Omega + 700\ \Omega} = 327.1\ \Omega$$

The Thevenin equivalent circuit will be as in Fig. 10-34.

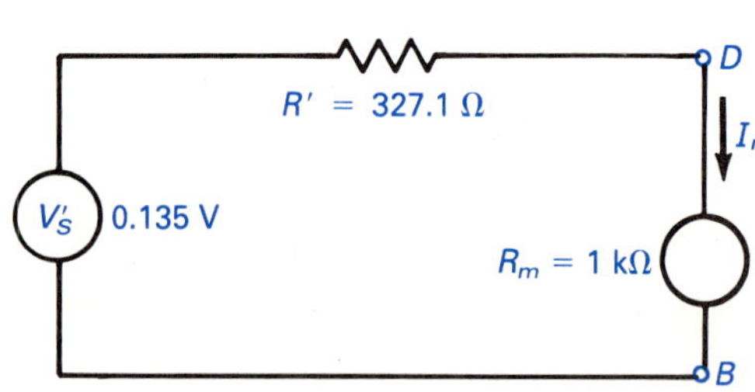

Fig. 10-34. Example 10-8: Thevenin equivalent circuit of Fig. 10-31.

$$I_m = \frac{V'}{R' + R_m} = \frac{0.135\text{ V}}{327.1\ \Omega + 1\text{ k}\Omega}$$

$$= \mathbf{0.1017\ mA} \quad \text{or} \quad \mathbf{101.7\ \mu A}$$

EXAMPLE 10-9 What power is actually delivered to load R_3 in the three-wire circuit in Fig. 10-35? The nominal load powers are: P_1 = 2.5 kW at 120 V, P_2 = 1.5 kW at 120 V, and P_3 = 3, 6, 9, or 12 kW at 240 V. Assume that the load resistances do not change appreciably if the load voltages are lower than the rated values.

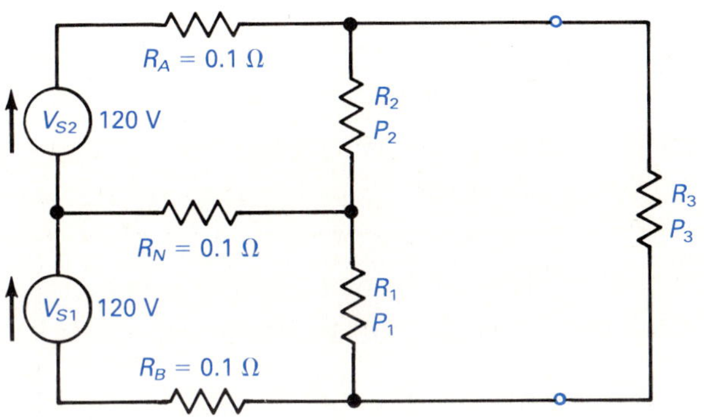

Fig. 10-35. Example 10-9: Given circuit.

Solution Calculating the values of the load resistances using the rated load voltages,

$$P = \frac{V^2}{R} \quad \text{or} \quad R = \frac{V^2}{P}$$

$$R_1 = \frac{V_1^2}{P_1} = \frac{(120\ \text{V})^2}{2.5\ \text{kW}} = 5.76\ \Omega$$

$$R_2 = \frac{V_2^2}{P_2} = \frac{(120\ \text{V})^2}{1.5\ \text{kW}} = 9.6\ \Omega$$

$$R_3 = \frac{V_3^2}{P_3} = \frac{(240\ \text{V})^2}{3\ \text{kW}} = 19.2\ \Omega \quad \text{or} \quad \frac{(240\ \text{V})^2}{6\ \text{kW}} = 9.6\ \Omega$$

$$\text{or} \quad \frac{(240\ \text{V})^2}{9\ \text{kW}} = 6.4\ \Omega \quad \text{or} \quad \frac{(240\ \text{V})^2}{12\ \text{kW}} = 4.8\ \Omega$$

Removing the load R_3 to obtain the Thevenin voltage gives the circuit in Fig. 10-36.

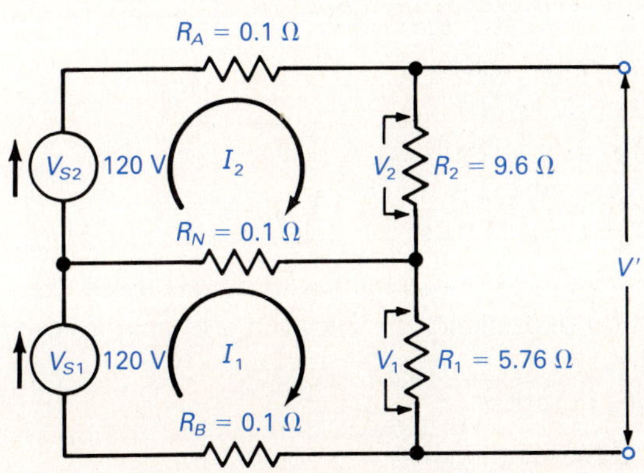

Fig. 10-36. Thevenin voltage in Fig. 10-35.

Loop analysis can be used in determining the Thevenin voltage. The loop equations are

$$5.96I_1 - 0.1I_2 = 120$$
$$-0.1I_1 + 9.8I_2 = 120$$

The solutions to these simultaneous equations are

$$I_1 = 20.34\ \text{A} \quad \text{and} \quad I_2 = 12.45\ \text{A}$$

The Thevenin voltage will be

$$\begin{aligned} V_S' &= V_1 + V_2 = I_1R_1 + I_2R_2 \\ &= 20.34\ \text{A} \times 5.76\ \Omega + 12.45\ \text{A} \times 9.6\ \Omega \\ &= 236.7\ \text{V} \end{aligned}$$

Removing the load R_3 and both generators results in the circuit in Fig. 10-37*a*. Converting the π network $R_AR_2R_N$ into an equivalent T network gives the circuit in Fig. 10-37*b*.

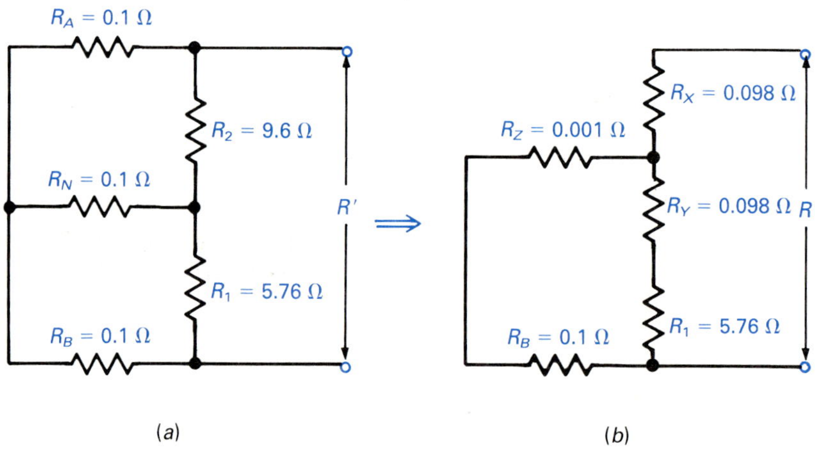

Fig. 10-37. *(a)* Thevenin resistance in Fig. 10-35; *(b)* first reduction of Fig. 10-37*a* by converting the π network R_A, R_2, R_N to the T network R_X, R_Y, R_Z.

$$\Sigma R_\nabla = R_A + R_2 + R_N = 0.1\ \Omega + 9.6\ \Omega + 0.1\ \Omega = 9.8\ \Omega$$

$$R_X = \frac{R_AR_2}{\Sigma R_\nabla} = \frac{0.1\ \Omega \times 9.6\ \Omega}{9.8\ \Omega} = 0.098\ \Omega$$

$$R_Y = \frac{R_2R_N}{\Sigma R_\nabla} = \frac{9.6\ \Omega \times 0.1\ \Omega}{9.8\ \Omega} = 0.098\ \Omega$$

$$R_Z = \frac{R_AR_N}{\Sigma R_\nabla} = \frac{0.1\ \Omega \times 0.1\ \Omega}{9.8\ \Omega} = 0.001\ \Omega$$

Figure 10-37*b* can now be simplified to the circuit in Fig. 10-38, from which the value of the Thevenin resistance R' can be computed.

R_P is R_Z and R_B in series:

$$R_P = R_Z + R_B = 0.001\ \Omega + 0.1\ \Omega = 0.101\ \Omega$$

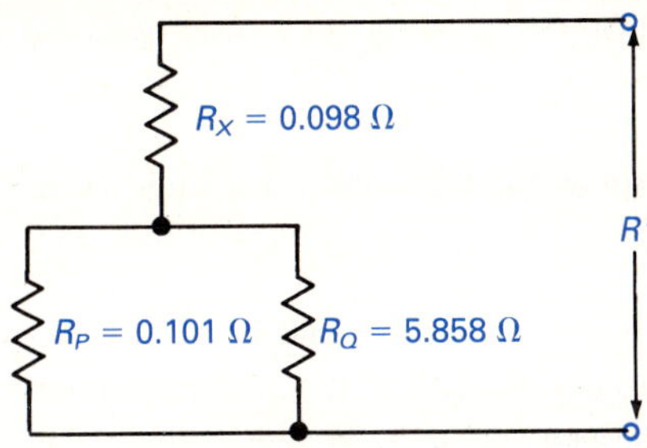

Fig. 10-38. Second reduction of Fig. 10-37*a* by converting the series groups $R_Z R_B$ and $R_Y R_1$ to elements R_P and R_Q.

R_Q is R_Y and R_1 in series:

$$R_Q = R_Y + R_1 = 0.098\ \Omega + 5.76\ \Omega = 5.858\ \Omega$$

$$R' = R_X + \frac{R_P R_Q}{R_P + R_Q} = 0.098\ \Omega + \frac{0.101\ \Omega \times 5.858\ \Omega}{0.101\ \Omega + 5.858\ \Omega}$$

$$= 0.1973\ \Omega$$

The Thevenin equivalent circuit will be as in Fig. 10-39.

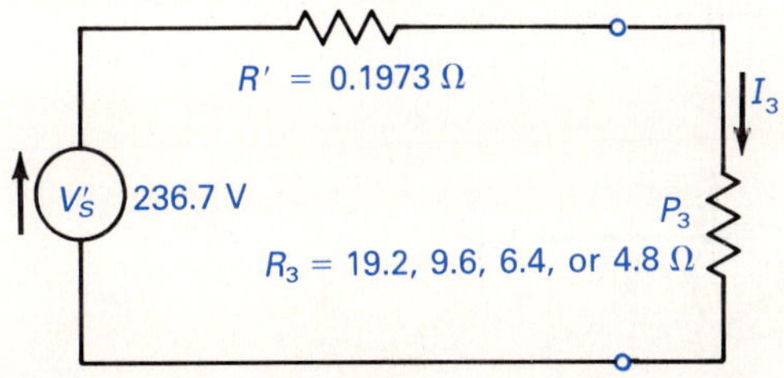

Fig. 10-39. Example 10-9: Thevenin equivalent of Fig. 10-35.

When $R_3 = 19.2\ \Omega$,

$$I_3 = \frac{V'_S}{R' + R_3} = \frac{236.7\ \text{V}}{0.1973\ \Omega + 19.2\ \Omega} = 12.2\ \text{A}$$

$$P_3 = I_3^2 R_3 = (12.2\ \text{A})^2 \times 19.2\ \Omega = \mathbf{2.858\ kW}$$

When $R_3 = 9.6\ \Omega$,

$$I_3 = \frac{V'_S}{R' + R_3} = \frac{236.7\ \text{V}}{0.1973\ \Omega + 9.6\ \Omega} = 24.16\ \text{A}$$

$$P_3 = I_3^2 R_3 = (24.16\ \text{A})^2 \times 9.6\ \Omega = \mathbf{5.603\ kW}$$

When $R_3 = 6.4\ \Omega$,

$$I_3 = \frac{V'_S}{R' + R_3} = \frac{236.7\ \text{V}}{0.1973\ \Omega + 6.4\ \Omega} = 35.88\ \text{A}$$

$$P_3 = I_3^2 R_3 = (35.88\ \text{A})^2 \times 6.4\ \Omega = \mathbf{8.239\ kW}$$

When $R_3 = 4.8\ \Omega$,

$$I_3 = \frac{V'_S}{R' + R_3} = \frac{236.7\ \text{V}}{0.1973\ \Omega + 4.8\ \Omega} = 47.37\ \text{A}$$

$$P_3 = I_3^2 R_3 = (47.37\ \text{A})^2 \times 4.8\ \Omega = \mathbf{10.77\ kW}$$

In the solution of Example 10-9 it took a lot of work to get to the Thevenin equivalent circuit. For a single-value load it would not be worthwhile, as loop or nodal analysis would be faster. With a number of different load values, four in this case, the work pays off, as the Thevenin circuit only has to be found once; this is the point of the Thevenin approach. With any analysis method that does not involve an equivalent circuit, the whole solution has to be repeated for each load value.

You will see that, because of the line resistances, each of the calculated powers is lower than the rated value. You should also note that the percentage power loss becomes greater as the load power increases. When the rated load power is 3 kW, 95.3 percent of this value is developed. When the power is intended to be 12 kW, only 89.8 percent of this rated value appears at the load.

EXAMPLE 10-10 Figure 10-40 shows part of an electronic circuit. What is the base-to-emitter voltage V_{BE} of the transistor if its base-to-emitter resistance R_{BE} is 800 Ω?

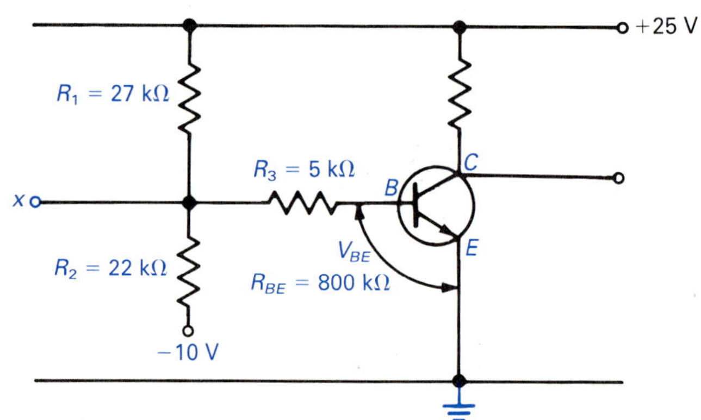

Fig. 10-40. Example 10-10: Given circuit.

Solution From the viewpoint of the required information, the circuit can be redrawn as in Fig. 10-41.

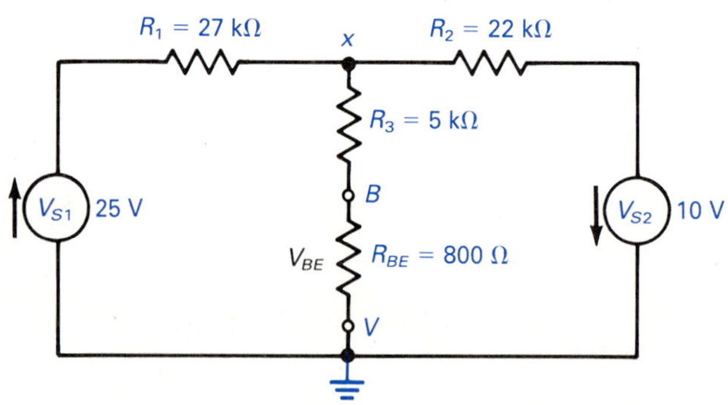

Fig. 10-41. Equivalent base-to-emitter circuit for Fig. 10-40.

Disconnecting R_3 and R_{BE} gives the circuit in Fig. 10-42.

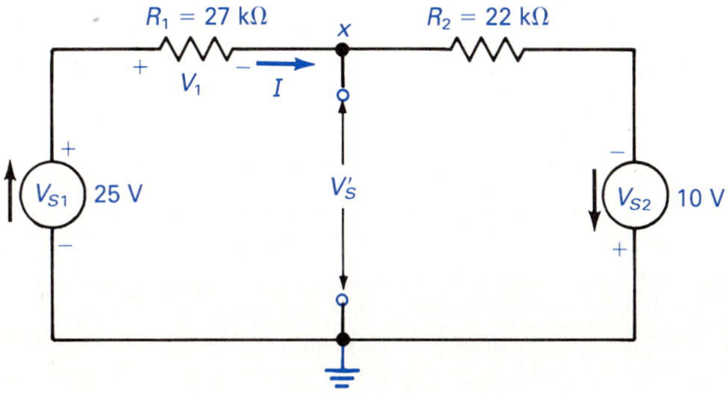

Fig. 10-42. Thevenin voltage in Fig. 10-40.

The two generator voltages aid each other, so the current I in Fig. 10-42 will be

$$I = \frac{V_{S1} + V_{S2}}{R_1 + R_2} = \frac{25\text{ V} + 10\text{ V}}{27\text{ k}\Omega + 22\text{ k}\Omega} = 0.7143\text{ mA}$$

The voltage across R_1 will be

$$V_1 = IR_1 = 0.7143\text{ mA} \times 27\text{ k}\Omega = 19.29\text{ V}$$

The Thevenin voltage is the voltage between point x and ground.

$$V'_S = V_{S1} - V_1 = 25\text{ V} - 19.29\text{ V} = 5.71\text{ V}$$

With both generators removed, the Thevenin resistance between point x and ground will be R_1 and R_2 in parallel.

$$R' = \frac{R_1R_2}{R_1 + R_2} = \frac{27\text{ k}\Omega \times 22\text{ k}\Omega}{27\text{ k}\Omega + 22\text{ k}\Omega} = 12.12\text{ k}\Omega$$

The Thevenin equivalent circuit will therefore be as in Fig. 10-43.

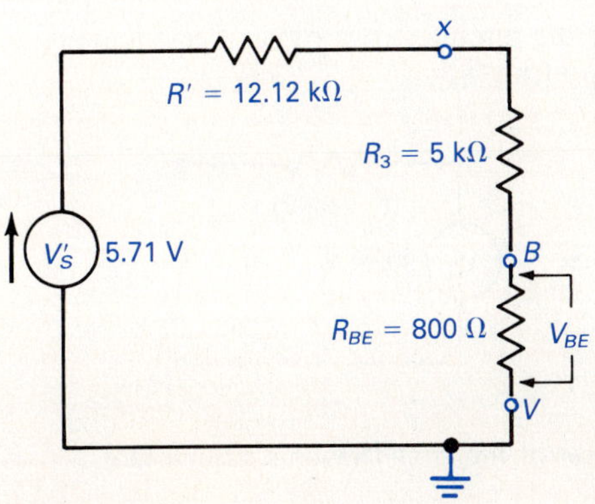

Fig. 10-43. Example 10-10: Thevenin equivalent of Fig. 10-40.

The base-to-emitter voltage will be

$$V_{BE} = \frac{V_S' R_{BE}}{R' + R_3 + R_{BE}} = \frac{5.71\text{ V} \times 800\ \Omega}{12.12\text{ k}\Omega + 5\text{ k}\Omega + 800\ \Omega}$$
$$= \mathbf{0.2549\ V}$$

Successive Use of Thevenin's Theorem

To introduce the Thevenin approach to network analysis, single elements have been isolated for attention. It is not necessary, however, to be so restrictive; complete networks within a circuit can also be isolated. Any two points in a circuit can be chosen, and then all of the circuit on one side of these points can be detached as a unit. The Thevenin equivalent of the circuit on the other side of the points is then determined. By this means, complex circuits can be dealt with in sections to minimize the analysis work.

This successive Thevenin technique is demonstrated by a numerical example.

EXAMPLE 10-11 What is the value of the current in R_6 in the circuit in Fig. 10-44?

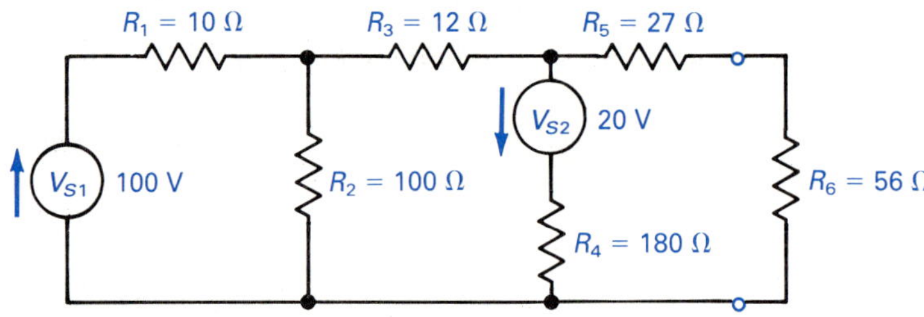

Fig. 10-44. Example 10-11: Given circuit.

Solution It is quite feasible to apply Thevenin's theorem directly; resistor R_6 can be removed, and the value of the Thevenin voltage and resistance calculated. The following technique, while requiring a lengthy explanation, is easily applied because the individual calculations are simple. The steps are:

1. Disconnect all of the circuit to the right of R_2, leaving a network as in Fig. 10-45*a*.

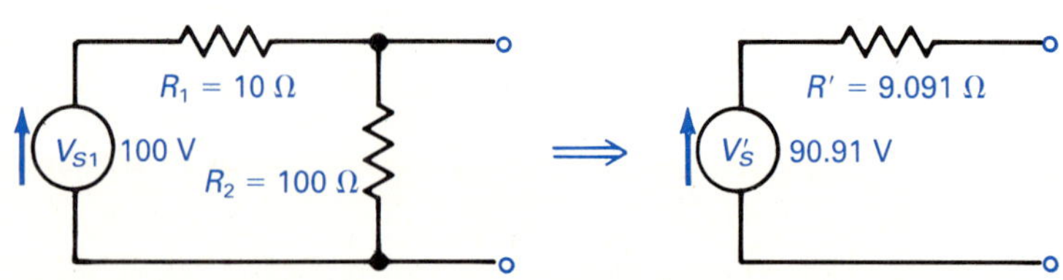

Fig. 10-45. Example 10-11: Derivation of the first Thevenin equivalent circuit.

Figure 10-45*b* is the Thevenin equivalent of Fig. 10-45*a*.

$$V_S' = V_{S1}\frac{R_2}{R_1 + R_2} = 100\text{ V}\frac{100\ \Omega}{10\ \Omega + 100\ \Omega} = 90.91\text{ V}$$

$$R' = \frac{R_1R_2}{R_1 + R_2} = \frac{10\ \Omega \times 100\ \Omega}{10\ \Omega + 100\ \Omega} = 9.091\ \Omega$$

2. Add to the first Thevenin circuit elements R_3, R_4, and V_{S2}. Leave the network to the right of R_4 and V_{S2} disconnected. The circuit in Fig. 10-46*a* results.

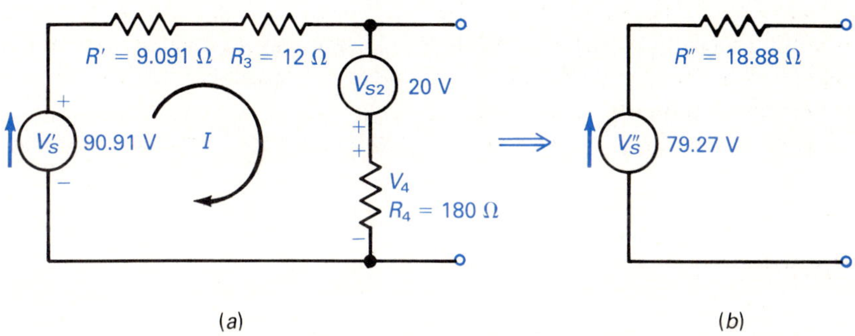

Fig. 10-46. Example 10-11: Derivation of the second Thevenin equivalent circuit.

The two generators in Fig. 10-46*a* are aiding each other, so the total emf is $V_S' + V_{S2}$. The current in the circuit is given by

$$I = \frac{V_S' + V_{S2}}{R' + R_3 + R_4} = \frac{90.91\text{ V} + 20\text{ V}}{9.091\ \Omega + 12\ \Omega + 180\ \Omega} = 0.5515\text{ A}$$

The voltage across R_4 is

$$V_4 = IR_4 = 0.5515\text{ A} \times 180\ \Omega = 99.27\text{ V}$$

V_4 and V_{S2} are of opposing polarity, so that the second Thevenin voltage is the difference between them.

$$V_S'' = V_4 - V_{S2} = 99.27\text{ V} - 20\text{ V} = 79.27\text{ V}$$

The second Thevenin resistance is the resistance between the terminals in Fig. 10-46*a*, with both generators removed. It is the series-parallel combination of R_4 in parallel with the series group R', R_3.

$$R'' = \frac{R_4(R' + R_3)}{R_4 + R' + R_3} = \frac{180\ \Omega(9.091\ \Omega + 12\ \Omega)}{180\ \Omega + 9.091\ \Omega + 12\ \Omega} = 18.88\ \Omega$$

3. Connect R_5 to the second Thevenin circuit, Fig. 10-46*b*. This connection gives the circuit in Fig. 10-47*a*. When R_6

is added, as in Fig. 10-47*b*, the Thevenin equivalent of the entire original circuit in Fig. 10-44 results.

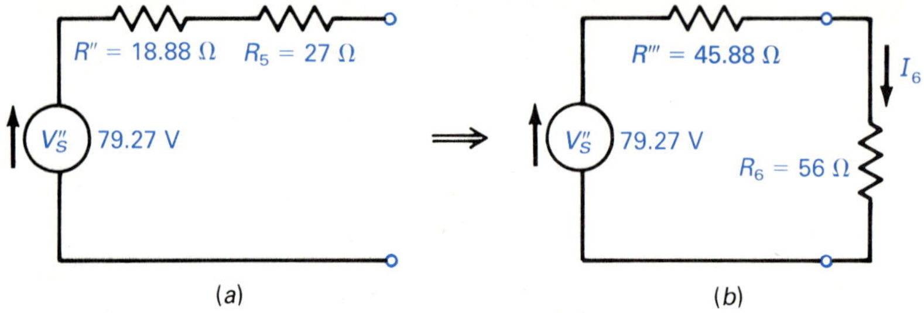

Fig. 10-47. Example 10-11: *(a)* Second Thevenin emf and resistance together with R_5; *(b)* final Thevenin equivalent of Fig. 10-44.

$$R''' = R'' + R_5 = 18.88\ \Omega + 27\ \Omega = 45.88\ \Omega$$

$$I_6 = \frac{V_S''}{R''' + R_6} = \frac{79.27\text{ V}}{45.88\ \Omega + 56\ \Omega} = \mathbf{0.7781\ A}$$

REVIEW QUESTIONS FOR SECTION 10-2

1. Of what elements does a Thevenin equivalent consist?
2. When a circuit is reduced to its Thevenin equivalent, to what voltage is the Thevenin emf V' equal?
3. When a circuit is reduced to its Thevenin equivalent, to what is the Thevenin resistance R' equal?
4. When determining the Thevenin emf and resistance of a circuit, the element about which information is required is left connected while calculations or measurements are being made. Correct?
5. The validity of the Thevenin approach can be proven using a T network. Why is this sufficient proof that Thevenin methods apply to more complex networks?
6. Thevenin analysis may be used even though the details of a circuit are not known. In such situations, how are the Thevenin emf and resistance values found?
7. When a number of values associated with a circuit element are to be verified, how is the Thevenin approach more advantageous than loop or nodal analysis?
8. How can a complex network in an electronic system be most easily reduced to its Thevenin equivalent circuit?

EXERCISE PROBLEMS FOR SECTION 10-2

1. Rework Prob. 1, Sec. 9-1 (page 268), by reducing the circuits to their Thevenin equivalents to find I_L.

(*ans. a.* 68.04 mA; *b.* 1.107 mA; *c.* 1.25 A; *d.* 6.5 μA)

2. Rework Prob. 2, Sec. 9-1 (page 268), by reducing the circuits to their Thevenin equivalents to find I_L.

3. Reduce Fig. 9-8*a* and *b* (page 269) to equivalent Thevenin circuits and compare. (*ans*. same circuit)

4. Using Thevenin's theorem, find the magnitude and direction of the meter current in Example 10-8 in the text if R_4 is changed to 200 Ω.

5. Reduce the circuit in Fig. 9-48 (page 296) to its Thevenin equivalent and find the value of I_R. (*ans*. 5.31 mA)

6. Reduce the circuit in Fig. 9-49 (page 296) to its Thevenin equivalent and find the value of the meter current.

7. Using Thevenin's theorem, find the power dissipation of R_2 in Example 10-11 (page 336) in the text. (*ans*. 63.3 W)

8. Using the ammeter as the element of interest in Fig. 9-47 (page 296), find

a. The Thevenin equivalent circuit.

b. The meter current.

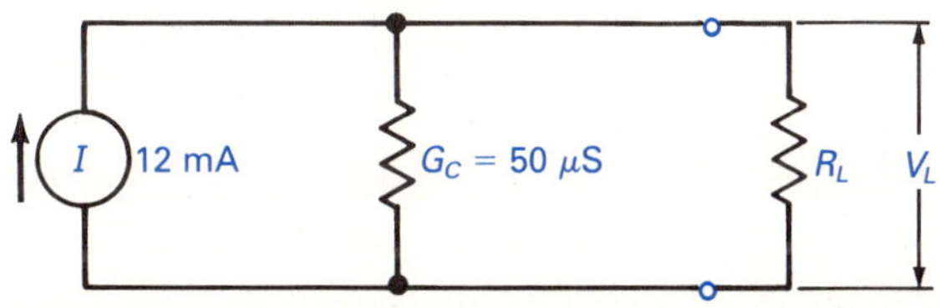

Fig. 10-48. Problem 9: Circuit.

9. Figure 10-48 shows the output equivalent circuit of a transistor amplifier. Convert the circuit to a Thevenin equivalent and find V_L when

a. R_L = 470 Ω (*ans*. 5.51 V)

b. R_L = 1000 Ω (*ans*. 11.42 V)

c. R_L = 1500 Ω (*ans*. 16.74 V)

d. R_L = 2700 Ω (*ans*. 28.55 V)

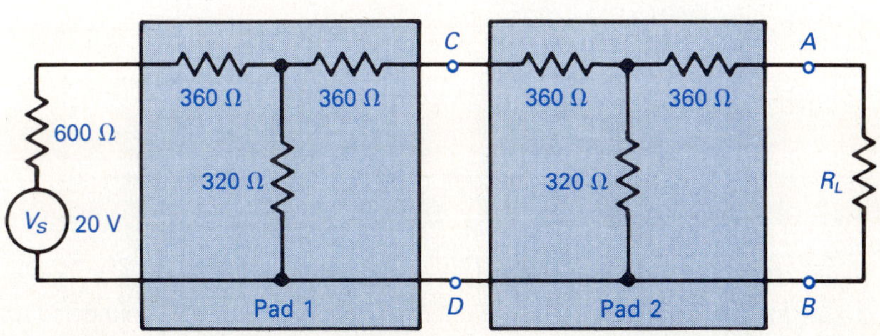

Fig. 10-49. Problem 10: Circuit.

10. The circuit in Fig. 10-49 shows a 600-Ω generator followed by two T-pad attenuators.

a. Find the Thevenin equivalent for the circuit at points *C* and *D* with pad 2 disconnected.

b. Find the Thevenin equivalent for the circuit at points *A* and *B* with both pads connected.

c. With R_L connected between points *A* and *B*, find the voltage developed across the load when
   i. $R_L = 300\ \Omega$
   ii. $R_L = 600\ \Omega$

d. What is the maximum power obtainable at points *A* and *B* with both pads connected?

11. The open-circuit voltage output of a generator is 20 V. When a 1000-Ω resistor is connected across the output, the terminal voltage falls to 15.4 V. Find the terminal voltage when a 470-Ω resistor is connected across the output. (*ans*. 12.24 V)

12. If an input signal of −1 V is applied through a 1000-Ω resistance to the input in Fig. 10-76 (page 354), find the base current when the base-to-emitter resistance is 200 Ω.

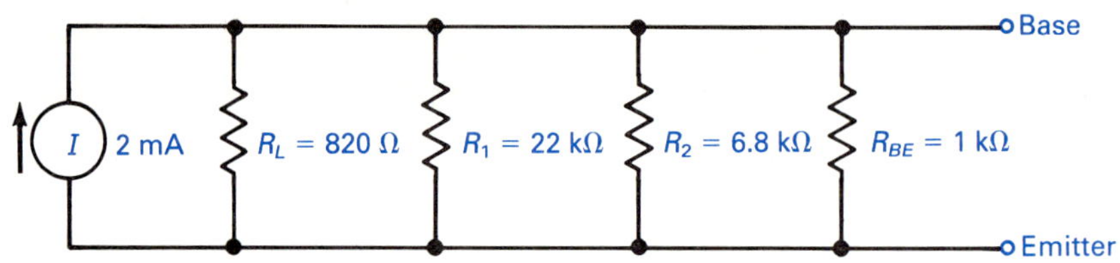

Fig. 10-50. Problem 13: Circuit.

13. The circuit in Fig. 10-50 represents the interstage coupling between two transistor amplifiers. Find the Thevenin equivalent of the circuit with R_{BE} as the element of interest. Find the signal current through R_{BE}.

(*ans*. 1.42 V, 708 Ω, 829 μA)

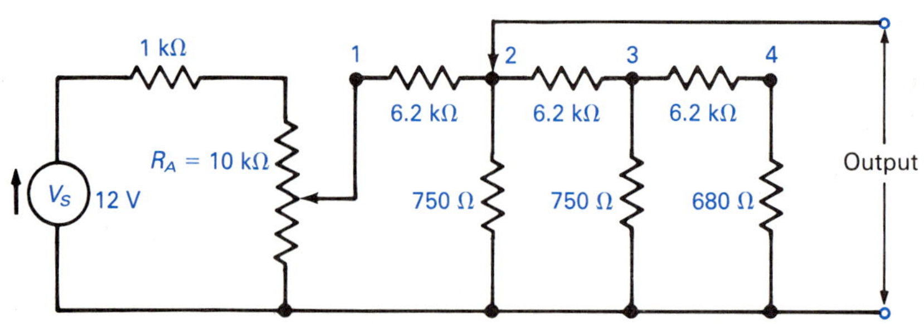

Fig. 10-51. Problem 14: Circuit.

14. Figure 10-51 is the output equivalent circuit of a generator. With R_A set to 5 kΩ, find the equivalent Thevenin circuit when the output is connected to
a. Tap 2.
b. Tap 3.

10-3 MAXIMUM POWER TRANSFER

In a circuit containing only steady direct currents, maximum power is transferred into an element when its resistance equals the Thevenin equivalent resistance of the remainder of the circuit.

This statement is the dc version of the maximum power transfer theorem or principle.

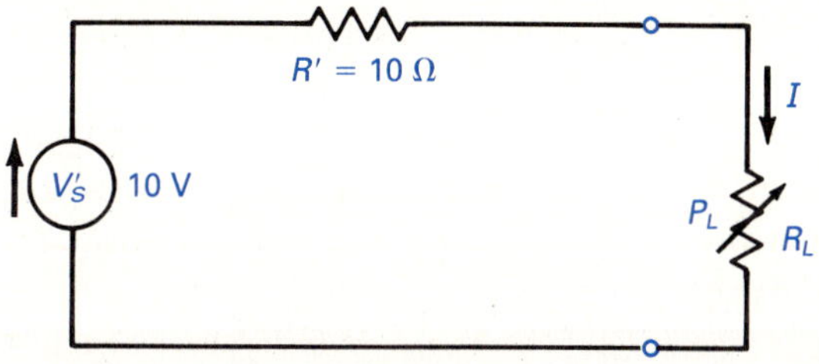

Fig. 10-52. Maximum power transfer. Thevenin equivalent of a circuit with its load resistance variable.

In Fig. 10-52 the Thevenin voltage V_S' and resistance R' should be considered to have been derived from a circuit which was either simple or complex. The load R_L is the element whose power we wish to maximize:

$$I_L = \frac{V_S'}{R' + R_L}$$

$$P_L = I^2 R_L = \frac{(V_S')^2 R_L}{(R' + R_L)^2} = \frac{(10\text{ V})^2 R_L}{(10\ \Omega + R_L)^2}$$

Computing values of load power related to various values of load resistance R_L gives the results shown in Table 10-1.

TABLE 10-1

R_L (Ω)	$10^2 R_L$	$10 + R_L$	$(10 + R_L)^2$	$P_L = \frac{10^2 R_L}{(10 + R_L)^2}$ (W)
0	0	10	100	0
5	500	15	225	2.222
7	700	17	289	2.422
8	800	18	324	2.469
9	900	19	361	2.493
9.5	950	19.5	380.25	2.498
10	1000	20	400	2.5
10.5	1050	20.5	420.25	2.499
11	1100	21	441	2.494
12	1200	22	484	2.479
20	2000	30	900	2.222
50	5000	60	3600	1.389

It can be seen from Table 10-1 that the load power P_L is greatest when the load resistance R_L is 10 Ω, the same as the Thevenin resistance R'. Figure 10-53 is a graph using the data in Table 10-1 to show how the power rises to a maximum and then drops off, peaking at the 10-Ω value of R_L.

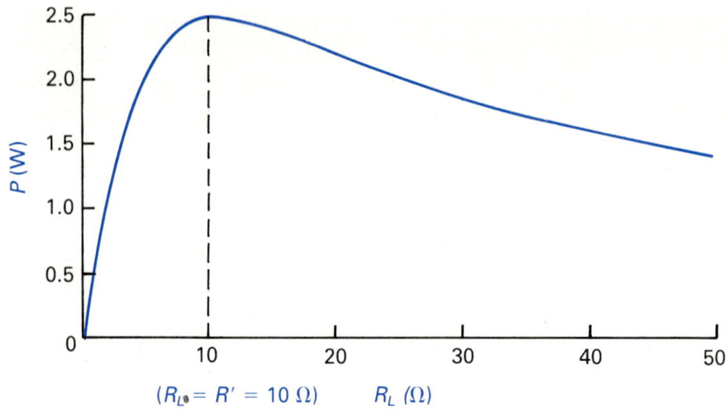

Fig. 10-53. Graph of load power with respect to load resistance in the circuit in Fig. 10-52.

It is of interest to review what influences the load power as the load resistance varies. Referring to Fig. 10-54, which shows the variations in current and load voltage with changing load resistance,

1. When R_L is zero, the circuit current has maximum value. However large this current value may be, there can be no voltage across a 0-Ω resistance, so no power can be developed in it.
2. As R_L increases from zero, the current drops but the load voltage rises. The rise of the load voltage from zero causes the power to rise.
3. When R_L becomes equal to R', the current will have dropped to 50 percent of its initial value, and the load voltage will have risen to 50 percent of the emf V'_S. This equality of generator and load resistances results in maximum load power.
4. As R_L becomes progressively greater than R', the current continues to drop and the load voltage to rise. The rate of change in both current and voltage is then low. The power does not change much, but it does steadily decrease until, if R_L were to become infinity (open circuit), the current and the power would be zero.

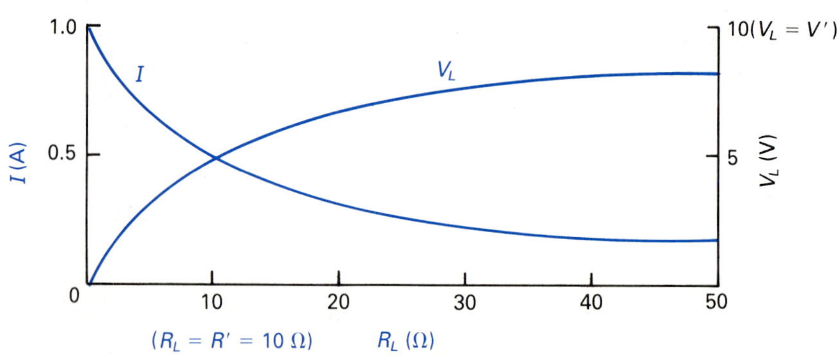

Fig. 10-54. Graphs of current and load voltage with respect to load resistance in the circuit in Fig. 10-52.

The validity of the dc maximum power transfer principle can be demonstrated by a proof which involves some elementary calculus.

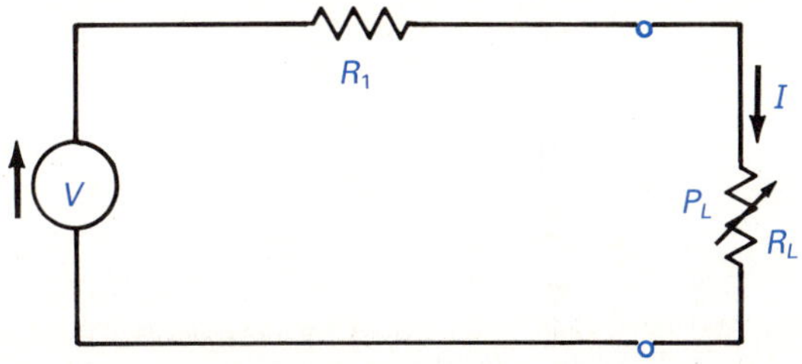

Fig. 10-55. Maximum power transfer. Circuit for a general derivation of conditions.

In reference to Fig. 10-55, the power in the load P_L is determined as follows:

$$I = \frac{V}{R_1 + R_L}$$

$$P_L = I^2 R_L = \frac{V^2 R_L}{(R_1 + R_L)^2}$$

Differentiating with respect to the variable R_L,

$$\frac{dP_L}{dR_L} = V^2 \frac{(R_1 + R_L)^2 - 2R_L(R_1 + R_L)}{(R_1 + R_L)^4}$$

$$\frac{dP_L}{dR_L} = V^2 \frac{R_1 - R_L}{(R_1 + R_L)^3}$$

The load power P_L will be maximum when this derivative equals zero.

$$V^2 \frac{R_1 - R_L}{(R_1 + R_L)^3} = 0$$

$$R_1 - R_L = \frac{(R_1 + R_L)^3}{V^2} \times 0$$

The right side of the equation is zero, leaving

$$R_1 - R_L = 0$$

or

$$R_1 = R_L$$

EXAMPLE 10-12 What is the value of load resistance R_L which will result in maximum load power in the circuit in Fig. 10-56?

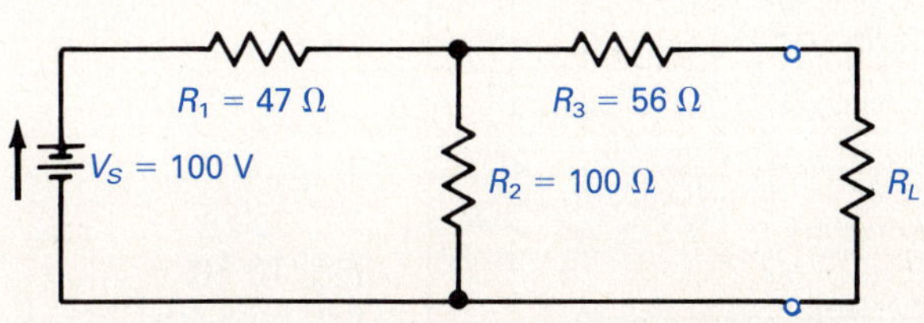

Fig. 10-56. Example 10-12: Given circuit.

Solution The Thevenin equivalent circuit is as in Fig. 10-57.

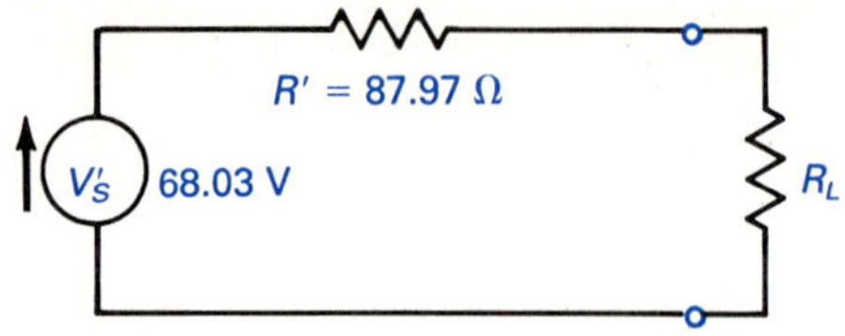

Fig. 10-57. Thevenin equivalent of Fig. 10-56.

For maximum load power,

$$R_L = R' = \mathbf{87.97\ \Omega}$$

Because maximum load power exists when the load resistance equals the Thevenin equivalent resistance of the remainder of the circuit, it naturally follows that in these circumstances,

$$\text{The load voltage } V_L = \frac{V_S'}{2}$$

$$\text{The load current } I_L = \frac{V_S'}{2R_L}$$

$$\text{The load power } P_L = V_L I_L = \frac{(V_S')^2}{4R_L}$$

EXAMPLE 10-13 An electronic power supply with an internal resistance of 0.5 Ω delivers 40 V to its output terminals when no load is connected between them. What are *(a)* the value of the load resistance which results in maximum load power, *(b)* the value of the maximum load power, *(c)* the load voltage and current when maximum load power is developed, and *(d)* the total power developed within the power supply?

Solution Whatever its internal circuit is, the power supply can be depicted electrically as the circuit in Fig. 10-58, which is its Thevenin equivalent.

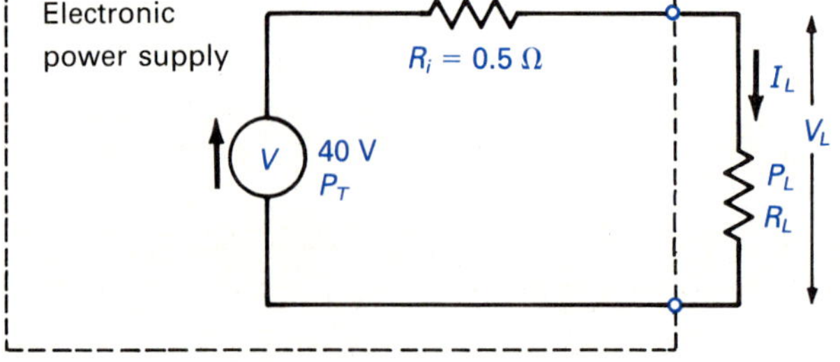

Fig. 10-58. Example 10-13: Circuit.

a. For maximum power transfer to the load,

$$R_L = R_i = \mathbf{0.5\ \Omega}$$

b.
$$P_L = \frac{V^2}{4R_L} = \frac{(40\text{ V})^2}{4 \times 0.5\ \Omega} = \mathbf{800\ W}$$

c.
$$V_L = \frac{V}{2} = \frac{40\text{ V}}{2} = \mathbf{20\ V}$$

$$I_L = \frac{V}{2R_L} = \frac{40\text{ V}}{2 \times 0.5\ \Omega} = \mathbf{40\ A}$$

d.
$$P_T = VI_L = 40\text{ V} \times 40\text{ A} = \mathbf{1600\ W}$$

As might be expected, Example 10-13 shows that, when maximum power is transferred to a load, this power is one-half the total power developed within the energy source, the load voltage being one-half the source emf. It is thus evident that criteria other than maximum power transfer must be applied when power efficiency is important.

For purposes such as the distribution of electrical energy, maximum load power is not the condition sought. The distribution system internal resistance is much lower than the load resistance, so as to maintain good voltage regulation (constant output voltage).

The maximum power condition is usually applied in electronic systems in which, to gain certain system performance characteristics, low power efficiency can be tolerated.

REVIEW QUESTIONS FOR SECTION 10-3

1. Under what conditions is maximum power transferred to a circuit element?
2. Under maximum power transfer conditions, what is the value of the load voltage in terms of the Thevenin equivalent voltage?
3. What percentage of the total circuit power is dissipated by a load under maximum power transfer conditions?

EXERCISE PROBLEMS FOR SECTION 10-3

1. An amplifier has an open-circuit output voltage of 24 V and an output resistance of 8 Ω. How can four 8-Ω speakers be connected to the amplifier so as to obtain maximum power output to the speakers? How much power is delivered to each speaker? (*ans*. series-parallel, 4.5 W)
2. The open-circuit voltage of a generator is 60 V, and the short-circuit current is 200 mA. What is the maximum power this generator can deliver to a load?
3. Find the maximum power that the circuit in Fig. 9-7 (page 268) can deliver to its load G_L. What must be the value of G_L for this maximum power transfer? (*ans*. 371.1 μW, 16.71 mS)

4. For the circuit in Prob. 14, Sec. 10-2, when the variable control R_A is set for maximum voltage, what is the maximum power the generator can deliver to a load from
   a. Tap 2?
   b. Tap 3?

10-4 NORTON'S THEOREM

Norton's theorem is the dual of Thevenin's theorem. It too is used to find the values of such quantities as current or voltage associated with one element in a circuit. It too involves the reduction of a network to a two-element equivalent circuit. Norton's theorem is concerned with a current-generator equivalent circuit rather than the voltage-generator equivalent circuit of Thevenin's theorem.

Norton's theorem states that the current in any conductance G, connected to two terminals of a network, is the same as it would be if G were connected to a current source where,

1. The source current is the short-circuit current at the terminals of G.
2. The internal conductance is the conductance at the terminals of G, with all other network sources replaced by conductances equal to their internal conductances.

Norton's theorem is stated in the given form to maintain the practice of regarding current sources and parallel networks from the conductance viewpoint, this view being consistent with the duality of series and parallel circuits. Should you wish to adopt a resistance approach to Norton's theorem, you can of course make the usual substitutions: resistance for conductance, and R for G, in the statements above; $1/R$ for G and reciprocal values in equations and calculations.

Norton Equivalent Circuit

In a network, shown in block form in Fig. 10-59*a*, there is an element G about which data are sought. To evaluate these data, the Norton equivalent circuit in Fig. 10-59*b* is developed.

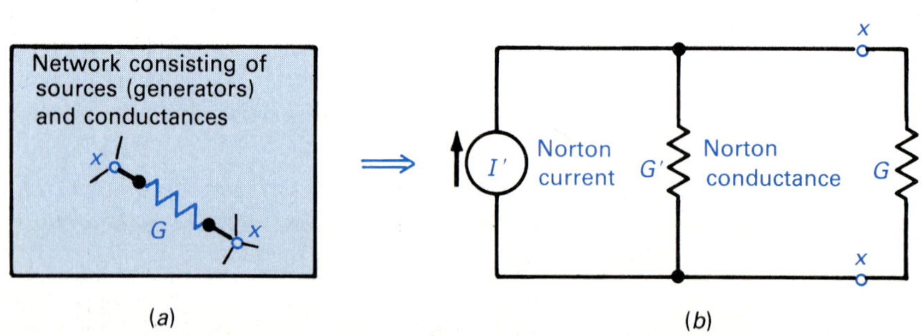

Fig. 10-59. *(a)* Network of unspecified type; *(b)* basic Norton circuit.

The *Norton current* I' is the short-circuit current at the element terminals (the current which would exist between the x,x terminals if they were connected by a good conductor).

The *Norton conductance* G' is the conductance at the x,x terminals with those terminals open-circuited (the element G removed), all sources in the remaining network being removed and replaced by their internal conductances. Note that the Norton conductance is defined in the same manner as the Thevenin resistance. The Norton conductance reciprocal (Norton resistance) is exactly the same as the Thevenin resistance.

Figure 10-60 shows a current source supplying a load through a π network. A π network is representative of all circuits which can be reduced to an equivalent π network, so by applying the Norton approach to Fig. 10-60 all such circuits are covered.

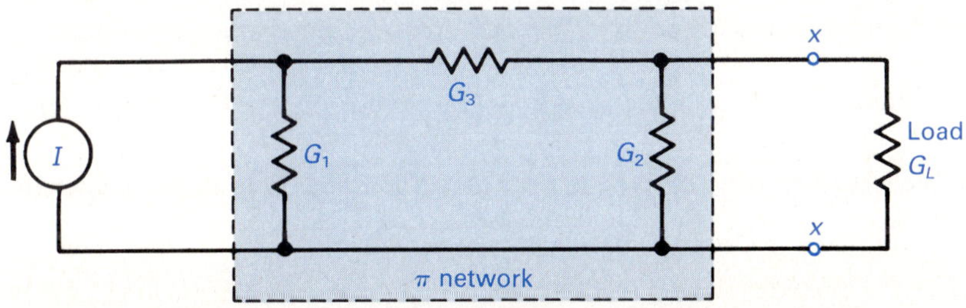

Fig. 10-60. Generator connected to a load through a π network.

To obtain the value of the Norton current I',

1. Short-circuit G_L at the x,x terminals.
2. Determine the current in the short circuit.

When the load terminals are short-circuited, the result is as shown in Fig. 10-61 (short-circuiting G_L also short-circuits G_2).

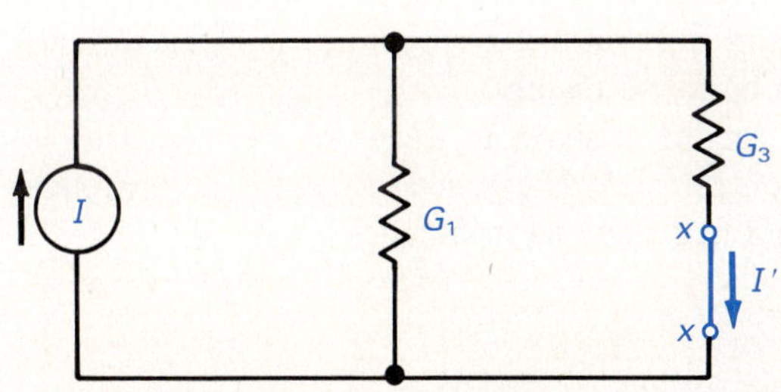

Fig. 10-61. Derivation of the Norton current for the circuit in Fig. 10-60.

The current in G_3 is the current in the short circuit x,x, and is thus the Norton equivalent current.

Applying the current-divider principle,

$$I' = I\frac{G_3}{G_1 + G_3} \tag{10-12}$$

To obtain the value of the Norton equivalent conductance G',

1. Disconnect the load G_L at the x,x terminals.
2. Remove the current source I. Since no internal conductance is shown for this source, it is assumed to have zero value, to be an open circuit.
3. Determine the conductance between the terminals.

With G_L and the current source removed, the circuit across the terminals will be series-parallel, as in Fig. 10-62.

$$G' = G_2 + \frac{G_1 G_3}{G_1 + G_3} \tag{10-13}$$

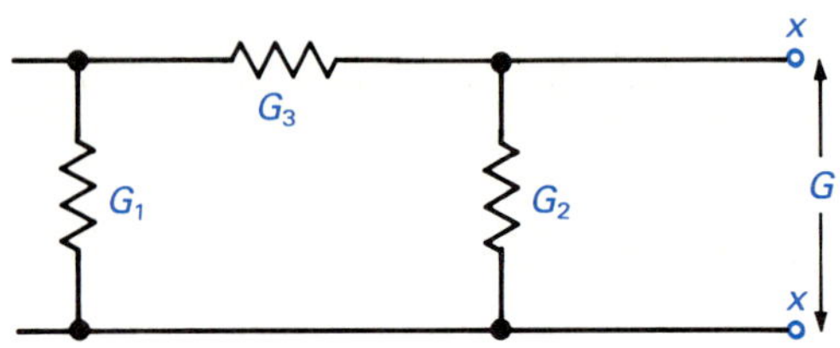

Fig. 10-62. Derivation of the Norton conductance for the circuit in Fig. 10-60.

The Norton equivalent of the circuit in Fig. 10-60 is therefore the one shown in Fig. 10-63.

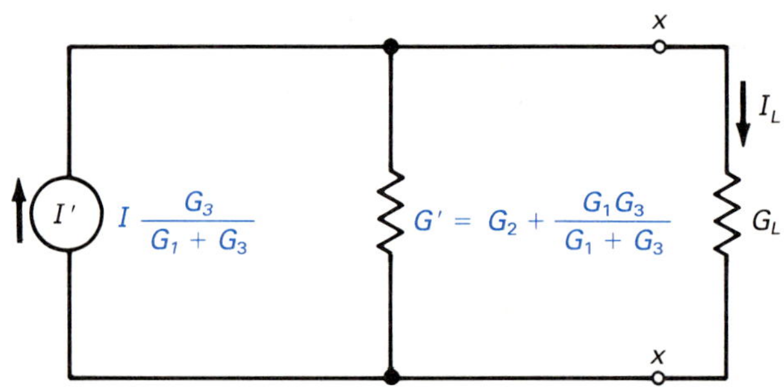

Fig. 10-63. Norton equivalent of the circuit in Fig. 10-60.

In any application, after the Norton circuit has been drawn, the load current, voltage, power, and energy can be found readily.

Norton's theorem is popular for the analysis of transistor circuits. This is particularly true when the Norton principle is extended from the dc approach established in this section to high-frequency ac techniques.

Guidelines

The suitability of circuits for the application of Norton analysis can be judged by whether they conform to some or all of the following guidelines.

1. Data for one element are required.
2. The conductance of that element has various alternative values.
3. There are more current sources than voltage sources.
4. The circuit details are not known, but the Norton current and conductance can be measured.

EXAMPLE 10-14 The circuit in Fig. 10-64 is made up of preferred-value resistors. What is the load current?

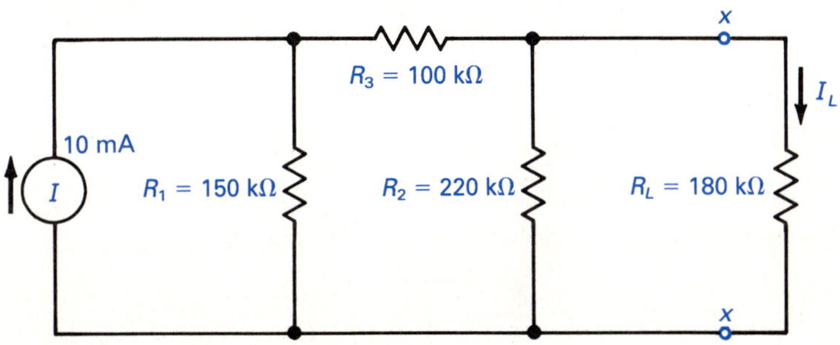

Fig. 10-64. Example 10-14: Circuit.

Solution Redrawing the circuit with the resistances converted to conductances gives Fig. 10-65.

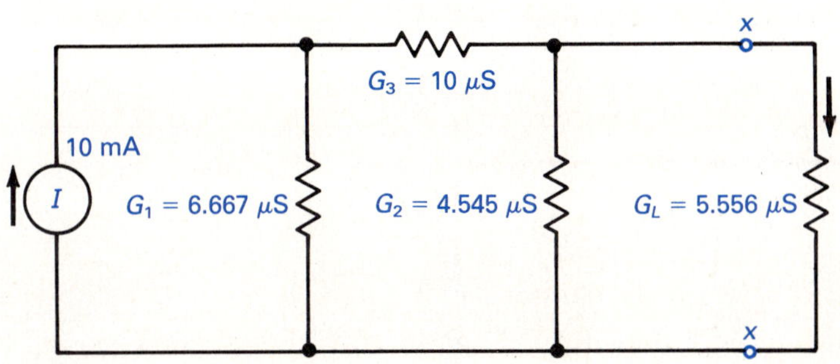

Fig. 10-65. Figure 10-64 with conductance values replacing resistance values.

With the x, x terminals short-circuited, the circuit becomes that in Fig. 10-66.

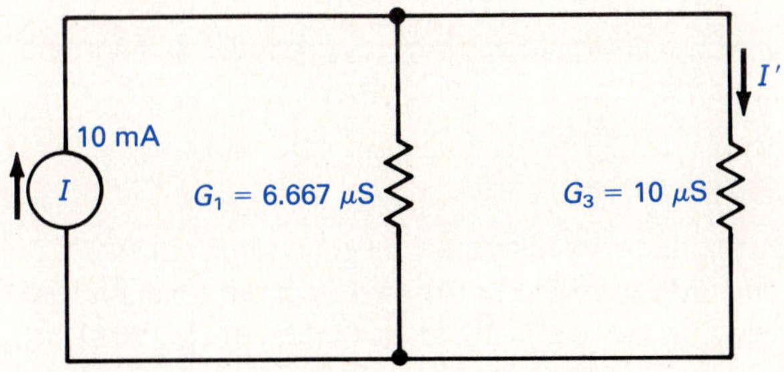

Fig. 10-66. Derivation of the Norton current for the circuit in Example 10-14.

The Norton current will be

$$I' = I\frac{G_3}{G_1 + G_3} = 10\text{ mA}\frac{10\ \mu\text{S}}{6.667\ \mu\text{S} + 10\ \mu\text{S}} = 6\text{ mA}$$

With the load and the generator removed, the circuit will be that in Fig. 10-67.

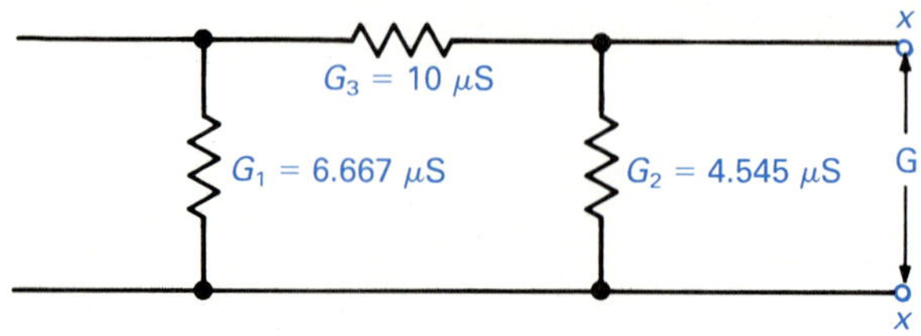

Fig. 10-67. Derivation of the Norton conductance for the circuit in Example 10-14.

The Norton conductance will be

$$G' = G_2 + \frac{G_1 G_3}{G_1 + G_3} = 4.545\ \mu S + \frac{6.667\ \mu S \times 10\ \mu S}{6.667\ \mu S + 10\ \mu S}$$
$$G' = 8.545\ \mu S$$

The Norton circuit equivalent to that in Fig. 10-64 is therefore as shown in Fig. 10-68.

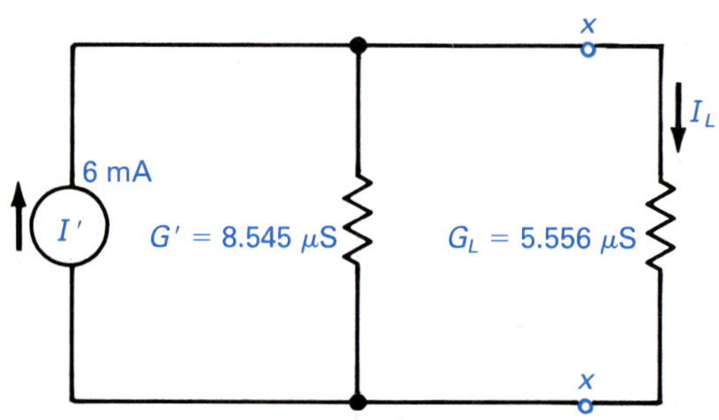

Fig. 10-68. Example 10-14: Norton equivalent circuit.

The load current is

$$I_L = I' \frac{G_L}{G' + G_L} = 6\ \text{mA}\ \frac{5.556\ \mu S}{8.545\ \mu S + 5.556\ \mu S} = \mathbf{2.364\ mA}$$

EXAMPLE 10-15 A multielement network reduced to an equivalent π network is shown in Fig. 10-69 connected between a current generator and a set of loads. If the load resistances can be 10 Ω, 100 Ω, or 1 kΩ, what are the related load currents?

Solution The Norton current is

$$I' = I \frac{G_3}{G_1 + G_3} = 1\ \text{A}\ \frac{8.3\ \text{mS}}{10\ \text{mS} + 8.3\ \text{mS}} = 0.4536\ \text{A}$$

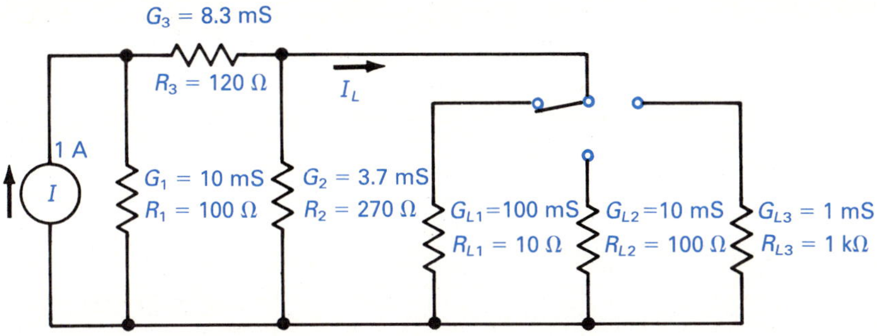

Fig. 10-69. Example 10-15: Circuit.

The Norton conductance is

$$G' = G_2 + \frac{G_1G_3}{G_1 + G_3} = 3.7 \text{ mS} + \frac{10 \text{ mS} \times 8.3 \text{ mS}}{10 \text{ mS} + 8.3 \text{ mS}} = 8.2 \text{ mS}$$

The Norton equivalent circuit therefore is that in Fig. 10-70.

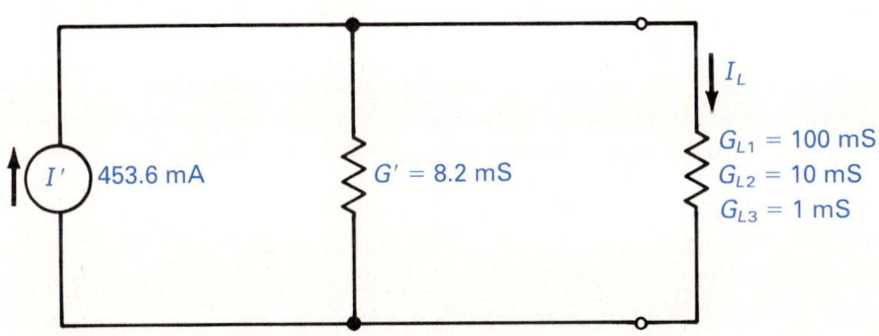

Fig. 10-70. Example 10-15: Norton equivalent circuit.

When G_L is 100 mS (or R_L = 10 Ω),

$$I_L = I' \frac{G_{L1}}{G_{L1} + G'} = 453.6 \text{ mA} \frac{100 \text{ mS}}{100 \text{ mS} + 8.2 \text{ mS}} = \mathbf{419.2\ mA}$$

When G_L is 10 mS (or R_L = 100 Ω),

$$I_L = I' \frac{G_{L2}}{G_{L2} + G'} = 453.6 \text{ mA} \frac{10 \text{ mS}}{10 \text{ mS} + 8.2 \text{ mS}} = \mathbf{249.2\ mA}$$

When G_L is 1 mS (or R_L = 1 kΩ),

$$I_L = I' \frac{G_{L3}}{G_{L3} + G'} = 453.6 \text{ mA} \frac{1 \text{ mS}}{1 \text{ mS} + 8.2 \text{ mS}} = \mathbf{49.3\ mA}$$

You will see from Example 10-15 that, as in the case of the Thevenin approach, Norton's theorem permits the ready calculation of the current in a network element when this element has a range of different resistance values. The Norton current and conductance need be computed only once.

EXAMPLE 10-16 An electronic power supply has the following measured characteristics at its output terminals: with the terminals connected through a low-resistance ammeter, the meter indicates 0.75 A; with the power switched off, the resistance between the terminals is 100 Ω. What current will this power supply deliver to a 50-Ω load?

Solution Figure 10-71*a* and *b* shows the measured conditions. The current I' and the resistance R' are the values required for a Norton equivalent circuit.

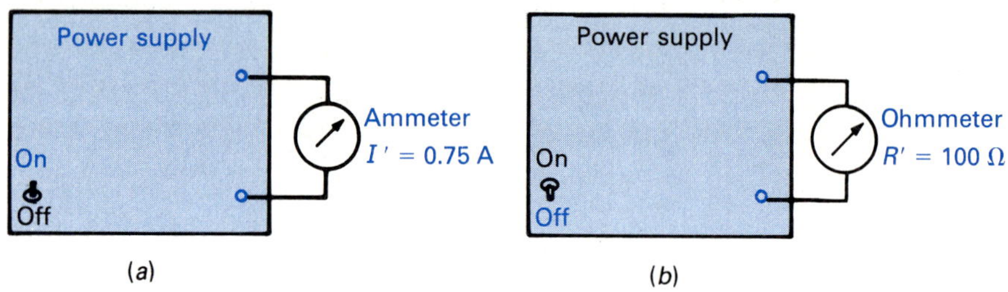

Fig. 10-71. Example 10-16: Measuring the Norton characteristics at the power-supply terminals: *(a)* Norton current; the power supply is ON. *(b)* Norton resistance; the power supply is OFF.

This circuit, with the 50-Ω load, is given in Fig. 10-72.

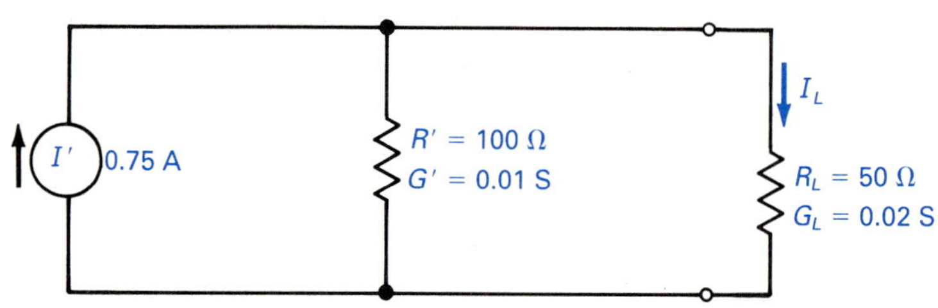

Fig. 10-72. Example 10-16: Norton equivalent circuit.

$$I_L = I' \frac{G_L}{G_L + G'} = 0.75\text{ A}\,\frac{0.02\text{ S}}{0.02\text{ S} + 0.01\text{ S}} = \mathbf{0.5\ A}$$

In Example 10-16, it was unnecessary to know anything about the power-supply circuit to determine the load current. If the short-circuit current and the open-circuit resistance at the terminals, as defined by Norton's theorem, can be measured, the current in any element connected between the terminals can be evaluated.

EXAMPLE 10-17 Two electronic power supplies, each with the same measured values as in Example 10-16, are connected in parallel with each other. The combined output is then applied to a 50-Ω load. What is the load current?

Solution Figure 10-73 shows the initial circuit. Since all elements are parallel-connected, the circuit can be redrawn as in Fig. 10-74.

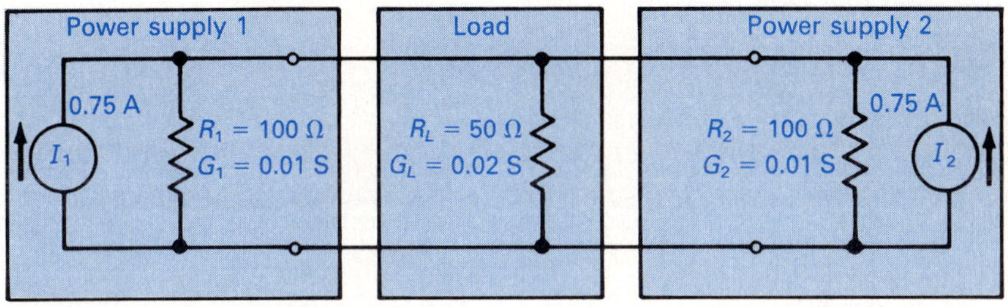

Fig. 10-73. Example 10-17: Circuit.

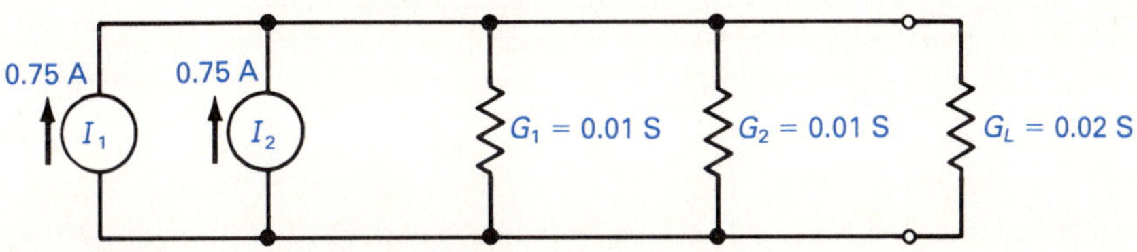

Fig. 10-74. Rearrangement of the components in Fig. 10-73.

The circuit can be further reduced to the Norton equivalent in Fig. 10-75.

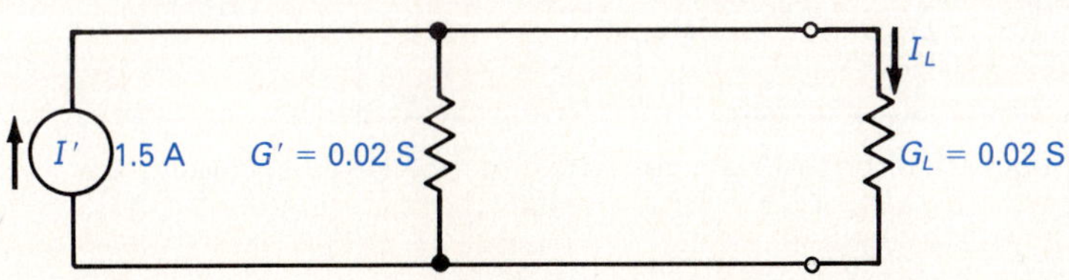

Fig. 10-75. Example 10-17: Norton equivalent circuit.

$$I' = I_1 + I_2 = 1.5 \text{ A}$$

$$G' = G_1 + G_2 = 0.02 \text{ S}$$

$$I_L = I' \frac{G_L}{G_L + G'} = 1.5 \text{ A} \frac{0.02 \text{ S}}{0.02 \text{ S} + 0.02 \text{ S}} = \mathbf{0.75 \text{ A}}$$

In Example 10-17 two similar energy sources connected in parallel supply more current to a load than one of the sources alone. This is always true if the sources have the same characteristics.

EXAMPLE 10-18 What is the base current I_B in the transistor switching circuit in Fig. 10-76? The base-to-emitter resistance R_{BE} can have any value between 125 and 200 Ω.

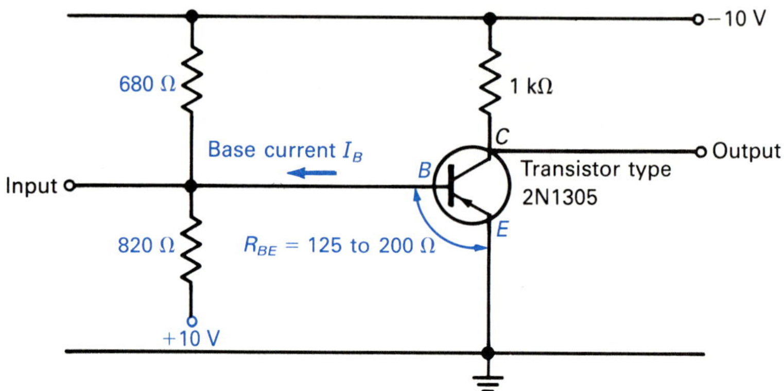

Fig. 10-76. Example 10-18: Given circuit.

Solution The equivalent circuit, from the viewpoint of the base-to-emitter resistance, can be redrawn in current-generator form as in Fig. 10-77.

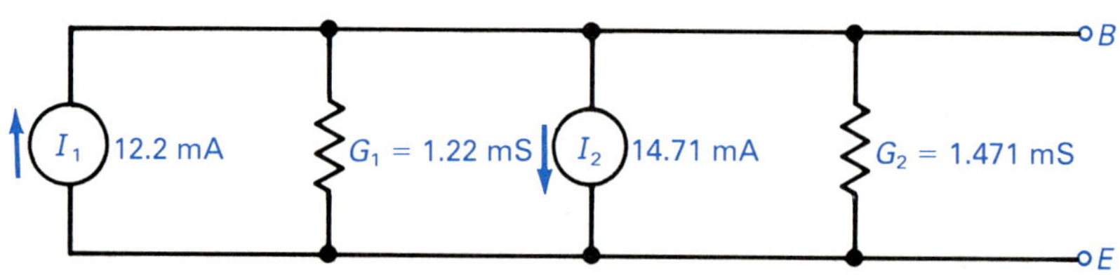

Fig. 10-77. Equivalent base-to-emitter circuit in Fig. 10-76 excluding the transistor.

The positive 10-V supply and the 820-Ω resistor are shown as I_1 and G_1.

$$I_1 = \frac{10\text{ V}}{820\ \Omega} = 12.2\text{ mA}$$

$$G_1 = \frac{1}{820\ \Omega} = 1.22\text{ mS}$$

The negative 10-V supply and the 680-Ω resistor are shown as I_2 and G_2.

$$I_2 = \frac{10\text{ V}}{680\ \Omega} = 14.71\text{ mA}$$

$$G_2 = \frac{1}{680\ \Omega} = 1.471\text{ mS}$$

The Norton equivalent of Fig. 10-77 is shown in Fig. 10-78.

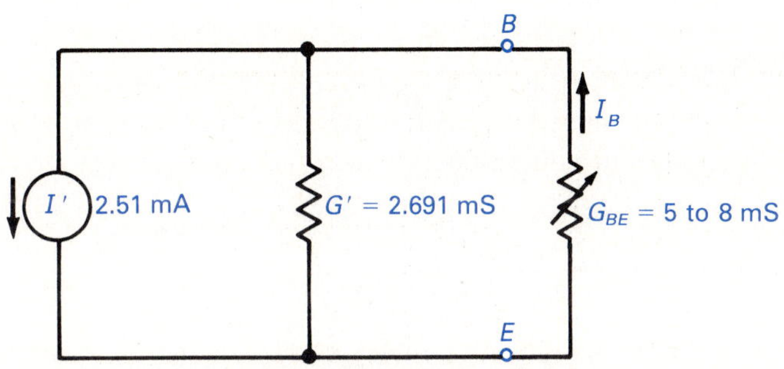

Fig. 10-78. Example 10-18: Norton equivalent circuit.

The currents I_1 and I_2 in Fig. 10-77 are in opposite directions, so that

$$I' = I_2 - I_1 = 14.71\text{ mA} - 12.2\text{ mA} = 2.51\text{ mA}$$
$$G' = G_1 + G_2 = 1.22\text{ mS} + 1.471\text{ mS} = 2.691\text{ mS}$$

When the base-to-emitter resistance is 125 Ω, $G_{BE} = 8$ mS,

$$I_B = I' \frac{G_{BE}}{G_{BE} + G'} = 2.51\text{ mA} \frac{8\text{ mS}}{8\text{ mS} + 2.691\text{ mS}} = \mathbf{1.878\ mA}$$

When the base-to-emitter resistance is 200 Ω, $G_{BE} = 5$ mS,

$$I_B = I' \frac{G_{BE}}{G_{BE} + G'} = 2.51\text{ mA} \frac{5\text{ mS}}{5\text{ mS} + 2.691\text{ mS}} = \mathbf{1.632\ mA}$$

In Example 10-18 the direction of the base current is from emitter to base, making the base negative with respect to the emitter. This polarity is correct for conduction in the PNP transistor shown in Fig. 10-76.

Successive Use of Norton's Theorem

Where it is considered advantageous, as, for instance, in circuits containing a number of current sources separated by series elements, Norton equivalent circuits can be derived in successive steps to obtain the overall Norton current and resistance.

EXAMPLE 10-19 What is the load current in the circuit in Fig. 10-79?

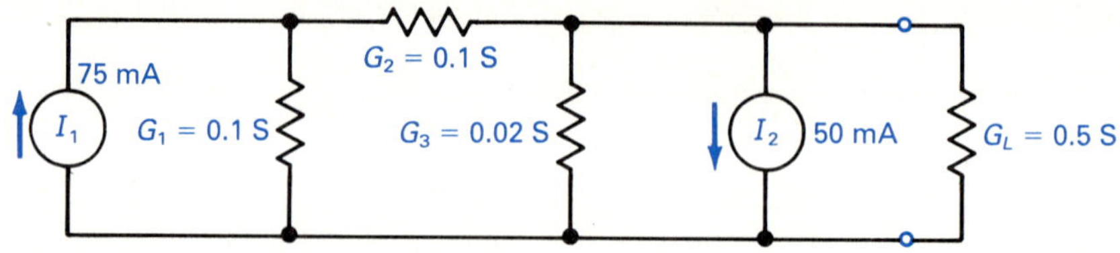

Fig. 10-79. Example 10-19: Circuit.

Solution The steps are:

1. Remove all of the circuit to the right of G_2, leaving the network in Fig. 10-80*a*.

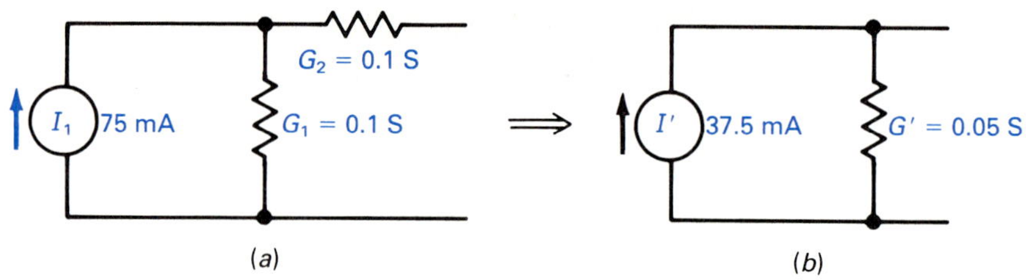

Fig. 10-80. Example 10-19: Derivation of the first Norton equivalent circuit.

Figure 10-80*b* is the Norton equivalent of Fig. 10-80*a*. G_1 and G_2, of equal value, will be in parallel when the terminals are short-circuited, so

$$I' = \frac{I}{2} = \frac{75\ \text{mA}}{2} = 37.5\ \text{mA}$$

G_1 and G_2 will be in series when the terminals are open-circuited and the current source removed. Therefore,

$$G' = \frac{G_1}{2} = \frac{G_2}{2} = \frac{0.1\ \text{S}}{2} = 0.05\ \text{S}$$

2. Add to the first Norton circuit the elements G_3, I_2, and G_L. The circuit in Fig. 10-81 results.

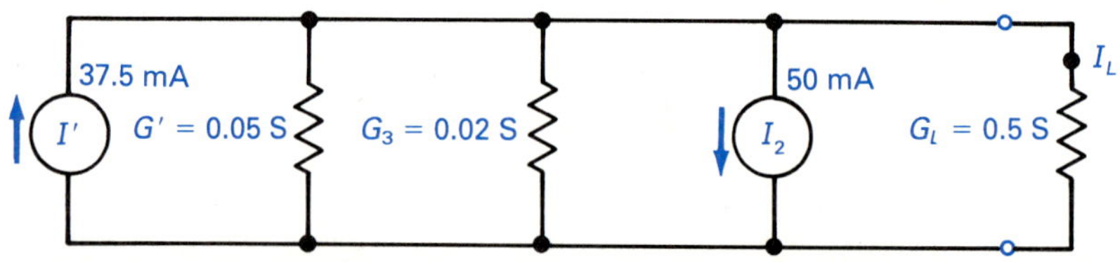

Fig. 10-81. Example 10-19: Circuit in Fig. 10-79 with the first Norton circuit replacing I_1, G_1, and G_2.

All the elements in Fig. 10-81 are in parallel with I_1 and I_2. The directions of I_1 and I_2 are opposite. The overall Norton equivalent circuit is as shown in Fig. 10-82.

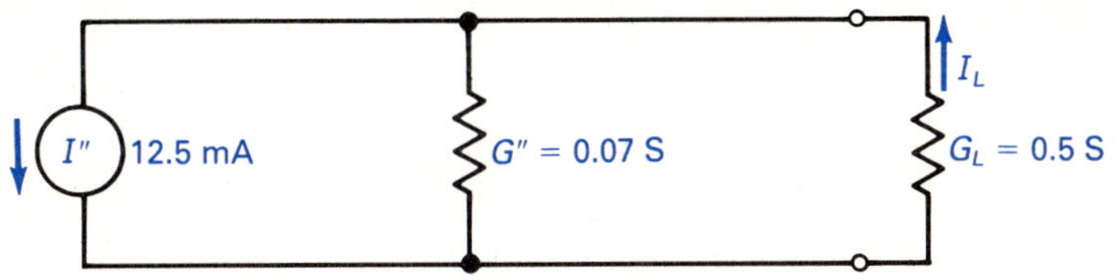

Fig. 10-82. Example 10-19: The second Norton equivalent circuit gives a final equivalent for the original circuit in Fig. 10-79.

$$I'' = I_2 - I' = 50 \text{ mA} - 37.5 \text{ mA} = 12.5 \text{ mA}$$
$$G'' = G' + G_3 = 0.05 \text{ S} + 0.02 \text{ S} = 0.07 \text{ S}$$
$$I_L = I'' \frac{G_L}{G'' + G_L} = 12.5 \text{ mA} \frac{0.5 \text{ S}}{0.07 \text{ S} + 0.5 \text{ S}} = \mathbf{10.96 \text{ mA}}$$

REVIEW QUESTIONS FOR SECTION 10-4

1. In what way is a Norton equivalent circuit similar to a Thevenin equivalent?

2. In what way(s) does a Norton equivalent circuit differ from a Thevenin equivalent?

3. To what is the Norton current I' equal?

4. To what is the Norton conductance G' equal?

5. When a circuit contains one or more current sources separated by series elements, how might the reduction to a Norton equivalent be simplified?

6. When converting a Thevenin circuit to its Norton equivalent, to what are the following equal?
   a. The Norton current I'.
   b. The Norton conductance G'.

7. When converting a Norton circuit to its Thevenin equivalent, to what are the following equal?
   a. The Thevenin voltage V'.
   b. The Thevenin resistance R'.

EXERCISE PROBLEMS FOR SECTION 10-4

1. Convert the circuit in Fig. 10-83 to its Norton equivalent when
   a. $I = 100$ mA, $G_g = 1.67$ mS, $G_1 = 1$ mS, $G_2 = 890\ \mu$S, $G_3 = 1$ mS (*ans*. 25 mA, 1.667 mS)
   b. $I = 100$ mA, $R_g = 820\ \Omega$, $R_1 = 390\ \Omega$, $R_2 = 470\ \Omega$, $R_3 = 560\ \Omega$ (*ans*. 36 mA, 3.125 mS)

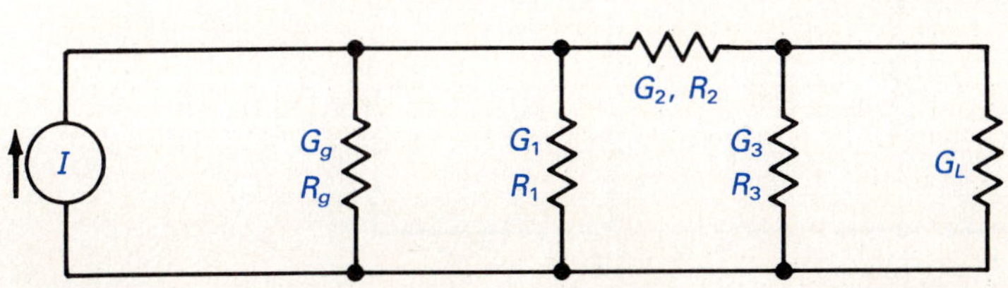

Fig. 10-83. Problem 1: Circuit.

2. Using Fig. 10-60 (page 347), find the voltage across G_L, using Norton's theorem, when

a. $I = 50$ mA, $G_1 = 0.01$ S, $G_2 = 0.01$ S, $G_3 = 0.01$ S, $G_L = 0.01$ S

b. $I = 100$ mA, $G_1 = 1.66$ mS, $G_2 = 3.33$ mS, $G_3 = 3.33$ mS, $G_L = 1.66$ mS

3. For the output equivalent circuit in Fig. 10-84, find the voltage developed across R_L when

a. $R_L = 820\ \Omega$ (*ans.* 3.84 V)

b. $R_L = 1800\ \Omega$ (*ans.* 7.83 V)

c. $R_L = 2700\ \Omega$ (*ans.* 11.02 V)

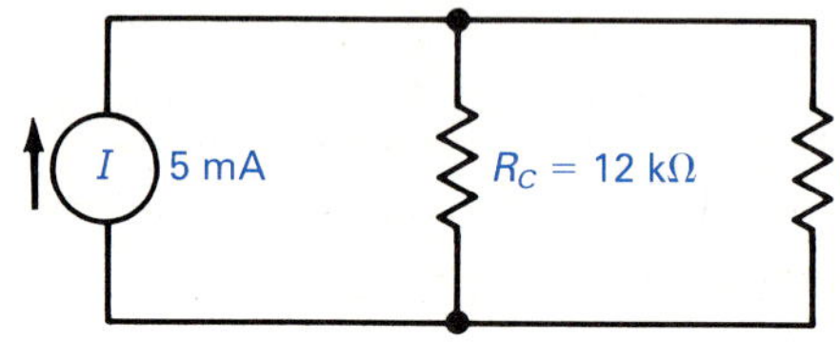

Fig. 10-84. Problem 3: Circuit.

4. Figure 10-85 is the interstage equivalent circuit between two transistor amplifiers. Using Norton's theorem, find I_{b2}, the input current to the second stage, when $h_{fe} = 65$ and $I_b = 6\ \mu$A.

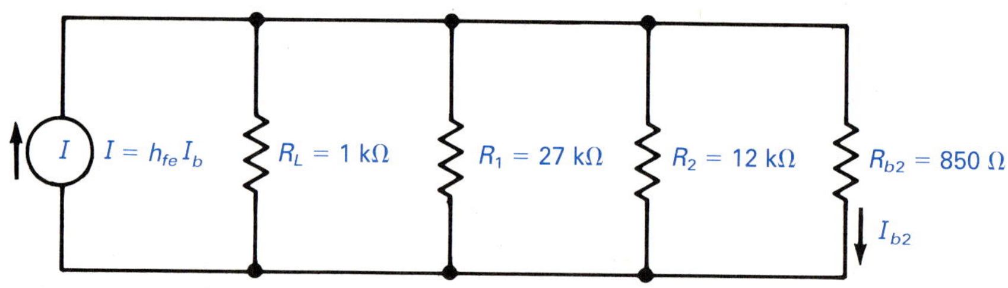

Fig. 10-85. Problem 4: Circuit.

5. Under a given set of conditions, the output circuit of a transistor amplifier can be drawn as in Fig. 10-86. Find the Norton equivalent circuit and the current through R_{b2}, the input resistance of the next amplifier, when $I_b = 30\ \mu$A and $h_{fe} = 50$. (*ans.* 1.5 mA, 1.025 mS, 873.5 μA)

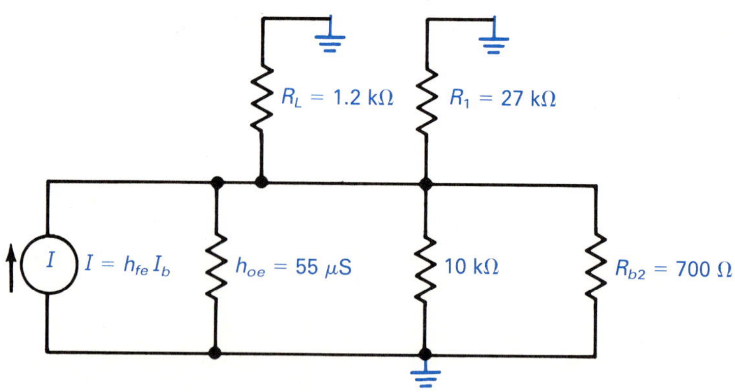

Fig. 10-86. Problem 5: Circuit.

6. The short-circuit current of a generator is 0.5 A. The open-circuit voltage is 35 V. Find the Norton equivalent circuit.
7. The short-circuit current of a generator is 250 mA. If the generator develops 60 V across a 720-Ω load, find the Norton equivalent circuit of the generator. (*ans*. 250 mA, 2.778 mS)

10-5 RELATIONSHIP BETWEEN THEVENIN AND NORTON CIRCUITS

By applying the definitions of the two theorems (which in fact really means applying the basic principles of voltage and current generators), Thevenin and Norton circuits can be interchanged. Figure 10-87*a* shows a Thevenin circuit and Fig. 10-87*b* a Norton circuit. It is implied that both circuits are derived from the same initial circuit and are thus equivalent to each other.

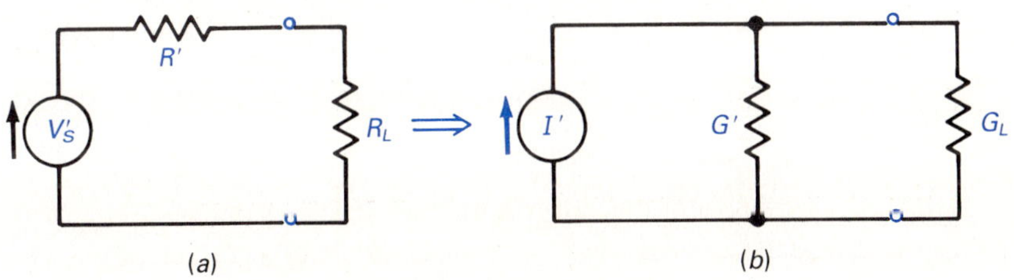

Fig. 10-87. Thevenin and Norton equivalence.

From the definition of the Norton current, it must equal the current in the short-circuited terminals of either circuit in Fig. 10-87.

In Fig. 10-87*a*, the short-circuit current equals V'_S/R'. The Norton current and conductance in Fig. 10-87*b* can be written using the symbols of the Thevenin circuit in Fig. 10-87*a*:

$$I' = \frac{V'_S}{R'} \qquad \text{and} \qquad G' = \frac{1}{R'}$$

From the definition of the Thevenin voltage, it must equal the voltage across the open-circuited terminals of either circuit in Fig. 10-87.

In Fig. 10-87*b*, the open-circuit voltage equals I'/G'. The Thevenin voltage and resistance in Fig. 10-87*a* can be written using the symbols of the Norton circuit in Fig. 10-87*b*.

$$V'_S = \frac{I'}{G'} \qquad \text{and} \qquad R' = \frac{1}{G'}$$

These relationships for Thevenin and Norton circuits are the same as those for voltage and current generator equivalence given in Sec. 7-4.

EXAMPLE 10-20 What are the Norton equivalent values of the Thevenin circuit in Fig. 10-88*a*?

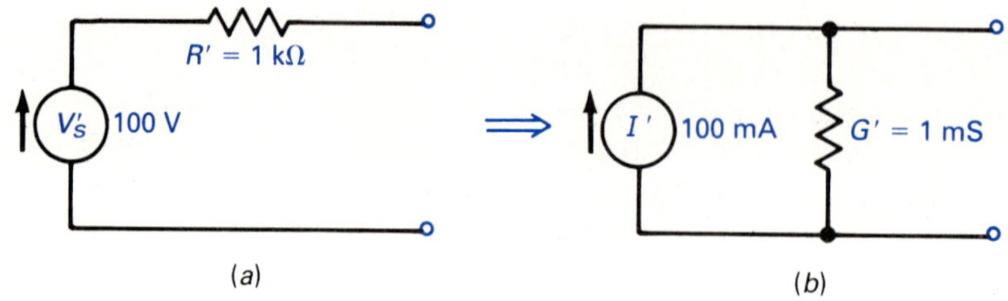

Fig. 10-88. Example 10-20: *(a)* Given Thevenin circuit; *(b)* Norton equivalent.

Solution

$$I' = \frac{V'_S}{R'} = \frac{100\text{ V}}{1\text{ k}\Omega} = \mathbf{100\ mA}$$

$$G' = \frac{1}{R'} = \frac{1}{1\text{ k}\Omega} = \mathbf{1\ mS}$$

EXAMPLE 10-21 What are the Thevenin equivalent values of the Norton circuit in Fig. 10-89*a*?

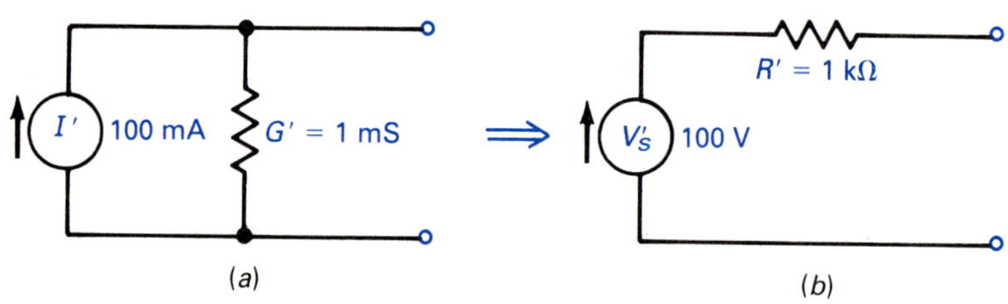

Fig. 10-89. Example 10-21: *(a)* Given Norton circuit; *(b)* Thevenin equivalent.

Solution

$$V'_S = \frac{I'}{G'} = \frac{100\text{ mA}}{1\text{ mS}} = \mathbf{100\ V}$$

$$R' = \frac{1}{G'} = \frac{1}{1\text{ mS}} = \mathbf{1\ k\Omega}$$

The results in Example 10-20 are the data given for Example 10-21. The results in Example 10-21 are the same as the data given for Example 10-20. Thus, commencing with the Thevenin voltage source in Fig. 10-88*a*, converting to a Norton current source (Figs. 10-88*b* and 10-89*a*), and then converting again to a Thevenin voltage source (Fig. 10-89*b*) leads back to the original data. This return to the starting conditions after two conversions validates the methods of conversion between the Thevenin and Norton source circuits.

10-6 CONSECUTIVE USE OF THEVENIN AND NORTON CIRCUITS

With some circuits, particularly those which consist of a long chain of elements connecting a load to a generator, it can be helpful to use the successive equivalent-circuit technique and apply Thevenin and Norton methods alternately. This can reduce the amount of computing effort, because the calculations can be carried out mentally, a calculator being used for converting resistance to conductance values.

One numerical example is given showing how to use this analysis method.

EXAMPLE 10-22 What is the current delivered to the load in the circuit in Fig. 10-90*a* if the load resistance is correct for maximum power transfer?

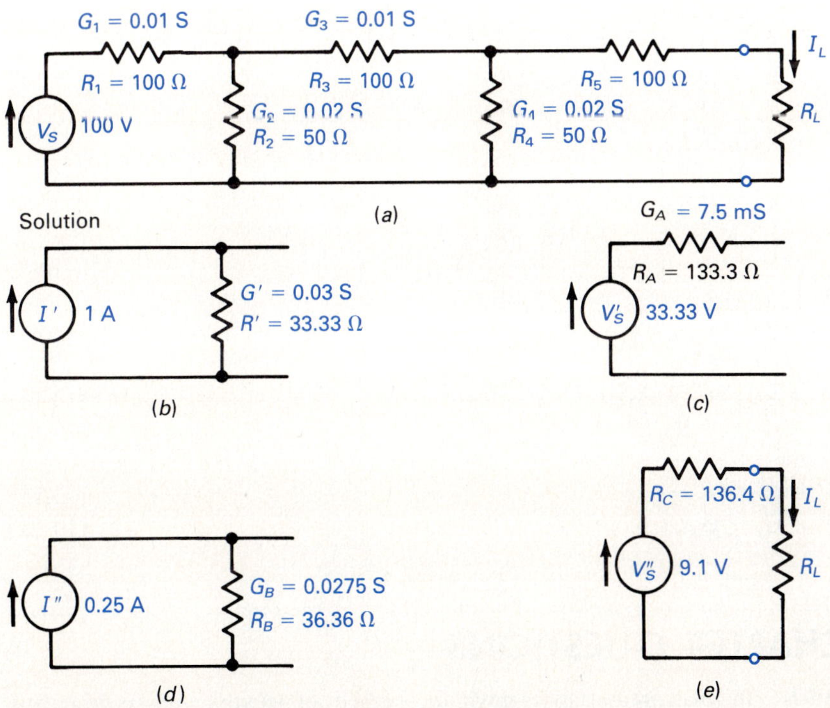

Fig. 10-90. Example 10-22: *(a)* Given circuit; *(b)* to *(e)* steps in deriving the Thevenin equivalent of the circuit.

Solution The steps are:

1. Change V_S and R_1 into current source form. Add R_2 written in conductance form. Draw the equivalent Norton circuit (Fig. 10-90*b*).

2. Change the Norton circuit into a Thevenin circuit. Add R_3 to the Thevenin resistance and label the sum R_A (Fig. 10-90*c*).
3. Change the Thevenin circuit into a Norton circuit. Express R_4 and R_A in conductance terms and add G_4 to G_A, calling the sum G_B (Fig. 10-90*d*).
4. Change the Norton circuit into a Thevenin circuit. Add R_5 to the Thevenin resistance, calling the sum R_C (Fig. 10-90*e*). Add the load R_L to complete the overall Thevenin equivalent of the original network in Fig. 10-90*a*. Since $R_C = 136.4\ \Omega$, for maximum power transfer, R_L should also be 136.4 Ω. The load current will be

$$I_L = \frac{V_S''}{R_C + R_L} = \frac{9.1\text{ V}}{136.4\ \Omega + 136.4\ \Omega} = \mathbf{33.36\ mA}$$

SUMMARY OF EQUATIONS

$\pi \leftrightarrow$ T conversions

$R_1 + R_2 = (R_BR_C + R_CR_A)/\Sigma R_\nabla$ (10-1)

$R_2 + R_3 = (R_AR_B + R_CR_A)/\Sigma R_\nabla$ (10-2)

$R_3 + R_1 = (R_AR_B + R_BR_C)/\Sigma R_\nabla$ (10-3)

$R_1 = R_BR_C/\Sigma R_\nabla$ (10-4)

$R_2 = R_CR_A/\Sigma R_\nabla$ (10-5)

$R_3 = R_AR_B/\Sigma R_\nabla$ (10-6)

$R_A = (R_1R_2 + R_2R_3 + R_3R_1)/R_1$

$= \Sigma R_Y/R_1$ (10-7)

$R_B = \Sigma R_Y/R_2$ (10-8)

$R_C = \Sigma R_Y/R_3$ (10-9)

Thevenin's theorem

$V' = VR_3/(R_1 + R_3)$ (10-10)

$R' = R_2 + R_1R_3/(R_1 + R_3)$ (10-11)

Norton's theorem

$I' = IG_3/(G_1 + G_3)$ (10-12)

$G' = G_2 + G_1G_3/(G_1 + G_3)$ (10-13)

CHAPTER QUESTIONS

10-1. In what circumstances might the methods of network analysis described in Chap. 9 be unsuitable?

10-2. Why is it that problems involving finding conditions in a *load* often occur in circuit analysis?

10-3. When it is said that two networks are *equivalent* to each other, what is meant?

10-4. In a subject that may be difficult to understand, why add to the possible confusion by giving two names to each of two basic networks: T and Y, and π and Δ?

10-5. Why are the Thevenin and Norton approaches to circuits regarded as very important?

10-6. What is the relationship between circuit reduction, using T $\leftrightarrow$ π conversions, and the Thevenin and Norton analysis methods?

10-7. Compare the methods used in Thevenin and Norton analysis.

10-8. Which is preferred for application to high-resistance circuits, the Thevenin or Norton method?

10-9. When would the *successive* methods of applying Thevenin or Norton techniques be appropriate?

10-10. How is it that maximum power can be transferred to a load from a source when the load voltage is only 50 percent of the source emf?

10-11. Could the Norton and Thevenin circuit analysis methods each be used with either resistances or conductances?

CHAPTER PROBLEMS

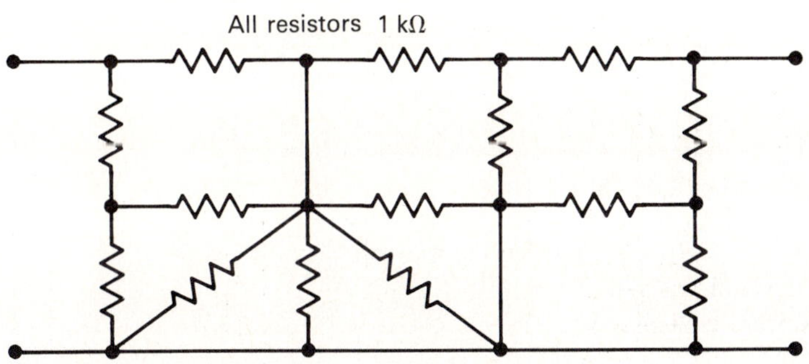

Fig. 10-91. Problems 10-1 and 10-2: Circuit.

10-1. In Fig. 10-91, reduce the network to a single T form without using T-to-π conversions (use only π-to-T conversions and series- or parallel-circuit principles). (*ans.* 689.4 Ω, 723.6 Ω, 38.59 Ω)

10-2. Reduce the network in Fig. 10-91 to a single π form without using π-to-T conversions (use only T-to-π conversions and series- or parallel-circuit principles).

10-3. Refer to Fig. 10-92. Find the current in resistor R_3 by applying

a. Thevenin analysis.
b. Norton analysis.
} (*ans.* 16.95 μA)

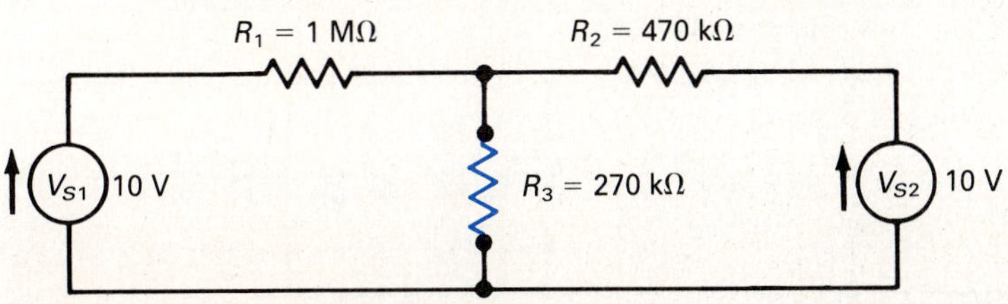

Fig. 10-92. Problem 10-3: Circuit.

10-4. Refer to Fig. 10-93. Find the current in resistor R_3 by using
a. Thevenin analysis.
b. Norton analysis.

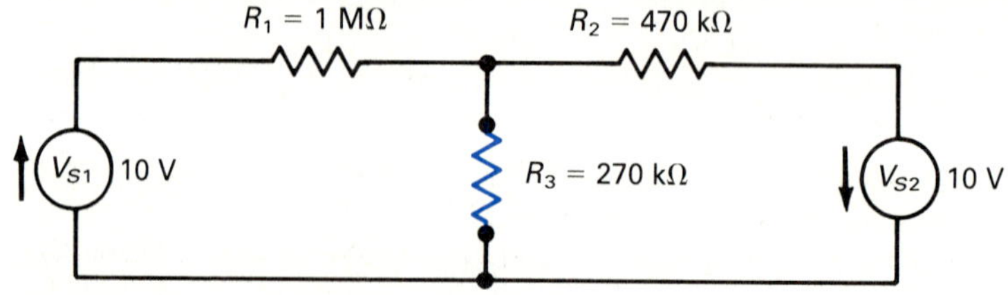

Fig. 10-93. Problem 10-4: Circuit.

10-5. Refer to Fig. 10-94. Find the current in resistor R_3 by applying
a. Thevenin analysis.
b. Norton analysis. } (*ans*. 10.92 A)

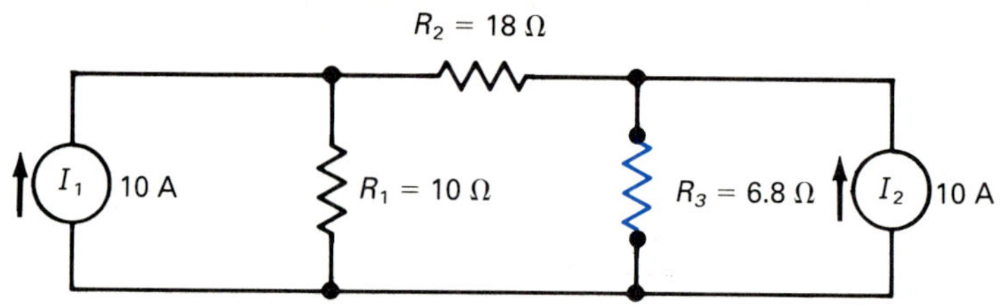

Fig. 10-94. Problem 10-5: Circuit.

10-6. Refer to Fig. 10-95. Find the current in resistor R_3 with
a. Thevenin analysis.
b. Norton analysis.

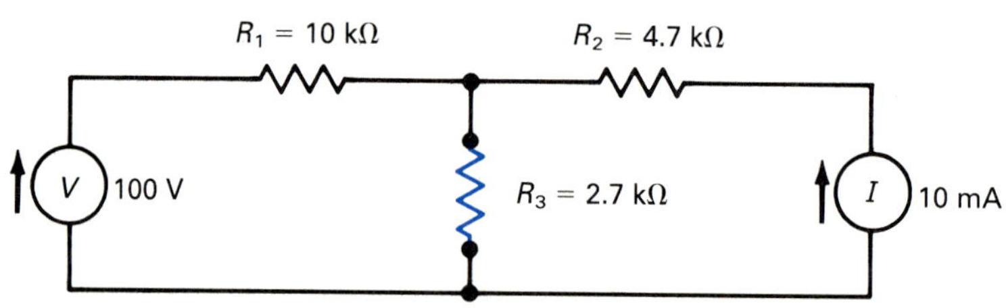

Fig. 10-95. Problem 10-6: Circuit.

10-7. Refer to Fig. 10-96. Find the current in resistor R_3 by applying
a. Thevenin analysis.
b. Norton analysis. } (*ans*. 3.225 mA)

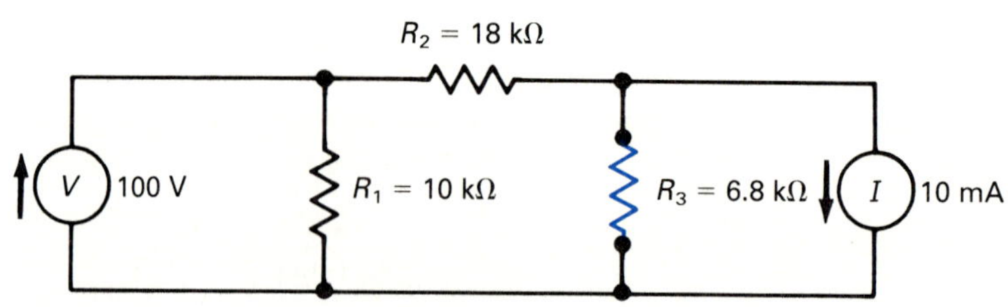

Fig. 10-96. Problem 10-7: Circuit.

10-8. An electric generator puts out 440 V when its terminals are open-circuit. It gives 100 A to a short circuit at its terminals.

a. What is the maximum power it can deliver to a load?

b. What are the load voltage and current under maximum load power conditions?

10-9. An electronic power supply delivers 2 A to any value of load resistance. The internal conductance of the supply is 0.8 S. What is the load voltage range when the load resistance varies between 1 Ω and 25 Ω? (*ans*. 11.11 V to 23.81 V)

10-10. To obtain the low voltages needed to measure some transistor characteristics, a voltage divider is connected across the output terminals of the power supply described in Prob. 10-9. Figure 10-97 shows the network.

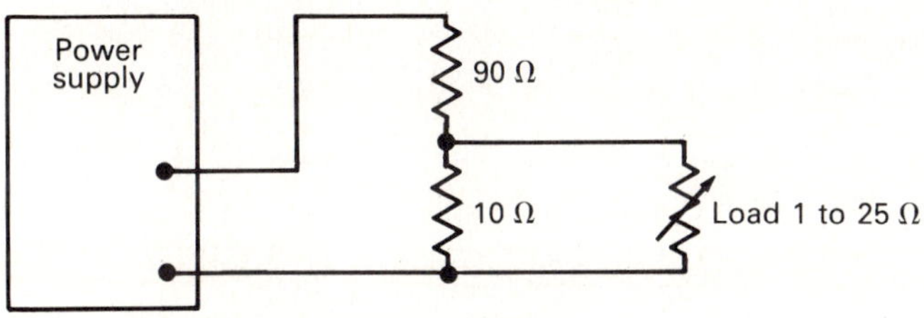

Fig. 10-97. Problem 10-10: Network.

What is the available voltage across a load that may vary in resistance between 1 Ω and 25 Ω?

10-11. Figure 10-98 shows part of the circuit of a transistor amplifier. If the

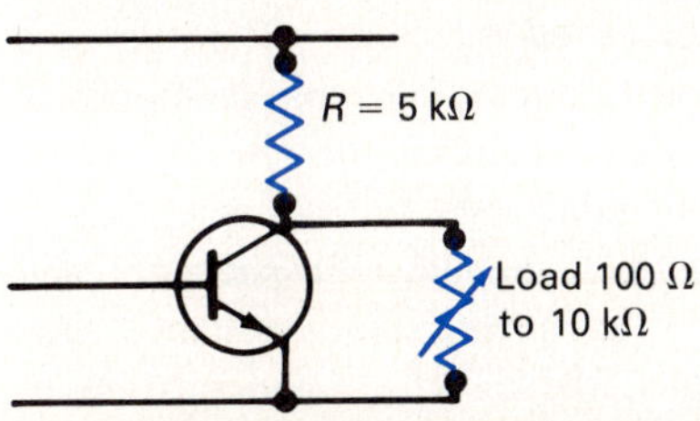

Fig. 10-98. Problem 10-11: Circuit.

transistor acts as a generator of 10 V open-circuit emf and 1 kΩ internal resistance, what range of voltages is delivered to a load that may have any resistance between 100 Ω and 10 kΩ? The resistor *R* should be considered as in parallel with the load.

(*ans*. 896.1 mV to 7.695 V)

11

DC TRANSIENTS

The circuits discussed so far have each contained one type of element; they have been made up of *all* resistance, *all* capacitance, or *all* inductance. That approach has been necessary in order to introduce you to circuit behavior as simply as possible. But it is not completely practical, as devices and circuits always contain all three elements, resistance, capacitance, *and* inductance. For example, a resistor is predominately resistive, but it has a small amount of capacitance and inductance. Similarly, capacitors and inductors each have some resistance; they dissipate power. In these circuit components, the effects of the mixture of characteristics are usually undesirable but not great enough to be important. However, these same effects can be useful and deliberately built into circuits.

So what is the difference between R, C, and L separately and combined? In general, the difference is the effect of *time*.

Resistance, capacitance, or inductance *alone* reacts to a change of voltage or current immediately; there is no time lag between the change of voltage and the change of current.

Where there is more than one element type present (R and C, R and L, etc.), a change of voltage may not occur at the same time as a change of current. For instance, if the voltage applied to an RL circuit changes suddenly, the current does not change suddenly. This time lag applies to a small degree to all resistors, capacitors, and inductors, but it is more noticeable when the amounts of the elements are intentionally made significant.

This chapter looks at the effects of mixing R and C or R and L with respect to time when the supply source is steady dc. Later chapters look at the effects of the mixed elements with ac supplies.

11-1 CHARGING CAPACITANCE THROUGH RESISTANCE

Figure 11-1*a* through *d* shows a sequence of events in a series *RC* circuit.

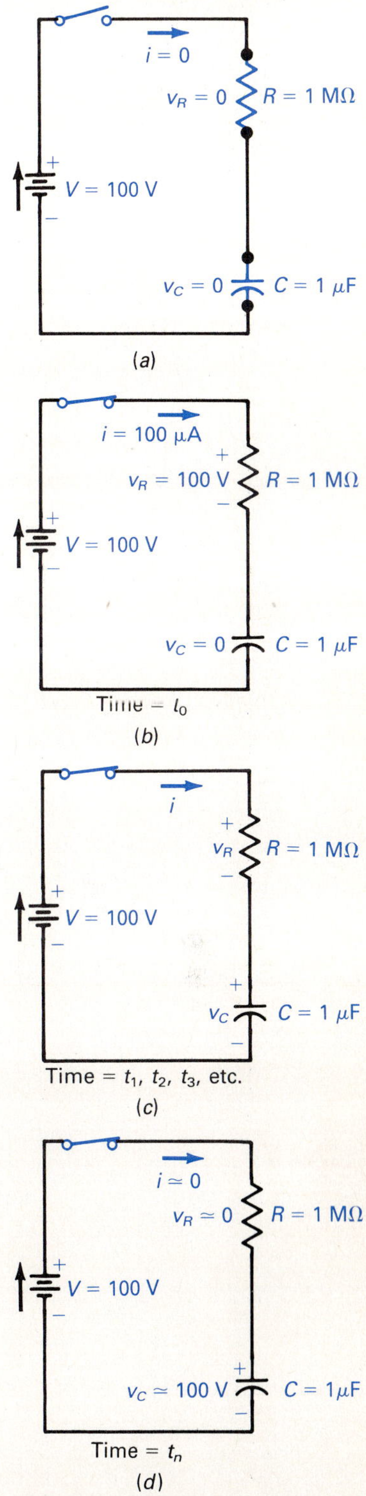

Fig. 11-1. Charging capacitor through resistor.

Figure 11-1*a* The capacitance is initially uncharged. With the switch open, the current i, the capacitive voltage v_C, and the resistive voltage v_R are all zero.

Figure 11-1*b* When the switch is first closed, the capacitance is still uncharged, or v_C is still zero, since capacitance opposes an instantaneous change in the voltage across its terminals. But Kirchhoff's voltage law must prevail in the closed loop.

$$V = v_R + v_C$$

or

$$v_R = V - v_C = 100\ \text{V} - 0\ \text{V} = 100\ \text{V}$$

Ohm's law must also be satisfied, so there must be current in the resistance.

$$i = \frac{v_R}{R} = \frac{100\ \text{V}}{1\ \text{M}\Omega} = 100\ \mu\text{A}$$

This is the initial condition on completing the circuit. The events are considered to occur at time t_0, the instant the switch is closed and the emf V is applied.

Figure 11-1*c* The current causes the capacitance to charge, so that, after the initial condition, the voltage v_C will have a value above zero. At this moment we have insufficient information to determine what that value is, except that v_C is greater than zero but must be less than the 100-V emf (the maximum possible voltage in this loop). The situation can be summarized:

$$0 < v_C < 100\ \text{V} \qquad (v_C \text{ is greater than zero but less than 100 V})$$

$$v_R = V - v_C = 100\ \text{V} - v_C$$

or

$$100\ \text{V} > v_R > 0 \qquad (v_R \text{ is less than 100 V but greater than zero})$$

$$i = \frac{v_R}{R} = \frac{100\ \text{V} - v_C}{1\ \text{M}\Omega} = 100\ \mu\text{A} - \frac{v_C}{1\ \text{M}\Omega}$$

or

$$100\ \mu\text{A} > i > 0 \qquad (i \text{ is less than } 100\ \mu\text{A but greater than zero})$$

These conditions are to be considered representative of times t_1, t_2, t_3, etc., between the initial time t_0 in Fig. 11-1*b* and the final time t_n in Fig. 11-1*d*.

Figure 11-1*d* As time goes on, the charge on the capacitance increases, as does its voltage v_C. Ultimately v_C becomes virtually equal to V, which is the maximum possible circuit voltage. This situation can be summarized:

$$v_C \simeq V \simeq 100\ \text{V}$$

$$v_R = V - v_C \simeq 100\ \text{V} - 100\ \text{V} \simeq 0\ \text{V}$$

$$i = \frac{v_R}{R} \simeq \frac{0\ \text{V}}{1\ \text{M}\Omega} \simeq 0\ \text{A}$$

This final event, we say, happens at a time t_n, implying a long time after time t_0.

RC Circuit Charge Curves

Graphical plots of the events in Fig. 11-1 are given in Fig. 11-2.

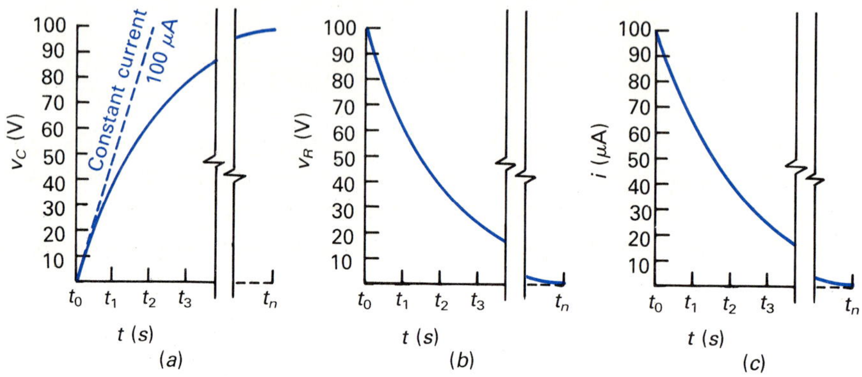

Fig. 11-2. For the *RC* circuit in Fig. 11-1, charge curves with respect to time: *(a)* Capacitor voltage; *(b)* resistor voltage; *(c)* current.

Why Are the Curves Curved?

With a constant charging current the curve of capacitive voltage with respect to time is a straight line. In Fig. 11-2*a* the v_C curve really is curved. Obviously, then, the current is not constant. Figure 11-2*c* shows the current decreasing as time progresses. Why does the current decrease?

To understand why the current decreases, let us take a sequence of events similar to those in Figs. 11-1 and 11-2, but with the assumption that the current stays constant for equal periods of time. In this way the action is frozen, so that calculations can be made without using calculus.

1. At time t_0, the conditions are

$$v_C = 0 \text{ V}$$
$$v_R = 100 \text{ V}$$
$$i = 100 \text{ } \mu\text{A}$$

The current is now assumed to stay at 100 μA for 0.2 s. The charge in the capacitance was zero at time t_0. An amount of charge Δq is now added. From Eq. (2-9),

$$\Delta q = i \, \Delta t$$
$$\Delta q = 100 \text{ } \mu\text{A} \times 0.2 \text{ s} = 20 \text{ } \mu\text{C}$$

The capacitive voltage will rise from zero by an amount Δv_C. From Eq. (4-3),

$$\Delta v_C = \frac{\Delta q}{C}$$

$$\Delta v_C = \frac{20\ \mu\text{C}}{1\ \mu\text{F}} = 20\ \text{V}$$

Therefore, at the end of the first 0.2-s period, the capacitive voltage is

$$v_{C1} = 0 + \Delta v_C = 20\ \text{V}$$

2. At time t_0 + 0.2 s, the conditions are

$$v_{C1} = 20\ \text{V}$$

$$v_{R1} = V - v_{C1} = 100\ \text{V} - 20\ \text{V} = 80\ \text{V}$$

$$i_1 = \frac{v_{R1}}{R} = \frac{80\ \text{V}}{1\ \text{M}\Omega} = 80\ \mu\text{A}$$

The conditions at time t_0 are compared with those at the end of the first 0.2-s period in Fig. 11-3. As shown in Fig. 11-3*b*, the polarity of the 20 V of v_{C1} acts in opposition to the 100 V of V, causing the net voltage across R to become 80 V. The circuit current is now reduced to 80 μA.

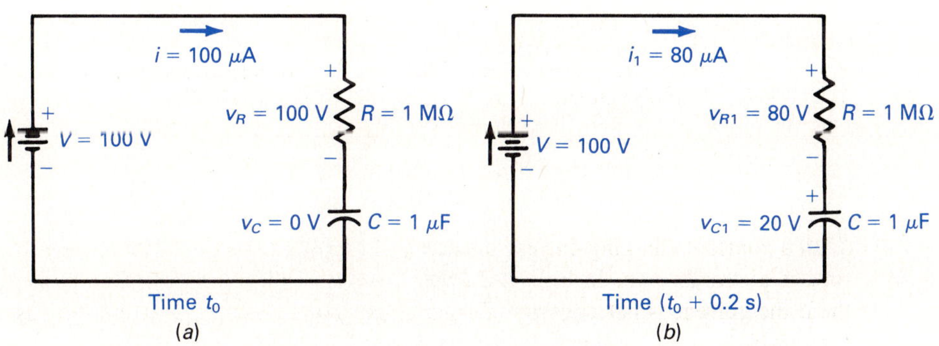

Fig. 11-3. *RC* circuit in Fig. 11-1 *(a)* at time t_0; *(b)* at 0.2 s after time t_0.

Assume that the current now stays at 80 μA for 0.2 s.

$$\Delta q = i_1\ \Delta t = 80\ \mu\text{A} \times 0.2\ \text{s} = 16\ \mu\text{C}$$

$$\Delta v = \frac{\Delta q}{C} = \frac{16\ \mu\text{C}}{1\ \mu\text{F}} = 16\ \text{V}$$

At the end of the second 0.2-s period, the total capacitive voltage will be

$$v_{C2} = v_{C1} + \Delta v = 20\ \text{V} + 16\ \text{V} = 36\ \text{V}$$

3. At time t_0 + 0.4 s, the conditions are

$$v_{C2} = 36\ \text{V}$$

$$v_{R2} = V - v_{C2} = 100\ \text{V} - 36\ \text{V} = 64\ \text{V}$$

$$i_2 = \frac{v_{R2}}{R} = \frac{64\ \text{V}}{1\ \text{M}\Omega} = 64\ \mu\text{A}$$

Figure 11-4*a* shows the situation at the end of the second 0.2-s period. Assuming that the current stays at 64 μA for 0.2 s,

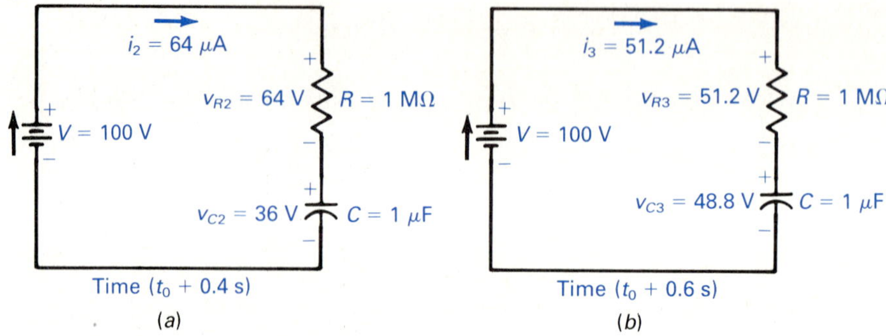

Fig. 11-4. *RC* circuit in Fig. 11-1 at *(a)* 0.4 s after time t_0 and *(b)* 0.6 s after time t_0.

$$\Delta q = i\,\Delta t = 64\ \mu A \times 0.2\ s = 12.8\ \mu C$$

$$\Delta v = \frac{\Delta q}{C} = \frac{12.8\ \mu C}{1\ \mu F} = 12.8\ V$$

At the end of the third 0.2-s period, the total capacitive voltage will be

$$v_{C3} = v_{C2} + \Delta v = 36\ V + 12.8\ V = 48.8\ V$$

4. At time t_0 + 0.6 s, the conditions are

$$v_{C3} = 48.8\ V$$

$$v_{R3} = V - v_{C3} = 100\ V - 48.8\ V = 51.2\ V$$

$$i_3 = \frac{v_{R3}}{R} = \frac{51.2\ V}{1\ M\Omega} = 51.2\ \mu A$$

Figure 11-4*b* shows the current and voltages at the end of the third 0.2-s period.

The voltages and currents for ten 0.2-s periods after time t_0 are shown in Table 11-1.

The values of Table 11-1 are plotted as dots in Fig. 11-5.

TABLE 11-1 Voltages and Current in Charging *RC* Circuit

t (s)	0	0.2	0.4	0.6	0.8	1.0	1.2	1.4	1.6	1.8	2.0
v_C (V)	0	20	36	48.8	59	67.2	73.8	79	83.2	86.6	89.3
v_R (V)	100	80	64	51.2	41	32.8	26.2	21	16.8	13.4	10.7
i (μA)	100	80	64	51.2	41	32.8	26.2	21	16.8	13.4	10.7

The values in Table 11-1 and Fig. 11-5 are not quite correct. They assume that the current remains constant for 0.2-s periods of time when it actually changes continuously. However, if lines are drawn through the dots, three curves result which are sufficiently similar to those in Fig. 11-2 to indicate that the original curves represent the true variations in voltage and current with respect to time.

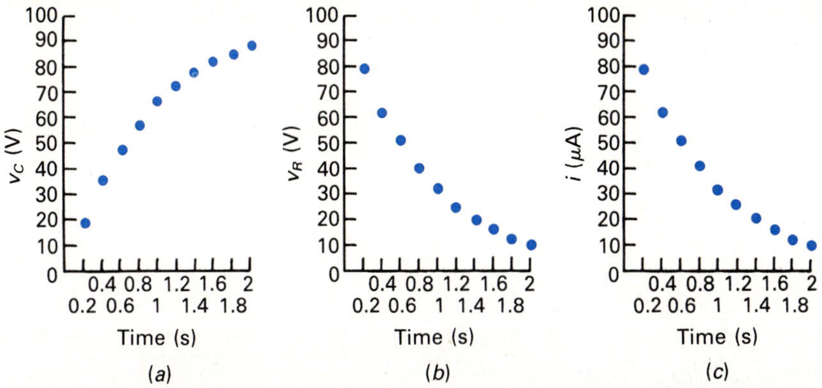

Fig. 11-5. Plots of the voltage and current values given in Table 11-1. *(a)* Capacitor voltage; *(b)* resistor voltage; *(c)* current.

Accurate relationships for the *RC* circuit can be obtained only by applying calculus methods to the loop equations. These methods take account of the continuous current change.

Equations for Capacitor Charging

The basic loop equation for the circuit in Figs. 11-1, 11-3, and 11-4 is

$$V = v_C + v_R \qquad \textbf{(11-1)}$$

An expression for v_C can be obtained from Eq. (4-1)

$$i = C\frac{dv_C}{dt}$$

As explained in Chap. 4, dv_C/dt is a way of writing "the rate of change of capacitor voltage" for ease of handling in equations. This term may be broken down when convenient, so that

$$dv_C = \frac{i}{C}\,dt$$

From this expression we can obtain an equation for the current in a charging *RC* circuit:

$$i = \frac{V}{R}e^{-t/RC} \qquad \textbf{(11-2)}$$

where i = instantaneous current, A
V = applied emf, V
R = resistance, Ω
e = base of natural logarithms[1] (a numerical constant ≃ 2.718)
t = time from t_0, s
C = capacitance, F

[1]The standard symbol for the base of natural logarithms is e. An acceptable alternative is ϵ, the lowercase Greek letter epsilon.

The value of the resistive voltage at any time can be found by substituting the expression for the current in Ohm's law:

$$v_R = iR = \left(\frac{V}{R} e^{-t/RC}\right) R$$

$$v_R = Ve^{-t/RC} \qquad \textbf{(11-3)}$$

Substituting for v_R in Eq. (11-1) gives an expression for v_C.

$$v_C = V - v_R \qquad (11\text{-}1)$$

$$= V - Ve^{-t/RC}$$

$$v_C = V(1 - e^{-t/RC}) \qquad \textbf{(11-4)}$$

In Eqs. (11-3) and (11-4),

v_R = resistive voltage, V
v_C = capacitive voltage, V
V = applied emf, V
e = base of natural logarithms ≃ 2.718
t = time from t_0, s
R = resistance, Ω
C = capacitance, F

EXAMPLE 11-1 In the circuit in Figs. 11-1 to 11-4, what are the correct instantaneous values of current and voltage 1 s after time t_0? It has already been noted that the values given in Table 11-1 are not quite correct.

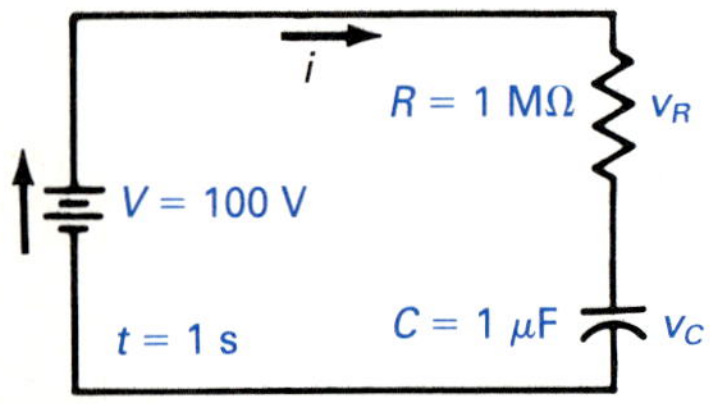

Fig. 11-6. Example 11-1: Circuit.

Solution (See Fig. 11-6.)

$$i = \frac{V}{R} e^{-t/RC} \qquad (11\text{-}2)$$

$$i = \frac{100 \text{ V}}{1 \text{ M}\Omega} e^{-1\text{s}/(1 \text{ M}\Omega \times 1\ \mu\text{F})} = 100\ \mu\text{A}\ e^{-1} = \frac{100\ \mu\text{A}}{2.718} = \mathbf{36.79\ \mu A}$$

$$v_R = iR = 36.79\ \mu\text{A} \times 1 \text{ M}\Omega = \mathbf{36.79\ V}$$

$$v_C = V - v_R = 100 \text{ V} - 36.79 \text{ V} = \mathbf{63.21\ V}$$

REVIEW QUESTIONS FOR SECTION 11-1

1. A series circuit consists of a voltage source V, a resistor R, and a capacitor C. When the circuit is first closed, what are the values of i, v_R, and v_C?
2. When the circuit of Question 1 has stabilized, after a long period of time, what are the values of i, v_R, and v_C?
3. Graphical plots of voltages and currents during charge and discharge of an RC circuit show curved lines. Why are the graphs curved?

4. What is the basic loop equation for the circuit of Question 1?
5. What is the equation for the current i in a charging RC circuit?
6. What is the equation for the voltage across the resistance v_R in a charging RC circuit?
7. What is the equation for the voltage across the capacitance v_C in a charging RC circuit?

EXERCISE PROBLEMS FOR SECTION 11-1

1. Using Fig. 11-7 with $V = 50$ V, $R = 1$ kΩ, and $C = 10$ μF, find i, v_R, and v_C using Eqs. (11-2) to (11-4) at the following times after the switch is closed.
   a. 0 s (*ans*. 50 mA, 50 V, 0 V)
   b. 3 ms (*ans*. 37 mA, 37 V, 13 V)
   c. 7000 μs (*ans*. 24.8 mA, 24.8 V, 25.2 V)
2. Repeat Prob. 1 with $V = 12$ V, $R = 100$ kΩ, and $C = 470$ pF, finding i, v_R, and v_C at
   a. 5 μs
   b. 68 μs
   c. 0.12 ms

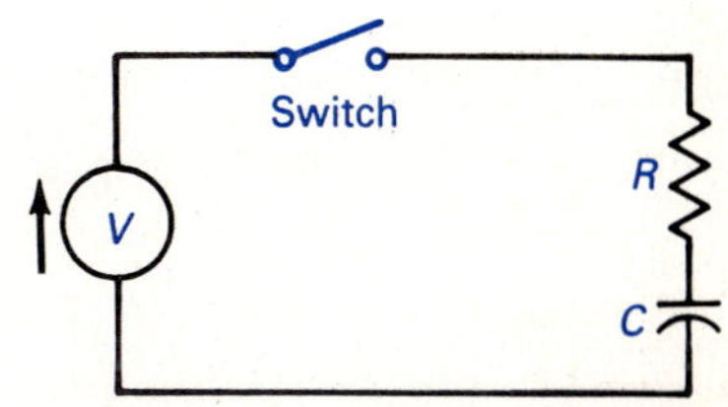

Fig. 11-7. Problems 1 and 2: Circuit.

11-2 CAPACITIVE TIME CONSTANT

In Example 11-1 the values form the basis of a principle relating resistance and capacitance to time. In the example they were

$$t = 1 \text{ s} \qquad R = 1 \text{ M}\Omega \qquad \text{and} \qquad C = 1\ \mu\text{F}$$

giving the magnitude of the exponent of e as

$$\frac{t}{RC} = 1 \qquad \text{or} \qquad t = RC$$

During this charging time, the capacitor became charged to 63.21 percent of the applied voltage. This would apply to any values of resistance and capacitance, their product $R \times C$ being called the *time constant*.

The time constant of a resistance-capacitance combination is the time taken for the charging capacitor's voltage to rise from zero to 63.21 percent of the source voltage.

The time constant is used frequently in electric circuit work and has the symbol τ, the lowercase Greek letter tau.

$$\tau = RC \tag{11-5}$$

where τ = time constant, s
R = resistance, Ω
C = capacitance, F

Table 11-1 can now be reproduced with correct values in terms of the circuit time constant ($\tau = RC = 1\ \text{M}\Omega \times 1\ \mu\text{F} = 1$ s) as shown in Table 11-2. The values are given to one decimal place, as this accuracy is usually good enough in practical time-constant situations.

TABLE 11-2 Universal Time Constants

t (s)	0	0.2τ	0.4τ	0.6τ	0.8τ	1.0τ	1.2τ	1.4τ	1.6τ	1.8τ	2.0τ
v_C (V)	0	18.2	33	45.1	55.1	63.2	69.9	75.3	79.8	83.5	86.5
v_R (V)	100	81.8	67	54.9	44.9	36.8	30.1	24.7	20.2	16.5	13.5
i (μA)	100	81.8	67	54.9	44.9	36.8	30.1	24.7	20.2	16.5	13.5

When, as in Table 11-2, the values of current and voltage are listed in terms of the time constant τ, rather than in absolute time units, the result is a *universal time-constant* table. Table 11-2 gives the percentage values of v_C, v_R, and i for *any RC* circuit. For example, if the circuit had $R = 47\ \text{k}\Omega$ and $C = 560$ pF, the time constant τ would be 26.32 μs. In this case Table 11-2 can be used where $0.2\tau = 5.264\ \mu$s, $0.4\tau = 10.53\ \mu$s, $0.6\tau = 15.79\ \mu$s, etc.

Charge Equations in Time-Constant Terms

Equations (11-2) to (11-4) can be rewritten with the time constant τ substituted for the product RC.

Equation (11-2) becomes

$$i = \frac{V}{R} e^{-t/\tau} \tag{11-6}$$

Equation (11-3) becomes

$$v_R = Ve^{-t/\tau} \tag{11-7}$$

Equation (11-4) becomes

$$v_C = V(1 - e^{-t/\tau}) \tag{11-8}$$

Universal Time-Constant Curves

From the data in Table 11-2 (extended to $t = 5\tau$) the universal time-constant curves in Fig. 11-8 can be drawn. The curves have been labeled $y = e^{-x}$ and $y = 1 - e^{-x}$, so that they can be used for any charge or discharge situation. The symbol y represents the current or voltage to be found, and x is the value of t/τ.

Universal time-constant curves can be used to determine approximate relationships. The horizontal axis x represents time expressed in terms of the time constant. The vertical axis y gives the fraction of the maximum value of current or

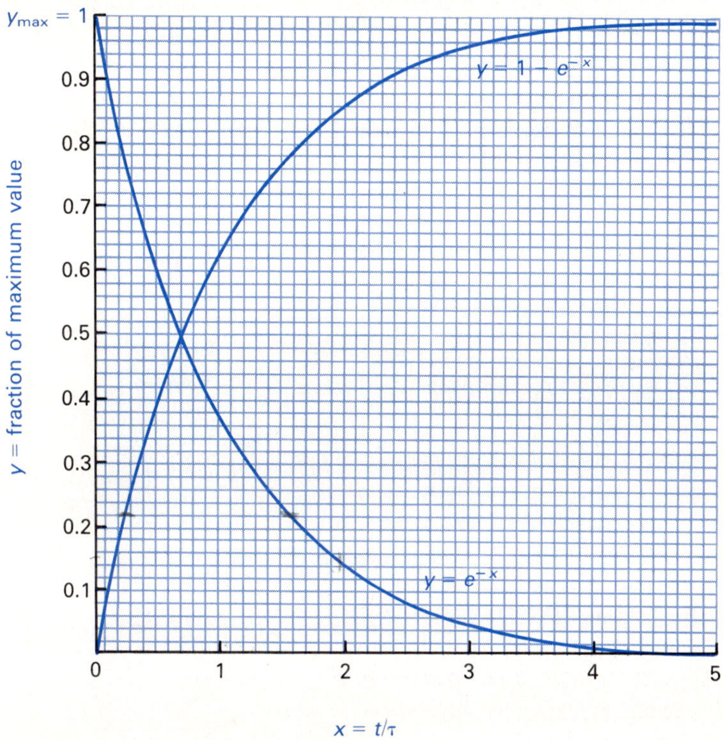

Fig. 11-8. Universal time-constant curves.

voltage attained at a specified time. To use the curves, it is first necessary to decide which curve applies to the situation. It is then a simple matter of projection to find the required data. The following examples show how to apply the curves during capacitive charge.

EXAMPLE 11-2 What is the voltage across the resistor in the circuit in Fig. 11-9 at a time 2.5 ms after the capacitor starts to charge?

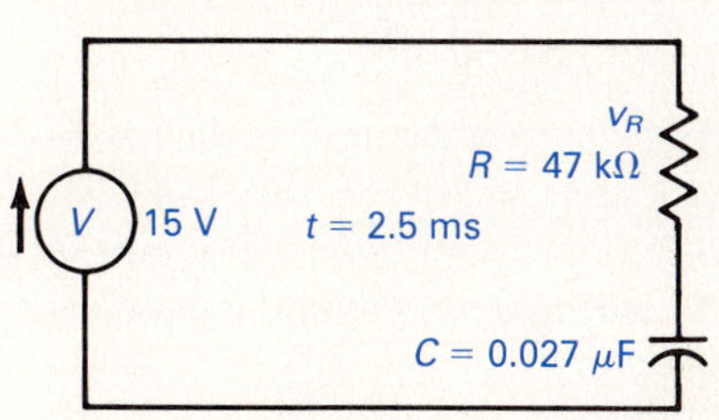

Fig. 11-9. Example 11-2: Circuit.

Solution The steps adopted are:

1. *Calculate the time constant:*

$$\tau = RC \qquad (11\text{-}5)$$
$$= 47\ \text{k}\Omega \times 0.027\ \mu\text{F} = 1.27\ \text{k}\Omega \cdot \mu\text{F}$$
$$\tau = 1.27\ \text{ms*}$$

2. *Select the curve:* During charge, the resistive voltage is given by Eq. (11-3), $v_R = Ve^{-t/RC}$, so the decreasing-value curve labeled $y = e^{-x}$ is used.
3. *Project from axis to axis:*

$$x = \frac{t}{\tau} = \frac{2.5\ \text{ms}}{1.27\ \text{ms}} = 1.97$$

From the point on the horizontal axis $x = 1.97$, project onto the curve $y = e^{-x}$ and then onto the vertical axis. The value of y becomes

$$y \simeq 0.14 \text{ of the maximum value of } v_R$$

The maximum value of v_R is $V = 15$ V. Therefore, 2.5 ms after the commencement of charge,

$$v_R \simeq 0.14 \text{ of } 15\ \text{V} \simeq \mathbf{2.1\ V}$$

An electronic calculator with exponential function capability makes time-constant problems easy to solve. For the data of Example 11-2, the method would be

$$\tau = RC = 47\ \text{k}\Omega \times 0.027\ \mu\text{F} = 1.269\ \text{ms}$$
$$\frac{t}{\tau} = \frac{2.5\ \text{ms}}{1.269\ \text{ms}} = 1.97$$

Changing the sign gives

$$-\frac{t}{\tau} = -1.97$$

Pressing the e^x button[1] gives

$$e^{-t/\tau} = 0.1394$$

Multiplying by the generator voltage, 15 V, results in the correct answer. To four significant figures this is 2.092 V.

EXAMPLE 11-3 Figure 11-10*b* is the equivalent circuit of the input of the transistor pulse amplifier shown in outline in Fig. 11-10*a*. If a 0.5-V pulse is applied to the transistor input, how long does it take for the base-to-emitter voltage to reach 90 percent of the pulse amplitude?

*Three significant figures are adequate when curves such as those in Fig. 11-8 are used.

[1] Refer to the operating instructions for your calculator if its buttons are labeled with other symbols for exponential functions.

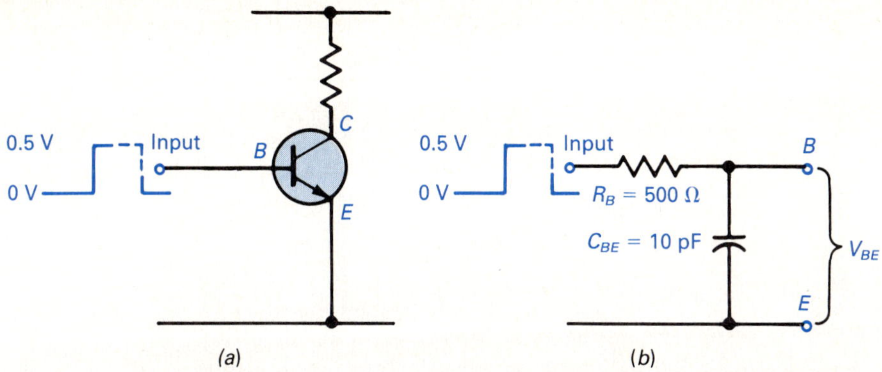

Fig. 11-10. Example 11-3: *(a)* Given circuit; *(b)* equivalent base-to-emitter circuit.

Solution The input pulse rising instantaneously from 0 to 0.5 V is equivalent to applying a 0.5-V battery via a switch. Thus the base-to-emitter voltage V_{BE} is the voltage across the base-to-emitter capacitance C_{BE} while it is charging from 0 V toward a maximum of 0.5 V.

1. *Calculating the time constant:*

$$\tau = R_B C_{BE}$$
$$= 0.5\ \text{k}\Omega \times 10\ \text{pF}$$
$$\tau = 5\ \text{ns}$$

2. *Selecting the curve:* During charge, the capacitive voltage is given by Eq. (11-4), $v_C = V(1 - e^{-t/RC})$, so that the increasing-value curve $y = 1 - e^{-x}$ is used.
3. *Projecting from axis to axis:*

$$y = 0.9 \text{ of maximum}$$

From the 0.9 point on the vertical axis, projecting onto the increasing-value curve and then onto the horizontal axis gives

$$x = \frac{t}{\tau} \simeq 2.3$$

Therefore the time taken for the base-to-emitter voltage V_{BE} to reach 0.9 of the pulse amplitude will be

$$t \simeq 2.3\tau \simeq 2.3 \times 5\ \text{ns} \simeq \mathbf{11.5\ ns}$$

In Example 11-3, the 11.5 ns taken for the base-to-emitter voltage to reach 0.9 of the pulse amplitude may seem to be a very small time period. Actually, in many applications of pulse amplifiers, such as high-speed computers, times of this order are much too long. A transistor with the specified input characteristics would be quite unacceptable.

This example can be viewed as an instance of waveform distortion.

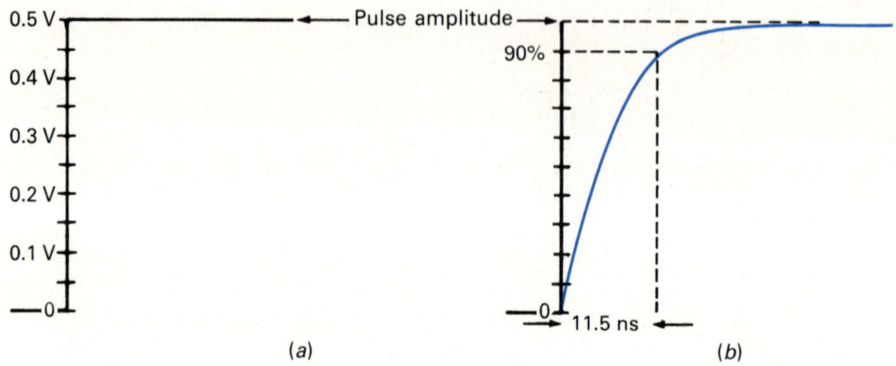

Fig. 11-11. Example 11-3: *(a)* 0.5-V pulse applied to the amplifier input. *(b)* The pulse as it appears between the base and the emitter of the transistor.

Figure 11-11*a* shows the input pulse rise. This pulse is assumed to increase from 0 to 0.5 V in almost zero time. Figure 11-11*b* shows the shape of the waveform actually appearing between the base and emitter of the transistor.

EXAMPLE 11-4 The variable resistance and fixed-value capacitance given in Fig. 11-12 are used to determine the time during which voltage is applied to the electrodes of a welding machine. If this time is that taken for the capacitance to charge from 0 to 10 V, what is the time range covered by the extremes of the resistance values?

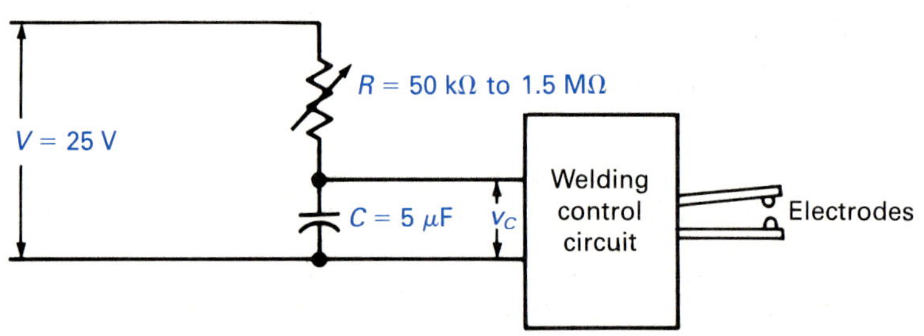

Fig. 11-12. Example 11-4: Data.

Solution The voltage required to determine the end of the welding time is

$$v_C = 10\ \text{V}$$

The supply voltage $V = 25$ V is the maximum possible value of v_C; therefore

$$v_C = \frac{10\ \text{V}}{25\ \text{V}} = 0.4 \text{ of maximum}$$

Projecting from $y = 0.4$ onto the $y = 1 - e^{-x}$ curve in Fig. 11-8 and then onto the horizontal axis gives

$$x = \frac{t}{\tau} \simeq 0.5 \qquad \text{or} \qquad t \simeq 0.5\tau$$

When $R = 50\ \text{k}\Omega$,

$$\tau = RC \tag{11-5}$$
$$\tau = 50\ \text{k}\Omega \times 5\ \mu\text{F} = 250\ \text{ms}$$
$$t \simeq 0.5\tau \simeq 125\ \text{ms}$$

When $R = 1.5\ \text{M}\Omega$,

$$\tau = RC = 1.5\ \text{M}\Omega \times 5\ \mu\text{F} = 7.5\ \text{s}$$
$$t \simeq 0.5\tau \simeq 3.75\ \text{s}$$

The welding arc time range can therefore be varied from **0.125** to **3.75 s** by varying the resistance value throughout its range.

REVIEW QUESTIONS FOR SECTION 11-2

1. What is the time constant of an *RC* circuit?
2. What is the symbol used for time constant?
3. In what unit is the time constant expressed?
4. There are two universal time-constant curves, one rising exponentially, the other falling exponentially. State the generalized equation for each curve.
5. Express the equations of Questions 5, 6, and 7, Sec. 11-1, in terms of time.

EXERCISE PROBLEMS FOR SECTION 11-2

1. Repeat Prob. 1, Sec. 11-1, using the universal time-constant curves, at the following times:
   a. 190 μs (*ans.* 2.1 μA, 0.21 V, 11.8 V)
   b. 15 μs (*ans.* 87 μA, 8.7 V, 3.3 V)
   c. 75 μs (*ans.* 24 μA, 2.4 V, 9.6 V)
2. Repeat Prob. 2, Sec. 11-1, using the universal time-constant curves, at the following times:
   a. 2000 μs *b.* 15 ms *c.* 8 ms

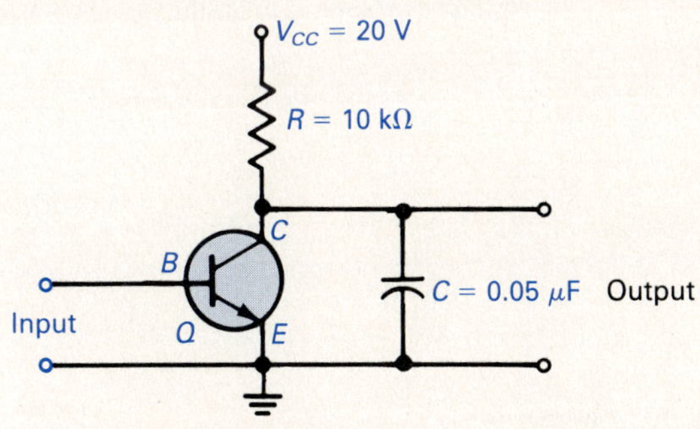

Fig. 11-13. Problem 3: Circuit.

3. Figure 11-13 shows the circuit of a simple ramp generator. *Q* is a transistor switch. When *Q* is on, it is the equivalent of a short circuit across *C*; when *Q* is off, it is in effect an open circuit and *C* is permitted to charge

through R. The transistor Q is initially on, and then is switched off. Plot the output voltage at 10-μs intervals for 50 μs after Q is switched off.

(*ans.* 0.396 V, 0.784 V, 1.165 V, 1.538 V, 1.9 V)

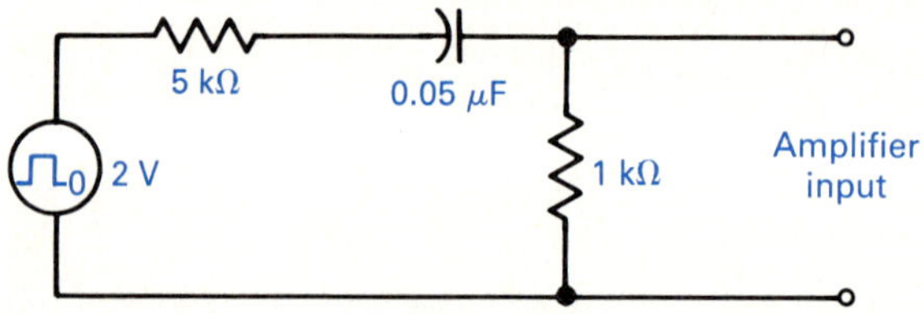

Fig. 11-14. Problems 4 and 5: Circuit.

4. The circuit in Fig. 11-14 represents a capacitive coupling between a pulse generator and the input of an amplifier. With a 2-V 100-μs pulse applied, find the voltage at the amplifier input at
   a. The beginning of the pulse.
   b. The end of the pulse.
5. For the circuit in Fig. 11-14 and Prob. 4, to what must the value of C be changed so that the voltage at the input falls to only 90 percent of its initial value by the end of the 100-μs pulse? (*ans.* 0.158 μF)

11-3 EVALUATING TIME

Example 11-4 is typical of situations in which it is desired to evaluate the time period elapsed during capacitance charge. The results given are not very accurate.

To find a value of time accurately, Eqs. (11-6) to (11-8) can be rearranged to make the time t the subject. It is then a simple job with an electronic calculator.

Equation (11-6):

$$i = \frac{V}{R} e^{-t/\tau}$$

becomes

$$t = \tau \ln \frac{V}{iR} \qquad \text{or} \qquad \tau \ln \frac{I_{\max}}{i}$$

(ln means "natural logarithm") **(11-9)**

Equation (11-7):

$$v_R = Ve^{-t/\tau}$$

becomes

$$t = \tau \ln \frac{V}{v_R} \tag{11-10}$$

Equation (11-8):

$$v_C = V(1 - e^{-t/\tau})$$

becomes

$$t = \tau \ln \frac{V}{V - v_C} \tag{11-11}$$

Using Eqs. (11-9) to (11-11), charge-condition times can be found easily to any order of accuracy within the number of display digits of your calculator. To four significant figures, the solutions of Examples 11-3 and 11-4 are

Example 11-3:

$$t = \tau \ln \frac{V}{V - v_C} \tag{11-11}$$

$$= 5 \text{ ns} \times \ln \frac{0.5 \text{ V}}{0.5 \text{ V} - 0.45 \text{ V}} = 5 \text{ ns} \times \ln 10$$

$$= 5 \text{ ns} \times 2.303$$

$$t = \mathbf{11.51 \text{ ns}}$$

Example 11-4:
When $R = 50\ \text{k}\Omega$,

$$t = \tau \ln \frac{V}{V - v_C} \tag{11-11}$$

$$= 250 \text{ ms} \times \ln \frac{25 \text{ V}}{25 \text{ V} - 10 \text{ V}} = 250 \text{ ms} \times \ln 1.667$$

$$= 250 \text{ ms} \times 0.5108$$

$$t = \mathbf{127.7 \text{ ms}}$$

When $R = 1.5\ \text{M}\Omega$,

$$t = 7.5 \text{ s} \times 0.5108$$

$$t = \mathbf{3.831 \text{ s}}$$

EXERCISE PROBLEMS FOR SECTION 11-3

1. Using Fig. 11-7 with $V = 25$ V, $R = 470\ \Omega$, and $C = 100$ pF, and Eq. (11-8), find the time required for v_C to reach the following values after the switch is closed:
   a. 20 V (*ans.* 75.6 ns)
   b. 9 V (*ans.* 21 ns)
   c. 2.5 V (*ans.* 4.95 ns)
   d. 15.8 V (*ans.* 47 ns)
2. The rise time of a waveform is the time required for the waveform to rise from 10 to 90 percent of its final amplitude. By finding the time to rise to 90 percent and subtracting the time required to rise to 10 percent, find a general expression for the rise time of an exponential waveform. Your answer should be in the form of a constant times a time constant.
3. Find the time constant and the rise time of v_C for the circuit in Prob. 1, Sec. 11-1. (*ans.* 10 ms, 22 ms)
4. Find the time constant and the rise time of v_C for the circuit in Prob. 2, Sec. 11-1.

11-4 CHARGING FROM A PARTIALLY CHARGED STATE

A capacitance may charge from an initial condition where it is not fully discharged, where the voltage at time t_0 is not zero. An example will show how this type of situation can be dealt with using the techniques already developed.

EXAMPLE 11-5 If, in Example 11-4, the welding control circuit does not fully discharge the capacitor after a weld is completed but leaves v_C equal to 2.5 V, what will be the maximum weld time during the next welding cycle?

Solution Figure 11-15*a* shows the conditions existing before the weld cycle starts: $v_C = 2.5$ V. In Fig. 11-15*b* the welding cycle has been completed, and $v_C = 10$ V. We wish to know how long it takes for v_C to rise from 2.5 to 10 V.

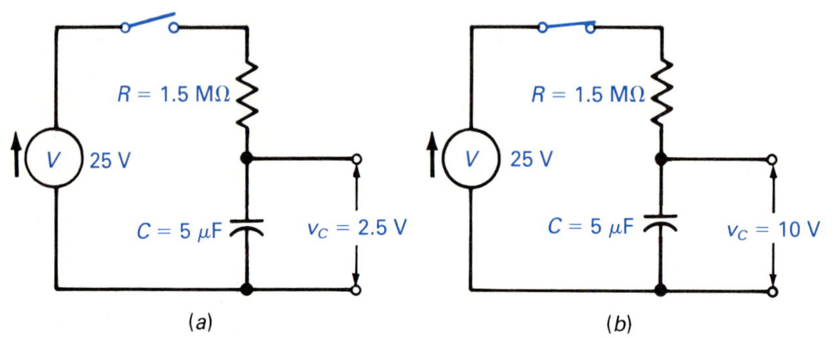

Fig. 11-15. Example 11-5: Circuit. *(a)* Before the weld cycle starts; *(b)* at completion of weld cycle.

The maximum weld time occurs when $R = 1.5\ \text{M}\Omega$ or $\tau = 7.5$ s. The time taken for the capacitor to charge from $v_C = 2.5$ V to $v_C = 10$ V will be the difference between the times taken to charge from a fully discharged state to $v_C = 10$ and 2.5 V, respectively. This statement can be written symbolically:

$$\Delta t = t_{10} - t_{2.5}$$

t_{10} is the time in Example 11-4, 3.831 s.

$$t_{2.5} = \tau \ln \frac{V}{V - v_C} \qquad (11\text{-}11)$$

$$= 7.5\ \text{s} \times \ln \frac{25\ \text{V}}{25\ \text{V} - 2.5\ \text{V}} = 7.5\ \text{s} \times 0.1054$$

$$t_{2.5} = 0.7902\ \text{s}$$

The time taken for the capacitor to charge from 2.5 V to 10 V will be

$$\Delta t = 3.831\ \text{s} - 0.7902\ \text{s} = \mathbf{3.041\ s}$$

11-5 DISCHARGING CAPACITANCE THROUGH RESISTANCE

Figure 11-16*a* through *d* shows a series of events in which a charged capacitor is discharged through a resistor.

During charge (Fig. 11-1) the capacitor is part of the energy source load. During discharge the capacitor *is* the energy source, and the resistor is its load. The discharge current direction follows the normal source-load convention: From capacitor positive through the resistor to capacitor negative.

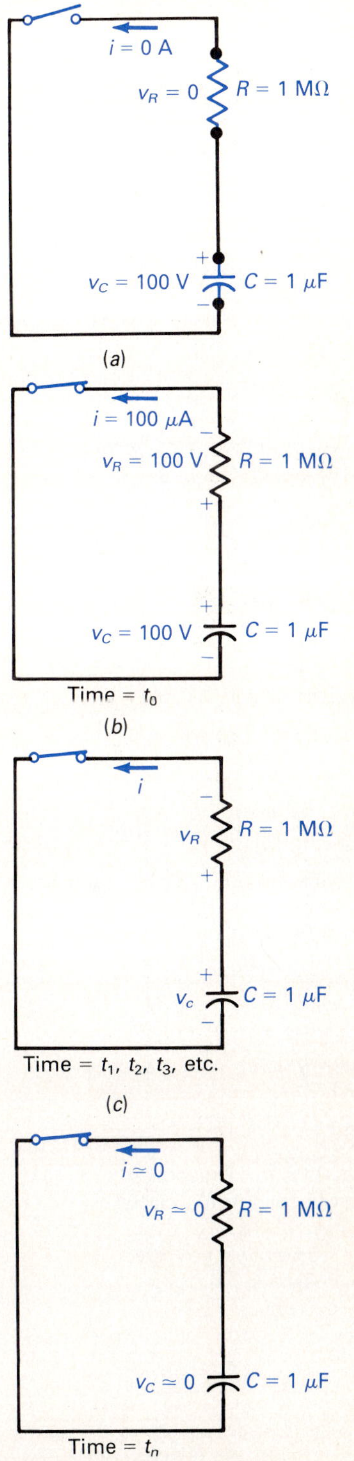

Fig. 11-16. Discharging capacitance through a resistor.

Figure 11-16*a* The capacitor is assumed to have been charged until $v_C = 100$ V. Since the switch is open, there is no current and therefore no resistive voltage.

Figure 11-16*b* At time t_0, when the switch is first closed, the initial conditions are

$$v_C = 100 \text{ V}$$
$$v_R = v_C = 100 \text{ V}$$
$$i = \frac{v_R}{R} = \frac{100 \text{ V}}{1 \text{ M}\Omega} = 100\ \mu\text{A}$$

Figure 11-16*c* In supplying current to the circuit, the energy imparted to the capacitor during its charge is extracted. This energy is transferred to the resistor, which dissipates it as heat. As energy leaves the capacitor, its voltage drops toward zero ($W = V^2C/2$ or $V \propto \sqrt{W}$). Since v_R equals v_C and $i = v_R/R$ during discharge, at successive time intervals t_1, t_2, t_3, etc., after the switch is closed (time t_0), both voltages v_C and v_R and the current i decrease toward zero. The voltage and current values can be summarized as follows:

At times t_1, t_2, t_3, etc.,

$$\left.\begin{aligned} 100 \text{ V} > v_C > 0 \\ 100 \text{ V} > v_R > 0 \\ 100\ \mu\text{A} > i > 0 \end{aligned}\right\} \text{Each quantity decreases from its value at time } t_0 \text{ toward zero}$$

Figure 11-16*d* After a period greater than about 5τ, the voltages and current become virtually zero (at 5τ they will each be 0.67 percent of their initial values). Thus at time t_n, a considerable time after t_0, the conditions become

$$v_C \simeq 0 \text{ V}$$
$$v_R \simeq 0 \text{ V}$$
$$i \simeq 0 \text{ A}$$

Comparison of Charge and Discharge

When Fig. 11-16 is compared with Fig. 11-1, the following points can be seen:

1. *a.* v_C increases from zero during charge.
b. v_C decreases toward zero during discharge.
c. The polarity of v_C is the same during charge and discharge.

2. *a.* v_R decreases toward zero during charge and discharge.
b. The polarity of v_R during discharge is opposite that during charge.

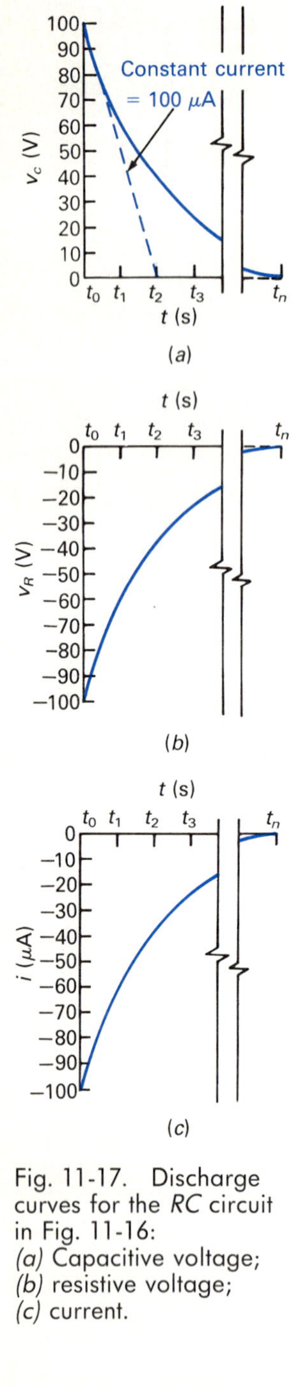

Fig. 11-17. Discharge curves for the *RC* circuit in Fig. 11-16:
(a) Capacitive voltage;
(b) resistive voltage;
(c) current.

3. *a.* i decreases toward zero during charge and discharge.
b. The direction of i during discharge is opposite that during charge.

RC Circuit Discharge Curves

Figure 11-17 shows the discharge curves of v_C, v_R, and i with respect to time. The reversals of polarity and direction relative to charge are indicated by the negative values of v_R and i.

Curves which follow the same law from a maximum value toward zero are the same shape, whether they represent positive or negative values. The three discharge curves are the same shape. Furthermore, they are the same shape as the decreasing-value charge curves (Fig. 11-2*b* and *c*).

The logic explaining the curvature of the discharge curves is essentially the same as in the case of charging: As the capacitor discharges, its voltage drops. This voltage is the circuit emf during discharge, so the current decreases. The decreasing current slows the rate at which the discharge takes place, thus producing the curvature.

The broken straight line in Fig. 11-17*a* shows the manner in which the capacitive voltage would decrease if the discharge current stayed constant at its initial value of 100 μA. The time it would take for v_C to reach zero at a constant 100-μA discharge current is $t_2 = \tau$ s. A general conclusion follows.

The initial charge held by the capacitor is

$$Q = CV_C \tag{4-3}$$

but

$$Q = it \tag{2-9}$$

Therefore

$$it = CV_C$$

$$t = C\frac{V_C}{i}$$

V_C is the initial capacitive voltage at the start of discharge; i is the discharge current, here assumed to be constant at a value V_C/R (or $V_C/i = R$). Thus

$$t = RC = \tau$$

This relationship also applies to the straight broken line in Fig. 11-2*a*.

Another way of defining time-constant results is: The *time constant* τ is the time it would take for the capacitive voltage to equal the applied emf during charge, or zero during discharge, if the current were maintained constant at its initial value.

Equations for Capacitor Discharging

The loop equation for the circuit in Fig. 11-16 is

$$v_C = v_R \qquad \text{or} \qquad v_C - v_R = 0 \tag{11-12}$$

An equation for the current in a discharging *RC* circuit can be obtained:

$$i = \frac{V_0}{R} e^{-t/RC} \tag{11-13}$$

where i = instantaneous circuit current, A
V_0 = initial circuit voltage (the voltage to which the capacitor was charged initially, or v_C at time t_0), V
e = base of natural logarithms $\simeq$ 2.718
R = resistance, Ω
C = capacitance, F
t = time, s

Applying Ohm's law and Eq. (11-13) gives

$$v_R = iR = \left(\frac{V_0}{R} e^{-t/RC}\right) R$$

$$v_R = V_0 e^{-t/RC} \qquad \textbf{(11-14)}$$

Substituting for v_R in Eq. (11-14),

$$v_C = v_R = V_0 e^{-t/RC} \qquad \textbf{(11-15)}$$

By comparing Eqs. (11-13) to (11-15) with Eqs. (11-2) and (11-3), it can be seen that all five are of the same form (they all refer to decreasing values of current or voltage). Therefore the calculation techniques used for i and v_R in the charging circuit can be used for i, v_R, and v_C under discharge conditions. They include the use of

1. The time constant ($\tau = RC$).
2. A universal time-constant table (e.g., Table 11-2). When this table is employed in the discharge situation, the v_R values are used for v_R and v_C.
3. A universal time-constant curve (e.g., Fig. 11-8). Since only decreasing values of current and voltage exist in the discharge circuit, the curve $y = e^{-x}$ is used for i, v_R, and v_C.
4. A calculator.

EXAMPLE 11-6 What are the approximate values of current and voltage in the circuit in Fig. 11-18 at t = 5 μs after the switch is closed? The capacitor is initially charged until v_C = 15 V.

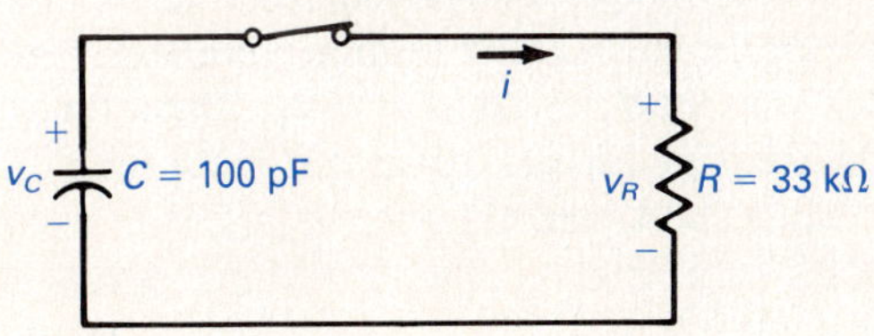

Fig. 11-18. Example 11-6: Circuit.

Solution

$$\tau = RC \qquad (11\text{-}5)$$

$$\tau = 33\ \text{k}\Omega \times 100\ \text{pF} = 3300\ \text{ns} = 3.3\ \mu\text{s}$$

$$t = 5\ \mu\text{s} = \frac{5\ \mu\text{s}}{3.3\ \mu\text{s}}\,\tau = 1.52\tau$$

Using the $y = e^{-x}$ curve in Fig. 11-8, projection gives

$$v_C \simeq 22 \text{ percent of } V_0 \simeq 22 \text{ percent of } 15 \text{ V}$$
$$v_C \simeq \mathbf{3.3\ V}$$
$$v_R = v_C \simeq \mathbf{3.3\ V}$$
$$i = \frac{v_R}{R} \simeq \frac{3.3\text{ V}}{33\text{ k}\Omega} \simeq \mathbf{100\ \mu A}$$

EXAMPLE 11-7 A 10-μF capacitor is charged to 100 V. A 100-V 20-kΩ/V voltmeter is connected across the capacitor, and the time of discharge through the voltmeter resistance is measured with a stopwatch. How much time elapses between the voltmeter readings of 90 and 10 V?

Solution (See Fig. 11-19.)

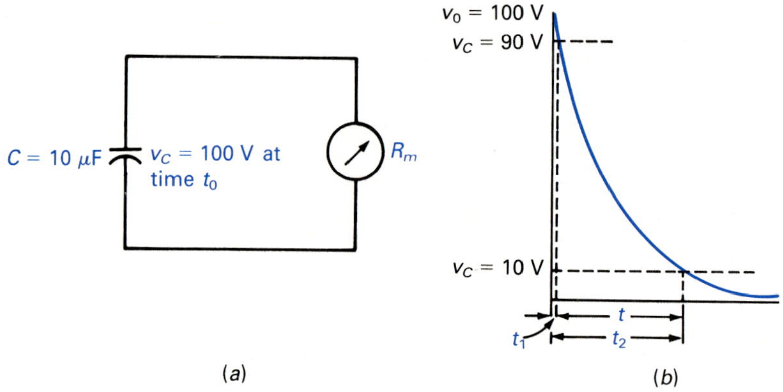

Fig. 11-19. Example 11-7: *(a)* Circuit; *(b)* curve of v_c with respect to time.

$$R_m = 100\text{ V} \times 20\text{ k}\Omega/\text{V} = 2\text{ M}\Omega$$
$$\tau = R_m C \qquad (11\text{-}5)$$
$$\tau = 2\text{ M}\Omega \times 10\ \mu\text{F} = 20\text{ s}$$

The time taken for the voltage to drop from 90 V to 10 V is the difference between the times taken to drop from 100 V to 10 V and from 100 V to 90 V, or $\Delta t = t_2 - t_1$. These times can be calculated by adapting Eq. (11-10) as it is in the form of a time version of Eqs. (11-14) and (11-15),

$$t = \tau \ln \frac{V}{v_R} \qquad (11\text{-}10)$$

Applied to this example, it becomes

$$t = \tau \ln \frac{v_0}{v_R}$$

$$\Delta t = t_2 - t_1 = 20\text{ s} \times \ln \frac{100\text{ V}}{10\text{ V}} - 20\text{ s} \times \ln \frac{100\text{ V}}{90\text{ V}}$$

$$= 46.05 \text{ s} - 2.107 \text{ s}$$
$$\Delta t = \mathbf{43.94 \text{ s}}$$

Example 11-7 could have been solved more rapidly by assuming an initial voltage value $v_C = 90$ V and determining the time taken for v_C to decrease to 10 V.

$$t = 20 \text{ s} \times \ln (90 \text{ V}/10 \text{ V}) = 20 \text{ s} \times \ln 9$$
$$t = 20 \text{ s} \times 2.197 = \mathbf{43.94 \text{ s}}$$

EXAMPLE 11-8 Figure 11-20*a* is the circuit of a relaxation oscillator used for generating a pulse waveform. As the capacitor C charges through R_1, the voltage v_e rises. When v_e becomes 6 V, the unijunction transistor internal resistance R_e drops suddenly from a very high value to 400 Ω, discharging the capacitor. At this time an output voltage pulse v_o is produced. When v_e drops to 2 V, R_e again becomes very high, allowing the capacitor to recharge. The action automatically repeats itself, so that a series of output pulses occurs at regular time intervals. The resulting waveforms of v_e and v_o are shown in Fig. 11-20*b*. What is the time t between consecutive pulses?

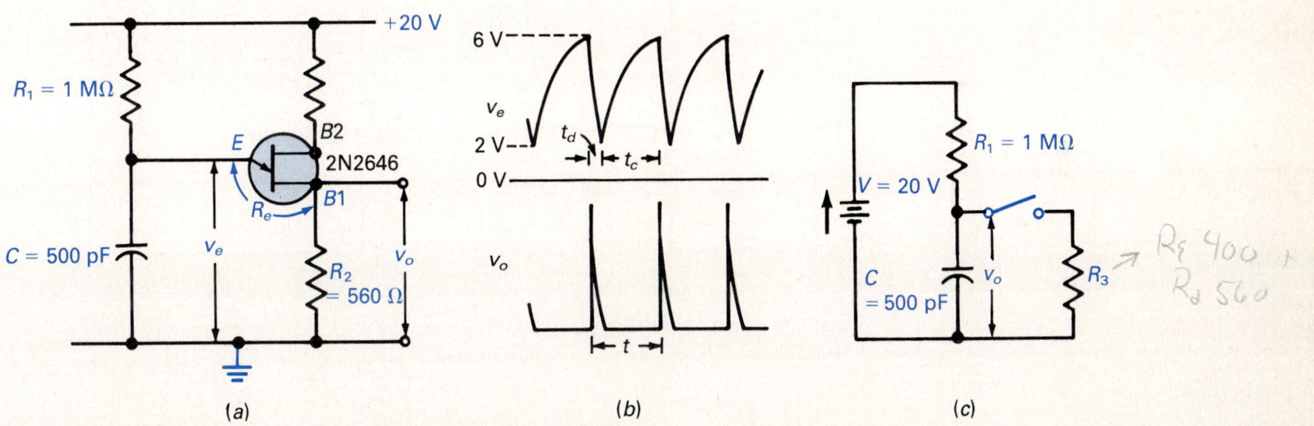

Fig. 11-20. Example 11-8: *(a)* Given circuit; *(b)* waveforms of v_e and v_o; *(c)* equivalent circuit.

Solution The complexity of Fig. 11-20*a* can be eliminated for this example by considering the timing circuit equivalent in Fig. 11-20*c*. The switch simulates the transistor action. When the capacitor C is charging to 6 V, the switch is open; for discharging from 6 to 2 V, the switch is closed, connecting resistance R_3 across the capacitor. R_3 is the series equivalent of R_e (400 Ω during capacitor discharge), and $R_2 = 560$ Ω.

While the capacitor C is charging between 2 and 6 V from the 20-V supply (switch open in Fig. 11-20c), the charging resistor is R_1. The charging time constant is

$$\tau_c = R_1 C \qquad (11\text{-}5)$$
$$\tau_c = 1\ \text{M}\Omega \times 500\ \text{pF} = 500\ \mu\text{s}$$

The charging time is the difference between the times for charging to 6 V and to 2 V,

$$t_c = t_6 - t_2$$

Applying Eq. (11-11),

$$t_c = \tau \ln \frac{V}{V - 6\ \text{V}} - \tau \ln \frac{V}{V - 2\ \text{V}}$$

As subtracting logarithms corresponds to dividing their original numbers,

$$t_c = \tau \ln \frac{V - 6\ \text{V}}{V - 2\ \text{V}} = 500\ \mu\text{s} \times \ln \frac{(20 - 2)\ \text{V}}{(20 - 6)\ \text{V}}$$
$$t_c = 125.7\ \mu\text{s}$$

While the capacitor is discharging between 6 and 2 V (switch closed in Fig. 11-20c), the equivalent circuit is that of Fig. 11-21a.

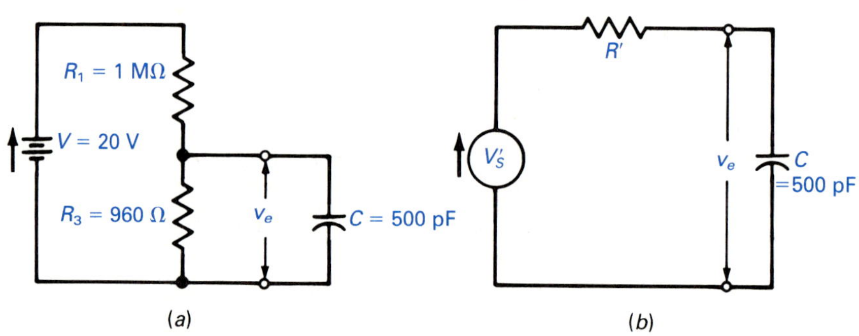

Fig. 11-21. *(a)* Discharge circuit in Fig. 11-20; *(b)* Thevenin equivalent of discharge circuit.

The series-parallel circuit in Fig. 11-21a must be converted into a series circuit for direct evaluation of discharge time. The Thevenin principle used with the capacitor as the load can be applied to perform the conversion. Figure 11-21b shows the resulting Thevenin equivalent circuit.

$$V_S' = V \frac{R_3}{R_1 + R_3} = 20\ \text{V} \frac{960\ \Omega}{1\ \text{M}\Omega + 960\ \Omega} = 0.0192\ \text{V}$$

$$R' = \frac{R_1 R_3}{R_1 + R_3} = \frac{1\ \text{M}\Omega \times 960\ \Omega}{1\ \text{M}\Omega + 960\ \Omega} = 959.1\ \Omega$$

Since the Thevenin voltage $V'_S \simeq 0$ V and the Thevenin resistance $R' \simeq R_3$, for practical purposes it may be considered that C discharges directly into the 960 Ω of R_3. The discharge extends from $v_e = 6$ V to $v_e = 2$ V.

$$\tau_d \simeq R_3C \simeq 960\ \Omega \times 500\ \text{pF} \simeq 480\ \text{ns} \simeq 0.48\ \mu\text{s}$$

Applying Eq. (11-10) in its discharge form,

$$t_d = \tau \ln \frac{v_o}{v_e} = 0.48\ \mu\text{s} \times \ln \frac{6\ \text{V}}{2\ \text{V}} = 0.6654\ \mu\text{s}$$

The time between pulses is the sum of the charge and discharge times:

$$t = t_c + t_d = 125.7\ \mu\text{s} + 0.6654\ \mu\text{s} = \mathbf{126.4\ \mu s}$$

EXAMPLE 11-9 What will be the current i_C in the circuit in Fig. 11-22*a* at $t = 10$ ms after the switch is closed if the capacitor is initially uncharged?

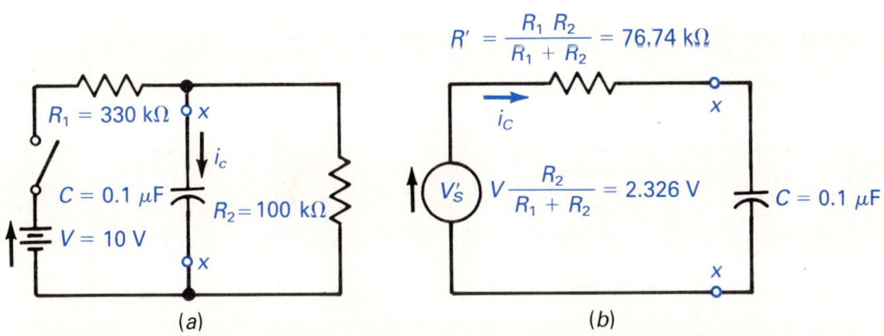

Fig. 11-22. Example 11-9: *(a)* Given circuit; *(b)* Thevenin equivalent.

Solution Figure 11-22*a* is transposed into the Thevenin series equivalent in Fig. 11-22*b* for ease of handling as a time-constant circuit.

$$\tau = R'C = 76.74\ \text{k}\Omega \times 0.1\mu\text{F} = 7.674\ \text{ms}$$

From Eq. (11-6),

$$i = \frac{V'_S}{R'} e^{-t/\tau} = \frac{2.326\ \text{V}}{76.74\ \text{k}\Omega} e^{-10\ \text{ms}/7.674\ \text{ms}}$$

$$i = \mathbf{0.0083\ mA} \qquad \text{or} \qquad \mathbf{8.3\ \mu A}$$

A review of Example 11-9 from a different approach can aid in understanding these circuits.

1. *Maximum capacitive voltage*. The maximum possible voltage which could be developed across the capacitor is the voltage across R_2 in Fig. 11-23 with the capacitor disconnected:

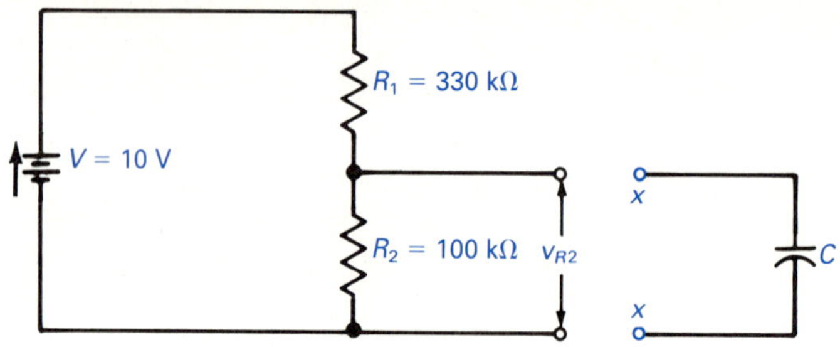

Fig. 11-23. Figure 11-22*a* rearranged.

$$V_{R2} = V \frac{R_2}{R_1 + R_2} = 10\ \text{V} \frac{100\ \text{k}\Omega}{330\ \text{k}\Omega + 100\ \text{k}\Omega} = 2.326\ \text{V}$$

Thus the voltage of the capacitor during its charge increases from zero toward 2.326 V.

2. *Charging resistance*. With or without R_2, the available initial charging current will be the same, because $v_C = 0$ V. This is shown in Fig. 11-24.

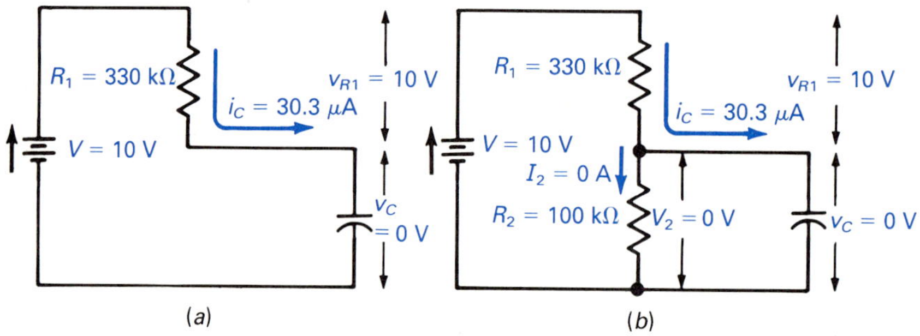

Fig. 11-24. Initial charging conditions for Figure 11-23: *(a)* Without R_2; *(b)* with R_2.

If the charge were to be completed ($t \gg 5\tau$), the conditions without and with R_2 would be as shown in Fig. 11-25.

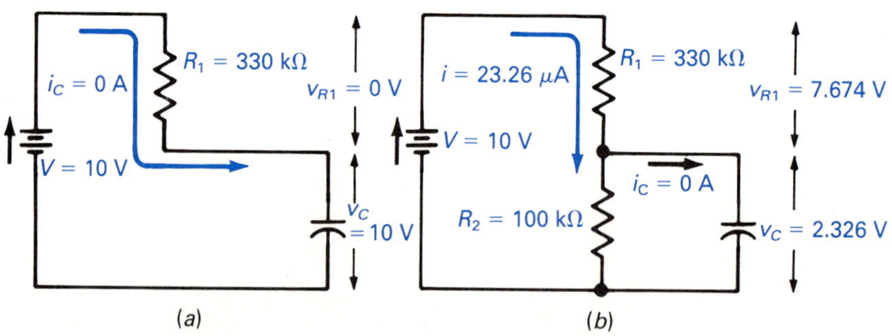

Fig. 11-25. Conditions at completion of charge for Figure 11-23: *(a)* Without R_2; *(b)* with R_2.

From the viewpoint of the capacitor terminals, the situation for R_2 will be:

1. Figure 11-24*b*: With the capacitor fully discharged, $v_C = 0$ V and the circuit provides an initial charging current of 0.0303 mA.
2. Figure 11-25*b*: With the capacitor fully charged, $i_C = 0$ A and the capacitive voltage is 2.326 V.

Item 2 confirms that the equivalent charging emf is 2.326 V. Items 1 and 2 together lead to the conclusion that the equivalent charging resistance is equal to 2.326 V/30.3 μA = 76.74 kΩ. This resistance value is that obtained by considering R_1 and R_2 to be effectively in parallel.

REVIEW QUESTIONS FOR SECTION 11-5

1. At the instant a circuit is closed to discharge a capacitor through a resistor, what are the values of the current i and the voltage across the resistor v_R in terms of capacitor voltage v_C?
2. What is the direction of current in an *RC* circuit during the discharge period as compared with the direction during the charging period?
3. What is the polarity of the voltage across the resistor v_R during the discharge period of an *RC* circuit, as compared with the polarity during charge?
4. What is the polarity of the voltage across the capacitor v_C during the discharge of an *RC* circuit as compared with the polarity during charge?
5. Define capacitive time constant in terms of applied emf and initial current value.
6. State the equations for i, v_R, and v_C for a discharging *RC* circuit.
7. When a capacitor is being charged through a circuit involving both series and shunt resistors, what approach can be taken to establish the time constant and final voltage?

EXERCISE PROBLEMS FOR SECTION 11-5

1. Find the time required for the given capacitor to discharge through the resistor from the voltage V_0 to the voltage v_C in each of the following cases:
   a. $V_0 = 45$ V, $v_C = 25$ V, $R = 4.7$ kΩ, $C = 0.001$ μF (*ans*. 2.763 μs)
   b. $V_0 = 10$ V, $v_C = 0.5$ V, $R = 22$ kΩ, $C = 0.02$ μF (*ans*. 1.318 ms)
   c. $V_0 = 100$ V, $v_C = 50$ V, $R = 10$ kΩ, $C = 0.2$ μF (*ans*. 1.386 ms)
2. Find the voltage across the capacitor when it has been discharging through the resistor from voltage V_0 for the given time t.
   a. $V_0 = 45$ V, $R = 4.7$ kΩ, $C = 0.001$ μF, $t = 2$ μs
   b. $V_0 = 10$ V, $R = 22$ kΩ, $C = 0.02$ μF, $t = 1$ ms
   c. $V_0 = 100$ V, $R = 10$ kΩ, $C = 0.2$ μF, $t = 600$ μs

11-6 *RC* CIRCUIT WITH APPLIED SQUARE WAVES

Within the broken-line box in Fig. 11-26*a* is a basic square-wave generator.

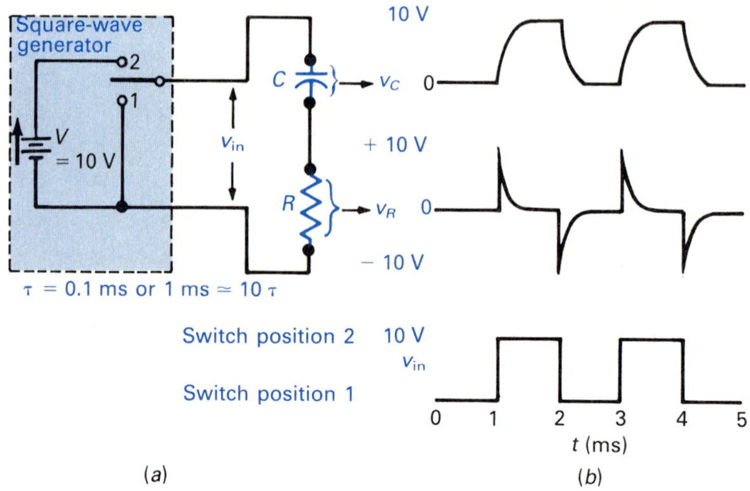

Fig. 11-26. *RC* circuit with applied square waves.

When the switch is at position 1, $v_{in} = 0$ V and the *RC* network is short-circuited so that any charge on the capacitor will be dissipated in the resistor. When the switch is at position 2, $v_{in} = 10$ V is supplied to the *RC* network as a charging voltage. The resulting network supply voltage is shown as the square waveform at the bottom of Fig. 11-26*b*, it being assumed that the switch is automatically operated so that it stays in each position for 1 ms and rapidly switches between the positions.

The time constant $\tau = RC$ is chosen to be about 0.1 ms (1 ms $\simeq 10\tau$), so that the resistive and capacitive voltages have time to stabilize between switch operations.

The waveforms of v_R and v_C shown in Fig. 11-26*b* are the result of successive charging and discharging. They are thus composites of Figs. 11-2*a* and 11-17*a* for v_C, and Figs. 11-2*b* and 11-17*b* for v_R.

Differentiation

When the applied voltage changes value suddenly, the resistive voltage v_R also changes value suddenly. When the applied voltage suddenly increases from 0 to +10 V (a 10-V positive swing), v_R also swings 10 V in a positive direction. Conversely, when v_{in} drops from +10 to 0 V (a 10-V negative swing), v_R swings 10 V in a negative direction.

If the applied voltage does not change (as, for instance, during the 1-ms periods in Fig. 11-26), the resistive voltage is due only to the capacitive charge conditions stabilizing.

In general, the resistive voltage is determined by the rate and direction of change in the applied voltage. This concept is similar to differentiation in math, in which rates of change are significant. The resistive voltage v_R is called the *differential voltage*. If the output of an *RC* series circuit is taken from the resistance, the circuit is called a *differentiator,* the time constant usually being not more than 10 percent of the minimum available charge or discharge time.

Integration

While the applied voltage is at the 10-V level, each increment of capacitive voltage adds to the existing voltage as the capacitor charges. This summation is related to integration in mathematics. The capacitive voltage v_C is called the *integral voltage*. If the output of an *RC* series circuit is taken from the capacitor, the circuit is called an *integrator,* the time constant typically being chosen to be at least 10 times the maximum available charge and/or discharge time.

EXAMPLE 11-10 The rectangular wave v_{in} in Fig. 11-27*b* is applied to the input of the differentiator (Fig. 11-27*a*). What is the reason for the shape of the output waveform v_{out}?

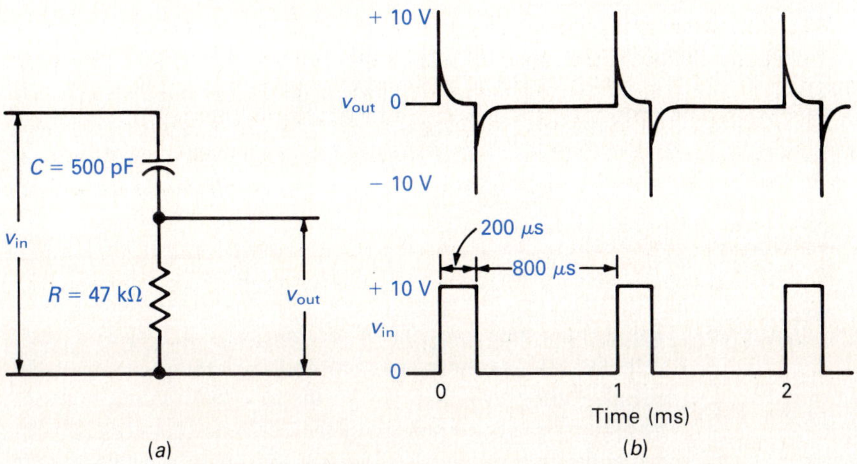

Fig. 11-27. Example 11-10: *(a)* Circuit; *(b)* input and output voltage waveforms.

Solution The rectangular wave v_{in} has unequal times for the pulse duration and the period between the pulses. Since the 23.5-μs time constant of the differentiator circuit (47 kΩ × 500 pF) is short compared with the 200-μs pulse duration and with the 800-μs interval between pulses, the capacitor always has time to charge before the input pulse changes level. Thus the output voltage v_{out} is always close to zero prior to an input level change. When the input level changes from 0 to 10 V, a positive swing, the output rises suddenly from 0 to +10 V. When the input level drops from 10 to 0 V, a negative swing, the output *rises* from 0 to −10 V. The output voltage v_{out} consists of spikes which rise almost instantaneously in either a positive or negative direction on an input level change, dropping virtually to zero before another input level change occurs.

EXAMPLE 11-11 The rectangular wave v_{in} in Fig. 11-28*b* is applied to the input of the integrator (Fig. 11-28*a*). What is the reason for the shape of the output waveform v_{out}?

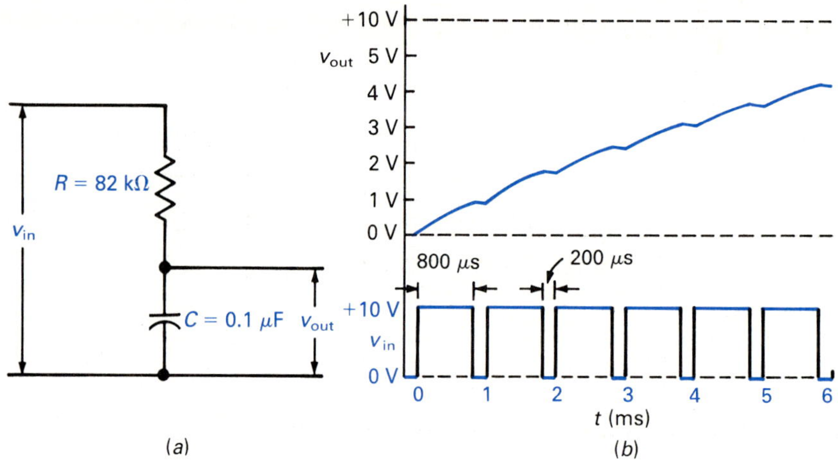

Fig. 11-28. Example 11-11: *(a)* Circuit; *(b)* input and output voltage waveforms.

Solution The 8.2-ms time constant of the integrator is long compared with the 800-μs pulse duration and the 200 μs between the pulses. During an 800-μs period, the capacitor voltage increases by approximately 10 percent of the difference between its voltage at the start of the period and its maximum possible voltage 10 V. During a 200-μs period, the capacitor loses about 2.4 percent of its voltage. Therefore the output voltage v_{out} gradually builds up toward the maximum of 10 V.

Examples 11-10 and 11-11 typify the operation of circuits in a television set which separate the synchronizing signals to maintain a steady picture.

Short-duration pulses constitute part of the transmitted signal. These enter a short-time-constant differentiator and leave as a series of sharp spikes, as in Example 11-10. These voltage spikes synchronize the horizontal sweep circuit.

Also included with the transmitted signal are groups of long pulses. Each time one of these groups of pulses enters a long-time constant integrator, it leaves as one built-up pulse, the action being similar to that in Example 11-11. The output pulses synchronize the vertical sweep circuit. The vertical synchronizing waveform of Fig. 11-28*b* may be simultaneously applied to the differentiator of Fig. 11-27*a*. The resulting series of pulses, shown in Fig. 11-29, would be used to maintain horizontal synchronization during the vertical synchronization period.

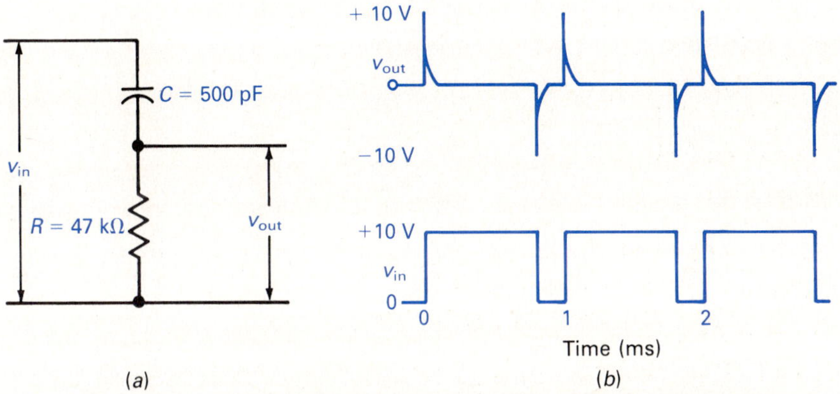

Fig. 11-29. Differentiator of Example 11-10. (*a*) Circuit; (*b*) waveforms when the input is the same as in Example 11-11.

REVIEW QUESTIONS FOR SECTION 11-6

1. If the input voltage across an *RC* circuit changes very rapidly, across which element of the circuit will this same voltage change appear?
2. If an *RC* circuit is connected so as to give the voltage across the resistor as its output, what is the circuit called?
3. If an *RC* circuit is connected so as to give the voltage across the capacitor as its output, what is the circuit called?
4. If a square wave is applied to a very-short-time-constant *RC* circuit, what voltage waveform will appear across
   a. The resistor?
   b. The capacitor?
5. If a square wave is applied to a very-long-time-constant *RC* circuit, what voltage waveform will appear across the capacitor?

11-7 CHARGING INDUCTANCE THROUGH RESISTANCE

Figure 11-30 shows a sequence of events in a series *RL* circuit during the charge of the inductance.

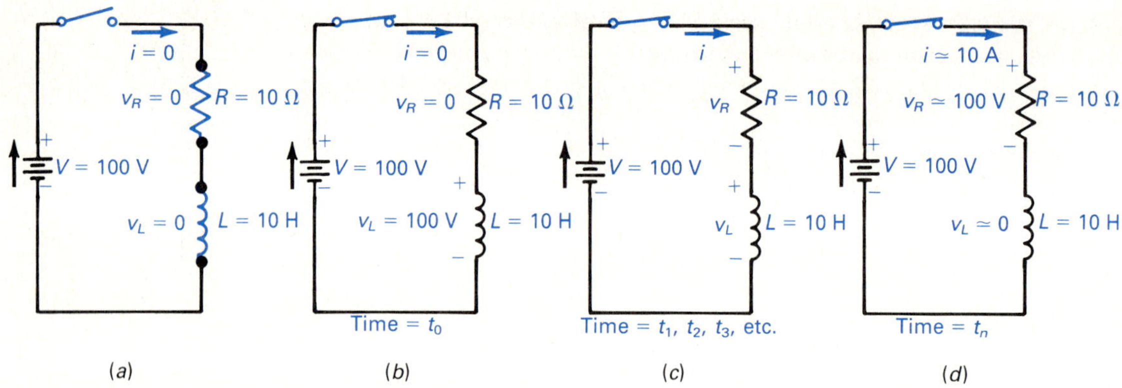

Fig. 11-30. Charging inductance through resistor.

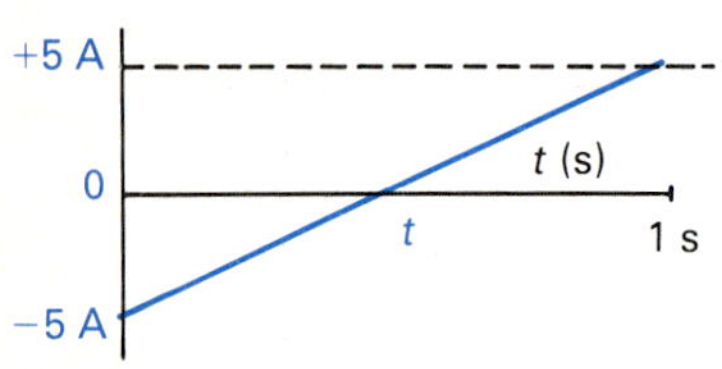

Fig. 11-31. Rate of change in current of 10 A/s when the current magnitude is zero.

Figure 11-30*a* With the switch open, the current i, the inductive voltage v_L, and the resistive voltage v_R are all zero.

Figure 11-30*b* There is no current at the instant the switch is closed, but there is a maximum rate of change in current. This concept needs some explanation.

Consider a current which, in a period of 1 s, changes from 5 A in one direction to 5 A in the opposite direction. This change is shown graphically in Fig. 11-31.

The current is changing at the rate of 10 A/s for the whole 1-s period. Therefore it is changing at this rate at time t. At this time, the current value is zero.

In Fig. 11-30*b*, at time t_0, it might be said that the current is *trying* to increase as rapidly as possible from zero value (which certainly existed before the switch was closed) to a maximum value of $V/R = 10$ A. In doing so, it achieves a high rate of increase. This current increase causes an induced emf in the inductance, which opposes the current change (Lenz' law). The initial effect is that the magnitude of the induced emf becomes equal to that of the applied emf and so prevents current. Of course, the zero current value exists only for an immeasurably small time period, since the current must change to produce a voltage.

At time t_0,

$$v_L = V = 100 \text{ V}$$
$$v_R = V - v_L = 100 \text{ V} - 100 \text{ V} = 0 \text{ V}$$
$$i = 0 \text{ A}$$

Figure 11-30*c* As the current increases, the resistive voltage also increases from zero. Its polarity is such as to reduce the total circuit voltage applied across the inductance. Thus the rate of current rise decreases. As time progresses and v_R becomes higher, the current rise tapers off almost to zero.

The current and voltage values at times t_1, t_2, t_3, etc., are

$$100 \text{ V} > v_L > 0$$
$$0 < v_R < 100 \text{ V}$$
$$0 < i < 10 \text{ A}$$

Figure 11-30d Time t_n is to be regarded as sufficiently long after time t_0 for conditions virtually to have become stabilized.

$$v_L \simeq 0 \text{ V}$$
$$v_R = V - v_L \simeq 100 \text{ V} - 0 \text{ V} \simeq 100 \text{ V}$$
$$i = \frac{v_R}{R} \simeq \frac{100 \text{ V}}{10 \ \Omega} \simeq 10 \text{ A}$$

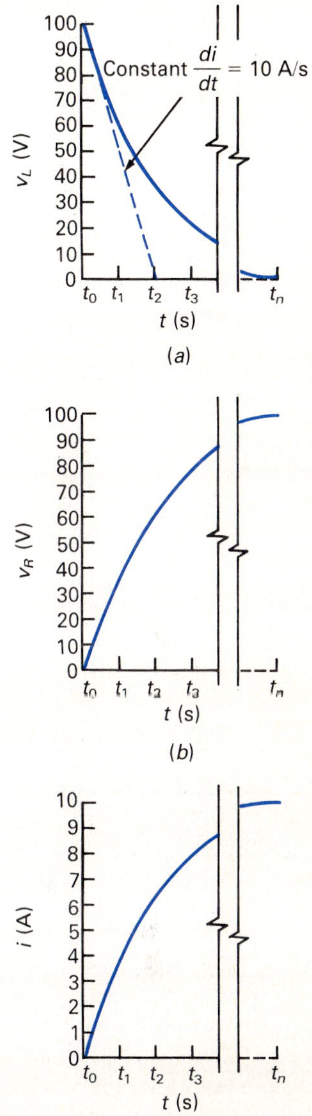

Fig. 11-32. Charge curves with respect to time for *RL* circuit in Fig. 11-30: *(a)* Inductive voltage; *(b)* resistive voltage; *(c)* current.

RL Circuit Charge Curves

The curves in Fig. 11-32 summarize the events in Fig. 11-30.

Equations for Inductor Charge

The loop equation for the circuit in Fig. 11-30 is

$$V = v_L + v_R \qquad \textbf{(11-16)}$$

or

$$V = L\frac{di}{dt} + iR$$

From this expression an equation can be obtained for the current in an *RL* circuit during charge:

$$i = \frac{V}{R}(1 - e^{-tR/L}) \qquad \textbf{(11-17)}$$

Applying Ohm's law and Eq. (11-17) gives

$$v_R = iR = V(1 - e^{-tR/L}) \qquad \textbf{(11-18)}$$

Substituting for v_R in Eq. (11-16),

$$v_L = V - v_R = V - V(1 - e^{-tR/L})$$
$$v_L = Ve^{-tR/L} \qquad \textbf{(11-19)}$$

In Eqs. (11-17) to (11-19),

i = instantaneous current, A
v_R = instantaneous resistive voltage, V
v_L = instantaneous inductive voltage, V
V = supply emf, V
R = resistance, Ω
L = inductance, H
t = time, s
e = base of natural logarithms ≃ 2.718

REVIEW QUESTIONS FOR SECTION 11-7

1. A series circuit consists of a voltage source V, a resistor R, and an inductor L. When the circuit is first closed, what are the values of i, v_R, and v_L?
2. When the circuit of Question 1 has stabilized, after a long period of time, what are the values of i, v_R, and v_L?

3. At what value of current does the current have the greatest rate of change when charging an *RL* circuit?
4. What is the basic loop equation for the circuit of Question 2?
5. What are the equations for current i, voltage across the resistor v_R, and voltage across the inductor v_L for a charging *RL* circuit?

EXERCISE PROBLEMS FOR SECTION 11-7

1. Using Fig. 11-33 with $V = 50$ V, $R = 100\ \Omega$, and $L = 100$ mH, and Eqs. (11-17) to (11-19), find i, v_R, and v_L at the following times after the switch is closed:
   a. $t = 0+$ (a short time after t_0) (*ans.* 0 A, 0 V, 50 V)
   b. $t = 1$ ms (*ans.* 316 mA, 31.6 V, 18.4 V)
   c. $t = 500\ \mu$s (*ans.* 196.7 mA, 19.67 V, 30.33 V)

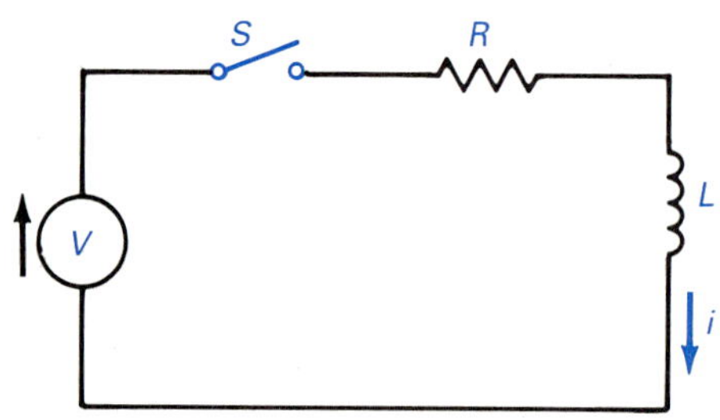

Fig. 11-33. Problems 1 to 3: Circuit.

2. Repeat Prob. 1, using the universal time-constant curves, at the following times after the switch is closed:
   a. $t = 200\ \mu$s
   b. $t = 3$ ms
3. Using Fig. 11-33 with $V = 10$ V, $R = 22\ \Omega$, and $L = 400\ \mu$H, find the time required to reach the following voltages and currents after the switch is closed:
   a. $v_R = 9$ V (*ans.* 41.86 μs)
   b. $v_L = 9$ V (*ans.* 1.92 μs)
   c. $i = 200$ mA (*ans.* 10.54 μs)

11-8 DISCHARGING INDUCTANCE THROUGH RESISTANCE

The sequence of events depicted in Fig. 11-34 applies to an inductor when its magnetic field is collapsing and its energy is being dissipated in a resistor. No switch is shown in this circuit because the sequence cannot start with the conditions that exist when a charged inductor is connected to an open circuit. Under these conditions the magnetic field would collapse in an indeterminate time.

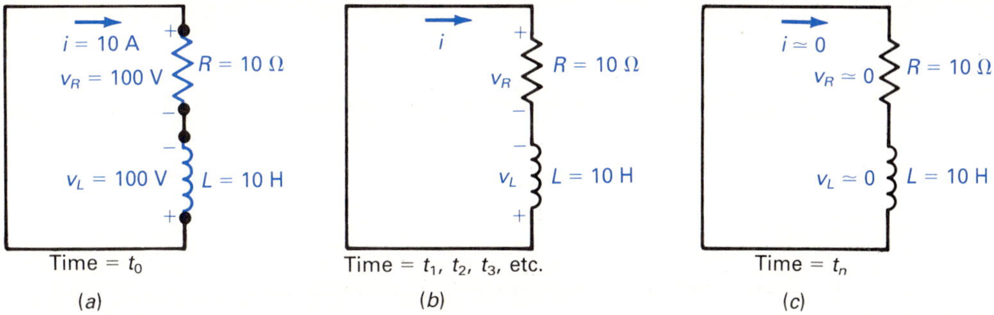

Fig. 11-34. Discharging inductance through resistor.

One way in which the transition from charge to discharge (Fig. 11-30*d* to Fig. 11-34*a*) can take place without the circuit continuity being broken is shown in Fig. 11-35.

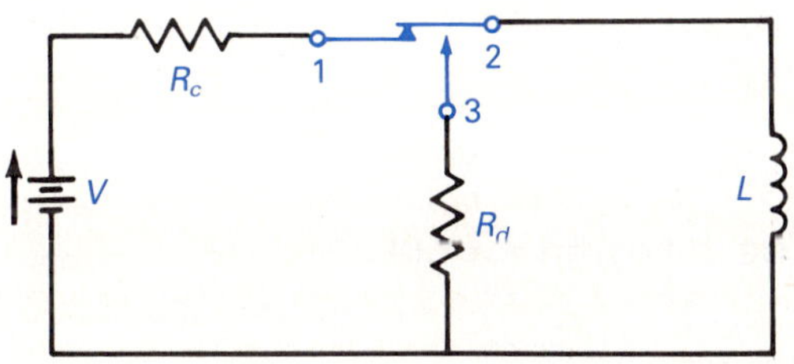

Fig. 11-35. Charge-discharge circuit with make-before-break contacts.

In Fig. 11-35 a make-before-break switch is shown. The circuit by which the inductor L is charged through resistor R_C is completed through contacts 1 and 2. When the switch is operated, contacts 2 and 3 connect together before contacts 1 and 2 open. Thus the discharge circuit through resistor R_d is completed before the charge circuit is broken. The charge and discharge circuits are both connected across the inductor during the time the switch is completing its operation. Therefore the switch operation should be rapid to avoid significant interference with the discharge conditions.

Figure 11-34*a* The inductance is assumed to have its magnetic field developed as in Fig. 11-30*d*. Since the circuit is transformed into that in Fig. 11-34*a* without first becoming open, the collapsing field produces an induced emf of 100 V. The polarity of this emf is opposite to that caused by the increasing field during charge. This polarity results in a circuit current in the same direction as that for charge. The current and voltage values at the instant these conditions apply will be, at time t_0,

$$v_L = 100 \text{ V}$$

$$v_R = v_L = 100 \text{ V}$$

$$i = \frac{v_R}{R} = \frac{100 \text{ V}}{10 \ \Omega} = 10 \text{ A}$$

Figure 11-34*b* As the energy of the magnetic field is dissipated in the resistor, the circuit current and voltages decrease toward zero. At times t_1, t_2, t_3, etc.,

$$100\ \text{V} > v_L > 0$$
$$100\ \text{V} > v_R > 0$$
$$10\ \text{A} > i > 0$$

Figure 11-34*c* When almost all the field energy has been dissipated, the current and voltages will all be approximately zero. At time t_n,

$$v_L \simeq 0\ \text{V}$$
$$v_R \simeq 0\ \text{V}$$
$$i \simeq 0\ \text{A}$$

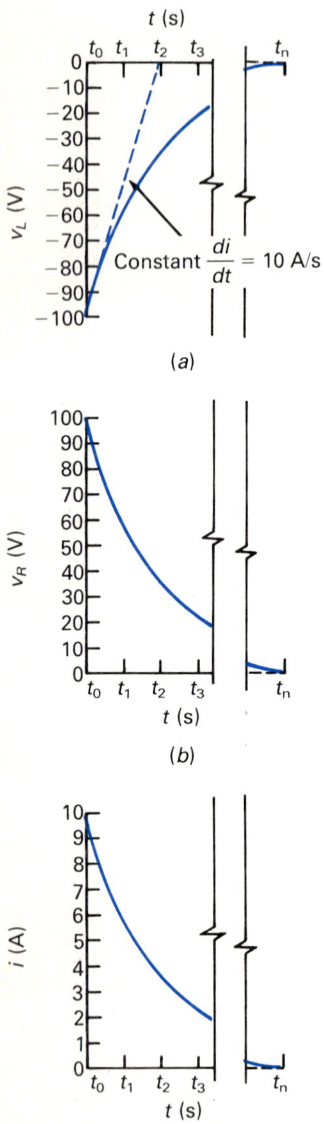

Fig. 11-36. Discharge curves with respect to time for *RL* circuits in Fig. 11-35. *(a)* Inductive voltage; *(b)* resistive voltage; *(c)* current.

RL Circuit Discharge Curves

Figure 11-36 shows how the current and voltages associated with the discharge circuit in Fig. 11-34 vary with respect to time.

Equations for Inductor Discharging Circuit

The loop equation for the *RL* discharge circuit in Fig. 11-34 is

$$v_L = v_R \tag{11-20}$$

or

$$L\frac{di}{dt} - iR = 0$$

An equation for the discharge current can be obtained:

$$i = \frac{V_0}{R}e^{-tR/L} \quad \text{or} \quad I_0e^{-tR/L} \tag{11-21}$$

Substituting the expression for i in Ohm's law,

$$v_R = iR = V_0e^{-tR/L} \tag{11-22}$$

Substituting for v_R in Eq. (11-20) gives

$$v_L = v_R = V_0e^{-tR/L} \tag{11-23}$$

In Eqs. (11-21) to (11-23),

i = instantaneous current, A
v_R = instantaneous resistive voltage, V
v_L = instantaneous inductive voltage, V
V_0 = initial induced emf produced by the collapsing field, V
L = inductance, H
R = resistance, Ω
t = time elapsed, s
e = base of natural logarithms ≃ 2.718

REVIEW QUESTIONS FOR SECTION 11-8

1. The series RL circuit of Question 1, Sec. 11-7, has been completely charged. If it is then discharged through a short circuit, what would be the initial values of i, v_R, and v_L?
2. During the discharge of the RL circuit of Question 1, what is the direction of i, polarity of v_R, and polarity of v_L as compared with when the circuit was charging?
3. State the equations for i, v_R, and v_L for a discharging RL circuit.

11-9 INDUCTIVE TIME CONSTANT

The meaning of the time constant in RL circuits is essentially the same as in RC circuits. The evaluation of τ is obtained by equating the exponent tR/L to unity:

$$\frac{tR}{L} = 1$$

$$t = \frac{L}{R}$$

or

$$\tau = \frac{L}{R} \tag{11-24}$$

where τ = time constant, s
R = resistance, Ω
L = inductance, H

As in the case of a resistance-capacitance circuit, the time constant of a resistance-inductance circuit can be stated in terms of voltage change during charge.

The time constant of a resistance-inductance combination is the time taken for the charging inductor's voltage to drop to 36.79 percent of the source voltage.

Referring back to Figs. 11-32 and 11-36, it can be seen that straight broken lines are shown representing a constant rate of current change of V/R = 100 V/10 Ω = 10 A/s. This rate of change in current is the initial value when charge or discharge commences. The time constant for an RL circuit can be expressed in terms of a linear current change, in a manner similar to that adopted for the RC circuit.

The time constant τ of an RL series circuit is the time it would take for an inductive current to equal the maximum possible value (V/R) during charge, or zero during discharge, if the rate of change in current were maintained constant at its initial value.

REVIEW QUESTIONS FOR SECTION 11-9

1. What is the time constant of an RL circuit?
2. Express the equations of Question 5, Sec. 11-7, in terms of time.
3. Express the equations of Question 3, Sec. 11-8, in terms of time.

11-10 CALCULATIONS FOR *RL* CIRCUITS

Equations (11-17) to (11-19) and Eqs. (11-21) to (11-23) indicate that the shapes of the curves for the *RL* circuit are the same as for the *RC* circuit. Any of the techniques previously applied to *RC* networks can be used for *RL* network problems. The time constant L/R is substituted for RC. The methods include:

1. The time constant ($\tau = L/R$).
2. A universal time-constant table (e.g., Table 11-2). Apply this table as follows:
   a. Inductance charging:
   For v_L, use v_R row.
   For v_R, use v_C row.
   For i, use v_C row.
   b. Inductance discharging:
   For v_L, use v_R row.
   For v_R and i, use as labeled.
3. Universal time-constant curves (e.g., Fig. 11-8).
   a. Inductance charging:
   For v_L, use the $y = e^{-x}$ curve.
   For v_R and i, use the $y = (1 - e^{-x})$ curve.
   b. Inductance discharging:
   For v_L, v_R, and i, use the $y = e^{-x}$ curve.
4. A calculator to solve the equations.

EXERCISE PROBLEMS FOR SECTIONS 11-8, 11-9, AND 11-10

1. In the circuit in Fig. 11-37, switch S is a make-before-break switch. If switch contacts 1 and 2 are closed, allowing the inductor to charge for time t_A, then contacts 2 and 3 are closed, allowing discharge for time t_B. Find i and v_L at the end of time t_B when
   a. $t_A = 0.5$ s, $t_B = 0.5$ s (*ans*. 2.39 A, 23.9 V)
   b. $t_A = 3$ s, $t_B = 1$ s (*ans*. 3.5 A, 35 V)
   c. $t_A = 1$ s, $t_B = 3$ s (*ans*. 0.315 A, 3.15 V)

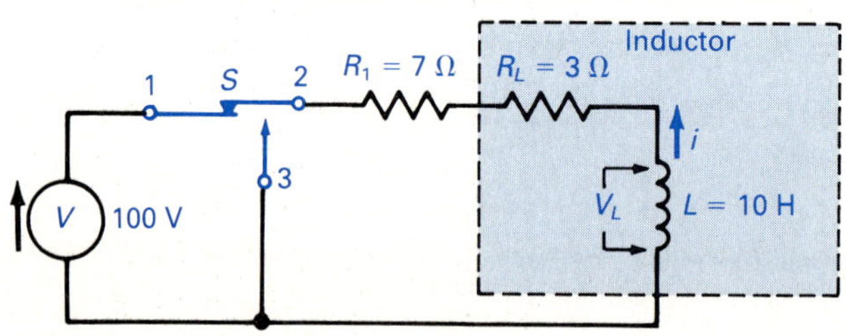

Fig. 11-37. Problem 1: Circuit.

2. Repeat all the conditions in Prob. 1, using Fig. 11-38.

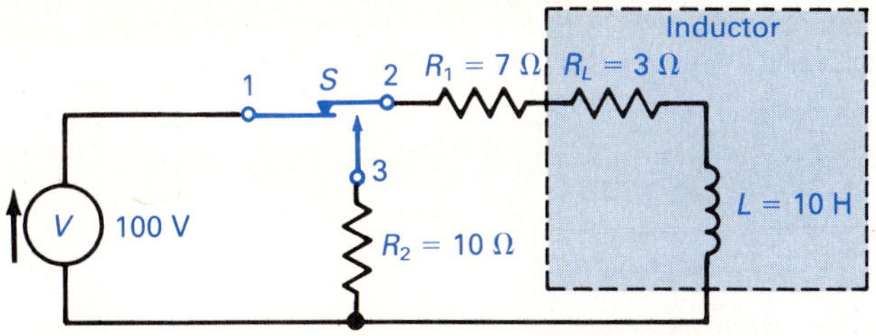

Fig. 11-38. Problem 2: Circuit.

3. Repeat all the conditions in Prob. 1, using Fig. 11-39.
(*ans*. *a*. 3.39 A, 10.16 V; *b*. 7.04 A, 21.11 V; *c*. 2.57 A, 7.71 V)

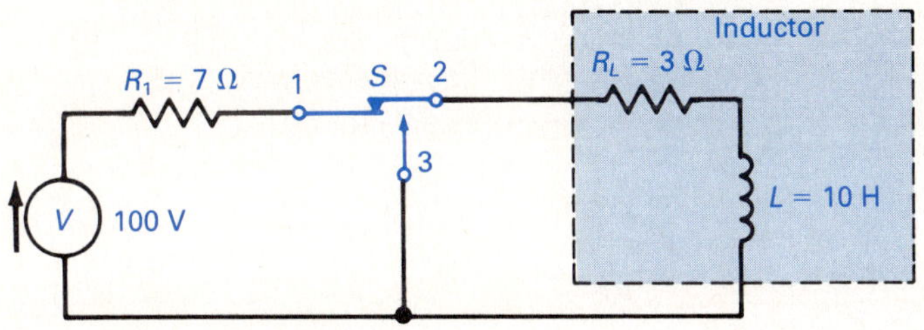

Fig. 11-39. Problem 3: Circuit.

4. A large inductor is to be checked with a 6.6-V battery and an ammeter. Thirty seconds after the battery is connected, the current indicated by the ammeter is 1.2 A. If the final current is 1.65 A, find the inductance of the inductor.
5. If the inductor in Prob. 4 is discharged by placing a short circuit across the winding and then removing the battery, find the time required for the current to reach the accepted zero level. (*ans*. 115.5 s)
6. The battery across the inductor in Prob. 4 is disconnected while the inductor is fully charged, and the current through the ensuing arc falls to zero, at a linear rate, in 5 ms. Find the voltage across the open terminals of the inductor during that period.

11-11 COMPARISON OF CHARGE AND DISCHARGE OF *RL* CIRCUITS

By comparing Figs. 11-32 and 11-36, the following can be seen:

1. *a*. v_L decreases toward zero during charge and discharge.
   b. The polarity of v_L during discharge is opposite that during charge.
2. *a*. v_R increases from zero during charge.
   b. v_R decreases toward zero during discharge.
   c. The polarity of v_R is the same during charge and discharge.
3. *a*. i increases from zero during charge.
   b. i decreases toward zero during discharge.
   c. The direction of i is the same during charge and discharge.

EXAMPLE 11-12 An inductor has a resistance of 5 Ω and an inductance of 1.6 mH. What are the current, resistive voltage, and inductive voltage 1 ms after applying 25 V across the inductor?

The equivalent circuit of the inductor is shown in Fig. 11-40.

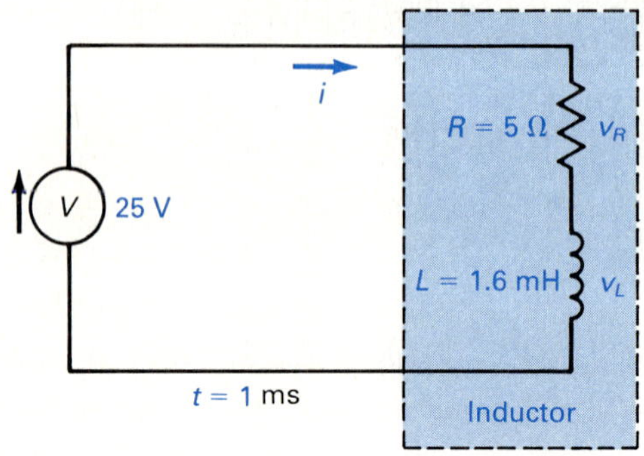

Fig. 11-40. Example 11-12: Equivalent circuit of the inductor.

Solution

$$\tau = \frac{L}{R} \tag{11-24}$$

$$\tau = \frac{1.6 \text{ mH}}{5 \ \Omega} = 0.32 \text{ ms}$$

From Eq. (11-17),

$$i = \frac{V}{R}(1 - e^{-t/\tau}) = \frac{25 \text{ V}}{5 \ \Omega}(1 - e^{-1 \text{ ms}/0.32 \text{ ms}}) = \mathbf{4.78 \text{ A}}$$

$$v_R = iR = 4.78 \text{ A} \times 5 \ \Omega = \mathbf{23.9 \text{ V}}$$

$$v_L = V - v_R = 25 \text{ V} - 23.9 \text{ V} = \mathbf{1.1 \text{ V}}$$

Unlike capacitors, which can be made highly efficient, practical inductors usually have substantial energy losses. These losses are commonly represented as equivalent resistances in series with the inductance. Depending on the application, this resistance can materially affect performance. The time constant of the inductor in Example 11-12 was 0.32 ms. To increase this time constant with the same inductance would mean decreasing the resistance. This solution could be an expensive one, requiring a larger wire size and greater bulk.

EXAMPLE 11-13 A fully charged power-supply filter inductor's resistance is 25 Ω and its inductance is 30 H with a 0.1 A current. What is the current 1 s after a short circuit is placed across the inductor?

Solution If the fully charged inductor carries a current of 0.1 A, the charged conditions will be

$$i = 0.1 \text{ A}$$

$$v_R = iR = 0.1 \text{ A} \times 25 \ \Omega = 2.5 \text{ V}$$

$$v_L = 0 \text{ V}$$

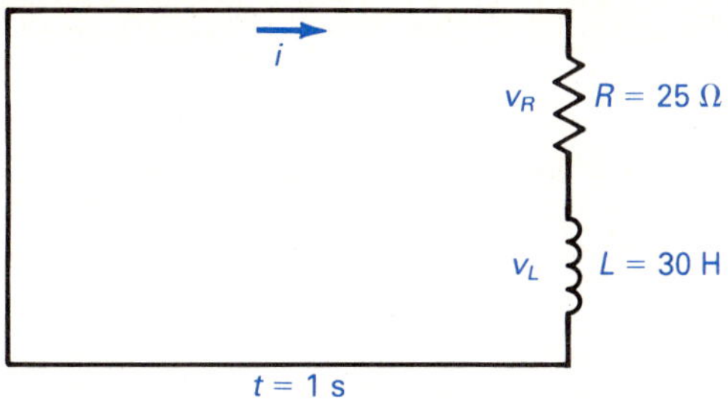

Fig. 11-41. Example 11-13: Schematic diagram of the inductor with a short circuit.

If, while these conditions exist, a short circuit is connected across the inductor, as in Fig. 11-41, the initial inductive voltage will equal the resistive voltage in the charged state. Therefore the initial discharge current will equal the final charge current.

$$I_0 = 0.1 \text{ A}$$

$$\tau = \frac{L}{R} \qquad (11\text{-}24)$$

$$\tau = \frac{30 \text{ H}}{25 \ \Omega} = 1.2 \text{ s}$$

$$t = 1 \text{ s} = \frac{1 \text{ s}}{1.2 \text{ s}} \tau \simeq 0.83\tau$$

Using the universal time-constant curves in Fig. 11-8, the inductive discharge current follows the law $y = e^{-x}$; so with $t \simeq 0.83\tau$,

$$i \simeq 0.44 I_0 \simeq 0.44 \times 0.1 \text{ A} \simeq \mathbf{44 \ mA}$$

EXAMPLE 11-14 A telephone relay has a resistance of 100 Ω and an inductance of 0.75 H. It has a 100-Ω resistor connected in parallel with it. How long will it take for the relay current to become 400 mA *(a)* when 48 V is applied across it and *(b)* if the 48 V is disconnected after the inductance has become fully charged?

Solution (See Fig. 11-42.)

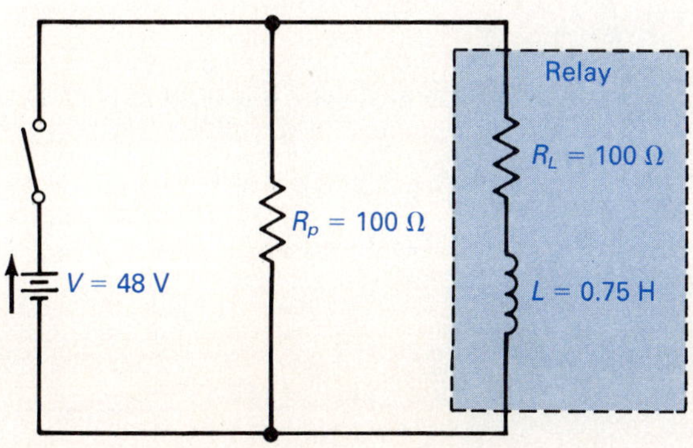

Fig. 11-42. Example 11-14: Circuit.

a. When the switch is closed, 48 V is connected directly across the relay (R_L and L in series), charging the inductance.

The charge time constant is

$$\tau_c = \frac{L}{R_L} \qquad (11\text{-}24)$$

$$\tau_c = \frac{0.75\ \text{H}}{100\ \Omega} = 7.5\ \text{ms}$$

The maximum charge current $I_{max} = 48\ \text{V}/100\ \Omega = 480\ \text{mA}$.

Equation (11-17) relates current and time in an RL circuit. This equation can be put into a form with time as the subject:

$$i = \frac{V}{R}(1 - e^{-tR/L}) \quad \text{or} \quad \frac{V}{R}(1 - e^{-t/\tau}) \qquad (11\text{-}17)$$

$$t = \tau \ln \frac{I_{max}}{I_{max} - i}$$

The time taken for the current to become 400 mA can be found by substituting these values:

$$t = 7.5\ \text{ms} \times \ln \frac{480\ \text{mA}}{480\ \text{mA} - 400\ \text{mA}} = \mathbf{13.44\ ms}$$

b. Assuming that the charge of the inductance is first completed, when the switch is opened the current will drop from the value

$$I_0 = \text{charging } I_{max} = 480\ \text{mA}$$

The circuit through which the inductance discharges consists of R_L and R_P in series. The discharge time constant is, from Eq. (11-24),

$$\tau_d = \frac{L}{R_L + R_P} = \frac{0.75\ \text{H}}{200\ \Omega} = 3.75\ \text{ms}$$

Using the time form of Eq. (11-21) will give the time taken for the relay's current to drop from its initial discharge value of 480 mA to 400 mA.

$$i = \frac{v_0}{R} e^{-tR/L} \quad \text{or} \quad I_{max}\, e^{-tR/L} \qquad (11\text{-}21)$$

$$t = \tau \ln \frac{I_{max}}{i} = 3.75\ \text{ms} \times \ln \frac{480\ \text{mA}}{400\ \text{mA}}$$

$$= \mathbf{0.6837\ ms}$$

SUMMARY OF EQUATIONS

RC circuits

$V = v_C + v_R$	(11-1)	Charge
$i = (V/R)e^{-t/RC}$	(11-2)	

$v_R = Ve^{-t/RC}$	(11-3)	Charge
$v_C = V(1 - e^{-t/RC})$	(11-4)	
$\tau = RC$	(11-5)*	
$i = (V/R)e^{-t/\tau}$	(11-6)	Charge
$v_R = Ve^{-t/\tau}$	(11-7)	
$v_C = V(1 - e^{-t/\tau})$	(11-8)	
$t = \tau \ln I_{max}/i$	(11-9)	
$t = \tau \ln V/v_R$	(11-10)	
$t = \tau \ln V/(V - v_C)$	(11-11)	
$v_C = v_R$ or $v_C - v_R = 0$	(11-12)	Discharge
$i = (V_0/R)e^{-t/RC}$	(11-13)	
$v_R = V_0e^{-t/RC}$	(11-14)	
$v_C = v_R = V_0e^{-t/RC}$	(11-15)	
RL circuits		
$V = v_L + v_R$	(11-16)	Charge
$i = (V/R)(1 - e^{-tR/L})$	(11-17)	
$v_R = iR = V(1 - e^{-tR/L})$	(11-18)	
$v_L = Ve^{-tR/L}$	(11-19)	
$v_L = v_R$	(11-20)	Discharge
$i = (V_0/R)e^{-tR/L}$	(11-21)	
$v_R = iR = V_0e^{-tR/L}$	(11-22)	
$v_L = v_R = V_0e^{-tR/L}$	(11-23)	
$\tau = L/R$	(11-24)*	

*The capacitive and inductive time constants are the same for both charge and discharge.

CHAPTER QUESTIONS

11-1. Why is it that capacitive currents and inductive voltages do not usually remain constant in practical circuits?

11-2. The time taken for *RC* and *RL* circuit voltages and currents to reach specified values can be determined accurately. Why is this so?

11-3. Give two examples of the way in which the relationships between time and other quantities in *RC* and *RL* circuits may be undesirable.

11-4. What could account for the fact that the currents and voltages in *RC* or *RL* circuits may not change at the same time or at the same rate?

11-5. In what way are the straight line and the curve related in Figs. 11-2*a* and 11-17*a*?

11-6. In solving *RC* and *RL* circuit problems, in what circumstances might the universal time-constant curves be of doubtful value?

11-7. What is required of an electronic calculator if it is to be of use in solving *RC* or *RL* circuit problems?

11-8. Compare the behavior of an *RC* circuit during charge and discharge with respect to

a. Current direction and change of magnitude.
b. Capacitor voltage polarity and change of magnitude.
c. Resistor voltage polarity and change of magnitude.

11-9. Compare the behavior of an *RL* circuit during charge and discharge
a. With respect to current direction and change of magnitude.
b. With respect to inductor voltage polarity and change of magnitude.
c. With respect to resistor voltage polarity and change of magnitude.

11-10. Why can it be helpful to use a Thevenin approach to *RC* and *RL* circuit problem solutions?

11-11. What is the main difference between the characteristics of *RC* and *RL* circuits when they are opened while charged?

CHAPTER PROBLEMS

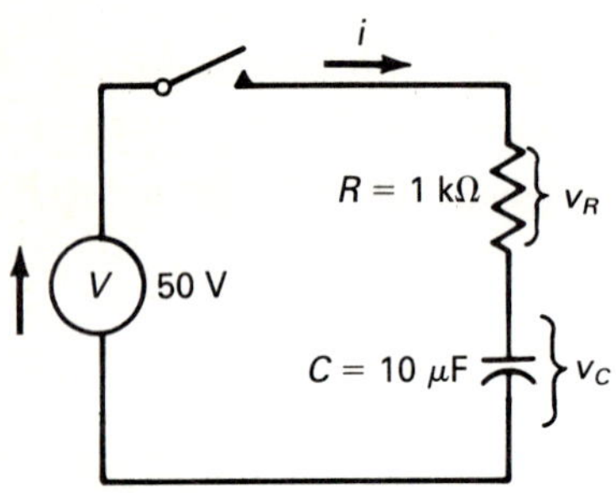

Fig. 11-43. Problem 11-1: Circuit.

11-1. Refer to Fig. 11-43. What are the values of current, capacitor voltage, and resistor voltage 20 ms after the switch is closed?
(*ans.* 6.77 mA, 6.77 V, 43.23 V)

11-2. Refer to Fig. 11-44. What are the values of current, capacitor voltage, and resistor voltage 50 ms after the switch is closed?

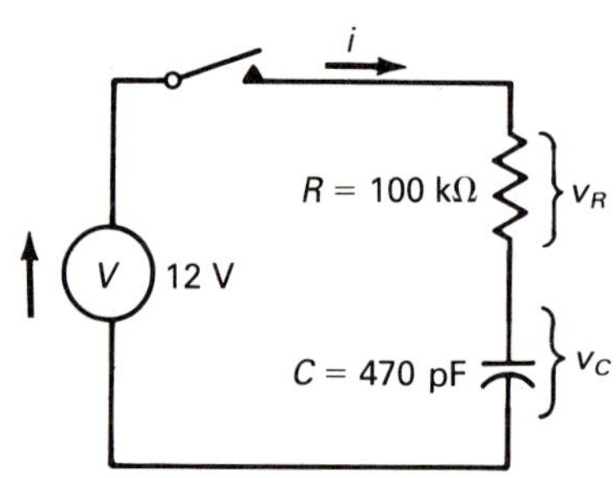

Fig. 11-44. Problem 11-2: Circuit.

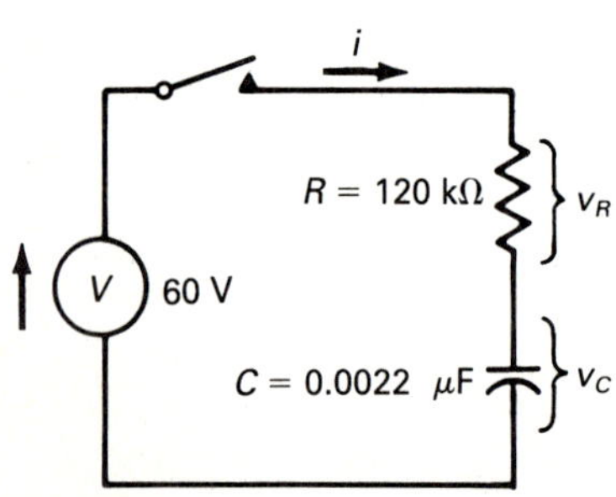

Fig. 11-45. Problem 11-3: Circuit.

11-3. Refer to Fig. 11-45. How long after the switch is closed does it take for the resistor voltage to become 6 V? (*ans.* 607.9 μs)

11-4. Refer to Fig. 11-46. How long after the switch is closed does it take for the current to become 15 mA?

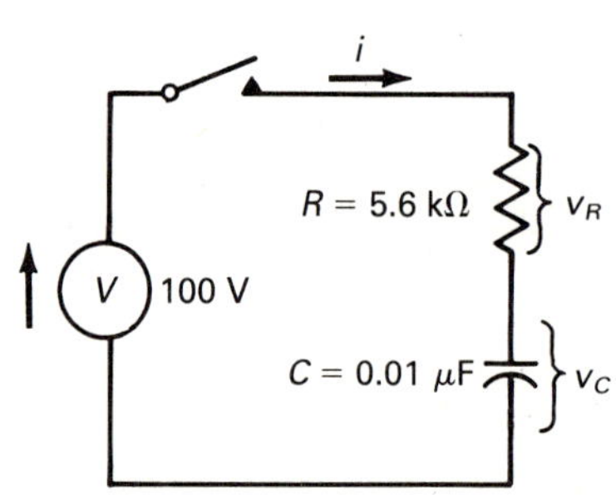

Fig. 11-46. Problem 11-4: Circuit.

11-5. What is the voltage across a 1000-pF capacitor, initially charged to 20 V, when it is discharged through a 100-kΩ resistor for 75 μs?
(*ans.* 9.447 V)

11-6. What is the time needed for a 1000-pF capacitor, initially charged to 20 V, to discharge to 5 V through a 100-kΩ resistor?

11-7. Refer to Fig. 11-47. With the capacitor initially uncharged, the switch is moved to position *A* and held there for 5 ms. It is then moved to position *B* for 10 ms. What are the final resistor and capacitor voltages? (*ans.* −2.008 V, 1.303 V)

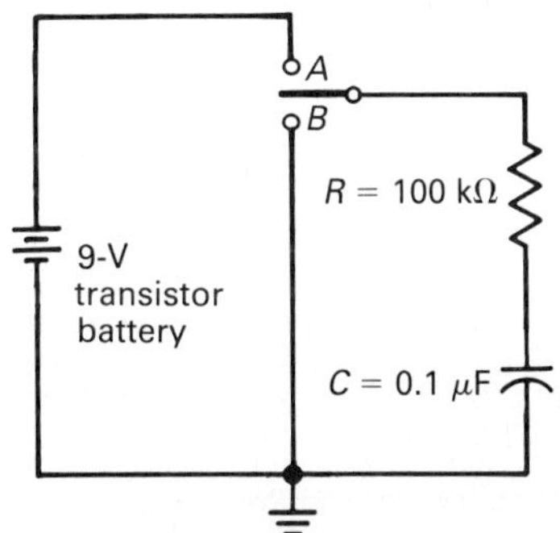

Fig. 11-47. Problems 11-7 and 11-8: Circuit.

11-8. If a 100-kΩ resistor is connected between point *B* and ground in the circuit of Fig. 11-47, but the other conditions remain the same as those of Prob. 11-7, what will then be the voltages?

11-9. Refer to Fig. 11-48. The switch is closed for 5 ms and then opened. What are the resistor and capacitor voltages 10 ms after the switch opens? (*ans.* 0 V, 2.093 V, 2.093 V)

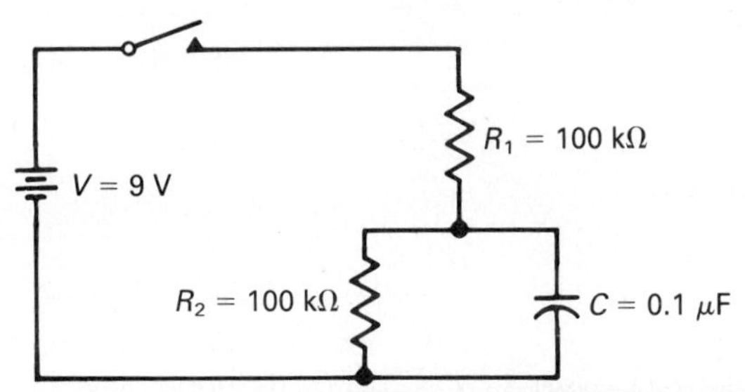

Fig. 11-48. Problem 11-9: Circuit.

11-10. Figure 11-49 shows the main items affecting the input characteristics of an oscilloscope (an electronic measuring instrument). What is the overall rise time of this circuit, the rise time being the time taken for the amplifier voltage to rise from 10 to 90 percent of the level of a voltage pulse that rises instantaneously?

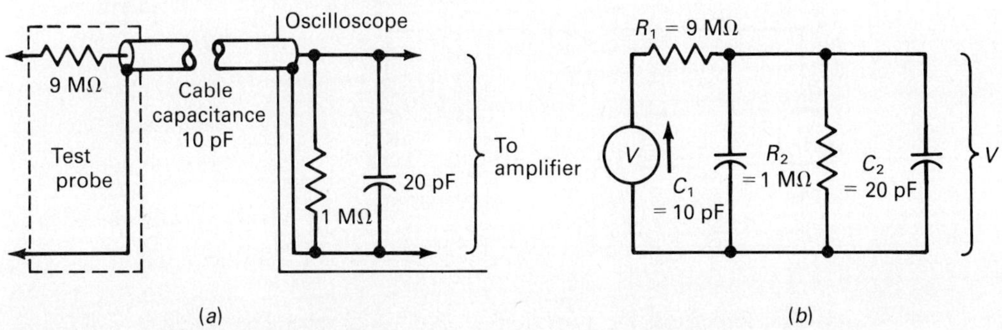

Fig. 11-49. Problem 11-10: Circuit.

11-11. A square-wave generator circuit is shown in Fig. 11-50. The frequency f of the output is related to the time constant of resistor R and capacitor C such that $f = \frac{1}{2}\tau$. If the resistor has a value of 120 kΩ, what approximate range of capacitor values is needed for the generator to cover the frequency range 20 Hz to 20 kHz (the unit of frequency is the hertz, Hz)? (*ans.* 200 pF to 0.2 μF)

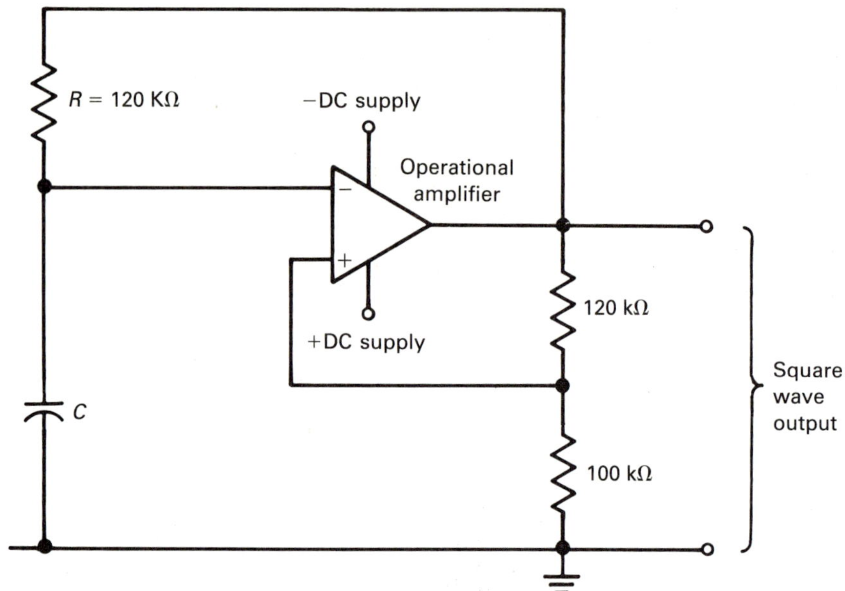

Fig. 11-50. Problem 11-11: Circuit.

11-12. Using Fig. 11-51 and starting with the capacitor completely discharged, the switch is first placed to position A and the capacitor charges for time t_A. The switch is then moved to position B, and the capacitor discharges for time t_B. Find the capacitor voltage v_C at the end of time t_B when

a. $t_A = 1$ ms, $t_B = 1$ ms
b. $t_A = 4$ ms, $t_B = 2$ ms
c. $t_A = 2$ ms, $t_B = 4$ ms

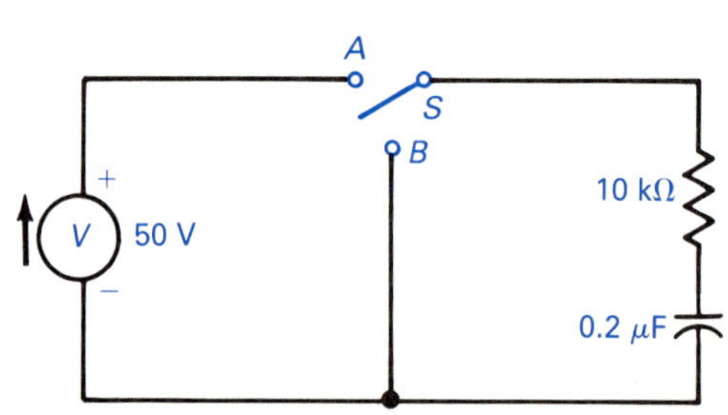

Fig. 11-51. Problem 11-12: Circuit.

11-13. Repeat all the conditions in Prob. 12, using Fig. 11-52.
(*ans. a.* 15.32 V; *b.* 26.22 V; *c.* 11.63 V)

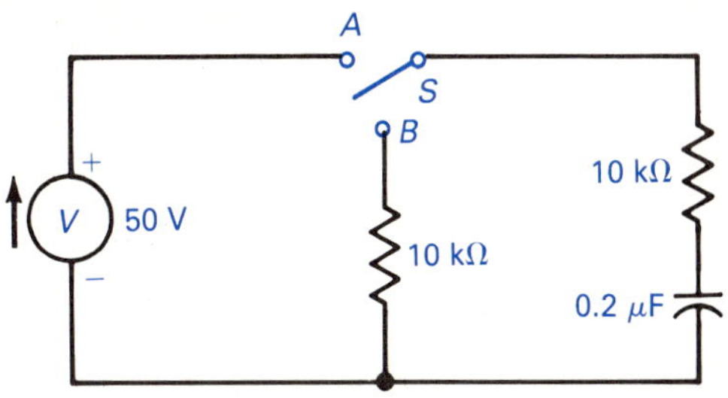

Fig. 11-52. Problem 11-13: Circuit.

11-14. Repeat all the conditions in Prob. 1, using Fig. 11-53.

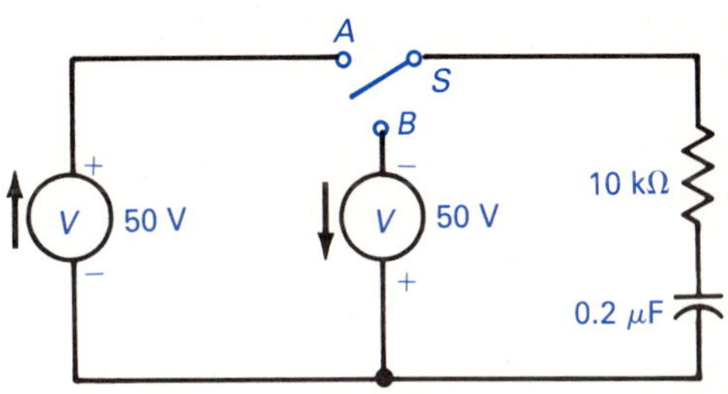

Fig. 11-53. Problem 11-14: Circuit.

11-15. In the circuit in Fig. 11-54, find the time constant of the equivalent *RC* circuit, the rise time (10 to 90 percent), and the final v_C for each of the following cases:

a. $R_1 = 1000\ \Omega$, $R_2 = 1000\ \Omega$ (*ans.* 5 μs, 11 μs, 50 V)
b. $R_1 = 5\ \text{k}\Omega$, $R_2 = 500\ \Omega$ (*ans.* 4.545 μs, 10 μs, 9.09 V)
c. $R_1 = 500\ \Omega$, $R_2 = 5\ \text{k}\Omega$ (*ans.* 4.545 μs, 10 μs, 90.9 V)

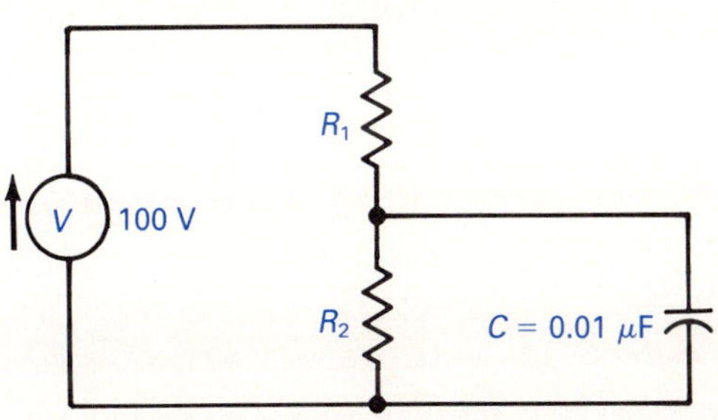

Fig. 11-54. Problem 11-15: Circuit.

11-16. Figure 11-55 represents the input of an oscilloscope fed from a 50-Ω pulse generator *V*. Find the time constant of the equivalent *RC* circuit and the rise time.

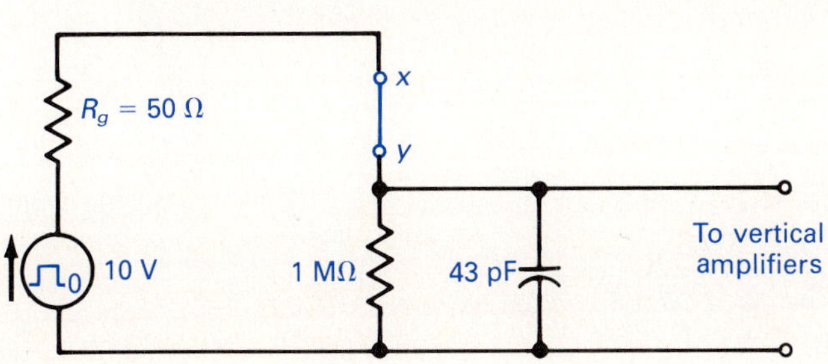

Fig. 11-55. Problems 11-16 and 11-17: Circuit.

11-17. In Prob. 16, if a resistor of 1 MΩ is placed between points *x* and *y* to reduce the signal to the oscilloscope input by 50 percent, find

a. The new time constant of the circuit (*ans*. 21.5 μs)

b. The new rise time. (*ans*. 47.3 μs)

c. The final voltage to the input of the vertical amplifiers. (*ans*. 5 V)

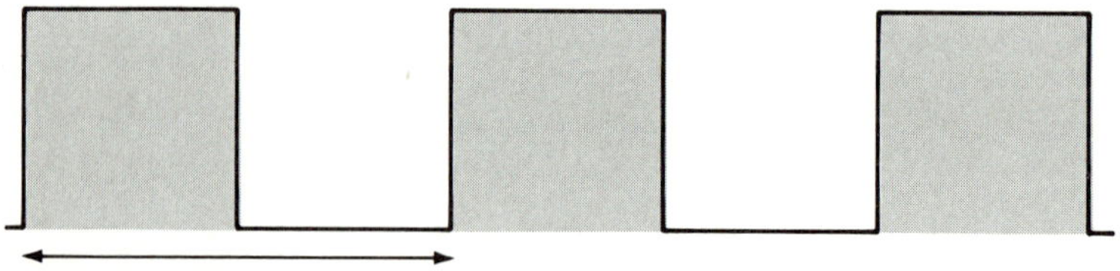

Time $t = 1/f$

At a frequency of 20 Hz, $1/f = 50$ ms

At a frequency of 20 kHz, $1/f = 50$ μs

Fig. 11-56. Problem 11-18.

11-18. Figure 11-56 shows the square voltage output from the generator of Fig. 11-50. The output of the circuit of Fig. 11-49 is the voltage fed to the oscilloscope amplifier and is therefore the waveform seen on the scope screen. Sketch the waveform seen on the screen when the square-wave generator is connected to the oscilloscope through the circuit of Fig. 11-49. Consider the generator to operate at 20 Hz and 20 kHz ($1/f = 50$ ms and 50 μs).

12

ALTERNATING CURRENT AND VOLTAGE

AC quantities (alternating voltage, current, and power) are time-varying; their magnitudes change as time progresses.

Except in Chap. 11, we have previously discussed only circuits in which voltages and currents did not vary; they were steady-state. In Chap. 11 too, the circuits were supplied from unvarying dc sources, although the internal circuit voltages and currents did change with respect to time. The time-varying effects made it necessary to adopt approaches different from those used for steady-state dc. The approaches need to be different again where, as in ac circuits, the voltage and current sources are continuously variable.

As in the case of dc transients, all the basic laws and principles developed for steady dc conditions apply to ac circuits, but the methods of applying them can be different.

In this chapter we will see how ac differs from dc and meet some language special to ac situations. The following chapters cover principles and techniques that refer particularly to ac circuits.

12-1 DC AND AC VOLTAGE AND CURRENT

''*DC*'' and ''*ac*'' are electrical terms meaning ''*direct*'' and ''*alternating*.'' They may refer to voltage, current, or power. This means that, although ''dc'' literally stands for ''direct current,'' it is more usual to say ''dc current.'' In the electrical and electronics industry, *ac current* is understood to mean *alternating current* and *ac voltage* to mean *alternating voltage*.

A dc voltage always has the same polarity and a dc current always has the same direction. They do not always have the same magnitude or value. On a time-related basis, dc can be shown by waveforms such as those in Fig. 12-1.

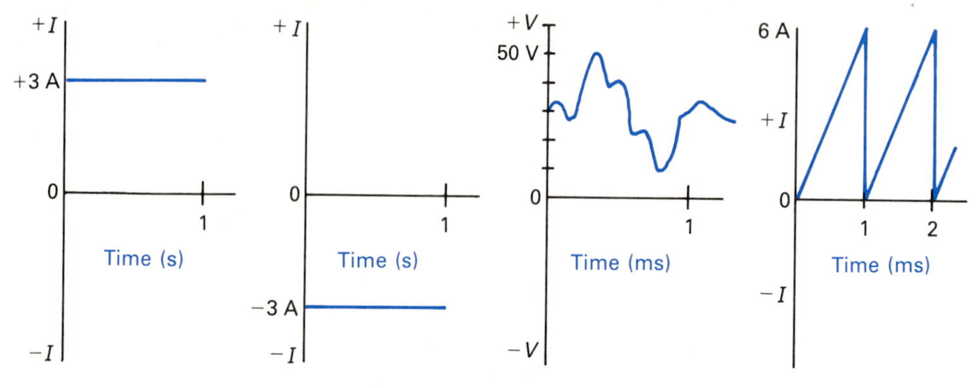

(a) (b) (c) (d)

Fig. 12-1. Waveforms: *(a)* and *(b)* Steady direct currents of opposite direction; *(c)* randomly varying direct voltage; *(d)* sawtooth direct current.

Figure 12-1*a* shows a steady dc current. Its value is constant at 3 A throughout the 1-s period indicated on the horizontal *time* axis.

Figure 12-1*b* also shows a steady 3-A direct current. Its value is shown as negative, while that in Fig. 12-1*a* is shown as positive. This change means that the currents represented by the two curves are of the same magnitude but have opposite directions. There is no fundamental meaning behind a positive current; it is an arbitrary choice. If, in a particular circuit situation, it is decided to call clockwise currents positive, then counterclockwise currents will be negative.

The Fig. 12-1*c* voltage always has the same polarity, because it is shown as positive for the whole 1-ms period. But its value starts at 30 V and then varies between 10 and 50 V in an apparently random manner. This too is a dc voltage.

The waveform in Fig. 12-1*d* depicts a dc current because, although its minimum value is zero, it never changes direction. It is shown as positive, rising in a linear manner from 0 to 6 A and then dropping back to 0 A. It repeats this movement at regular time intervals of 1 ms, thus constituting a periodic dc waveform.

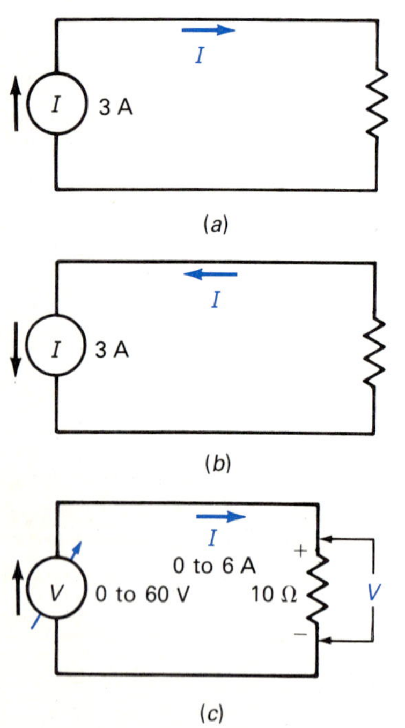

Fig. 12-2. Circuits for producing the dc waveforms in Fig. 12-1. *(a)* Steady 3-A clockwise current; *(b)* steady 3-A counterclockwise current; *(c)* varying direct voltage or current.

The circuits in Fig. 12-2 cover all the situations shown in Fig. 12-1.

In Fig. 12-2*a* a constant-current source produces a clockwise current of 3 A, as in Fig. 12-1*a*.

In Fig. 12-2*b* a constant-current source causes a 3-A current in a counterclockwise direction. Relative to Fig. 12-2*a* this is a negative current. If Fig. 12-1*a* represents a clockwise current, Fig. 12-1*b* represents a counterclockwise current.

The variable voltage source in Fig. 12-2*c* can result in either of the waveforms in Fig. 12-1*c* or *d*. The voltage applied to the resistance could vary in a random manner, as in Fig. 12-1*c*, or it could be controlled in a sawtooth manner to result in the current waveform in Fig. 12-1*d*.

Ac can be *periodic*, where alternations occur at regular time intervals, or *random*, where alternations do not occur at regular time intervals. Only periodic ac is discussed in this text, as it is of the greatest practical importance.

The polarity of a periodic ac voltage reverses at regular time intervals. The direction of a periodic ac current reverses at regular time intervals. Between alternations (reversals), the magnitude of an ac voltage or current can be constant or varying.

Figure 12-3 shows some typical ac waveforms.

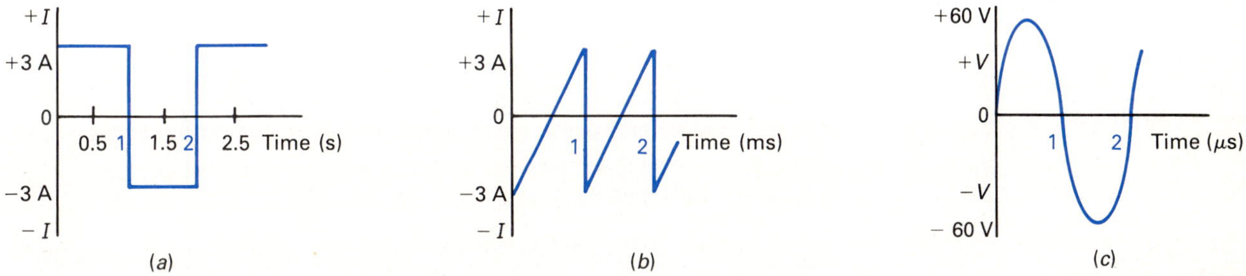

Fig. 12-3. Waveforms: *(a)* Square-wave alternating current; *(b)* sawtooth alternating current; *(c)* sine-wave alternating voltage.

Figure 12-3*a* shows a square-wave alternating current. Its magnitude stays constant at +3 A for 1 s, suddenly changes to −3 A, maintains this value for 1 s, and then changes back to +3 A. Thus this waveform represents a current which changes its direction at 1-s intervals but, between these reversals, has a constant magnitude of 3 A. The 1-s constant values of +3 A and −3 A, taken individually, are the same as the direct currents in Fig. 12-1*a* and *b*.

The sawtooth waveform in Fig. 12-3*b* represents a current which is the ac equivalent of the dc sawtooth in Fig. 12-1*d*. They both refer to currents which rise in a linear manner through a range of 6 A and then revert to the initial value. The ac sawtooth current starts at 3 A in one direction, goes to 0 A, and then continues to 3 A in the opposite direction.

Figure 12-3*c* shows a sine-wave ac voltage. This waveform is of such special interest in electrical and electronic technology that almost all ac examples and problems in this text refer to sine waves. The sine wave is treated in detail in Secs. 12-4 to 12-6.

The waveforms in Fig. 12-3 can be related to the respective circuits in Fig. 12-4.

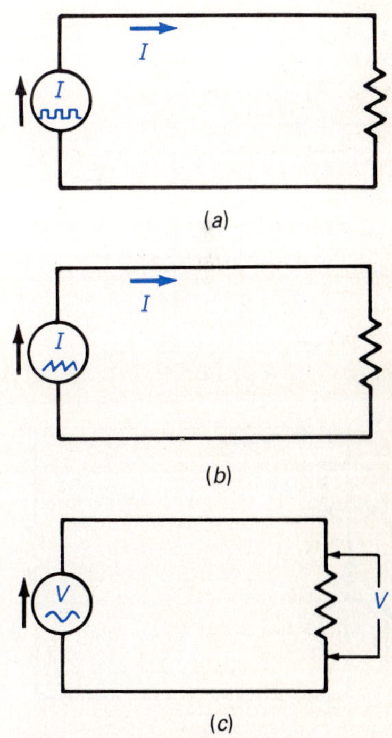

Fig. 12-4. Circuits for producing the ac waveforms in Fig. 12-3. *(a)* Square-wave current; *(b)* sawtooth current; *(c)* sine-wave voltage.

Source Symbols

There appear to be no generally accepted standard symbols for energy sources. As for the dc sources in previous chapters, we adopt simple, clear symbols for ac sources. Current sources are shown in Fig. 12-4 as circles labeled *I*. Voltage sources are shown as circles labeled *V*. Both symbols show the type of waveform supplied by the source.

Arrows on AC Diagrams

Not only do ac voltage polarities and current directions change individually at regular time intervals, but, if inductance or capacitance is involved, the polarities and directions of the various voltages and currents in an ac circuit change with respect to each other. Therefore arrows on ac circuit diagrams do not necessarily represent simple directional relationships as they do on dc circuit diagrams. In general, an arrow on an ac circuit diagram indicates only the *presence* of a current or a voltage, not its direction or its polarity.

Certain conditions in ac circuits are sometimes indicated by arrows on diagrams. We shall show later that it can be convenient to use arrows of opposite direction in each branch of a diagram for an ac parallel inductance-capacitance circuit. These arrows indicate the fact that, at a certain time instant, the currents in each branch are of opposite direction.

No values of voltage or current are given in the circuit diagrams in Fig. 12-4. This omission is deliberate, as the magnitude values for ac quantities have to be stated in special terms because of their varying nature.

REVIEW QUESTIONS FOR SECTION 12-1

1. What is meant by ''time-varying quantity''?
2. A current waveform starts at zero, rises to 6 A, falls to zero, and then repeats this sequence at regular time intervals. Is this waveform ac?
3. What symbol is used in this book to represent a sine-wave ac generator?

12-2 WAVEFORM VALUES

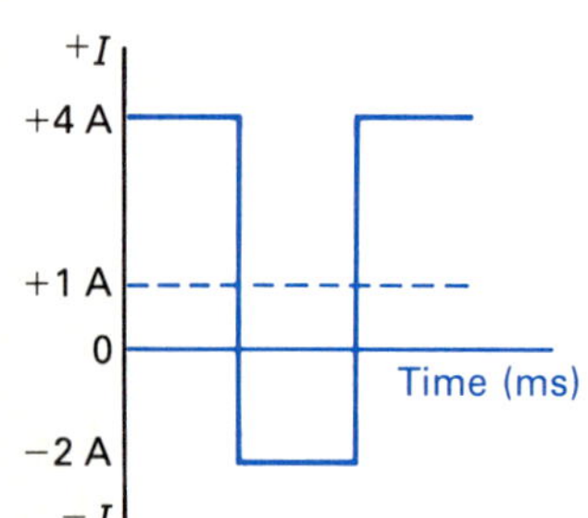

(*a*)

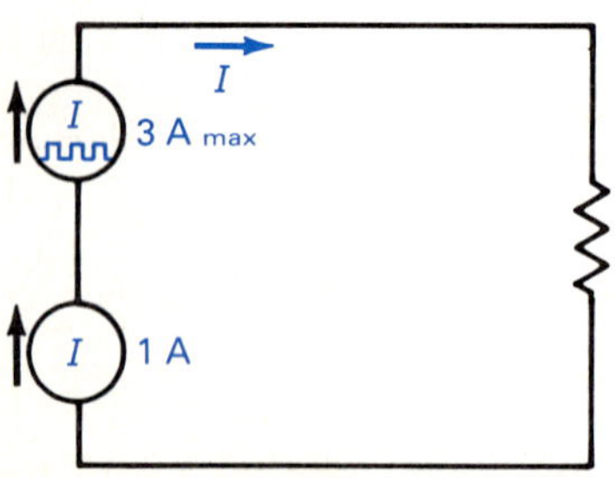

(*b*)

Fig. 12-5. *(a)* Asymmetrical square-wave current; *(b)* circuit for producing the asymmetrical square-wave current.

Waveform shape influences the effects produced by a current or voltage in a given circuit. Measurements are also influenced by waveshape. To satisfy the demands of different functional and measuring needs, current and voltage magnitudes can be specified in terms of one or another of several values, each with its own area of application.

Maximum or Peak

The *maximum* or *peak value* of an ac current or voltage is the maximum *magnitude* reached during a positive or negative excursion. If the waveform is symmetrical, as in Fig. 12-3, the positive and negative peak values are the same. The subscript ''max'' is used to denote a maximum or peak value.

Referring to Fig. 12-3*a* and *b*,

$$I_{\max} = 3 \text{ A}$$

Referring to Fig. 12-3*c*,

$$V_{\max} = 60 \text{ V}$$

Figure 12-5*a* shows a square-wave current which has the same overall 6-A swing as that in Fig. 12-3*a*.

The waveform in Fig. 12-5*a* is asymmetrical, the negative and positive peaks being of different magnitude. In such cases it is necessary to name the peaks referred to:

Positive peak: $I_{max} = 4$ A
Negative peak: $I_{max} = 2$ A

This square-wave current can be considered to be made up of two components: a symmetrical ac square wave of 3 A maximum value superimposed on a 1-A steady direct current. From this viewpoint, the current swings 3 A positive and 3 A negative with respect to +1 A. The current is said to have a + 1-A dc level. The circuit in Fig. 12-5*b* shows this dc level as a 1-A dc source in series with a square-wave ac source of 3 A maximum value.

The minimum degree of insulation needed for the protection of wiring, components, equipment, and personnel is determined by the peak value of ac voltage applied to the circuit, since at least this value will be present at certain instants of time.

A peak voltage value may be measured by displaying the waveform on a calibrated oscilloscope and then measuring the amount by which the waveform peak is displaced from the horizontal zero axis. The positive and negative peak values can be individually measured in this manner.

A peak current value can be measured by passing the current through a known resistance, thus developing a voltage of the same waveform as the current and of magnitude proportional to that of the current. Alternatively a current probe, a special transformer, can be used to produce the voltage. The current-derived voltage obtained by either method is applied to an oscilloscope for measurement and interpretation in terms of the known current voltage relationship.

Peak-to-Peak

The magnitude of an ac waveform taken from its most positive value to its most negative value is called the *peak-to-peak value*. The subscript "p-p" is used to identify peak-to-peak values.

In Fig. 12-3*a* and *b* the peak-to-peak value of current is

$$I_{\text{p-p}} = 6 \text{ A} \qquad \text{from } +3 \text{ to } -3 \text{ A}$$

In Fig. 12-3*c* the peak-to-peak value of the voltage is

$$V_{\text{p-p}} = 120 \text{ V} \qquad \text{from } +60 \text{ to } -60 \text{ V}$$

The peak-to-peak value of the waveform in Fig. 12-5*a* is the same as that in Fig. 12-3*a* and *b*:

$$I_{\text{p-p}} = 6 \text{ A}$$

In the case of Fig. 12-5*a*, the 6 A is from +4 to −2 A.

The concept of peak-to-peak value can be usefully applied to a varying dc waveform by finding the peak-to-peak value of its ac component. The ac component of a varying dc current or voltage is the actual variation in magnitude.

With reference to the random dc voltage in Fig. 12-1*c*, it can be seen that a 40-V variation takes place between +50 and +10 V. This waveform has a 40-$V_{\text{p-p}}$ ac component.

The sawtooth current in Fig. 12-1*d* is direct current and has the same peak-to-peak value as the ac currents in Figs. 12-3 and 12-5,

$$I_{\text{p-p}} = 6 \text{ A}$$

Although it is only of academic interest, perhaps it should be mentioned that steady direct current, as in Fig. 12-1*a* and *b*, has a zero peak-to-peak value.

Peak-to-peak values take no account of dc levels. Thus these values should not be stated when dc levels are important.

A peak-to-peak value of voltage or current is most readily measured with an oscilloscope. The peak-to-peak value is determined by measuring the displacement between the most positive and most negative peaks of the display and relating this displacement to the oscilloscope vertical range setting.

The ease with which peak-to-peak values associated with complex waveforms can be measured with an oscilloscope is the common justification for the extensive use of these values in the electronics industry.

Instantaneous

For any waveform an *instantaneous value* is a value of magnitude at a specified instant. The term ''specified instant'' means a selected point on the horizontal axis of the relevant waveform curve. Usually this axis represents time, but it may represent angle or distance. Figures 12-1, 12-3, and 12-5 all have time as the horizontal axis variable.

It is usual to imply that a specified instant occurs at the stated time after a reference time t_0. Time t_0 can have any of several meanings, such as:

1. The time a switch is closed.
2. The time observation begins.
3. The curve origin. This refers to a graphical presentation of the waveform, as in Figs. 12-1, 12-3, and 12-5. The origin is the point at which the horizontal and vertical axes intercept, the common zero point.
4. The time at which the waveform has zero magnitude. When this meaning is used, it is important to specify which zero is meant. For example, in Fig. 12-3*c*, at times $t = 0\ \mu s$ and $t = 2\ \mu s$, the voltage is zero swinging positive, while at time $t = 1\ \mu s$ the voltage is zero swinging negative.

Instantaneous values are given in lowercase symbols. Referring to Fig. 12-3*a*, and taking time t_0 as the curve origin:
At time $t = 0.5\ \mu s$,

$$i = +3 \text{ A}$$

At time $t = 1.5\ \mu s$,

$$i = -3 \text{ A}$$

At time $t = 1\ \mu s$,

$$i = \pm 3 \text{ A}$$

The last example needs explanation. If the current changes from +3 to −3 A at a given time instant, then the magnitude must have both these values at that time. Although it is not physically possible for the current to change value in zero time, it can change rapidly enough for it to be considered so. For example, if the current actually takes 1 ns to change from +3 to −3 A, this time is virtually zero when compared with the 1 s between changes. If it is desired to specify an

instantaneous value immediately prior to or after a rapid change, it can be shown as:
At the instant immediately prior to time $t = 1$ s,

$$i_{1s-} = +3\ \text{A}$$

At the instant immediately after time $t = 1$ s,

$$i_{1s+} = -3\ \text{A}$$

Average

The *average value* of an alternating current or voltage is the mean magnitude of the positive or negative part of its waveform. If the waveform is symmetrical, the positive and negative parts will be the same and their mean magnitudes will be equal. An asymmetrical waveform will have unequal positive and negative parts, and thus either the positive average value or the negative average value should be referred to.

It is important to distinguish between the average (practical) and the true average (theoretical) values.

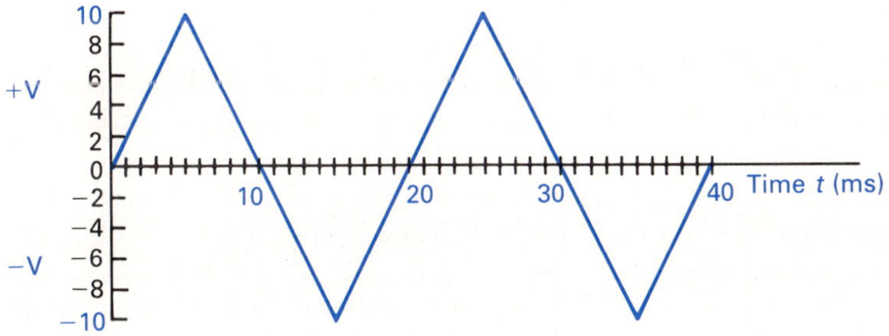

Fig. 12-6. Triangular-wave voltage.

The waveform in Fig. 12-6 is that of a symmetrical triangular voltage. During any 20-ms time interval the voltage magnitude varies in a linear manner between +10 V and −10 V. In determining the average value over a 20-ms period, all positive instantaneous magnitudes are in effect canceled by equal and opposite negative instantaneous magnitudes, thus giving a true average value of 0 V. This zero result would be obtained for any symmetrical voltage waveform averaged over an interval between instants when the magnitudes have the same value and direction of change (swinging positive or swinging negative). Therefore the true average value is of no practical use in specifying the relative magnitudes of symmetrical ac voltages and currents.

Again with reference to the symmetrical voltage in Fig. 12-6, the mean magnitude over a time interval during which the voltage is either always positive or always negative is related to the peak value. Hence this interpretation of average value is of use in comparing the magnitudes of waveforms which have the same shape but differ in peak value. Over the 10-ms intervals between zero values, for instance, from $t = 0$ to $t = 10$ ms, the average magnitude is +5 V. This 5 V is the average value of the waveform as recognized in practical electrical work. The subscript "av" is used for average value.

Referring to Fig. 12-3*a*,

$$I_{av} = 3 \text{ A}$$

Referring to Fig. 12-3*b*,

$$I_{av} = 1.5 \text{ A}$$

The average value of a sine wave, as in Fig. 12-3*c*, is discussed in Sec. 12-6.

Almost all meters used for the measurement of current and voltage are based on a *moving-coil* meter movement. This type of movement requires direct current if it is to result in a meaningful indication. It indicates zero with ordinary symmetrical alternating current. The moving-coil movement is fitted with a rectifier to adapt it for measuring alternating current and voltage. The rectifier causes the movement current to be of constant direction. When the movement is fitted with a rectifier, the deflection of its pointer is proportional to the practical average value of the ac current or voltage.

Although ac meters are usually calibrated in rms values, the fact that the moving-coil type is the most popular makes measurement the most common application of ac average values.

Root Mean Square (RMS)

The rms value of an ac current or voltage is the value most widely used in practice. It is the value used to equate all waveforms, ac, steady dc, or varying dc. It relates waveforms of current or voltage on a power or energy basis. As it is concerned with the effect of energy transfer, such as occurs when heat is produced in resistance, rms value is also called *effective value. Unless otherwise stated, ac currents or voltages are understood to be given in rms values.*

The *rms value* of an ac current or voltage is that value which is equivalent, from a power or energy viewpoint, to steady direct current of the same magnitude.

If a steady direct current of 1 A will light a lamp, then an alternating current of 1 A rms will light the lamp to the same brilliance.

The concept of rms is arrived at as follows:

1. Power is expressed in terms of current or voltage for a given resistance value:

$$P = I^2R \qquad \text{or} \qquad P = \frac{V^2}{R}$$

Power is proportional to the *square* of either current or voltage.

2. In steady dc terms the power is constant, but where the current and voltage vary, the instantaneous power changes too. The steady equivalent of a varying power is represented by its average or *mean* value. When ac power is expressed in current or voltage terms, rms values are implied:

$$P_{av} = I_{rms}^2 R \qquad \text{or} \qquad \frac{V_{rms}^2}{R}$$

$$I_{rms}^2 = \frac{P_{av}}{R} \qquad \text{or} \qquad V_{rms}^2 = P_{av}R$$

3. To obtain the rms value of current or voltage, a square *root* operation must be performed:

$$I_{\text{rms}} = \sqrt{\frac{P_{\text{av}}}{R}} \qquad \text{or} \qquad V_{\text{rms}} = \sqrt{P_{\text{av}}R}$$

Items 1 to 3 refer to square, mean, and root. The term rms therefore aptly describes the overall operation of taking the square *root* of the *mean* value of the current or voltage *squared*.

In Fig. 12-3*a* the current is constant at +3 A for 1 s. The square of +3 A constant is $+9\ \text{A}^2$, also constant. For the next 1 s, the current is constant at −3 A. The square of −3 A is $+9\ \text{A}^2$ too. In this example, the square of the current is always $+9\ \text{A}^2$, so the mean square of the current is $9\ \text{A}^2$. The square root of $9\ \text{A}^2$ is 3 A. Thus the rms value of a square wave is the same as its peak value, because the magnitude is constant. The fact that $(-3\ \text{A})^2$ is the same as $(+3\ \text{A})^2$ is not just mathematically correct, since it is logical that power is dissipated equally well by current in either direction (or voltage of either polarity).

Other waveforms do not share the simplicity of the square wave in the derivation of their rms values. Even the apparently simple sawtooth wave in Fig. 12-3*b* requires a calculus solution to obtain its rms value. The principle does not change, however; the rms value is the root of the mean of the squares of all the instantaneous values.

The rms values of sine-wave currents and voltages are dealt with in Sec. 12-6.

As ac voltage and current magnitudes are usually stated in rms terms, meters are usually required to indicate rms values. Most meters actually measure the average value of an ac waveform. Therefore the meter dial must be calibrated to register in rms values. The relationship between rms and average values is a function of the waveshape, thus making the measurements reliable only for the waveshape used in calibration. Most commercial ac meters are calibrated to indicate rms values of sine-wave currents and voltages, although meters which register the rms values of current or voltage waveforms of any shape are available.

EXAMPLE 12-1 For the current waveform in Fig. 12-7, what are *(a)* the positive peak value, *(b)* the negative peak value, *(c)* the rms value, and *(d)* the instantaneous value 2.5 ns after time t_0?

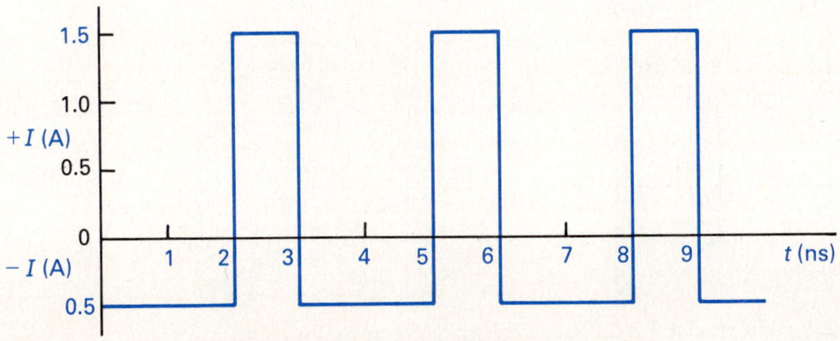

Fig. 12-7. Example 12-1: Current waveform.

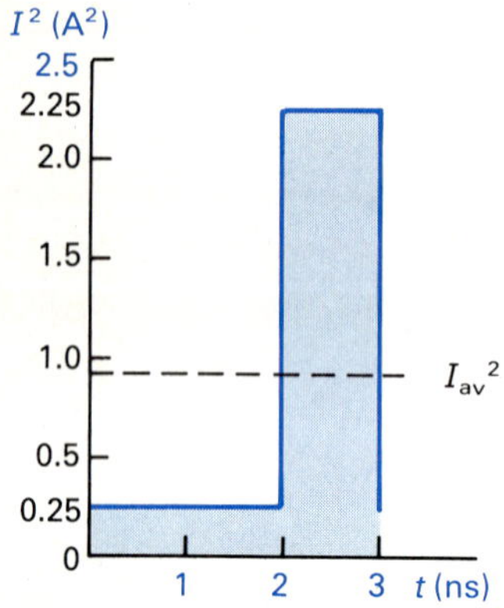

Fig. 12-8. Portion of the waveform in Fig. 12-7 with all instantaneous current magnitudes squared.

Solution Figure 12-7 shows a rectangular wave which is asymmetrical about the horizontal axis. Therefore the positive and negative peaks must be dealt with separately.

a. Positive I_{max} = **1.5 A**

b. Negative I_{max} = **0.5 A**

c. Mean of positive values squared $= (1.5\text{ A})^2 = 2.25\text{ A}^2$

Mean of negative values squared $= (-0.5\text{ A})^2 = 0.25\text{ A}^2$

Figure 12-8 shows the graph of I^2 for a 3-ns period (a complete alternation). The mean magnitude of this I^2 curve will be that value obtained by taking the total area between the curve and the horizontal axis and spreading it out evenly over the whole 3-ns time occupied by the cycle.

$$\text{Total area under curve} = 0.25\text{ A}^2 \times 2\text{ ns} + 2.25\text{ A}^2 \times 1\text{ ns}$$

$$\text{Total area} = 2.75\text{ A}^2 \cdot \text{ns}$$

$$\text{Average magnitude} = \frac{\text{average area}}{\text{total time}}$$

$$I_{av}^2 = \frac{2.75\text{ A}^2 \cdot \text{ns}}{3\text{ ns}} = 0.9167\text{ A}^2$$

$$I_{rms} = \sqrt{I_{av}^2} = \sqrt{0.9167\text{ A}^2} = \mathbf{0.9574\text{ A}}$$

d. The instantaneous value at time $t = 2.5$ ns is

$$i = \mathbf{+1.5\text{ A}}$$

Figure 12-7 can be called a *pulse waveform* or *pulse train* because it consists of a sequence (or train) of relatively short-duration pulses of current in the positive direction with longer time intervals between the pulses. The pulse train in Fig. 12-7 consists of positive pulses, since the short-duration pulses are in a positive direction.

Pulse voltage waveforms and pulse current waveforms have similar characteristics.

EXAMPLE 12-2 What is the average value of the trapezoidal waveform in Fig. 12-9?

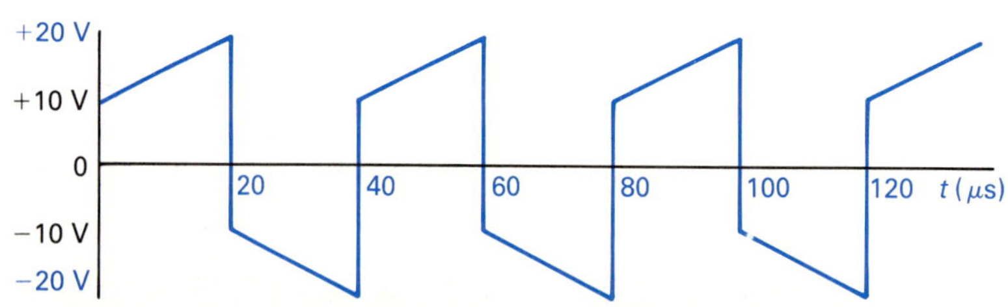

Fig. 12-9. Example 12-2: Trapezoidal voltage waveform.

Solution Using the conventional meaning of the average value of a waveform, either the positive or the negative part of the symmetrical waveform in Fig. 12-9 can be selected to obtain the solution.

The waveform in Fig. 12-9 may be regarded as being equivalent to square waves of 10-V magnitude with triangular waves of 10-V peak magnitude superimposed on them. Figure 12-10 shows how this concept can be applied to a positive section of the original waveform.

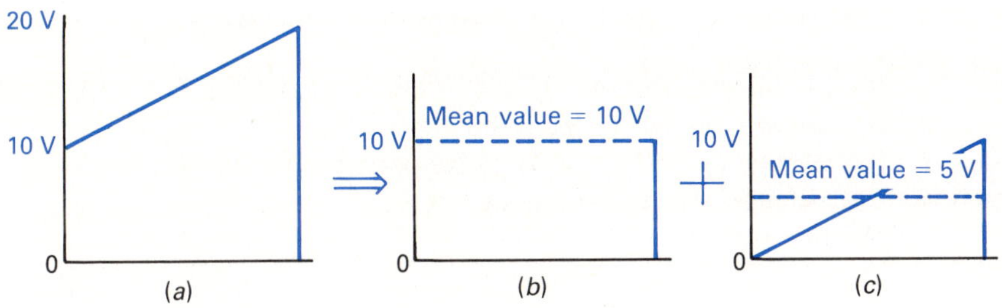

Fig. 12-10. Components of the waveform in Fig. 12-9.

Figure 12-10*a* is represented as the sum of Fig. 12-10*b* and *c*. The mean value of the square-wave section in Fig. 12-10*b* is 10 V. The mean value of the triangular-wave section in Fig. 12-10*c* is 5 V. The average value of the waveform in Fig. 12-9 is the sum of these two mean values:

$$V_{av} = 10\text{ V} + 5\text{ V} = \mathbf{15\ V}$$

REVIEW QUESTIONS FOR SECTION 12-2

1. Referring to a voltage or current waveform, what is meant by
   a. Peak value?
   b. Peak-to-peak value?
   c. Instantaneous value?
   d. RMS value?
2. What symbol is used for
   a. Peak voltage?
   b. Peak-to-peak voltage?
   c. Instantaneous current?
   d. RMS current?
3. Can peak-to-peak values apply to dc voltages and currents?
4. What common measuring technique justifies the use of peak-to-peak values?
5. Why are rms values widely used?
6. How does rms value in ac compare with steady dc value?
7. What do the letters "rms" stand for?

EXERCISE PROBLEMS FOR SECTION 12-2

1. A square-wave voltage has positive and negative maxima of 10 V, and the positive and the negative periods are each 1 ms. Find

a. The peak-to-peak voltage. (*ans*. 20 V)

b. The average voltage for the positive period. (*ans*. 10 V)

c. The average voltage for the negative period. (*ans*. 10 V)

d. The average voltage over one cycle. (*ans*. 10 V)

e. The rms voltage. (*ans*. 10 V)

2. A current square wave has a positive maximum of 5 A, a negative maximum of 10 A, a positive period of 1 ms, and a negative period of 2 ms. Find

a. The peak-to-peak current.

b. The average current for the positive period.

c. The average current for the negative period.

d. The average current over one cycle.

e. The rms current.

12-3 CYCLE, PERIOD, FREQUENCY, AND WAVELENGTH

Cycle

The term ''cycle'' is used extensively in ac circuit work. A *cycle* is a complete alternation; it is the waveform of current or voltage between equivalent magnitudes. By ''equivalent magnitude'' is meant the same magnitude varying in the same direction.

Figure 12-11 shows a triangular waveform. Points $t = 0$, 8, and 16 ms are equivalent points, because here the waveform has zero magnitude and is changing in a positive direction. Points $t = 4$, 12, and 20 ms are also equivalent points; here the waveform has zero value and is changing in a negative direction. Points $t = 0$, 4, 8, 12, 16, and 20 ms are not all equivalent, even though they all represent zero values; in some the value is changing in a positive direction, and in others negative. Points $t = 2$, 10, and 18 ms are all positive peaks and are equivalent. The negative peaks, points $t = 6$ and 14 ms, are equivalent too.

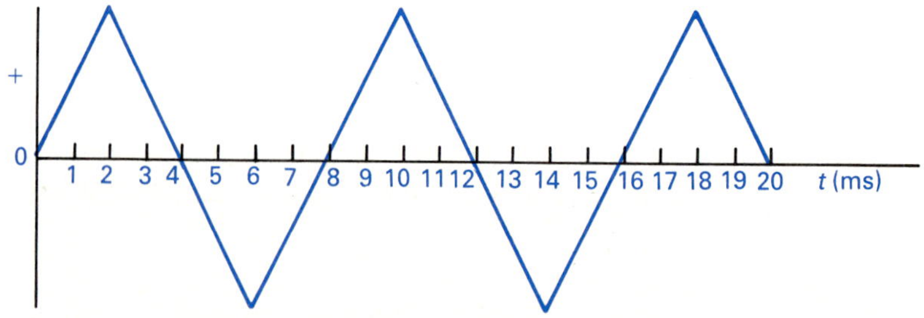

Fig. 12-11. Triangular waveform.

Points $t = 1$, 9, and 17 ms are equivalent, while points $t = 1$, 3, and 5 ms are not.

With reference to Fig. 12-11, typical cycles are the waveform existing between points

$t = 0$ and 8 ms
$t = 1$ and 9 ms
$t = 2$ and 10 ms
$t = 5$ and 13 ms

In Sec. 12-2 there were several references to part of the waveform. In each case this phrase meant a half-cycle. Although a half-cycle, as its name implies, is any half of a cycle, it is common to use the term in a limited sense to mean the waveform between adjacent zero values of magnitude. This usage results in the terms "positive half-cycle" and "negative half-cycle." Again with reference to Fig. 12-11, the positive half-cycles are the waveform between points

$t = 0$ and 4 ms
$t = 8$ and 12 ms
$t = 16$ and 20 ms

The negative half-cycles are the waveform between points

$t = 4$ and 8 ms
$t = 12$ and 16 ms

While this limited meaning of half-cycle is common, it must be remembered that the waveform between, say, points $t = 1$ and 5 ms is also a half-cycle.

The term "cycle" is concerned with the waveform itself. It is not a measurement of amplitude or time; it has no symbol or unit.

Period

The time occupied by one complete cycle is the *period* of the waveform. The symbol for period is T, and the unit for period is the second, symbol s.

In Fig. 12-11 the period is the time from point $t = 0$ ms to point $t = 8$ ms, or between any other two points representing one cycle. Thus the period T is 8 ms.

Frequency

The number of complete cycles occuring each second is the *frequency* of the waveform. The symbol for frequency is f, and the unit for frequency is the hertz, symbol Hz.

Referring to Fig. 12-11, there is one complete cycle in 8 ms; there will be 125 cycles in 1 s. Therefore the frequency f of this waveform is 125 Hz.

There will be f cycles each second, so every cycle will occupy a time of $1/f$ second. This time is the period T.

$$T = \frac{1}{f} \qquad \text{or} \qquad f = \frac{1}{T} \tag{12-1}$$

where T = period, s
f = frequency, Hz

The two forms of Eq. (12-1), show that period and frequency are reciprocal quantities.

Wavelength

Sometimes the horizontal axis of a waveform is scaled in terms of physical distance. This is done when the waveform pertains to the voltage or current associated with propagated energy, such as sound or radio waves. There is no conflict in using time or distance as the horizontal axis quantity for an ac waveform. Distance is directly related to time and is used when it provides a convenient reference.

The distance occupied by one complete cycle of a propagated wave is a *wavelength*. The symbol for wavelength is λ (the lowercase Greek letter lambda). The unit of wavelength is the meter, symbol m.

Since one cycle occupies λ meters, f cycles will occupy $f\lambda$ meters. The number of cycles in 1 s is f. A propagated wave will travel at $f\lambda$ meters per second, so that $f\lambda$ is its velocity of propagation. This gives the equation

$$v = f\lambda \qquad \textbf{(12-2)}$$

where f = frequency, Hz
λ = wavelength, m
v = velocity of propagation, m/s

From Eq. (12-2) it can be seen that frequency and wavelength are reciprocal:

$$f = \frac{v}{\lambda} \qquad \text{or} \qquad \lambda = \frac{v}{f}$$

This reciprocity is compatible with Eq. (12-1), since period and wavelength are both horizontal axis measures of one cycle.

In the case of electromagnetic radiation, such as radio waves, the velocity of propagation is the speed of light c, which is approximately 3×10^8 m/s. This velocity gives

$$c = f\lambda \qquad \text{or} \qquad f\lambda = 3 \times 10^8 \text{ m/s} \qquad \textbf{(12-3)}$$

EXAMPLE 12-3 What is the frequency of the current in Example 12-1?

Solution The time occupied by one complete cycle is 3 ns. Therefore the period is

$$T = 3 \text{ ns}$$

The frequency is the reciprocal of the period:

$$f = \frac{1}{T} \tag{12-1}$$

$$f = \frac{1}{3 \text{ ns}} = \frac{1}{3 \times 10^{-9} \text{ s}} = \mathbf{333.3 \text{ MHz}}$$

EXAMPLE 12-4 If in Example 12-3 the waveform represents a current pulse train propagated along a transmission line, what would be the scaling of the horizontal axis in Fig. 12-7 to represent distance? Assume that the pulse train is propagated along the transmission line at the speed of light.

Solution From Example 12-3,

$$f = 333.3 \text{ MHz}$$

The wavelength will be

$$\lambda = \frac{c}{f} \tag{12-3}$$

$$\lambda = \frac{3 \times 10^8 \text{ m/s}}{333.3 \text{ MHz}} = 0.9 \text{ m} \quad \text{or} \quad 90 \text{ cm}$$

The physical length of one complete cycle will be 90 cm. Therefore the scaling of the horizontal axis will be **30 cm per division,** where one division in Fig. 12-7 is 1 ns.

The solution to Example 12-4 indicates that, on a propagation distance basis, Fig. 12-12 gives the waveform of the current pulse train.

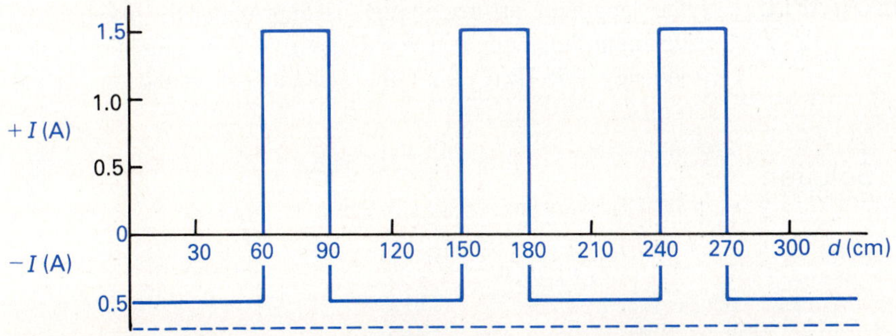

Fig. 12-12. Current waveform in Fig. 12-7 in terms of the scaling in Example 12-4.

EXAMPLE 12-5 What is the frequency of an AM radio transmission if its wavelength is 350 m?

Solution

$$f = \frac{c}{\lambda} = \frac{3 \times 10^8 \text{ m/s}}{350 \text{ m}} = \mathbf{857.1\ kHz}$$

EXAMPLE 12-6 The FM radio band covers the frequency range 88 to 108 MHz. What is the equivalent wavelength range?

Solution

$$\lambda = \frac{c}{f} = \frac{3 \times 10^8 \text{ m/s}}{f \text{ Hz}} \quad \text{or} \quad \frac{300 \text{ Mm/s}}{f \text{ MHz}}$$

At the lowest-frequency end of the FM band,

$$\lambda = \frac{300 \text{ Mm/s}}{88 \text{ MHz}} = \mathbf{3.409\ m}$$

At the highest-frequency end of the band,

$$\lambda = \frac{300 \text{ Mm/s}}{108 \text{ MHz}} = \mathbf{2.778\ m}$$

For the efficient reception of very-high-frequency radio transmission, the length of an antenna is made approximately one-half the wavelength of the signal. Therefore the length of an antenna for the reception of FM-broadcast radio is typically of the order of 1.5 m (5 ft), one-half the average of 3.409 and 2.778 m, the results for Example 12-6.

EXAMPLE 12-7 If the length of an organ pipe is approximately one-quarter wavelength, how long a pipe will be needed for a middle-C tone (frequency 256 Hz)? Take the velocity of sound in air as 336 m/s.

Solution

$$\lambda = \frac{v}{f} \tag{12-2}$$

$$\lambda = \frac{336 \text{ m/s}}{256 \text{ Hz}} = 1.313 \text{ m}$$

The organ pipe is to be approximately $\lambda/4$ in length:

$$\text{Pipe length} \simeq \frac{\lambda}{4} \simeq \frac{1.313\ \text{m}}{4} \simeq \mathbf{0.3282\ m} \qquad \text{(approximately 1 ft)}$$

EXAMPLE 12-8 The waveform in Fig. 12-13 is displayed on an oscilloscope screen. The horizontal sweep speed is 2 ms per division. What is the waveform frequency?

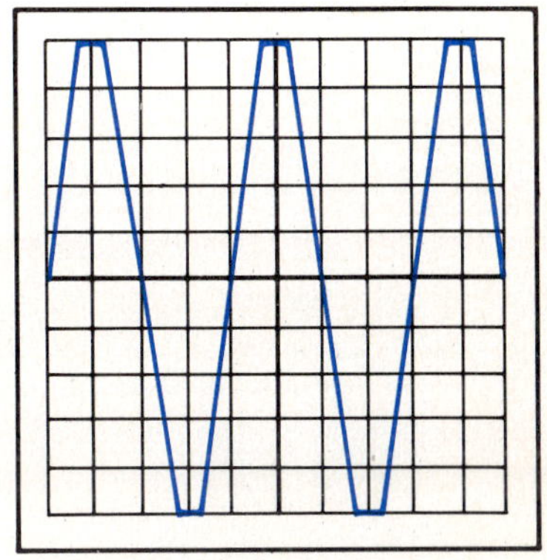

Fig. 12-13. Example 12-8: Oscilloscope display.

Solution Figure 12-13 shows a display of 2.5 cycles of a trapezoidal waveform. One cycle occupies four divisions of the horizontal scale. Four divisions at 2 ms per division represent 8 ms. The period of the waveform T is the time occupied by one cycle:

$$T = 8\ \text{ms}$$

The frequency of the waveform f is the reciprocal of the period T:

$$f = \frac{1}{T} \tag{12-1}$$

$$f = \frac{1}{8\ \text{ms}} = \mathbf{125\ Hz}$$

The solution to Example 12-8 could have been obtained more directly as follows:

$$f = \frac{\text{number of cycles displayed}}{\text{total time occupied}}$$

$$f = \frac{2.5\ \text{cycles}}{20\ \text{ms}} = 125\ \text{Hz}$$

REVIEW QUESTIONS FOR SECTION 12-3

1. Referring to ac waveforms, what is meant by
   a. Cycle?
   b. Period?
   c. Frequency?
   d. Wavelength?

2. What symbols are used for
   a. Cycle?
   b. Period?
   c. Frequency?
   d. Wavelength?

3. What is usually meant by "half-cycle"?

4. State the equation relating
   a. Frequency and period.
   b. Frequency and wavelength.
5. What is the speed of radio waves?

EXERCISE PROBLEMS FOR SECTION 12-3

1. Find the period of the waveform when the frequency is
   a. 10 kHz (*ans*. 100 μs)
   b. 60 Hz (*ans*. 16.67 ms)
   c. 30 MHz (*ans*. 33.3 ns)
2. Find the frequency of the waveform when the period is
   a. 2 μs
   b. 16.67 ms
   c. 50 ns
3. The velocity of radio waves is given as 3×10^8 m/s. Find the wavelengths for the following radio frequencies
   a. 500 kHz (*ans*. 600 m)
   b. 6 GHz (*ans*. 5 cm)
   c. 1.2 MHz (*ans*. 250 m)
4. The velocity of sound in air is given as 336 m/s. Find the wavelengths for the following sound waves
   a. 20 Hz
   b. 15 kHz
   c. 1000 Hz
5. The length of a dipole antenna for a given frequency is $\lambda/2$. Find the length of the antenna if the frequency is
   a. 500 kHz (*ans*. 300 m)
   b. 2170 kHz (*ans*. 69.12 m)
   c. 88 Mhz (*ans*. 1.7 m)
   d. 450 MHz (*ans*. 33.3 cm)
6. Find the frequency of electromagnetic radiations with the following wavelengths:
   a. 2.5 m
   b. 3 cm
   c. 80 m

12-4 THE IMPORTANCE OF THE SINE WAVE

The *sine wave* is the waveform represented by the equation $y = k \sin x$. This rather uninspiring definition gives no clue to the reason for the dominance of the sine wave over all other waveforms in ac circuits. Before considering the details of sine-wave characteristics, we should obtain an idea of the significance of the sine wave in electrical technology.

The sine wave is important in ac circuits because:

1. It is the easiest waveform to generate.
2. It is the easiest waveform to distribute and use.
3. All other periodic waveforms can be synthesized from sine-wave components.

Generation of Sine Waves

Electrical energy used in industry, commerce, and the home is produced by electromagnetic generators. The functioning of these generators is based on electromagnetic induction, which was dealt with in Chap. 5.

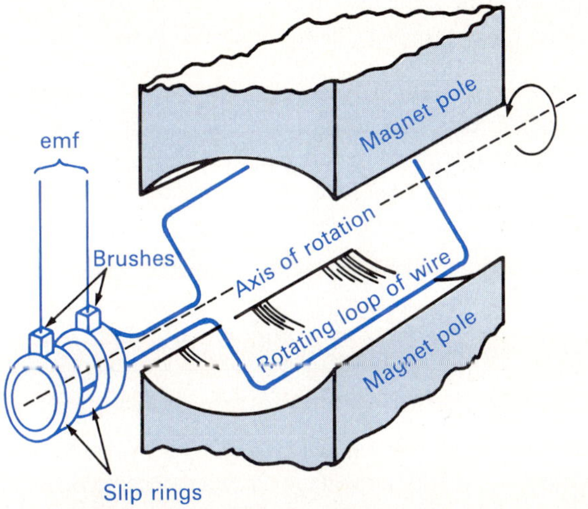

Fig. 12-14. Basic details of an ac generator or alternator.

Figure 12-14 shows a basic electromagnetic generator. It consists of a conducting loop of wire which is rotated within a magnetic field. The movement of the conducting loop sides across the field results in an emf being induced between the ends of the loop. Each loop end is connected to a metal slip ring. A conducting brush makes contact with each slip ring, and so connects the emf to an external network at all angular positions of the loop.

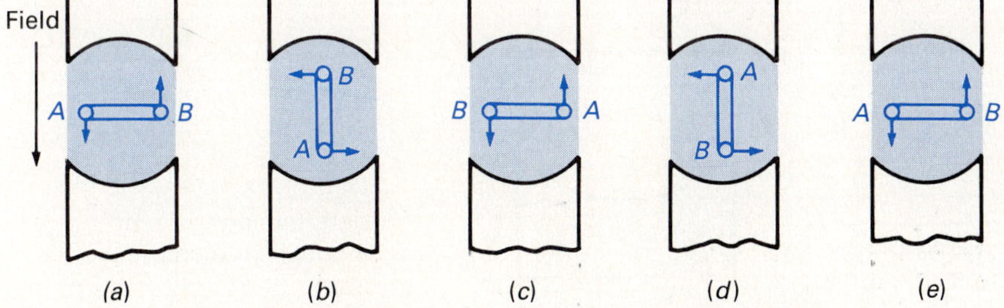

Fig. 12-15. Positions of the sides of a wire loop at 90° intervals.

Figure 12-15 shows a sequence of events which occurs while the loop is rotating. Although it is free to rotate either way, the loop is shown rotating in a counterclockwise direction. The small circles *A* and *B* represent sectional views of the loop sides.

Figure 12-15*a* At this instant the motion of the loop sides is parallel to the vertical direction of the magnetic field. There is no emf because there is no relative motion between the loop sides and the field; the conductor is not cutting the field.

Figure 12-15*b* Now the motion of each loop side is horizontal, perpendicular to the field direction. This is maximum relative motion which causes maximum emf to be generated.

Figure 12-15*c* Here the situation is the same as that in Fig. 12-15*a*, except that the sides of the loop, *A* and *B*, have changed places. The emf is again zero.

Figure 12-15*d* In this position the emf will have maximum magnitude, as in Fig. 12-5*b*. The loop side *A* is now moving to the left, whereas in Fig. 12-15*b* it was moving to the right. Similarly, side *B* has reversed its direction across the field. The effect of this reversal of direction of relative motion is a reversal of emf polarity.

Figure 12-15*e* The loop is now back in the same position as in Fig. 12-15*a*. The emf is zero again.

The complete rotation of the loop shown in Figs. 12-14 and 12-15 represents one complete cycle of alternation. A cycle of ac voltage has been generated. The voltage magnitudes have gone from zero to a maximum with one polarity, to zero, to a maximum with the opposite polarity, and then back to the original zero.

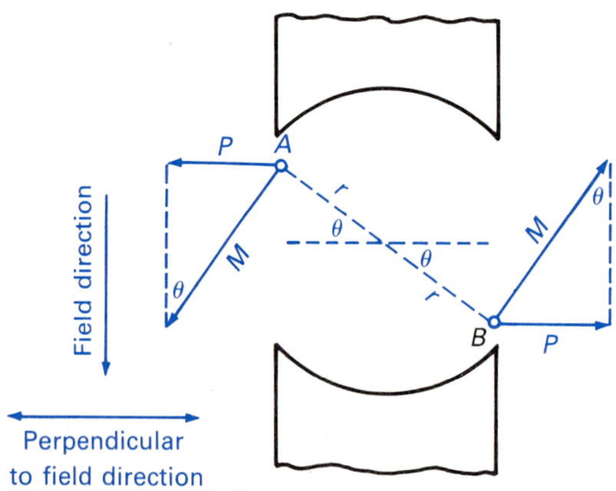

Fig. 12-16. Factors determining the emf induced in a rotating wire loop.

The diagram in Fig. 12-16 typifies the situation which exists when the generated voltage is neither zero nor maximum. The plane of the loop (a line through sides *A* and *B*) is at an angle θ to the horizontal. The instantanteous direction of a loop side is perpendicular to the radius of rotation and at an angle of θ to the vertical.

The generated emf is always proportional to the velocity of the component of conductor motion perpendicular to the field direction.

In Fig. 12-16 the actual directions of the instantaneous motion of each side of the loop (A and B) are shown by the arrows M. The directions of the components of motion perpendicular to the field direction are shown by the arrows P. The lengths of these arrows indicate the respective velocities.

The relationship between the actual and perpendicular velocities is

$$P \propto M \sin \theta$$

where θ is the angle between the instantaneous direction of a loop side and the direction of the field. θ can be specified in degrees or radians.

The generated emf v at any instant is proportional to the perpendicular velocity component P:

$$v \propto P \propto M \sin \theta$$

The actual velocity M of the conductors which form the loop sides A and B is the peripheral velocity of the loop and is related to the radius of rotation r by the equation

$$M = \omega r$$

where ω = angular velocity of rotation (the speed at which the loop turns), in radians per second.

Thus M is directly proportional to r, and the relationship between the generated emf and the radius of rotation is

$$v \propto r \sin \theta \qquad \textbf{(12-4)}$$

For our present purposes, the significant conclusion to be drawn from Eq. (12-4) is: The emf generated by an electromagnetic generator has a sine waveform.

Electronic generators of ac which incorporate inductance-capacitance resonant circuits are used extensively. They are called *oscillators* and typically are found as subsystems in radio transmitters and receivers and in measuring instruments. Resonant circuits naturally tend to develop sine-wave currents and voltages, and so oscillators incorporating resonant circuits are sine-wave generators.

Distribution and Use of Sine-Wave AC

The distribution and use of ac always involves inductance and capacitance. This is particularly true of inductance; it is an inherent characteristic of distribution transformers, electric motors, and other electrical equipment. Unless sine waves are used, inductance and capacitance will result in waveform changes that prevent effective operation of a system. These waveform changes occur because the relationships between inductive or capacitive voltages and currents are dependent on rate of change ($v = L\, di/dt$ and $i = C\, dv/dt$). A sine wave has a rate-of-change curve that is also a sine wave; once a sine wave, always a sine wave.

The shape of a rate-of-change curve may seem to be a detail of purely theoretical interest, but actually it is of the utmost importance. It is so important that, even if sine waves were difficult to generate, they would have to be generated anyway. Why are they of such importance? Generating sine waves and supplying them to an electric network results in the same waveform for all currents and voltages throughout the network; there will be sine waves throughout.

Let us see what could happen if other current waveforms were used in an inductive circuit.

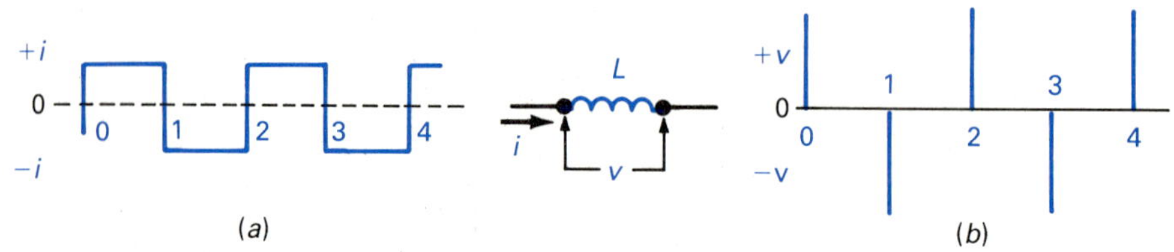

Fig. 12-17. Square-wave current in an inductor results in a pulse voltage waveform.

A square-wave alternating current, as in Fig. 12-17, in an inductor will produce a voltage across the inductor only during current reversals; there will be no voltage the remainder of the time because the current is constant.

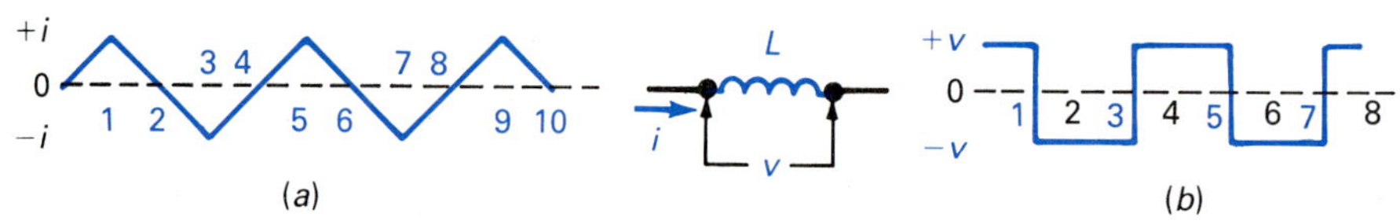

Fig. 12-18. Triangular-wave current in an inductor results in a square voltage waveform.

A triangular-wave current, as in Fig. 12-18, in an inductor will cause a constant magnitude of voltage to be developed across the inductor. The polarity of the voltage will reverse at each current peak.

Figures 12-17 and 12-18 show just two possibilities. In a complex network involving many points of distribution and utilization of electrical energy, the waveforms will be difficult to predict and will render the system useless unless sine waves are used.

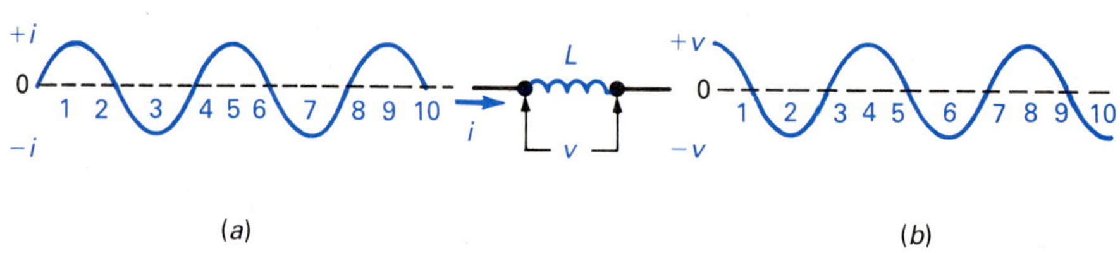

Fig. 12-19. Sine-wave current in an inductor results in a sine-wave voltage.

If an inductor carries the sine-wave current in Fig. 12-19*a*, the voltage across the inductor will have the waveform in Fig. 12-19*b*. Figure 12-19*b* is a cosine waveform, but sine and cosine waves are curves of the same shape displaced relative to each other on the horizontal time axis. Any linear network supplied with energy from a sine-wave ac source will contain only sinusoidal currents and voltages. These currents and voltages will all have the same frequency; their amplitudes and time relationships will depend on the resistance, inductance, and capacitance values within the network.

Nonsinusoidal Waveforms

Any periodic wave can be reduced to a dc component and a number of sine-wave components of various frequencies, magnitudes, and time relationships. The mathematics of Fourier analysis are needed to fully prove that this is so, but the objectives of this text do not justify such a treatment. However, if you can be convinced that the principle is true, it will help you to understand the importance of sine waves. A demonstration that one non-sine wave can be synthesized from sine waves should help you to accept the fact that a general principle is involved.

A square wave, such as those in Figs. 12-17*a* and 12-18*b*, can be produced by adding together an infinite number of sine waves. The sine waves must start in phase; they must all have zero value and be swinging positive at the time instant that the square wave is to be zero swinging positive. If the frequency of the square wave is to be f Hz, then the frequencies of the sine-wave components must be odd multiples of f Hz; they must be f Hz, $3f$ Hz, $5f$ Hz, $7f$ Hz, etc. If the maximum value of the f-Hz sine wave is 1 V, the maximum values of the other sine-wave components must be $\frac{1}{3}$ V, $\frac{1}{5}$ V, $\frac{1}{7}$ V, etc.

The addition of sine waves can be shown graphically by adding the heights of the curves at each interval of time.

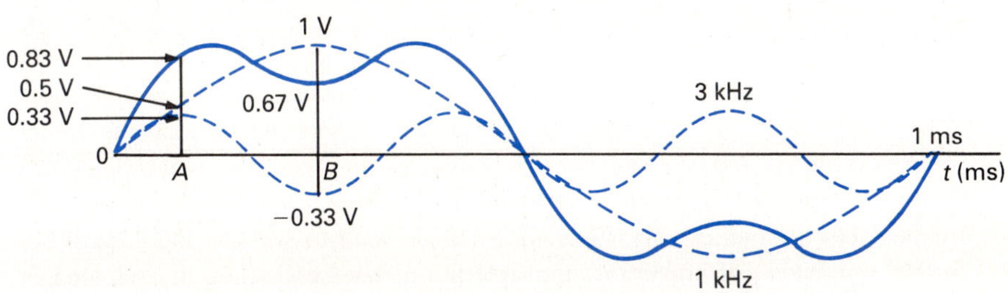

Fig. 12-20. A sine wave and its third harmonic added.

Figure 12-20 shows three waveforms. Two of these are sine waves and are shown as broken lines; the frequency of one is 1 kHz with a maximum value of 1 V, and the other has a frequency of 3 kHz and a maximum value of $\frac{1}{3}$ V. The solid-line curve is the result of adding the two sine waves. At each instant of time from 0 to 1 ms, the height of this solid curve is the sum of the heights of the two sine-wave curves. At time A the sine waves have values of 0.5 and 0.33 V, so the addition curve has the value of 0.83 V. At time B, the sine waves have values of +1 and −0.33 V, so the addition curve has the value 0.67 V.

The solid-line curve in Fig. 12-20 shows the beginning of the transformation from a 1-kHz sine wave to a 1-kHz square wave.

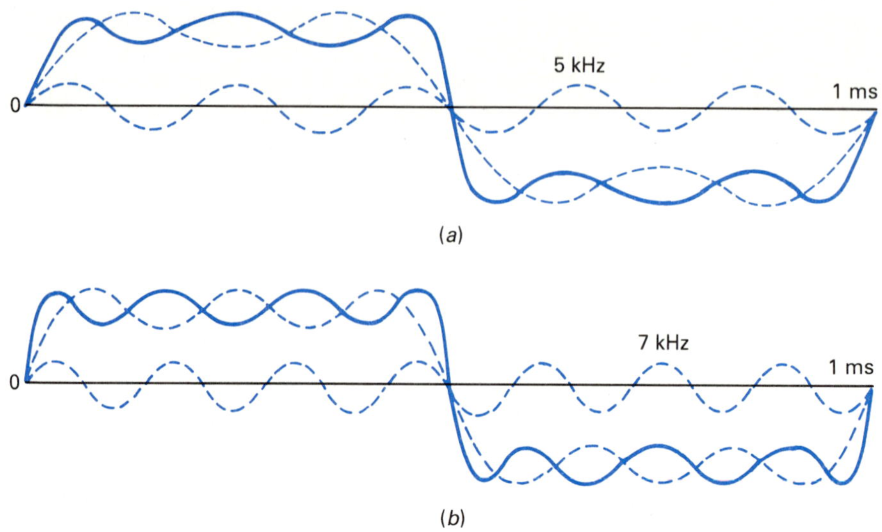

Fig. 12-21. *(a)* The addition waveform in Fig. 12-20 added to a fifth harmonic; *(b)* the addition waveform in Fig. 12-21*a* added to a seventh harmonic.

The solid-line curve in Fig. 12-20, representing the sum of the 1- and 3-kHz sine waves, is shown as a broken line in Fig. 12-21*a*. Also shown as a broken line is a 5-kHz sine wave of maximum value $\frac{1}{5}$ V. The solid-line curve is the sum of the two broken-line curves, and hence represents the sum of the 1-, 3-, and 5-kHz sine waves.

In Fig. 12-21*b*, the solid-line curve in Fig. 12-21*a* is shown as a broken line, as is a 7-kHz $\frac{1}{7}$-V maximum-value sine wave. The sum of these curves is the solid-line curve. This latter curve represents the addition of the 1-kHz (1-V), 3-kHz ($\frac{1}{3}$-V), 5-kHz ($\frac{1}{5}$-V), and 7-kHz ($\frac{1}{7}$-V) sine waves.

By comparing Figs. 12-20 and 12-21*a* and *b*, it can be seen that, as the higher frequencies are added, the total curve becomes closer to a square wave. The sides become steeper; the tops and bottoms become flatter, the ripples being more frequent but of smaller magnitude. It is not difficult to see that the addition of further sine waves of higher odd-multiple frequencies will result in even steeper sides and flatter tops and bottoms. The end result, if the number of sine waves added were to become infinite, would be the vertical sides and flat top and bottom of the perfect square wave.

The addition of sine waves of all multiple frequencies, odd and even, gives a sawtooth wave if the amplitudes are again reciprocals of the frequency multiples (for example, the amplitude of the $4f$ sine wave is one-quarter that of the f sine wave). If sine waves which are even multiples of a given frequency are added, the result can be a sharp pulse of current or voltage.

The principle that all periodic waveforms can be regarded as sine waves added together is used in circuit analysis and design. For analysis, the sine-wave components may be considered to be produced by separate generators, the total circuit effect being determined by the superposition method. Transient waveforms such as square and sawtooth waves, containing very-high-frequency sine-wave components, require circuit designs which take these frequencies into account.

REVIEW QUESTIONS FOR SECTION 12-4

1. What is the general equation for a sine wave?
2. Give three important reasons for the use of sine waves in electrical and electronic work.
3. What is unique about the rate of change of a sine wave?
4. Why is the rate of change of a sine wave important?
5. What is the relationship between a sine wave and a non-sine-wave periodic waveform?

12-5 THE SINE WAVE

The emf produced by a simple electromagnetic generator, at any time instant, is proportional to the sine of an angle. Basic trigonometric principles can be used to draw a sine waveform.

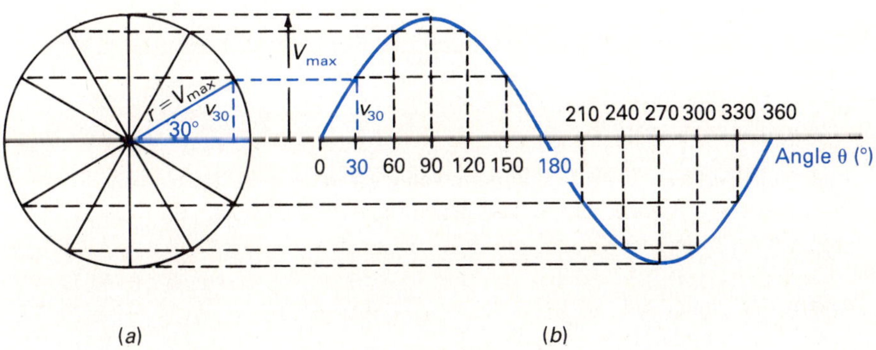

Fig. 12-22. Drawing a sine wave by projection from a circle.

In Fig. 12-22*a* a circle of radius V_{max} is divided into 12 equal segments, so that each segment represents an angle of 30° or $\pi/6$ rad. Figure 12-22*b* has a horizontal axis scaled so that each 30° angle is allocated an equal distance. This axis extends from 0 to 360°, which is one complete revolution (2π rad), or one cycle of ac voltage.

For any angle to the horizontal in Fig. 12-22*a*, the vertical projection from the circumference to the horizontal gives a line of length representing the instantaneous magnitude at that angle. Thus the line v_{30} represents the magnitude at 30°.

The waveform in Fig. 12-22*b* is drawn by erecting verticals at each 30° point and making their lengths equal to the vertical projections obtained from Fig. 12-22*a*. Since a vertical projection is equal to the circle radius multiplied by the sine of the relevant angle (for example, $v_{30} = v_{max} \sin 30°$), the resulting waveform is a sine wave representing all instantaneous voltage values for one ac cycle.

The maximum value of the ac voltage in Fig. 12-22 is V_{max} volts. The instantaneous value at any angle θ from the zero point will be

$$v = V_{max} \sin \theta \qquad \textbf{(12-5)}$$

Equation (12-5) is the fundamental sine-wave ac relationship; it applies equally well to currents:

$$i = I_{\max} \sin \theta \tag{12-5}$$

The shaft of an ac generator, or alternator, normally rotates at a constant rate. The circle in Fig. 12-22, together with its radius lines, should be regarded as rotating at a constant rate.

Angular Velocity

The rate of rotation is the *angular velocity,* the rate at which the angle changes. The symbol for angular velocity is ω (the lowercase Greek letter omega). The unit for angular velocity is the radian per second, symbol rad/s.

An ac frequency of f hertz can be related to Fig. 12-22 by saying that the radius rotates f revolutions per second. During each revolution the radius traverses 2π radians. Therefore the angular velocity or *angular frequency* can be expressed as

$$\omega = 2\pi f \tag{12-6}$$

where ω = angular velocity or angular frequency, rad/s
π = the numerical constant $\simeq 3.142$
f = frequency, Hz

Angular velocity is the angle traversed each second, so that

$$\omega = \frac{\theta}{t} \quad \text{or} \quad \theta = \omega t \quad \text{or} \quad t = \frac{\theta}{\omega} \tag{12-7}$$

where θ = angle, rad
ω = angular velocity, rad/s
t = time, s

Equation (12-7) gives the important link between angle and time. The sine wave in Fig. 12-22 can be redrawn with time as the horizontal axis by substituting $t = \theta/\omega$.

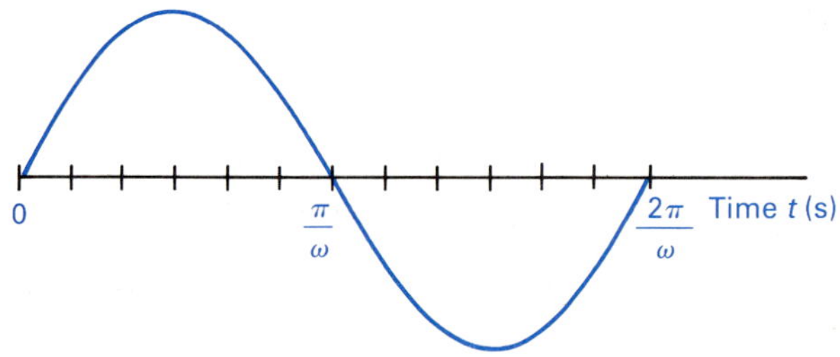

Fig. 12-23. Sine wave with respect to time.

Figure 12-23 is the *same* as Fig. 12-22*b*; it is not merely the same shape. Appreciating this fact will help you to avoid confusion when faced with the various methods of expressing ac quantities used in electrical work. In the two figures the horizontal axis represents angle or time. From Eqs. (12-6) and (12-7) it can be seen that these two quantities are related by $2\pi f$, which is a numerical constant for any given periodic wave.

As an example, consider an angle of π radians (180°), which is equal to one half-cycle of a sine wave:

$$\theta = \pi$$

but $$\theta = \omega t \qquad (12\text{-}7)$$

so $$\omega t = \pi$$

or $$t = \frac{\pi}{\omega}$$

EXAMPLE 12-9 A wall outlet delivers 60-Hz sine-wave ac at 170-V maximum value to domestic lamps and appliances. What is the instantaneous voltage for the following angles after angle θ_0: *(a)* $\pi/4$ rad, *(b)* $3\pi/4$ rad, *(c)* $5\pi/4$ rad, and *(d)* $7\pi/4$ rad?

Solution If it is not otherwise specified, it is normally assumed that angle θ_0 is the point on the horizontal axis at which the voltage magnitude is zero and is swinging in a positive direction. Figure 12-24 is drawn on this assumption.

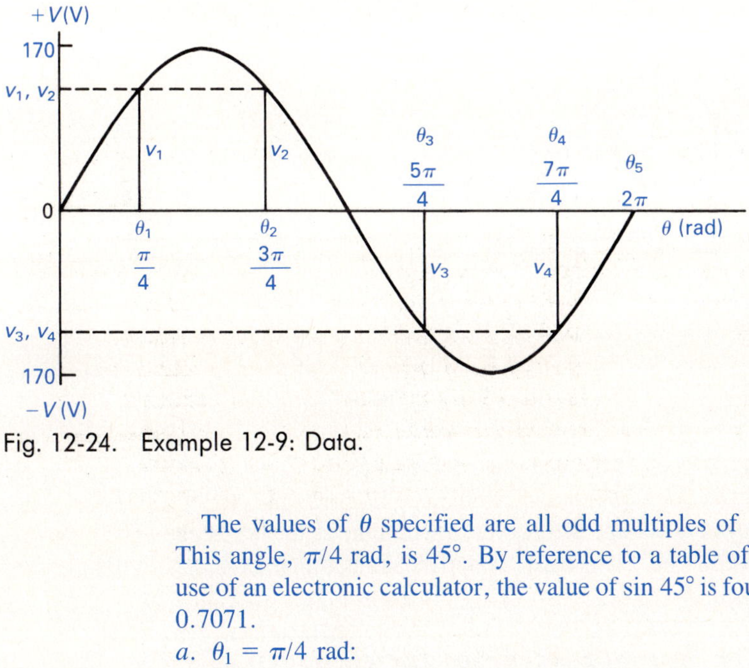

Fig. 12-24. Example 12-9: Data.

The values of θ specified are all odd multiples of $\pi/4$ rad. This angle, $\pi/4$ rad, is 45°. By reference to a table of sines or use of an electronic calculator, the value of sin 45° is found to be 0.7071.

a. $\theta_1 = \pi/4$ rad:

$$v_1 = V_{max} \sin \theta_1 \qquad (12\text{-}5)$$

$$v_1 = 170\text{ V} \times \sin\ 45° = 170\text{ V} \times 0.7071 \simeq \mathbf{120\ V}$$

It is clear from Fig. 12-24 that the magnitudes of the voltages at all the specified angles are the same. At $\pi/4$ and $3\pi/4$ rad both voltages have the same positive polarity. At $5\pi/4$ and $7\pi/4$ rad both voltages have negative polarity. Therefore,

b. $\theta_2 = 3\pi/4$ rad:

$$v \simeq \mathbf{120\ V}$$

c. $\theta_3 = 5\pi/4$ rad:

$$v_3 \simeq \mathbf{-120\ V}$$

d. $\theta_4 = 7\pi/4$ rad:

$$v_4 \simeq \mathbf{-120\ V}$$

It is important to note that for angles from 0 to 180° (0 to π rad) the sines are positive, while for angles between 180 and 360° (π to 2π rad) the sines are negative. If this is not clear to you, a review of basic trigonometry is necessary. Sines, cosines, tangents, and arctangents will be met frequently in the discussions on ac circuits.

In the solutions to Example 12-9, each of the voltages has the magnitude $V_{max} \times 0.7071$, which is also the rms value of the voltage. This is because the sine of 45° and the relationship between rms and peak values for a sine wave are both $1/\sqrt{2}$.

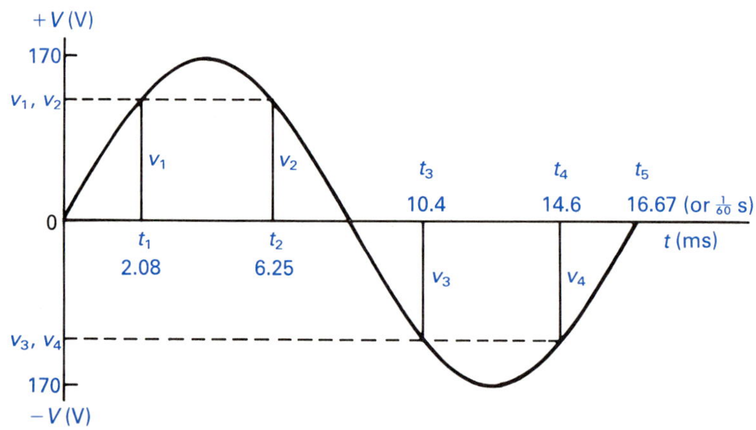

Fig. 12-25. Figure 12-24 with time reference.

The sine-wave voltage of Fig. 12-24 with time as the horizontal axis is shown in Fig. 12-25.

EXERCISE PROBLEMS FOR SECTION 12-5

1. For a 60-Hz sine wave (starting at zero and going positive), find the angle θ, in radians and in degrees, at the following times after t_0:

a. 1 ms (*ans.* 0.377 rad, 21.6°)

b. 8.33 ms (*ans.* π rad, 180°)

c. 0.55 s (*ans*. 0 rad, 0°)

2. Find the angular velocities for the following frequencies:
a. 60 Hz *c*. 1.5 MHz
b. 1000 Hz *d*. 2.8 GHz

12-6 SINE WAVE VALUES

The methods by which the magnitude of an ac wave can be specified apply to sine-wave voltages and currents. For the reasons given in Sec. 12-4, sine-wave alternating current is so common that the various means for specifying its magnitude deserve separate treatment.

In the following paragraphs describing sine-wave values, only voltage waves are shown; current waves follow the same principles.

Maximum or Peak

The *maximum* or *peak value* of a sine wave is the greatest magnitude either side of zero.

A sine wave is symmetrical, its positive and negative half-cycles being the same. Thus its positive and negative peaks have the same magnitude. Figure 12-26 shows the maximum values of a sine-wave emf.

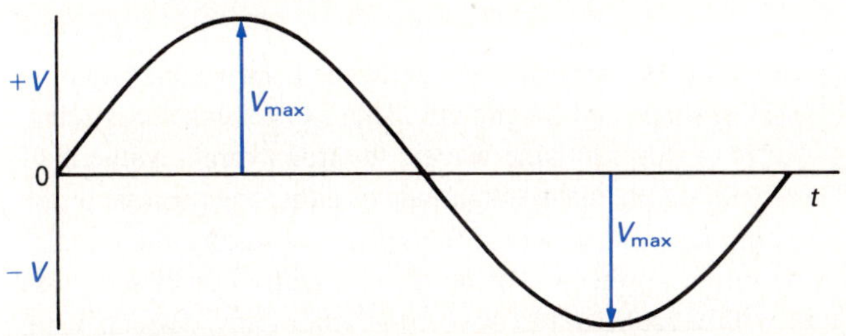

Fig. 12-26. Sine-wave maximum or peak value.

Peak-to-Peak

The *peak-to-peak value* of a sine wave is the total waveform magnitude taken from the positive peak to the negative peak.

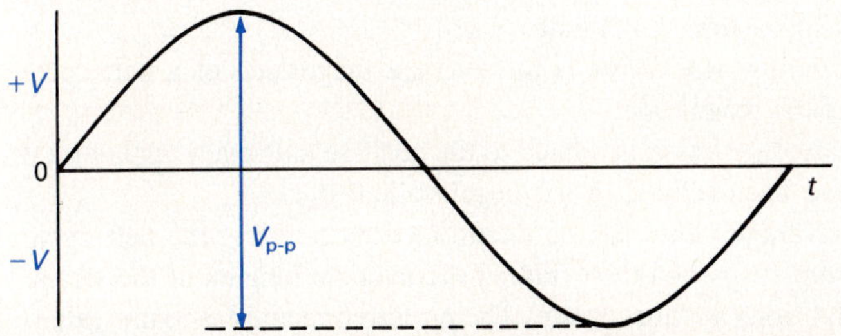

Fig. 12-27. Sine-wave peak-to-peak value.

As you can see from Fig. 12-27, because of the symmetry of the sine wave,

$$V_{\text{p-p}} = 2V_{\max} \tag{12-8}$$

Instantaneous

The *instantaneous value* of a sine wave is the magnitude at a specified instant of time after a reference time. The reference time is usually called t_0, the time at which the magnitude is zero and is swinging in a positive direction.

Sines are expressed in terms of angles. The relationship between angle and time is given in Eq. (12-7) as

$$\theta = \omega t \tag{12-7}$$

Applying this relationship to Eq. (12-5) ($v = V_{max} \sin \theta$) gives an expression for the instantaneous value of the emf shown in Fig. 12-28:

$$v = V_{max} \sin \omega t \tag{12-9}$$

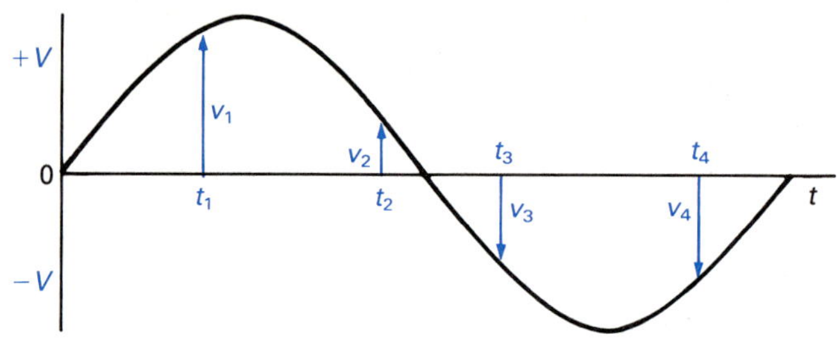

Fig. 12-28. Sine-wave instantaneous values.

Average

The true average or mean value of a sine wave is zero, the positive and negative half-cycles being identical in shape and magnitude. Thus, as a means of specifying the relative magnitudes of different sine waves, the true average value is of no practical use. By determining the mean magnitude of either the positive or the negative half-cycle, a value is obtained which is useful for establishing the significance of the reading on an ac meter. An ac meter usually consists of a dc meter movement fitted with a rectifier to convert the alternating current being measured into dc half-cycles. The meter pointer indicates the average magnitude of these half-cycles.

All references to the average value of a sine wave in this text have the conventional meaning: the average magnitude of a positive or a negative half-cycle.

A half-cycle can be specified as beginning at any point in a cycle. To ensure that a full positive or negative half-cycle is chosen, it is usual to specify that it commences at a point of zero magnitude.

The *average value* of a sine wave is the average magnitude of a half-cycle which begins with zero magnitude.

In Fig. 12-29 the average value is related to the positive half-cycle, although it could as readily have been related to the negative half-cycle.

Graphically, the average value can be determined by dividing the half-cycle into many equal-width strips and then finding the average heights of the strips. This procedure is laborious and inaccurate. The preferred method is to use calculus which gives the result:

$$V_{av} = \frac{2V_{max}}{\pi} \quad \text{or} \quad 0.6366V_{max} \tag{12-10}$$

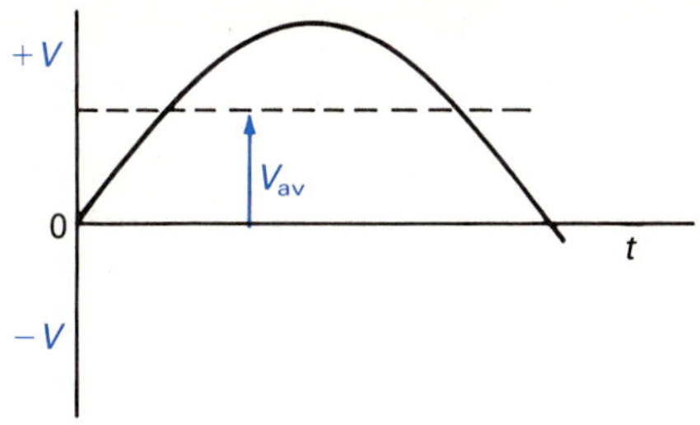

Fig. 12-29. Sine-wave average value.

RMS

A sine wave of voltage or current in a network will result in a certain power level. The steady dc voltage or current which will result in the same power is the *rms* or *effective value* of the sine wave. Figure 12-30 shows how the level of the rms value is related to a sine waveform of voltage.

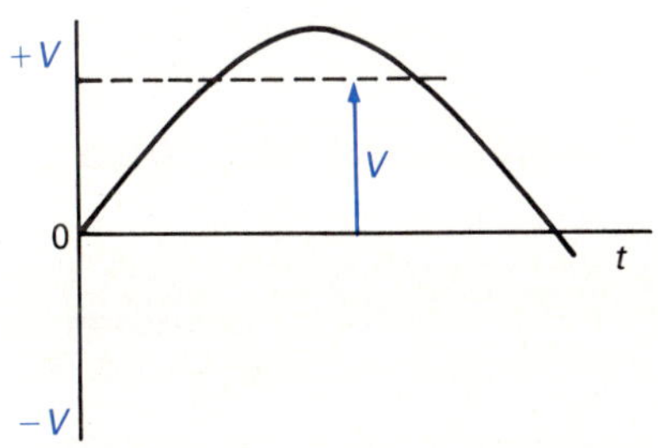

Fig. 12-30. Sine-wave rms value.

The derivation of the rms value can be shown graphically. A curve is drawn through all the points representing the squares of the instantaneous values. The average height of this curve is then found. The square root of this height is the rms value. The calculus method is faster and more accurate. It gives the rms value of a sine-wave voltage as

$$V = \frac{V_{max}}{\sqrt{2}} \quad \text{or} \quad 0.7071V_{max} \tag{12-11}$$

Note that the symbol V (no subscript) implies rms value.

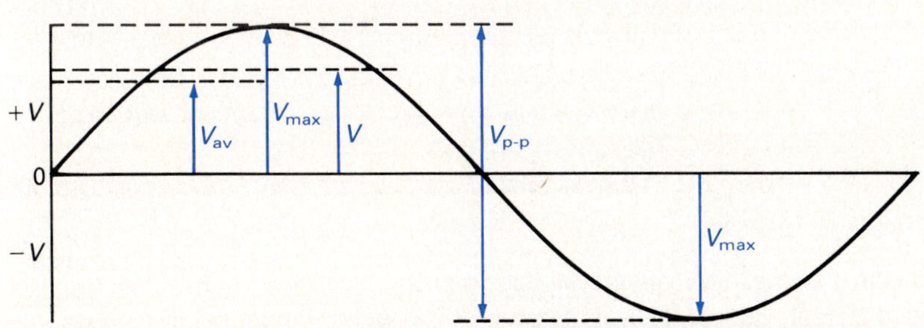

Fig. 12-31. Comparison of sine-wave magnitude values.

Figure 12-31 shows one cycle of a sine-wave emf with maximum or peak, peak-to-peak, average, and rms values as defined above. The comparative magnitudes of these values can be clearly seen. Instantaneous values have been omitted because they can have any magnitude, depending on the specified point in the cycle.

EXAMPLE 12-10 For a domestic electricity supply rated at 120 V 60 Hz, what are *(a)* the maximum value, *(b)* the average value, and *(c)* the instantaneous value 10 ms after time t_0?

Solution Domestic electricity supplies are sine-wave ac for the reasons given in Sec. 12-4 (ease of generation, distribution, and use). A supply is rated in rms values, unless otherwise stated. Therefore the supply in this example can be more completely described as 120 V rms 60 Hz sine-wave ac. One cycle of this sine wave is shown in Fig. 12-32.

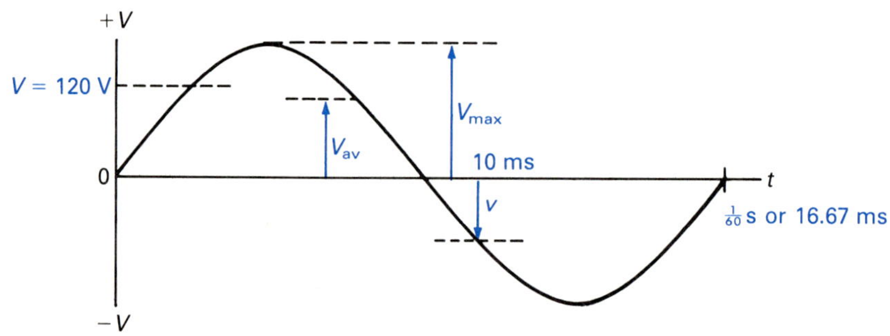

Fig. 12-32. Example 12-10: Waveform data.

a. From Eq. (12-11),

$$V_{max} = \sqrt{2}V$$
$$V_{max} = 1.414 \times 120\ \text{V} = \mathbf{169.7\ V}$$

b. $$V_{av} = \frac{2V_{max}}{\pi} \qquad (12\text{-}10)$$

$$V_{av} = 0.6366 \times 169.7\ \text{V} = \mathbf{108\ V}$$

c. $$\omega = 2\pi f \qquad (12\text{-}6)$$

$$\omega = 2\pi \times 60\ \text{Hz} = 377\ \text{rad/s}$$
$$v = V_{max} \sin \omega t = 169.7\ \text{V} \times \sin 377\ \text{rad/s} \times 10\ \text{ms}$$
$$= 169.7\ \text{V} \times \sin 3.77\ \text{rad} \quad \text{or} \quad 169.7\ \text{V} \times \sin 216°$$
$$v = 169.7\ \text{V} \times (-0.5878) = \mathbf{-99.75\ V}$$

In North America the voltage of the domestic electricity supply has one of several different values, 110, 115, and 117 V being common. The value of 120 V, as given in Example 12-11, is the accepted standard.

It should be noted that the peak value of the ordinary household supply is about 170 V. The insulation of wiring and equipment must be able to safely withstand this peak voltage.

EXAMPLE 12-11 The instantaneous value of a 5-kHz sine-wave current is 10 mA, 40° after angle θ_0. What are the *(a)* peak, *(b)* peak-to-peak, and *(c)* rms values of the current?

Solution See Fig. 12-33.

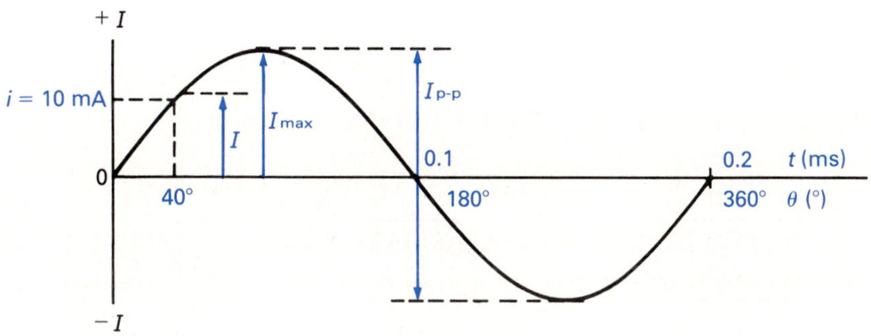

Fig. 12-33. Example 12-11: Waveform data.

a. From Eqs. (12-5) or (12-9),

$$I_{max} = \frac{i}{\sin\theta} = \frac{10\text{ mA}}{\sin 40°} = \frac{10\text{ mA}}{0.6428} = \mathbf{15.56\ mA}$$

b. $$I_{p\text{-}p} = 2I_{max} \qquad (12\text{-}8)$$

$$I_{p\text{-}p} = 2 \times 15.56\text{ mA} = \mathbf{31.12\ mA}$$

c. $$I = 0.7071I_{max} \qquad (12\text{-}11)$$

$$I = 0.7071 \times 15.56\text{ mA} = \mathbf{11\ mA}$$

EXAMPLE 12-12 A symmetrical ac sine-wave voltage displayed on the screen of an oscilloscope appears as in Fig. 12-34. The vertical calibration of the oscilloscope is 5 mV per division. What is *(a)* the peak-to-peak voltage, *(b)* the peak voltage, and *(c)* the rms voltage?

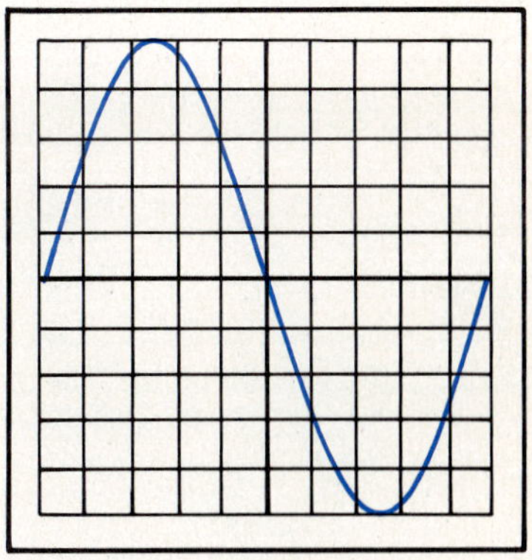

Fig. 12-34. Example 12-12: Oscilloscope display.

Solution *a*. The total height of the display is 10 divisions. The peak-to-peak voltage is the height multiplied by the vertical calibration.

$$V_{p\text{-}p} = 10\text{ divisions} \times 5\text{ mV per division} = \mathbf{50\ mV}$$

b. From Eq. (12-8),

$$V_{max} = \frac{V_{p\text{-}p}}{2} = \frac{50\text{ mV}}{2} = \mathbf{25\ mV}$$

c. From Eq. (12-11),

$$V = \frac{V_{max}}{\sqrt{2}} = \frac{25\text{ mV}}{\sqrt{2}} = \mathbf{17.68\ mV}$$

The solutions given for Example 12-12 are valid only if it is known that the waveform displayed is a sine wave and symmetrical about the zero magnitude level. The display does not necessarily show whether the waveform is symmetrical or asymmetrical, or even whether it is alternating or direct current. To know, one must be aware of the vertical location of the zero voltage level on the display. Since we are told that Fig. 12-34 represents a symmetrical ac sine-wave voltage, we know that the zero voltage level is located at the central horizontal line.

When an oscilloscope is being used, it often happens that the nature of the waveform and the position of the zero voltage level are not known, resulting in situations similar to those depicted in Fig. 12-35.

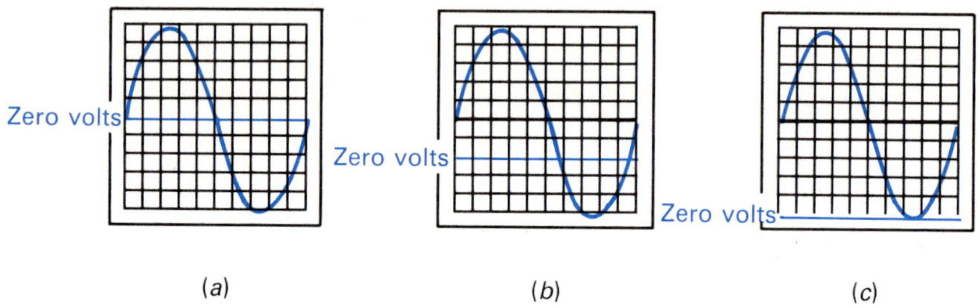

Fig. 12-35. Oscilloscope displays: *(a)* Symmetrical ac voltage; *(b)* asymmetrical ac voltage; *(c)* varying dc voltage.

The displays in Fig. 12-35*a*, *b*, and *c* are identical.

In Fig. 12-35*a* the zero voltage level is on the central horizontal line, making this display representative of a symmetrical ac voltage of 25 mV peak value, as in Fig. 12-34.

Figure 12-35*b* shows the zero voltage level two divisions, or 10 mV, below the center line. This display thus represents an asymmetrical ac voltage with a positive peak value of 35 mV and a negative peak value of 15 mV.

The display in Fig. 12-35*c* has its zero voltage level five divisions, or 25 mV, below the center line. The voltage represented is therefore varying direct current with a 50-mV peak-to-peak ac component and a dc level of +25 mV.

EXAMPLE 12-13 The *form factor* of a waveform is the ratio of its rms value to its average value. What is the form factor of a sine-wave current?

Solution From Eq. (12-10),

$$I_{av} = \frac{2I_{max}}{\pi}$$

From Eq. (12-11),

$$I = \frac{I_{max}}{\sqrt{2}}$$

$$\text{Form factor} = \frac{I}{I_{av}} = \frac{I_{max}}{\sqrt{2}} \div \frac{2I_{max}}{\pi} = \frac{1}{\sqrt{2}} \times \frac{\pi}{2} = \mathbf{1.111}$$

REVIEW QUESTIONS FOR SECTION 12-6

1. Referred to a sine wave's peak value, what is the
   a. Peak-to-peak value?
   b. Instantaneous value?
   c. RMS value?
2. What is the usual reference time t_0 for a sine wave?
3. How is the rms value of a sine wave related to dc voltage or current?
4. If the household electricity voltage is 120 V, what is
   a. The peak voltage?
   b. The rms voltage?
   c. The peak-to-peak voltage?

EXERCISE PROBLEMS FOR SECTION 12-6

1. A sinusoidal current has a frequency of 1000 Hz and a maximum value of 5 A. Find the instantaneous current at the following times after t_0:
   a. 1 ms (*ans*. 0 A)
   b. 200 μs (*ans*. 4.76 A)
   c. 2.25 ms (*ans*. 5 A)
   d. 0.125 ms (*ans*. 3.536 A)
   e. 35.625 ms (*ans*. −3.536 A)
   f. 0.75 ms (*ans*. −5 A)
2. A 10-kHz sinusoidal voltage has a maximum value of 25 V. Find V_{av}.
3. A 10-kHz sinusoidal voltage has an average value of 25 V. Find $V_{p\text{-}p}$. (*ans*. 78.6 V)
4. A 10-kHz sinusoidal voltage has an instantaneous value of 25 V, 12.5 μs after t_0. Find V_{av}.
5. A 10-kHz sinusoidal voltage has an average value of 25 V. Find the instantaneous value of voltage at $t = 65$ μs. (*ans*. −31.8 V)
6. Find the effective value of current when the peak value of an alternating current is 30 mA.
7. A signal generator is to be set to 5 V rms. If the generator is to be monitored by an oscilloscope, what peak-to-peak voltage should be read on the oscilloscope? (*ans*. 14.14 V)
8. If a 120-V 60-Hz line voltage is viewed on an oscilloscope, how many volts peak-to-peak will be indicated?
9. What is the average value of a 120-V 60-Hz line voltage? (*ans*. 108 V)
10. A 400-Hz generator has an output voltage of 200 V. Find the instantaneous voltage at the following times after t_0:
    a. 1 ms
    b. 2 ms
    c. 3 ms

11. $v = 212 \sin 377t$ describes a voltage.
    a. What is the rms value? (*ans*. 150 V)
    b. What is the frequency? (*ans*. 60 Hz)
12. The peak-to-peak voltage read across a 100-Ω resistor by an oscilloscope is 28.3 V. Find
    a. The effective current in the resistor.
    b. The power dissipated by the resistor.
13. An ac ammeter in series with a 470-Ω resistor reads 35 mA. Find the peak voltage across the resistor. (*ans*. 23.26 V)
14. Write the equation for the instantaneous voltage when $V = 50$ V and $f = 1000$ Hz.
15. An oscilloscope, using a current probe, indicates a peak-to-peak current of 30 mA and a wave period of 833 μs. If the circuit being measured contains a 3300-Ω resistor, find the instantaneous voltage across the resistor at 300 μs. (*ans*. 38.1 V)
16. A current probe on an oscilloscope is used to read the current in a circuit consisting of a generator and a 220-Ω resistor. Find V when the peak-to-peak value of current is 18 mA.
17. An oscilloscope is switched to a vertical range of 1 V per division and 5 μs per horizontal division when the display is as shown in Fig. 12-36. What are the following waveform values?
    a. Peak-to-peak. (*ans*. 10 V)
    b. RMS. (*ans*. 3.535 V)
    c. Average. (*ans*. 3.183 V)
    d. Frequency. (*ans*. 20 kHz)

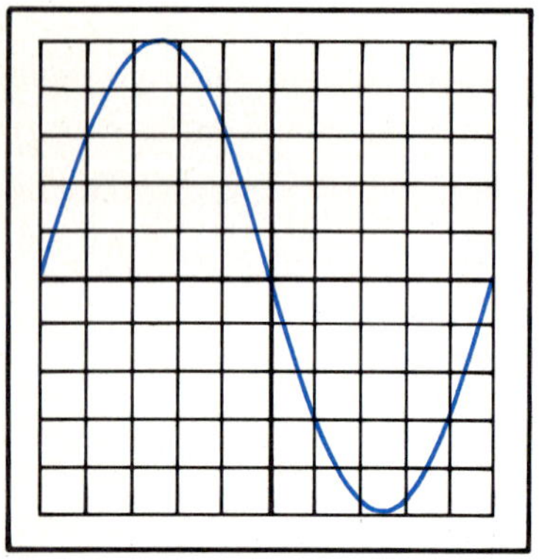

Fig. 12-36. Problem 17: Oscilloscope display.

12-7 PHASE

When two or more waveforms of the same frequency are present simultaneously in a circuit, the time relationship between them is called *phase*. When this relationship is expressed as an angle, it is called the *phase angle*.

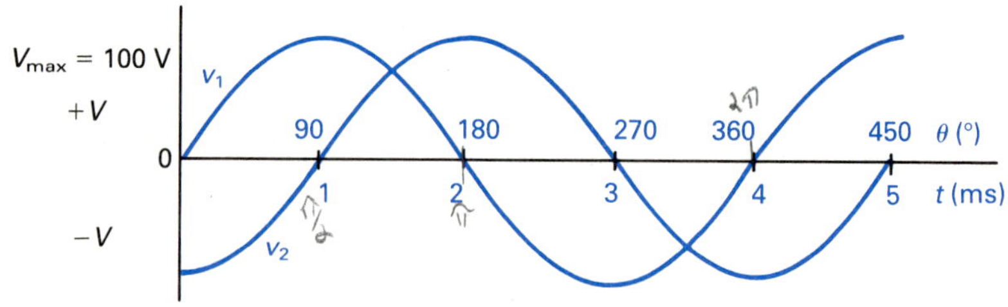

Fig. 12-37. Two out-of-phase sine-wave voltages.

Figure 12-37 shows two sine-wave voltages. They both have a peak magnitude of 100 V and a frequency of 250 Hz (the period is 4 ms). The equivalent magnitudes of each voltage have a time separation of 1 ms. The voltages have a 1-ms phase difference. As the 1-ms phase difference of v_1 and v_2 represents $\frac{1}{4}$ cycle, the phase angle between the two voltages is 90° (or $\pi/2$ rad).

The symbol for phase angle is ϕ or θ (lowercase Greek letter phi or theta). The more commonly used symbol is ϕ. The SI unit for phase angle is the radian (rad), although at present the degree (°) is more frequently used.

Lead and Lag

In a waveform diagram, such as Fig. 12-37, movement to the right means successive time intervals or *elapsed* time. Time 1 (1 ms) occurs after time 0 (*reference* time); time 2 (2 ms) occurs after time 1.

Each point on the waveform of v_2 occurs after the equivalent point on the waveform of v_1. For instance, the positive peak of v_1 occurs at time 1, whereas the positive peak of v_2 occurs 1 ms later at time 2. It is said that voltage v_2 *lags* voltage v_1.

Each point on the waveform of v_1 occurs before the equivalent point on the waveform of v_2. Voltage v_1 *leads* voltage v_2.

A leading phase relationship is regarded as positive, and a lagging phase as negative. In Fig. 12-37 the phase angle of v_1 with respect to v_2 is $+\phi$ ($+90°$ or $+\pi/2$ rad). The phase angle of v_2 with respect to v_1 is $-\phi$ ($-90°$ or $-\pi/2$ rad).

It is important to realize that the phase between two ac quantities is meaningful only if they have the same frequency. If they have different frequencies, the time between equivalent points on the respective waveforms will continually change, and a value cannot be specified.

In Phase, Antiphase, and Phase Quadrature

Three phase relationships encountered often in circuits are each given specific names. They are shown in Fig. 12-38.

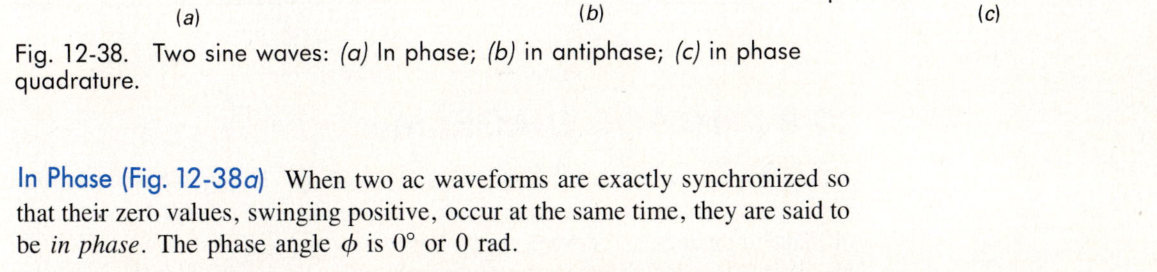

Fig. 12-38. Two sine waves: *(a)* In phase; *(b)* in antiphase; *(c)* in phase quadrature.

In Phase (Fig. 12-38*a*) When two ac waveforms are exactly synchronized so that their zero values, swinging positive, occur at the same time, they are said to be *in phase*. The phase angle ϕ is 0° or 0 rad.

Antiphase (Fig. 12-38*b*) If two waveforms have their zero values occurring at the same time, but one swings positive when the other swings negative, they are said to be in *antiphase*. The phase angle ϕ is 180° or π rad. Another common way of expressing antiphase is *180° out of phase*.

Phase Quadrature (Fig. 12-38*c*) If two waveforms have their positive-going zero values displaced by one quarter-cycle, they are said to be in *phase quadrature*. The phase angle ϕ is 90° or $\pi/2$ rad.

The voltages in Fig. 12-37 are in phase quadrature, with v_1 leading v_2.

In waveform diagrams for in-phase and antiphase conditions (refer to Fig. 12-38*a* and *b*), there is no clear lag or lead. For all other phase angles between 0° and 180°, lag and lead show clearly. A phase angle of over 180° is confusing in waveform diagrams. For example, the diagram for 270° lead is the same as for 90° lag. It must be made clear that these statements refer only to waveform diagrams, and not to other ways of depicting phase relationships, such as phasor diagrams or trigonometric expressions.

Circuit Phase Angle

The angle between the *total voltage* and the *total current,* with voltage as the reference, is the *circuit phase angle*.

If the *current lags* the *voltage,* the circuit is said to have a *lagging phase angle*. Conversely, a *leading phase angle* occurs when the *current leads* the *voltage*.

REVIEW QUESTIONS FOR SECTION 12-7

1. If two waves have a phase difference of $\frac{1}{4}$ cycle, what is their phase angle?
2. What is meant by phase
   a. Lead?
   b. Lag?
3. What is meant by
   a. In phase?
   b. In antiphase?
   c. In phase quadrature?
4. To what does the term "circuit phase angle" refer?

12-8 PHASOR DIAGRAM

Although ac quantities can be drawn as waveforms, for circuit analysis this procedure is tedious. Furthermore, if several quantities are involved, it becomes difficult to distinguish between them. For these reasons it is convenient to use phasor diagrams to present ac relationships.

A phasor diagram consists of a number of straight lines, one for each quantity represented. Each line is drawn from a common origin. The length of each line is proportional to the magnitude of the quantity it represents. As the purpose of this scaling is to give a visual comparison of magnitudes, the values must be equivalent for each quantity. Rms values are usually adopted. The angles between lines represent the phase angles between the quantities.

The common current in a series circuit and the common voltage in a parallel circuit are typical reference quantities. The phasor lines for such reference quantities are usually placed in a horizontal position and point to the right.

It is a standard convention that counterclockwise rotation from one phasor line to another is regarded as positive motion and represents a leading phase. Conversely, clockwise rotation is negative motion representing a lagging phase. Figure 12-39 shows some phasor conventions.

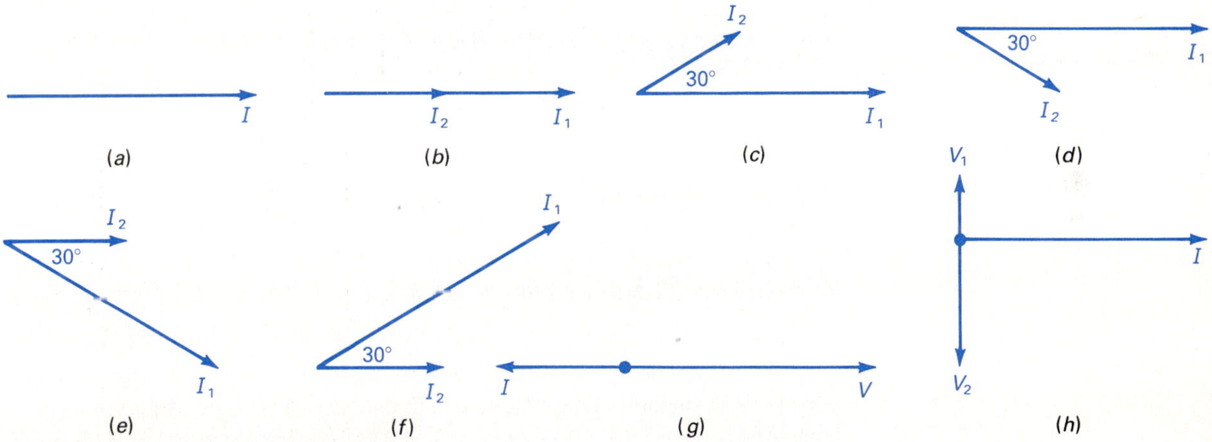

Fig. 12-39. Typical phasor diagrams.

Figure 12-39*a* shows a phasor for a current I. Its horizontal position with the arrowhead pointing to the right identifies it as being in the *reference* position.

In Fig. 12-39*b* two current phasors I_1 and I_2 are both in the reference position. They are in phase. The rms value of I_1 is twice that of I_2.

In Fig. 12-39*c* the current I_1 again has twice the magnitude of I_2, but now I_1 is a reference phasor with I_2 leading it by 30° or $\pi/6$ rad.

Figure 12-39*d* shows current I_1 with twice the magnitude of I_2. I_2 lags I_1 by 30°.

Figure 12-39*e* and *f* shows the same circuit conditions as Fig. 12-39*c* and *d*, respectively. I_2 is shown as the reference phasor in Fig. 12-39*e* and *f*. It may be said that Fig. 12-39*e* is Fig. 12-39*c* rotated 30° clockwise, and that Fig. 12-39*f* is Fig. 12-39*d* rotated 30° counterclockwise.

In Fig. 12-39*g* a voltage and a current are shown in antiphase.

Two voltages, each in phase quadrature with a current, are shown in Fig. 12-39*h*.

Ratios or products of voltage and current magnitudes (V/I, I/V, or VI) cannot be determined from phasor diagrams such as Fig. 12-39*g* and *h*, unless the magnitude scalings for voltage and current are specified.

REVIEW QUESTIONS FOR SECTION 12-8

1. In a phasor diagram, what is indicated by
   a. The length of a line?
   b. The angle between two lines?
2. In a phasor diagram, what is the usual reference phasor for a
   a. Series circuit?
   b. Parallel circuit?
3. How does a phasor diagram distinguish between lead and lag?
4. How does a phasor diagram show
   a. In phase?
   b. Antiphase?

12-9 POLAR AND RECTANGULAR

A phasor diagram is drawn if it is desired to present a visual impression of circuit conditions. Whether or not the diagram is drawn, the phasor concept is used in circuit work by expressing the length and angular position of phasors in either polar or rectangular form.

Phasors in Polar Form

A phasor is represented in a diagram as a line of length proportional to the quantity magnitude, positioned so that its angle with respect to the reference position is the relative phase angle of the quantity. This type of presentation is known as *polar*.

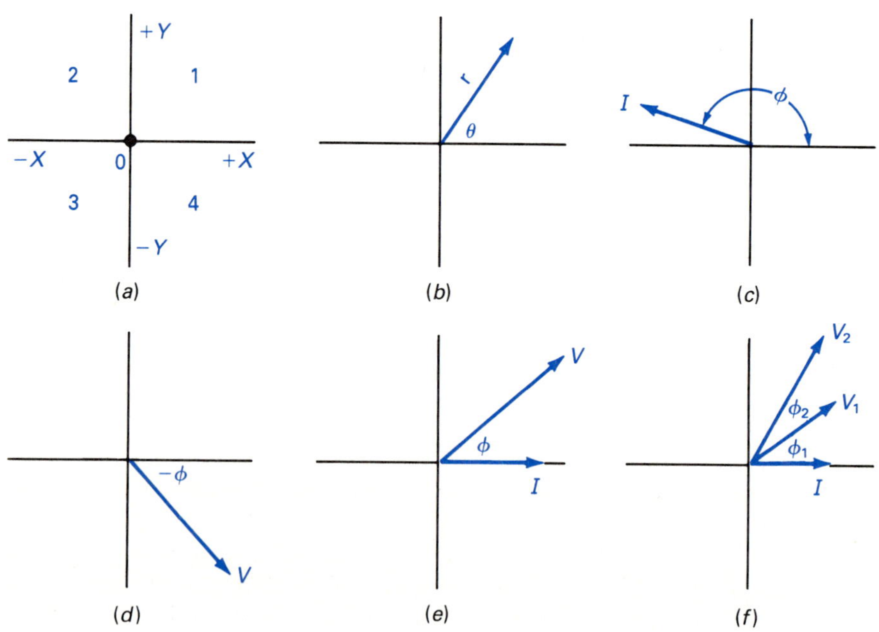

Fig. 12-40. Polar presentation of phasors.

Figure 12-40*a* shows the basic polar coordinate system. The horizontal, or *X*, axis extends to the right (positive) and to the left (negative) of the origin 0. The vertical, or *Y*, axis extends up (positive) and down (negative) from the origin. The quadrants so formed are given the numbers 1 through 4 in a counterclockwise rotation from the reference position.

Figure 12-40*b* shows a typical phasor. Its magnitude is represented by the length *r*, and its relative phase by the angle θ. The phasor is called $r\underline{/\theta}$ (magnitude *r*, angle θ).

Figure 12-40*c* shows a current phasor $I\underline{/\phi}$.

The voltage phasor in Fig. 12-40*d* is $V\underline{/-\phi}$.

The two phasors in Fig. 12-40*e* are $I\underline{/0}$ and $V\underline{/\phi}$.

The three phasors in Fig. 12-40*f* are $I\underline{/0}$, $V_1\underline{/\phi_1}$, and $V_2\underline{/\phi_1 + \phi_2}$.

It should be noted that all phasor magnitudes are measured from the common origin, and all phase angles from the reference position.

Scalar Value

A phasor quantity is completely expressed in its polar form $r\underline{/\theta}$. In many applications of ac, only the current and voltage magnitudes are of interest. For this reason, special notation is used to denote the magnitude of a phasor quantity while disregarding the phase angle. For the phasor quantity $r\underline{/\theta}$, the magnitude alone can be expressed as $|r|$. Alternative terms for magnitude are "scalar value" and "modulus."

The practice of writing a quantity symbol between vertical lines to indicate a scalar value is not always rigidly adhered to. When, in a given situation, it is clear that magnitudes are referred to, the vertical lines can be omitted. This technique is adopted in the following chapters.

Complex Quantities

Phasors expressed in rectangular form use complex numbers to identify horizontal and vertical components. Complex quantities are not complicated. They are called "complex" because they have so-called real and imaginary parts. Before explaining how the rectangular form of phasor presentation is derived, an introduction to the *j* operator will be given. The *j*-operator concept is the basis of complex quantities and is the only difficult idea involved.

Signs like +, −, ×, ÷, and $\sqrt{\ }$ are *operators,* because they convey instructions to carry out operations. The operator *j* gives the instruction, "rotate 90° or $\pi/2$ rad." Often it means, "change the direction from horizontal to vertical."

Unlike most operators, *j* has a numerical value, $\sqrt{-1}$, and can be operated on by other operators or even by *j* itself. To use *j* in ac circuit work it is necessary to get an idea of how to arrive at its numerical value.

In Fig. 12-41*a* the $+n$-unit phasor lies in the reference position. To move this phasor into the negative position to the left of the origin, it must be rotated through an angle of π rad (180°). It then becomes a $-n$-unit phasor. As $-n = (-1) \times (+n)$, it can be said that the original $+n$ phasor has been rotated through 180° or multiplied by -1.

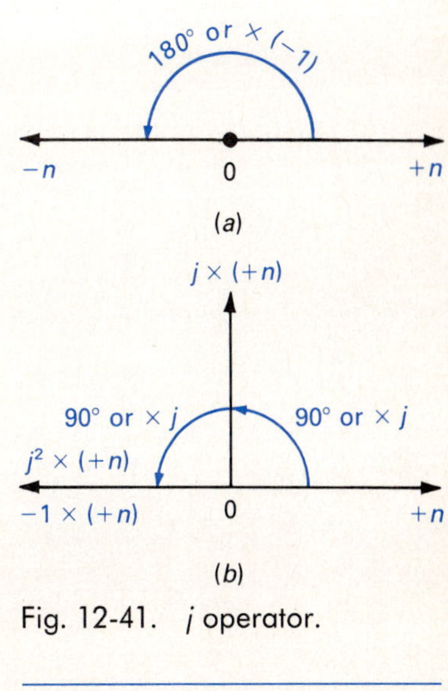

Fig. 12-41. *j* operator.

In Fig. 12-41b the $+n$ phasor has been moved into the $-n$ position in two stages. First it was moved into the vertical position by a 90° rotation, and then it was rotated a further 90°. Each 90° rotation is defined as multiplying by j. To go from $+n$ to $-n$ two multiplications by j are needed, thus implying multiplication by j^2.

Combining the results of Fig. 12-41a and b, we see that rotation through 180° is the same as multiplying by -1 or multiplying by j^2. Therefore,

$$j^2 = -1 \qquad \text{or} \qquad j = \sqrt{-1} \tag{12-12}$$

Because it is not a rational quantity, j is called *imaginary*. However, the use of this term is unfortunate because it gives the impression that j is not valid for use in calculating, which it certainly is, since j^2 is a real number, -1.

Listing the effect of successive rotations through 90° ($\pi/2$ rad) leads to the rectangular coordinate system.

Reference Position	$+n$
90° (π/2-rad) rotation	jn or $\sqrt{-1}n$
180° (π-rad) rotation	j^2n or $-1n$ or $-n$
270° (3π/2-rad) rotation	j^3n or $-\sqrt{-1}n$ or $-jn$
360° (2π-rad) rotation	j^4n or $(-1)^2n$ or $+n$

Phasors in Rectangular Form

Figure 12-42a shows rectangular coordinates with j notation. In Fig. 12-42b to e phasors are shown in each of the four quadrants. They are

Figure 12-42b:	$a + jb$
Figure 12-42c:	$-a + jb$
Figure 12-42d:	$-a - jb$
Figure 12-42e:	$a - jb$

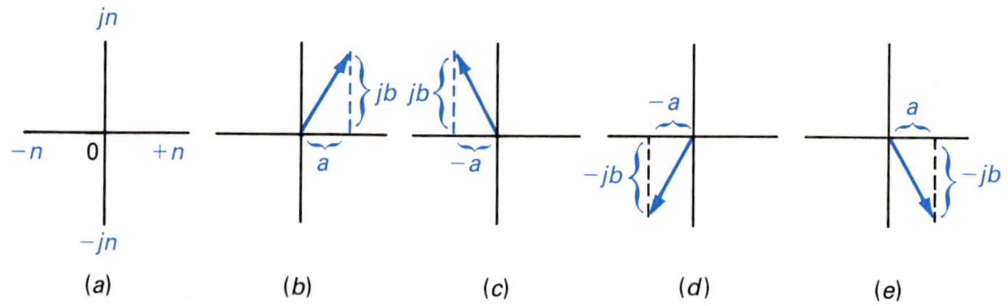

Fig. 12-42. Rectangular presentation of phasors.

Conversion between Polar and Rectangular Forms

A phasor expressed in polar form can be converted into its equivalent rectangular form by applying trigonometry.

The length r of the phasor is the hypotenuse of a right triangle of which the horizontal component a is the adjacent side and the vertical component jb is the opposite side. These symbols are shown in Fig. 12-43b.

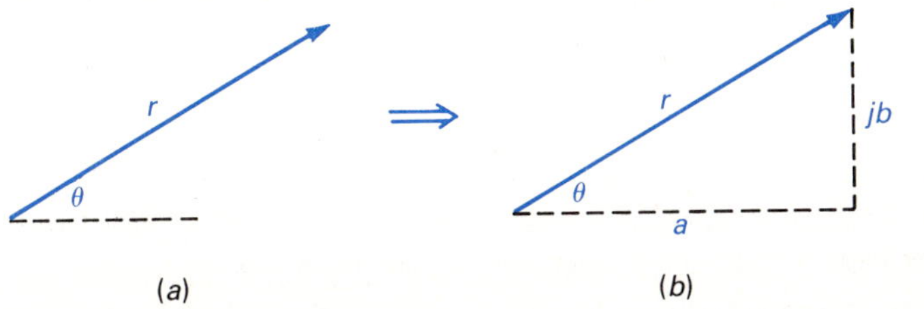

Fig. 12-43. Polar-to-rectangular conversion.

The horizontal component is sometimes called the real part of the phasor. Its magnitude is given by

$$|a| = r \cos \theta$$

The vertical component is sometimes called the imaginary part of the phasor. Its magnitude is given by

$$|b| = r \sin \theta$$

The rectangular expression for the phasor is therefore

$$a + jb = r \cos \theta + jr \sin \theta$$

Combining this with the polar expression, we obtain the equation for conversion from polar to rectangular form,

$$r \angle \theta = r \cos \theta + jr \sin \theta \qquad \textbf{(12-13)}$$

A phasor expressed in rectangular form can be converted into polar form by applying Pythagoras' theorem and trigonometry.

The magnitudes of the phasor and its rectangular components are r, a, and b, as shown in Fig. 12-44a.

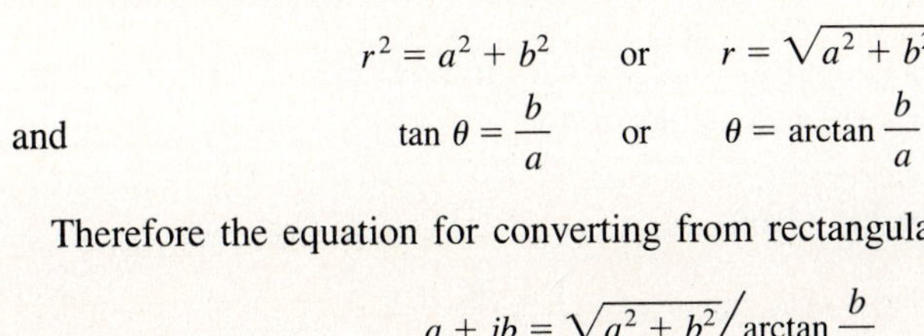

Fig. 12-44. Rectangular-to-polar conversion.

$$r^2 = a^2 + b^2 \qquad \text{or} \qquad r = \sqrt{a^2 + b^2}$$

and

$$\tan \theta = \frac{b}{a} \qquad \text{or} \qquad \theta = \arctan \frac{b}{a}$$

Therefore the equation for converting from rectangular to polar form is

$$a + jb = \sqrt{a^2 + b^2} \angle \arctan \frac{b}{a} \qquad \textbf{(12-14)}$$

EXAMPLE 12-14 In a three-element series network the current I is 1 A. The voltages across the elements are $V_1 = 10$ V in phase with I; $V_2 = 20$ V in phase quadrature, leading I; and $V_3 = 25$ V in phase quadrature, lagging I. Draw to scale *(a)* a waveform diagram and *(b)* a phasor diagram.

Solution

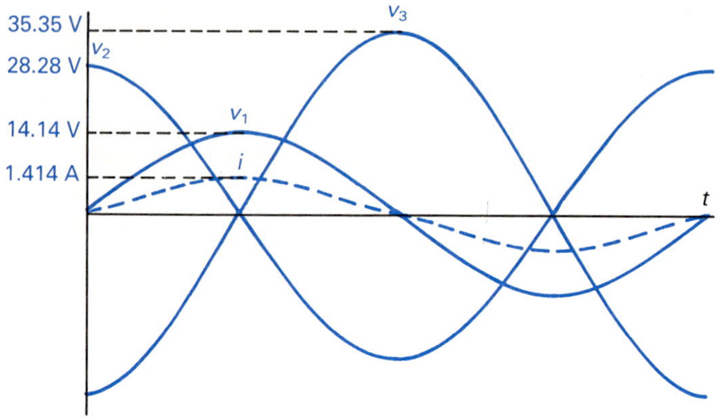

Fig. 12-45. Example 12-14: Waveforms.

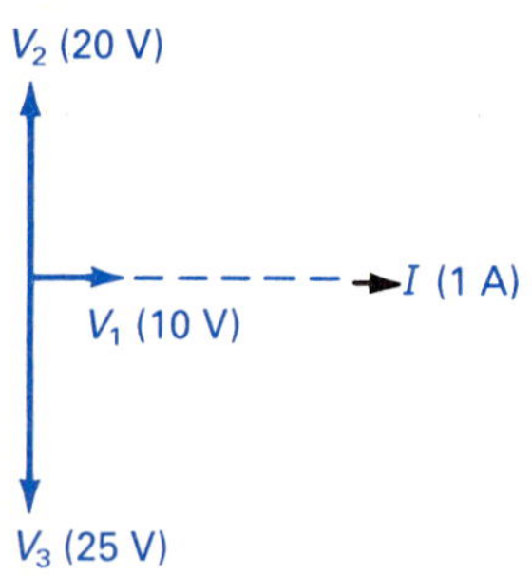

Fig. 12-46. Example 12-14: Phasor diagram.

a. See Fig. 12-45. The given current and voltage symbols I, V_1, V_2, and V_3 indicate rms values. The related maximum values are shown on the vertical axis in Fig. 12-45. The curves are labeled with instantaneous-value symbols. The current curve is shown as a broken line, and the voltage curves as solid lines.

b. See Fig. 12-46.

In waveform or phasor diagrams the relative scaling for voltage and current has no significance, as two different units are involved. Any convenient scaling can be chosen. In Fig. 12-45, the current magnitude is shown smaller than that of all the voltages. In Fig. 12-46, the current magnitude is shown greater. In both Figs. 12-45 and 12-46 the voltage magnitudes have been shown to scale, that is, in their correct proportions.

EXAMPLE 12-15 Express the current and voltages in Example 12-14 in *(a)* polar form and *(b)* rectangular form.

Solution Refer to Fig. 12-46.

a.

$$I = \mathbf{1\ A}\ \underline{/0^\circ}$$
$$V_1 = \mathbf{10\ V}\ \underline{/0^\circ}$$

$$V_2 = \mathbf{20\ V\ \underline{/90°}}$$
$$V_3 = \mathbf{25\ V\ \underline{/-90°}}$$

b.

$$I = \mathbf{(1 + j0)\ A}$$
$$V_1 = \mathbf{(10 + j0)\ V}$$
$$V_2 = \mathbf{(0 + j20)\ V}$$
$$V_3 = \mathbf{(0 - j25)\ V}$$

EXAMPLE 12-16 The voltage across a parallel network is 50 mV. The current in one branch is 100 mA and leads the voltage by 60°. The current in a second branch is 150 mA and lags the voltage by 45°. Draw a phasor diagram for this network.

Solution See Fig. 12-47.

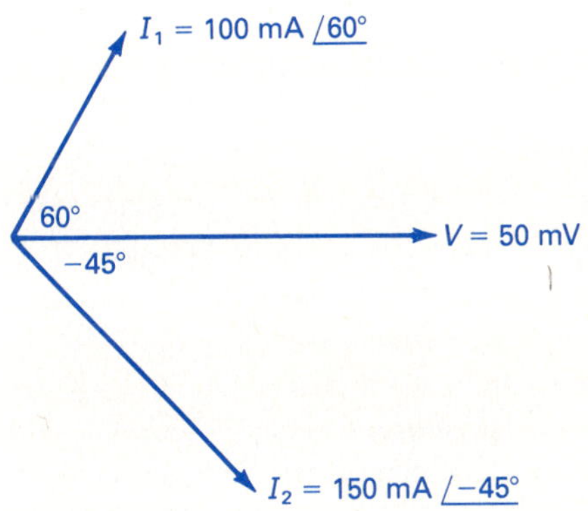

Fig. 12-47. Examples 12-16 and 12-17: Phasor diagram.

EXAMPLE 12-17 Express the voltage and the currents in Example 12-16 in *(a)* polar form and *(b)* rectangular form.

Solution Refer to Fig. 12-47.

a.

$$V = \mathbf{50\ mV\ \underline{/0°}}$$
$$I_1 = \mathbf{100\ mA\ \underline{/60°}}$$
$$I_2 = \mathbf{150\ mA\ \underline{/-45°}}$$

b. Apply Eq. (12-13) ($r\,\underline{/\theta} = r \cos\theta + j \sin\theta$):

$$V = 50 \cos 0° + j \sin 0° = \mathbf{(50 + j0)\ mV}$$
$$I_1 = 100 \cos 60° + j100 \sin 60° = 100 \times 0.5 + j100 \times 0.866$$
$$I_1 = \mathbf{(50 + j86.6)\ mA}$$
$$I_2 = 150 \cos(-45°) + j150 \sin(-45°)$$
$$= 150 \times 0.7071 + j150 \times (-0.7071)$$
$$I_2 = \mathbf{(106 - j106)\ mA}$$

EXAMPLE 12-18 A network is connected to a 120-V 60-Hz commercial supply. Relative to this emf, there is a network pd of $(75 - j30)$ V and a network current of $(-2.5 + j3)$ A. Express the emf and the network pd and current in polar form.

Solution The supply voltage is, in a case like this, regarded as the reference. A phasor diagram helps to visualize the conditions. (See Fig. 12-48.)

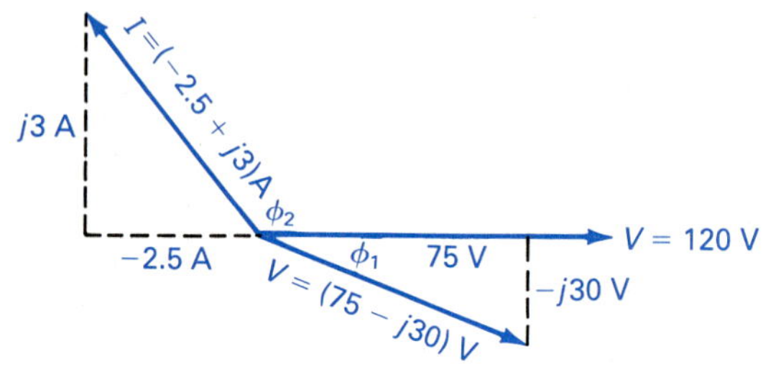

Fig. 12-48. Example 12-18: Phasor diagram.

$$V = 120 + j0 \text{ V} = \mathbf{120\ V\ \underline{/0^\circ}}$$

Apply Eq. (12-14):

$$V = \sqrt{75^2 + 30^2}\ \underline{/\arctan(-30)/75} = \mathbf{80.78\ V\ \underline{/-21.8^\circ}} \quad \text{or} \quad \mathbf{\underline{/-0.3805\ rad}}^*$$

$$I = \sqrt{2.5^2 + 3^2}\ \underline{/\arctan 3/(-2.5)} = \mathbf{3.905\ A\ \underline{/129.8^\circ}} \quad \text{or} \quad \mathbf{\underline{/2.265\ rad}}$$

It was previously mentioned that a knowledge of trigonometry is necessary for ac circuit work. This knowledge should include familiarity with the tangent forms for each quadrant.

A phasor diagram will give the location of a phasor's angle, as is evident from Fig. 12-48. Angle ϕ_1 [arctan $(-30)/75$] is in the fourth quadrant, and its value is thus between 270 and 360° (between $3\pi/2$ and 2π rad), or between -90 and 0°. Angle ϕ_2 [arctan $3/(-2.5)$] is in the second quadrant, its value being between 90 and 180° (between $\pi/2$ and π rad).

It is likely that you will interconvert between tangents and angles by using tables or an electronic calculator and that the angles will be in degrees. To convert degrees to radians, the SI unit for angles, divide by 57.3. For instance, in Example 12-18, the phase angle of the current is 129.8°, which, divided by 57.3, is 2.265 rad.

If you are not accustomed to using the radian, it is recommended that you make an effort to develop some familiarity with this unit. At present the degree is used more often than the radian, and there is little evidence that the radian will become dominant in the near future. However, the radian is the SI unit by international agreement, and its use will probably increase.

*Angles are given in both degrees and radians. For example, in this case, $\underline{/-21.8^\circ}$ and $\underline{/-0.3805\text{ rad}}$ are the same angle.

REVIEW QUESTIONS FOR SECTION 12-9

1. What are the components of a complex quantity called?
2. Give the rectangular equivalent of the phasor $I\,\angle\theta$.
3. Give the polar equivalent of the phasor $(a + jb)$.
4. What is the conversion factor to convert degrees to radians?

EXERCISE PROBLEMS FOR SECTION 12-9

1. Two devices A and B are connected in series across a 120-V 60-Hz energy source. If the phase angle of the emf with respect to the current is $-51°$ and the voltage across device A is $95\text{ V}\,\angle 50°$, what is the voltage across device B? (*ans*. 166.6 V, $-85°$)
2. Express the current and voltages shown in Fig. 12-49 in
   a. Polar form.
   b. Rectangular form.
3. For Fig. 12-50, express V with respect to I in
   a. Polar form. (*ans*. 88.6 V, $-73.6°$)
   b. Rectangular form. (*ans*. $(25 - j85)$ V)

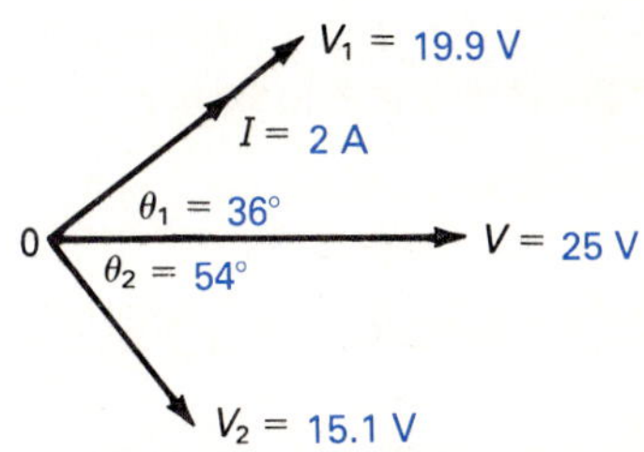

Fig. 12-49. Problem 2.

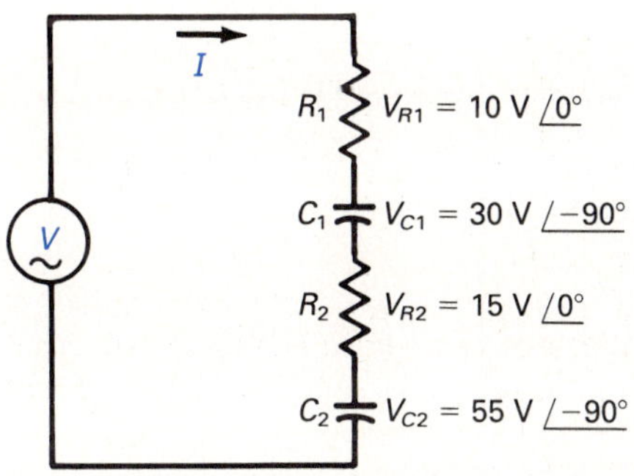

Fig. 12-50. Problem 3.

4. Two inductors A and B are connected in series across a source of emf. $V_{L_A} = 75\text{ V}\,\angle 80°$ and $V_{L_B} = 65\text{ V}\,\angle 61°$. Find the magnitude of the emf and its phase angle with respect to the current.

12-10 PHASOR ARITHMETIC

Addition, subtraction, multiplication, division, and other arithmetic operations on phasor quantities are all encountered in ac circuit calculations. By the use of rectangular and polar phasor forms, any of these operations can be carried out with comparative ease. The following techniques should be memorized.

Addition and Subtraction *Use the rectangular phasor form*. Using ordinary arithmetic methods, add or subtract the real parts of each phasor, and then add or subtract the imaginary (j) parts. Keep the sums or differences of each part separate.

The general operations are

Addition: $$(a + jb) + (c + jd) = (a + c) + j(b + d) \tag{12-15}$$
Subtraction: $$(a + jb) - (c + jd) = (a - c) + j(b - d) \tag{12-16}$$

a, b, c, and d are the magnitudes of the component parts of the phasors and may be either positive or negative.

Multiplication and Division

Use the polar phasor form. Multiply or divide the magnitudes, and add or subtract the phase angles.

The general operations are

Multiplication: $$r_1 \underline{/\theta_1} \times r_2 \underline{/\theta_2} = r_1 r_2 \underline{/\theta_1 + \theta_2} \tag{12-17}$$
Division: $$r_1 \underline{/\theta_1} \div r_2 \underline{/\theta_2} = r_1 / r_2 \underline{/\theta_1 - \theta_2} \tag{12-18}$$

Powers and Roots

Raising to a power or extracting a root is rarely used in phasor work. However, when it is needed, the use of exponents can be regarded as an extension of multiplication.

$$(r \underline{/\theta})^n = r^n \underline{/n\theta} \tag{12-19}$$

Squaring and taking a square root are good examples of the use of exponents.

$$(r \underline{/\theta})^2 = r^2 \underline{/2\theta}$$
$$\sqrt{r \underline{/\theta}} = (r \underline{/\theta})^{1/2} = \sqrt{r} \underline{/\theta/2}$$

Reciprocal of a Phasor

The reciprocal of phasor $r \underline{/\theta}$ is $(1/r) \underline{/-\theta}$. This can be determined as follows:

$$\frac{1}{r \underline{/\theta}} = (r \underline{/\theta})^{-1} = r^{-1} \underline{/-1(\theta)}$$
$$\frac{1}{r \underline{/\theta}} = \frac{1}{r} \underline{/-\theta} \tag{12-20}$$

Similarly, $$\frac{1}{r \underline{/-\theta}} = \frac{1}{r} \underline{/+\theta} \tag{12-21}$$

Reciprocal of j

The reciprocal of j is $-j$ (or the reciprocal of $-j$ is j). This can be shown to be true by applying the phasor reciprocal principle above or by *rationalizing*. Rationalizing is based on the idea that multiplying any number by 1 will not change the number. Any number divided by itself equals 1; this includes $j/j = 1$.

The reciprocal of j is $1/j$,

$$\frac{1}{j} = \frac{1}{j} \times \frac{j}{j} = \frac{j}{j^2} = \frac{j}{-1}$$
$$\frac{1}{j} = -j \tag{12-22}$$

The reciprocal of $-j$ is $1/(-j)$,

$$\frac{1}{-j} = \frac{1}{-j} \times \frac{j}{j} = \frac{j}{-j^2} = \frac{j}{+1}$$

$$\frac{1}{-j} = j \qquad \textbf{(12-23)}$$

EXAMPLE 12-19 There are two voltages in a series circuit, $V_1 = (10 + j15)$ V and $V_2 = (7 - j5)$ V. What is the total circuit voltage V_T?

Solution In phasor form, Kirchhoff's voltage law becomes,

$$V_T = V_1 + V_2 = (10 + j15) + (7 - j5) = (10 + 7) + j(15 - 5)$$
$$V_T = \mathbf{(17 + j10)\ V}$$

EXAMPLE 12-20 The total current I_T in a two-branch series-parallel circuit is $(250 - j150)$ mA. One of the branch currents I_1 is $(-100 + j100)$ mA. What is the other branch current I_2?

Solution Apply Kirchhoff's current law,

$$I_T = I_1 + I_2 \qquad \text{or} \qquad I_2 = I_T - I_1$$

Use phasor subtraction,

$$\begin{aligned} I_2 &= (250 - j150) - (-100 + j100) \\ &= [250 - (-100)] + j[(-150) - 100] \\ &= (250 + 100) + j(-150 - 100) \\ I_2 &= \mathbf{(350 - j250)\ mA} \end{aligned}$$

EXAMPLE 12-21 The ratio of the voltage to the current in a two-branch parallel network is given by

$$\frac{V}{I} = \frac{(10 + j20)(5 + j3)}{(10 + j20) + (5 + j3)}$$

Express this ratio as a polar phasor.

Solution The numerator N involves multiplication, so each phasor should be converted to polar form.

$$10 + j20 = \sqrt{10^2 + 20^2}\ \underline{/\arctan \tfrac{20}{10}} = 22.36\ \underline{/63.43^\circ} \text{ or } \underline{/1.107 \text{ rad}}$$
$$5 + j3 = \sqrt{5^2 + 3^2}\ \underline{/\arctan \tfrac{3}{5}} = 5.831\ \underline{/30.96^\circ} \text{ or } \underline{/0.5403 \text{ rad}}$$
$$N = (10 + j20)(5 + j3) = 22.36\ \underline{/63.43^\circ} \times 5.831\ \underline{/30.96^\circ}$$

Using phasor multiplication,

$$N = 22.36 \times 5.831\ \underline{/63.43^\circ + 30.96^\circ}$$
$$N = 130.4\ \underline{/94.39^\circ} \qquad \text{or} \qquad \underline{/1.647 \text{ rad}}$$

The denominator D involves addition, so the rectangular form is used as given:

$$D = (10 + j20) + (5 + j3) = 15 + j23$$

For division D should be converted to polar form:

$$D = \sqrt{15^2 + 23^2}\ \underline{/\arctan \tfrac{23}{15}} = 27.45\ \underline{/56.89^\circ} \quad \text{or} \quad \underline{/0.9928 \text{ rad}}$$

$$\frac{V}{I} = \frac{N}{D} = \frac{130.4\ \underline{/94.39^\circ}}{27.45\ \underline{/56.89^\circ}} \quad \text{or} \quad \frac{\underline{/1.647 \text{ rad}}}{\underline{/0.9928 \text{ rad}}}$$

Using phasor division,

$$\frac{V}{I} = \frac{130.4}{27.45}\ \underline{/94.39^\circ - 56.89^\circ} \quad \text{or} \quad \underline{/1.647 \text{ rad} - 0.9928 \text{ rad}}$$

$$\frac{V}{I} = \mathbf{4.75\ \underline{/37.5^\circ}} \quad \text{or} \quad \mathbf{\underline{/0.6542 \text{ rad}}}$$

REVIEW QUESTIONS FOR SECTION 12-10

1. What is the preferred form for phasor quantities that are to be
a. Added?
b. Subtracted?
c. Multiplied?
c. Divided?

2. What is the reciprocal of
a. $V\ \underline{/\phi}$?
b. ja

EXERCISE PROBLEMS FOR SECTION 12-10

1. Evaluate the following:
a. $(10 + j10) + (15 + j15)$ (*ans.* $25 + j25$)
b. $(5 + j25) + (20 - j10)$ (*ans.* $25 + j15$)
c. $(15 - j30) + (-10 - j20)$ (*ans.* $5 - j50$)

2. Evaluate the following:
a. $(+j10)(+j20)$
b. $(+j15)(-j25)$
c. $(-j50)(-j30)$

3. Evaluate the following:
a. $(14\ \underline{/35^\circ})(8\ \underline{/15^\circ})$ (*ans.* $112\ \underline{/50^\circ}$)
b. $(25\ \underline{/75^\circ})(6\ \underline{/-40^\circ})$ (*ans.* $150\ \underline{/35^\circ}$)
c. $(21\ \underline{/-24^\circ})(11\ \underline{/-12^\circ})$ (*ans.* $231\ \underline{/-36^\circ}$)

4. Evaluate the following:
a. $14\ \underline{/35^\circ} + 8\ \underline{/15^\circ}$
b. $25\ \underline{/75^\circ} + 6\ \underline{/-40^\circ}$
c. $22\ \underline{/-24^\circ} + 11\ \underline{/-12^\circ}$

5. Evaluate the following:
   a. $(10 + j10)(15 + j15)$ (*ans.* $300 \angle 90°$)
   b. $(5 + j25)(20 - j10)$ (*ans.* $570 \angle 52.13°$)
   c. $(15 - j30)(-10 - j20)$ (*ans.* -750)
6. Evaluate the following:
   a. $(10 + j10)/(15 + j15)$
   b. $(5 + j25)/(20 - j10)$
   c. $(15 - j30)/(-10 - j20)$
7. Convert the following quantities to rectangular form and draw each result as a phasor.
   a. $120 \text{ V} \angle -6°$ [*ans.* $(119.3 - j12.54)$ V]
   b. $30 \text{ mA} \angle 65°$ [*ans.* $(12.68 + j27.19)$ mA]
   c. $240 \text{ V} \angle 48°$ [*ans.* $(160.6 + j178.4)$ V]
8. Convert the following quantities to polar form:
   a. $(707 + j707)$ V
   b. $(40 - j13)$ A
   c. $(40 - j360)$ V
9. Two inductors *A* and *B* are connected in series across a 120-V 60-Hz generator. The angle of total voltage with respect to current is 80°. Voltage $V_{L_A} = 75 \text{ V} \angle 80°$. Find V_{L_B} in polar form. (*ans.* $45 \text{ V} \angle 80°$)
10. Two motors are connected in parallel across a 120-V 60-Hz source. Find the total current with its phase angle referenced to voltage when the current drawn by the motors are $2.3 \text{ A} \angle -28°$ and $3.3 \text{ A} \angle -38°$.

12-11 IMPEDANCE AND ADMITTANCE

"Impedance" and "admittance" are general terms for the characteristics of an element or a network which determine current-voltage ratios. Although they can vary, impedance and admittance are normally regarded as parameters, as it is usual to be primarily interested in the current-voltage behavior of networks with impedances and admittances of constant value.

The words "impedance" and "admittance" are self-explanatory, referring to the ability to impede or restrict current and the ability to admit or permit current, respectively. Both impedance and admittance apply to all types of voltage and current, and so phasor values are relevant

Impedance

The *impedance* of an element or a network is the voltage-current ratio. The symbol for impedance is *Z*, and the unit for impedance is the ohm, symbol Ω.

$$Z = \frac{V}{I} \quad \text{or} \quad V = ZI \tag{12-24}$$

Admittance

The *admittance* of an element or a network is the current-voltage ratio. The symbol for admittance is Y, and the unit for admittance is the siemens, symbol S.

$$Y = \frac{I}{V} \qquad \text{or} \qquad I = YV \tag{12-25}$$

Because they are general, the terms "impedance" Z and "admittance" Y can be used for any type of individual element or any combination of elements. However, it is usual to employ specific terms and symbols whenever it is convenient.

Specific Impedances and Admittances

Specific terms are used to refer to the voltage-current behavior of resistance, which dissipates energy, and of capacitance and inductance, which store energy.

Resistance *Resistance* is the impedance of an element or a network in which there can be no energy storage. This view of resistance is the same as that adopted in the dc circuit sections of this text. The use of the word "impedance" does not conflict with, but adds a new dimension to, the concept of resistance. The symbol for resistance is R.

Reactance *Reactance* is the impedance of an element or a network in which all the energy can be stored in electric or magnetic fields. The general reactance symbol is X, the inductive reactance symbol is X_L, and the capacitive reactance symbol is X_C.

Conductance *Conductance* is the admittance of an element or a network in which there can be no storage of energy. The symbol for conductance is G.

Susceptance *Susceptance* is the admittance of an element or a network in which all the energy can be stored in electric or magnetic fields. The general susceptance symbol is B, the inductive susceptance symbol is B_L, and the capacitive susceptance symbol is B_C.

Reciprocal Parameters

All impedances are voltage-current ratios, while all admittances are current-voltage ratios. Thus they are *reciprocal* parameters.

$$Z = \frac{1}{Y} \qquad \text{or} \qquad R = \frac{1}{G} \qquad \text{or} \qquad X_C = \frac{1}{B_C} \qquad \text{or} \qquad X_L = \frac{1}{B_L} \tag{12-26}$$

EXAMPLE 12-22 A three-element series network carries a current of 2 A. The elements are *(a)* a resistance of 100 Ω, *(b)* a capacitive reactance of 50 Ω, and *(c)* an inductive reactance of 75 Ω. What is the magnitude of the voltage across each element?

Solution *a.*

$$R = 100\ \Omega$$
$$V = IR = 2\text{ A} \times 100\ \Omega = \mathbf{200\ V}$$

b.

$$X_C = 50\ \Omega$$
$$V = IX_C = 2\text{ A} \times 50\ \Omega = \mathbf{100\ V}$$

c.

$$X_L = 75\ \Omega$$
$$V = IX_L = 2\text{ A} \times 75\ \Omega = \mathbf{150\ V}$$

EXAMPLE 12-23 A parallel network consisting of the same elements as those in Example 12-22 is supplied with 120-V 60-Hz ac. What is the magnitude of the current in each element?

Solution *a.*

$$G = \frac{1}{R} = \frac{1}{100\ \Omega} = 0.01\text{ S}$$
$$I = VG = 120\text{ V} \times 0.01\text{ S} = \mathbf{1.2\ A}$$

b.

$$B_C = \frac{1}{X_C} = \frac{1}{50\ \Omega} = 0.02\text{ S}$$
$$I = VB_C = 120\text{ V} \times 0.02\text{ S} = \mathbf{2.4\ A}$$

c.

$$B_L = \frac{1}{X_L} = \frac{1}{75\ \Omega} = 0.0133\text{ S}$$
$$I = VB_L = 120\text{ V} \times 0.0133\text{ S} = \mathbf{1.6\ A}$$

Example 12-23 could have been dealt with by using the $I = V/Z$ form of Ohm's law, thus avoiding separate computations for conductance and susceptance. In simple cases, using this technique is probably the sensible thing to do. Adopting the $I = VY$ form of Ohm's law is the start of a deliberate effort to get you thinking in terms of current being proportional to admittance. Such thinking develops an appreciation of parallel-circuit characteristics and forms the basis of some techniques commonly used in advanced circuit work.

REVIEW QUESTIONS FOR SECTION 12-11

1. What is meant by
   a. Impedance?
   b. Admittance?
2. What are the symbols and units for
   a. Impedance?
   b. Admittance?

3. What are other terms for the impedance of
   a. A resistive circuit?
   b. An inductive circuit?
   c. A capacitive circuit?
4. What are other terms for the admittance of
   a. A resistive circuit?
   b. An inductive circuit?
   c. A capacitive circuit?

EXERCISE PROBLEMS FOR SECTION 12-11

1. A series circuit with a current of 15 mA has a resistance of 1500 Ω $\underline{/0^\circ}$, a capacitive reactance of 2000 Ω $\underline{/-90^\circ}$, and an inductive reactance of 3500 Ω $\underline{/90^\circ}$. Find the voltage across each element, the total voltage, and the phase angle of the total voltage with respect to the current.
   (*ans*. 22.5 V, 30 V, 52.5 V, 31.82 V, +45°)
2. A three-element series circuit has an emf of 120 V $\underline{/-12^\circ}$. Element *A* is a 100-Ω $\underline{/0^\circ}$ resistance with a pd of 85 V $\underline{/0^\circ}$. Element *B* has a pd of 75 V $\underline{/-90^\circ}$. Find the pd across the third element and its impedance.
3. A parallel circuit, fed from a 120-V 60-Hz source, consists of resistance of 60 Ω $\underline{/0^\circ}$, a motor with an impedance of $(200 + j120)$ Ω, and a capacitor with an impedance of 135 Ω $\underline{/-90^\circ}$. Find the magnitude of the current in each element, the total current, and the phase angle of the total current with respect to the voltage.
   (*ans*. 2 A, 0.5145 A $\underline{/-30.96^\circ}$, 0.8889 A $\underline{/90^\circ}$, 2.519 A $\underline{/14.3^\circ}$)
4. The current drawn from a 240-V 60-Hz source by two motors connected in parallel is $(4.6 - j2.9)$ A. If one of the motors draws 1.9 A $\underline{/-36^\circ}$, find the current drawn by the other motor.

SUMMARY OF EQUATIONS

$T = 1/f$ or $f = 1/T$ (12-1)

$v = f\lambda$ (12-2)

$c = f\lambda$ or $f\lambda = 3 \times 10^8$ m/s (12-3)

$v \propto r \sin \theta$ (12-4)

$v = V_{max} \sin \theta$ or $i = I_{max} \sin \theta$ (12-5)

$\omega = 2\pi f$ (12-6)

$\omega = \theta/t$ or $\theta = \omega t$ or $t = \theta/\omega$ (12-7)

$V_{p\text{-}p} = 2V_{max}$ (12-8)

$v = V_{max} \sin \omega t$ (12-9)

$V_{av} = 2V_{max}/\pi$ or $0.6366\ V_{max}$ (12-10)

$V = V_{max}/\sqrt{2}$ or $0.7071\ V_{max}$ (12-11)

$j^2 = -1$ or $j = \sqrt{-1}$ (12-12)

$r\ \underline{/\theta} = r \cos \theta + jr \sin \theta$ (12-13)

$a + jb = \sqrt{a^2 + b^2}\ \underline{/\arctan b/a}$ (12-14)

$$(a + jb) + (c + jd) = (a + c) + j(b + d) \quad (12\text{-}15)$$

$$(a + jb) - (c + jd) = (a - c) + j(b - d) \quad (12\text{-}16)$$

$$r_1 \angle\theta_1 \times r_2 \angle\theta_2 = r_1 r_2 \angle\theta_1 + \theta_2 \quad (12\text{-}17)$$

$$r_1 \angle\theta_1 \div r_2 \angle\theta_2 = r_1/r_2 \angle\theta_1 - \theta_2 \quad (12\text{-}18)$$

$$(r \angle\theta)^n = r^n \angle n\theta \quad (12\text{-}19)$$

$$1/(r \angle\theta) = (1/r) \angle-\theta \quad (12\text{-}20)$$

$$1/(r \angle-\theta) = (1/r) \angle+\theta \quad (12\text{-}21)$$

$$1/j = -j \quad (12\text{-}22)$$

$$-1/j = j \quad (12\text{-}23)$$

$$Z = V/I \quad \text{or} \quad V = ZI \quad (12\text{-}24)$$

$$Y = I/V \quad \text{or} \quad I = YV \quad (12\text{-}25)$$

$$Z = 1/Y \quad \text{or} \quad R = 1/G \quad \text{or} \quad X_C = 1/B_C \quad \text{or} \quad X_L = 1/B_L \quad (12\text{-}26)$$

CHAPTER QUESTIONS

12-1. Compare the terms "steady dc," "varying dc," "dc transient," and "ac."

12-2. What are the meanings of arrows shown beside the symbols for a dc generator and an ac generator?

12-3. How is the distinction made between *pure ac* and *ac with a dc level* in
a. A curve diagram?
b. A schematic diagram?

12-4. Why is it necessary to distinguish between
a. The *average value* and the *rms value* of a sine wave?
b. The *true average value* and the *practical average value* of a sine wave?

12-5. Why is the term "cycle" not regarded as referring to a quantity (as is "*frequency*")?

12-6. What is the difference between the *period* of a wave and its *wavelength?*

12-7. Why is sine-wave ac power used in all electric systems for domestic and industrial use throughout the world?

12-8. Why does the study of electricity and electronics almost always refer to the sine-wave type of ac?

12-9. *Angular velocity* is sometimes referred to as *angular frequency*. Why is that so?

12-10. What does the *form factor* of a voltage tell about its waveform?

12-11. What is a phasor quantity?

12-12. Why use a line to represent a phasor quantity?

12-13. Why is it usual to have phasor line lengths represent *rms* values in phasor diagrams?

12-14. A *complex* quantity is not necessarily *complicated*. To what, then, does the word "complex" refer?

12-15. Which is the better way to represent phasors, rectangular form or polar form?

12-16. Why is the *j* operator

a. Assigned the numerical value $\sqrt{-1}$?

b. Represented by a change of position of a phasor line by 90°?

12-17. What words are used for

a. The ratio V/I?

b. The ratio I/V?

CHAPTER PROBLEMS

12-1. The voltage waveform seen on an oscilloscope screen is shown in Fig. 12-51. What are the voltage's values,

a. Peak-to-peak? (*ans*. 20 V)

b. Average over one complete cycle? (*ans*. 0 V)

c. RMS? (*ans*. 10 V)

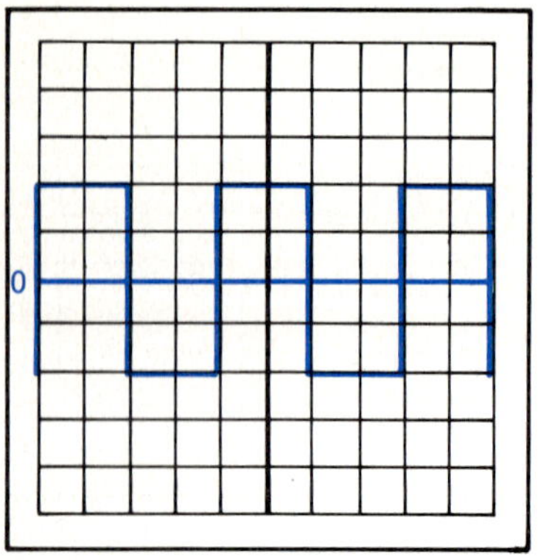

Vertical 5 V/division
Horizontal 500 μs/division

Fig. 12-51. Problem 12-1: Waveform.

12-2. The waveform seen on an oscilloscope screen is shown in Fig. 12-52. What are the values of the displayed voltage,

a. Peak-to-peak?

b. Average over one complete cycle?

c. RMS?

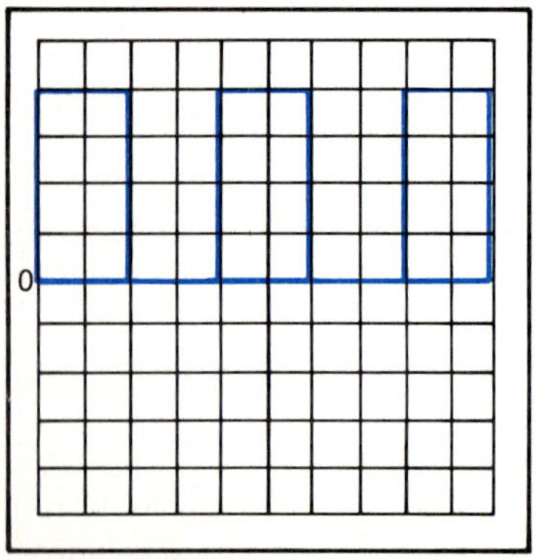

Vertical 5 V/division
Horizontal 500 μs/division

Fig. 12-52. Problem 12-2: Waveform.

12-3. What are the values of the voltage, as seen on an oscilloscope, displayed in Fig. 12-53,

a. Peak-to-peak? (*ans*. 20 V)

b. Average over one complete cycle? (*ans*. 15 V)

c. RMS? (*ans*. 18.03 V)

12-4. Figure 12-54 shows an oscilloscope display of a voltage. What are its values,

a. Peak-to-peak?

b. Average over one complete cycle?

c. RMS?

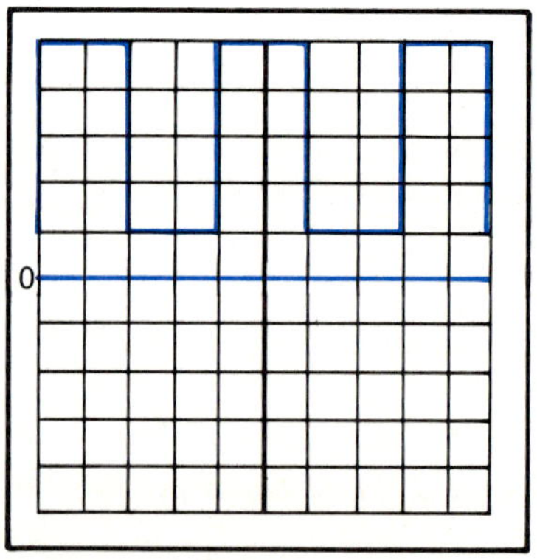

Vertical 5 V/division
Horizontal 500 μs/division

Fig. 12-53. Problem 12-3: Waveform.

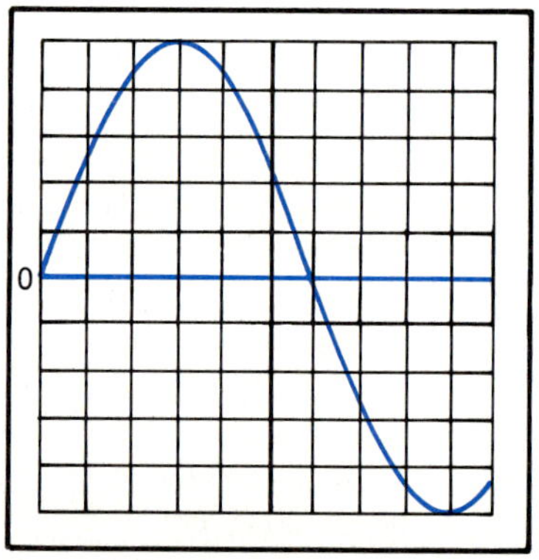

Vertical 10 mV/division
Horizontal 10 ns/division

Fig. 12-54. Problems 12-4 to 12-7, 12-9 to 12-10: Waveform.

12-5. What is the "practical average" value of the voltage in Fig. 12-54? What would this value become if the negative half-cycles were absent (called *half-wave rectification*)? Or if they were made positive so that the waveform consisted of positive half-cycles, each following the other (called *full-wave rectification*)? (*ans.* 15.92 V, 7.958 V, 15.92 V)

12-6. What are the frequencies and periods of each voltage in Figs. 12-51 to 12-54?

12-7. If the voltage waveform of Fig. 12-54 is radiated as a radio wave, what is its wavelength? (*ans.* 36 m)

12-8. The waveforms of Fig. 12-55 are oscilloscope displays of the overall voltage and current of a circuit. What are the circuit's

a. Phase angle?

b. Impedance?

c. Admittance?

Draw the circuit phasor diagram to scale.

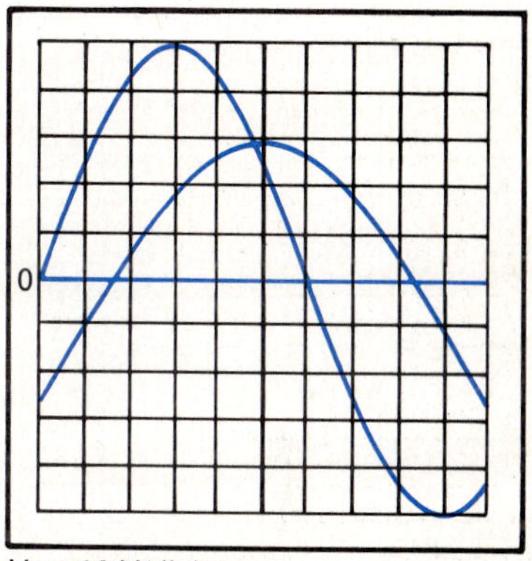

V at 10 V/divison
I at 50 mA/division

Fig. 12-55. Problem 12-8: Waveforms.

12-9. Refer to Fig. 12-54. How would this voltage be expressed in polar and rectangular forms? [*ans.* 35.36 mV $\angle 0$, $(35.36 + j0)$ mV]

12-10. What are the instantaneous values of the voltage of Fig. 12-54

a. At an angle of 50° after the voltage is zero, going positive?

b. 40 ns after the voltage is zero, going negative?

12-11. Figure 12-56 shows two voltages out of phase with each other. With respect to V_1, express each voltage

a. In polar form. (*ans.* 141.4 V $\angle 0°$, 282.8 V $\angle -80°$)

b. In rectangular form. [*ans.* $(141.4 + j0)$ V, $(49.11 - j278.5)$ V]

12-12. The two voltages of Fig. 12-56 are measured across two circuit elements connected in series. The total voltage across both elements is the phasor sum of the two voltages. Draw a phasor diagram that shows the original voltages and their phasor sum. Express the sum in polar and rectangular forms.

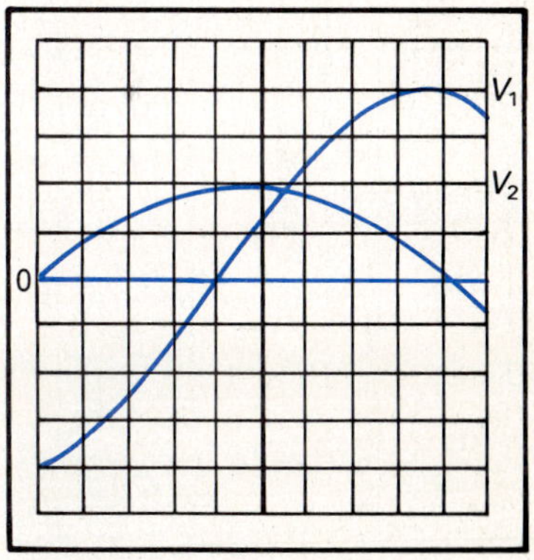

Vertical 100 V/division

Fig. 12-56. Problems 12-11 and 12-12: Waveforms.

12-13. With the use of a current probe, an oscilloscope measures the current in a 5.6-kΩ resistor as 20 mA peak-to-peak. How much power does this resistor dissipate? (*ans.* 280 mW)

12-14. The minimum voltage rating of a capacitor is the peak value of any voltage for which it is used. What is the minimum voltage rating for a capacitor to be used in a 240-V 60-Hz circuit?

12-15. An electric motor drawing 1.7 A $\angle -42°$ and two 100-W light bulbs whose currents are in phase with the 120-V 60-Hz supply are all connected in parallel. What are the total current and its phase with respect to the supply voltage? Draw a phasor diagram for the conditions. (*ans.* 3.143 A $\angle -21.2°$)

12-16. The motor of Prob. 12-15 is connected in parallel with a capacitor across the 120-V supply. What are the total current and its phase with respect to the supply voltage if the capacitor's current is 0.5 A and leads the voltage by 90°? Draw a phasor diagram of this situation.

12-17. What is the total admittance of the circuit of Prob. 12-16? (*ans.* 11.79 mS $\angle -26.8°$)

12-18. What is the total impedance of the circuit of Prob. 12-15?

13

AC EFFECTS OF RESISTANCE, CAPACITANCE, AND INDUCTANCE

Throughout Chaps. 1 to 10, when dealing with steady dc circuits, resistance was the important parameter. It was the only thing controlling current, voltage, power, and energy. Resistance is important in circuits with varying dc and those with ac too. However, little needs to be added, as resistance affects these circuits in the same manner as it does steady dc circuits. Resistance always opposes current flow; it always dissipates energy. Except when combined with capacitance or inductance, resistance does not affect phase or timing. The waveform of a current in a resistance always has the same shape as the waveform of the voltage across it. Ohm's and Kirchhoff's laws always apply to resistive circuits without special conditions.

Capacitance and inductance have been mentioned before, but no attempt has been made to explain them. Now that we are concerned with currents and voltages that change value, these two parameters must be considered. Capacitance and inductance are quite different from resistance in their circuit effects. For most of the things said about resistance, the opposite would have to be said about capacitance and inductance. For example, resistance dissipates energy, while capacitance and inductance store it; resistance has no phase effects, capacitance and inductance do.

13-1 RESISTANCE

The relationship between voltage and current in any resistive circuit is given by Ohm's law ($V = IR$). In an ac resistive circuit this law applies for the values of voltage and current at any time instant (including peak values), for rms values, and for average values. The voltage and current magnitudes have a constant ratio (equal to the resistance) which is not a function of time.

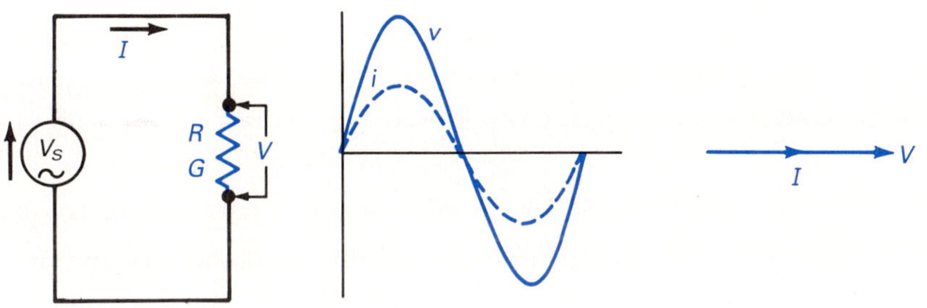

(a) (b) (c)

Fig. 13-1. Resistance in the ac circuit.

In Fig. 13-1*a* a sine-wave ac source has an emf of V_S volts rms. This emf is applied across a resistor of R ohms (a conductance of G siemens). The circuit current is I amperes rms, and the resistive pd is V volts rms.

Ohm's law, expressed in rms values, for a resistive ac circuit is the same as for a dc circuit:

$$V = IR \qquad \text{or} \qquad I = VG$$

At any instant of time, Ohm's law in a resistive ac circuit is

$$v = iR \qquad \text{or} \qquad i = vG$$

The waveforms of current and voltage at all time instants in one cycle are shown in Fig. 13-1*b*. A phasor diagram is given in Fig. 13-1*c*.

The characteristics of a resistive ac circuit are:

1. The current and the voltage are in phase ($\phi = 0$).
2. The impedance is the resistance ($Z = R$).
3. The admittance is the conductance ($Y = G = 1/R$).
4. The circuit dissipates average power ($P = VI = I^2R = V^2/R$).

REVIEW QUESTIONS FOR SECTION 13-1

1. In what manner does the current in a resistive ac circuit depend on its voltage?
2. If sine-wave ac is applied to a resistive circuit, what is the current value when the voltage is zero?
3. How is the impedance of a resistive circuit expressed?
4. What is the admittance of a resistive circuit?
5. What is the phase angle of a resistive circuit?

6. Draw the phasor diagram of a resistive circuit.
7. How is the power dissipated by a resistive circuit expressed?

13-2 CAPACITANCE

The relationship between voltage and current in a capacitive circuit at any time instant is

$$i = C\frac{dv}{dt}$$

As the instantaneous value of a sine wave is $V_{max} \sin \omega t$,*

$$i = C\frac{dV_{max} \sin \omega t}{dt}$$

From this equation we can obtain another that leads to a form of Ohm's law:

$$i = \omega C V_{max} \sin\left(\omega t + \frac{\pi}{2}\right) \qquad \textbf{(13-1)}$$

Analyzing Eq. (13-1), we see that

1. The current is of sine waveform.
2. The phase angle of the circuit is $\pi/2$ rad (90°). This conclusion is reached from sin $(\omega t + \pi/2)$. The plus sign indicates a leading phase angle; the current leads the voltage.
3. The instantaneous value of the current is proportional to the maximum value.
4. The factor linking the magnitudes of the current and the voltage is ωC.

By replacing $\omega t + \pi/2$ with θ, Eq. (13-1) can be written:

$$i = \omega C V_{max} \sin \theta$$

This can be rewritten in rms form:

$$I = \omega C V \qquad \textbf{(13-2)}$$

Equation (13-2) is the Ohm's law of the capacitive circuit.

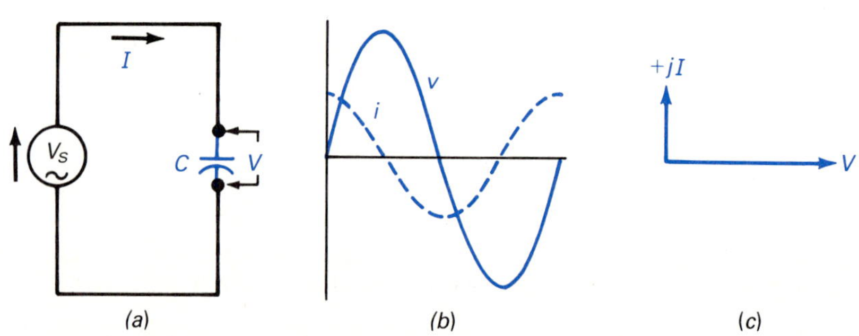

Fig. 13-2. Capacitance in the ac circuit.

*Recall that ω is angular frequency in rad/s and that $\omega = 2\pi f$, where f is the frequency in Hz.

Figure 13-2 shows a capacitive circuit with waveform and phasor diagrams. The phasor diagram in Fig. 13-2*c* can be shown with different symbols (Fig. 13-3).

Figure 13-3*a* is the same as Fig. 13-2*c*, except that the current phasor is labeled in terms of voltage, as in Eq. (13-2). The upward vertical position of the current phasor is indicated by $+j$.

In Fig. 13-3*b* the phasor diagram is shown with current as the reference phasor. It can be said that Fig. 13-3*b* is Fig. 13-3*a* rotated clockwise. Also, the voltage phasor in Fig. 13-3*b* is labeled in terms of current, from Eq. (13-2). The downward vertical position of the voltage phasor is indicated by $-j$.

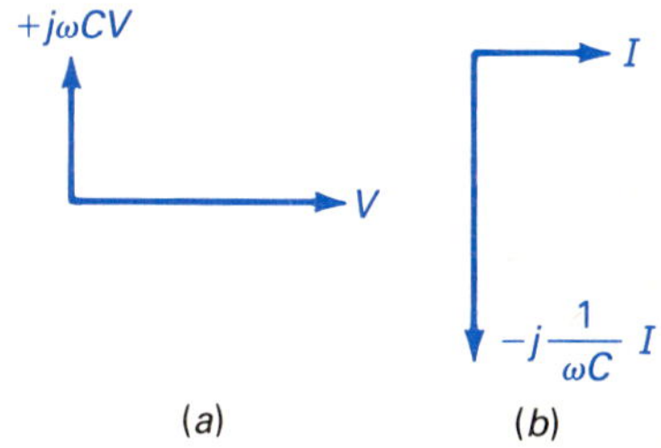

Fig. 13-3. Phasor diagram of a capacitive circuit: *(a)* In admittance form; *(b)* in impedance form.

From Fig. 13-3*a* it can be seen that

$$I = +j\omega CV \qquad \text{or} \qquad \frac{I}{V} = +j\omega C$$

In Sec. 12-11 it was stated that the ratio I/V is admittance, the ability of a circuit to permit current. In the capacitive circuit, the admittance is capacitive susceptance B_C.

$$B_C = +j\omega C \tag{13-3}$$

From Fig. 13-3*b* it can be seen that

$$V = -j\frac{1}{\omega C}I \qquad \text{or} \qquad \frac{V}{I} = -j\frac{1}{\omega C}$$

The ratio V/I is impedance, the ability of a circuit to restrict current. The impedance of a capacitive circuit is its capacitive reactance X_C.

$$X_C = -j\frac{1}{\omega C} \tag{13-4}$$

The sine-wave ac characteristics of a capacitive circuit can be summarized:

1. The current and the voltage are in phase quadrature, the current leading the voltage.
2. The circuit phase angle (the phase angle of the current with respect to the voltage) is $+90°$ or $+\pi/2$ rad.
3. The admittance is the capacitive susceptance ($Y = B_C = +j\omega C$ siemens).
4. The impedance is the capacitive reactance ($Z = X_C = -j1/\omega C$ ohms).
5. The circuit does not dissipate average power ($P = 0$).

REVIEW QUESTIONS FOR SECTION 13-2

1. In what manner does the current in a capacitive ac circuit depend on its voltage?
2. If sine-wave ac is applied to a capacitive circuit, what is the current value when the voltage is zero?
3. How is the impedance of a capacitive circuit expressed?
4. What is the admittance of a capacitive circuit?
5. What is the phase angle of a capacitive circuit?
6. Draw the phasor diagram of a capacitive circuit.
7. How is the power dissipated by a capacitive circuit expressed?

13-3 INDUCTANCE

In a circuit which is purely inductive, the voltage and current at any instant of time are related by the equation

$$v = L\frac{di}{dt}$$

If the current is of sine waveform, this equation becomes

$$v = L\frac{dI_{\text{max}} \sin \omega t}{dt}$$

This equation can give another, from which an inductive version of Ohm's law can be obtained:

$$v = \omega L I_{\text{max}} \sin\left(\omega t + \frac{\pi}{2}\right) \qquad \textbf{(13-5)}$$

Equation (13-5) tells us that

1. The voltage is of sine waveform.
2. The phase angle of the circuit is 90° ($\pi/2$ rad). The plus sign in $\sin(\omega t + \pi/2)$ indicates that the voltage leads the current or that the current lags the voltage. Therefore the circuit has a lagging phase angle (−90°).
3. The instantaneous value of the voltage is proportional to the maximum value.
4. The factor linking the magnitudes of the voltage and the current is ωL.

Letting $\omega t + \pi/2 = \theta$, Eq. (13-5) can be written as

$$v = \omega L I_{\text{max}} \sin \theta$$

which can be rewritten in rms form as

$$V = \omega L I \qquad \textbf{(13-6)}$$

Equation (13-6) is the Ohm's law of the inductive circuit.

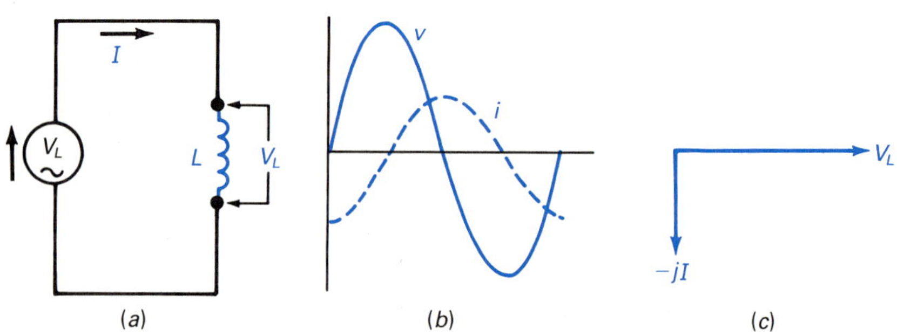

Fig. 13-4. Inductance in the ac circuit.

An inductive circuit with its waveform and phasor diagrams is shown in Fig. 13-4. The phasor diagram in Fig. 13-4*c* can be shown with different symbols (Fig. 13-5).

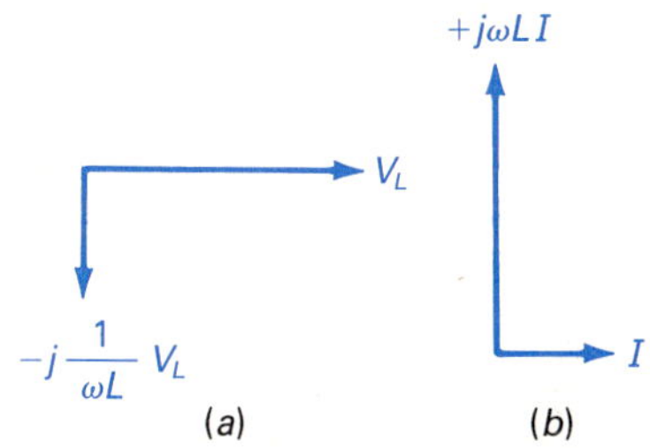

Fig. 13-5. Phasor diagram of an inductive circuit: *(a)* In admittance form; *(b)* in impedance form.

Figure 13-5*a* is the same as Fig. 13-4*c*, except that the current phasor is labeled in terms of voltage, from Eq. (13-6). The downward vertical position of the current phasor is indicated by $-j$.

In Fig. 13-5*b* the phasor diagram is shown with current as the reference phasor. The voltage phasor is labeled in terms of the current, as in Eq. (13-6). The upward vertical position of the voltage phasor is indicated by $+j$.

From Fig. 13-5*a*,

$$I = -j\frac{1}{\omega L}V_L \qquad \text{or} \qquad \frac{I}{V_L} = -j\frac{1}{\omega L}$$

The ratio I/V_L is admittance, the ability of a circuit to permit current. In the inductive circuit, the admittance is the inductive susceptance B_L.

$$B_L = -j\frac{1}{\omega L} \tag{13-7}$$

From Fig. 13-5*b*,

$$V_L = +j\omega LI \qquad \text{or} \qquad \frac{V_L}{I} = +j\omega L$$

The ratio V_L/I is impedance, the ability of a circuit to restrict current. The impedance of an inductive circuit is its inductive reactance X_L.

$$X_L = +j\omega L \tag{13-8}$$

The sine-wave characteristics of a circuit which is purely inductive can be summarized:

1. The current and the voltage are in phase quadrature, the current lagging the voltage.
2. The circuit phase angle (the phase angle of the current with respect to the voltage) is $-90°$ or $-\pi/2$ rad.
3. The admittance is the inductive susceptance ($Y = B_L = -j\,1/\omega L$ siemens).
4. The impedance is the inductive reactance ($Z = X_L = j\omega L$ ohms).
5. The circuit does not dissipate average power ($P = 0$).

REVIEW QUESTIONS FOR SECTION 13-3

1. In what manner does the voltage in an inductive circuit depend on its current?
2. If sine-wave ac is applied to an inductive circuit, what is the current value when the voltage is zero?
3. How is the impedance of an inductive circuit expressed?
4. What is the admittance of an inductive circuit?
5. What is the phase angle of an inductive circuit?
6. Draw the phasor diagram of an inductive circuit.
7. How is the power dissipated by an inductive circuit expressed?

EXERCISE PROBLEMS FOR SECTIONS 13-2 AND 13-3

1. Find the capacitive susceptance of the following capacitors at the frequencies stated.

a. 10 μF, 20 Hz (*ans.* 1.257 mS)
b. 10 μF, 60 Hz (*ans.* 3.77 mS)
c. 10 μF, 1000 Hz (*ans.* 62.8 mS)
d. 0.047 μF, 100 kHz (*ans.* 29.53 mS)
e. 33 pF, 100 kHz (*ans.* 20.73 μS)
f. 33 pF, 20 MHz (*ans.* 4.147 mS)
g. 0.0015 μF, 20 MHz (*ans.* 188.5 mS)
h. 220 pF, 20 MHz (*ans.* 27.65 mS)

2. Find the capacitive reactance of the capacitors in Prob. 1 at the frequencies stated.

3. Find the inductive susceptance of the following inductors at the frequencies stated.

a. 5 H, 20 Hz (*ans.* 1.592 mS)
b. 5 H, 100 Hz (*ans.* 318.3 μS)
c. 5 H, 1000Hz (*ans.* 31.83 μS)
d. 200 mH, 50 kHz (*ans.* 15.92 μS)
e. 100 mH, 100 kHz (*ans.* 15.92 μS)
f. 5.6 μH, 10 MHz (*ans.* 2.842 mS)
g. 35 μH, 10 MHz (*ans.* 454.5 μS)
h. 160 μH, 10 MHz (*ans.* 99.5 μS)

4. Find the inductive reactance of the inductors in Prob. 3 at the frequencies stated.

5. Find the current in a circuit consisting of a 1000-Hz generator with an output of 50 V when each of the following capacitors is individually connected:

a. 5 μF (*ans.* 1.571 A)
b. 0.033 μF (*ans.* 10.37 mA)
c. 1000 pF (*ans.* 314.2 μA)
d. 0.0022 μF (*ans.* 691.2 μA)

6. Find the voltage across each of the following inductors when they are carrying a current of 30 mA at 10 kHz:

a. 0.2 H *b.* 50 mH *c.* 100 μH *d.* 1 mH

7. An audio amplifier uses a 0.1-μF capacitor to couple two stages together. Find the reactance of the capacitor at the following audio frequencies:

a. 50 Hz (*ans.* 31.83 kΩ)
b. 1000 Hz (*ans.* 1.592 kΩ)
c. 15 kHz (*ans.* 106.1 Ω)

8. A 0.47-μF capacitor and a 0.27-μF capacitor are connected in parallel across a 120-V 400-Hz supply. Find

a. The admittance of each capacitor.
b. The total admittance.
c. The reactance of each capacitor.
d. The total reactance.
e. The total equivalent capacitance.
f. The total circuit current.

9. A 0.47-μF capacitor and a 0.27-μF capacitor are connected in series across a 120-V 400-Hz supply. Find
a. The total admittance. (*ans*. 431 μS)
b. The total reactance. (*ans*. 2.32 kΩ)
c. The total equivalent capacitance. (*ans*. 0.1715 μF)
d. The current in the circuit. (*ans*. 51.72 mA)
e. The pd across each capacitor. (*ans*. 43.78 V, 76.22 V)

10. A radio-frequency choke (inductor) permits the passage of low frequencies in preference to high frequencies. Find the reactance of a 300-μH radio-frequency choke at
a. 2 MHz
b. 400 Hz

11. Two relay coils, one 0.5 H and the other 1.0 H, are connected in parallel across a 120-V 800-Hz system. Find
a. The admittance of each coil. (*ans*. 398 μS, 199 μS)
b. The total admittance. (*ans*. 597 μS)
c. The reactance of each coil. (*ans*. 2.513 kΩ, 5.027 kΩ)
d. The total reactance. (*ans*. 1.676 kΩ)
e. The total equivalent inductance. (*ans*. 0.333 H)
f. The total current in the circuit. (*ans*. 71.61 mA)

12. A 20-mH inductor and a 100-mH inductor are connected in series across a 25-V 1000-Hz generator. Find
a. The total admittance.
b. The total reactance.
c. The total equivalent inductance.
d. The current in the circuit.
e. The pd across the 100-mH inductor.

13-4 EXAMPLES ON SECTIONS 13-1, 13-2, AND 13-3

EXAMPLE 13-1 What is the resistance of a 2-kW electric stove hot plate when it is operating? The supply is 120-V 60-Hz ac.

Solution When ac voltages and currents are stated in rms values, a resistive element operating on ac is treated in exactly the same manner as if it were dc-operated:

$$P = \frac{V^2}{R}$$

$$R = \frac{V^2}{P} = \frac{(120\ \text{V})^2}{2\ \text{kW}} = \mathbf{7.2\ \Omega}$$

EXAMPLE 13-2 The current in a 500-pF capacitor is 1 mA at 1 MHz. What is the magnitude of the voltage across it?

Solution The impedance of the capacitor is its capacitive reactance X_C. From Eq. (13-4),

$$X_C = \frac{1}{\omega C} = \frac{1}{2\pi f C} = \frac{1}{2\pi \times 1\ \text{MHz} \times 500\ \text{pF}} = 318.3\ \Omega$$

Apply Ohm's law and omit j, as it is not important when only magnitudes are of interest. From Eqs. (13-2) and (13-4),

$$V = IX_C = 1\ \text{mA} \times 318.3\ \Omega = \mathbf{318.3\ mV}$$

EXAMPLE 13-3 What is the magnitude of the voltage across a pure inductance of 20 mH when its current is 80 mA at 1 kHz?

Solution The impedance of the inductance is its inductive reactance X_L. From Eq. (13-8),

$$X_L = \omega L = 2\pi f L = 2\pi \times 1\ \text{kHz} \times 20\ \text{mH} = 125.7\ \Omega$$

Apply Ohm's law. From Eqs. (13-6) and (13-8),

$$V = IX_L = 80\ \text{mA} \times 125.7\ \Omega = \mathbf{10.06\ V}$$

The term "pure inductance" is used in Example 13-3 to indicate that, if a practical inductor is being considered, its resistance is negligible. This distinction is important because, unlike practical capacitors, real inductors often have so much resistance that it cannot be ignored. More will be said in later chapters about these resistive effects in inductors. Meanwhile it is assumed that, if the resistance of an inductor is not specifically mentioned, it is not significant to the circuit conditions.

EXAMPLE 13-4 What is the magnitude of the current in a 50-μF capacitor when 50 V at 60 Hz is developed across it?

Solution The admittance of the capacitor is its capacitive susceptance B_C. From Eq. (13-3),

$$B_C = \omega C = 2\pi f C = 2\pi \times 60\ \text{Hz} \times 50\ \mu\text{F} = 0.01885\ \text{S}$$

Apply the admittance form of Ohm's law. From Eqs. (13-2) and (13-3),

$$I = VB_C = 50\ \text{V} \times 0.01885\ \text{S} = \mathbf{0.9425\ A}$$

EXAMPLE 13-5 An electronic generator (an oscillator) applies 1 V at 5 MHz across a radio tuning coil. What is the inductance of the coil if its current magnitude is 0.5 mA?

Solution Ohm's law in admittance form is, from Eqs. (13-6) and (13-7),

$$I = VB_L$$

The admittance of the inductor is its inductive susceptance B_L:

$$B_L = \frac{I}{V} = \frac{0.5\text{ mA}}{1\text{ V}} = 0.5\text{ mS}$$

From Eq. (13-7),

$$B_L = \frac{1}{\omega L}$$

$$L = \frac{1}{\omega B_L} = \frac{1}{2\pi f B_L} = \frac{1}{2\pi \times 5\text{ MHz} \times 0.5\text{ mS}}$$

$$L = \mathbf{0.06366\text{ mH}} \quad \text{or} \quad \mathbf{63.66\ \mu H}$$

EXAMPLE 13-6 The inductance of a motor starting relay is 1 H. Assuming that its resistance is negligible, what is the phasor value of its current when the relay is supplied with 120-V 60-Hz ac?

Solution

$$X_L = j\omega L \qquad (13\text{-}8)$$

$$X_L = j2\pi fL = j2\pi \times 60\text{ Hz} \times 1\text{ H} = j377\ \Omega$$

From Eqs. (13-6) and (13-8),

$$I = \frac{V}{X_L}$$

$$I = \frac{120\text{ V}}{j377\ \Omega} = \mathbf{-j0.3183\text{ A}}$$

EXAMPLE 13-7 One of the components in the tuning circuit of a radio is a 35- to 300-pF variable capacitor. When the capacitor is set at 35 pF, the radio is tuned to the high end of the AM band, 1600 kHz. When the radio is tuned to the low end of the band, 550 kHz, the capacitor is in its 300-pF position. If the signal voltage across the capacitor is constant at 100 μV, what is the phasor value of its current at each end of the AM band?

Solution At the high-frequency end of the band, $f = 1600$ kHz and $C = 35$ pF. The phasor value of the capacitor's impedance is given by Eq. (13-4):

$$X_C = -j\frac{1}{\omega C} = -j\frac{1}{2\pi fC}$$

$$= -j\frac{1}{2\pi \times 1600\text{ kHz} \times 35\text{ pF}}$$

$$= -j2842\ \Omega$$

From Eqs. (13-2) and (13-4),

$$I = \frac{V}{X_C} = \frac{100\ \mu\text{V}}{-j2842\ \Omega} = \mathbf{j0.0352\ \mu A} \qquad \text{or} \qquad \mathbf{j35.2\ nA}$$

At the low-frequency end of the band, $f = 550$ kHz and $C = 300$ pF:

$$X_C = -j\frac{1}{\omega C} = -j\frac{1}{2\pi \times 550\ \text{kHz} \times 300\ \text{pF}} = -j964.6\ \Omega$$

$$I = \frac{V}{X_C} = \frac{100\ \mu\text{V}}{-j964.6\ \Omega} = \mathbf{j0.1037\ \mu A} \qquad \text{or} \qquad \mathbf{j103.7\ nA}$$

The solutions of Examples 13-6 and 13-7 used the reciprocal of j principle developed in Sec. 12-10. This technique of interchanging j and $1/j$ with a change in sign is used often in ac work.

In Example 13-7, the current at the low-frequency end of the band could have been obtained by applying proportionality to the high-frequency result as follows:

Current is proportional to susceptance ($I \propto B$).
Capacitive susceptance is proportional to the product of the angular frequency and the capacitance ($B_C \propto \omega C$).
Angular frequency is proportional to frequency ($\omega \propto f$).

$$I \propto B_C \propto \omega C \propto fC$$

These proportionalities can be used to produce the following relationship between the low-frequency current and the high-frequency current (I_{550} and I_{1600}):

$$I_{550} = I_{1600} \times \frac{f_{550}C_{550}}{f_{1600}C_{1600}} = j35.2\ \text{nA} \times \frac{550\ \text{kHz} \times 300\ \text{pF}}{1600\ \text{kHz} \times 35\ \text{pF}}$$

$$I_{550} = j103.7\ \text{nA}$$

13-5 AC POWER IN *R, C,* AND *L*

Instantaneous Power

Whenever there is voltage *and* current, there is power. For sine-wave alternating current in any type of circuit, the power level will vary from instant to instant, sometimes being zero in magnitude.

Figure 13-6*a* shows that, for a resistive circuit, whenever there is voltage, there is also current. Power is zero when both i and v are zero.

In Fig. 13-6*b* and *c* it can be seen that, for inductive and capacitive circuits, there are instants when there is voltage with zero current and other instants when there is current with zero voltage. These are the instants of zero power.

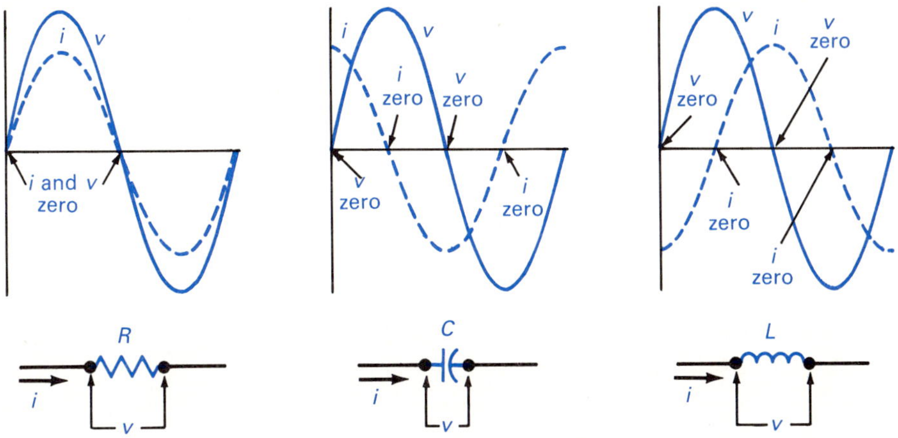

(a) (b) (c)

Fig. 13-6. Waveforms in ac circuits: *(a)* Resistive; *(b)* capacitive; *(c)* inductive.

Instantaneous power ($p = vi$) can be determined for resistive, inductive, or capacitive circuits for any point in the cycle. The maximum instantaneous power (peak power) is important in some electronics applications, but apart from this, instantaneous powers are rarely of interest. The average power over an extended period of time is of importance in most electric and electronic systems.

Average Power

In ac circuit terminology, ''power'' usually refers to average power, the true average taken over a complete cycle.

The average power in a resistive circuit carrying ac is the same as if the circuit carried steady dc of magnitude equal to the ac rms value. This is to be expected, since rms values are defined in terms of the power equivalence of steady direct current.

The power equations (2-14), (3-6), and (3-7) can be used for resistive ac circuits:

$$P = VI \quad \text{or} \quad I^2R \quad \text{or} \quad \frac{V^2}{R}$$

where P = power, W
V = emf or pd, rms V
I = current, rms A
R = resistance, Ω

The average power in a capacitive or inductive circuit is zero.

The power statements for resistance, capacitance, and inductance are derived by mathematics. Here it is sufficient to refer to the basic logic of ac power.

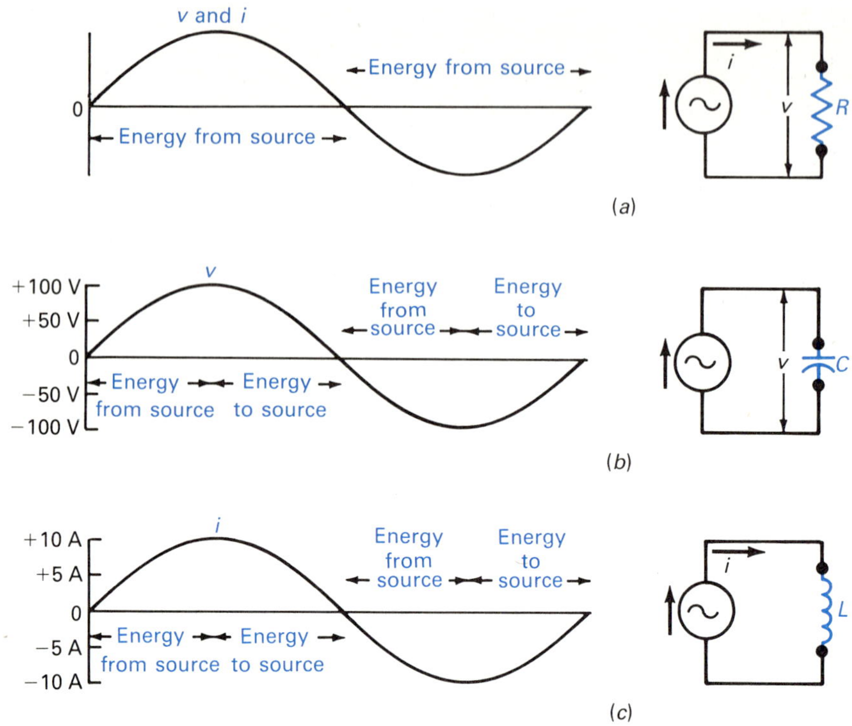

Fig. 13-7. Energy distribution per cycle: *(a)* Resistive circuit; *(b)* capacitive circuit; *(c)* inductive circuit.

As indicated in Fig. 13-7*a*, energy is supplied to a resistor all the time alternating current is present. The energy leaves the circuit, mainly as heat, and cannot be returned to the source.

Energy is supplied to a capacitor whenever the voltage across its terminals increases. It is returned to the source whenever the terminal voltage decreases. With reference to Fig. 13-7*b*, if the capacitance is 10 μF, the energy stored in the capacitor ($V^2C/2$) when the voltage across it is 0, 50, and 100 V will be 0 J, 12.5 mJ, and 50 mJ, respectively. Thus, during the first quarter-cycle, the energy will rise from 0 to 12.5 to 50 mJ, this energy coming *from* the source. During the second quarter-cycle, the energy stored in the capacitor will drop from 50 to 12.5 to 0 mJ, this energy being returned *to* the source.

Energy is supplied to an inductor whenever the current through it increases. It is returned to the source whenever the current through the inductor decreases. The reasoning is essentially the same as for a capacitor. With reference to Fig. 13-7*c*, during the first quarter-cycle, when the current is rising from zero to 10 A, if the inductance is 1 mH, the energy *from* the source will rise from 0 to 50 mJ ($I^2L/2$). During the second quarter-cycle, this 50 mJ of energy will be returned *to* the source while the current is dropping from 10 to 0 A.

When a capacitor or an inductor is connected to an ac source, no energy leaves the circuit; it is repeatedly transferred between the source and the element. No net energy is taken from the source, and hence there is no net or average power dissipation by the circuit.

REVIEW QUESTIONS FOR SECTION 13-5

1. What must be the magnitude of a steady dc voltage to produce the same average power as a given ac voltage?
2. How much power is dissipated in an ac circuit that contains only
   a. Resistance?
   b. Capacitance?
   c. Inductance?
3. In an ac circuit, what happens to the energy in a capacitor when its voltage
   a. Increases?
   b. Decreases?
4. In an ac circuit, what happens to the energy in an inductor when its current
   a. Increases?
   b. Decreases?
5. How is energy stored in an inductor when its voltage is zero or in a capacitor when its current is zero?
6. How is energy stored in a resistor?

EXERCISE PROBLEMS FOR SECTION 13-5

1. A 100-Ω resistance is fed from a 120-V 60-Hz source. Reference time t_0 is when the voltage is zero going positive.
   a. Find the total power consumed by the circuit. (*ans*. 144 W)
   b. Find the instantaneous power at the following times:
      i. t = 2.083 ms (*ans*. 144 W)
      ii. t = 4.167 ms (*ans*. 289 W)
      iii. t = 6.25 ms (*ans*. 144 W)
      iv. t = 10.42 ms (*ans*. 144 W)
      v. t = 12.5 ms (*ans*. 289 W)
      vi. t = 14.58 ms (*ans*. 144 W)
2. A 20-μF capacitor is fed from a 120-V 60-Hz source. Reference time t_0 is when the voltage is zero going positive.
   a. Find the total power consumed by the circuit.
   b. Find the instantaneous power at the following times:
      i. t = 2.083 ms
      ii. t = 4.167 ms
      iii. t = 6.25 ms
      iv. t = 10.42 ms
      v. t = 12.5 ms
      vi. t = 14.58 ms
3. A circuit consists of a 5-μF capacitor connected across a 120-V 400-Hz generator. Find the energy stored in the electric field of the capacitor at the following times after t_0:
   a. 0.2 ms (*ans*. 16.68 mJ)
   b. 0.625 ms (*ans*. 72 mJ)
   c. 0.937 ms (*ans*. 36 mJ)
   d. 1.56 ms (*ans*. 35.5 mJ)
   e. 1.875 ms (*ans*. 72 mJ)
   f. 2.5 ms (*ans*. 0 J)
4. A circuit consists of a 5-H inductor connected across a 120-V 400-Hz generator. Find the energy stored in the magnetic field of the inductor at the times after t_0 listed for Prob. 3.

13-6 FREQUENCY, IMPEDANCE, AND ADMITTANCE

The impedance and admittance of elements at different frequencies are often of importance in the operation of circuits. In some applications varying frequency characteristics are vital to the functioning of the equipment. An example is radio tuning, which is carried out by controlling reactance. In other situations frequency effects can limit the usefulness of equipment. For instance, stray capacitive and inductive reactance can render equipment useless at high frequencies.

Impedance

Figure 13-8 shows the relationships between frequency and the magnitudes of resistance, inductive reactance, and capacitive reactance.

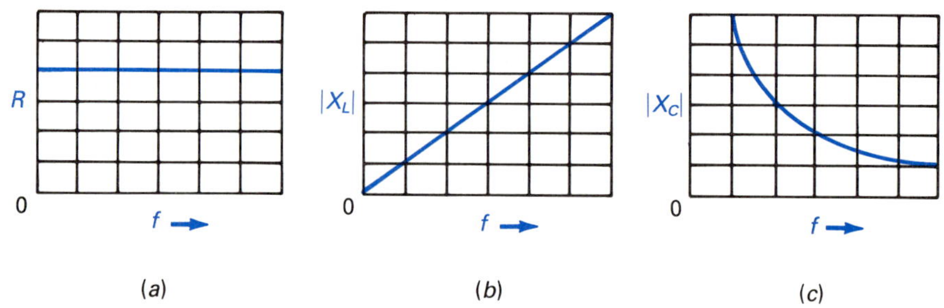

Fig. 13-8. Impedance magnitude with respect to frequency: *(a)* Resistance; *(b)* inductive reactance; *(c)* capacitive reactance.

Resistance is the same at all frequencies, as shown by the horizontal line in Fig. 13-8*a*.

The magnitude of inductive reactance ($|X_L| = 2\pi fL$) is directly proportional to frequency, as shown by the inclined line in Fig. 13-8*b*.

The magnitude of capacitive reactance ($|X_C| = 1/2\pi fC$) is inversely proportional to frequency. The exponential curve in Fig. 13-8*c* shows this relationship.

Figure 13-8*b* and *c* shows only the magnitudes of X_L and X_C. To depict the relative phases of inductive and capacitive circuits, it is usual to adopt a frequency-magnitude diagram which distinguishes between $+j$ and $-j$.

In Fig. 13-9 the frequency–inductive-reactance curve is shown above the horizontal axis to be related to $+j$. The frequency–capacitive-reactance curve is shown in the $-j$ location, below the horizontal axis.

The magnitude of impedance at any frequency in diagrams such as Figs. 13-8 and 13-9 is indicated by the distance of the curve above or below the horizontal axis. There are no reactance differences between Figs. 13-8 and 13-9. Any change in magnitude away from zero is an increase. Any change in magnitude toward zero is a decrease. In Fig. 13-9 the magnitude of X_L is shown increasing with increasing frequency, as in Fig. 13-8*b*, while the magnitude of X_C is shown decreasing, as in Fig. 13-8*c*.

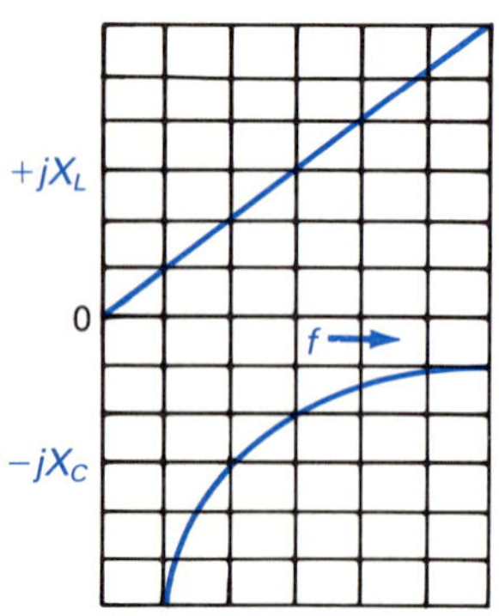

Fig. 13-9. Inductive and capacitive reactance curves showing the relative phases.

Admittance

Admittance is the reciprocal of impedance, so the curves of admittance with respect to frequency are the reciprocals of Figs. 13-8 and 13-9.

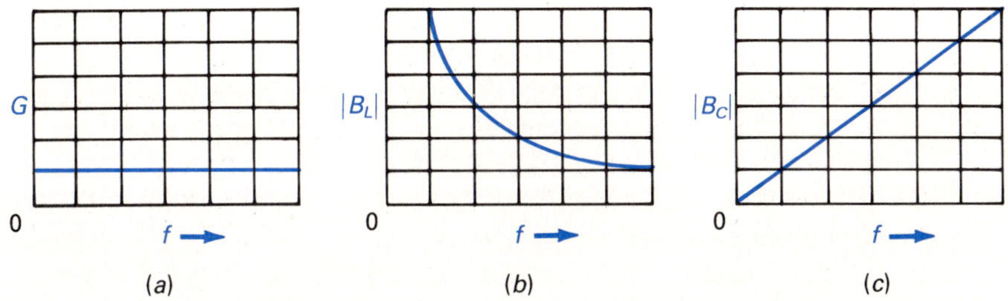

Fig. 13-10. Admittance magnitude with respect to frequency: *(a)* Conductance; *(b)* inductive susceptance; *(c)* capacitive susceptance.

Conductance is independent of frequency, as shown in Fig. 13-10*a*.

The magnitude of inductive susceptance ($|B_L| = 1/2\pi fL$) is inversely proportional to frequency, as indicated in Fig. 13-10*b*.

In Fig. 13-10*c* the magnitude of capacitive susceptance ($|B_C| = 2\pi fC$) is directly proportional to frequency.

Figure 13-11 shows the capacitive and inductive susceptance curves in their respective $+j$ and $-j$ positions.

It is evident that there are considerable differences in the behavior of resistive, inductive, and capacitive elements from a frequency point of view. These differences are among the most important of the resources at the disposal of the ac circuit designer.

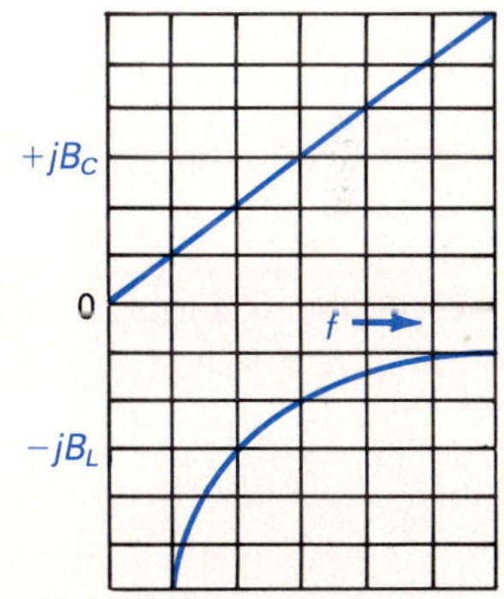

Fig. 13-11. Inductive and capacitive susceptance curves showing the relative phases.

REVIEW QUESTIONS FOR SECTION 13-6

1. Why is the effect of ac frequency different for a resistor than for a capacitor or an inductor?
2. When frequency rises, what happens to the impedance of
   a. A capacitor?
   b. An inductor?
3. When frequency rises, what happens to the admittance of
   a. A capacitor?
   b. An inductor?
4. When frequency rises, what happens to the phase angle of
   a. A capacitor?
   b. An inductor?

EXERCISE PROBLEMS FOR SECTION 13-6

1. A 0.001-μF capacitor and a 5-mH inductor are connected in series across a variable-frequency source. Find the reactance of each at the following frequencies:

a. 0 Hz (*ans.* $\infty\ \Omega$, $0\ \Omega$)
b. 1 kHz (*ans.* $-j159.2\ \mathrm{k\Omega}$, $+j31.42\ \Omega$)
c. 20 kHz (*ans.* $-j7.958\ \mathrm{k\Omega}$, $+j628.3\ \Omega$)
d. 100 kHz (*ans.* $-j1.592\ \mathrm{k\Omega}$, $+j3.142\ \mathrm{k\Omega}$)
e. 1 MHz (*ans.* $-j159.2\ \Omega$, $+j31.42\ \mathrm{k\Omega}$)
f. 50 MHz (*ans.* $-j3.183\ \Omega$, $+j1.571\ \mathrm{M\Omega}$)

2. When impedances are connected in series, the total impedance equals the phasor sum of the individual impedances. Find the total impedance for the circuit in Prob. 1 at each frequency.

3. Find the frequency at which $X_L = X_C$ in the circuit in Prob. 1. (*ans.* 71.18 kHz)

4. A 1500-pF capacitor and a 2-mH inductor are connected in parallel across a variable-frequency source. Find the susceptance of each at the following frequencies:
a. 0 Hz *d.* 100 kHz
b. 1 kHz *e.* 1 MHz
c. 20 kHz *f.* 50 MHz

5. When admittances are connected in parallel, the total admittance equals the phasor sum of the individual admittances. Find the total admittance for the circuit in Prob. 4 at each frequency.
(*ans. a.* ∞ S; *b.* $-j97.57$ mS; *c.* $-j3.791$ mS; *d.* $+j146.7\ \mu$S; *e.* $+j9.345$ mS; *f.* $+j471.2$ mS)

6. Find the frequency at which $B_L = B_C$ for the circuit in Prob. 4.

SUMMARY OF EQUATIONS

$i = \omega C V_{max} \sin(\omega t + \pi/2)$	(13-1)	$v = \omega L I_{max} \sin(\omega t + \pi/2)$	(13-5)
$I = \omega C V$	(13-2)	$V = \omega L I$	(13-6)
$B_C = +j\omega C$	(13-3)	$B_L = -j1/\omega L$	(13-7)
$X_C = -j1/\omega C$	(13-4)	$X_L = +j\omega L$	(13-8)

CHAPTER QUESTIONS

13-1. Compare the function of resistance in an ac circuit with its function in a dc circuit?

13-2. Why are inductance and capacitance generally of greater importance in ac circuits than in dc circuits?

13-3. What precaution needs to be taken in applying Ohm's law in an ac resistive circuit?

13-4. What precautions need to be taken in applying Ohm's law in ac inductive and capacitive circuits?

13-5. What is Ohm's law as used when dealing with resistive, inductive, and capacitive circuits?

13-6. Why is it said that inductive reactance is a $+j$ quantity, while capacitive reactance is a $-j$ quantity?

13-7. Why is it said that inductive reactance is a $+j$ quantity, while inductive susceptance is a $-j$ quantity?

13-8. Compare *average power* and *instantaneous power* as they apply in resistive, capacitive, and inductive circuits.

13-9. Why is it sometimes necessary to distinguish between *inductor* and *inductance* when dealing with ac circuits, but rarely necessary to make the same distinction between *capacitor* and *capacitance?*

13-10. How do Figs. 13-9 and 13-11 show the phase angles of reactance and susceptance?

CHAPTER PROBLEMS

13-1. Refer to Example 13-5. It shows that the inductance needs to be 63.66 μH to give a susceptance of 0.5 mS (reactance of 2 kΩ) at 5 MHz. What capacitance would give the same value at that frequency? (*ans.* 15.92 pF)

13-2. If the frequency of Prob. 13-1 is raised to 10 MHz, but the inductance and capacitance remain the same, what happens to the inductive reactance?

13-3. What would the capacitance have to be for the frequency of Prob. 13-1 to be raised to 10 MHz and both inductive and capacitive reactances to be the same? (*ans.* 7.96 pF)

13-4. What would the inductance have to be for the frequency of Prob. 13-1 to be raised to 10 MHz, the capacitance to remain the same, and both inductive and capacitive susceptances to be the same?

13-5. At a frequency of 1 MHz, what is the impedance of
a. A 10-kΩ resistor? (*ans.* 10 kΩ)
b. A 1-mH inductor? (*ans.* 6.28 kΩ)
c. A 10-pF capacitor? (*ans.* 15.92 kΩ)

13-6. What frequency changes would be needed in Prob. 13-5 to make the admittances of the inductor and the capacitor the same as that of the resistor?

13-7. A transistor radio is tuned to 1.6 MHz, the highest frequency of the AM radio band, by the circuit of Fig. 13-12. If the inductive reactance has to be equal to the capacitive reactance for tuning, what value of inductance is needed? (*ans.* approximately 200 μH)

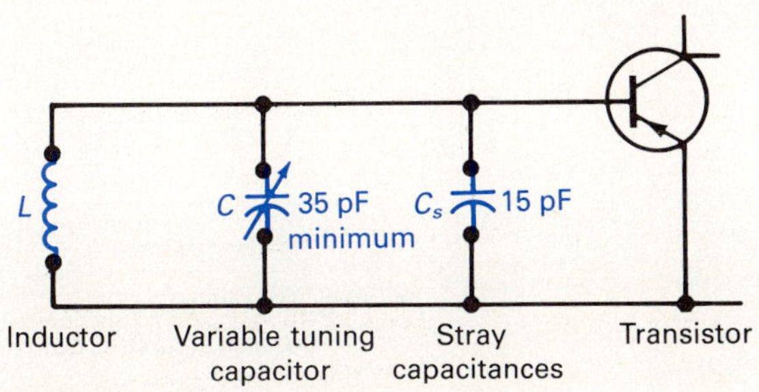

Fig. 13-12. Problems 13-7 and 13-8: Circuit.

13-8. In Prob. 13-7, the tuning capacitor is variable, its maximum value being 425 pF. What is the frequency range of the AM band over which the radio may be tuned? The values of inductance and stray capacitance are to be the same as in Fig. 13-12.

13-9. An example in a later chapter (Example 15-14) refers to *power-factor improvement* as it may apply to an industrial electrical situation. The capacitor required to give the improvement will draw 254.8 A from a 440-V 60-Hz supply. What should be the capacitor's value? If you wish, look at the example *after* you have attempted this problem? (*ans*. 1536 μF)

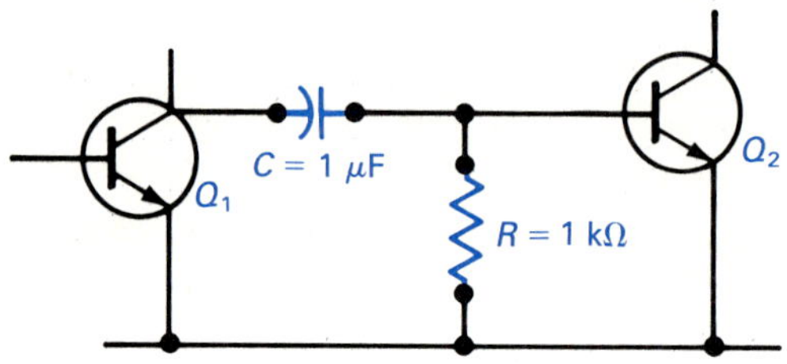

Fig. 13-13. Problem 13-10: Circuit.

13-10. In an electronic amplifier circuit, one stage is connected to the next by the network shown in Fig. 13-13. How do the impedances of the capacitor and the resistor compare over the audio- (sound-) frequency range, 20 Hz to 20 kHz?

14

AC CIRCUITS WITH *R* AND *C* OR *R* AND *L*

Before detailing the behavior of ac circuits containing resistance and inductance or capacitance, a preview of some general characteristics of ac circuits will be helpful.

1. All the principles and laws of dc circuits, including Ohm's and Kirchhoff's laws and Thevenin's and Norton's theorems, apply to ac circuits.
2. Where a circuit contains both resistive and nonresistive components, there will be phase differences between voltages, between currents, and between voltages and currents. In these circumstances, the *magnitudes of voltage or current occurring at a specified instant of time can be added arithmetically, but peak, peak-to-peak, average, and rms values cannot.*
3. When *Kirchhoff's laws* are applied to *ac circuits, voltages or currents are added,* these usually being expressed *in rms terms*. The *additions must be carried out using phasor methods.*
4. *Power is dissipated only in the resistive components of a circuit.* It is important to distinguish between *the rate at which energy leaves the circuit from resistive components (called active power), the rate at which energy is stored in and discharged from reactive components (reactive power),* and *the product of total voltage and total current (apparent power).* The term "apparent power" refers to its similarity to dc circuit power, both

being the product of total voltage and total current. But *the real ac equivalent of dc power is active power, the actual power dissipated.*

5. Often the production of active power is of particular significance in an ac circuit application; for example, in an electric motor the active power is responsible for the mechanical output power at the motor shaft. But apparent power can never be ignored. The apparent power determines the overall effect of the circuit on the supply source, the amount of current drawn for a given supply voltage.
6. The importance of both active power and apparent power has led to the adoption of a special term for the ratio between them. The *power factor is the ratio of active power to apparent power;* the power factor describes the amount of power dissipated by the circuit in relation to the voltage and current supplied to the circuit.
7. Often electrical utilities base their rates on apparent power, since this determines the sizes of conductors, transformers, and other equipment. In such cases, the power factor tells the user how much useful power is available.

Addition of Phasors

Where an ac circuit contains inductance or capacitance, the currents and voltages do not rise and fall together; there are time lag or lead relationships between them, necessitating the use of phasor techniques in analysis.

As a common example arising in ac circuit work, consider the addition of two sine waveforms such as is involved in applying Kirchhoff's laws. Figure 14-1 shows two sine-wave voltages.

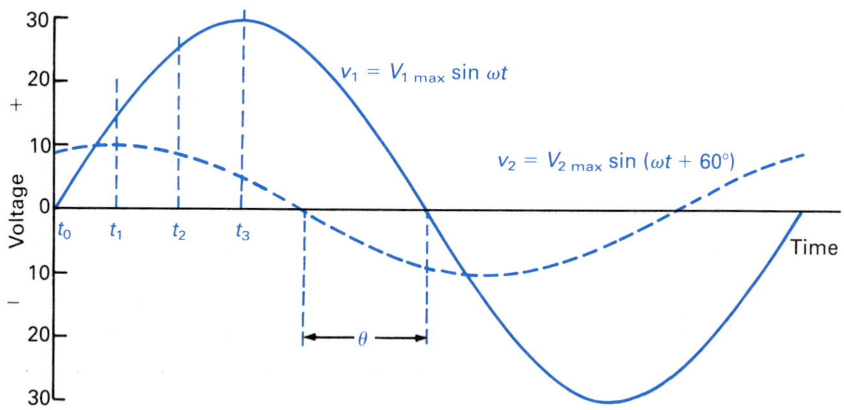

Fig. 14-1. Two sine waves differing in phase.

In Fig. 14-1 waveform v_1 (solid line) has three times the peak value of waveform v_2 (broken line), or $V_{1\,max} = 3\,V_{2\,max}$. Also, waveform v_2 leads waveform v_1 by 60°. At any given time instant, the sum of the magnitudes of v_1 and v_2 is given by arithmetic addition: $v_T = v_1 + v_2$. Examples are:

At time t_0, $v_1 = 0$, $v_2 = +8.66$, and $v_T = 0 + 8.66 = 8.66$.
At time t_1, $v_1 = +15$, $v_2 = +10$, and $v_T = 15 + 10 = 25$.

At time t_2, $v_1 = +26$, $v_2 = +8.66$, and $v_T = 26 + 8.66 = 34.66$.
At time t_3, $v_1 = +30$, $v_2 = +5$, and $v_T = 30 + 5 = 35$.

At time t_1, v_1 is approaching its maximum value, while v_2 is at its maximum value. At time t_3, v_1 is at its maximum value, but v_2 has passed its maximum value. At no time do the peaks of v_1 and v_2 occur simultaneously; thus *it cannot be said that* $V_{T\,\max} = V_{1\,\max} + V_{2\,\max}$.

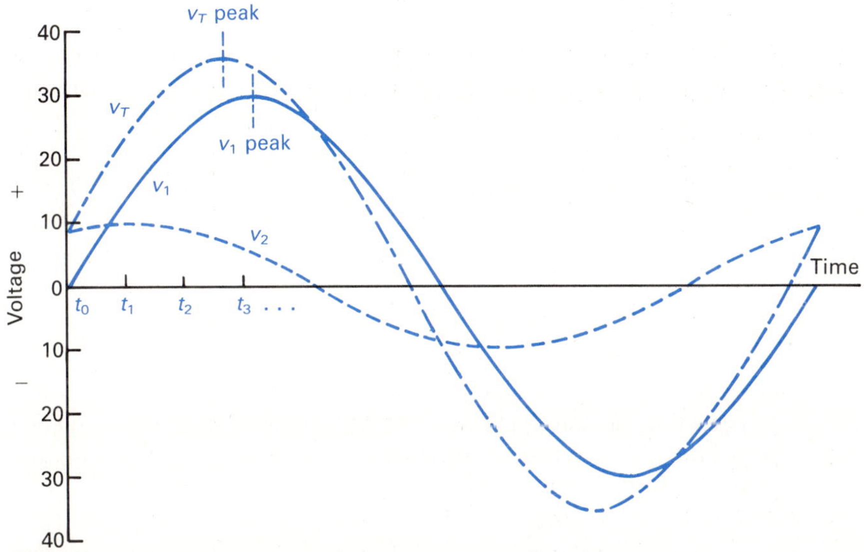

Fig. 14-2. Addition of two sine waves.

Arithmetic addition of each pair of instantaneous values v_1 and v_2 throughout the cycle in Fig. 14-1 results in the formation of a third sine wave v_T which represents the phasor sum of the two original sine waves. The three waveforms v_1, v_2, and v_T are shown in Fig. 14-2. You will see that the peak of v_T occurs slightly ahead of the peak of v_1. A phasor diagram helps visualize and determine the values of magnitude and phase angle of waveform v_T relative to those of v_1 and v_2. To keep the treatment general, the phase angle between V_1 and V_2 is initially given the symbol θ, and the phase angle between V_T and V_1 is given the symbol ϕ.

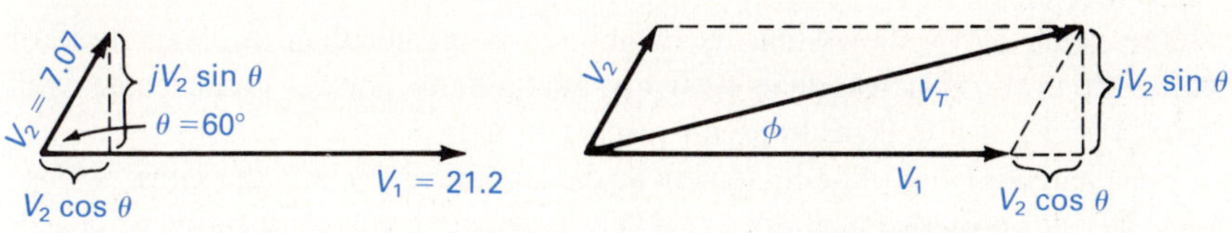

Fig. 14-3. *(a)* Sine waves in Fig. 14-2 as phasors; *(b)* adding the phasors.

Figure 14-3*a* shows the phasors V_1 and V_2, with V_1 as the reference phasor. The rectangular expressions for the phasors are

$$V_1 = V_1 + j0$$
$$V_2 = V_2 \cos\theta + jV_2 \sin\theta$$

The two phasors and their sum, the phasor V_T, are shown in Fig. 14-3*b*. The projection of V_T onto the horizontal reference axis is $V_1 + V_2 \cos\theta$. The vertical projection of V_T is $jV_2 \sin\theta$. Thus the phasor V_T is expressed in rectangular form as

$$V_T = (V_1 + V_2 \cos\theta) + jV_2 \sin\theta$$

The magnitude of V_T is

$$|V_T| = \sqrt{(V_1 + V_2 \cos\theta)^2 + (V_2 \sin\theta)^2}$$
$$|V_T| = \sqrt{V_1^2 + V_2^2 + 2V_1V_2 \cos\theta}$$

The phase angle of V_T with respect to V_1 is

$$\phi = \arctan \frac{V_2 \sin\theta}{V_1 + V_2 \cos\theta}$$

Therefore the polar form of the sum phasor appears as

$$V_T = \sqrt{V_1^2 + V_2^2 + 2V_1V_2 \cos\theta} \quad \angle \arctan \frac{V_2 \sin\theta}{V_1 + V_2 \cos\theta}$$

The derivations of the rectangular and polar expressions for a phasor which is the sum of two other phasors are given to show how the basic principle of adding voltages or currents is applied in ac terms. This apparently simple task can involve rather extensive mathematics. In practice, matters are usually much simpler than this, since the original values are known and can be used throughout a calculation. As an example, consider the situation in Figs. 14-1 to 14-3, where the rms values of V_1 and V_2 are 21.2 and 7.07, respectively, and the phase angle between them is 60°. The sum of these phasors is found in the following manner:

$$V_1 = 21.2 \angle 0° = 21.2 + j0$$
$$V_2 = 7.07 \angle 60° = 7.07(\cos 60° + j \sin 60°) = 3.54 + j6.12$$
$$V_T = V_1 + V_2 = (21.2 + j0) + (3.54 + j6.12)$$
$$V_T = 24.74 + j6.12$$

or, in polar form,

$$V_T = \sqrt{24.74^2 + 6.12^2} \quad \angle \arctan \frac{6.12}{24.74}$$
$$V_T = 25.48 \angle 13.9°$$

The value of V_T shows that its magnitude is greater than the magnitude of either V_1 or V_2, but less than their arithmetic sum, and its phase angle with respect to V_1 is less than that with respect to V_2.

The characteristics of multielement ac circuits are essentially the same as those of steady dc, provided a phasor viewpoint is adopted. The application of phasor principles to ac circuits is introduced most simply by restricting the initial discussion to networks which contain only *R* and *C* or *R* and *L*. In adopting a practical approach to ac circuits, *rms values of voltage and current are used unless otherwise stated.*

14-1 SERIES *RC* CIRCUIT

The basic characteristics of any series ac circuit are:

1. The current in each element is the same.
2. The total circuit voltage is the phasor sum of the voltages of the individual elements (Kirchhoff's voltage law).

A two-element *RC* series circuit is shown in Fig. 14-4.

Figure 14-4*a* is a schematic diagram in which the current I is common to both elements. It is convenient to make this common current the reference phasor $I\ \underline{/0^\circ}$ or $I + j0$.

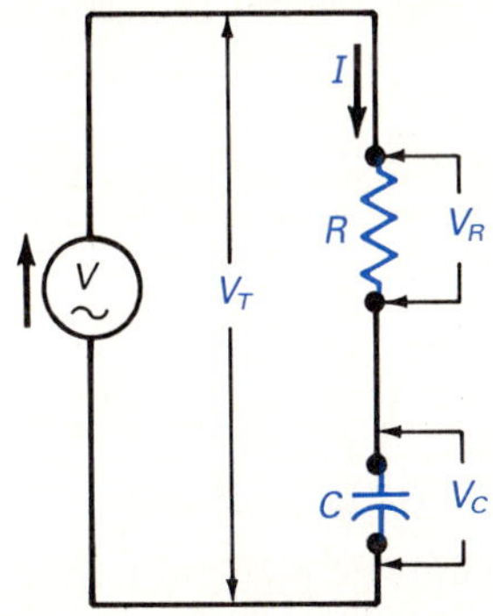

(*a*)

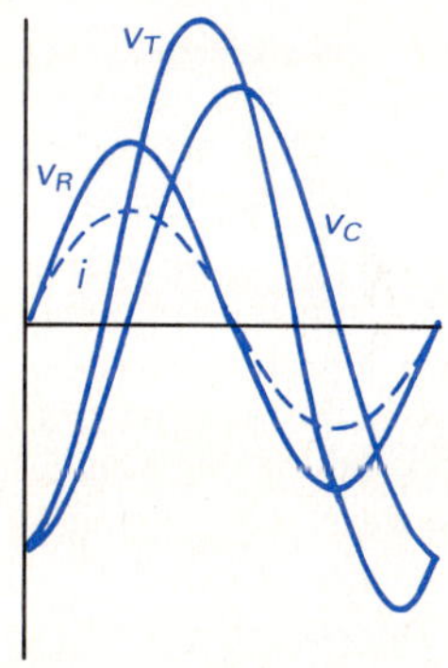

(*b*)

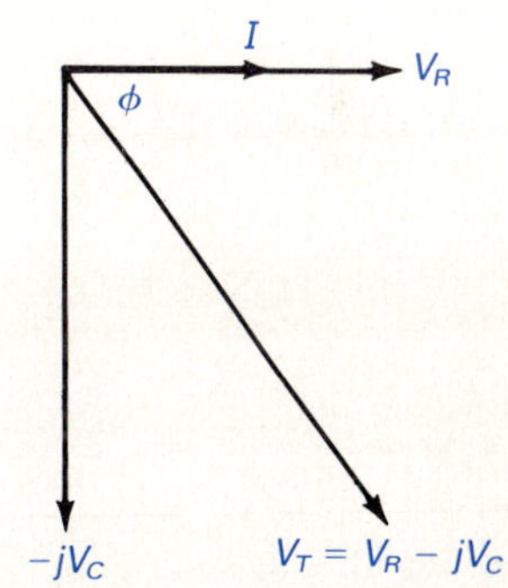

(*c*)

Fig. 14-4. Series *RC* circuit diagrams: (*a*) Schematic; (*b*) waveform; (*c*) phasor.

Voltages

The current is shown as the reference phasor in the waveform diagram in Fig. 14-4*b* and in the phasor diagram in Fig. 14-4*c*.

The resistive voltage V_R is in phase with the current, so it has the phasor value $V_R\ \underline{/0^\circ}$ or $V_R + j0$. The capacitive voltage V_C lags the current by 90° or $\pi/2$ rad, making its phasor value $V_C\ \underline{/-90^\circ}$ or $0 - jV_C$.

The total circuit voltage V_T is the phasor sum of the resistive and capacitive voltages:

$$V_T = V_R + V_C = (V_R + j0) + (0 - jV_C)$$

$$V_T = V_R - jV_C \qquad \textbf{(14-1)}$$

Impedances

By applying Ohm's law to each voltage term in Eq. (14-1), it can be rewritten

$$IZ = IR - jIX_C$$

By dividing each term of this equation by the common current I, an impedance equation for the series *RC* circuit is obtained:

$$Z = R - jX_C \qquad \textbf{(14-2)}$$

Powers

Each voltage term in Eq. (14-1) can be multiplied by the common current I:

$$V_TI = V_RI - jV_CI \qquad \textbf{(14-3)}$$

In ac circuits containing reactance (X_C or X_L), the product of the total voltage and total current V_TI is greater than the average power dissipated by the circuit.

V_TI is called *apparent power*. From the viewpoint of the voltage-current demand on the supply source it appears that the circuit uses this much power. The symbol for apparent power is S, the initial letter of the word "sum," apparent power being the phasor sum of two other powers. The unit for apparent power is the voltampere (VA).

Average power dissipation takes place only in resistances, as explained in Sec. 13-4. In Eq. (14-3) the term $V_R I$ represents the power actively dissipated by the circuit and is thus called *active power*. The symbol for active power is P, the general power symbol. The unit for active power is the watt (W), the general power unit.

The quantity $V_C I$ in Eq. (14-3) is not a true average power because, as we have seen in Chap. 13, the average power dissipated in a capacitor is zero. However, since $V_C I$ is a voltage-current product associated with a reactance, it is called *reactive power*. The symbol for reactive power is Q. This is the initial letter of the word "quadrature," reactive power being the quadrature component of apparent power. The unit of reactive power is the voltampere reactive (var).

There is international agreement that power relationships in ac circuits of all types will be considered on a voltage-reference basis, so that the resulting phase angle is the *circuit phase angle: the angle of the total current with respect to the total voltage*.

In Fig. 14-4*c* current is the reference phasor, and the phase angle is $-\phi$. If this phasor diagram is rotated counterclockwise through the angle ϕ, the phasor V_T will be in the reference position, I will be conventionally related to it, and the phase angle will be $+\phi$, the accepted phase angle of a capacitive circuit. This rotation is shown in Fig. 14-5. Figure 14-5*a* is in the same position as Fig. 14-4*c*, while Fig. 14-5*b* is Fig. 14-5*a* rotated counterclockwise through the angle ϕ.

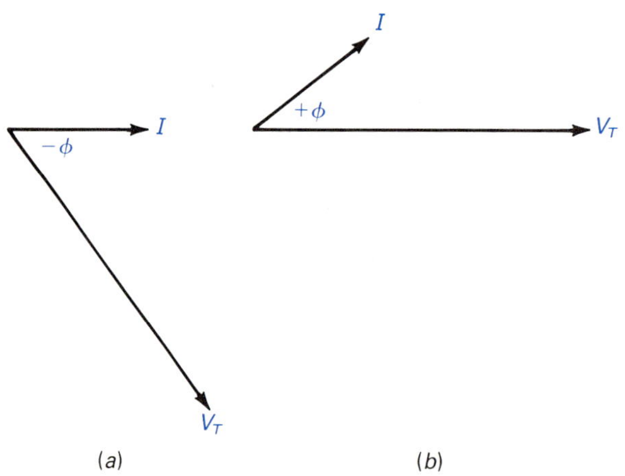

Fig. 14-5. *(a)* Current and voltage phasors in Fig. 14-4*c*; *(b)* phasors rotated $\phi°$ counterclockwise.

In Fig. 14-6*a* the phasor diagram in Fig. 14-5*b* is repeated with the addition of the horizontal and vertical components of the current.

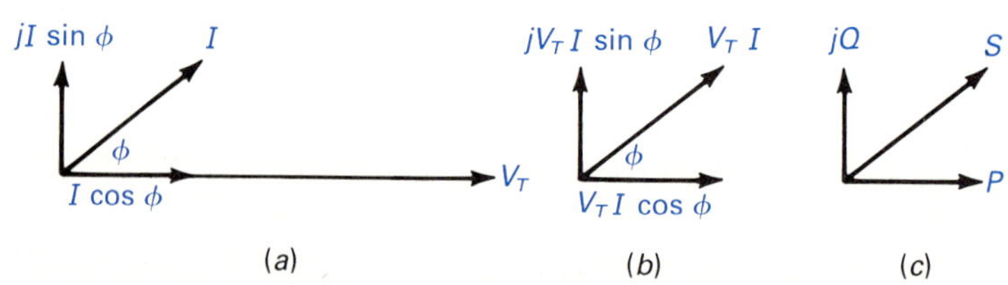

Fig. 14-6. Series *RC* circuit. *(a)* Phasor diagram with voltage reference; *(b)* current component phasors multiplied by the total voltage; *(c)* power phasors.

The current I and its horizontal and vertical components, $I \cos \phi$ (resistive) and $jI \sin \phi$ (reactive), may each be multiplied by the total circuit voltage V_T to obtain the three powers in current terms. The results of this operation are

Apparent power:	$S = V_T I$		**(14-4)**
Active power:	$P = V_T I \cos \phi$	or $S \cos \phi$	**(14-5)**
Reactive power:	$Q = V_T I \sin \phi$	or $S \sin \phi$	**(14-6)**

The powers given in Eqs. (14-4) to (14-6) are shown as phasors in Fig. 14-6*b* and *c*. In Fig. 14-6 the phase angle has a positive sign, which is regarded as correct for a capacitive circuit. From this conventional point of view, the sign of the reactive power is positive, thus producing the usual form of the power equation for a capacitive circuit:

$$S = P + jQ \qquad \textbf{(14-7)}$$

In Eqs. (14-1) to (14-7).

V_T = total network voltage, V
V_R = resistive voltage, V
V_C = capacitive voltage, V
ϕ = phase angle, ° or rad
Z = total circuit impedance, Ω
R = resistance, Ω
X_C = capacitive reactance, Ω
I = common circuit current, A
S = apparent power, VA
P = active power, W
Q = reactive power, var

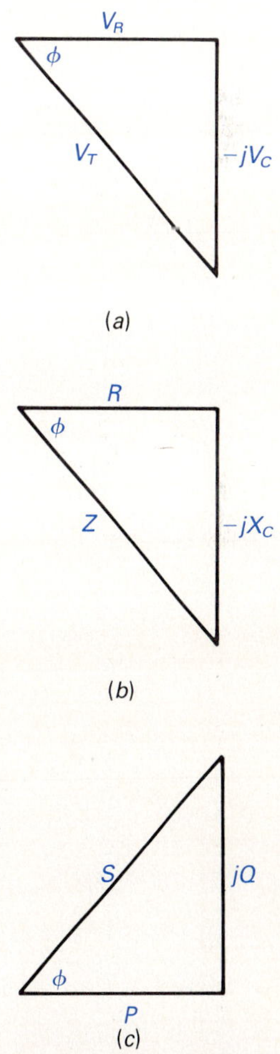

Fig. 14-7. Series *RC* circuit. (*a*) Voltage triangle; (*b*) impedance triangle; (*c*) power triangle.

Voltage, Impedance, and Power Triangles

When shown diagrammatically in the form of triangles, Eqs. (14-1), (14-2), and (14-7) become easier to analyze into polar and rectangular coordinates.

The relationships applicable to a series *RC* circuit can be summarized by reference to Eqs. (14-1), (14-2), and (14-7), and Fig. 14-7*a* to *c*.

Voltage triangle (Fig. 14-7*a*):

$$V_T = V_R - jV_C \qquad (14\text{-}1)$$

$$|V_T| = \sqrt{V_R^2 + V_C^2}$$

$$\phi = -\arctan \frac{V_C}{V_R}$$

$$V_R = V_T \cos \phi$$

$$V_C = V_T \sin \phi$$

Impedance triangle (Fig. 14-7*b*):

$$Z = R - jX_C \qquad (14\text{-}2)$$

$$|Z| = \sqrt{R^2 + X_C^2}$$

$$\phi = -\arctan \frac{X_C}{R}$$

$$R = Z \cos \phi$$
$$X_C = Z \sin \phi$$

Power triangle (Fig. 14-7*c*):

$$S = P + jQ \tag{14-7}$$
$$|S| = \sqrt{P^2 + Q^2}$$
$$\phi = \arctan \frac{Q}{P}$$
$$P = S \cos \phi = V_T I \cos \phi$$
$$Q = S \sin \phi = V_T I \sin \phi$$

Power Factor

A useful figure of merit for electric components or systems is the *power factor*, the ratio of active power to apparent power.

For industrial electric systems it is usually desirable to obtain a required level of active power for minimum cost. This objective is achieved by designing the system so that its apparent power demand is as small as possible consistent with its useful active power. Another way of expressing this objective is to say that the aim is to make the apparent power equal to the active power, or to produce *unity* power factor (power factor = 1).

For some other applications, power factors less than unity may be the aim. For capacitors (and inductors) it is desirable that there be no power dissipation; there should be zero active power, or the apparent power should equal the reactive power. Thus these reactive devices should ideally possess zero power factor.

The power factor is the ratio P/S. Referring to Fig. 14-7*c*, it can be seen that this ratio is the cosine of the phase angle ϕ. The standard symbol for the power factor is $\cos \phi$. Since it is a ratio of powers, the power factor is a dimensionless quantity; it has no unit.

$$\cos \phi = \frac{P}{S} \tag{14-8}$$

From Fig. 14-7*a* and *b*, the power factor can be expressed in terms of voltage or impedance:

$$\cos \phi = \frac{V_R}{V_T} \quad \text{or} \quad \frac{R}{Z} \tag{14-9}$$

In Eqs. (14-8) and (14-9),

$\cos \phi$ = power factor
P = active power, W
S = apparent power, VA
V_R = resistive voltage, V
V_T = total circuit pd, V
R = resistance, Ω
Z = impedance, Ω

EXAMPLE 14-1 A series circuit consists of a resistor of 1 kΩ and a capacitor of 0.022 μF connected to a 10-V 5-kHz source. What is *(a)* the voltage across the resistor, *(b)* the voltage across the capacitor, and *(c)* the phase angle of the circuit?

Solution See Fig. 14-8.

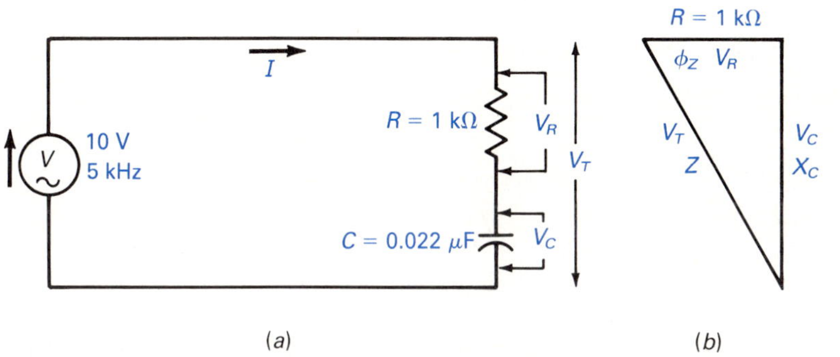

Fig. 14-8. Example 14-1: *(a)* Circuit; *(b)* impedance or voltage triangle.

$$\omega = 2\pi f = 2\pi \times 5\text{ kHz} = 31.42\text{ krad/s}$$

$$X_C = \frac{1}{\omega C} = \frac{1}{31.42\text{ krad/s} \times 0.022\ \mu\text{F}} = 1.447\text{ k}\Omega$$

$$Z = R - jX_C \qquad (14\text{-}2)$$

$$Z = (1 - j1.447)\text{ k}\Omega = \sqrt{1^2 + 1.447^2}\text{ k}\Omega \,\angle \arctan\frac{-1.447\text{ k}\Omega}{1\text{ k}\Omega}$$

$$Z = 1.759\text{ k}\Omega \angle -55.35^\circ \quad \text{or} \quad \angle -0.966\text{ rad}$$

$$I = \frac{V}{Z} = \frac{10\text{ V} \angle 0^\circ}{1.759\text{ k}\Omega \angle -55.35^\circ} = 5.685\text{ mA} \angle 55.35^\circ \quad \text{or} \quad \angle 0.966\text{ rad}$$

a. $V_R = IR = 5.685\text{ mA} \angle 55.35^\circ \times 1\text{ k}\Omega = \mathbf{5.685\ V \angle 55.35^\circ} \quad \text{or} \quad \mathbf{\angle 0.966\ rad}$

b. $V_C = I(-jX_C) = 5.685\text{ mA} \angle 55.35^\circ \times 1.447\text{ k}\Omega \angle -90^\circ$

$= 5.685\text{ mA} \times 1.447\text{ k}\Omega \angle 55.35^\circ - 90^\circ \quad \text{or} \quad \angle (0.966 - \pi/2)\text{ rad}$

$V_C = \mathbf{8.226\ V \angle -34.65^\circ} \quad \text{or} \quad \mathbf{\angle -0.604\ rad}$

c. From the phasor expression for the current *I* above,

$$\phi = \mathbf{55.35^\circ = 0.966\ rad}$$

It should be noted from Example 14-1 that, if the emf or total voltage is made the reference phasor $V \angle 0^\circ$ or $V_T \angle 0^\circ$, the current phasor derived from it automatically gives the circuit phase angle. In the example the phase angle is a positive quantity, 55.35°, which is in agreement with the concept that a leading current phase angle exists in a capacitive circuit.

The phasor diagram for Example 14-1, with the total voltage as reference, is given in Fig. 14-9.

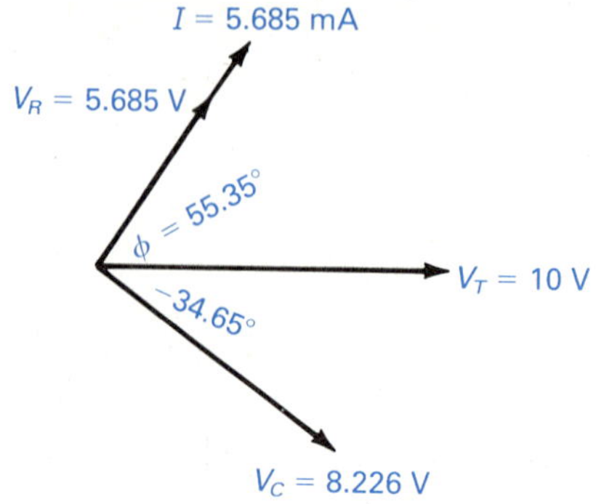

Fig. 14-9. Example 14-1: Phasor diagram.

EXAMPLE 14-2 A 50-pF capacitor has a power factor of 0.0001 at 10 MHz. What is its series equivalent resistance?

Solution

$$\omega = 2\pi f = 2\pi \times 10 \text{ MHz} = 62.83 \text{ Mrad/s}$$

$$X_C = \frac{1}{\omega C} = \frac{1}{62.83 \text{ Mrad/s} \times 50 \text{ pF}} = 318.3 \ \Omega$$

$$\cos\phi = \frac{R_s}{Z} \tag{14-9}$$

$$\cos\phi = \frac{R_s}{\sqrt{R_s^2 + X_C^2}} = \frac{R_s}{\sqrt{R_s^2 + (318.3 \ \Omega)^2}}$$

$$R_s^2 = \cos^2\phi \,[R_s^2 + (318.3 \ \Omega)^2]$$

$$R_s^2 = 10^{-8}[R_s^2 + (318.3 \ \Omega)^2]$$

$$R_s^2 = 10^{-8} R_s^2 + 10^{-8}(318.3 \ \Omega)^2$$

$$R_s \simeq \sqrt{(318.3 \ \Omega)^2 \times 10^{-8}} \simeq 318.3 \ \Omega \times 10^{-4} \simeq \mathbf{0.0318 \ \Omega}$$

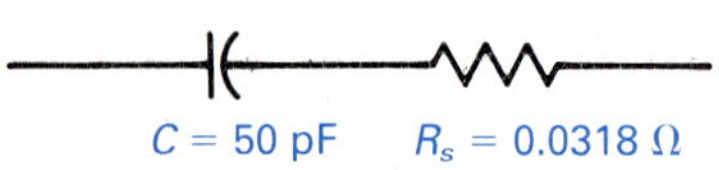

Fig. 14-10. Example 14-2: Equivalent series circuit of the capacitor.

From the results for Example 14-2, the capacitor appears to have a series equivalent circuit as shown in Fig. 14-10.

The term "equivalent" is used to qualify both the resistance and the circuit:

1. R_s is that value of resistance which, if connected in series with a theoretically perfect 50-pF capacitor, would result in the same power dissipation as occurs within this real imperfect capacitor. The components contributing to the power dissipation of the capacitor (connecting wire resistance, dielectric losses, radiation, etc.) may not all be series elements. However, as will be shown in Sec. 14-5, the series equivalent of any parallel element can be found, so that the grouping of all power-dissipating elements into a single equivalent series resistance is all right.
2. A parallel circuit for the capacitor can be derived by a method described in Sec. 14-5. Both the series and parallel circuits simulate the behavior of the capacitor, but neither can be said to be its exact replica. Each circuit is the electrical equivalent of the capacitor.

EXAMPLE 14-3 A 500-W 120-V heater is to be operated from a 230-V 50-Hz supply. Assuming that the resistance of the heater remains constant with varying temperature, what value of capacitor could be used to drop the unwanted voltage?

Solution For the heater alone on 120 V (Fig. 14-11*a*),

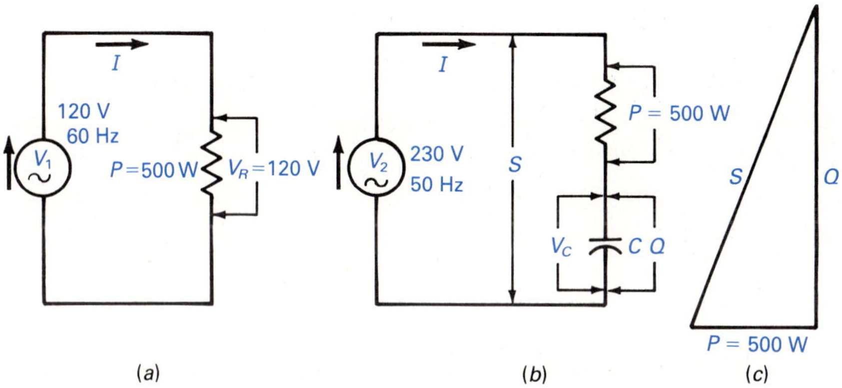

Fig. 14-11. Example 14-3: *(a)* Circuit of the heater operating on 120 V; *(b)* circuit of the heater operating on 230 V; *(c)* phasor diagram of the heater operating on 230 V.

$$P = V_R I$$

$$I = \frac{P}{V_R} = \frac{500\text{ W}}{120\text{ V}} = 4.167\text{ A}$$

For the heater and the capacitor on 230 V (Fig. 14-11*b* and *c*),

$$S = V_2 I = 230\text{ V} \times 4.167\text{ A} = 958.4\text{ VA}$$

$$Q = \sqrt{S^2 - P^2} = \sqrt{(958.4\text{ VA})^2 - (500\text{ W})^2} = 817.6\text{ var}$$

The reactive power Q is $V_C I$. By applying Ohm's law, $V_C = IX_C$, Q can be expressed as I^2X_C:

$$Q = I^2 X_C$$

$$X_C = \frac{Q}{I^2} = \frac{817.6\text{ var}}{(4.167\text{ A})^2} = 47.09\ \Omega$$

$$X_C = \frac{1}{\omega C} = \frac{1}{2\pi f C}$$

$$C = \frac{1}{2\pi f X_C} = \frac{1}{2\pi \times 50\text{ Hz} \times 47.09\ \Omega} = \mathbf{67.6\ \mu F}$$

Example 14-3 reveals some important ac circuit principles:

1. Capacitive reactance can be used as a series impedance to operate resistive equipment from an ac voltage higher than that for which it is designed. This result is achieved without dissipating extra power. In the example the

heater is intended to operate on 120-V 60-Hz North American supply but is to be used on the 230-V 50-Hz supply common in Europe and other countries.

2. The frequency is significant only as far as the reactance is concerned. A change in frequency does not affect resistive equipment such as heaters.
3. The constancy of resistance with temperature referred to in the example is important. The resistance of a heater, or other resistive equipment, can rise considerably between the instant it is switched on cold and when it is operating hot. The reactance of the capacitor will not change appreciably with temperature. Any calculation, such as the example solution, is valid only for a given value of resistance, and hence for a given temperature.
4. If a reactive element is added in series with a resistive element to operate from a higher voltage, the active power can remain unchanged but the apparent power will increase (or the power factor decrease).

EXAMPLE 14-4 A 100-V 400-Hz generator has an internal impedance equivalent to a resistance of 5 Ω. If it supplies a synchronous motor whose equivalent impedance is a 40-Ω capacitive reactance in series with a 10-Ω resistance, what is the magnitude of the motor's voltage? Refer to the diagram in Fig. 14-12.

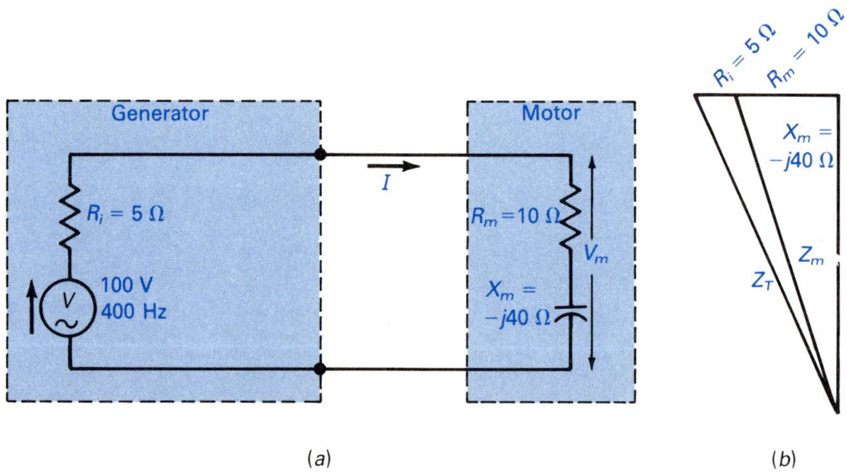

Fig. 14-12. Example 14-4: *(a)* Circuit; *(b)* impedance diagram.

Solution The impedance of the motor is, from Eq. (14-2),

$$Z_m = R_m - jX_m = (10 - j40)\ \Omega \qquad \text{or} \qquad |Z_m| = 41.23\ \Omega$$

The impedance of the generator is its internal resistance:

$$Z_g = R_i = (5 + j0)\ \Omega$$

The total impedance of the circuit is the phasor sum of the motor and generator impedances:

$$Z_T = Z_g + Z_m$$
$$= (5 + j0)\ \Omega + (10 - j40)\ \Omega = (15 - j40)\ \Omega \qquad \text{or} \qquad |Z_T| = 42.72\ \Omega$$
$$|I| = \frac{|V|}{|Z_T|} = \frac{100\ \text{V}}{42.72\ \Omega} = 2.34\ \text{A}$$
$$|V_m| = |I||Z_m| = 2.34\ \text{A} \times 41.23\ \Omega = \mathbf{96.48\ V}$$

In Example 14-4 the load receives 96.48 percent of the available emf. If this were a dc circuit in which the generator resistance was 5 Ω and the load resistance was 10 Ω + 40 Ω = 50 Ω, the load would receive only 90.9 percent of the available emf. This difference is inherent in the difference between ac phasor methods and dc algebraic methods of calculation.

EXAMPLE 14-5 The network in Fig. 14-13*a* is part of a transistor stereo amplifier circuit. If the ac voltage delivered to the *RC* circuit by transistor Q_1 is 5 V at any audio frequency, what voltage is passed on to transistor Q_2 at 20 Hz, 1 kHz, and 20 kHz? Assume that the input impedance to transistor Q_2 is very high, so that there is no loading effect across resistor *R*.

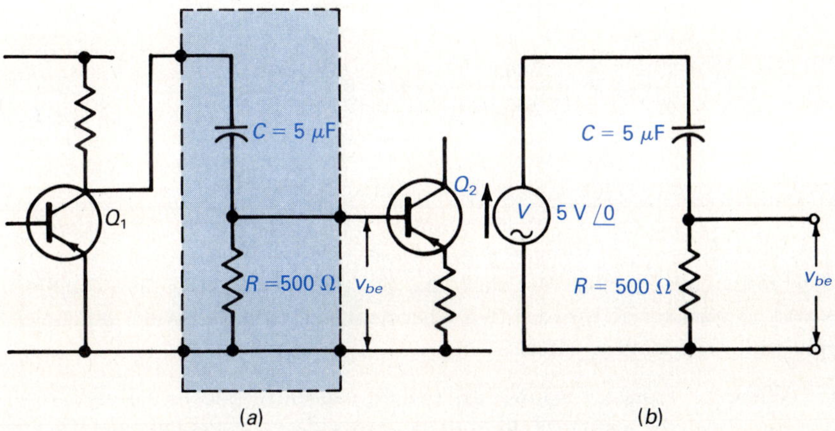

Fig. 14-13. Example 14-5: *(a)* Given circuit; *(b)* equivalent circuit.

Solution Figure 14-13*b* shows the equivalent circuit of Fig. 14-13*a* from the viewpoint of the series *RC* circuit. Transistor Q_1 is represented by the 5-V emf generator. v_{be} is the voltage supplied to transistor Q_2 if loading effects are ignored.

In this type of situation it is not necessary to calculate the current. The unloaded voltage-divider principle applies to ac if a phasor approach is adopted:

$$v_{be} = V\frac{R}{Z}$$

a. When $f = 20$ Hz,

$$\omega = 2\pi f = 2\pi \times 20\text{ Hz} = 125.7\text{ rad/s}$$

$$X_C = \frac{1}{\omega C} = \frac{1}{125.7\text{ rad/s} \times 5\ \mu\text{F}} = 1.59\text{ k}\Omega$$

$$Z = R - jX_C = (0.5 - j1.59)\text{ k}\Omega = 1.667\text{ k}\Omega\ \underline{/-72.53^\circ} \quad \text{or} \quad \underline{/-1.265\text{ rad}}$$

$$v_{be} = V\frac{R}{Z} = 5\text{ V}\ \underline{/0^\circ}\ \frac{500\ \Omega\ \underline{/0^\circ}}{1.667\text{ k}\Omega\ \underline{/-72.53^\circ}} = \mathbf{1.5\ V}\ \underline{/\mathbf{72.53^\circ}} \quad \text{or} \quad \underline{/\mathbf{1.265\ rad}}$$

b. When $f = 1$ kHz,

$$\omega = 2\pi f = 2\pi \times 1\text{ kHz} = 6.283\text{ krad/s}$$

$$X_C = \frac{1}{\omega C} = \frac{1}{6.283\text{ krad/s} \times 5\ \mu\text{F}} = 31.83\ \Omega$$

$$Z = R - jX_C = (500 - j31.83)\ \Omega = 501\ \Omega\ \underline{/-3.65^\circ} \quad \text{or} \quad \underline{/-0.0637\text{ rad}}$$

$$v_{be} = V\frac{R}{Z} = 5\text{ V}\ \underline{/0^\circ}\ \frac{500\ \Omega\ \underline{/0^\circ}}{501\ \Omega\ \underline{/-3.65^\circ}}$$

$$v_{be} = \mathbf{4.99\ V}\ \underline{/\mathbf{3.65^\circ}} \quad \text{or} \quad \underline{/\mathbf{0.0637\ rad}}$$

c. When $f = 20$ kHz,

$$\omega = 2\pi \times 20\text{ kHz} = 125.7\text{ krad/s}$$

$$X_C = \frac{1}{\omega C} = \frac{1}{125.7\text{ krad/s} \times 5\ \mu\text{F}} = 1.59\ \Omega$$

$$Z = R - jX_C = (500 - j1.59)\ \Omega \simeq 500\ \Omega\ \underline{/0^\circ}$$

$$v_{be} = V\frac{R}{Z} \simeq 5\text{ V}\ \underline{/0^\circ}\ \frac{500\ \Omega\ \underline{/0^\circ}}{500\ \Omega\ \underline{/0^\circ}} \simeq \mathbf{5\ V}\ \underline{/\mathbf{0^\circ}}$$

The results for Example 14-4 show that, for a series *RC* circuit as a coupling network to transmit ac between two other networks, voltage-divider principles apply with the following effects:

1. When the frequency is low, such that the capacitive reactance is high compared with the resistance, the voltage across the resistor is low, and it leads the applied voltage by an appreciable angle.
2. For high frequencies, where the capacitive reactance is low compared with the resistance, the voltage across the resistor is almost equal to the supply voltage and in phase with it.

When the value of v_{be} and its phase angle relative to V are plotted over the complete audio-frequency spectrum from 20 Hz to 20 kHz, the results are as shown in Fig. 14-14.

In Fig. 14-14 the horizontal frequency axis is logarithmic (equal distances represent equal frequency ratios). This method of display is commonly used so that all parts of the spectrum are shown with equal clarity. For instance, the

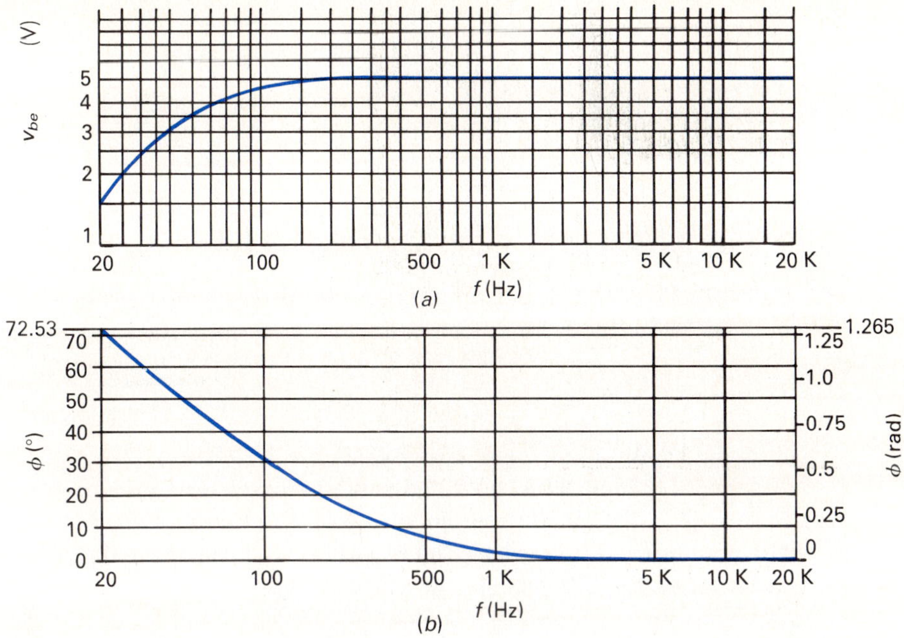

Fig. 14-14. Example 14-5: Frequency response: *(a)* Magnitude; *(b)* phase.

frequency range 100 to 500 Hz occupies the same space as the 1 to 5 kHz range. In Fig. 14-14*a* the vertical voltage axis is also shown as logarithmic for the same basic reason.

In the case of audio-frequency equipment there is another reason for using logarithmic scaling for frequency and magnitude. The human ear tends to respond to changes in frequency (pitch) and sound level in a logarithmic manner. There is no equivalent relationship for phase angle magnitude, and it is usual to employ linear scaling, as shown on the vertical axis in Fig. 14-14*b*.

REVIEW QUESTIONS FOR SECTION 14-1

1. What is the most convenient reference phasor in a series ac circuit problem?
2. In a series *RC* circuit, what is
   a. The phase relationship between V_R and I?
   b. The phase relationship between V_C and I?
   c. The phase relationship between V_R and V_C?
   d. The total voltage with respect to the element voltages?
   e. The impedance with respect to the resistance and the capacitance?
3. In an ac circuit, what is the product of current and total voltage called? What are the symbol and the unit for this quantity?
4. In an ac circuit, what is the product of current and resistive voltage called? What are the symbol and the unit for this quantity?
5. In an ac circuit, what is the product of current and reactive voltage called? What are the symbol and the unit for this quantity?

6. What is the phase angle between a circuit's current and its voltage called?
7. What is the power factor of a circuit in terms of its phase angle?
8. What would be an advantage of using a capacitor, rather than a resistor, as a voltage-dropping element?
9. What is the power factor of a perfect capacitor?

EXERCISE PROBLEMS FOR SECTION 14-1

1. Calculate V_R, V_C, Z, I, and ϕ for series circuits having the values
   a. $V_T = 25$ V, $R = 1.2$ kΩ, $C = 0.01$ μF, $f = 10$ kHz
   (*ans.* 15.05 V, 19.97 V, 1.994 kΩ, 12.54 mA, 52.99°)
   b. $V_T = 6$ V, $R = 100$ Ω, $C = 10$ μF, $f = 60$ Hz
   (*ans.* 2.117 V, 5.615 V, 283.5 Ω, 21.17 mA, 69.35°)
   c. $V_T = 50$ mV, $R = 2.2$ kΩ, $C = 100$ pF, $f = 1.2$ MHz
   (*ans.* 42.82 mV, 25.81 mV, 2.569 kΩ, 19.46 μA, 31.08°)
2. Find S, Q, and P for each of the sets of values given in Prob. 1.
3. Use each of the three methods shown in the text to determine the power factor for each of the sets of values given in Prob. 1.
   (*ans. a.* 0.602; *b.* 0.3527; *c.* 0.9146)
4. A 20-μF capacitor with a power factor of 0.001 is connected across a 120-V 60-Hz supply. Find
   a. The equivalent series resistance.
   b. The current in the capacitor.
   c. The apparent power.
   d. The reactive power.
   e. The active power.
5. A 1-μF capacitor is connected across the output of a signal generator. If the generator has an internal resistance of 600 Ω and is set for an open-circuit voltage of 10 V, find the circuit current, the voltage across the capacitor, and the phase angle of V_C with respect to I at each of the following frequencies:
   a. 50 Hz (*ans.* 3.087 mA, 9.826 V, −79.32°)
   b. 500 Hz (*ans.* 14.72 mA, 4.685 V, −27.95°)
   c. 15 kHz (*ans.* 16.66 mA, 0.1768 V, −1.01°)
6. The voltage across the 680-Ω resistor of a series *RC* circuit is 4.6 V. The current in the circuit leads the emf by 0.91 rad. If the frequency is 1000 Hz, find
   a. The voltage across the capacitor.
   b. The total voltage of the circuit.
   c. The value of the capacitor.
7. Figure 14-15 shows the circuit of a high-pass *RC* filter. Find the value of *C* and the phase relationship of the output voltage to the input voltage. (*ans.* 2122 pF, 36.87°)

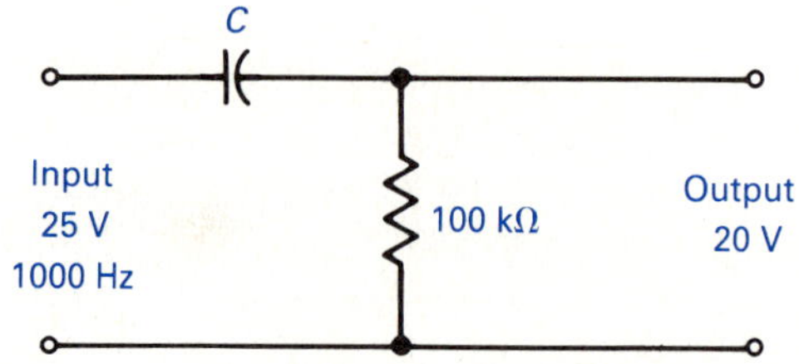

Fig. 14-15. Problems 7 to 9: Circuit.

8. For Prob. 7, find the output voltages and the phase angle of the output voltage with respect to the input voltage when the frequency is changed to
a. 100 Hz
b. 10 kHz

9. For Prob. 7, what is the minimum input frequency if the output voltage is to be greater than 95 percent of the input voltage? (*ans*. 2.281 kHz)

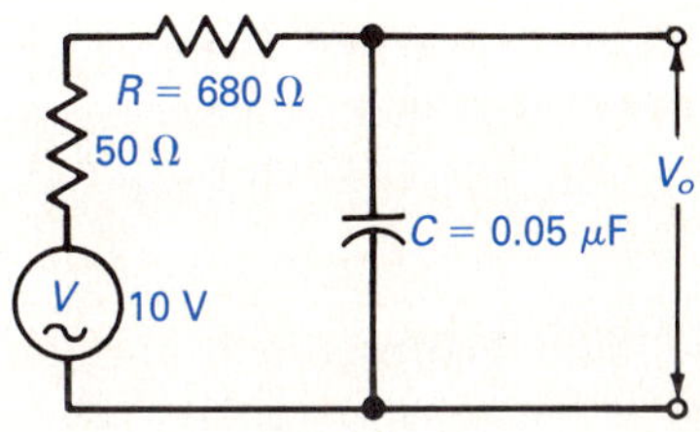

Fig. 14-16. Problems 10 and 11: Circuit.

10. For the low-pass filter circuit shown in Fig. 14-16, the cutoff frequency f_0 is reached when the total resistance equals the total capacitive reactance.
a. Find f_0.
b. Find the magnitude of the output voltage V_0 and its phase relative to the input voltage V at the frequency f_0.

11. For Prob. 10, find the output voltage and phase angle at
a. 500 Hz (*ans*. 9.934 V, 83.46°)
b. 20 kHz (*ans*. 2.13 V, 12.3°)

14-2 SERIES *RL* CIRCUIT

The series *RL* circuit behaves in a manner which is the dual of the series *RC* circuit. Phasor quantities such as total voltage and impedance, which have real and imaginary parts, are conjugate[1] for each type of circuit. For example, the impedance of a series *RL* circuit is $R + jX$, while the impedance of a series *RC* circuit is $R - jX$. Thus the outstanding distinction between the two types of circuits is the sign of the imaginary part of the related rectangular expressions (or the sign of the phase angle).

The general characteristics of the series *RL* circuit are so similar to those of the series *RC* circuit that the treatment of the series *RL* circuit is essentially the same as that given in Sec. 14-1. Thus the following discussion summarizes the characteristics of the *RL* series circuit and notes the points of difference between it and the series *RC* circuit.

The common current I in the circuit in Fig. 14-17*a* is made the reference phasor $I\,\underline{/0°}$ or $I + j0$ in Fig. 14-17*c*.

[1] *Conjugate* quantities are equal except for the sign of the imaginary part. Thus $a + jb$ and $a - jb$ are conjugate, as are $5 + j4$ and $5 - j4$.

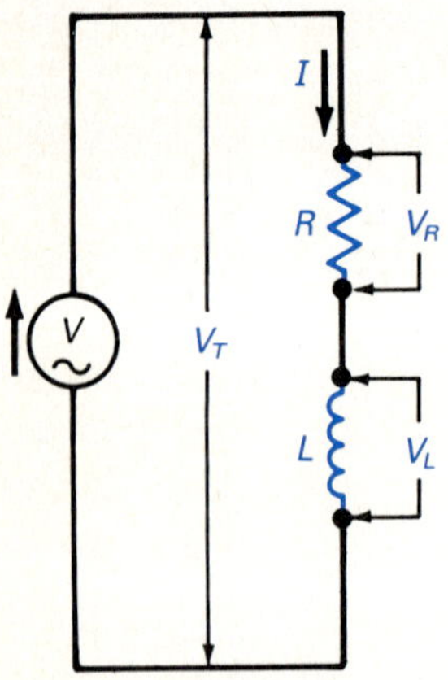

(a)

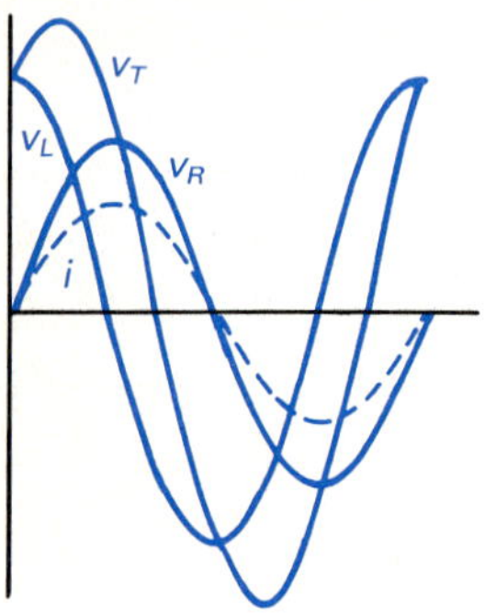

(b)

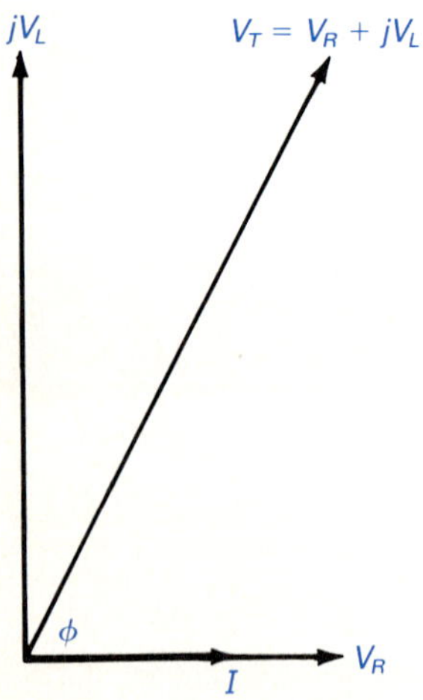

(c)

Fig. 14-17. Series *RL* circuit diagrams: *(a)* Schematic; *(b)* waveform; *(c)* phasor.

Voltages

The resistive voltage V_R is in phase with the circuit current I and thus has the phasor value $V_R \angle 0°$ or $V_R + j0$.

The inductive voltage V_L leads the current by 90°, giving a phasor value of $V_L \angle 90°$ or $0 + jV_L$.

The total circuit pd V_T is equal to the phasor sum of the resistive and inductive voltages:

$$V_T = V_R + V_L = (V_R + j0) + (0 + jV_L)$$

$$V_T = V_R + jV_L \tag{14-10}$$

Impedances

The application of Ohm's law to each voltage in Eq. (14-10) gives:

$$IZ = IR + jIX_L$$

Dividing each term of this equation by the common factor I results in an equation for the series *RL* circuit in terms of impedances:

$$Z = R + jX_L \tag{14-11}$$

Powers

If each voltage in Eq. (14-10) is multiplied by the common current I, a power equation for the series *RL* circuit is obtained:

$$V_T I = V_R I + jV_L I \tag{14-12}$$

These power equations for the *RL* circuit have been derived from the phasor relationships in Fig. 14-17*c*, which refers voltages to the common current. The conventional approach to ac powers is based on the circuit phase angle, which refers current to voltage. Rotating Fig. 14-17*c* clockwise through an angle of ϕ degrees will place the total-voltage phasor in the required reference position. This rotation is shown in Fig. 14-18*a* and *b*.

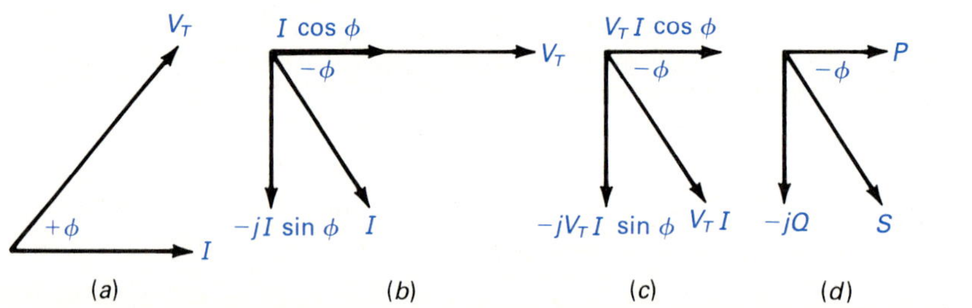

Fig. 14-18. Series *RL* circuit phasor diagrams: *(a)* and *(b)* Voltage and current; *(c)* and *(d)* power.

Figure 14-18*b* shows the current I lagging the total voltage V_T by the angle ϕ. This view gives the phase angle $-\phi$ which is usually associated with an inductive circuit. The rectangular components of the current, $I \cos \phi$ (resistive) and $-jI \sin \phi$ (reactive), are also given in Fig. 14-18*b*.

Multiplying each current phasor in Fig. 14-18*b* by the total voltage gives expressions for each of the circuit powers. The resulting power phasors are shown in Fig. 14-18*c* and *d*.

The individual power equations for the series *RL* circuit, as obtained from Fig. 14-18*c* and *d*, are

Apparent power: $S = V_T I$ **(14-13)**

Active power: $P = V_T I \cos \phi = S \cos \phi$ **(14-14)**

Reactive power: $Q = V_T I \sin \phi = S \sin \phi$ **(14-15)**

Figure 14-18*d* uses the standard ac power symbols and shows that, from the conventional viewpoint, the power equation for a series *RL* circuit is

$$S = P - jQ \qquad \textbf{(14-16)}$$

In Eqs. (14-10) to (14-16),

V_T = total network voltage, V
V_R = resistive voltage, V
V_L = inductive voltage, V
ϕ = phase angle, ° or rad
Z = total circuit impedance, Ω
R = resistance, Ω
X_L = inductive reactance, Ω
I = common circuit current, A
S = apparent power, VA
P = active power, W
Q = reactive power, var

Voltage, Impedance, and Power Triangles

Figure 14-19*a* to *c* shows Eqs. (14-10), (14-11), and (14-16) in triangle form. The series *RL* circuit relationships can conveniently be summarized from these triangles:

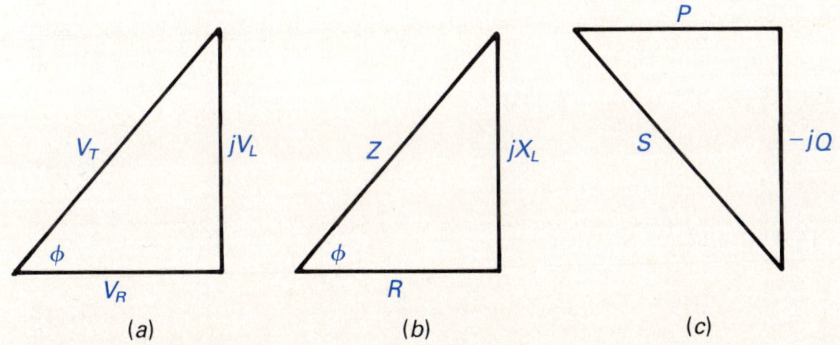

Fig. 14-19. Series *RL* circuit: *(a)* Voltage triangle; *(b)* impedance triangle; *(c)* power triangle.

Voltage triangle (Fig. 14-19*a*):

$$V_T = V_R + jV_L \qquad (14\text{-}10)$$

$$|V_T| = \sqrt{V_R^2 + V_L^2}$$

$$\phi = \arctan \frac{V_L}{V_R}$$

$$V_R = V_T \cos \phi$$
$$V_L = V_T \sin \phi$$

Impedance triangle (Fig. 14-19*b*):

$$Z = R + jX_L \tag{14-11}$$
$$|Z| = \sqrt{R^2 + X_L^2}$$
$$\phi = \arctan \frac{X_L}{R}$$
$$R = Z \cos \phi$$
$$X_L = Z \sin \phi$$

Power triangle (Fig. 14-19*c*):

$$S = P - jQ \tag{14-16}$$
$$|S| = \sqrt{P^2 + Q^2}$$
$$\phi = -\arctan \frac{Q}{P}$$
$$P = S \cos \phi = V_T I \cos \phi$$
$$Q = S \sin \phi = V_T I \sin \phi$$

Power Factor and *Q* Factor

The power factor $\cos \phi$ of any circuit is the ratio of active power to apparent power, P/S. Electric motors, transformers, and electromagnetic controls constitute the bulk of industrial electrical equipment. This equipment is usually inductive, and thus the total industrial load includes inductive reactive power, resulting in a lagging power factor. To obtain the best service rates from an electric supply utility, the reactive power should be zero, with the active power and the apparent power equal. This would give unity power factor. Industrial circuits usually deviate widely from this ideal situation, often justifying the power-factor improvement techniques referred to in Chap. 15.

From the above, or from Fig. 14-19*c*,

$$\cos \phi = \frac{P}{S} \tag{14-17}$$

From Fig. 14-19*a* and *b*,

$$\cos \phi = \frac{V_R}{V_T} = \frac{R}{Z} \tag{14-18}$$

Inductors are used for a number of purposes in electrical and electronic equipment. Here the objective may be to obtain zero power factor, or reactive power without active power. This result cannot be achieved in practice, because inductors which do not dissipate power cannot be designed and manufactured. A frequently used figure of merit for an inductor is its Q factor. This term has been used for many years and originally meant quality. Today it is best to regard it as being related to the Q, or quadrature, component of power. An inductor with a perfect Q factor (Q = infinity) does not dissipate power; it has zero active power P. Its reactive power Q is the only power component and accounts for all the apparent power. Relating the quality factor to quadrature power in this manner should help to dispel some of the confusion caused by the fact that the symbol Q has more than one meaning.

Q factor is defined as the ratio of stored energy to dissipated energy.

$$Q^* = \frac{W_{\text{stored}}}{W_{\text{dissipated}}} = \frac{\text{reactive power} \times \text{time}}{\text{active power} \times \text{time}} = \frac{Qt}{Pt}$$

$$Q = \frac{\text{reactive power}}{\text{active power}} = \frac{Q}{P} = \frac{I^2X_L}{I^2R}$$

$$Q = \frac{X_L}{R} = \tan\phi \tag{14-19}$$

Substituting ωL for X_L in Eq. (14-19),

$$Q = \frac{\omega L}{R} \tag{14-20}$$

In Eqs. (14-17) to (14-20),

$\cos\phi$ = circuit power factor
P = active power, W
S = apparent power, VA
V_R = resistive voltage, V
V_T = total circuit voltage, V
R = resistance, Ω
X_L = inductive reactance, Ω
Z = total circuit impedance, Ω
Q = circuit quality or quadrature factor
or Q = reactive power, var
W = energy, J
t = time, s
I = circuit current, A
$\tan\phi$ = tangent of the circuit phase angle
ω = angular velocity or angular frequency, rad/s
L = inductance, H

EXAMPLE 14-6 A series circuit consists of a resistor of 1 kΩ and an inductor of 50 mH connected to a 10-V 5-kHz source. What are *(a)* the voltage across the resistor, *(b)* the voltage across the inductor, and *(c)* the phase angle of the circuit?

Solution See Fig. 14-20.

*Q on the left side of the following equations is Q factor. On the right side of the equations Q is reactive power, one of the unfortunate results of too few available symbols.

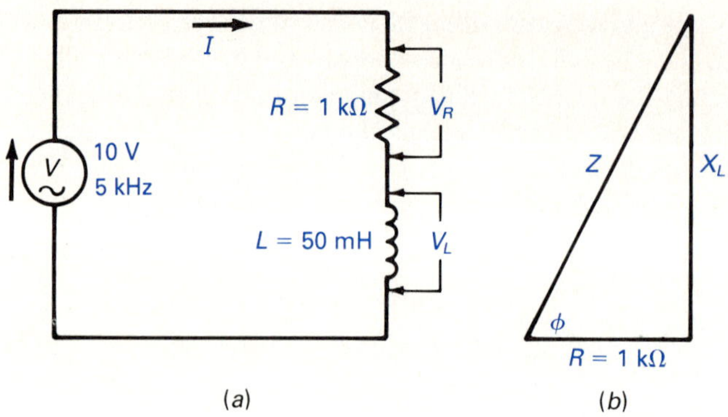

Fig. 14-20. Example 14-6: *(a)* Circuit; *(b)* impedance triangle.

$$\omega = 2\pi f = 2\pi \times 5 \text{ kHz} = 31.42 \text{ krad/s}$$

$$X_L = \omega L = 31.42 \text{ krad/s} \times 50 \text{ mH} = 1.571 \text{ k}\Omega$$

$$Z = R + jX_L \qquad (14\text{-}11)$$

$$= (1 + j1.571) \text{ k}\Omega = \sqrt{1^2 + 1.571^2} \text{ k}\Omega \;\underline{/\arctan \tfrac{1.571}{1}}$$

$$Z = 1.862 \text{ k}\Omega \;\underline{/57.5^\circ} \quad \text{or} \quad \underline{/1.003 \text{ rad}}$$

$$I = \frac{V}{Z} = \frac{10 \text{ V } \underline{/0^\circ}}{1.862 \text{ k}\Omega \;\underline{/57.5^\circ}}$$

$$I = 5.37 \text{ mA } \underline{/-57.5^\circ} \quad \text{or} \quad \underline{/-1.003 \text{ rad}}$$

a. $V_R = IR = 5.37 \text{ mA } \underline{/-57.5^\circ} \times 1 \text{ k}\Omega$

$$V_R = \mathbf{5.37\ V}\ \underline{/\mathbf{-57.5^\circ}} \quad \text{or} \quad \underline{/\mathbf{-1.003\ rad}}$$

b. $V_L = I(jX_L) = 5.37 \text{ mA } \underline{/-57.5^\circ} \times 1.57 \text{ k}\Omega \;\underline{/90^\circ}$

$$= 5.37 \text{ mA} \times 1.571 \text{ k}\Omega \;\underline{/-57.5^\circ + 90^\circ}$$

$$V_L = \mathbf{8.43\ V}\ \underline{/\mathbf{32.5^\circ}} \quad \text{or} \quad \underline{/\mathbf{0.5678\ rad}}$$

c. From the phasor expression for the current I above,

$$\phi = \mathbf{-57.5^\circ} = \mathbf{-1.003\ rad}$$

The phasor diagram for Example 14-6 with the total voltage as the reference phasor is shown in Fig. 14-21.

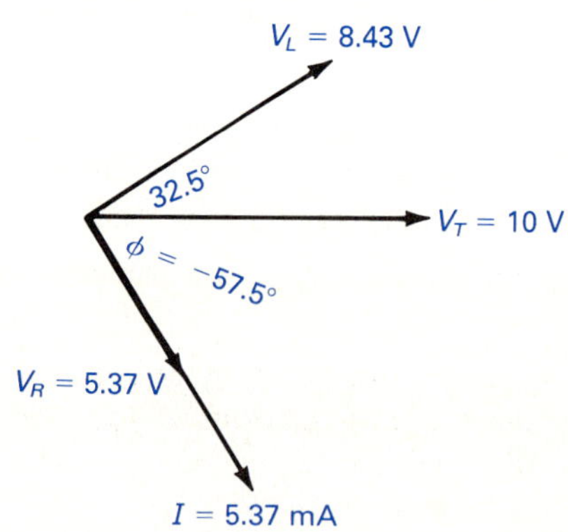

Fig. 14-21. Example 14-6: Phasor diagram.

EXAMPLE 14-7 A 17.5-μH radio tuning coil has a Q of 200 at 5 MHz. What are its *(a)* series equivalent resistance and *(b)* power factor?

Solution

$$\omega = 2\pi f = 2\pi \times 5 \text{ MHz} = 31.42 \text{ Mrad/s}$$

$$X_L = \omega L = 31.42 \text{ Mrad/s} \times 17.5\ \mu\text{H} = 549.9\ \Omega$$

a.

$$Q = \frac{X_L}{R} \tag{14-19}$$

$$R = \frac{X_L}{Q} = \frac{549.9\ \Omega}{200} = \mathbf{2.749\ \Omega}$$

b.

$$Q = \tan \phi \tag{14-19}$$

$$\phi = \arctan Q = \arctan 200 = 89.7° = 1.565 \text{ rad}$$

$$\cos \phi = \mathbf{\cos 89.7° = 0.0052}$$

In Example 14-7 the power factor is almost equal to the reciprocal of Q ($\frac{1}{200} = 0.005$). In an efficient inductor, the equivalent series resistance is so low compared with the reactance that the impedance is approximately the same as the reactance, giving this result.

For most purposes a practical inductor will have a Q value of at least 10. If an inductor is at least that good, the power factor will equal the reciprocal of the Q within 0.5 percent.

Unless otherwise specified, the resistance of an inductor is its total effective resistance. This resistance accounts for the many different forms in which power dissipation occurs in a practical inductor. They are:

1. An inductor is basically a coil of wire. The ordinary resistance of the wire is called the inductor's *ohmic resistance* or its *dc resistance*.
2. If the frequency is high, current tends to concentrate on the surface of a conductor rather than being evenly distributed throughout its cross section. This is called *skin effect* and acts to increase the resistance of an inductor's coil to a value higher than the ohmic value.
3. If the inductor has an iron core, currents will be developed within the core because it is a conductor in a varying magnetic field. These currents absorb energy from the inductor. This energy absorption is called *eddy-current loss*.
4. The magnetization of an iron core does not exactly follow the variations in coil current. This lag between cause and effect is called *hysteresis* and constitutes a dissipation of power within an iron-core inductor.
5. If the frequency is very high, the insulating materials used in the coil and the core can dissipate power in the form of *dielectric loss*.
6. At high frequencies energy may be emitted in the form of radio waves. This emission is called *radiation loss*.

The various forms of power dissipation which may be encountered in an inductor are listed to give reasons why there can be a large disparity between the resistance measured by an ohmmeter (dc) and the resistance determined by the phase angle (ac).

EXAMPLE 14-8 Measurements made on an iron-core inductor gave the following results:

	60 Hz	1 kHz
V	5 V	5 V
I	80 mA	5 mA
$\cos\phi$	0.27	0.34

What are the effective resistance and inductance of the inductor at 60 Hz and at 1 kHz?

Solution

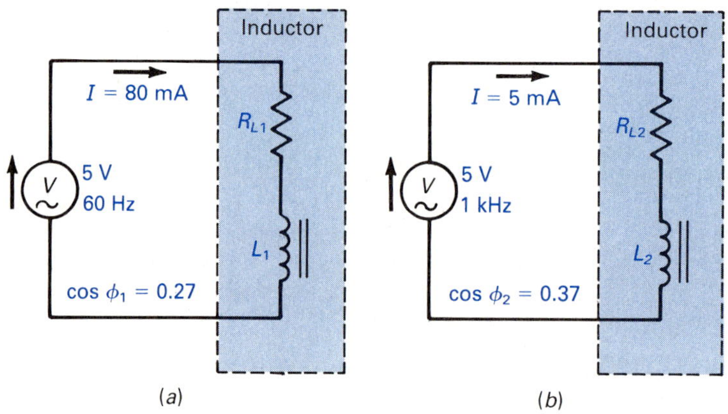

Fig. 14-22. Example 14-8: *(a)* Circuit with 60-Hz values; *(b)* circuit with 1-kHz values.

a. At 60 Hz (Fig. 14-22*a*),

$$Z_1 = \frac{V}{I} = \frac{5\text{ V}}{80\text{ mA}} = 62.5\ \Omega$$

$$\cos\phi_1 = \frac{R_{L1}}{Z_1} \qquad (14\text{-}18)$$

$$\frac{R_{L1}}{62.5\ \Omega} = 0.27$$

$$R_{L1} = \mathbf{16.88\ \Omega}$$

$$\omega_1 = 2\pi f_1 = 2\pi \times 60\text{ Hz} = 377\text{ rad/s}$$

$$Z_1 = \sqrt{R_{L1}{}^2 + X_{L1}{}^2} = 62.5\ \Omega$$

$$R_{L1}{}^2 + (\omega_1 L_1)^2 = (62.5\ \Omega)^2$$

$$(16.88\ \Omega)^2 + (377\text{ rad/s})^2 L_1{}^2 = (62.5\ \Omega)^2$$

$$L_1 = \sqrt{\frac{(62.5\ \Omega)^2 - (16.88\ \Omega)^2}{(377\text{ rad/s})^2}}$$

$$= 0.1596\text{ H} \simeq \mathbf{0.16\ H}$$

b. At 1 kHz (Fig. 14-22*b*),

$$Z_2 = \frac{V}{I} = \frac{5\ \text{V}}{5\ \text{mA}} = 1\ \text{k}\Omega$$

$$\cos \phi_2 = \frac{R_{L2}}{Z_2}$$

$$\frac{R_{L2}}{1000\ \Omega} = 0.34$$

$$R_{L2} = \mathbf{340\ \Omega}$$

$$\omega_2 = 2\pi f_2 = 2\pi \times 1000\ \text{Hz} = 6283\ \text{rad/s}$$

$$Z_2 = \sqrt{R_{L2}{}^2 + X_{L2}{}^2} = 1000\ \Omega$$

$$R_{L2}{}^2 + (\omega_2 L_2)^2 = (1000\ \Omega)^2$$

$$(340\ \Omega)^2 + (6283\ \text{rad/s})^2 L_2{}^2 = (1000\ \Omega)^2$$

$$L_2 = \sqrt{\frac{(1000\ \Omega)^2 - (340\ \Omega)^2}{(6283\ \text{rad/s})^2}}$$

$$= 0.1497\ \text{H} \simeq \mathbf{0.15\ H}$$

The data given for Example 14-8 are typical of the values to be expected from an iron-core inductor. They show the substantial increase in effective resistance as the frequency becomes higher.

Inductance is usually considered to be constant over a wide frequency spectrum. Some reduction in inductance can occur at higher frequencies as the result of a decrease in the incremental permeability of the core material and a change in effective wire dimensions caused by skin effect.

EXAMPLE 14-9 A 500-W 120-V heater is to be operated from a 230-V 50-Hz supply. Assuming that the resistance of the heater remains constant with varying temperature, what value of inductance could be used to drop the unwanted voltage?

Solution This example is the same as Example 14-3, except for the use of an inductor instead of a capacitor. Therefore most of the data are the same. See Fig. 14-23.

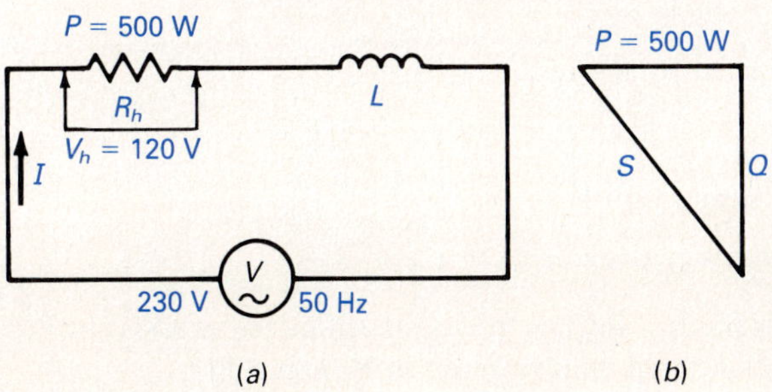

Fig. 14-23. Example 14-9: *(a)* Circuit; *(b)* power triangle.

$$X_L = 47.09\ \Omega$$
$$X_L = \omega L = 2\pi f L$$
$$L = \frac{X_L}{2\pi f} = \frac{47.09\ \Omega}{2\pi \times 50\ \text{Hz}} = 0.1498\ \text{H} \simeq \mathbf{0.15\ H}$$

It might appear that the inductor in Example 14-8 could be used as the voltage-dropping element in Example 14-9. However, if the effective resistance of the inductor is taken into account, the power dissipated by the heater will be less than its rated 500 W. This power reduction is shown in the following calculation, which assumes that the inductor's effective resistance is approximately the same at 50 and at 60 Hz. (See Fig. 14-24.)

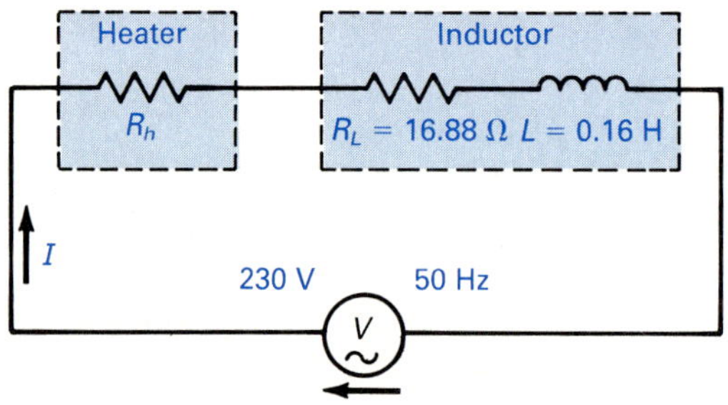

Fig. 14-24. Example 14-9: Heater circuit including the inductor in Example 14-8.

Using the heater data in Example 14-9,

$$P = \frac{V_h^2}{R_h}$$
$$R_h = \frac{V_h^2}{P} = \frac{(120\ \text{V})^2}{500\ \text{W}} = 28.8\ \Omega$$

Using data from solution *(a)* for Example 14-8,

$$X_L = 2\pi f L = 2\pi \times 50\ \text{Hz} \times 0.16\ \text{H} = 50.26\ \Omega$$
$$Z = R_h + R_L + jX_L = (28.8 + 16.88)\ \Omega + j50.26\ \Omega$$
$$Z = 67.92\ \Omega$$
$$I = \frac{V}{Z} = \frac{230\ \text{V}}{67.92\ \Omega} = 3.386\ \text{A}$$

The actual power dissipated by the heater will be

$$P = I^2 R_h = (3.386\ \text{A})^2 \times 28.8\ \Omega = 330.2\ \text{W}$$

The voltage across the heater will be

$$V_h = IR_h = 3.386\ \text{A} \times 28.8\ \Omega = 97.52\ \text{V}$$

Thus the inductor in Example 14-8 has far too much effective resistance to permit it to be used for the voltage-dropping job in Example 14-9.

EXAMPLE 14-10 Figure 14-25*a* shows the basic circuit of an electronic light-dimming control. Increasing the value of the variable resistor R decreases the lagging phase of the voltage V_g applied to the gate of the silicon controlled rectifier relative to the supply voltage V. The decreased phase angle causes the average lamp current to increase, thereby increasing the intensity of the light from the lamp. If the resistance range of R is 50 to 500 Ω, what is the phase angle range? Assume that the transformer T does not introduce phase shift.

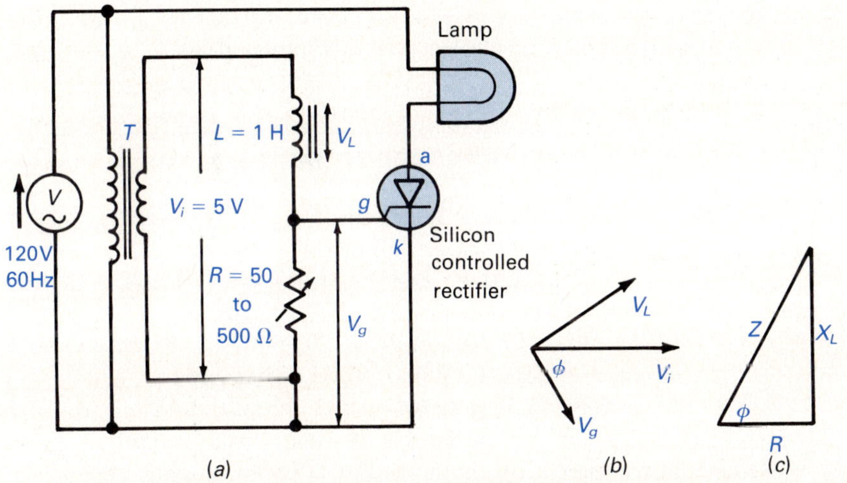

Fig. 14-25. Example 14-10: *(a)* Circuit; *(b)* phasor diagram and *(c)* impedance triangle for the silicon controlled rectifier gate circuit.

Solution It is assumed that V_i, the 5-V ac voltage at 60 Hz applied to the RL series circuit, is in phase with the supply emf V. We are thus concerned with the phase angle ϕ of the voltage V_g (developed across the variable resistor R) with respect to V_i. A voltage phasor diagram of the RL circuit is given in Fig. 14-25*b*. The value of ϕ is obtained most readily from the impedance triangle (Fig. 14-25*c*):

$$X_L = \omega L = 2\pi fL = 2\pi \times 60 \text{ Hz} \times 1 \text{ H} = 377 \ \Omega$$

When $R = 50 \ \Omega$,

$$\phi = \arctan \frac{X_L}{R} = \arctan \frac{377 \ \Omega}{50 \ \Omega} = 82.45° = 1.438 \text{ rad}$$

When $R = 500 \ \Omega$,

$$\phi = \arctan \frac{X_L}{R} = \arctan \frac{377 \ \Omega}{500 \ \Omega} = 37.01° = 0.6458 \text{ rad}$$

Thus the range of phase angle obtainable by varying R over its resistance range of 50 to 500 Ω is **−37.01° (−0.6458 rad)** to **−82.45° (−1.438 rad).**

REVIEW QUESTIONS FOR SECTION 14-2

1. In a series *RL* circuit, what is
   a. The phase relationship between V_R and I?
   b. The phase relationship between V_L and I?
   c. The phase relationship between V_R and V_L?
   d. The total voltage with respect to the element voltages?
   e. The impedance with respect to the resistance and the inductance?
2. Does a series *RL* circuit have a leading phase angle?
3. What factor, other than power factor, is commonly used as a figure of merit for an inductor?
4. If the factor referred to in Question 3 is greater than about 10, what is the approximate relationship between it and the power factor?

EXERCISE PROBLEMS FOR SECTION 14-2

1. Calculate V_R, V_L, Z, I, and ϕ for series circuits having the following values:
   a. $V_T = 300$ V, $R = 1000\ \Omega$, $L = 50$ mH, $f = 800$ Hz
   (*ans*. 291 V, 73.13 V, 1.031 kΩ, 291 mA, −14.11°)
   b. $V_T = 50$ V, $R = 20$ kΩ, $L = 0.1$ H, $f = 10$ kHz
   (*ans*. 47.7 V, 14.98 V, 20.96 kΩ, 2.385 mA, 17.44°)
   c. $V_T = 50\ \mu$V, $R = 1500\ \Omega$, $L = 150\ \mu$H, $f = 1$ MHz
   (*ans*. 42.33 μV, 26.6 μV, 1772 Ω, 0.0282 μA, 32.14°)
2. Find *S*, *Q*, and *P* for each of the sets of values given for Prob. 1.
3. Use each of the three methods shown in the text to determine the power factor for each of the sets of values given for Prob. 1.
   (*ans*. *a*. 0.9698, *b*. 0.954, *c*. 0.8468)
4. A 0.5-H inductor with a *Q* factor of 12 is connected across a 120-V 60-Hz supply. Find
   a. The equivalent series resistance.
   b. The current in the inductor.
   c. The apparent power.
   d. The reactive power.
   e. The active power.
5. A 120-V 60-Hz electric motor draws 2.5 A. The voltage leads the current by 60°. Find the impedance of the motor in
   a. Rectangular form. [*ans*. $(24 + j41.57)\ \Omega$]
   b. Polar form. (*ans*. 48 Ω, 60°)
6. For Prob. 5, find
   a. The apparent power.
   b. The reactive power.
   c. The active power.
7. For Prob. 5 with time referenced to 0 V going positive, find the instantaneous current and instantaneous power at times
   a. $t = 3.7$ ms (*ans*. 1.203 A, 201 W)
   b. $t = 9.5$ ms (*ans*. 2.018 A, −146.1 W)
8. Figure 14-26 shows the circuit of a low-pass *RL* filter. Find

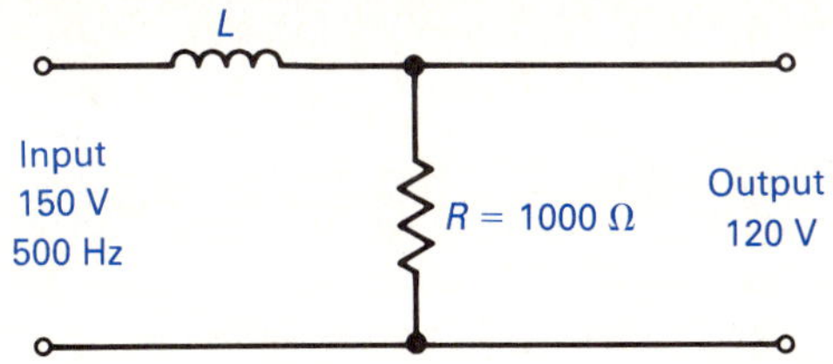

Fig. 14-26. Problems 8 and 9: Circuit.

a. The value of L.
b. The phase relationship of output voltage to input voltage.

9. For the circuit in Fig. 14-26, find the output voltage and phase relationship of output to input at
a. 50 Hz (*ans*. 149.6 V, 4.29° lag)
b. 5 kHz (*ans*. 19.8 V, 82.41° lag)

10. A 240-V 60-Hz electric motor has an impedance of $(55 + j35)\ \Omega$.
a. Express the impedance in polar form.
b. Find the current drawn by the motor.
c. Find the energy consumed by the motor in 1 h.

11. An electromagnetic relay, when connected across a 120-V 60-Hz supply, consumes 207 J of energy per minute. If the current is 48 mA, find the series equivalent circuit. (*ans*. 1497 Ω, 5.31 H)

12. An inductor with an inductance of 0.16 H has a dc resistance of 4 Ω. When connected across a 120-V 60-Hz source, the phase angle is 82.5°. Find
a. The ac resistance of the inductor at 60 Hz.
b. The current in the inductor.

14-3 PARALLEL *RC* CIRCUIT

The basic characteristics of the parallel *RC* circuit are:

1. The voltage across each element is the same.
2. The total circuit current is the phasor sum of the currents in the individual elements (Kirchhoff's current law).

In the *RC* parallel circuit in Fig. 14-27*a*, V is the common circuit voltage; it is equal to the emf of the source V_S, and it also equals the pds across the resistance and the capacitance. It is convenient to make this common voltage the reference phasor $V\ \underline{/0^\circ}$ or $V + j0$.

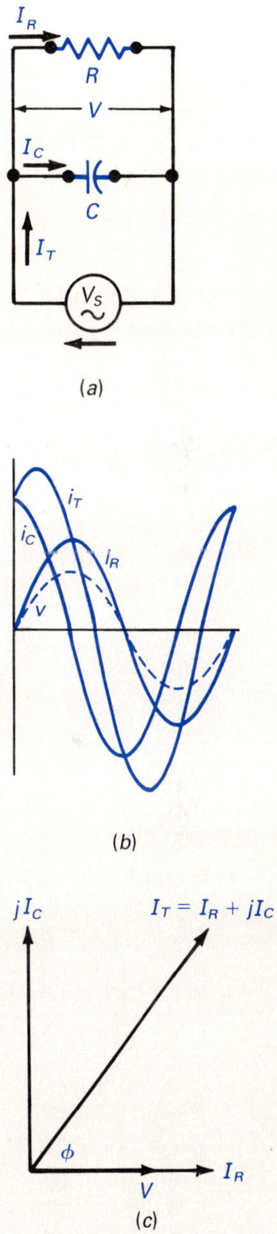

Fig. 14-27. Parallel *RC* circuit. *(a)* Schematic diagram; *(b)* waveform diagram; *(c)* phasor diagram.

Currents

The resistive current I_R is in phase with the voltage. Its phasor value is $I_R\ \underline{/0^\circ}$ or $I_R + j0$.

The capacitive current I_C leads the voltage by 90° ($\pi/2$ rad). Its phasor value is $I_C\ \underline{/90^\circ}$ or $0 + jI_C$.

The total circuit current I_T is the phasor sum of the resistive and capacitive currents:

$$I_T = I_R + I_C = (I_R + j0) + (0 + jI_C)$$
$$I_T = I_R + jI_C \tag{14-21}$$

Figure 14-27*b* and *c* shows the voltage and the currents as waveforms and as phasors.

Admittances

By applying the admittance form of Ohm's law ($I = VY$) to each current in Eq. (14-21), it can be written:

$$VY = VG + jVB_C$$

If each term of this equation is now divided by the common factor V, an admittance equation for the RC parallel circuit is obtained:

$$Y = G + jB_C \tag{14-22}$$

Powers

A power equation for the parallel RC circuit results from multiplying each term of the current equation (14-21) by the common voltage V:

$$VI_T = VI_R + jVI_C \tag{14-23}$$

In Eq. (14-23) the product of the voltage and the total current VI_T is the apparent power of the circuit S. The product of the voltage and the resistive current VI_R is the active power P dissipated by the circuit. The product of the voltage and the capacitive current VI_C is the reactive power Q. Substituting the symbols S, P, and Q into Eq. (14-23) gives:

$$S = P + jQ \tag{14-24}$$

Equations (14-21) to (14-24) are based on the use of voltage as the reference phasor in Fig. 14-27*c*, the currents in that figure being referred to the common voltage. Therefore the conventional view of phase angle, current referred to voltage, is implicit in this approach to parallel ac circuits. There is no need to rotate the phasor diagram, as was done for the series circuits, with the resultant change in sign of the reactive power component of apparent power.

Equations (14-21), (14-22), and (14-24) can be displayed as phasor diagrams, as shown in Fig. 14-28*a*, *b*, and *c*, respectively.

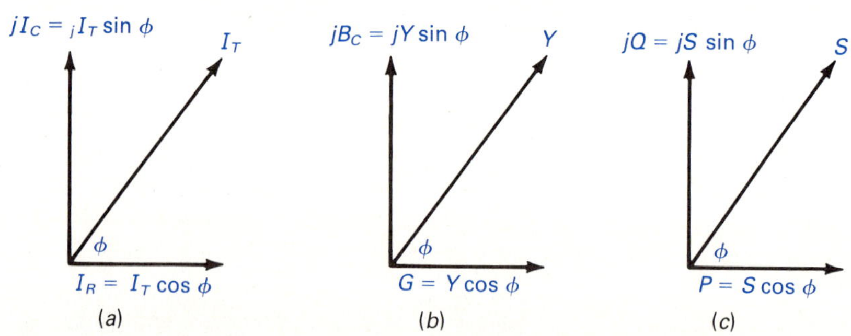

Fig. 14-28. Parallel *RC* circuit phasor diagrams: *(a)* Current; *(b)* admittance; *(c)* power.

From Eqs. (14-23) and (14-24) and Fig. 14-28*c*, equations for each of the individual powers of a parallel *RC* circuit can be obtained:

Apparent power:	$S = VI_T$	**(14-25)**
Active power:	$P = S\cos\phi = VI_T\cos\phi$	**(14-26)**
Reactive power:	$Q = S\sin\phi = VI_T\sin\phi$	**(14-27)**

In Eqs. (14-21) to (14-27),

I_T = total circuit current, A
I_R = resistive current, A
I_C = capacitive current, A
ϕ = circuit phase angle, ° or rad
Y = total circuit admittance, S
G = conductance, S
B_C = capacitive susceptance, S
V = common circuit voltage, V
S = apparent power, VA
P = active power, W
Q = reactive power, var

It is interesting to note that the power relationships [Eqs. (14-7) and (14-24)] are exactly the same for the series *RC* circuit and the parallel *RC* circuit. This equality is a fortunate outcome of standardizing the requirement for drawing the power triangles: that the circuit phase angle (total current referred to total voltage) be shown in its correct orientation. In the case of a capacitive circuit, a leading phase, or a positive phase angle, is shown regardless of the circuit configuration.

Current, Admittance, and Power Triangles

Figure 14-29 shows the current, admittance, and power triangles for the *RC* parallel circuit. These triangles can be used for visualizing the circuit relationships.

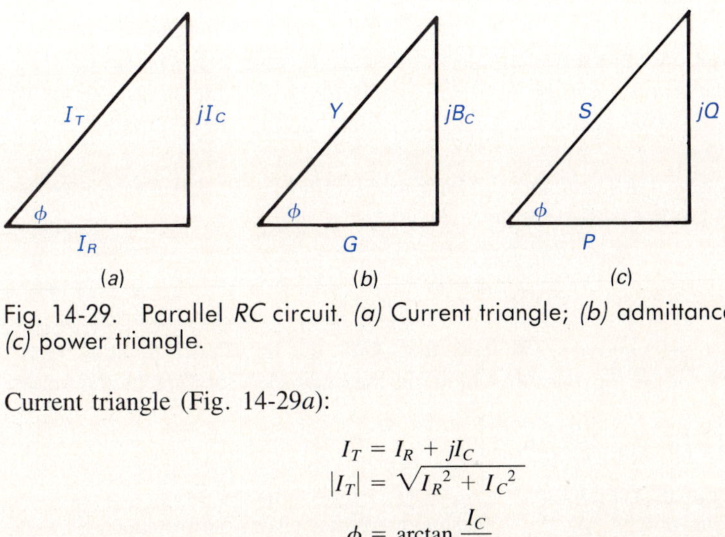

Fig. 14-29. Parallel *RC* circuit. *(a)* Current triangle; *(b)* admittance triangle; *(c)* power triangle.

Current triangle (Fig. 14-29*a*):

$$I_T = I_R + jI_C \qquad (14\text{-}21)$$

$$|I_T| = \sqrt{I_R^2 + I_C^2}$$

$$\phi = \arctan\frac{I_C}{I_R}$$

$$I_R = I_T\cos\phi$$

$$I_C = I_T\sin\phi$$

Admittance triangle (Fig. 14-29*b*):

$$Y = G + jB_C \tag{14-22}$$

$$|Y| = \sqrt{G^2 + B_C{}^2}$$

$$\phi = \arctan \frac{B_C}{G}$$

$$G = Y \cos \phi$$

$$B_C = Y \sin \phi$$

Power triangle (Fig. 14-29*c*):

$$S = P + jQ \tag{14-24}$$

$$|S| = \sqrt{P^2 + Q^2}$$

$$\phi = \arctan \frac{Q}{P}$$

$$P = S \cos \phi = VI_T \cos \phi$$

$$Q = S \sin \phi = VI_T \sin \phi$$

Power Factor

The power factor of a circuit is cos ϕ, or the ratio of the active power to the apparent power. In the parallel *RC* circuit the power factor can be expressed in terms of the sides of the triangles in Fig. 14-29:

$$\cos \phi = \frac{P}{S} = \frac{I_R}{I_T} = \frac{G}{Y} \tag{14-28}$$

If Eq. (14-28) is compared with Eqs. (14-8) and (14-9), another aspect of the duality of parallel and series circuits is revealed.

From the power ratio viewpoint [Eqs. (14-8) and (14-28)], the power factors of series and parallel circuits are the same.

From Eqs. (14-9) and (14-28) it is seen that the power factor can be expressed as a voltage ratio for the series circuit, or as a current ratio for the parallel circuit.

An impedance ratio can be used to express the power factor of a series ac circuit [Eq. (14-9)], while an admittance ratio applies to a parallel circuit [Eq. (14-28)].

By substituting $1/R$ for G and $1/Z$ for Y in Eq. (14-28), the power factor for the parallel *RC* circuit can be written

$$\cos \phi = \frac{Z}{R} \tag{14-29}$$

Equations (14-9) and (14-29) show that, from the impedance point of view, the power factor of the parallel circuit is the reciprocal of that of the series circuit.

In Eqs. (14-28) and (14-29),

$\cos \phi$ = circuit power factor
P = active power, W
S = apparent power, VA
I_R = resistive current, A
I_T = total circuit current, A
G = conductance, S
Y = total circuit admittance, S
Z = total circuit impedance, Ω
R = resistance, Ω

EXAMPLE 14-11 A parallel network consisting of a 1-kΩ resistor and a 0.022-μF capacitor is connected to a 10-V 5-kHz source. What are *(a)* the current in the resistor, *(b)* the current in the capacitor, *(c)* the total circuit current, and *(d)* the circuit phase angle?

Solution See Fig. 14-30.

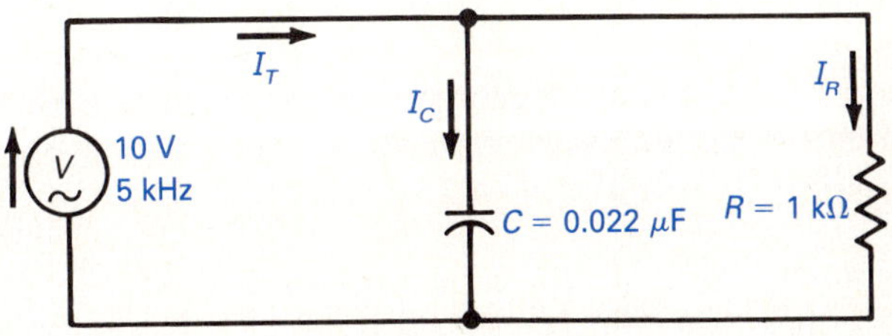

Fig. 14-30. Example 14-11: Circuit.

$$\omega = 2\pi f = 2\pi \times 5\text{ kHz} = 31.42\text{ krad/s}$$

$$B_C = \omega C = 31.42\text{ krad/s} \times 0.022\ \mu\text{F} = 0.6912\text{ mS}$$

$$G = \frac{1}{R} = \frac{1}{1\text{ k}\Omega} = 1\text{ mS}$$

a. $I_R = VG = 10\text{ V} \times 1\text{ mS} = \mathbf{10\ mA}$

b. $I_C = VB_C = 10\text{ V}\ \underline{/0^\circ} \times 0.6912\text{ mS}\ \underline{/90^\circ}$

$I_C = \mathbf{6.912\ mA\ \underline{/90^\circ}}$ or $\mathbf{\underline{/\pi/2\ rad}}$

c. $I_T = I_R + jI_C$ (14-21)

$I_T = \mathbf{(10 + j6.912)\ mA}$

or $I_T = \mathbf{12.16\ mA\ \underline{/34.65^\circ}}$ or $\mathbf{\underline{/0.6047\ rad}}$

d. From the answer to part *c*,

$$\phi = \mathbf{34.65^\circ = 0.6047\ rad}$$

In Example 14-11 the answer to part *a* could have been calculated as easily by using the resistance form of Ohm's law:

$$I_R = \frac{V}{R} = \frac{10\text{ V}}{1\text{ k}\Omega} = 10\text{ mA}$$

The conductance form was used for consistency in the adoption of an admittance view of parallel ac circuits.

You will have noticed that both the polar and the rectangular forms were given for the answer to part *c*. Either of these forms is acceptable, since each one specifies the circuit current as a complete phasor expression.

Parallel View of a Practical Capacitor

In Sec. 14-1 it was implied that a practical capacitor is shown as an equivalent series *RC* circuit. While this view is quite in order, it is just as valid to regard the capacitor as having an equivalent parallel circuit. From a logical approach, it is not easy to say for sure whether the various power losses within a capacitor should be represented as a series resistance component or as a resistance in parallel with a capacitance. We shall see later in this chapter that it is a fairly simple matter to convert a series equivalent circuit into parallel form or vice versa. Thus there is no technical difficulty in handling these alternative views of the practical capacitor.

EXAMPLE 14-12 A 50-pF capacitor has a power factor of 0.0001 at 10 MHz. What is its parallel equivalent resistance?

Solution

$$\omega = 2\pi f = 2\pi \times 10\ \text{MHz} = 62.83\ \text{Mrad/s}$$

$$B_C = \omega C = 62.83\ \text{Mrad/s} \times 50\ \text{pF} = 3.142\ \text{mS}$$

$$\cos\phi = \frac{G_p}{Y} \qquad (14\text{-}28)$$

$$10^{-4} = \frac{G_p}{\sqrt{G_p^2 + B_C^2}}$$

$$10^{-8}(G_p^2 + B_C^2) = G_p^2$$

$$10^{-8}B_C^2 \simeq G_p^2$$

$$G_p \simeq 10^{-4}B_C \simeq 3.142\ \text{mS} \times 10^{-4} \simeq 0.3142\ \mu\text{S}$$

$$R_p = \frac{1}{G_p} \simeq \frac{1}{0.3142\ \mu\text{S}} \simeq \mathbf{3.183\ M\Omega}$$

The capacitor in Example 14-12 is the same as that in Example 14-2. Thus a 50-pF capacitor with a power factor of 0.0001 can be considered to have either a series equivalent resistance of 0.0318 Ω or a parallel equivalent resistance of 3.183 MΩ.

Figure 14-31 shows the series and parallel versions of this 50-pF, 0.0001–power factor capacitor.

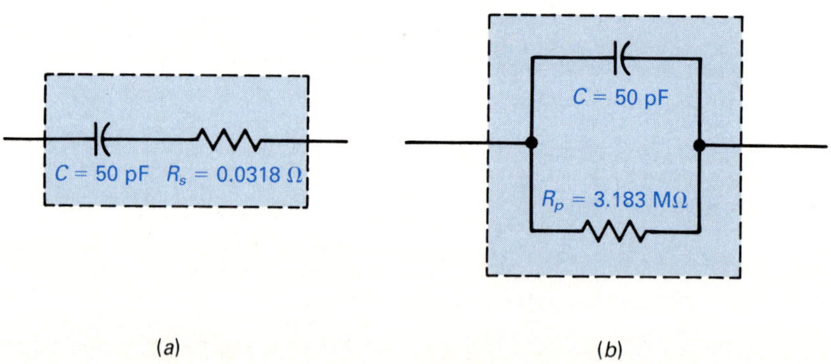

Fig. 14-31. Examples 14-2 and 14-12: Capacitor series and parallel equivalent circuits.

As a check on the validity of the series and parallel views of equivalent resistance, consider the power dissipation within the capacitor if an emf of 1 V at 10 MHz is applied across it.

In the series circuit in Example 14-2 (Fig. 14-31*a*), an applied emf of 1 V will result in a current of 3.142 mA. The power dissipated [$I^2R = (3.142 \text{ mA})^2 \times 0.0318\ \Omega$] will be 0.314 μW.

In the parallel circuit in Example 14-12 (Fig. 14-31*b*), an applied emf of 1 V will give a power dissipation [$V^2/R = (1 \text{ V})^2/3.183 \text{ M}\Omega$] of 0.314 μW.

Thus, although the series and parallel equivalent resistances differ by a ratio of 10^8, the power dissipation is the same.

It is interesting to think of the effect of the series and parallel resistances as being the same. If the parallel resistance were the same as the series resistance (0.0318 Ω), the result would be a virtual short circuit. On the other hand, if the series resistance were the same as the parallel resistance (3.183 MΩ), the result would be a virtual open circuit.

EXAMPLE 14-13 A constant-current source supplies 10 mA at 250 kHz to a network which consists of a 330-pF capacitor connected in parallel with a 3.3-kΩ resistor. What are *(a)* the circuit voltage, *(b)* the current through the resistor and *(c)* the current through the capacitor? Give the values with reference to the source current.

Solution See Fig. 14-32.

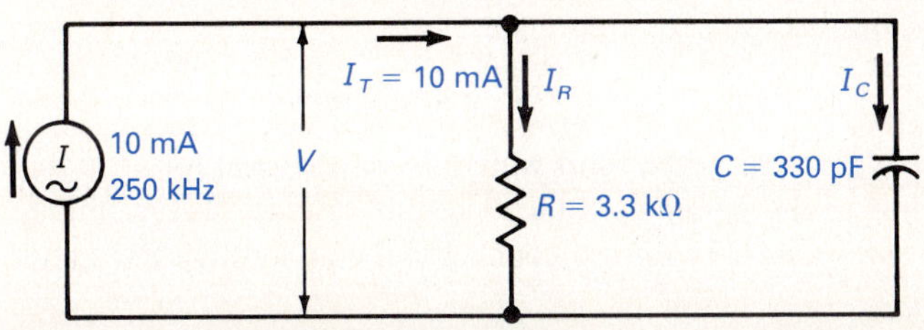

Fig. 14-32. Example 14-13: Circuit.

$$G = \frac{1}{R} = \frac{1}{3.3\text{ k}\Omega} = 0.303\text{ mS} \quad \text{or} \quad 303\ \mu\text{S}$$

$$\omega = 2\pi f = 2\pi \times 250\text{ kHz} = 1.57\text{ Mrad/s}$$

$$B_C = \omega C = 1.57\text{ Mrad/s} \times 330\text{ pF} = 518.4\ \mu\text{S}$$

$$Y = G + jB_C \qquad (14\text{-}22)$$

$$Y = (303 + j518.4)\ \mu\text{S} = 600.5\ \mu\text{S}\ \underline{/59.7^\circ} \quad \text{or} \quad \underline{/1.042\text{ rad}}$$

$$a.\ V = \frac{I_T}{Y} = \frac{10\text{ mA}\ \underline{/0^\circ}}{600.5\ \mu\text{S}\ \underline{/59.7^\circ}} = \mathbf{16.65\ V}\ \underline{/\mathbf{-59.7^\circ}} \quad \text{or} \quad \underline{/\mathbf{-1.042\ rad}}$$

The current-divider principle used for dc parallel circuits can be applied to ac circuits provided that phasor techniques are adhered to.

b. $$I_R = I_T \frac{G}{Y} = 10 \text{ mA } \angle 0^\circ \frac{303\ \mu\text{S } \angle 0^\circ}{600.5\ \mu\text{S } \angle 59.7^\circ}$$

$$I_R = \mathbf{5.046 \text{ mA } \angle -59.7^\circ} \quad \text{or} \quad \mathbf{\angle -1.042 \text{ rad}}$$

c. $$I_C = I_T \frac{B_C}{Y} = 10 \text{ mA } \angle 0^\circ \frac{518.4\ \mu\text{S } \angle 90^\circ}{600.5\ \mu\text{S } \angle 59.7^\circ}$$

$$I_C = \mathbf{8.633 \text{ mA } \angle 30.3^\circ} \quad \text{or} \quad \mathbf{\angle 0.5288 \text{ rad}}$$

The phasor diagram for Example 14-13, drawn to scale and with reference to the source current, is given in Fig. 14-33*a*.

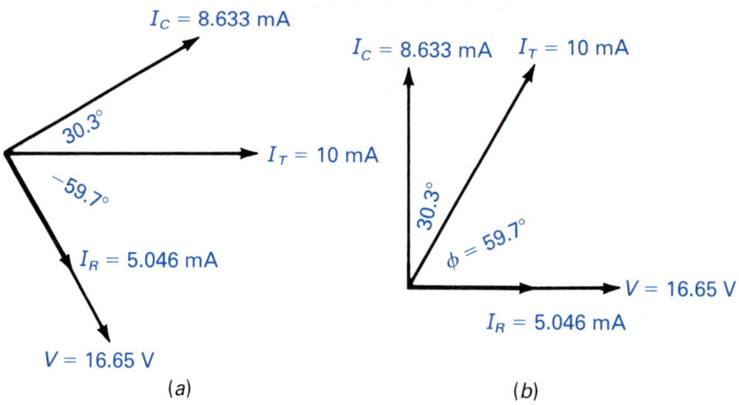

Fig. 14-33. Example 14-13: Phasor diagrams: *(a)* Source current I_T as reference; *(b)* common voltage V as reference.

Drawing the phasor diagram in the form which is conventional for a parallel circuit (with reference to the common voltage) results in Fig. 14-33*b*. Figure 14-33*b* is the same as Fig. 14-33*a* rotated counterclockwise through the circuit phase angle, 59.7°.

The common circuit voltage could have been calculated from either the resistive current or the capacitive current, once they had been obtained.

$$V = I_R R = 5.046 \text{ mA } \angle -59.7^\circ \times 3.3 \text{ k}\Omega \angle 0^\circ = 16.65 \text{ V } \angle -59.7^\circ$$

or $$V = \frac{I_C}{B_C} = \frac{8.633 \text{ mA } \angle 30.3^\circ}{518.4\ \mu\text{S } \angle 90^\circ} = 16.65 \text{ V } \angle -59.7^\circ$$

The individual element currents were found by applying the current-divider principle. One test of the validity of this method is to add these currents, in phasor form, to ensure that their sum equals the source current (Kirchhoff's current law):

$$I_R = 5.046 \text{ mA } \angle -59.7^\circ = 5.046 \cos(-59.7^\circ) + j5.046 \sin(-59.7^\circ)$$

$$I_R = (2.546 - j4.355) \text{ mA}$$

$$I_C = 8.633 \text{ mA } \angle 30.3^\circ = 8.633 \cos(30.3^\circ) + j8.633 \sin(30.3^\circ)$$

$$I_C = (7.454 + j4.355) \text{ mA}$$

$$I_R + I_C = (2.546 + 7.454) + j(4.355 - 4.355)$$
$$= (10 + j0) \text{ mA} \quad \text{or} \quad 10 \text{ mA} \angle 0^\circ$$

Therefore, $I_R + I_C = I_T$

EXAMPLE 14-14 If the parallel *RC* network in Example 14-13 were connected to a signal generator with a source current of 10 mA at 250 kHz and an internal conductance of 500 μS, what would be the current through the 3.3-kΩ resistor? (See Fig. 14-34.)

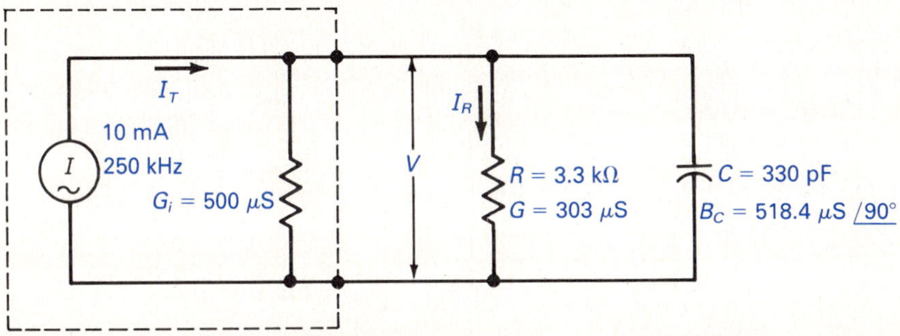

Fig. 14-34. Example 14-14: Circuit.

Solution From Example 14-13,

$$G = 303\ \mu\text{S}$$
$$B_C = 518.4\ \mu\text{S} \angle 90^\circ$$

The total admittance will be, from Eq. (14-22),

$$Y = (G_i + G) + jB_C = [(500 + 303) + j518.4]\ \mu\text{S}$$
$$Y = (803 + j518.4)\ \mu\text{S} = 955.8\ \mu\text{S} \angle 32.85^\circ \quad \text{or} \quad \angle 0.5733 \text{ rad}$$

The circuit voltage and the current through *R* can be determined by the admittance form of Ohm's law:

$$V = \frac{I_T}{Y} = \frac{10 \text{ mA} \angle 0^\circ}{955.8\ \mu\text{S} \angle 32.85^\circ} = 10.46 \text{ V} \angle -32.85^\circ$$
$$I_R = VG = 10.46 \text{ V} \angle -32.85^\circ \times 303\ \mu\text{S} \angle 0^\circ$$
$$I_R = \mathbf{3.17 \text{ mA} \angle -32.85^\circ} \quad \text{or} \quad \mathbf{\angle 0.5733 \text{ rad}}$$

Example 14-14 shows that the effect of the parallel internal conductance of a current source is to rob the load of current. This is analogous to the voltage robbing caused by the series internal resistance of a voltage source. The effect of each of these internal losses is to reduce the load voltage and current. A generator can be considered in relation to the voltage-source viewpoint (Thevenin) or in relation to the current-source viewpoint (Norton); the results are the same in either case.

EXAMPLE 14-15 What is the impedance at 2 MHz of a parallel circuit consisting of a 820-pF capacitor and a 47-Ω resistor?

Solution See Fig. 14-35.

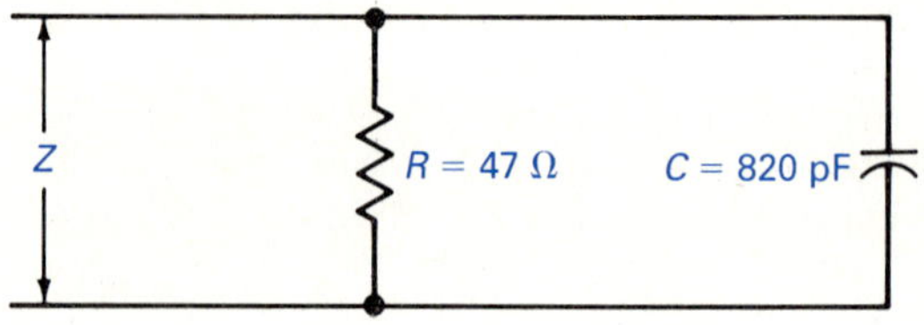

Fig. 14-35. Example 14-15: Circuit.

The principle for calculating the total resistance of two parallel resistors [$R_T = R_1R_2/(R_1 + R_2)$] can be applied to impedances if a phasor approach is used.

$$|X_C| = \frac{1}{\omega C} = \frac{1}{2\pi f C} = \frac{1}{2\pi \times 2\text{ MHz} \times 820\text{ pF}} = 97.05\ \Omega$$

$$X_C = -j97.05\ \Omega \quad \text{or} \quad 97.05\ \Omega\ \angle{-90^\circ}$$

$$Z = \frac{RX_C}{R + X_C} = \frac{47\ \Omega\ \angle 0^\circ \times 97.05\ \Omega\ \angle{-90^\circ}}{(47 - j97.05)\ \Omega}$$

$$= \frac{(47 \times 97.05)\ \Omega^2\ \angle 0^\circ + (-90^\circ)}{107.8\ \Omega\ \angle{-64.17^\circ}}$$

$$= \frac{4.561\text{ k}\Omega^2\ \angle{-90^\circ}}{107.8\ \Omega\ \angle{-64.17^\circ}} = (4561 \div 107.8)\ \Omega\ \angle{-90^\circ - (-64.17^\circ)}$$

$$Z = \mathbf{42.31\ \Omega\ \angle{-25.83^\circ}} \quad \text{or} \quad \mathbf{\angle{-0.4508\text{ rad}}} \quad \text{or} \quad \mathbf{(38.08 - j18.43)\ \Omega}$$

In Example 14-15 the solution showed that

1. The total impedance of the two elements in parallel is numerically less than the impedance of either element. This is to be expected for a normal parallel circuit.
2. The nature of the total impedance is directly related to the nature of the elements. Thus, where the elements are resistive and capacitive, the total impedance is similarly resistive and capacitive ($R - jX$ or $\angle{-\phi}$).

EXAMPLE 14-16 Figure 14-36*a* shows the input circuit of a transistor amplifying stage in a domestic AM-FM radio. The voltage generator is a simulation of the ac source which drives the transistor. The input resistance R_{be} of the transistor is constant at 1 kΩ. C is the stray capacitance of the input circuit. What is the voltage V_{be} at the transistor input at frequencies of 1 MHz on the AM band and 100 mHz on the FM band?

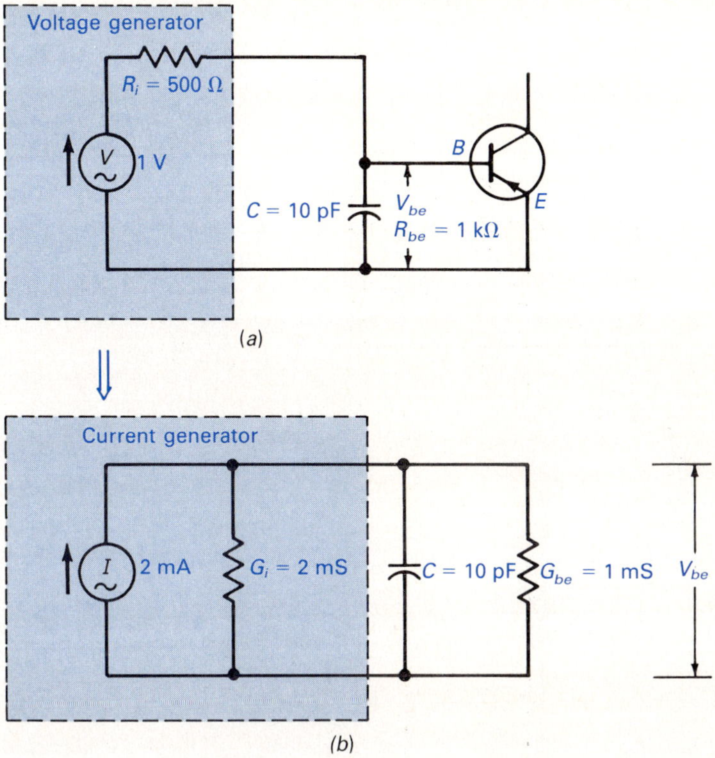

Fig. 14-36. Example 14-16: *(a)* Given circuit; *(b)* equivalent base-to-emitter circuit.

Solution The circuit in Fig. 14-36*b* is the equivalent of that in Fig. 14-36*a*. The voltage generator is converted to a current generator using the Thevenin-to-Norton conversion principles developed for dc circuits.

The source current I is given by

$$I = \frac{V}{R_i} = \frac{1\text{ V}}{500\ \Omega} = 2\text{ mA}$$

The source conductance G_i is

$$G_i = \frac{1}{R_i} = \frac{1}{500\ \Omega} = 2\text{ mS}$$

The transistor input conductance G_{be} is

$$G_{be} = \frac{1}{R_{be}} = \frac{1}{1\text{ k}\Omega} = 1\text{ mS}$$

Now that a straightforward parallel circuit has been obtained, the transistor input voltage V_{be} can be determined as the common circuit voltage.

When $f = 1$ MHz (AM band),

$$\omega = 2\pi f = 2\pi \times 1\ \text{MHz} = 6.283\ \text{Mrad/s}$$
$$B_C = \omega C = 6.283\ \text{Mrad/s} \times 10\ \text{pF} = 0.0628\ \text{mS}$$

The total admittance of the circuit is, from Eq. (14-22),

$$Y = (G_i + G_{be}) + jB_C$$
$$Y = (3 + j0.0628)\ \text{mS} \simeq 3\ \text{mS}\ \underline{/0^\circ}$$

Applying Ohm's law gives the common circuit voltage, which is the voltage developed at the transistor input.

$$V_{be} = \frac{I}{Y} \simeq \frac{2\ \text{mA}\ \underline{/0^\circ}}{3\ \text{mS}\ \underline{/0^\circ}} \simeq \mathbf{0.6667\ V}\ \underline{\mathbf{/0^\circ}}$$

When f = 100 MHz (FM band),

$$\omega = 2\pi f = 2\pi \times 100\ \text{MHz} = 628.3\ \text{Mrad/s}$$
$$B_C = \omega C = 628.3\ \text{Mrad/s} \times 10\ \text{pF} = 6.283\ \text{mS}$$

From Eq. (14-22),

$$Y = (G_i + G_{be}) + jB_C = (3 + j6.283)\ \text{mS} \quad \text{or} \quad 6.962\ \text{mS}\ \underline{/64.48^\circ}$$
$$V_{be} = \frac{I}{Y} = \frac{2\ \text{mA}\ \underline{/0^\circ}}{6.962\ \text{mS}\ \underline{/64.48^\circ}} = \mathbf{0.2873\ V}\ \underline{\mathbf{/-64.48^\circ}} \quad \text{or} \quad \underline{\mathbf{/-1.125\ rad}}$$

Example 14-16 demonstrates how a stray capacitance of 10 pF (which is a small value) can be significant at a high frequency but insignificant at a lower frequency.

If there were no stray capacitance, the transistor input voltage V_{be} would be 0.6667 V and in phase with the source current (or voltage in Fig. 14-36*a*).

When the frequency is relatively low, as on the AM radio band, the effect of the 10-pF stray capacitance is negligible, and the transistor input voltage is approximately the same as if there were no capacitance.

At the high FM frequencies, the susceptance of the 10-pF stray capacitance is great enough to affect the circuit admittance materially. The result is a reduced transistor input voltage which lags the source current (or voltage) by a substantial phase angle.

To summarize, we may say that the effect of shunt capacitance is determined by frequency; high frequencies have a greater effect than low frequencies. "High" and "low" are terms which must be related to the context. One megahertz is low when compared with 100 MHz, but would be high if compared with 1 kHz.

REVIEW QUESTIONS FOR SECTION 14-3

1. What is the most convenient quantity for a reference phasor for a parallel ac circuit?

2. In a parallel *RC* circuit, what is

a. The phase relationship between I_R and V?

b. The phase relationship between I_C and V?
c. The phase relationship between I_R and I_C?
d. The total current with respect to the element currents?

3. Why is it not necessary to rotate the phasor diagram of a parallel ac circuit to obtain the conventional form?

EXERCISE PROBLEMS FOR SECTION 14-3

1. A resistor R and a capacitor C are connected in parallel across a source of emf. Find I_R, I_C, I_T, Y, and ϕ when
a. $V = 25$ V, $R = 1.2$ kΩ, $C = 0.01$ μF, $f = 10$ kHz
(*ans*. 20.83 mA, 15.71 mA, 26.1 mA, 1.044 mS, 37.02°)
b. $V = 6$ V, $R = 100$ Ω, $C = 10$ μF, $f = 60$ Hz
(*ans*. 60 mA, 22.6 mA, 64.11 mA, 10.68 mS, 20.65°)
c. $V = 50$ mV, $R = 2.2$ kΩ, $C = 100$ pF, $f = 1.2$ MHz
(*ans*. 22.7 μA, 37.71 μA, 44.02 μA, 880.4 μS, 58.93°)

2. Find S, Q, and P for each of the sets of values given in Prob. 1.

3. A 20-μF capacitor with a power factor of 0.001 is connected across a 120-V 60-Hz supply. Find
a. The equivalent parallel resistance. (*ans*. 132.6 kΩ)
b. The total current in the circuit. (*ans*. 0.9048 A)
c. The apparent power. (*ans*. 108.6 VA)
d. The reactive power. (*ans*. 108.6 var)
e. The active power. (*ans*. 0.1086 W)
(Compare answers with those for Prob. 4, Sec. 14-1.)

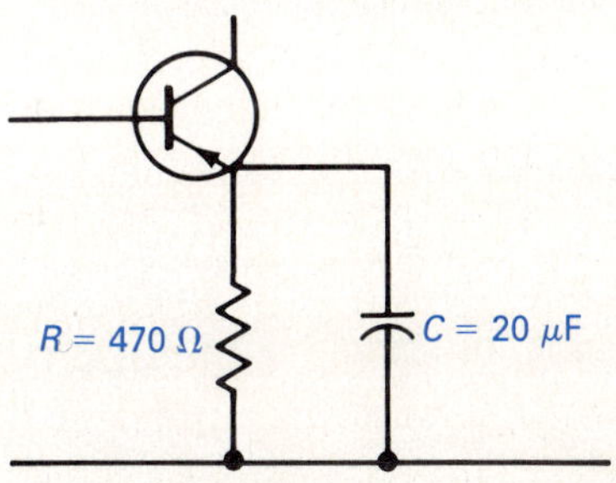

Fig. 14-37. Problems 4 and 5: Circuit.

4. Figure 14-37 shows a parallel RC circuit connected to the emitter circuit of a transistor audio amplifier. Find the admittance, in polar form, of the circuit at the following frequencies.
a. 10 kHz
b. 1 kHz
c. 20 Hz

5. Find the impedance, in polar and rectangular form, of the RC circuit in Prob. 4 at the frequencies stated.
[*ans*. *a*. 0.7956 Ω, −89.9°, (0.00139 − j0.7956) Ω;
b. 7.955 Ω, −89.03°, (0.1347 − j7.954) Ω;
c. 303.7 Ω, −49.74°, (196.2 − j231.7) Ω]

14-4 PARALLEL *RL* CIRCUIT

Except for circuit phase conditions (lag instead of lead), the parallel *RL* circuit is treated in the same manner as the parallel *RC* circuit.

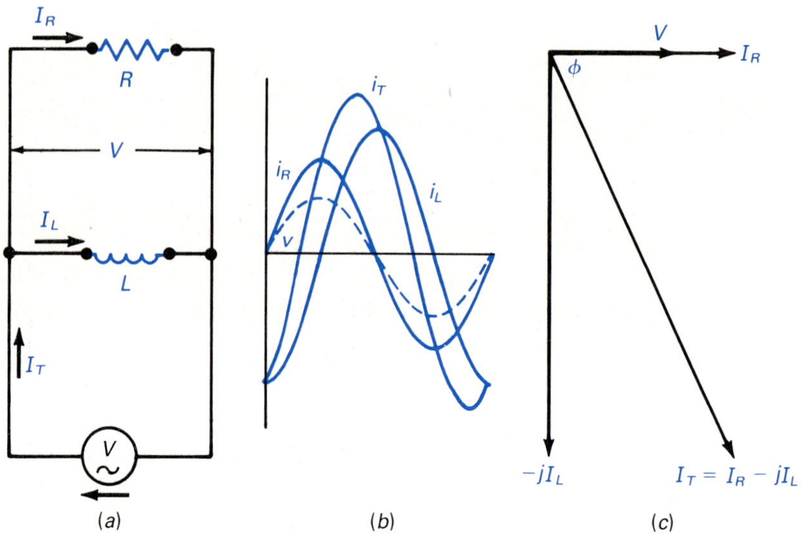

Fig. 14-38. Parallel *RL* circuit: *(a)* Schematic diagram; *(b)* waveform diagram; *(c)* phasor diagram.

In Fig. 14-38, V is the common circuit voltage; it is the pd across R and L as well as the source emf. It is the convenient choice for the reference phasor, so we make its phasor value $V \angle 0°$ or $V + j0$.

CURRENTS

The resistive current I_R is in phase with the voltage. Its phasor value is $I_R \angle 0°$ or $I_R + j0$.

The inductive current I_L lags the voltage by 90° or $\pi/2$ rad. Its phasor value is $I_L \angle -90°$ or $0 - jI_L$.

The phasor sum of the resistive and inductive currents is the total current I_T:

$$I_T = I_R + I_L = (I_R + j0) + (0 - jI_L)$$
$$I_T = I_R - jI_L \qquad \textbf{(14-30)}$$

Figure 14-38*a* shows the *RL* parallel circuit, while Fig. 14-38*b* and *c* show its waveform and phasor diagrams.

Admittances

Applying the admittance form of Ohm's law ($I = VY$) to each term in Eq. (14-30) gives

$$VY = VG - jVB_L$$

Dividing each term of this equation by the common factor V results in an admittance equation for the parallel circuit:

$$Y = G - jB_L \qquad \textbf{(14-31)}$$

Powers

Multiplying each term of the current equation (14-30) by the common voltage V gives a power equation for the parallel RL circuit:

$$VI_T = VI_R - jVI_L \qquad \textbf{(14-32)}$$

Equation (14-32) can be written in standard ac power symbol form:

$$S = P - jQ \qquad \textbf{(14-33)}$$

Since the currents have been referred to the common voltage, Eq. (14-33) expresses the power relationship for the RL parallel circuit in conventional form.

Using Eqs. (14-30), (14-31), and (14-33), we can draw phasor diagrams to show the currents, admittances, and powers in the parallel RL circuit. These diagrams are shown in Fig. 14-39*a*, *b*, and *c*, respectively.

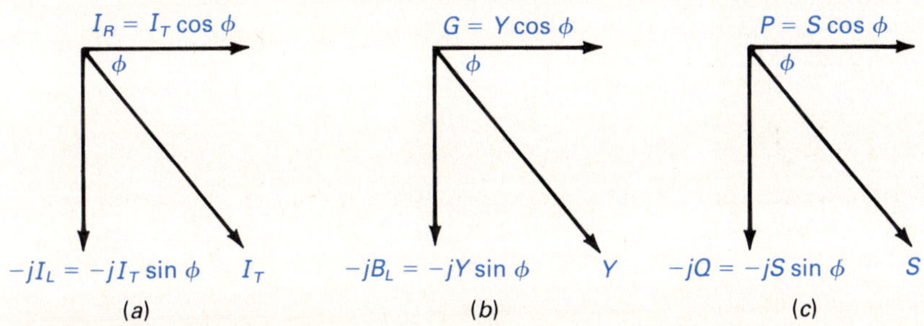

Fig. 14-39. Parallel *RL* circuit phasor diagrams: *(a)* Current; *(b)* admittance; *(c)* power.

From Eq. (14-32) and Fig. 14-39*c*, expressions for each of the individual powers of a parallel RL circuit can be obtained:

Apparent power:	$S = VI_T$	**(14-34)**
Active power:	$P = S\cos\phi = VI_T\cos\phi$	**(14-35)**
Reactive power:	$Q = S\sin\phi = VI_T\sin\phi$	**(14-36)**

In Eqs. (14-30) to (14-36),

I_T = total circuit current, A
I_R = resistive current, A
I_L = inductive current, A
ϕ = phase angle, ° or rad
Y = total circuit admittance, S
G = conductance, S
B_L = inductive susceptance, S
V = common circuit voltage, V
S = apparent power, VA
P = active power, W
Q = reactive power, var

As a result of the adoption of a standard convention for expressing ac power relationships (voltages as reference phasors), the power equations (14-16) and (14-33) for the series and parallel *RL* circuits are the same. In each case the lagging phase characteristic of inductive circuits is directly shown.

Current, Admittance, and Power Triangles

The current, admittance, and power triangles for the *RL* parallel circuit are shown in Fig. 14-40. The circuit relationships applicable to these triangles are the following:

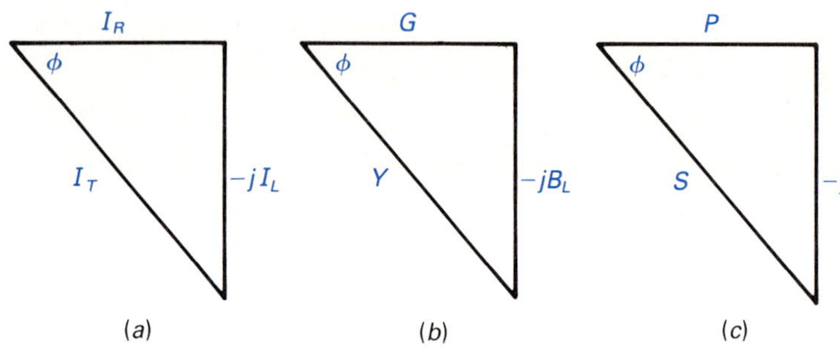

Fig. 14-40. Parallel *RL* circuit. *(a)* Current triangle; *(b)* admittance triangle; *(c)* power triangle.

Current triangle (Fig. 14-40*a*):

$$I_T = I_R - jI_L \tag{14-30}$$

$$|I_T| = \sqrt{I_R^2 + I_L^2}$$

$$\phi = \arctan \frac{I_L}{I_R}$$

$$I_R = I_T \cos \phi$$

$$I_L = I_T \sin \phi$$

Admittance triangle (Fig. 14-40*b*):

$$Y = G - jB_L \tag{14-31}$$

$$|Y| = \sqrt{G^2 + B_L^2}$$

$$\phi = \arctan \frac{B_L}{G}$$

$$G = Y \cos \phi$$

$$B_L = Y \sin \phi$$

Power triangle (Fig. 14-40*c*):

$$S = P - jQ \tag{14-33}$$

$$|S| = \sqrt{P^2 + Q^2}$$

$$\phi = \arctan \frac{Q}{P}$$

$$P = S \cos \phi = VI_T \cos \phi$$

$$Q = S \sin \phi = VI_T \sin \phi$$

Power Factor

The power factor of a parallel *RL* circuit is expressed in essentially the same terms as for the parallel *RC* circuit:

$$\cos\phi = \frac{P}{S} = \frac{I_R}{I_T} = \frac{G}{Y} = \frac{Z}{R} \tag{14-37}$$

Q Factor

The quality factor Q of an inductor can be obtained from its parallel equivalent circuit using the same definition as for the series equivalent. The *Q factor is the ratio of stored energy to dissipated energy:*

$$Q = \frac{W_{\text{stored}}}{W_{\text{dissipated}}} = \frac{\text{reactive power} \times \text{time}}{\text{active power} \times \text{time}} = \frac{Qt}{Pt}$$

$$Q = \frac{\text{reactive power}}{\text{active power}} = \frac{Q}{P}$$

The Q on the left side of the equations above represents *quality factor;* the Q on the right side of the equations represents *reactive power*.

By referring to Fig. 14-40*b* and *c*, which are similar triangles, it can be seen that B_L/G is the same as Q/P. Therefore,

$$Q = \frac{B_L}{G} = \tan\phi \tag{14-38}$$

Substituting $1/\omega L$ for B_L in Eq. (14-38),

$$Q = \frac{1}{\omega LG} \quad \text{or} \quad \frac{R}{\omega L} \quad \text{or} \quad \frac{R}{X_L} \tag{14-39}$$

In Eqs. (14-37) to (14-39),

P = active power, W
S = apparent power, VA
I_R = resistive current, A
I_T = total circuit current, A
G = conductance, S
B_L = inductive susceptance, S
Y = total circuit admittance, S
Z = total circuit impedance, Ω
R = resistance, Ω
X_L = inductive reactance, Ω
Q = quality or quadrature factor
or Q = reactive power, var
W = energy, J
t = time, s
$\tan\phi$ = tangent of the phase angle
ω = angular velocity or angular frequency, rad/s
L = inductance, H

A comparison of Eqs. (14-19), (14-20), (14-38), and (14-39) will show that:

1. In either series or parallel *RL* circuits, the Q factor is equal to the tangent of the phase angle.
2. In terms of resistance and reactance, the Q factors of the series and parallel circuits are reciprocal.

EXAMPLE 14-17 A parallel network consisting of a 1-kΩ resistor and a 50-mH inductor is connected to a 10-V 5-kHz source. What are (*a*) the current in the resistor, (*b*) the current in the inductor, (*c*) the total circuit current, and (*d*) the circuit phase angle?

Solution See Fig. 14-41.

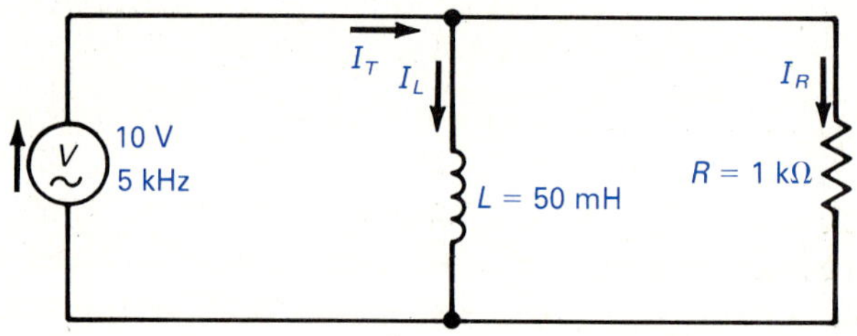

Fig. 14-41. Example 14-17: Circuit.

$$\omega = 2\pi f = 2\pi \times 5\text{ kHz} = 31.42\text{ krad/s}$$

$$B_L = \frac{1}{\omega L} = \frac{1}{31.42\text{ krad/s} \times 50\text{ mH}} = 0.6365\text{ mS}$$

$$G = \frac{1}{R} = \frac{1}{1\text{ k}\Omega} = 1\text{ mS}$$

a. $I_R = VG = 10\text{ V} \times 1\text{ mS} = \mathbf{10\text{ mA}}$

b. $I_L = VB_L = 10\text{ V}\ \underline{/0^\circ} \times 0.6365\text{ mS}\ \underline{/-90^\circ}$

$I_L = \mathbf{6.365\text{ mA}\ \underline{/-90^\circ}}$

c. $I_T = I_R - jI_L$ (14-30)

$I_T = \mathbf{(10 - j6.365)\text{ mA}}$

$I_T = \mathbf{11.85\text{ mA}\ \underline{/-32.41^\circ}}$ or $\mathbf{\underline{/-0.5656\text{ rad}}}$

d. From the answers to part *c*,

$$\phi = \mathbf{-32.41^\circ = -0.5656\text{ rad}}$$

EXAMPLE 14-18 A constant-current source supplies 10 mA at 250 kHz to a network consisting of a 2-mH inductor connected in parallel with a 3.3-kΩ resistor. What are (*a*) the circuit voltage, (*b*) the current through the resistor, and (*c*) the current through the inductor? Give the values with respect to the source current.

Solution See Fig. 14-42.

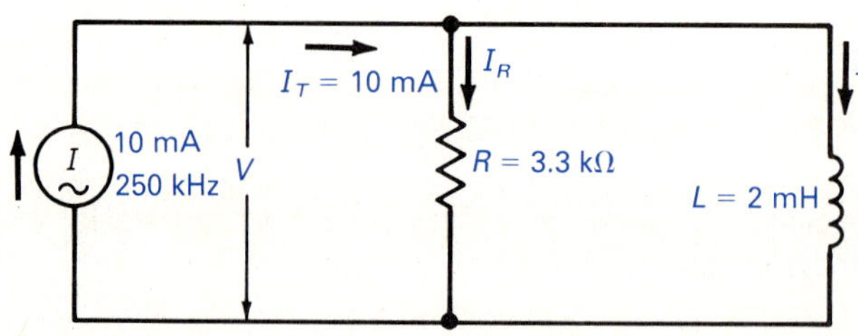

Fig. 14-42. Example 14-18: Circuit.

$$G = \frac{1}{R} = \frac{1}{3.3\ \text{k}\Omega} = 303\ \mu\text{S}$$

$$\omega = 2\pi f = 2\pi \times 250\ \text{kHz} = 1.57\ \text{Mrad/s}$$

$$B_L = \frac{1}{\omega L} = \frac{1}{1.57\ \text{Mrad/s} \times 2\ \text{mH}} = 318\ \mu\text{S}$$

The total admittance of the circuit is given by

$$Y = G - jB_L \qquad (14\text{-}31)$$

$$Y = (303 - j318)\ \mu\text{S} = 439.2\ \mu\text{S}\ \underline{/-46.4^\circ} \quad \text{or} \quad \underline{/-0.8098\ \text{rad}}$$

a. $$V = \frac{I_T}{Y} = \frac{10\ \text{mA}\ \underline{/0^\circ}}{439.2\ \mu\text{S}\ \underline{/-46.4^\circ}}$$

$$V = \mathbf{22.76\ V}\ \underline{/\mathbf{46.4^\circ}} \quad \text{or} \quad \underline{/\mathbf{0.8090\ rad}}$$

Using the current-divider principle,

b. $$I_R = I_T\frac{G}{Y} = 10\ \text{mA}\ \underline{/0^\circ}\ \frac{303\ \mu\text{S}\ \underline{/0^\circ}}{439.2\ \mu\text{S}\ \underline{/-46.4^\circ}}$$

$$I_R = \mathbf{6.899\ mA}\ \underline{/\mathbf{46.4^\circ}} \quad \text{or} \quad \underline{/\mathbf{0.8098\ rad}}$$

c. $$I_L = I_T\frac{B_L}{Y} = 10\ \text{mA}\ \underline{/0^\circ}\ \frac{318\ \mu\text{S}\ \underline{/-90^\circ}}{439.2\ \mu\text{S}\ \underline{/-46.4^\circ}}$$

$$I_L = \mathbf{7.24\ mA}\ \underline{/\mathbf{-43.6^\circ}} \quad \text{or} \quad \underline{/\mathbf{-0.761\ rad}}$$

The phasor diagram for Example 14-18, drawn to scale and with reference to the source current, is given in Fig. 14-43*a*.

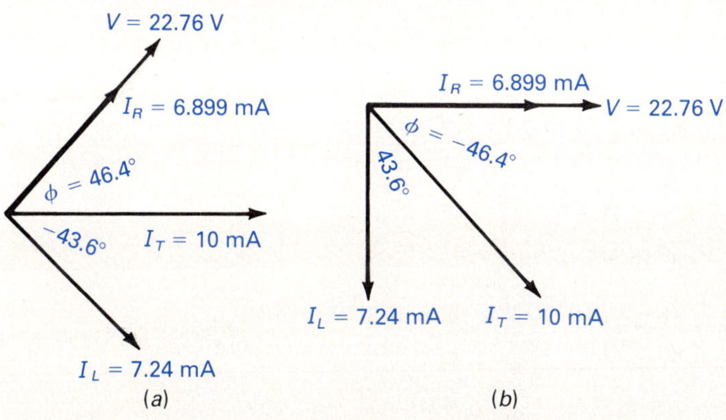

Fig. 14-43. Example 14-18: Phasor diagrams: *(a)* Source current I_T as reference; *(b)* common voltage V as reference.

Figure 14-43*b* shows the phasor diagram with the common voltage as the reference phasor. This is the conventional way of representing parallel-circuit phasors. Figure 14-43*b* can be considered to be Fig. 14-43*a* rotated clockwise through the circuit phase angle 46.4°.

The current-divider principle was used in the solutions for Example 14-18 as a demonstration. The currents could have been found by Ohm's law, perhaps more easily.

EXAMPLE 14-19 What is the impedance at 2 MHz of a parallel circuit consisting of a 47-Ω resistor and a 10-μH inductor?

Solution See Fig. 14-44.

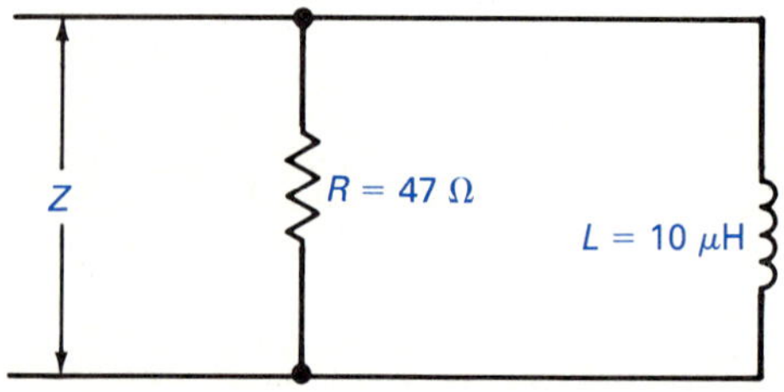

Fig. 14-44. Example 14-19: Circuit.

$$X_L = \omega L = 2\pi f L = 2\pi \times 2\text{ MHz} \times 10\ \mu\text{H} = 125.7\ \Omega$$

Applying the general relationship for the total impedance of two parallel impedances:

$$Z = \frac{Z_1 Z_2}{Z_1 + Z_2} = \frac{RX_L}{R + X_L} = \frac{47\ \Omega\ \underline{/0^\circ} \times 125.7\ \Omega\ \underline{/90^\circ}}{(47 + j125.7)\ \Omega}$$

$$= \frac{(47 \times 125.7)\ \Omega\ \underline{/0^\circ + 90^\circ}}{134.2\ \Omega\ \underline{/69.5^\circ}}$$

$$= \frac{5.908\text{ k}\Omega\ \underline{/90^\circ}}{134.2\ \Omega\ \underline{/69.5^\circ}} = (5908 \div 134.2)\ \Omega\ \underline{/90^\circ - 69.5^\circ}$$

$$Z = \mathbf{44.02\ \Omega\ \underline{/20.5^\circ}} \quad \text{or} \quad \mathbf{\underline{/0.3578\text{ rad}}}$$

EXAMPLE 14-20 Using data from the solutions for Examples 14-17 to 14-19, find the power factors of the circuits in Figs. 14-41, 14-42, and 14-44.

Solution From Example 14-17,

$$P = VI_R = 10\text{ V} \times 10\text{ mA} = 100\text{ mW}$$

$$Q = VI_L = 10\text{ V} \times 6.365\text{ mA} = 63.65\text{ mvar}$$

$$S = \sqrt{P^2 + Q^2} = \sqrt{(100\text{ mW})^2 + (63.65\text{ mvar})^2} = 118.5\text{ mVA}$$

$$\cos\phi = \frac{P}{S} = \frac{100\text{ mW}}{118.5\text{ mVA}} = \mathbf{0.8439}$$

From Example 14-18,

$$\cos\phi = \frac{I_R}{I_T} = \frac{6.899\text{ mA}}{10\text{ mA}} = \mathbf{0.6899}$$

From Example 14-19,

$$\cos\phi = \frac{Z}{R} = \frac{44.02\ \Omega}{47\ \Omega} = \mathbf{0.9366}$$

REVIEW QUESTIONS FOR SECTION 14-4

1. In a parallel *RL* circuit, what is
   a. The phase relationship between I_R and V?
   b. The phase relationship between I_L and V?
   c. The phase relationship between I_R and I_L?
   d. The total current with respect to the element currents?
2. How does a power triangle for a parallel *RL* circuit compare with that for a series *RL* circuit?
3. In a series *RL* circuit $Q = X/R$, while in a parallel *RL* circuit $Q = R/X$. How is this so if Q has the same meaning in both cases?

EXERCISE PROBLEMS FOR SECTION 14-4

1. Calculate I_R, I_L, I_T, Y, and ϕ for parallel circuits having the following values:
   a. $V = 300$ V, $R = 1000\ \Omega$, $L = 50$ mH, $f = 800$ Hz
   (*ans*. 0.3 A, 1.194 A, 1.231 A, 4.103 mS, 75.9°)
   b. $V = 50$ V, $R = 20$ kΩ, $L = 0.1$ H, $f = 10$ kHz
   (*ans*. 2.5 mA, 7.958 mA, 8.345 mA, 166.9 mS, 72.56°)
   c. $V = 50\ \mu$V, $R = 1500\ \Omega$, $L = 150\ \mu$H, $f = 1$ MHz
   (*ans*. 33.33 nA, 53.05 nA, 62.66 nA, 1.235 nS, 57.85°)
2. Find S, Q, and P for each of the sets of values given in Prob. 1.
3. A 240-V 60-Hz electric motor has an admittance of $(0.01 - j0.006)$ S.
   a. Express the admittance in polar form. (*ans*. $0.01166\ \text{S}\ \underline{/-30.96^\circ}$)
   b. Find the current drawn by the motor. (*ans*. 2.799 A)
   c. Find the power drawn by the motor. (*ans*. 576 W)
4. A 240-V 60-Hz electric motor has an impedance of $(55 + j35)\ \Omega$. Find
   a. The parallel equivalent circuit.
   b. The current drawn by the equivalent parallel resistance.
   c. The current drawn by the equivalent parallel inductance.
   d. The total current drawn by the equivalent parallel circuit.
5. For Prob. 4, find
   a. The value of capacitance that would have the same reactance as the equivalent parallel inductance. (*ans*. 21.8 μF)
   b. The total circuit current, and its phase angle, if the capacitor is connected across the equivalent parallel circuit. (*ans*. 3.106 A)

14-5 SERIES ↔ PARALLEL CONVERSION

In network analysis it is sometimes convenient to determine the parallel equivalent of a series ac circuit, or vice versa. For example, the series-parallel *resonant* network in Fig. 14-45*a* can be changed, as far as characteristics are concerned, to the totally parallel network in Fig. 14-45*b* by converting the series *RL* network L_sR_s into its parallel equivalent L_pR_p.

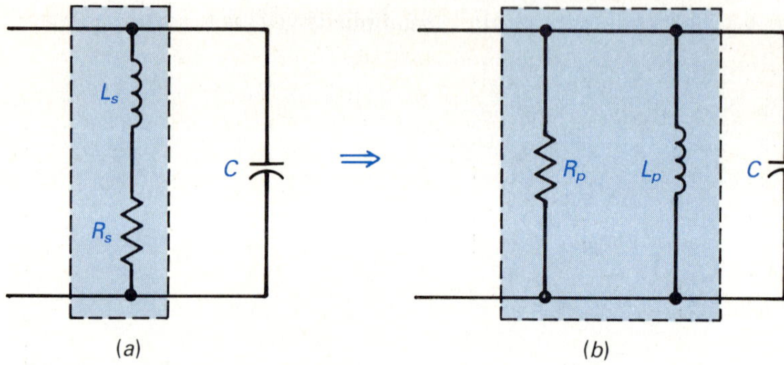

Fig. 14-45. *(a)* Series-parallel circuit; *(b)* parallel equivalent.

The circuit in Fig. 14-45*b* can be analyzed by direct parallel-circuit methods. The resulting calculations and phasor diagrams are simpler than those of the original series-parallel circuit. This simplification can be helpful in understanding circuit behavior.

To perform the conversions it is necessary only to equate impedances or admittances. This procedure involves rationalization, which will be dealt with first.

Rationalizing Complex Quantities

Rationalizing is used to separate the real and imaginary parts of an expression which has a complex denominator. Unless this separation is made, the expression lacks clarity. For example, $Z = [1/(0.03 + j0.04)]\ \Omega$ is not a clear way of expressing an impedance. By rationalizing, this expression can be converted to $Z = (12 - j16)\ \Omega$, which is in the standard rectangular form for impedance. A circuit diagram and an impedance triangle can be drawn, as in Fig. 14-46.

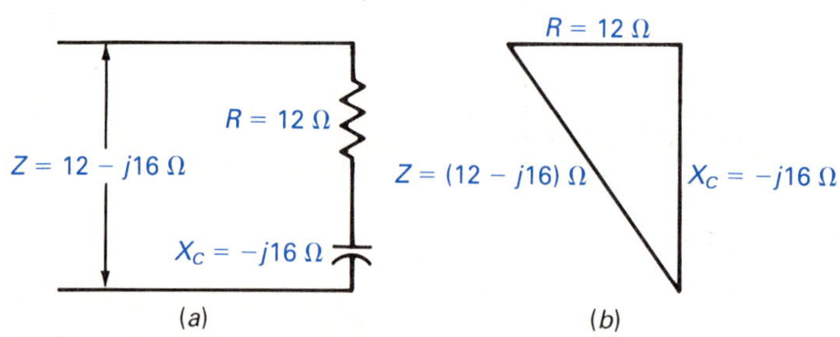

Fig. 14-46. The impedance $Z = [1/(0.03 + j0.04)]\ \Omega$: *(a)* As an equivalent network; *(b)* as an impedance triangle.

The technique of rationalization is based on the principle that a quantity is unchanged in value if it is multiplied by 1. The 1 can be any convenient quantity divided by itself. Another way of saying "multiplied by 1" is to say "numerator and denominator multiplied by the same quantity."

In complex number arithmetic, the quantity chosen for division by itself is the conjugate of the denominator in the original expression. When a complex number is multiplied by its conjugate, the imaginary parts are eliminated, leaving just a real number. For example, consider the expression $1/(R + jX)$. The denominator is $R + jX$, so the expression will be multiplied by $(R - jX)/(R - jX)$:

$$\frac{1}{R + jX} = \frac{1}{R + jX} \times \frac{R - jX}{R - jX} = \frac{R - jX}{(R + jX)(R - jX)}$$

Concentrating first on the final denominator as the product of two binomials, each a rectangular-form complex number,

$$(R + jX)(R - jX) = R^2 - jRX + jRX - j^2X^2$$

Since $j^2 = -1$, it can be seen that $-j^2 = +1$, so that

$$\text{Denominator} = R^2 + X^2$$

We started with a denominator $R + jX$ which contained both real and imaginary parts. We now have a denominator which has only real parts. The new denominator is an ordinary real number by which each part of the numerator can be divided. Therefore,

$$\frac{1}{R + jX} = \frac{R - jX}{R^2 + X^2} = \frac{R}{R^2 + X^2} - j\frac{X}{R^2 + X^2}$$

The overall result of this rationalization is to convert a fraction, $1/(R + jX)$, which has a complex denominator to a rectangular phasor form consisting of a real part, $R/(R^2 + X^2)$, and an imaginary part, $-jX/(R^2 + X^2)$.

You should check to see that, if the original denominator had been $R - jX$, the final denominator would also have been $R^2 + X^2$ but the imaginary part of the final expression would have been $+jX/(R^2 + X^2)$.

Summarizing the results of rationalizing a fraction with a complex denominator,

$$\frac{1}{R + jX} = \frac{R}{R^2 + X^2} - j\frac{X}{R^2 + X^2}$$

$$\frac{1}{R - jX} = \frac{R}{R^2 + X^2} + j\frac{X}{R^2 + X^2}$$

The numerical example given above and displayed in Fig. 14-46 is thus evaluated:

$$Z = \frac{1}{0.03 + j0.04} \times \frac{0.03 - j0.04}{0.03 - j0.04}$$

$$= \frac{0.03}{0.03^2 + 0.04^2} - j\frac{0.04}{0.03^2 + 0.04^2}$$

$$Z = (12 - j16)\ \Omega$$

The application of rationalization to series ↔ parallel conversion can now proceed.

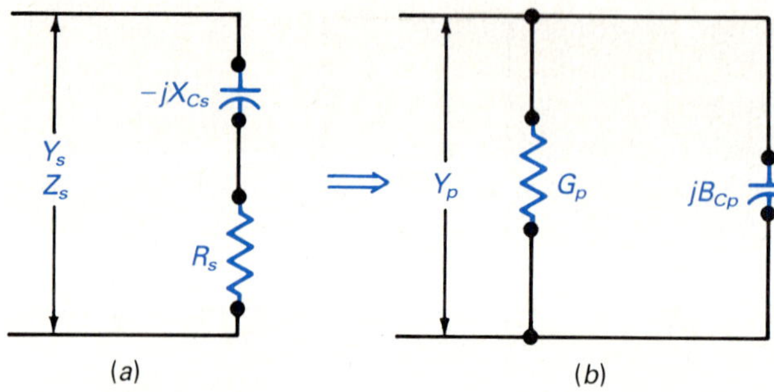

Fig. 14-47. *(a)* Series *RC* network; *(b)* its parallel equivalent.

Series-to-Parallel Conversion, *RC* Figure 14-47*a* shows a resistor R_s in series with a capacitive reactance $-jX_{Cs}$. The circuit impedance Z_s is given by

$$Z_s = R_s - jX_{Cs}$$

Its admittance Y_s will be

$$Y_s = \frac{1}{Z_s} = \frac{1}{R_s - jX_{Cs}}$$

Rationalizing, we obtain a rectangular expression for the admittance of the series circuit:

$$Y_s = \frac{R_s}{R_s^2 + X_{Cs}^2} + j\frac{X_{Cs}}{R_s^2 + X_{Cs}^2}$$

The admittance of the parallel circuit in Fig. 14-47*b* is

$$Y_p = G_p + jB_{Cp}$$

For the parallel circuit and the series circuit to be equivalent, their admittances must be equal. Therefore,

$$Y_p = Y_s$$

$$G_p + jB_{Cp} = \frac{R_s}{R_s^2 + X_{Cs}^2} + j\frac{X_{Cs}}{R_s^2 + X_{Cs}^2}$$

This equality can be valid only if the real parts are equal *and* the imaginary parts are equal. Thus,

$$G_p = \frac{R_s}{R_s^2 + X_{Cs}^2} \qquad \textbf{(14-40)}$$

$$jB_{Cp} = j\frac{X_{Cs}}{R_s^2 + X_{Cs}^2} \qquad \textbf{(14-41)}$$

Equations (14-40) and (14-41) give the necessary conditions for the conversion of the series circuit in Fig. 14-47*a* to its parallel equivalent, Fig. 14-47*b*.

Parallel-to-Series Conversion, *RC* With reference to Fig. 14-48,

$$Y_p = G_p + jB_{Cp}$$

$$Z_p = \frac{1}{Y_p} = \frac{1}{G_p + jB_{Cp}}$$

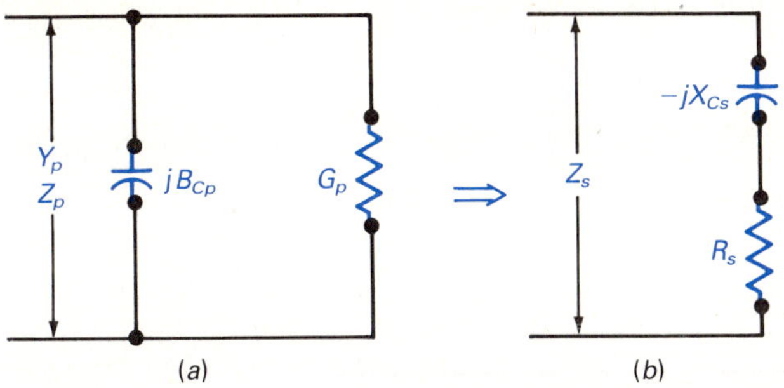

Fig. 14-48. *(a)* Parallel *RC* network; *(b)* its series equivalent.

Rationalizing,

$$Z_p = \frac{G_p}{G_p^2 + B_{Cp}^2} - j\frac{B_{Cp}}{G_p^2 + B_{Cp}^2}$$

To obtain expressions for series elements, impedances are equated:

$$Z_s = Z_p$$

$$R_s - jX_{Cs} = \frac{G_p}{G_p^2 + B_{Cp}^2} - j\frac{B_{Cp}}{G_p^2 + B_{Cp}^2}$$

The values of the series elements are found by equating the real parts and the imaginary parts of the impedances:

$$R_s = \frac{G_p}{G_p^2 + B_{Cp}^2} \tag{14-42}$$

$$-jX_{Cs} = -j\frac{B_{Cp}}{G_p^2 + B_{Cp}^2} \tag{14-43}$$

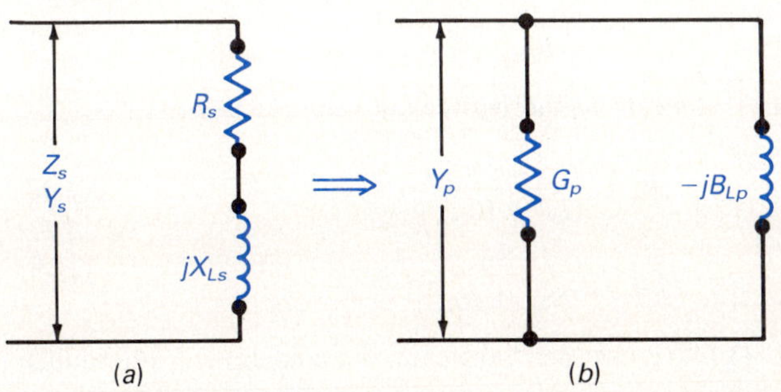

Fig. 14-49. *(a)* Series *RL* network; *(b)* its parallel equivalent.

Series-to-Parallel Conversion, *RL* With reference to Fig. 14-49,

$$Z_s = R_s + jX_{Ls}$$

$$Y_s = \frac{1}{Z_s} = \frac{1}{R_s + jX_{Ls}} = \frac{R_s}{R_s^2 + X_{Ls}^2} - j\frac{X_{Ls}}{R_s^2 + X_{Ls}^2}$$

Equating admittances,

$$Y_p = Y_s$$

$$G_p - jB_{Lp} = \frac{R_s}{R_s^2 + X_{Ls}^2} - j\frac{X_{Ls}}{R_s^2 + X_{Ls}^2}$$

Therefore, for Fig. 14-49*a* and *b* to be equivalent, the parallel elements must have the magnitudes

$$G_p = \frac{R_s}{R_s^2 + X_{Ls}^2} \qquad \textbf{(14-44)}$$

$$-jB_{Lp} = -j\frac{X_{Ls}}{R_s^2 + X_{Ls}^2} \qquad \textbf{(14-45)}$$

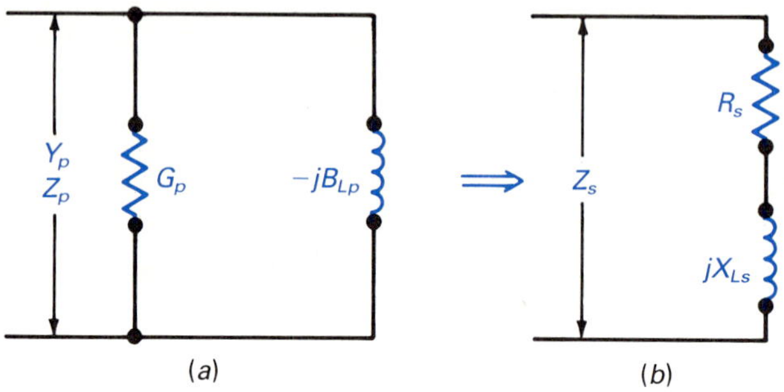

Fig. 14-50. *(a)* Parallel *RL* network; *(b)* its series equivalent.

Parallel-to-Series Conversion, *RL* With reference to Fig. 14-50,

$$Y_p = G_p - jB_{Lp}$$

$$Z_p = \frac{1}{G_p - jB_{Lp}} = \frac{G_p}{G_p^2 + B_{Lp}^2} + j\frac{B_{Lp}}{G_p^2 + B_{Lp}^2}$$

Equating impedances,

$$Z_s = Z_p$$

$$R_s + jX_{Ls} = \frac{G_p}{G_p^2 + B_{Lp}^2} + j\frac{B_{Lp}}{G_p^2 + B_{Lp}^2}$$

The magnitudes of the elements in the equivalent series circuit will be

$$R_s = \frac{G_p}{G_p^2 + B_{Lp}^2} \qquad \textbf{(14-46)}$$

$$jX_{Ls} = j\frac{B_{Lp}}{G_p^2 + B_{Lp}^2} \qquad \textbf{(14-47)}$$

Equations (14-40) to (14-47), for the values of impedance and admittance obtained as a result of series ↔ parallel conversion, can be summarized:

1. The nature of the elements does not change. For example, a capacitive element in the original circuit results in a capacitive element in the equivalent circuit.
2. If the conversion is series to parallel, the resulting elements are admittances. For a parallel-to-series conversion, the resulting elements are impedances.
3. The denominator for any derived element is the sum of the squares of the magnitudes of the original circuit elements.
4. The numerator for any derived element is the magnitude of the related

element in the original circuit. For instance, the magnitude of an original resistance becomes the numerator of a conductance.

5. The sign of an imaginary quantity follows the general conventions (examples are $-jX_C$ and $+jB_C$).

EXAMPLE 14-21 A series circuit consists of a 1-kΩ resistor and a 0.01-μF capacitor. What are the values of resistance and capacitance in the equivalent parallel circuit at a frequency of 16 kHz?

Solution See Fig. 14-51.

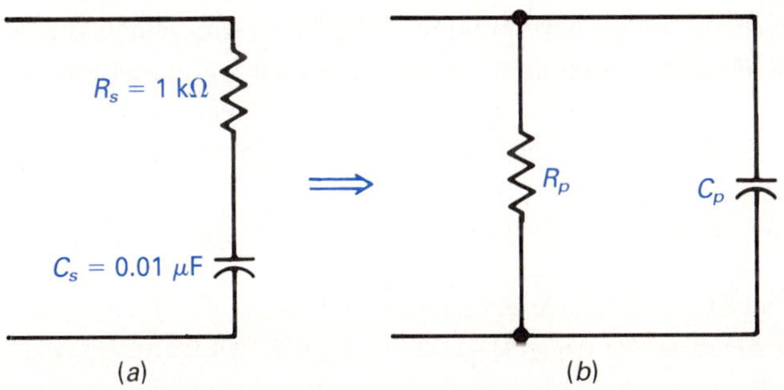

Fig. 14-51. Example 14-21: *(a)* Given network; *(b)* its parallel equivalent.

$$\omega = 2\pi f = 2\pi \times 16\text{ kHz} \simeq 100\text{ krad/s}$$

$$X_{Cs} = \frac{1}{\omega C_s} = \frac{1}{100\text{ krad/s} \times 0.01\ \mu\text{F}} = 1\text{ k}\Omega$$

$$R_s^2 + X_{Cs}^2 = (1\text{ k}\Omega)^2 + (1\text{ k}\Omega)^2 = 2\text{ M}\Omega^2$$

$$G_p = \frac{R_s}{R_s^2 + X_{Cs}^2} \qquad (14\text{-}40)$$

$$G_p = \frac{1\text{ k}\Omega}{2\text{ M}\Omega^2} = 0.5\text{ mS}$$

$$R_p = \frac{1}{G_p} = \frac{1}{0.5\text{ mS}} = \mathbf{2\ k\Omega}$$

$$jB_{Cp} = j\frac{X_{Cs}}{R_s^2 + X_{Cs}^2} \qquad (14\text{-}41)$$

$$B_{Cp} = \frac{1\text{ k}\Omega}{2\text{ M}\Omega^2} = 0.5\text{ mS}$$

$$B_{Cp} = \omega C$$

$$C = \frac{B_{Cp}}{\omega} = \frac{0.5\text{ mS}}{100\text{ krad/s}} = \mathbf{0.005\ \mu F}$$

The numbers in Example 14-21 were kept simple to aid in relating the original series circuit to its parallel equivalent.

It is important to be aware that the actual capacitance value changes in conversion. Only by this awareness can you hope to understand why the value of the capacitance of a capacitor is different for the series and parallel viewpoints. Instruments for measuring the characteristics of capacitors differ, some using the series view and others the parallel. This difference between the series and parallel capacitances of a practical capacitor is brought out in Example 14-22.

The series and parallel inductances of a practical inductor are also different.

You should note that the approximately equal sign ($\simeq$) is used only once in the solution to Example 14-21. Naturally, everything which follows is approximate if it depends on the value of ω. The only point in using the approximately equal sign is to indicate that the value given is close enough for our purposes. Therefore there is no need to repeat the sign.

Also note that the symbol distinction between scalar and phasor values was not made in Example 14-21.

It is not usual to maintain a rigid discipline in using the scalar value symbol unless clarity demands it. An example is $\sqrt{R^2 + X^2}$, which really ought to be written $\sqrt{|R|^2 + |X|^2}$. You will probably agree that the scalar symbol does nothing to clarify such a situation. Whether or not the scalar symbol should be used depends on whether or not the circumstances warrant it.

EXAMPLE 14-22 An electrolytic capacitor has a power factor of 0.05 and a capacitance of 25 μF when measured on a 1-kHz bridge (an instrument for measuring resistance, capacitance, and inductance). The bridge uses the parallel approach to capacitor measurement. What are the capacitor's *(a)* parallel resistance, *(b)* series resistance, and *(c)* series capacitance?

Solution See Fig. 14-52.

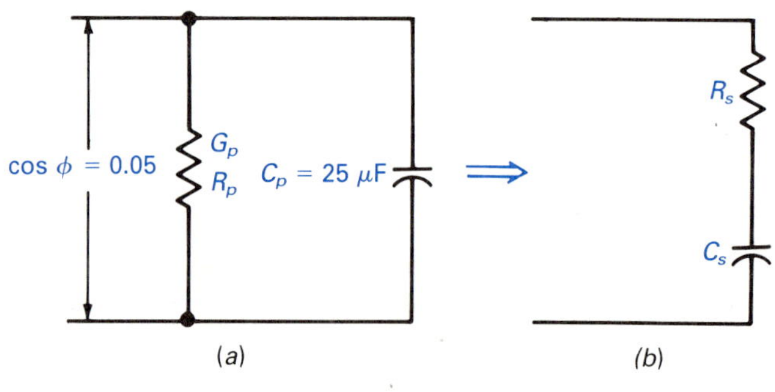

Fig. 14-52. Example 14-22: *(a)* Parallel equivalent network of the capacitor; *(b)* its series equivalent.

$$\omega = 2\pi f = 2\pi \times 1 \text{ kHz} = 6.283 \text{ krad/s}$$

$$B_{Cp} = \omega C = 6.283 \text{ krad/s} \times 25\ \mu\text{F} = 157.1 \text{ mS}$$

From Eq. (14-28),

$$\cos\phi = \frac{G_p}{Y_p} = \frac{G_p}{\sqrt{G_p^2 + B_{Cp}^2}}$$

$$(G_p \cos\phi)^2 + (B_{Cp} \cos\phi)^2 = G_p^2$$

$$G_p^2(1 - \cos^2\phi) = (B_{Cp} \cos\phi)^2$$

$$G_p^{\,2} = \frac{(B_{Cp}\cos\phi)^2}{1-\cos^2\phi} = \frac{(157.1\text{ mS} \times 0.05)^2}{1-0.05^2}$$

$$G_p \simeq 157.1\text{ mS} \times 0.05 \simeq 7.855\text{ mS}$$

a.
$$R_p = \frac{1}{G_p} \simeq \frac{1}{7.855\text{ mS}} \simeq \mathbf{127.3\ \Omega}$$

b.
$$G_p^{\,2} + B_{Cp}^{\,2} = (7.855\text{ mS})^2 + (157.1\text{ mS})^2 \simeq 24.74\text{ mS}^2$$

$$R_s = \frac{G_p}{G_p^{\,2} + B_{Cp}^{\,2}} \tag{14-42}$$

$$R_s = \frac{7.855\text{ mS}}{24.74\text{ mS}^2} = \mathbf{0.3175\ \Omega}$$

c.
$$-jX_{Cs} = -j\,\frac{B_{Cp}}{G_p^{\,2} + B_{Cp}^{\,2}} \tag{14-42}$$

$$-jX_{Cs} = -j\,\frac{157.1\text{ mS}}{24.74\text{ mS}^2} = -j6.35\ \Omega$$

$$X_{Cs} = \frac{1}{\omega C_s}$$

$$C_s = \frac{1}{\omega X_{Cs}} = \frac{1}{6.283\text{ krad/s} \times 6.35\ \Omega} = \mathbf{25.06\ \mu F}$$

The power factor of the capacitor in Example 14-22 is too high for many purposes. Such a high power factor could be tolerated only in an electrolytic capacitor used in high-power applications. This high value was given to show that it is possible to obtain a noticeably different capacitance value from the series and parallel viewpoints of the same capacitor. The difference between series and parallel capacitance is rarely an important matter, since the power factor of a capacitor is usually much lower than 0.05, resulting in the series and parallel equivalent capacitance values being virtually the same.

EXAMPLE 14-23 The series-parallel network in Fig. 14-53*a* has the values L_s = 100 μH, R_s = 50 Ω, and C = 400 pF. What are the element values in the equivalent parallel circuit in Fig. 14-53*b* at a frequency of 0.8 MHz?

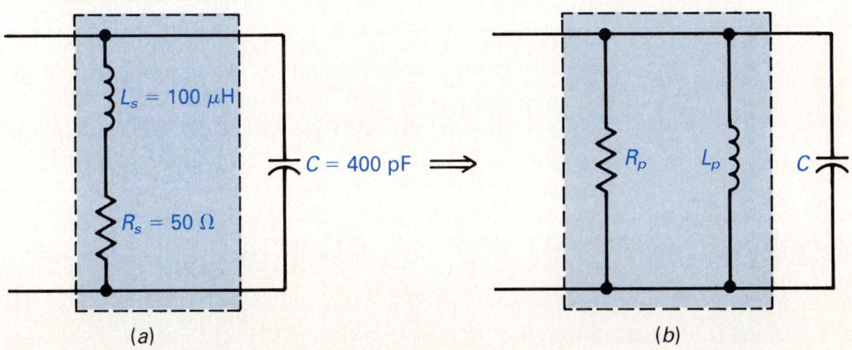

Fig. 14-53. Example 14-23: *(a)* Given network; *(b)* its parallel equivalent.

Solution Convert the series circuit L_sR_s to the parallel equivalent L_pR_p:

$$\omega = 2\pi f = 2\pi \times 0.8 \text{ MHz} = 5.027 \text{ Mrad/s}$$

$$X_{Ls} = \omega L = 5.027 \text{ Mrad/s} \times 100\ \mu\text{H} = 502.7\ \Omega$$

$$R_s^2 + X_{Ls}^2 = 255.2 \text{ k}\Omega^2$$

$$G_p = \frac{R_s}{R_s^2 + X_{Ls}^2} \qquad (14\text{-}44)$$

$$G_p = \frac{50\ \Omega}{255.2 \text{ k}\Omega^2} = 0.196 \text{ mS}$$

$$R_p = \frac{1}{G_p} = \frac{1}{0.196 \text{ mS}} = \mathbf{5.102\ k\Omega}$$

$$B_{Lp} = \frac{X_{Ls}}{R_s^2 + X_{Ls}^2} \qquad (14\text{-}45)$$

$$B_{Lp} = \frac{502.7\ \Omega}{255.2 \text{ k}\Omega^2} = 1.97 \text{ mS}$$

$$B_{Lp} = \frac{1}{\omega L_p}$$

$$L_p = \frac{1}{\omega B_{Lp}} = \frac{1}{5.027 \text{ Mrad/s} \times 1.97 \text{ mS}}$$

$$L_p = \mathbf{101\ \mu H}$$

The capacitance was not involved in the conversion process, so in the parallel circuit it is 400 pF, exactly the same as in the series circuit.

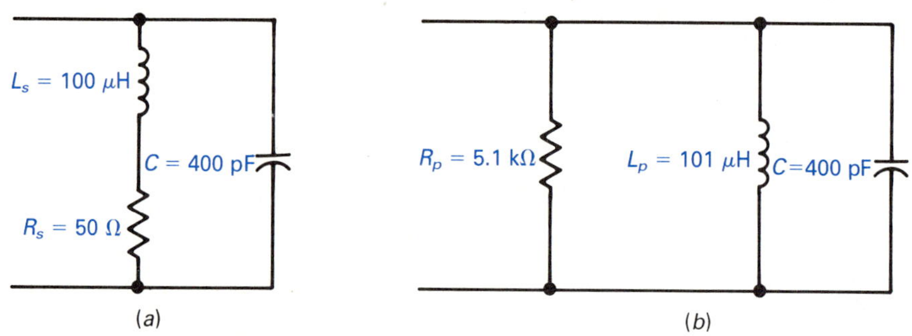

Fig. 14-54. Example 14-23: *(a)* Given network; *(b)* its parallel equivalent, with values.

Figure 14-54 shows the values of the circuit before and after the conversion from series-parallel to parallel. The parallel equivalent inductance differs from the original series value by only 1 percent, because the Q of the RL circuit is fairly high (a little over 10).

REVIEW QUESTIONS FOR SECTION 14-5

1. What is the purpose of rationalization?
2. What is the basic difference between the values of a series ac circuit and its parallel equivalent?

3. Compare the power factors of a series *RL* circuit with its parallel equivalent.
4. In what way is series ↔ parallel conversion related to frequency?
5. There is no dc version of series ↔ parallel conversion. Why is that so?

EXERCISE PROBLEMS FOR SECTION 14-5

1. Convert the series circuits in Prob. 1, Sec. 14-1, to their parallel equivalents, expressing the equivalent circuits in terms of resistance and capacitance.
(*ans. a.* 3.31 kΩ, 0.00638 μF; *b.* 803.6 Ω, 8.756 μF; *c.* 3 kΩ, 26.66 pF)
2. Convert the parallel circuits in Prob. 1, Sec. 14-3, to their series equivalents, expressing the equivalent circuits in terms of resistance and capacitance.
3. Convert the series circuits in Prob. 1, Sec. 14-2, to their parallel equivalents, expressing the equivalent circuits in terms of resistance and inductance.
(*ans. a.* 1.063 kΩ, 0.841 H; 21.97 kΩ, 1.113 H; *c.* 2.093 kΩ, 530 μH)
4. Convert the parallel circuits in Prob. 1, Sec. 14-4, to their series equivalents, expressing the equivalent circuits in terms of resistance and inductance.

14-6 SERIES-PARALLEL CIRCUITS (*RC* AND *RL*)

When the techniques for handling series and parallel ac circuits have been mastered, series-parallel circuits can be dealt with by adopting an approach similar to that used for dc circuits. This means analyzing the series-parallel circuit as series and parallel subcircuits.

As it is pointless to attempt to develop general rules which can be applied to the infinite variety of series-parallel networks, some numerical examples are given showing how to tackle typical problems. The techniques demonstrated are applied to practical situations in later chapters.

EXAMPLE 14-24 What is the impedance of the network in Fig. 14-55?

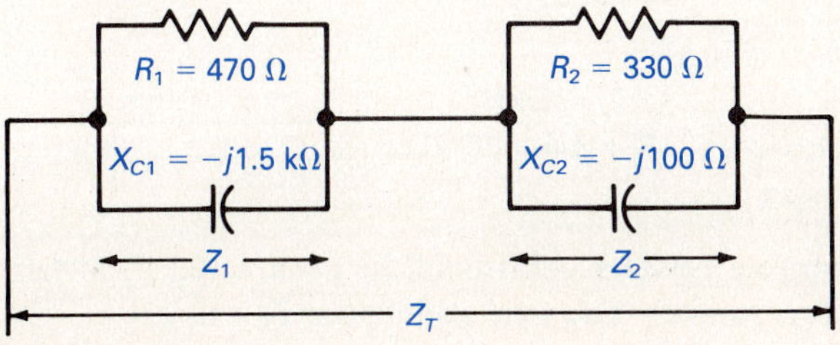

Fig. 14-55. Example 14-24: Network.

Solution Z_1 is the impedance of the parallel combination of R_1 and X_{C1}. Applying the principle for the determination of the total impedance of two parallel impedances $[Z = Z_b A_b/(Z_a + Z_b)]$,

$$Z_1 = \frac{R_1 X_{C1}}{R_1 + X_{C1}} = \frac{470\ \Omega(-j1.5\ \text{k}\Omega)}{(470 - j1.5)\ \text{k}\Omega} = \frac{705\ \text{k}\Omega^2 \angle -90^\circ}{1.572\ \text{k}\Omega \angle -72.6^\circ} = 448\ \Omega \angle -17.4^\circ$$

Z_2 is the impedance of the parallel combination of R_2 and X_{C2}:

$$Z_2 = \frac{R_2 X_{C2}}{R_2 + X_{C2}} = \frac{330\ \Omega(-j100\ \Omega)}{(330 - j100)\ \Omega} = \frac{33\ \text{k}\Omega^2 \angle -90^\circ}{334.8\ \Omega \angle -16.88^\circ} = 95.71\ \Omega \angle -73.12^\circ$$

The complete network consists of the group impedances Z_1 and Z_2 connected in series. Therefore the total impedance Z_T is

$$Z_T = Z_1 + Z_2 = 448\ \Omega \angle -17.4^\circ + 95.71\ \Omega \angle -73.12^\circ$$
$$= (427.5 - j134)\ \Omega + (27.79 - j91.58)\ \Omega$$
$$Z_T = \mathbf{(455.3 - j225.6)\ \Omega = 508.1\ \Omega \angle -26.36^\circ} \quad \text{or} \quad \mathbf{\angle -0.46\ rad}$$

EXAMPLE 14-25 What is the admittance of the network in Fig. 14-56?

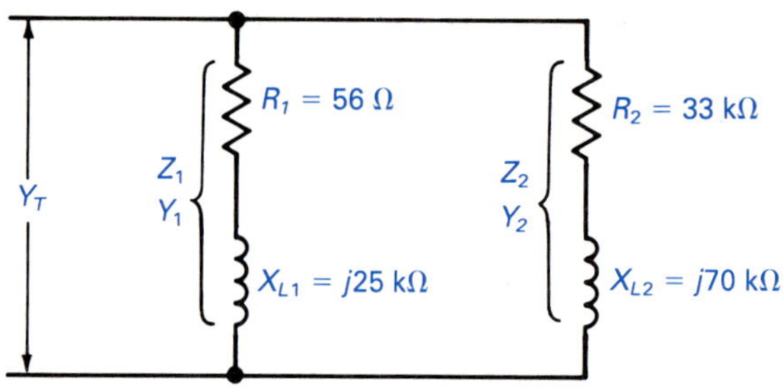

Fig. 14-56. Example 14-25: Network.

Solution Z_1 and Y_1 are the impedance and admittance of the series combination of R_1 and X_{L1}.

$$Z_1 = R_1 + jX_{L1} = (56 + j25)\ \text{k}\Omega$$
$$Y_1 = \frac{1}{Z_1} = \frac{1}{(56 + j25)\ \text{k}\Omega} = \frac{56\ \text{k}\Omega}{(56\ \text{k}\Omega)^2 + (25\ \text{k}\Omega)^2} - j\frac{25\ \text{k}\Omega}{(56\ \text{k}\Omega)^2 + (25\ \text{k}\Omega)^2}$$
$$Y_1 = (14.89 - j6.647)\ \mu\text{S}$$

Z_2 and Y_2 are the impedance and admittance of the series combination of R_2 and X_{L2}.

$$Z_2 = R_2 + jX_{L2} = (33 + j70)\ \text{k}\Omega$$
$$Y_2 = \frac{1}{Z_2} = \frac{1}{(33 + j70)\ \text{k}\Omega} = \frac{33\ \text{k}\Omega}{(33\ \text{k}\Omega)^2 + (70\ \text{k}\Omega)^2} - j\frac{70\ \text{k}\Omega}{(33\ \text{k}\Omega)^2 + (70\ \text{k}\Omega)^2}$$
$$Y_2 = (5.51 - j11.69)\ \mu\text{S}$$

The complete network consists of the admittances Y_1 and Y_2 connected in parallel. The total admittance Y_T will be

$$Y_T = Y_1 + Y_2 = (14.89 - j6.647)\ \mu S + (5.51 - j11.69)\ \mu S$$
$$Y_T = \mathbf{(20.4 - j18.34)\ \mu S = 27.43\ \mu S \angle -41.97^\circ} \quad \text{or} \quad \mathbf{\angle -0.7324\ rad}$$

EXAMPLE 14-26

What is the load voltage V_L in the circuit in Fig. 14-57?

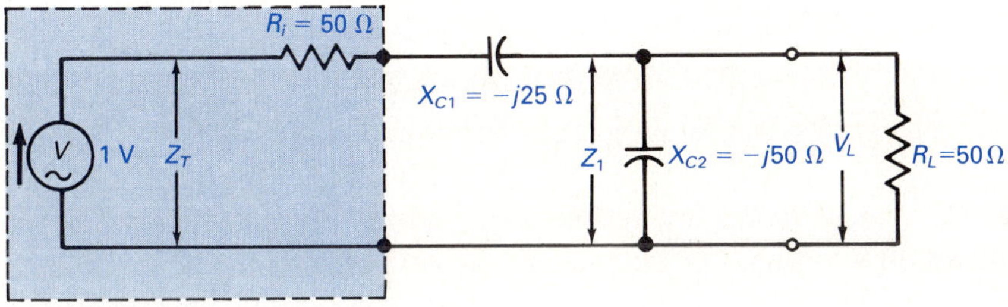

Fig. 14-57. Example 14-26: Circuit.

Solution Z_1 is the impedance of the parallel network $R_L X_{C2}$.

$$Z_1 = \frac{R_L X_{C2}}{R_L + X_{C2}} = \frac{50\ \Omega(-j50\ \Omega)}{(50 - j50)\ \Omega} = \frac{2500\ \Omega \angle -90^\circ}{70.71\ \Omega \angle -45^\circ}$$
$$Z_1 = 35.36\ \Omega \angle -45^\circ \quad \text{or} \quad (25 - j25)\ \Omega$$

Z_T is the total circuit impedance; it is Z_1 in series with R_i and X_{C1}.

$$Z_T = Z_1 + R_i + X_{C1} = (25 - j25)\ \Omega + (50 + j0)\ \Omega + (0 - j25)\ \Omega$$
$$Z_T = (75 - j50)\ \Omega \quad \text{or} \quad 90.14\ \Omega \angle -33.69^\circ$$

The voltage-divider principle can be applied to this example:

$$V_L = V\frac{Z_1}{Z_T} = 1\ V \angle 0^\circ \frac{35.36\ \Omega \angle -45^\circ}{90.14\ \Omega \angle -33.69^\circ}$$
$$V_L = \mathbf{0.3923\ V \angle -11.31^\circ} \quad \text{or} \quad \mathbf{\angle -0.1974\ rad}$$

EXAMPLE 14-27

What is the value of the load current I_L in the circuit in Fig. 14-58? Assume that no mutual inductance is present.

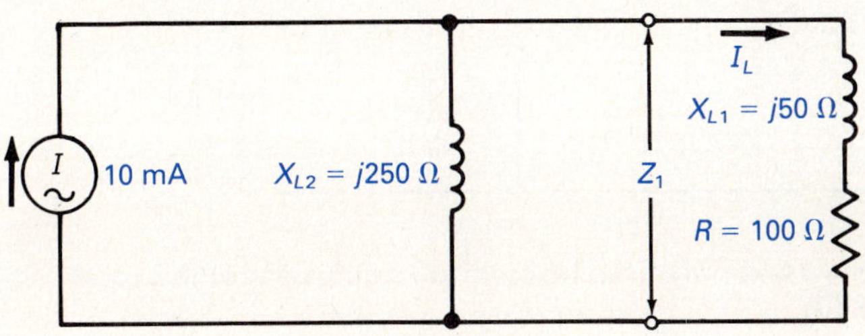

Fig. 14-58. Example 14-27: Circuit.

Solution The impedance of R and X_{L1} in series is Z_1.

$$Z_1 = R + jX_L = (100 + j50)\ \Omega$$

The current-divider principle can be applied to find the value of the current I_L:

$$I_L = I \frac{X_{L2}}{X_{L2} + Z_1} = 10\text{ mA }\angle 0^\circ \frac{250\ \Omega\ \angle 90^\circ}{j250\ \Omega + (100 + j50)\ \Omega}$$

$$= \frac{10\text{ mA }\angle 0^\circ \times 250\ \Omega\ \angle 90^\circ}{(100 + j300)\ \Omega}$$

$$I_L = \frac{10\text{ mA }\angle 0^\circ \times 250\ \Omega\ \angle 90^\circ}{316.2\ \Omega\ \angle 71.57^\circ} = \mathbf{7.906\ mA\ \angle 18.43^\circ} \quad \text{or} \quad \mathbf{\angle 0.3216\ rad}$$

EXAMPLE 14-28 Figure 14-59*a* is a schematic diagram of a two-stage transistor amplifier. An equivalent circuit of the network coupling transistors Q_1 and Q_2 is given in Fig. 14-59*b*. R_l is the effective load coupled by capacitor C_c and includes the effects of R_2 and R_3 together with the input resistance of Q_2. C_s is the stray capacitance of the input circuit of Q_2, including its wiring. What is the bandwidth of the network coupling transistors Q_1 and Q_2? The

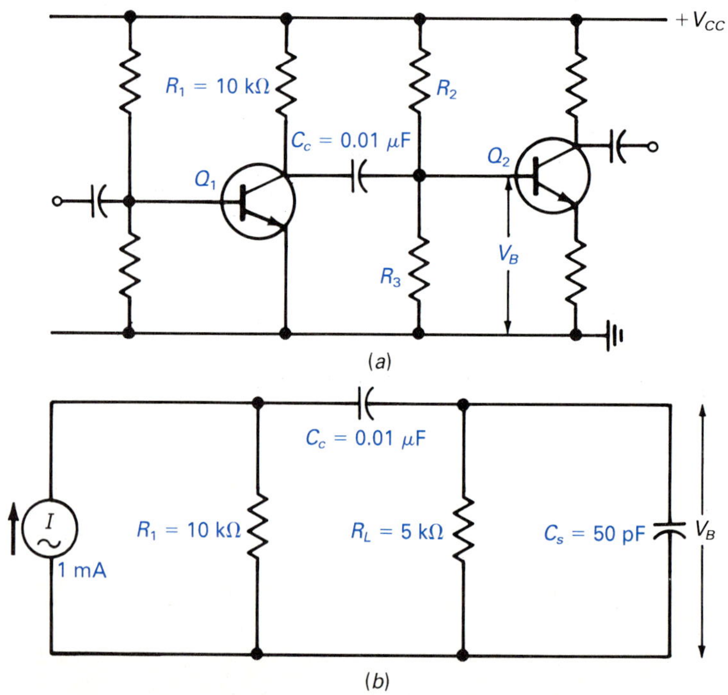

Fig. 14-59. Example 14-28: *(a)* Given circuit; *(b)* equivalent of the base-to-emitter circuit of Q_2.

bandwidth is defined as the difference in frequency between the points at which the magnitude of the voltage V_B fed to transistor Q_2 is $1/\sqrt{2}$ of its maximum value.

Solution The current generator in Fig. 14-59*b* represents transistor Q_1 in its effect on driving transistor Q_2. It is assumed that this source produces a constant alternating current of 1 mA. One milliampere is chosen as a convenient current level, since only the variation in voltage V_B, with respect to frequency, is of concern in this example.

Mid-frequency The maximum possible value of voltage V_B would occur if the effects of both capacitances C_c and C_s could be neglected. This condition does occur at a so-called mid-frequency when the series reactance of C_1 and the shunt susceptance of C_s are both small compared with the values of the other circuit elements. Figure 14-60*a* shows this situation.

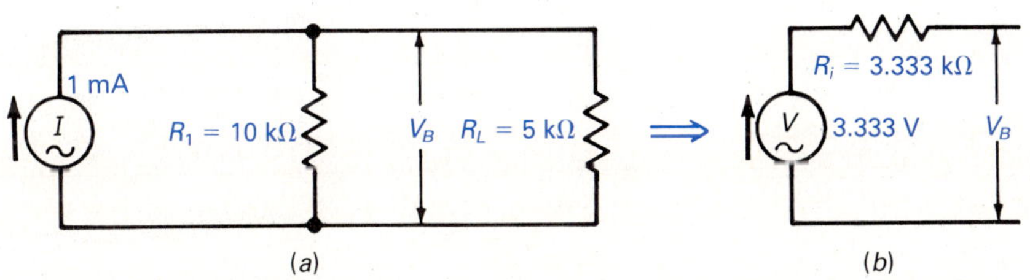

Fig. 14-60. Example 14-28: *(a)* Mid-frequency equivalent of Fig. 14-59*b*; *(b)* Thevenin equivalent.

The complete circuit in Fig. 14-60*a* can be regarded as a current generator with a source current of 1 mA and an internal conductance consisting of R_1 and R_L in parallel. This current generator can be converted to the voltage generator in Fig. 14-60*b*. Its internal resistance R_i is the parallel combination of R_1 and R_L:

$$R_i = \frac{R_1 L_L}{R_1 + R_L} = \frac{10\text{ k}\Omega \times 5\text{ k}\Omega}{10\text{ k}\Omega + 5\text{ k}\Omega} = 3.333\text{ k}\Omega$$

The voltage V_B will equal the source emf of the voltage generator:

$$V_B = V = IR_i = 1\text{ mA} \times 3.333\text{ k}\Omega = 3.333\text{ V}$$

This mid-frequency value of V_B (3.333 V) becomes our reference voltage level. The task now is to determine two frequencies, one below the mid-frequency and one above it, at which the value of V_B is 3.333 V $\div \sqrt{2}$ = 2.357 V.

Low Frequency At low frequencies the series reactance of C_c is significant, but the shunt susceptance of C_s is negligible. The effective circuit is that in Fig. 14-61*a*.

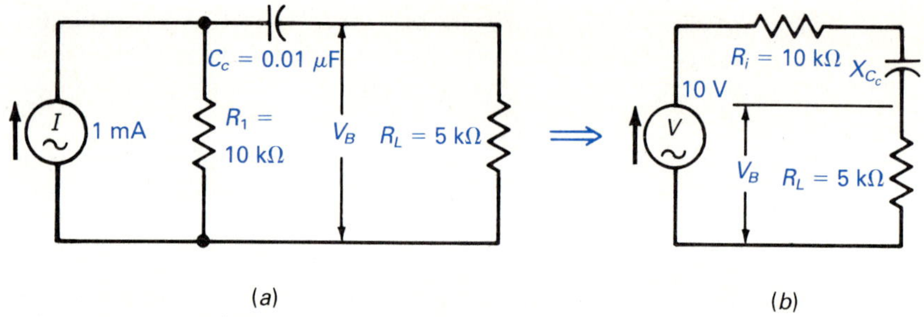

Fig. 14-61. Example 14-28: Low-frequency equivalent of Fig. 14-59*b*.

The current source and R_1 can be converted to a voltage source as indicated in Fig. 14-61*b*, where

$$R_i = R_1 = 10\ \text{k}\Omega$$

and

$$V = IR_1 = 1\ \text{mA} \times 10\ \text{k}\Omega = 10\ \text{V}$$

V_B can be determined by the voltage-divider principle:

$$V_B = V\frac{R_L}{Z_T} = V\frac{R_L}{R_L + R_i - jX_{Cc}} = 10\ \text{V}\frac{5\ \Omega}{5\ \text{k}\Omega + 10\ \text{k}\Omega - jX_{Cc}}$$

The low frequency of interest occurs when $V_B = 2.357$ V.

$$2.357\ \text{V} = 10\ \text{V}\frac{5\ \text{k}\Omega}{15\ \text{k}\Omega - jX_{Cc}}$$

$$15\ \text{k}\Omega - jX_{Cc} = 21.21\ \text{k}\Omega$$

$$X_{Cc} = \sqrt{(21.21\ \text{k}\Omega)^2 - (15\ \text{k}\Omega)^2}$$

$$X_{Cc} = 15\ \text{k}\Omega$$

$$\frac{1}{\omega C_c} = 15\ \text{k}\Omega$$

$$\omega = \frac{1}{0.01\ \mu\text{F} \times 15\ \text{k}\Omega} = 6.667\ \text{krad/s}$$

$$2\pi f = 6.667\ \text{krad/s}$$

$$f = 1.061\ \text{kHz}$$

High Frequency At high frequencies the shunt susceptance of C_s is significant, but the series reactance of C_c is negligible. The equivalent circuit is as in Fig. 14.62.

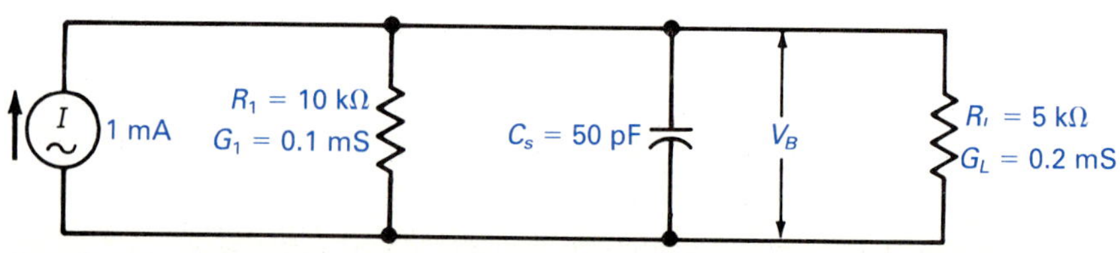

Fig. 14-62. Example 14-28: High-frequency equivalent of Fig. 14-59*b*.

The total load on the current source is

$$Y = G_1 + jB_{Cs} + G_L = 0.1\ \text{mS} + jB_{Cs} + 0.2\ \text{mS}$$

$$Y = 0.3\ \text{mS} + jB_{Cs}$$

The voltage V_B will be given by Ohm's law in admittance form:

$$V_B = \frac{I}{Y} = \frac{1\ \text{mA}}{0.3\ \text{mS} + jB_{Cs}}$$

The high frequency we wish to determine occurs when $V_B = 2.357$ V.

$$2.357\ \text{V} = \frac{1\ \text{mA}}{0.3\ \text{mS} + jB_{Cs}}$$

$$0.3\ \text{mS} + jB_{Cs} = 0.4242\ \text{mS}$$

$$B_{Cs} = \sqrt{(0.4242\ \text{mS})^2 - (0.3\ \text{mS})^2}$$

$$B_{Cs} = 0.3\ \text{mS}$$

$$\omega C = 0.3\ \text{mS}$$

$$\omega = \frac{0.3\ \text{mS}}{50\ \text{pF}} = 6\ \text{Mrad/s}$$

$$2\pi f = 6\ \text{Mrad/s}$$

$$f = \frac{6\ \text{Mrad/s}}{2\pi} = 0.955\ \text{MHz} = 955\ \text{kHz}$$

Therefore the bandwidth of the coupling network is the difference between the high frequency of 955 kHz and the low frequency of 1.061 kHz.

$$B = 955\ \text{kHz} - 1.061\ \text{kHz} = \mathbf{953.9\ kHz}$$

In Example 14-28 it can be seen that the lower bandwidth frequency occurs when X_{Cc} equals the net series resistance, and that the upper bandwidth frequency occurs when B_{Cs} equals the net shunt conductance. This is not a coincidence occurring in this particular example. The definition of bandwidth can be considered to be based on the 45° right triangles in Fig. 14-63.

In Fig. 14-63*a*,

$$R = X_{Cc} = \frac{Z}{\sqrt{2}}$$

In Fig. 14-63*b*,

$$G = B_{Cs} = \frac{Y}{\sqrt{2}}$$

Since the $1/\sqrt{2}$ points define bandwidth, the bandwidth frequencies of an *RC* coupling network can be described as occurring when

1. The series capacitive reactance equals the total equivalent series resistance of the coupling network and its load.
2. The shunt capacitive susceptance equals the total equivalent shunt conductance of the coupling network and its load.
3. The phase angle between the driving source and the coupled voltage is 45° or $\pi/4$ rad.

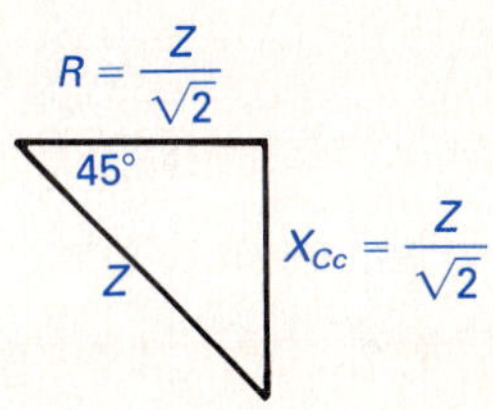

(*a*)

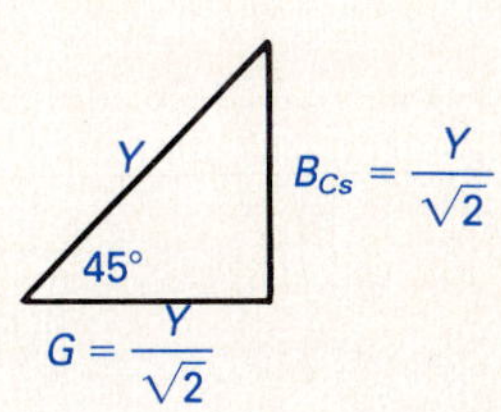

(*b*)

Fig. 14-63. Example 14-28: *(a)* Lower-bandwidth frequency impedance triangle; *(b)* upper-bandwidth frequency admittance triangle.

The terms "high" and "low" in the solution for Example 14-28 are relative terms. If one frequency is many times greater than another, "high" and "low" distinguishes between them. However, whether a frequency of 955 kHz is high depends on the circumstances.

EXAMPLE 14-29 An outline diagram of a Wein bridge oscillator is shown in Fig. 14-64*a*. If the gains of amplifiers $A1$ and $A2$ are adequate, the circuit will oscillate; that is, it will act as an electronic ac generator. The frequency of oscillation will be determined by the frequency at which the phase angle of V_{AB}, the input to amplifier $A1$, is zero with respect to the voltage V_o. This frequency is controlled by the values of R_1, R_2, C_1, and C_2. What is the frequency of oscillation in terms of these elements?

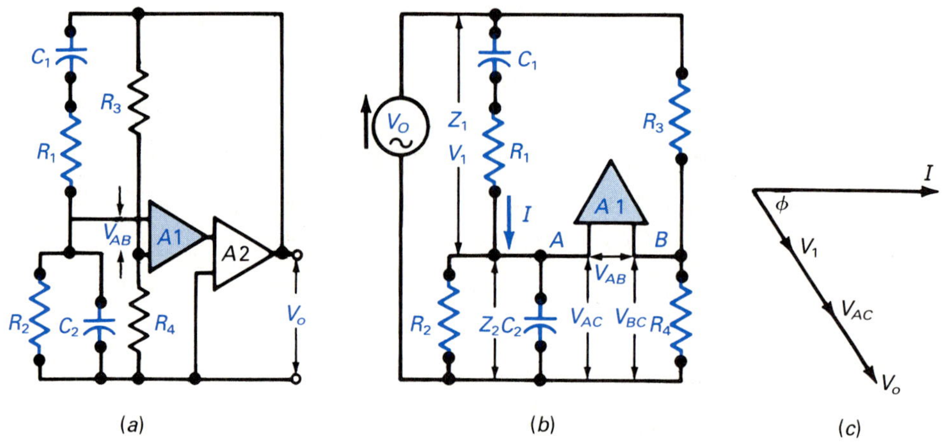

Fig. 14-64. Examples 14-29 and 14-30: *(a)* Given circuit; *(b)* part of the circuit rearranged with V_o as the input voltage; *(c)* phasor diagram.

Solution The phase-shift circuit controlling the input to amplifier $A1$ is given in Fig. 14-64*b*.

It is a general axiom that, if a number of networks, each with the same phase angle, are connected in series, the voltages across each network and across the complete series circuit are all in phase with each other. Figure 14-64*c* shows a phasor diagram in which the phase angles of the series network R_1C_1 and the parallel network R_2C_2 are each equal to $\phi°$. The current I common to both networks is given as the reference phasor. The voltage V_1 across the series network R_1C_1 lags the current by $\phi°$. The voltage V_{AC} across the parallel network R_2C_2 also lags the current by ϕ. The voltage V_o is the phasor sum of the in-phase voltages V_1 and V_{AC} and is in phase with each of them. Thus the voltage V_{AC} will be in phase with the voltage V_o if the phase angles of the series and parallel RC networks are equal. It will be obvious that, since the phase angles of R_3 and R_4 are both always zero, V_{BC} is in phase with V_o.

The input voltage V_{AB} of amplifier $A1$ is the phasor difference between V_{AC} and V_{BC}. Where V_{AB} and V_{BC} are each in phase with V_o, and hence in phase with each other, V_{AB} is in phase with V_o. The frequency of oscillation is the frequency at which these conditions exist, that is, the frequency at which the series network R_1C_1 has the same phase angle as the parallel network R_2C_2.

The phase angle of the series network R_1C_1 is given by

$$\phi_1 = \arctan \frac{X_{C1}}{R_1}$$

The phase angle of the parallel network R_2C_2 is given by

$$\phi_2 = \arctan \frac{B_{C2}}{G_2} = \arctan \frac{R_2}{X_{C2}}$$

The frequency of oscillation is the frequency at which ϕ_1 and ϕ_2 are equal, or when

$$\arctan \frac{X_{C1}}{R_1} = \arctan \frac{R_2}{X_{C2}}$$

or

$$\frac{X_{C1}}{R_1} = \frac{R_2}{X_{C2}}$$

$$\frac{1}{\omega C_1 \omega C_2} = R_1 R_2$$

$$\omega = \frac{1}{\sqrt{R_1 R_2 C_1 C_2}}$$

$$\mathbf{f = \frac{1}{2\pi\sqrt{R_1 R_2 C_1 C_2}}}$$

EXAMPLE 14-30 If in the circuit in Fig. 14-64 capacitors C_1 and C_2 are both 0.1 μF, and resistors R_1 and R_2 are both variable with a range of 1 to 10 kΩ and are ganged so that both have the same resistance value at all times, what is the oscillation frequency range?

Solution If $C_1 = C_2 = C$ and $R_1 = R_2 = R$, the frequency of oscillation will be

$$f = \frac{1}{2\pi\sqrt{R^2C^2}} = \frac{1}{2\pi RC}$$

When $R = 1$ kΩ,

$$f = \frac{1}{2\pi \times 1\ \text{k}\Omega \times 0.1\ \mu\text{F}} = 1.591\ \text{kHz}$$

When $R = 10\ \mathrm{k}\Omega$,

$$f = \frac{1}{2\pi \times 1\ \mathrm{k}\Omega \times 0.1\ \mu\mathrm{F}} = 159.1\ \mathrm{Hz}$$

The frequency of the oscillation is from **159.1 Hz** to **1.591 kHz.**

From the results for Example 14-30 it is clearly possible to make a Wien bridge oscillator of a chosen frequency range. In the example the capacitance value of 0.1 μF was selected for ease of calculation. Had the value been 0.1591 μF, the frequency range would have been 100 Hz to 1 kHz.

When capacitors of 0.01591, 0.1591, and 1.591 μF are connected into the circuit by selector switches, an oscillator with a frequency range of 10 Hz to 10 kHz will result. Such oscillators are commercially manufactured as signal generators for measuring and testing purposes. These generators are good examples of the application of ac circuit principles.

SUMMARY OF EQUATIONS

Series *RC* circuit

$V_T = V_R - jV_C$ (14-1)

$Z = R - jX_C$ (14-2)

$V_TI = V_RI - jV_CI$ (14-3)

$S = V_TI$ (14-4)

$P = V_TI \cos \phi$ or $S \cos \phi$ (14-5)

$jQ = jV_TI \sin \phi$ or $jS \sin \phi$ (14-6)

$S = P + jQ$ (14-7)

$\cos \phi = P/S$ (14-8)

$\cos \phi = V_R/V_T$ or R/Z (14-9)

Series *RL* circuit

$V_T = V_R + jV_L$ (14-10)

$Z = R + jX_L$ (14-11)

$V_TI = V_RI + jV_LI$ (14-12)

$S = V_TI$ (14-13)

$P = V_TI \cos \phi = S \cos \phi$ (14-14)

$-jQ = -jV_TI \sin \phi = -jS \sin \phi$ (14-15)

$S = P - jQ$ (14-16)

$\cos \phi = P/S$ (14-17)

$\cos \phi = V_R/V_T = R/Z$ (14-18)

$Q = X_L/R = \tan \phi$ (14-19)

$Q = \omega L/R$ (14-20)

Parallel *RC* circuit

$I_T = I_R + jI_C$ (14-21)

$Y = G + jB_C$ (14-22)

$VI_T = VI_R + jVI_C$ (14-23)

$S = P + jQ$ (14-24)

$S = VI_T$ (14-25)

$P = S \cos \phi = VI_T \cos \phi$ (14-26)

$jQ = jS \sin \phi = jVI_T \sin \phi$ (14-27)

$\cos \phi = P/S = I_R/I_T = G/Y$ (14-28)

$\cos \phi = Z/R$ (14-29)

Parallel *RL* circuit

$I_T = I_R - jI_L$ (14-30)

$Y = G - jB_L$ (14-31)

$VI_T = VI_R - jVI_L$ (14-32)

$S = P - jQ$ (14-33)

$S = VI_T$ (14-34)

$P = S \cos \phi = VI_T \cos \phi$ (14-35)

$-jQ = -jS \sin \phi = -jVI_T \sin \phi$ (14-36)

$\cos \phi = P/S = I_R/I_T = G/Y = Z/R$ (14-37)

$Q = B_L/G = \tan \phi$ (14-38)

$Q = 1/\omega LG$ or $R/\omega L$ or R/X_L (14-39)

Series ↔ parallel conversion

$G_P = R_S/(R_s^2 + X_{Cs}^2)$ (14-40)

$$jB_{Cp} = jX_{Cs}/(R_s^2 + X_{Cs}^2) \quad (14\text{-}41)$$
$$R_S = G_p/(G_p^2 + B_{Cp}^2) \quad (14\text{-}42)$$
$$-jX_{Cs} = -jB_{Cp}/(G_p^2 + B_{Cp}^2) \quad (14\text{-}43)$$
$$G_p = R_s/(R_s^2 + X_{Ls}^2) \quad (14\text{-}44)$$
$$-jB_{LP} = -jX_{Ls}/(R_s^2 + X_{Ls}^2) \quad (14\text{-}45)$$
$$R_s = G_p/(G_p^2 + B_{Lp}^2) \quad (14\text{-}46)$$
$$jX_{Ls} = jB_{Lp}/(G_p^2 + B_{Lp}^2) \quad (14\text{-}47)$$

The number of equations in Chap. 14 may look somewhat frightening. However, a review of them shows that many are very similar, differing only in the sign of the j term or in the symbols for the voltages and currents. Compare the following equations:

(14-1) to (14-9) with (14-10) to (14-18)
(14-21) to (14-28) with (14-30) to (14-37)
(14-40) to (14-43) with (14-44) to (14-47)

You will see that most of the 47 equations fall into groups of few really different types. There are only 10 basically different equations.

In *any* circuit,

1. The apparent power is the product of total voltage and total current, $S = VI$ [Eqs. (14-4), (14-13), (14-25), and (14-34)].
2. The active power is the product of the apparent power and the power factor, $P = S \cos \phi$ [Eqs. (14-5), (14-8), (14-14), (14-17), (14-26), (14-28), (14-35), and (14-37)].
3. The magnitude of the reactive power is the product of the apparent power and the sine of the phase angle, $|Q| = S \sin \phi$ [Eqs. (14-6), (14-15), (14-27), and (14-36)].
4. The apparent power is the phasor sum of the active and reactive powers, $S = P \pm jQ$ [Eqs. (14-7), (14-16), (14-24), and (14-33)].

In any *series* circuit,

5. The total voltage is the phasor sum of the resistive and reactive voltages, $V_T = V_R \pm jV_X$ [Eqs. (14-1) and (14-10)].
6. The total impedance is the phasor sum of the resistance and the reactance, $Z = R \pm jX$ [Eqs. (14-2) and (14-11)].
7. The power factor is the ratio of resistive voltage to total voltage or the ratio of resistance to impedance, $\cos \phi = V_R/V_T = R/Z$ [Eqs. (14-9) and (14-18)].

In any *parallel* circuit,

8. The total current is the phasor sum of the resistive and reactive currents, $I_T = I_R \pm jXI_B$ [Eqs. (14-21) and (14-30)].
9. The total admittance is the phasor sum of the conductance and the susceptance, $Y = G \pm jB$ [Eqs. (14-22) and (14-31)].
10. The power factor is the ratio of resistive current to total current, or the ratio of conductance to admittance, $\cos \phi = I_R/I_T = G/Y$ [Eqs. (14-28) and (14-37)].

In a *capacitive* circuit, series or parallel,

11. The power equation is $S = P + jQ$ [Eqs. (14-7) and (14-24)].

In an *inductive* circuit, series or parallel,

12. The power equation is $S = P - jQ$ [Eqs. (14-16) and (14-33)].
13. The quality factor is the tangent of the phase angle, $Q = \tan \phi$ [Eqs. (14-19) and (14-38)].

CHAPTER QUESTIONS

14-1. How is the treatment of dc and ac circuits similar?

14-2. How is the treatment of dc and ac circuits different?

14-3. Compare the characteristics of a series *RC* circuit with those of a series *RL* circuit if their impedance magnitudes are the same.

14-4. Compare the characteristics of a series *RC* circuit with those of a parallel *RC* circuit if their impedance magnitudes are the same.

14-5. Which is best, a high or low value of power factor?

14-6. Equations (14-9) and (14-18) give power factor as R/Z, whereas in Eqs. (14-29) and (14-37) it is given as Z/R. But in *any* circuit, power factor is always *active power* ÷ *apparent power*. What accounts for this situation?

14-7. What might be a disadvantage of using a capacitor as a voltage-dropping element rather than a resistor?

14-8. Why is the *impedance phase angle* of a series *RC* or *RL* circuit opposite in polarity (leading ↔ lagging) to its *circuit phase angle?*

14-9. Why is the *admittance phase angle* of a parallel *RC* or *RL* circuit the same as its *circuit phase angle?*

14-10. Why is it necessary to distinguish between active power, reactive power, and apparent power?

14-11. How might an electricity supply company (power company) justify charging an industrial user on the basis of apparent power rather than real or active power?

14-12. An electric motor has a lower power factor off load than when it is driving machinery. What could account for this?

CHAPTER PROBLEMS

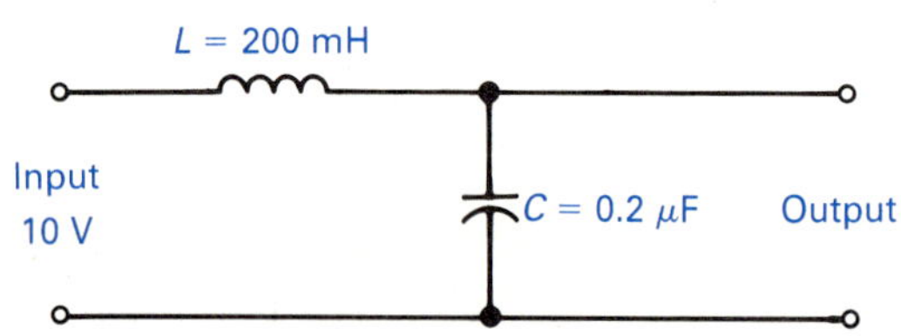

Figure 14-65. Problem 14-1: Circuit.

14-1. Figure 14-65 is the circuit of a low-pass *LC* filter. Find the output voltage at the following frequencies:

a. 500 Hz (*ans*. 16.52 V)
b. 5 kHz (*ans*. 0.26 V)
c. 20 kHz (*ans*. 0.0159 V)

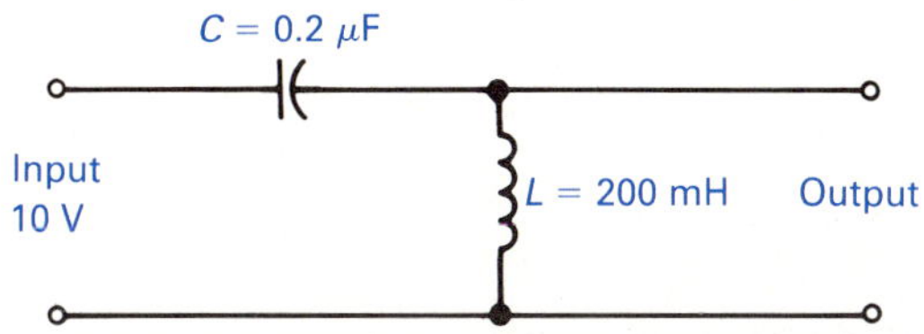

Fig. 14-66. Problem 14-2: Circuit.

14-2. Figure 14-66 is the circuit of a high-pass *LC* filter. Find the output voltage at the following frequencies:
a. 500 Hz
b. 5 kHz
c. 20 kHz

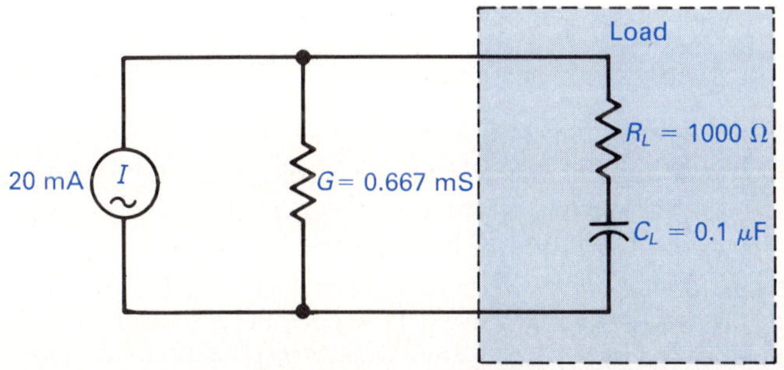

Fig. 14-67. Problems 14-3 and 14-4: Circuit.

14-3. Figure 14-67 shows the equivalent circuit of a transistor amplifier feeding a load consisting of a 1000-Ω resistor in series with a 0.1-μF capacitor. Find the voltage across the load and the phase angle of this voltage with respect to the total current at
a. 50 Hz (*ans*. 29.91 V, −2.69°)
b. 1000 Hz (*ans*. 19.02 V, −25.37°)
c. 5 kHz (*ans*. 12.49 V, −10.4°)

14-4. Repeat Prob. 14-3 when three similar loads are connected in parallel across the output of the amplifier.

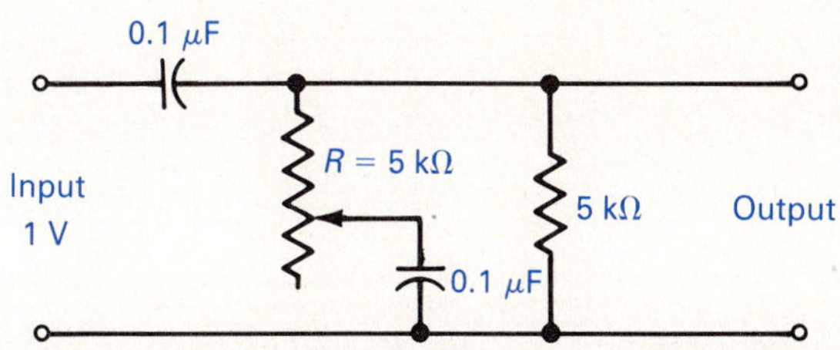

Fig. 14-68. Problem 14-5: Circuit.

14-5. Figure 14-68 shows the circuit of a tone control for an audio amplifier. When *R* is set at zero, find the output voltage of the tone control circuit at the following frequencies:

a. 100 Hz (*ans*. 0.266 V)
b. 1000 Hz (*ans*. 0.493 V)
c. 10 kHz (*ans*. 0.5 V)

14-6. Repeat Prob. 14-5 when R is set at 5000 Ω.

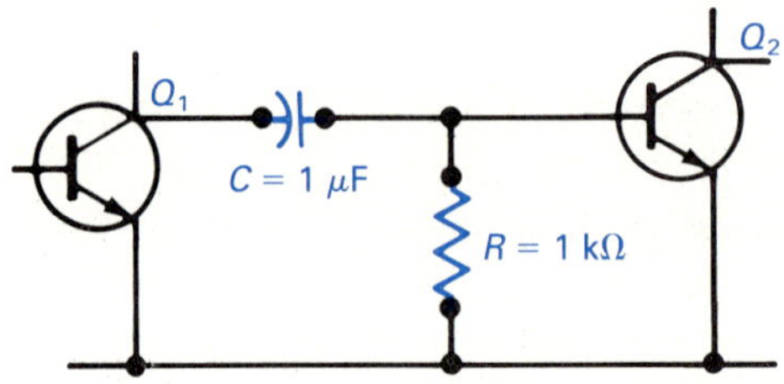

Fig. 14-69. Problems 14-7 to 14-9: Circuit.

14-7. Problem 13-10 refers to the transistor amplifer circuit of Fig. 13-13. This figure is reproduced as Fig. 14-69. What is the RC circuit impedance magnitude over the audio-frequency range, specifically at 20 Hz, 100 Hz, 500 Hz, 1 kHz, 5 kHz, 10 kHz, and 20 kHz?
(*ans*. 8.02 kΩ, 1.88 kΩ, 1.049 kΩ, 1.013 kΩ, 1.001 kΩ, 1 kΩ, and 1 kΩ)

14-8. In the circuit of Fig. 14-69, what are the values of voltage fed to the second transistor Q_2 if the first transistor Q_1 delivers 1 V to the RC network? Draw a curve that shows how the Q_2 transistor's voltage varies over the audio range from 20 Hz to 20 kHz, using the frequencies of Prob. 14-7.

14-9. Referring to the circuit of Fig. 14-69, how does the phase of the Q_2 voltage vary over the audio-frequency range, 20 Hz to 20 kHz? Take the lowest and highest frequency.
(*ans*. 1.45 rad to 0.008 rad or 82.8° to 0.456°)

14-10. A 240-V 60-Hz electric motor has a rating of 5 hp, an efficiency of 70 percent, and a lagging power factor of 0.6 on load. Draw a power triangle for this motor. What are its equivalent series resistance and inductance?

14-11. What would be the equivalent parallel conductance and inductance of the electric motor of Prob. 14-10? (*ans*. 59.21 mS, 59.74 mH)

14-12. What value of capacitance would have the same susceptance as the inductance in Prob. 14-11?

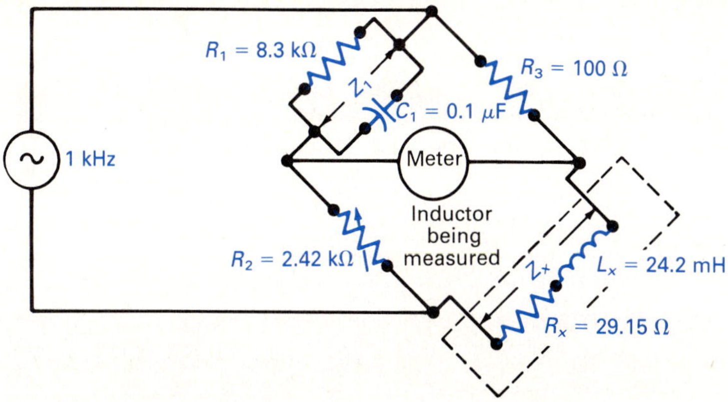

Fig. 14-70. Problem 14-13: Circuit of Maxwell bridge.

14-13. Figure 14-70 is the circuit of a Maxwell bridge, a measuring instrument. The values given are for the balance condition, from which the characteristics of the inductor being measured are found. Check that balance is achieved, that is, $|Z_x| \times |Z_1| = R_2 \times R_3$. Also check that the two impedances have phase angles of the same value but opposite sign.

15

NONRESONANT *RLC* CIRCUITS

The principles and techniques developed for ac circuits containing two types of elements (resistance and capacitance or resistance and inductance) can be applied to circuits which contain all three types of elements, resistance, capacitance, *and* inductance, often called *RLC* circuits. The only new point arising in dealing with the phasor relationships in *RLC* circuits is that, in addition to two phasors in quadrature (resistive and reactive), there is another reactive phasor. The two reactive phasors are in antiphase with respect to each other, so they partially or completely cancel in effect.

Partial cancellation of a circuit's phasors occurs when the inductive and capacitive components are unequal. This inequality leaves a net inductive or net capacitive condition. Together with the resistive (or conductive) component, the net inductive or net capacitive component in effect reduces the *RLC* circuit to an *RL* or an *RC* circuit (series or parallel). As these two-element circuits were fully covered in Chap. 14, we need consider only the cancellation of phasors to deal with *RLC* circuits.

Complete cancellation of the reactive components of a circuit's phasors is called *resonance,* and raises some rather special issues which justify a separate chapter. This chapter therefore is confined to the general techniques of handling *RLC* circuits, deliberately avoiding situations in which complete cancellation of reactive phasors occurs.

15-1 *RLC* SERIES CIRCUIT

A series circuit containing a resistor, a capacitor, and an inductor is shown in Fig. 15-1*a*.

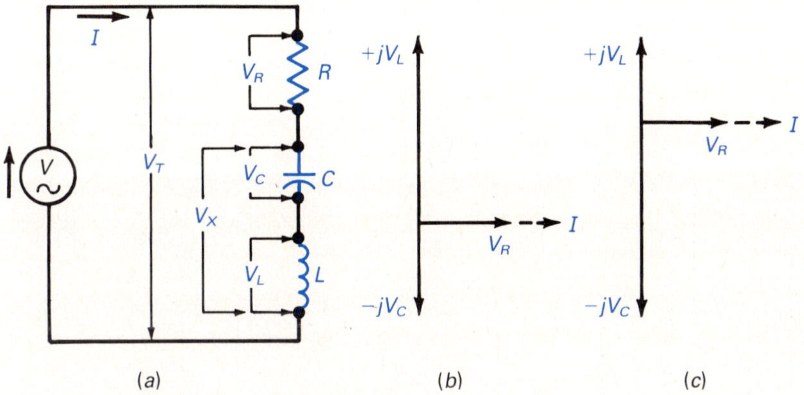

Fig. 15-1. *RLC* series circuit. *(a)* Schematic diagram; *(b)* partial phasor diagram when $V_L > V_C$; *(c)* partial phasor diagram when $V_L < V_C$.

Figure 15-1*b* and *c* shows partial phasor diagrams for the series *RLC* circuit in Fig. 15-1*a*. In each phasor diagram the common current I is shown as the reference phasor $I \angle 0°$ or $I + j0$. The resistive voltage V_R is in phase with the current and has the phasor value $V_R \angle 0°$ or $V_R + j0$. The inductive voltage leads the current by 90° and thus has the phasor value $V_L \angle 90°$ or $0 + jV_L$. The capacitive voltage V_C lags the current by 90° and has the phasor value $V_C \angle -90°$ or $0 - jV_C$.

As both V_L and V_C are quadrature components of the total voltage V_T, but are of opposite j sign, the magnitude of their phasor sum is the arithmetic difference between them. This phasor sum of the individual reactive voltages may be called the net reactive voltage V_X:

$$V_X = (+jV_L) + (-jV_C)$$

$$V_X = jV_L - jV_C \qquad \text{or} \qquad j(V_L - V_C) \tag{15-1}$$

If we concern ourselves only with situations in which inductive and capacitive components are unequal, there are just two alternatives for the series *RLC* circuit; V_L is either greater than or less than V_C. We will therefore consider each of these alternatives.

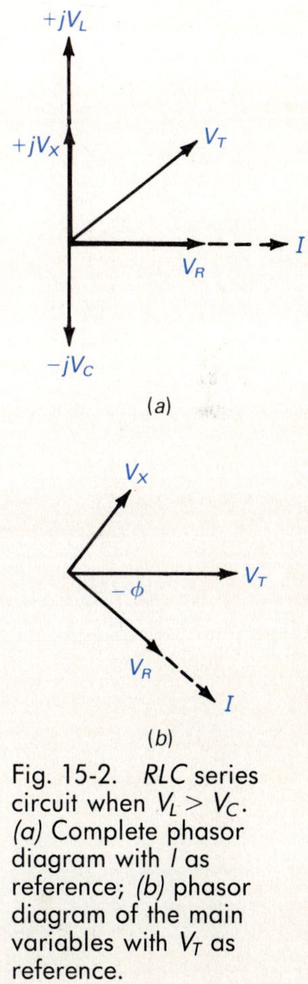

Fig. 15-2. *RLC* series circuit when $V_L > V_C$. *(a)* Complete phasor diagram with I as reference; *(b)* phasor diagram of the main variables with V_T as reference.

When $V_L > V_C$ The partial phasor diagram in Fig. 15-1*b* shows the phasor line for $+jV_L$ longer than that for $-jV_C$, indicating that the magnitude of V_L is greater than that of V_C. This partial phasor diagram leads to the full phasor diagram in Fig. 15-2*a*.

In Fig. 15-2*a* the net reactive voltage V_X is shown as a $+j$ quantity, since, with V_L greater than V_C, $V_L - V_C$ is a positive quantity.

The overall characteristics of the series *RLC* circuit can be determined by considering only the relationships between the current, the resistive voltage, the net reactive voltage, and the total circuit voltage. In Fig. 15-2*b* these four variables are shown in the conventional power position with the total voltage as the reference phasor.

From Fig. 15-2 it can be seen that, unless the characteristics of individual reactive elements are of interest, an *RLC* series circuit in which V_L is greater than V_C behaves as an *RL* series circuit.

All the characteristics of the *RL* series circuit referred to in Sec. 14-2 apply, with V_X substituted for V_L, X_T for X_L, etc.

When $V_L < V_C$ In Fig. 15-1*c* the phasor line for $+jV_L$ is shorter than that for $-jV_C$, indicating that the magnitude of V_L is less than that of V_C. A full phasor diagram for this situation is given in Fig. 15-3*a*.

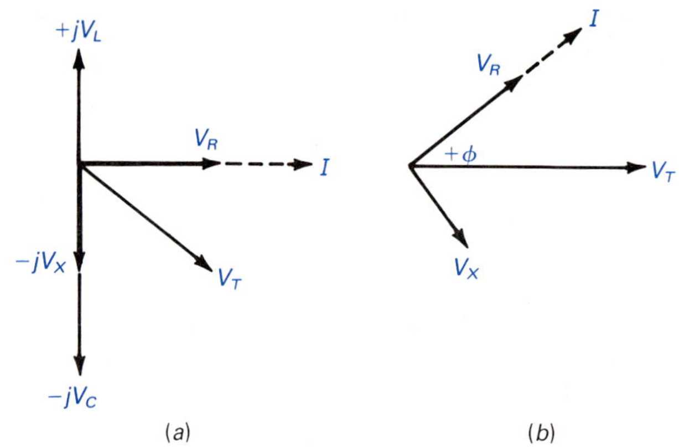

Fig. 15-3. *RLC* series when $V_L < V_C$. *(a)* Complete phasor diagram with I as reference; *(b)* phasor diagram of the main variables with V_T as reference.

When V_L is less than V_C, $V_L - V_C$ is a negative quantity; Fig. 15-3*a* illustrates this, V_X being shown as a $-j$ quantity. The common circuit current, the resistive voltage, the net reactive voltage, and the total circuit voltage are shown in Fig. 15-3*b* with the total voltage as the reference phasor, the conventional power viewpoint.

When V_L is less than V_C, the net reactive voltage is capacitive and the circuit has a leading phase angle. These characteristics are typical of an *RC* circuit.

All the characteristics of the *RC* series circuit referred to in Sec. 14-1 apply to the overall behavior of an *RLC* series circuit in which V_L is less than V_C. Substitute V_X for V_C, X_T for X_C, etc.

EXAMPLE 15-1 What is the value of the current in the circuit in Fig. 15-4*a*?

Solution As X_L is greater than X_C, for any value of current I, V_L will be greater than V_C. Therefore the circuit in Fig. 15-4*a* is net inductive, as shown in the equivalent circuit in Fig. 15-4*b*.

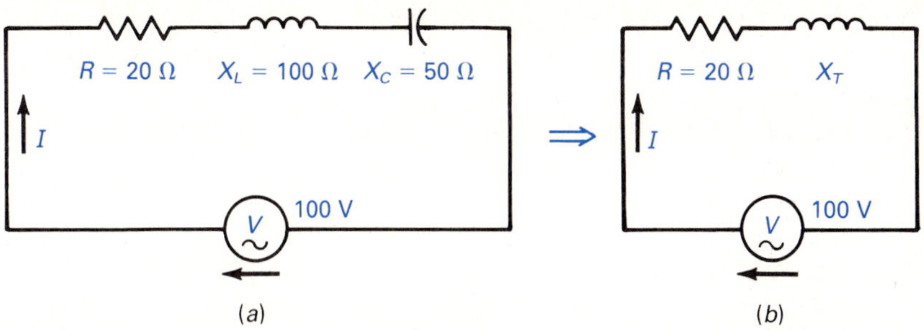

Fig. 15-4. Example 15-1: *(a)* Circuit; *(b)* circuit with X_L and X_C combined as X_T.

$$X_T = j(X_L - X_C) \qquad \text{extension of Eq. (15-1)}$$
$$X_T = j(100\ \Omega - 50\ \Omega) = j50\ \Omega$$
$$Z = R + jX_T = (20 + j50)\ \Omega = 53.85\ \Omega\ \underline{/68.2^\circ}$$

Taking the applied emf V as the reference phasor,

$$I = \frac{V}{Z} = \frac{100\text{ V}\ \underline{/0^\circ}}{53.85\ \Omega\ \underline{/68.2^\circ}} = \mathbf{1.857\ A}\ \underline{/\mathbf{-68.2^\circ}} \quad \text{or} \quad \underline{/\mathbf{-1.19\ rad}}$$

The result for Example 15-1 gives the expected nature of the current in an *RL* circuit; the current lags the emf by less than 90°. The effect of the capacitive reactance is offset by the greater inductive reactance.

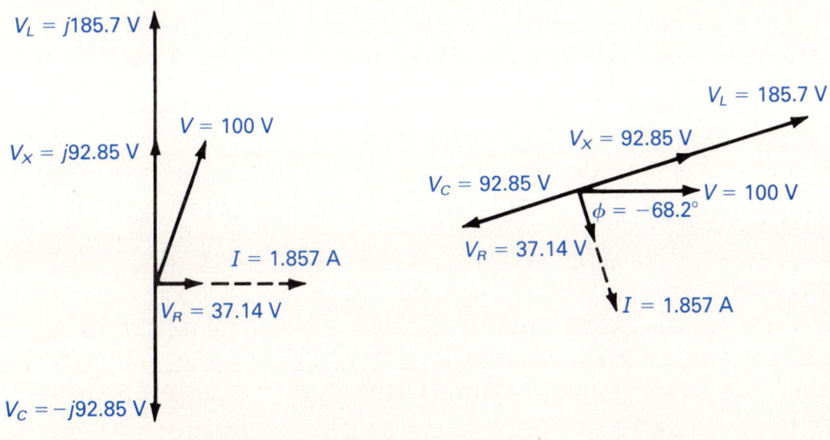

Fig. 15-5. Phasor diagrams. *(a)* With I as reference; *(b)* with V as reference.

Phasor diagrams for the circuit in Example 15-1 are given in Fig. 15-5. Figure 15-5*a* has the common current as the reference phasor, as is usual in deriving series circuit relationships, while Fig. 15-5*b* has the emf (or total circuit voltage) as the reference phasor, the conventional view for power considerations in any circuit.

Had the values of X_L and X_C in Example 15-1 been reversed ($X_L = 50\ \Omega$ and $X_C = 100\ \Omega$), the resulting current would have been numerically the same, but its phase angle would have been positive. A positive current phase angle (leading current) indicates a net capacitive circuit.

EXAMPLE 15-2 In the circuit in Fig. 15-6*a*, what is *(a)* each individual element voltage, *(b)* the net reactive voltage, and *(c)* the applied emf?

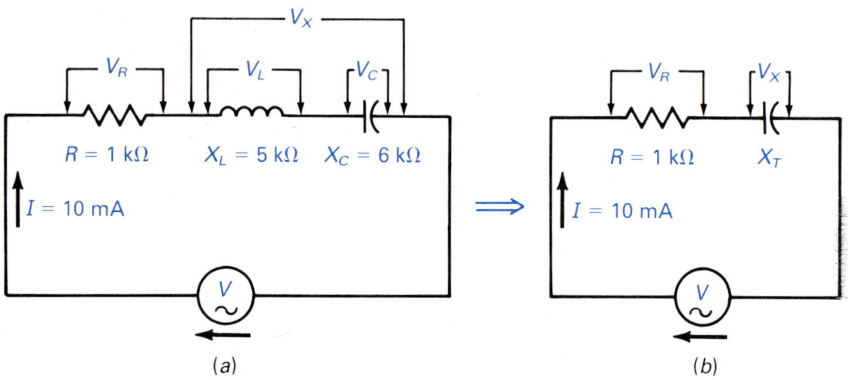

Fig. 15-6. Example 15-2: *(a)* Circuit; *(b)* circuit with X_L and X_C combined as X_T.

Solution As X_L is less than X_C, V_L is less than V_C. The circuit in Fig. 15-6*a* is thus net capacitive. Figure 15-6*b* shows the net reactance X_T to be capacitive.

a. Applying Ohm's law to each element with the common current I as the reference phasor,

$$V_R = IR = 10\ \text{mA}\ \underline{/0^\circ} \times 1\ \text{k}\Omega\ \underline{/0^\circ} = \mathbf{10\ V\ \underline{/0^\circ}}$$

$$V_C = IX_C = 10\ \text{mA}\ \underline{/0^\circ} \times 6\ \text{k}\Omega\ \underline{/-90^\circ} = \mathbf{60\ V\ \underline{/-90^\circ}} = \mathbf{-j60\ V}$$

$$V_L = IX_L = 10\ \text{mA}\ \underline{/0^\circ} \times 5\ \text{k}\Omega\ \underline{/+90^\circ} = \mathbf{50\ V\ \underline{/+90^\circ}} = \mathbf{+j50\ V}$$

b.
$$V_X = jV_L - jV_C \qquad (15\text{-}1)$$

$$V_X = j50\ \text{V} - j60\ \text{V} = \mathbf{-j10\ V} \quad \text{or} \quad \mathbf{10\ V\ \underline{/-90^\circ}}$$

c.
$$V = V_R + jV_X = 10\ \text{V}\ \underline{/0^\circ} + 10\ \text{V}\ \underline{/-90^\circ}$$

$$V = \mathbf{(10 - j10)\ V} = \mathbf{14.14\ V\ \underline{/-45^\circ}} \quad \text{or} \quad \mathbf{\underline{/-0.7853\ rad}}$$

The element voltages in Example 15-2*a* have the correct phase relationships with respect to their common current. The resistive voltage V_R is in phase with the current, the capacitive voltage V_C lags the current by 90°, and the inductive voltage V_L leads the current by 90°.

The capacitive reactance of this series circuit is greater than the inductive reactance; the net reactive voltage V_X is capacitive, lagging the circuit current by 90°. This conclusion is shown by the result for Example 15-2*b*.

The complete circuit in Example 15-2 behaves as a resistance-capacitance network, the emf lagging the current by less than 90° (or the current leading the emf by less than 90°). The result for part *c* substantiates this conclusion.

Phasor diagrams for the circuit in Example 15-2 are given in Fig. 15-7 with both current and total voltage references.

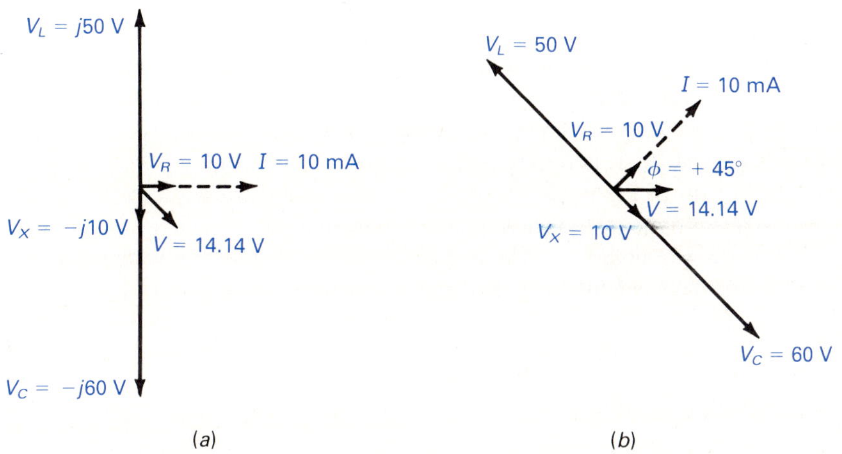

Fig. 15-7. Example 15-2: Phasor diagrams. *(a)* With *I* as reference; *(b)* with *V* as reference.

Voltage Polarities

In Example 15-1, $V_L = 185.7$ V, which is greater than the 100-V applied emf. In Example 15-2, $V_L = 50$ V and $V_C = 60$ V, both of which are greater than the 14.14-V applied emf. It is a unique characteristic of *RLC* series circuits that individual element voltages can be greater in magnitude than the emf applied to the circuit (or than the total voltage across the circuit). This characteristic is due to the tendency for series inductive and capacitive voltages to cancel.

The waveforms of the reactive voltages in Example 15-2 are shown in relation to each other and to the circuit current in Fig. 15-8.

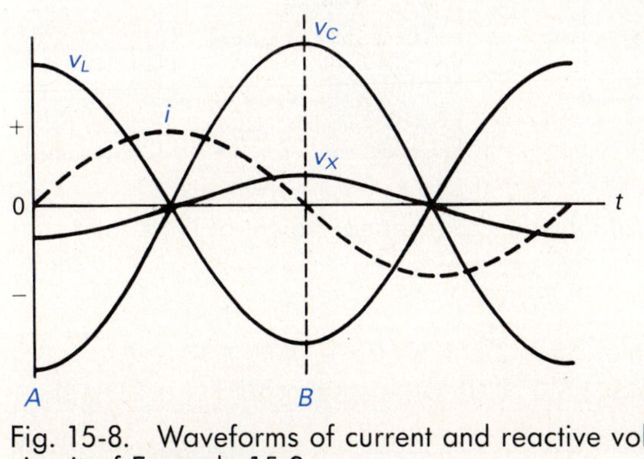

Fig. 15-8. Waveforms of current and reactive voltages in the series ac circuit of Example 15-2.

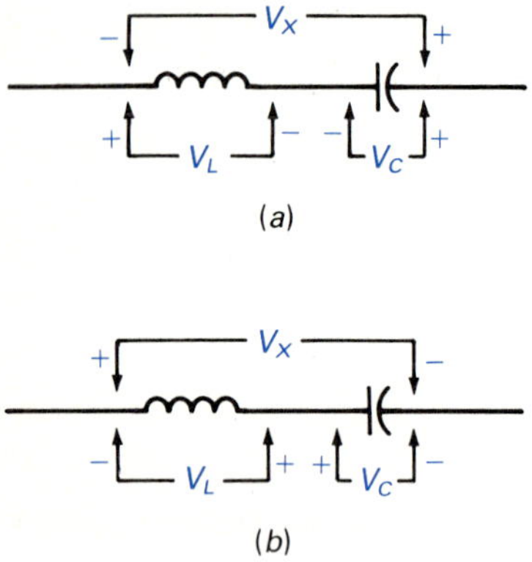

Fig. 15-9. Instantaneous reactive voltage polarities in a series ac circuit.

The waveform of the inductive voltage v_L leads the current i by 90°. The waveform of the capacitive voltage v_C lags the current by 90°. Therefore v_L and v_C have a relative phase of 180°; they are in antiphase; their polarities are always opposite.

Figure 15-9*a* shows the polarities of the reactive voltages at time *A* in Fig. 15-8. The inductive voltage v_L is at its positive maximum value, its polarity being indicated by + ⌇⌇⌇ −. The capacitive voltage v_C is at its negative maximum value, its polarity being shown as − ⊣⊢ +. Since the instantaneous polarities of v_L and v_C are opposite, the net reactive voltage v_X is the difference between them. The polarity of v_X is the same as that of v_C, the greater of the two individual reactive voltages; its relative magnitude is similarly a negative maximum.

One half-cycle later, at instant of time *B*, the voltage values are the same as at time *A* but both of the polarities are reversed, as shown in Fig. 15-9*b*.

At instants of time other than *A* or *B*, the inductive and capacitive voltages will still have opposite polarities. The magnitude of the net reactive voltage will be the difference between the capacitive and inductive voltages, and its polarity will be that of the capacitive voltage. The only exceptions are the time instants, two per cycle, when all three reactive voltages v_C, v_L, and v_X are zero.

EXAMPLE 15-3 Figure 15-10*a* shows two transistors coupled by an *LC* series circuit. Transistor Q_1 amplifies the video-frequency signal from the input and passes this amplified signal on to transistor Q_2 in the form of a voltage V_{be}. An equivalent circuit is given in Fig. 15-10*b*, where Q_1 is depicted as a voltage generator of 15 V emf and 200 Ω internal resistance. R_{be} is the total effective resistance between the base *B* of transistor Q_2 and ground. What is the value of the voltage V_{be} at a frequency of 2.5 MHz?

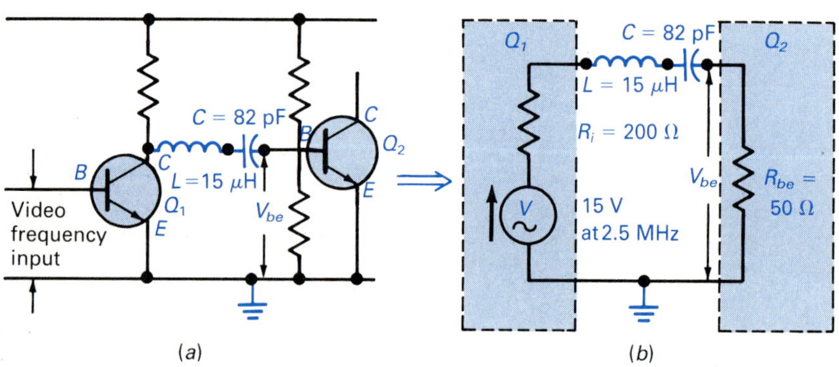

Fig. 15-10. Example 15-3: *(a)* Given circuit; *(b)* equivalent of the circuit connecting Q_1 to Q_2.

Solution

$$\omega = 2\pi f = 2\pi \times 2.5 \times 10^6 \text{ Hz} = 1.571 \times 10^7 \text{ rad/s}$$

$$X_L = \omega L = 1.571 \times 10^7 \text{ rad/s} \times 1.5 \times 10^{-5} \text{ H} = 235.6\ \Omega$$

$$X_C = \frac{1}{\omega C} = \frac{1}{1.571 \times 10^7 \text{ rad/s} \times 8.2 \times 10^{-11} \text{ F}} = 776.3\ \Omega$$

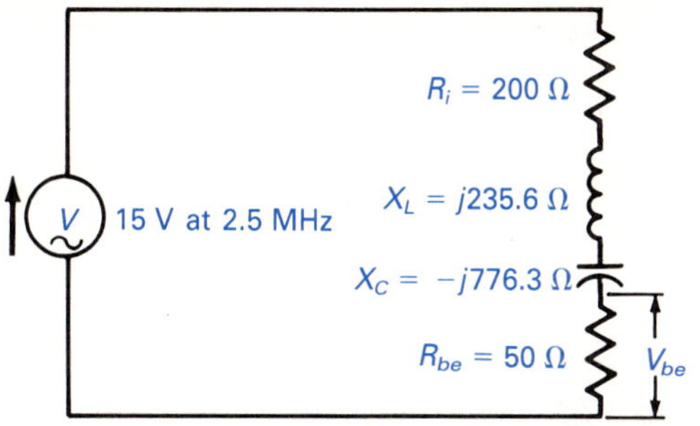

Fig. 15-11. Example 15-3: Fig. 15-10*b* redrawn.

The equivalent circuit can be redrawn as in Fig. 15-11.

The voltage-divider principle can be applied to any ac series circuit if phasor techniques are used.

First calculating the impedance of the circuit Z,

$$Z = (R_i + R_{be}) + j(X_L - X_C)$$
$$= [(200 + 50) + j(235.6 - 776.3)]\ \Omega$$
$$Z = (250 - j540.7)\ \Omega = 595.7\ \Omega\ \underline{/-65.19^\circ}$$

Now applying the voltage-divider principle,

$$V_{be} = V\frac{R_{be}}{Z} = 15\text{ V}\ \underline{/0^\circ} \times \frac{50\ \Omega\ \underline{/0^\circ}}{595.7\ \Omega\ \underline{/-65.19^\circ}}$$
$$V_{be} = \mathbf{1.259\ V}\ \underline{\mathbf{/65.19^\circ}} \qquad \text{or} \qquad \underline{\mathbf{/1.138\ rad}}$$

In Example 15-3 the capacitive reactance exceeds the inductive reactance. The total circuit is essentially a series resistance-capacitance network. Therefore the circuit current and the resistive voltages lead the applied emf. Thus the voltage V_{be} developed across the resistance R_{be} has a leading phase with respect to the emf. The magnitude of V_{be} is appreciably less than the emf, because R_{be} (50 Ω) is low compared with R_i (200 Ω) and the net reactance (540.7 Ω).

At a frequency of 4.5 MHz, X_C and X_L in Fig. 15-11 would be equal, resulting in the complete cancellation of reactive effects. At this frequency the circuit would behave as a resistive voltage divider, so that V_{be} would be the fraction $R_{be}/(R_{be} + R_i) = 0.2$ of the emf. This 3 V is the maximum possible value of V_{be}. Circuits such as that in Fig. 15-10 are used in television sets to separate sound (audio) and picture (video) signals. The 4.5-MHz sound intercarrier signal is directed to the audio section of the set. Picture signals, mainly below 2.5 MHz, are attenuated (reduced in level) and so discouraged from entering the audio section.

EXAMPLE 15-4

An electric motor designed to operate on a 60-Hz supply has equivalent series resistance and inductance of 2 Ω and 1 H, respectively. If the motor's power factor is 0.8 lagging, what is its equivalent series capacitance?

Solution Figure 15-12 is an equivalent series circuit of the motor connected to a 60-Hz supply.

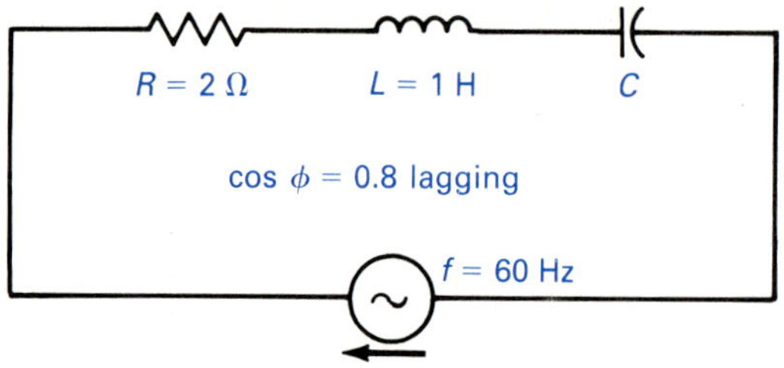

Fig. 15-12. Example 15-4: Equivalent circuit.

$$\omega = 2\pi f = 2\pi \times 60\ \text{Hz} = 377\ \text{rad/s}$$
$$X_L = \omega L = 377\ \text{rad/s} \times 1\ \text{H} = 377\ \Omega$$

Calculating the impedance magnitude,

$$\cos\phi = \frac{R}{Z} \quad \text{or} \quad Z = \frac{R}{\cos\phi}$$

The power factor $\cos\phi$ is 0.8, so

$$Z = \frac{2\ \Omega}{0.8} = 2.5\ \Omega$$

When the impedance magnitude is known, the net reactance can be found:

$$Z = \sqrt{R^2 + X_T{}^2}$$
$$Z^2 = R^2 + X_T{}^2$$
$$\text{or} \quad X_T{}^2 = Z^2 - R^2 = (2.5\ \Omega)^2 - (2\ \Omega)^2 = 2.25\ \Omega^2$$
$$X_T = \sqrt{2.25\ \Omega^2} = 1.5\ \Omega$$

From the net reactance and the inductive reactance, the capacitive reactance can be determined:

$$X_T = X_L - X_C \quad \text{extension of Eq. (15-1)}$$
$$\text{or} \quad X_C = X_L - X_T = 377\ \Omega - 1.5\ \Omega = 375.5\ \Omega$$

The value of the series capacitance of the motor can be obtained from the capacitive reactance:

$$X_C = \frac{1}{\omega C}$$
$$\text{or} \quad C = \frac{1}{\omega X_C} = \frac{1}{377\ \text{rad/s} \times 375.5\ \Omega} = \mathbf{7.06\ \mu F}$$

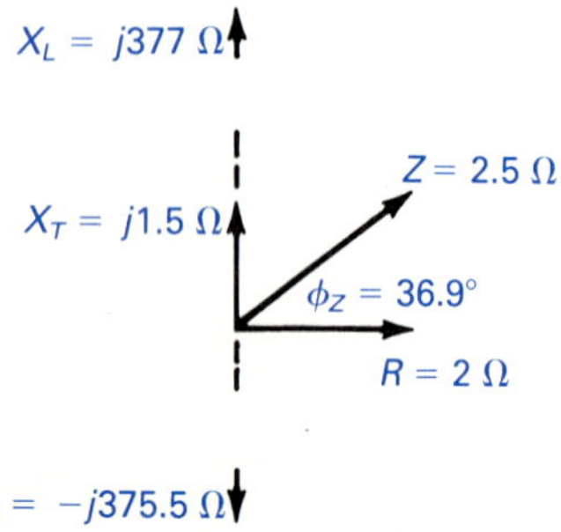

Fig. 15-13. Example 15-4: Impedance diagram.

For Example 15-4 an impedance diagram drawn in phasor diagram form may be helpful in visualizing the circuit conditions. Such an impedance diagram is shown in Fig. 15-13.

Figure 15-13 shows the phase angle of the impedance ϕ_Z as +36.9°. As impedances are proportional to voltages, the figure is basically the same as a voltage phasor diagram. A positive, or leading, voltage phase angle represents a lagging current phase angle. Thus the circuit in Example 15-4 has a lagging power factor.

EXAMPLE 15-5 In the circuit in Example 15-4, if the supply voltage is 240 V, what are *(a)* the apparent power, *(b)* the active power, and *(c)* the reactive power?

Solution Each component of power can be determined by first calculating the circuit current.

$$I = \frac{V}{Z} = \frac{240\ \text{V}}{2.5\ \Omega} = 96\ \text{A}$$

a. $S = I^2Z = (96\ \text{A})^2 \times 2.5\ \Omega = \mathbf{23.04\ kVA}$

b. $P = I^2R = (96\ \text{A})^2 \times 2\ \Omega = \mathbf{18.43\ kW}$

c. $Q = I^2X_T = (96\ \text{A})^2 \times 1.5\ \Omega = \mathbf{13.82\ kvar}$

A power triangle for Example 15-5 is given in Fig. 15-14.

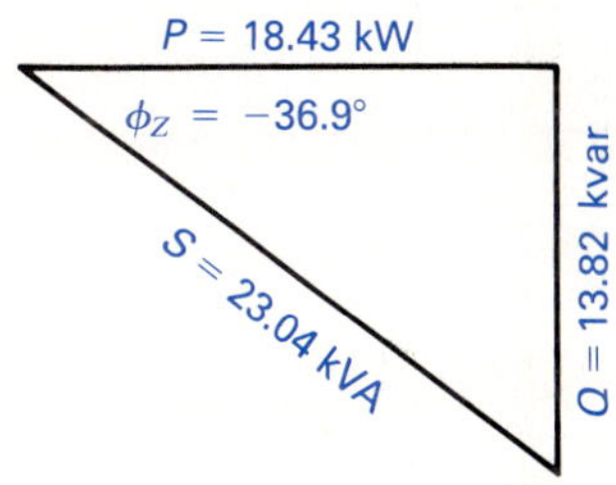

Fig. 15-14. Example 15-5: Power triangle.

REVIEW QUESTIONS FOR SECTION 15-1

1. In a series *RLC* circuit, which phasor is normally used as the reference?
2. In deriving the power equation of a series *RLC* circuit from current and voltage, the sign of the imaginary term is changed. Why?
3. In a series *RLC* circuit in which the inductive component of voltage is greater than the capacitive component, does the current lead the total voltage?
4. For the condition of Question 3, is the circuit phase angle leading?
5. What units are used for apparent power, reactive power, and active power in a series *RLC* circuit?
6. When the inductive reactance of a series *RLC* circuit is less than the capacitive reactance, is the net reactive voltage a $-j$ quantity?
7. When the inductive reactance of a series *RLC* circuit is less than the capacitive reactance, is the reactive power a $-j$ quantity?
8. In a series *RLC* circuit, a reactive voltage may be greater than the applied emf. How is this possible?
9. What is the phase relationship of v_L and v_C in a series *RLC* circuit?

EXERCISE PROBLEMS FOR SECTION 15-1

1. Using Fig. 15-1 (page 563), find the magnitudes of I, V_R, V_C, V_L, and the circuit phase angle, and show that the phasor sum of V_R, V_C, and V_L equals the emf when

a. $V = 100$ V, $R = 100\ \Omega$, $X_C = 250\ \Omega$, $X_L = 1000\ \Omega$
(*ans*. 132.2 mA, 13.22 V, 33.04 V, 132.2 V, 82.4° lag)
b. $V = 12$ V, $R = 1000\ \Omega$, $X_C = 1000\ \Omega$, $X_L = 3000\ \Omega$
(*ans*. 5.37 mA, 5.37 V, 5.37 V, 16.1 V, 63.43° lag)
c. $V = 200\ \mu$V, $R = 20\ \Omega$, $X_C = 300\ \Omega$, $X_L = 280\ \Omega$
(*ans*. 7.071 μA, 141.4 μV, 2.121 mV, 1.98 mV, 45° lead)
d. $V = 200\ \mu$V, $R = 2\ \Omega$, $X_C = 300\ \Omega$, $X_L = 280\ \Omega$
(*ans*. 9.95 μA, 19.9 μV, 2.985 mV, 2.786 mV, 84.29° lead)

2. Repeat Prob. 1 when $V = 1$ V, $R = 10\ \Omega$, $C = 500$ pF, $L = 200\ \mu$H, and
a. $f = 400$ kHz *c*. $f = 500$ kHz
b. $f = 450$ kHz *d*. $f = 550$ kHz

3. Using Fig. 15-1, find the magnitudes of I, V_C, and the phase angle, and show that the phasor sum of V_C, V_L, and V_R equals the emf when $V = 4$ mV, $R = 8\ \Omega$, $L = 250\ \mu$H, $f = 1$ MHz, and C is
a. 60 pF (*ans*. 3.693 μA, 9.8 mV, 89.58° lead)
b. 90 pF (*ans*. 20.22 μA, 35.87 mV, 87.67° lead)
c. 110 pF (*ans*. 32.19 μA, 46.58 mV, 86.31° lag)
d. 200 pF (*ans*. 5.16 μA, 4.106 mV, 89.41° lag)

4. Using Fig. 15-1, find the magnitude of I and the phase angle of the circuit when $V = 120$ V, $R = 40\ \Omega$, $L = 0.15$ H, $f = 60$ Hz, and C is
a. Short-circuited *b*. 200 μF *c*. 50 μF *d*. 20 μF

5. Find the apparent power, reactive power, and active power for each of the circuits in Prob. 15-4.
(*ans*. *a*. 208 VA, 169.8 var, 120.1 W; *b*. 244.3 VA, 179.4 var, 165.8 W; *c*. 358.6 VA, 31.23 var, 357.3 W; *d*. 167.5 VA, 148.3 var, 77.96 W)

15-2 *RLC* PARALLEL CIRCUIT

Figure 15-15*a* is a schematic diagram of an *RLC* parallel circuit.

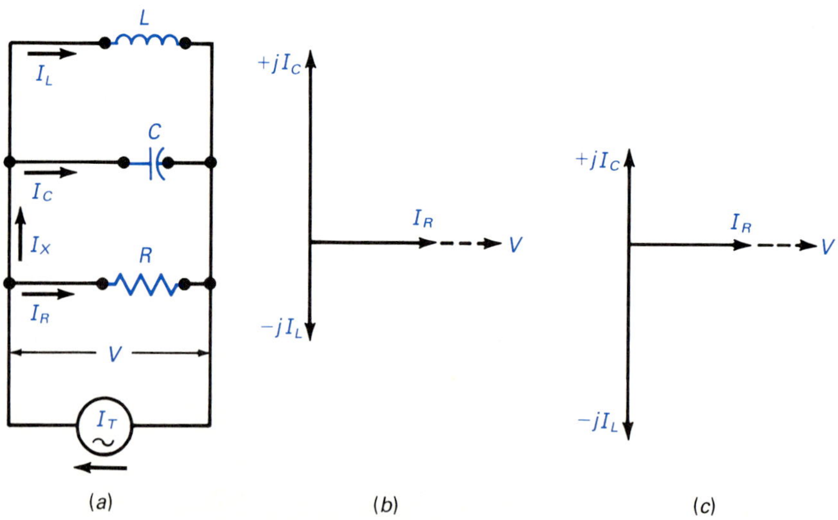

Fig. 15-15. *RLC* parallel circuit. *(a)* Schematic diagram; *(b)* partial phasor diagram when $I_C > I_L$; *(c)* partial phasor diagram when $I_C < I_L$.

Partial phasor diagrams for the parallel *RLC* circuit in Fig. 15-15*a* are given in Fig. 15-15*b* and *c*. In each of these phasor diagrams the common voltage V is made the reference phasor $V \underline{/0^\circ}$ or $V + j0$. The resistive current I_R is in phase with the voltage and has the phasor value $I_R \underline{/0^\circ}$ or $I_R + j0$. The capacitive current I_C leads the voltage by 90° and has the phasor value $I_C \underline{/90^\circ}$ or $0 + jI_C$. The inductive current I_L lags the voltage by 90° and has the phasor value $I_L \underline{/-90^\circ}$ or $0 - jI_L$.

The reactive currents I_C and I_L are of opposite j sign, so I_X, the net reactive current, is numerically equal to their difference:

$$I_X = jI_C - jI_L \qquad \text{or} \qquad j(I_C - I_L) \tag{15-2}$$

In this chapter we are concerned with situations in which reactive components are unequal. We will therefore consider only the *RLC* parallel circuit conditions in which I_C is greater than I_L and in which I_C is less than I_L.

When $I_C > I_L$ Figure 15-15*b* has the phasor line for $+jI_C$ longer than the line for $-jI_L$, illustrating the case in which I_C is greater than I_L. A full phasor diagram for a parallel *RLC* circuit in which I_C is greater than I_L is given in Fig. 15-16*a*.

When I_C is greater than I_L, $I_C - I_L$ is a positive quantity, so that the net reactive current I_X is a $+j$ quantity, as shown in Fig. 15-16*a*.

Figure 15-16*b* is essentially the same as Fig. 15-16*a* but concentrates on the variables which determine the overall characteristics of the *RLC* parallel circuit: the common voltage V, the resistive current I_R, the net reactive current I_X, and the total circuit current I_T. It can be seen that Fig. 15-16*b* is the phasor diagram of a parallel *RC* circuit. Unless individual element characteristics are of interest, the behavior of a parallel *RLC* circuit in which $I_C > I_L$ can be regarded as the same as that of an *RC* parallel circuit. Refer to Sec. 14-3 for a treatment of the parallel *RC* circuit. When this approach is used, I_X is substituted for I_C, B_T for B_C, etc.

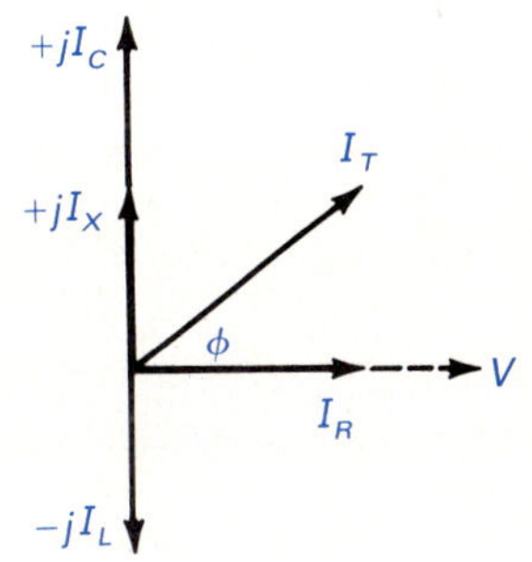

(*a*)

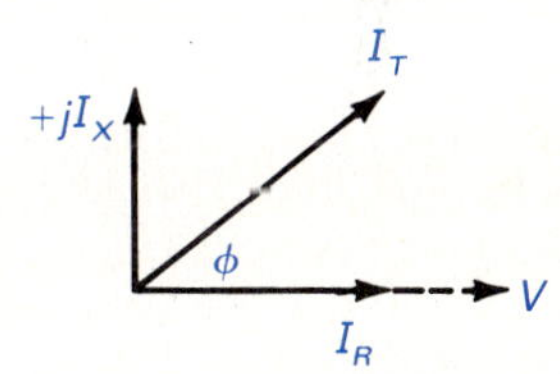

(*b*)

Fig. 15-16. *RLC* parallel-circuit phasor diagrams when $I_C > I_L$: (*a*) Complete; (*b*) main variables.

When $I_C < I_L$ The phasor line for $+jI_C$ in Fig. 15-15*c* is shorter than the line for $-jI_L$. This phasor diagram applies when I_C is less than I_L. Figure 15-17*a* is a full phasor diagram for the situation $I_C < I_L$.

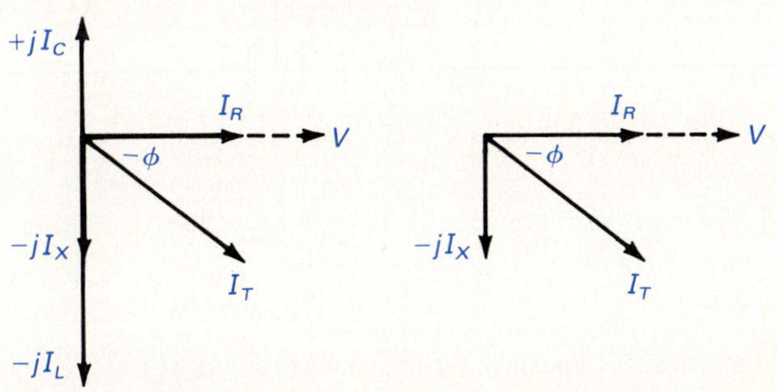

Fig. 15-17. *RLC* parallel-circuit phasor diagrams when $I_C < I_L$: (*a*) Complete; (*b*) main variables.

If I_C is less than I_L, $I_C - I_L$ is a negative quantity. Figure 15-17*a* shows the net reactive current I_X as a $-j$ quantity.

The main variables controlling the behavior of the total *RLC* parallel circuit are the common voltage *V*, the resistive current I_R, the net reactive current I_X, and the total circuit current I_T. These variables for the $I_C < I_L$ condition are shown in the phasor diagram in Fig. 15-17*b*. From this phasor diagram it can be seen that a parallel *RLC* circuit with $I_C < I_L$ may be treated in total as a parallel *RL* circuit. Therefore, if individual element characteristics are not of direct concern, the principles developed in Sec. 14-4 can be applied, with the substitution of I_X for I_L, B_T for B_L, etc.

It should be observed that in Figs. 15-16 and 15-17 no phasor diagram rotation was involved for the *a* and *b* diagrams, as was needed in Figs. 15-2 and 15-3. This was because the parallel-circuit approach is oriented toward a voltage reference, so that the conventional view for ac power considerations (the circuit phase angle is the angle of the total-current phasor with the total-voltage phasor as reference) is automatic when dealing with parallel circuits.

EXAMPLE 15-6 What is the value of the voltage *V* in the circuit in Fig. 15-18?

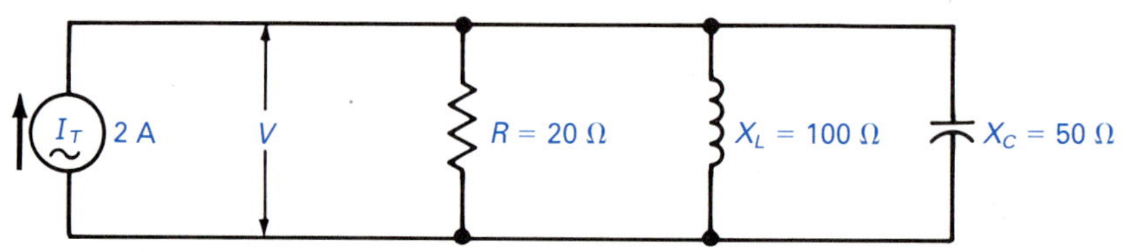

Fig. 15-18. Example 15-6: Circuit.

Solution Convert the impedances of the individual elements into admittances:

$$G = \frac{1}{R} = \frac{1}{20\ \Omega} = 0.05\ \text{S}$$

$$B_L = \frac{1}{jX_L} = -j\frac{1}{100\ \Omega} = -j0.01\ \text{S}$$

$$B_C = \frac{1}{-jX_C} = j\frac{1}{50\ \Omega} = j0.02\ \text{S}$$

$$B_T = j(B_C - B_L) \qquad \text{extension of Eq. (15-2)}$$

$$B_T = j(0.02\ \text{S} - 0.01\ \text{S}) = j0.01\ \text{S}$$

The total circuit admittance is obtained by the phasor addition of the conductance and the net susceptance:

$$Y = G + jB_T = 0.05\ \text{S} + j0.01\ \text{S}$$

$$Y = (0.05 + j0.01)\ \text{S} \qquad \text{or} \qquad 0.051\ \text{S}\ \underline{/11.31^\circ}$$

The common circuit voltage *V* (with respect to the total current I_T) can now be obtained by applying the admittance form of Ohm's law ($I = VY$):

$$V = \frac{I_T}{Y} = \frac{2\text{ A}\angle 0^\circ}{0.051\text{ S}\angle 11.31^\circ} = \mathbf{39.22\text{ V}\angle -11.31^\circ} \quad \text{or} \quad \mathbf{\angle -0.1974\text{ rad}}$$

The element values selected for Example 15-6 are the same as those for the series circuit in Example 15-1. Since the inductive reactance was greater than the capacitive reactance, the series circuit was predominantly inductive. The series-circuit current lagged the applied emf. With the voltage of the parallel circuit common, and the capacitive susceptance greater than the inductive susceptance, the capacitive current in Example 15-6 is greater than the inductive current. The parallel circuit is thus predominantly capacitive. The circuit voltage V lags the source current I_T, or the current leads the voltage.

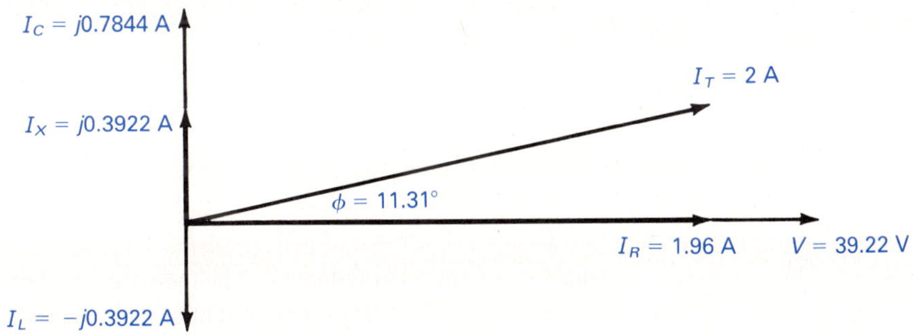

Fig. 15-19. Example 15-6: Phasor diagram.

A phasor diagram for the circuit in Example 15-6 is given in Fig. 15-19.

EXAMPLE 15-7 In the circuit in Figure 15-20, what is *(a)* each individual element current, *(b)* the net reactive current, and *(c)* the total circuit current?

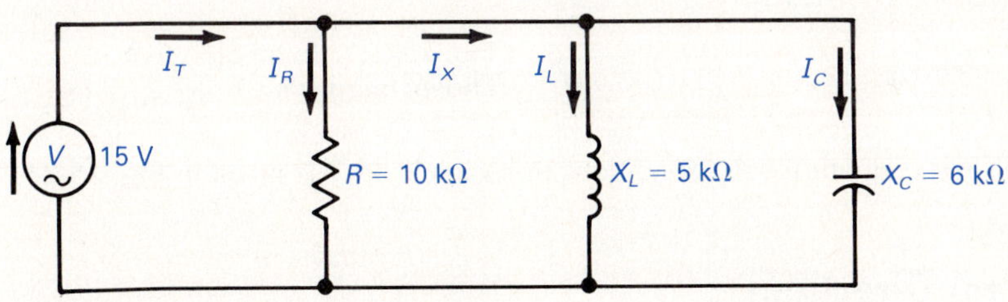

Fig. 15-20. Example 15-7: Circuit.

Solution Calculating the admittances of the individual elements,

$$G = \frac{1}{R} = \frac{1}{10\text{ k}\Omega} = 0.1\text{ mS}$$

$$B_L = \frac{1}{X_L} = \frac{1}{j5\text{ k}\Omega} = -j0.2\text{ mS}$$

$$B_C = \frac{1}{X_C} = \frac{1}{-j6\ \text{k}\Omega} = j0.1667\ \text{mS}$$

a. Applying Ohm's law,

$$I_R = VG = 15\ \text{V} \times 0.1\ \text{mS} = \mathbf{1.5\ mA}$$
$$I_L = VB_L = 15\ \text{V} \times (-j0.2\ \text{mS}) = \mathbf{-j3\ mA}$$
$$I_C = VB_C = 15\ \text{V} \times j0.1667\ \text{mS} = \mathbf{j2.5\ mA}$$

b. The net reactive current can be obtained by the phasor addition of the inductive and capacitive currents:

$$I_X = j(I_C - I_L) \qquad (15\text{-}2)$$
$$= j(2.5\ \text{mA} - 3\ \text{mA})$$
$$I_X = \mathbf{-j0.5\ mA = 0.5\ mA\ \underline{/-90°}} \quad \text{or} \quad \mathbf{\underline{/-\pi/2\ rad}}$$

c. The total circuit current is the phasor sum of the resistive current and the net reactive current:

$$I_T = I_R + jI_X$$
$$I_T = (1.5\ \text{mA} - j0.5)\ \text{mA} = \mathbf{1.58\ mA\ \underline{/-18.43°}} \quad \text{or} \quad \mathbf{\underline{/-0.3216\ rad}}$$

In the parallel *RLC* circuit in Example 15-7 the inductive reactance is less than the capacitive reactance, so that the circuit is net inductive. The inductive current is greater than the capacitive current, and the total circuit current lags the applied emf.

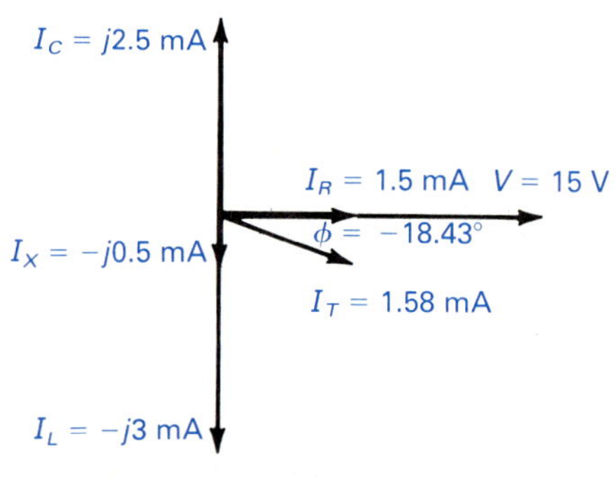

Fig. 15-21. Example 15-7: Phasor diagram.

A phasor diagram for the conditions in Example 15-7 is given in Fig. 15-21.

Current Directions

The total current in Example 15-7 (1.58 mA) is less than either its inductive component (3 mA) or its capacitive component (2.5 mA). It frequently happens that the values involved in an *RLC* parallel circuit result in branch currents which exceed the total circuit current in magnitude. This apparently illogical situation is due to the fact that, at any time instant, the inductive and capacitive currents have opposite directions. The waveform diagrams in Fig. 15-22 show the relationships between the reactive currents and the applied emf in the circuit in Example 15-7.

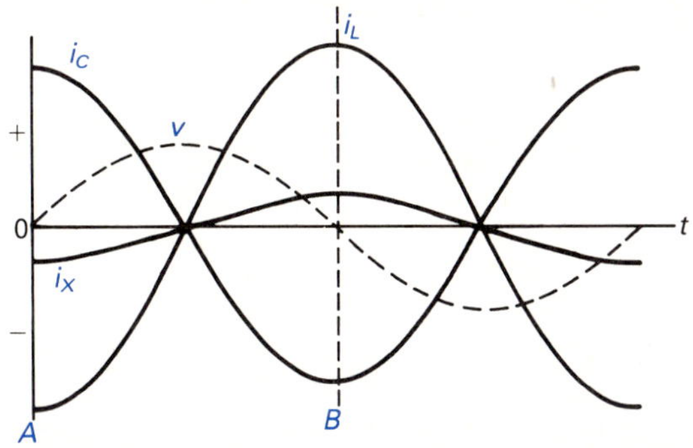

Fig. 15-22. Waveforms of the currents and the voltage in the parallel ac circuit of Example 15-7.

The waveform of the inductive current i_L lags that of the emf by 90°. The waveform of the capacitive current i_C leads that of the emf by 90°. Thus i_L and i_C are in antiphase (180° phase relationship). At any instant in time, one of the reactive currents is positive while the other is negative. These positive and negative signs indicate opposite current directions.

At time A in Fig. 15-22, the capacitive current i_C is at its positive peak value, while the inductive current i_L is at its negative peak value, their directions being opposite. The magnitude of i_L being greater than that of i_C results in the net reactive current i_X having a direction the same as that of the inductive current.

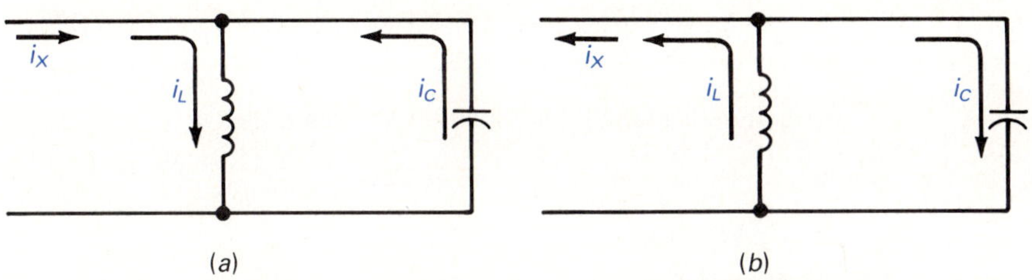

Fig. 15-23. Instantaneous current directions in a parallel LC circuit.

The conditions existing at instant A are shown in the circuit in Fig. 15-23a. Figure 15-23b shows the situation one half-cycle later, at instant B. At this time the current magnitudes are the same as in Fig. 15-23a, but their directions are reversed. At all instants (except when the currents are zero) the inductive and capacitive currents will be opposite in direction, the net reactive current direction being that of the inductive current.

It must be understood that the predominantly inductive conditions referred to above are related to specific circuit values such as those in Example 15-7.

EXAMPLE 15-8 A tape recorder head has an inductance of 100 mH, a self-capacitance of 250 pF, and a parallel internal resistance of 10 kΩ. The recorder amplifier delivers 1 mA to the head at all audible frequencies. What magnitude of current is developed in the coil's inductance at 20 Hz, 1 kHz, and 20 kHz?

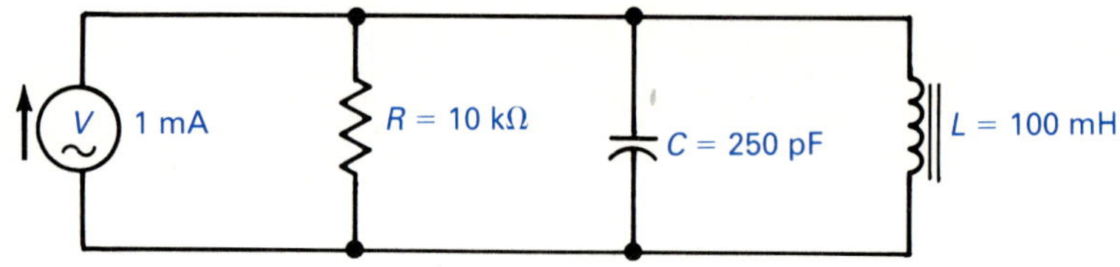

Fig. 15-24. Example 15-8: Circuit.

Solution (See Fig. 15-24.)

a. When $f_1 = 20$ Hz,

$$\omega = 2\pi f_1 = 2\pi \times 20 \text{ Hz} = 125.7 \text{ rad/s}$$
$$X_L = \omega L = 125.7 \text{ rad/s} \times 100 \text{ mH} = 12.57\ \Omega$$
$$X_C = \frac{1}{\omega C} = \frac{1}{125.7 \text{ rad/s} \times 2.5 \times 10^{-10}\text{ F}} = 31.82\text{ M}\Omega$$

At 20 Hz, X_L is so low compared with either R or X_C that the inductance can be considered to be the only significant element connected to the 10-mA source. Therefore,

$$I_L \simeq \mathbf{1\ mA}$$

b. When $f_2 = 1$ kHz, the ratio of the frequencies ($f_2/f_1 = 50$) can be used to compute the reactances at 1 kHz:

$$X_L \text{ at 1 kHz} = X_L \text{ at 20 Hz} \times 50 = 12.57\ \Omega \times 50 = 628.5\ \Omega$$
$$X_C \text{ at 1 kHz} = X_C \text{ at 20 Hz} \div 50 = 31.82\text{ M}\Omega \div 50 = 636\text{ k}\Omega$$

Again at 1 kHz, X_L is much lower than either R or X_C, so

$$I_L \simeq \mathbf{1\ mA}$$

c. When $f_3 = 20$ kHz, applying the ratio $f_3/f_1 = 1000$,

$$X_L \text{ at 20 kHz} = X_L \text{ at 20 Hz} \times 1000 = 12.57\ \Omega \times 1000 = 12.57\text{ k}\Omega$$
$$X_C \text{ at 20 kHz} = X_C \text{ at 20 Hz} \div 1000 = 31.82\text{ M}\Omega \div 1000 = 31.82\text{ k}\Omega$$

Thus at 20 kHz, X_L, X_C, and R are all of the same order of magnitude, and a detailed calculation of I_L is necessary.

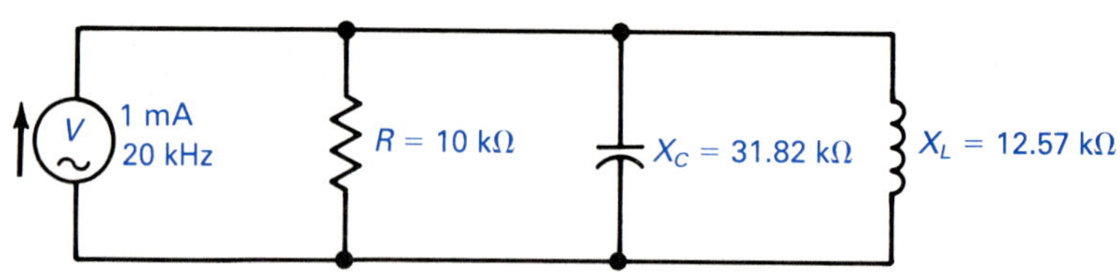

Fig. 15-25. Example 15-8: Circuit with reactance values at 20 kHz.

Figure 15-25 shows the values of the element impedances. Converting these impedances to admittances,

$$G = \frac{1}{R} = \frac{1}{10\text{ k}\Omega} = 100\ \mu\text{S}$$

$$B_C = \frac{1}{X_C} = \frac{1}{31.82\ \text{k}\Omega} = 31.4\ \mu\text{S}$$

$$B_L = \frac{1}{X_L} = \frac{1}{12.57\ \text{k}\Omega} = 79.6\ \mu\text{S}$$

$$B_T = j(B_C - B_L) \qquad \text{extension of Eq. (15-2)}$$

$$B_T = j(31.4\ \mu\text{S} - 79.6\ \mu\text{S}) = -j48.2\ \mu\text{S}$$

The total admittance Y is the phasor sum of the conductance and the net susceptance:

$$Y = G + jB_T = (100 - j48.2)\ \mu\text{S} = 111\ \mu\text{S}\ \underline{/-25.73^\circ}$$

The magnitudes of the common voltage V and the current I_L can both be determined by applying Ohm's law:

$$V = \frac{I}{Y} = \frac{1\ \text{mA}}{111\ \mu\text{S}} = 9.009\ \text{V}$$

$$I_L = VB_L = 9.009\ \text{V} \times 79.6\ \mu\text{S}$$

$$I_L = \mathbf{0.7171\ mA}$$

A tape recorder head consists of a coil of wire wound on an iron core. When the instrument is recording, the magnetic tape passes over the head while an alternating current in the coil produces an alternating magnetic field. This field magnetizes the tape by induction, resulting in a magnetic reproduction of the sound being recorded. The results for Example 15-8 show that, while the coil's self-capacitance and resistance play an insignificant role at lower frequencies, at the top of the audio spectrum, 20 kHz, these parameters of capacitance and resistance can noticeably reduce the inductive current. This current reduction will cause the high-frequency performance of the recording head to be lower than its performance at lower frequencies.

EXAMPLE 15-9 An industrial electric load has a power factor of 0.5 lagging and draws 5 kW from a 208-V 60-Hz supply. What value of parallel capacitor would be needed to raise the circuit power factor to 0.95 lagging?

Solution An equivalent parallel circuit of the load is shown in Fig. 15-26, where G and B_L are, respectively, load conductance and inductive susceptance. The active power is dissipated in this circuit entirely within the resistive component G.

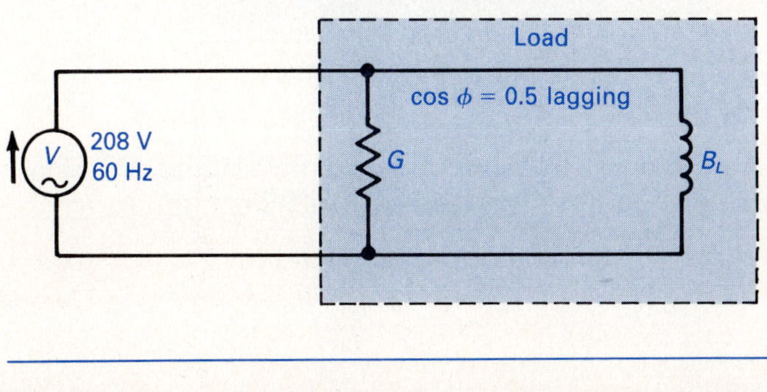

Fig. 15-26. Example 15-9: Circuit.

$$P = \frac{V^2}{R} \qquad \text{or} \qquad V^2G$$

$$G = \frac{P}{V^2} = \frac{5\ \text{kW}}{(208\ \text{V})^2} = 0.1156\ \text{S}$$

The power factor of a parallel circuit is the ratio of its conductance to its admittance:

$$\cos\phi = \frac{G}{Y}$$

$$Y = \frac{G}{\cos\phi} = \frac{0.1156\ \text{S}}{0.5} = 0.2312\ \text{S}$$

Figure 15-27 is an admittance triangle for the load.

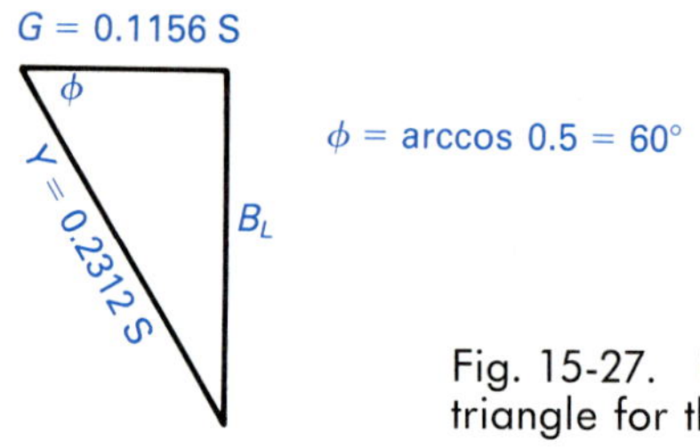

Fig. 15-27. Example 15-9: Admittance triangle for the load.

From the admittance triangle,

$$B_L = \sqrt{Y^2 - G^2} = \sqrt{(0.2312\ \text{S})^2 - (0.1156\ \text{S})^2} = 0.2002\ \text{S}$$

Thus the load has the characteristics of a parallel circuit consisting of a 0.1156-S conductance (8.651-Ω resistance) and a 0.2002-S inductive susceptance (4.995-Ω inductive reactance or 0.0132-H inductance).

When the capacitance is added in parallel, the circuit will appear as in Fig. 15-28.

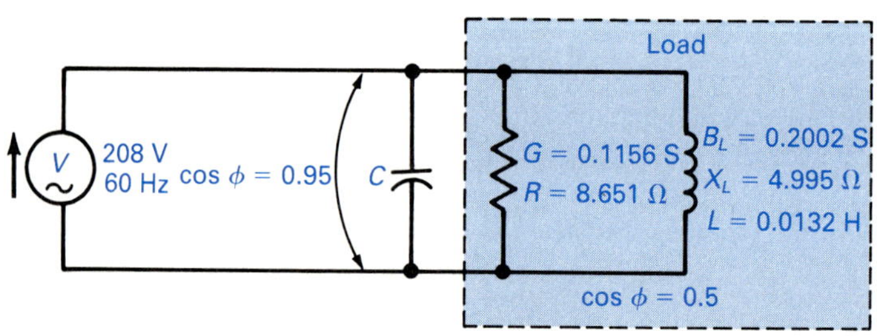

Fig. 15-28. Example 15-9: Circuit with added capacitor.

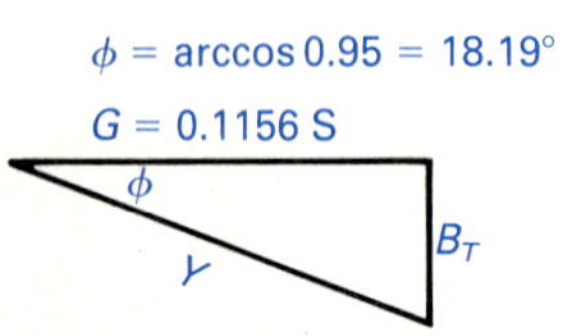

Fig. 15-29. Example 15-9: Admittance triangle with the capacitor in the circuit.

For the power factor $\cos\phi$ to equal 0.95 lagging, the admittance triangle must appear as in Fig. 15-29.

$$\cos\phi = \frac{G}{Y}$$

$$Y = \frac{G}{\cos\phi} = \frac{0.1156\text{ S}}{0.95} = 0.1217\text{ S}$$

The total susceptance B_T can now be found:

$$B_T = \sqrt{Y^2 - G^2} = \sqrt{(0.1217\text{ S})^2 - (0.1156\text{ S})^2} = 0.038\text{ S}$$

A complete admittance diagram, in phasor form, is shown in Fig. 15-30, and from it the value of the capacitive susceptance can be determined. The lagging power factor means that the net susceptance B_T is a $-j$ quantity (inductive). Therefore the capacitive susceptance, which is a $+j$ quantity, must be lower in value than the inductive susceptance of the load.

$$B_C = B_L - B_T = 0.2002\text{ S} - 0.038\text{ S} = 0.1622\text{ S}$$

The value of the capacitance C can now be calculated:

$$\omega = 2\pi f = 2\pi \times 60\text{ Hz} = 377\text{ rad/s}$$

$$B_C = \omega C$$

$$C = \frac{B_C}{\omega} = \frac{0.1622\text{ S}}{377\text{ rad/s}} = \mathbf{430.2\ \mu F}$$

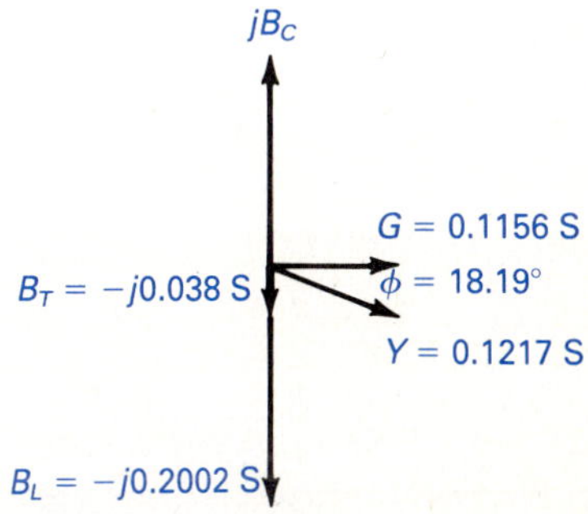

Fig. 15-30. Example 15-9: Admittance diagram.

The final circuit for Example 15-9 is given in Fig. 15 31.

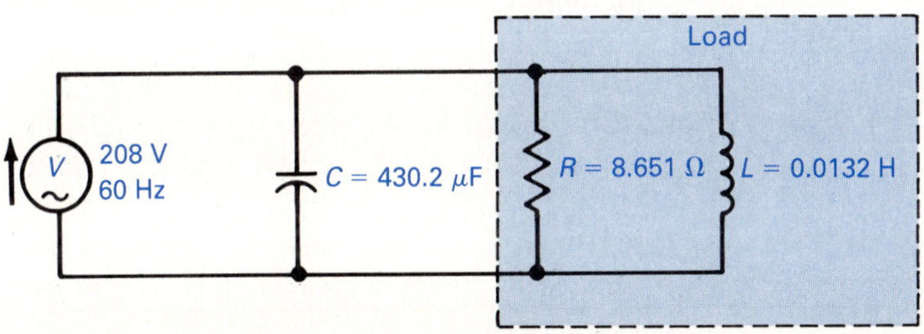

Fig. 15-31. Example 15-9: Final circuit with values.

The power dissipated by the total circuit in Fig. 15-31 is 5 kW, the same as that of the original load (in Fig. 15-26). The addition of the parallel capacitor has not affected the active power but has reduced the apparent power from 10 to 5.263 kVA. In some industrial situations, electric supply costs are directly related to apparent power. If this were so in the case of Example 15-9, a saving of 52.6 percent of the cost of supplying the load power could result from connection of the 430-μF capacitor. This supply cost reduction could justify the installation of the capacitor.

The connection of a reactive element in parallel with a load to raise the circuit power factor toward unity ($\cos\phi = 1$) is called *power-factor improvement.*

Power-factor improvement has been treated in Example 15-9 as an application of parallel ac circuits, the load being represented as a parallel *RL* network. This is in order, but it is more usual in practice to adopt a series view of such a load. When this is done, power-factor improvement becomes a series-parallel matter. This approach is seen in the next section.

REVIEW QUESTIONS FOR SECTION 15-2

1. In a parallel *RLC* circuit, which phasor is normally used as the reference?
2. In deriving the power equation for a parallel *RLC* circuit, it is not necessary to change the sign of the imaginary term as is done for the series *RLC* circuit. Why?
3. In a parallel *RLC* circuit in which the capacitive current is greater than the inductive current, does the circuit current lead the voltage?
4. For the conditions of Question 3, is the circuit phase angle leading?
5. When the inductive reactance of a parallel *RLC* circuit is less than the capacitive reactance, is the net reactive current a $-j$ quantity?
6. When the inductive susceptance of a parallel *RLC* circuit is less than the capacitive susceptance, is the reactive power a $-j$ quantity?
7. In parallel *RLC* circuits, reactive branch currents may be larger than the total current. How is this possible?
8. What is the phase relationship of I_C to I_L in a parallel *RLC* circuit?

EXERCISE PROBLEMS FOR SECTION 15-2

1. Using Fig. 15-15 (page 572), find the magnitudes of I_T, I_R, I_L, I_C, and the phase angle, and show that the phasor sum of I_R, I_L, and I_C equals I_T when
   a. $V = 120$ V, $R = 30\ \Omega$, $X_L = 60\ \Omega$, $X_C = 120\ \Omega$
   (*ans.* 4.123 A, 4 A, 2 A, 1 A, 14.04° lag)
   b. $V = 120$ V, $R = 60\ \Omega$, $X_L = 120\ \Omega$, $X_C = 60\ \Omega$
   (*ans.* 2.236 A, 2 A, 1 A, 2 A, 26.57° lead)
   c. $V = 1$ V, $R = 27$ kΩ, $X_L = 5$ kΩ, $X_C = 2$ kΩ
   (*ans.* 302.3 μA, 37 μA, 200 μA, 500 μA, 82.97° lead)
   d. $V = 10$ V, $R = 100$ kΩ, $X_L = 1$ kΩ, $X_C = 2$ kΩ
   (*ans.* 5.001 mA, 100 μA, 10 mA, 5 mA, 88.85° lag)
2. Repeat Prob. 1 when $V = 1$ V, $R = 40$ kΩ, $C = 500$ pF, $L = 200\ \mu$H, and
   a. $f = 400$ kHz
   b. $f = 450$ kHz
   c. $f = 500$ kHz
   d. $f = 550$ kHz
3. Using Fig. 15-15, find the magnitudes of I_T, I_L, and the phase angle, and show that the phasor sum of I_R, I_L, and I_C equals I_T when $V = 10$ mV, $R = 250$ kΩ, $L = 250\ \mu$H, $f = 1$ MHz, and C is
   a. 60 pF (*ans.* 2.596 μA, 6.365 μA, 89.12° lag)
   b. 90 pF (*ans.* 0.7101 μA, 6.365 μA, 86.77° lag)
   c. 110 pF (*ans.* 0.5474 μA, 6.365 μA, 85.81° lead)
   d. 200 pF (*ans.* 6.201 μA, 6.365 μA, 89.63° lead)
4. Using Fig. 15-15, find the magnitude of I_T and the phase angle of the circuit when $V = 120$ V, $R = 40\ \Omega$, $L = 0.15$ H, $f = 60$ Hz, and C is
   a. Open-circuited
   b. 200 μF
   c. 50 μF
   d. 20 μF
5. Find the apparent power, active power, and reactive power for each of the circuits in Prob. 4.
   (*ans.* *(a)* 441 VA, 360 W, 254.6 var *(b)* 906 VA, 360 W, 831.4 var *(c)* 360.4 VA, 360 W, 16.8 var *(d)* 388.4 VA, 360 W, 146 var)

15-3 *RLC* SERIES-PARALLEL CIRCUIT

Series-parallel *RLC* circuits are dealt with by:

1. Applying the technique of considering each subcircuit, series or parallel, and then combining the results
2. Adopting series-to-parallel (or parallel-to-series) conversion to produce a circuit which is simple series or parallel in configuration

Some examples will demonstrate how these alternative techniques may be used.

EXAMPLE 15-10 What is the total current in the circuit in Fig. 15-32?

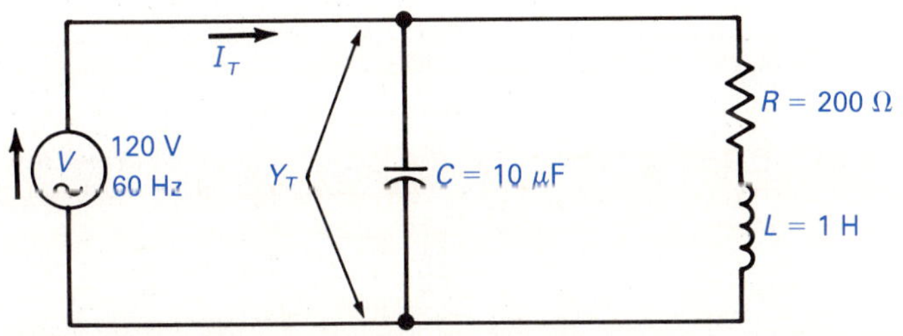

Fig. 15-32. Example 15-10: Circuit.

Solution

$$\omega = 2\pi f = 2\pi \times 60 \text{ Hz} = 377 \text{ rad/s}$$

$$X_C = \frac{1}{\omega C} = \frac{1}{377 \text{ rad/s} \times 10\ \mu\text{F}} = 265.3\ \Omega$$

$$B_C = \frac{1}{X_C} = \frac{1}{265.3\ \Omega} = 3.77 \text{ mS}$$

$$X_L = \omega L = 377 \text{ rad/s} \times 1 \text{ H} = 377\ \Omega$$

Finding the impedance of the *RL* series branch,

$$Z_{RL} = R + jX_L = (200 + j377)\ \Omega = 426.8\ \Omega\ \underline{/62.05^\circ}$$

Converting the *RL* branch impedance to its equivalent admittance,

$$Y_{RL} = \frac{1}{Z_{RL}} = \frac{1}{426.8\ \Omega\ \underline{/62.05^\circ}} = 2.343 \text{ mS}\ \underline{/-62.05^\circ}$$

$$Y_{RL} = (1.098 - j2.07) \text{ mS}$$

The total admittance Y_T of the series-parallel network can be obtained by the phasor addition of the admittances of the capacitive and inductive branches:

$$Y_T = Y_C + Y_{RL} = j3.77 \text{ mS} + (1.098 - j2.07) \text{ mS}$$
$$= [1.098 + j(3.77 - 2.07)] \text{ mS}$$
$$Y_T = (1.098 + j1.7) \text{ mS} \quad \text{or} \quad 2.024 \text{ mS} \angle 57.1°$$

The application of Ohm's law will give the total circuit current I_T:

$$I_T = VY_T = 120 \text{ V} \angle 0° \times 2.024 \text{ mS} \angle 57.1°$$
$$I_T = \mathbf{242.9 \text{ mA} \angle 57.1°} \quad \text{or} \quad \mathbf{\angle 0.9965 \text{ rad}}$$

EXAMPLE 15-11 Verify the result for Example 15-10 by series-to-parallel conversion.

Solution Figure 15-33*a* shows the series *RL* branch in Fig. 15-32 with the elements identified as series by the symbols R_s and X_{Ls}. In Fig. 15-33*b* the equivalent parallel-circuit elements have the symbols G_p and B_{Lp}. Applying the series-to-parallel relationships developed in Chap. 14,

$$G_p = \frac{R_s}{R_s^2 + X_{Ls}^2} \tag{14-44}$$

$$G_p = \frac{200 \ \Omega}{(200 \ \Omega)^2 + (377 \ \Omega)^2} = 1.098 \text{ mS}$$

$$B_{Lp} = -j \frac{X_{Ls}}{R_s^2 + X_{Ls}^2} \tag{14-45}$$

$$B_{Lp} = -j \frac{377 \ \Omega}{(200 \ \Omega)^2 + (377 \ \Omega)^2} = -j2.07 \text{ mS}$$

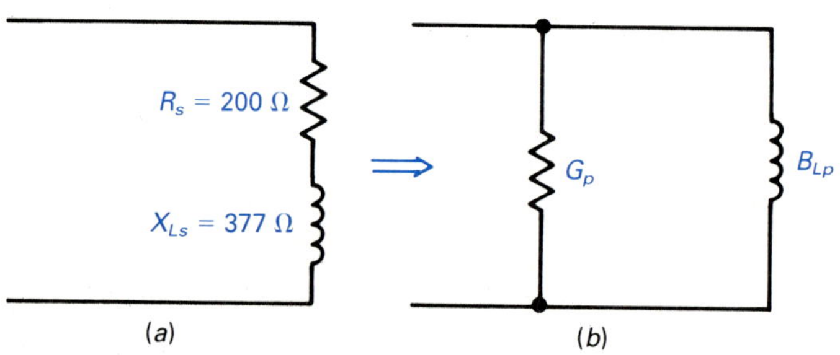

Fig. 15-33. Example 15-11: *(a)* The *RL* series circuit in Fig. 15-32; *(b)* its parallel equivalent.

The capacitive susceptance, calculated in the solution to Example 15-10, is $B_C = 3.77$ mS.

Figure 15-34 is the parallel equivalent of the series-parallel circuit in Fig. 15-32.

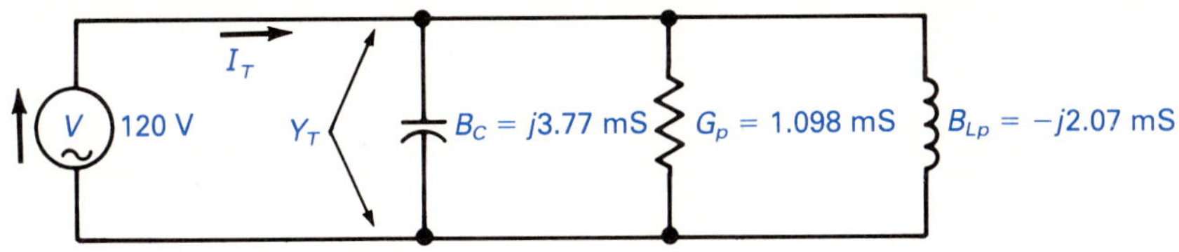

Fig. 15-34. Example 15-11: Parallel equivalent of Fig. 15-32.

The total admittance Y_T is the phasor sum of the element admittances B_C, G_p, and B_{Lp}:

$$Y_T = G_p + jB_T = G_p + j(B_C - B_L)$$
$$Y_T = 1.098 + j(3.77 - 2.07) \text{ mS} = (1.098 + j1.7) \text{ mS}$$

This expression for Y_T is the same as that obtained in Example 15-10, thus verifying the result for that example.

EXAMPLE 15-12 What is the active power in the circuit in Fig. 15-35?

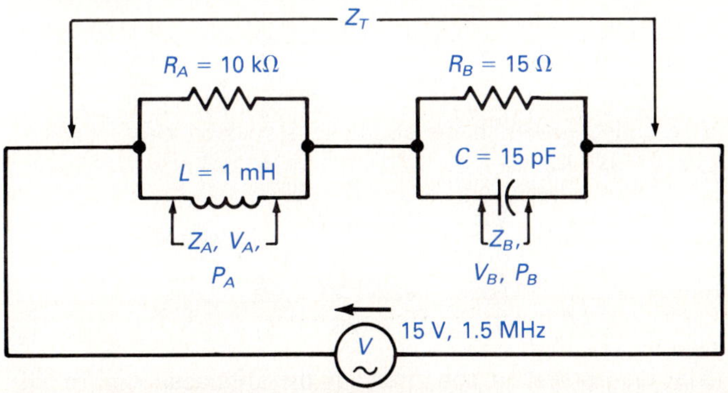

Fig. 15-35. Example 15-12: Circuit.

Solution

$$\omega = 2\pi f = 2\pi \times 1.5 \text{ MHz} = 9.425 \text{ Mrad/s}$$
$$X_L = \omega L = 9.425 \text{ Mrad/s} \times 1 \text{ mH} = 9.425 \text{ k}\Omega$$
$$X_C = \frac{1}{\omega C} = \frac{1}{9.425 \text{ Mrad/s} \times 15 \text{ pF}} = 7.073 \text{ k}\Omega$$

Calculating the impedance Z_A of the RL parallel group by applying the method for two parallel elements $[Z_T = Z_1Z_2/(Z_1 + Z_2)]$:

$$Z_A = \frac{Z_{R_A}Z_L}{Z_{R_A} + Z_L} = \frac{R_A \times jX_L}{R_A + jX_L} = \frac{jR_AX_L}{R_A + jX_L}$$

$$Z_A = \frac{j(10\ \text{k}\Omega \times 9.425\ \text{k}\Omega)}{10\ \text{k}\Omega + j9.425\ \text{k}\Omega} = \frac{94.25\ (\text{k}\Omega)^2\ \underline{/90^\circ}}{13.74\ \text{k}\Omega\ \underline{/43.3^\circ}}$$

$$Z_A = 6.86\ \text{k}\Omega\ \underline{/46.7^\circ}$$

Calculate the impedance Z_B of the *RC* parallel group:

$$Z_B = \frac{Z_{R_B} Z_C}{Z_{R_B} + Z_C} = \frac{R_B \times (-jX_C)}{R_B - jX_C} = \frac{-jR_B X_C}{R_B - jX_C}$$

$$= \frac{-j(15\ \text{k}\Omega \times 7.073\ \text{k}\Omega)}{15\ \text{k}\Omega - j7.073\ \text{k}\Omega} = \frac{106.1\ (\text{k}\Omega)^2\ \underline{/-90^\circ}}{16.58\ \text{k}\Omega\ \underline{/-25.25^\circ}}$$

$$Z_B = 6.399\ \text{k}\Omega\ \underline{/-64.75^\circ}$$

The group impedances Z_A and Z_B are in series. Therefore the total impedance Z_T is the phasor sum of Z_A and Z_B:

$$Z_T = Z_A + Z_B = 6.86\ \text{k}\Omega\ \underline{/46.7^\circ} + 6.399\ \text{k}\Omega\ \underline{/-64.75^\circ}$$

$$= (4.705 + j4.993)\ \text{k}\Omega + (2.73 - j5.788)\ \text{k}\Omega$$

$$= [(4.705 + 2.73) + j(4.993 - 5.788)]\ \text{k}\Omega$$

$$Z_T = (7.435 - j0.795)\ \text{k}\Omega \qquad \text{or} \qquad 7.477\ \text{k}\Omega\ \underline{/-6.1^\circ}$$

The voltages across each of the two parallel groups can be found by applying the voltage-divider principle:

$$V_A = V\frac{Z_A}{Z_T} = 15\ \text{V}\ \underline{/0^\circ}\ \frac{6.86\ \text{k}\Omega\ \underline{/46.7^\circ}}{7.477\ \text{k}\Omega\ \underline{/-6.1^\circ}} = 13.76\ \text{V}\ \underline{/52.8^\circ}$$

$$V_B = V\frac{Z_B}{Z_T} = 15\ \text{V}\ \underline{/0^\circ}\ \frac{6.399\ \text{k}\Omega\ \underline{/-64.75^\circ}}{7.477\ \text{k}\Omega\ \underline{/-6.1^\circ}} = 12.84\ \text{V}\ \underline{/-58.65^\circ}$$

The active power in the circuit is developed only in resistances R_A and R_B:

$$P_A = \frac{V_A{}^2}{R_A} = \frac{(13.76\ \text{V})^2}{10\ \text{k}\Omega} = 18.93\ \text{mW}$$

$$P_B = \frac{V_B{}^2}{R_B} = \frac{(12.84\ \text{V})^2}{15\ \text{k}\Omega} = 10.99\ \text{mW}$$

The total power in any circuit is the arithmetic sum of the individual powers:

$$P_T = P_A + P_B = 18.93\ \text{mW} + 10.99\ \text{mW} = \mathbf{29.92\ mW}$$

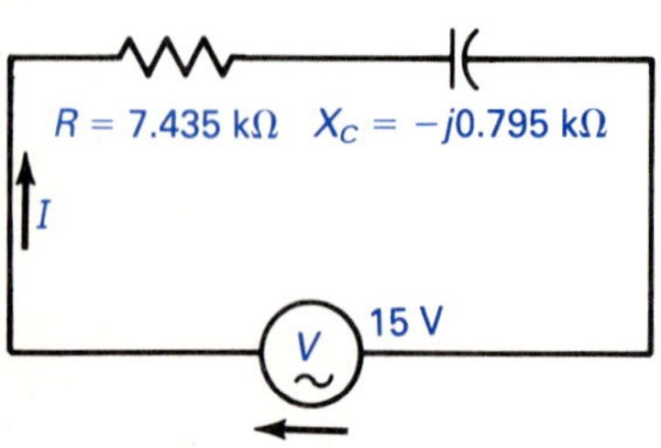

Fig. 15-36. Series circuit with the same overall characteristics as Fig. 15-35.

In Example 15-12 the active power for each parallel subnetwork was calculated and the resulting powers added to arrive at the total circuit power, as this technique is perhaps the easiest to visualize. Actually, the amount of calculation can be reduced appreciably by realizing what is implied by the rectangular expression for the total circuit impedance Z_T.

Figure 15-36 is a series circuit with impedance $Z_T = 7.435 - j0.795$ kΩ, the same as the impedance of the series-parallel circuit in Example 15-12. The power dissipated by the 7.435-kΩ resistance in Fig. 15-36 is the same as the total active power in Example 15-12:

$$|I| = \frac{V}{Z_T} = \frac{15\text{ V}}{7.477\text{ k}\Omega} = 2.006\text{ mA}$$

$$P_T = |I|^2 R = (2.006\text{ mA})^2 \times 7.435\text{ k}\Omega = \mathbf{29.92\text{ mW}}$$

EXAMPLE 15-13 A noise filter circuit in an audio amplifier is shown in Fig. 15-37. When the switch S is open, high audio frequencies are reduced in level relative to when the switch is closed. What is the magnitude of the ratio V_o/V_i at a frequency of 14 kHz when (*a*) S is open and (*b*) S is closed?

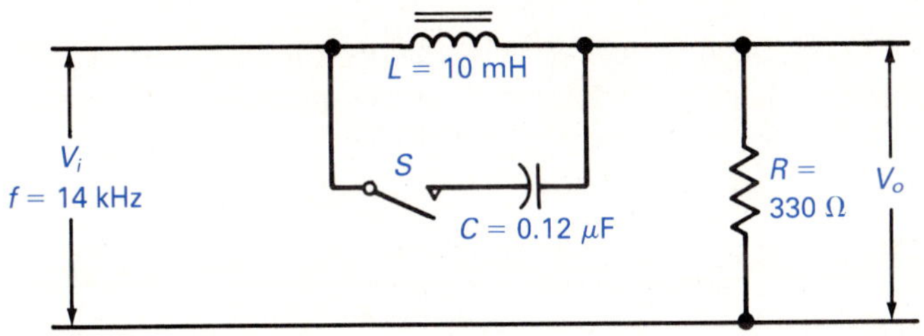

Fig. 15-37. Example 15-13: Circuit.

Solution *a.* When S is open, the equivalent circuit of Fig. 15-37 is that in Fig. 15-38.

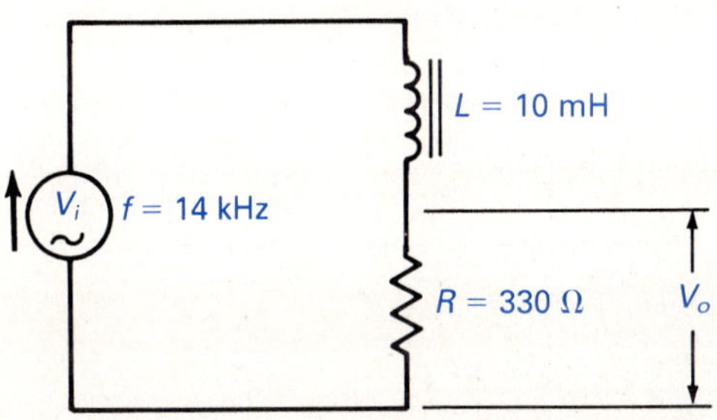

Fig. 15-38. Example 15-13: Equivalent of Fig. 15-37 when the switch is open.

$$\omega = 2\pi f = 2\pi \times 14\text{ kHz} = 87.96\text{ krad/s}$$

$$X_L = \omega L = 87.96\text{ krad/s} \times 10\text{ mH} = 879.6\ \Omega$$

The impedance of the series RL network is

$$Z = R + jX_L = (330 + j879.6)\ \Omega \qquad \text{or} \qquad 939.5\ \Omega\ \underline{/69.42^\circ}$$

The magnitude of the voltage V_o can be obtained by applying the voltage-divider principle:

$$V_o = V_i \frac{R}{Z} = V_i \frac{330\ \Omega}{939.5\ \Omega} = 0.3513\ V_i$$

Therefore, when the switch is open, the ratio V_o/V_i is

$$\frac{V_o}{V_i} = \mathbf{0.3513}$$

b. When the switch is closed, the equivalent circuit of Fig. 15-37 is that in Fig. 15-39.

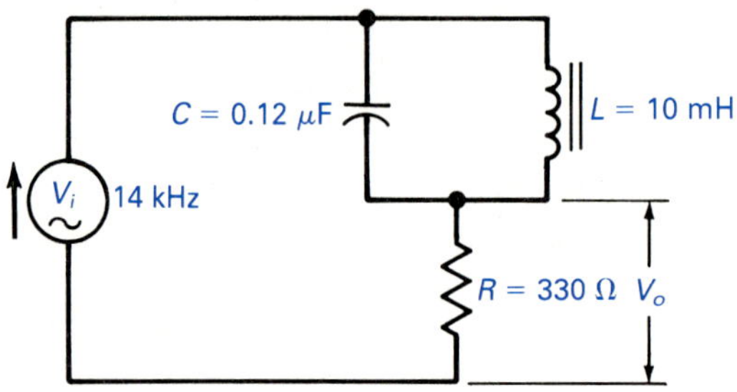

Fig. 15-39. Example 15-13: Equivalent of Fig. 15-37 when the switch is closed.

$$\omega = 87.96 \text{ krad/s}$$
$$X_L = 879.6\ \Omega$$
$$X_C = \frac{1}{\omega C} = \frac{1}{87.96 \text{ krad/s} \times 0.12\ \mu\text{F}} = 94.74\ \Omega$$

The impedance of the *LC* parallel circuit is

$$Z_{LC} = \frac{Z_L Z_C}{Z_L + Z_C} = \frac{jX_L(-jX_C)}{j(X_L - X_C)} = \frac{-j^2 X_L X_C}{j(X_L - X_C)} = -j\frac{X_L X_C}{X_L - X_C}$$

$$Z_{LC} = -j\frac{879.6\ \Omega \times 94.74\ \Omega}{879.6\ \Omega - 94.74\ \Omega} = -j106.2\ \Omega$$

The total impedance of the series-parallel network is

$$Z_T = Z_R + Z_{LC} = (330 - j106.2)\ \Omega \qquad \text{or} \qquad 346.7\ \Omega\ \underline{/-17.85^\circ}$$

The magnitude of the voltage V_o obtained by applying the voltage-divider principle is

$$V_o = V_i \frac{R}{Z_T} = V_i \frac{330\ \Omega}{346.7\ \Omega} = 0.9518\ V_i$$

Thus, when the switch is closed, the ratio V_o/V_i is

$$\frac{V_o}{V_i} = \mathbf{0.9518}$$

The results for Example 15-13 show that, with the circuit in Fig. 15-38, substantially less of the voltage V_i is transmitted through the circuit when the switch is open than when it is closed. Much of the irritating background noise in a poor-quality recording or radio transmission occurs at audio frequencies higher than about 12 kHz. A simple ac circuit, such as that in Fig. 15-38, can be used to reduce the level of this type of noise. The switch would be opened only when the background noise is objectionable. At other times the higher audio frequencies would be transmitted for good sound fidelity.

Power-Factor Improvement

A load on an electric supply system is usually inductive, its power factor being less than unity and lagging. This means that the apparent power is greater than the active power. For a given supply voltage the current drawn from the supply is greater than the current needed for the same useful power at unity power factor. On a domestic load the heavy load components, such as the kitchen stove and the clothes dryer, are resistive, so that, although the total current can be about 100 A, the power factor is close to unity. However, an industrial load may consist mainly of electric motors, fluorescent lights, and other equipment with a low power factor. Also the total current drawn by an industrial load may be several hundreds or even thousands of amperes. In such a situation the electric supply company may base its rates on the maximum current drawn by the load, since this determines the distribution transformer and conductor sizes. The result is that the cost of installing equipment which will raise the power factor toward unity may be more than offset by the savings in supply costs. Installation of equipment to increase the power factor is called *power-factor improvement* or *power-factor correction*. The additional equipment must have a leading power factor to oppose the lagging power factor of the load; it must be capacitive. Power-factor improvement in no way affects the performance characteristics of the original load. It is still connected across the supply, so its apparent, active, and reactive powers are not changed. Only the effect on the electric supply is changed.

EXAMPLE 15-14 An industrial load draws 500 A from a 440-V 60-Hz supply at a power factor of 0.6 lagging. What is *(a)* the reactive power of a capacitor which will raise the circuit power factor to 0.9 lagging, *(b)* the capacitor's current, and *(c)* its capacitance?

Solution (See Fig. 15-40.)

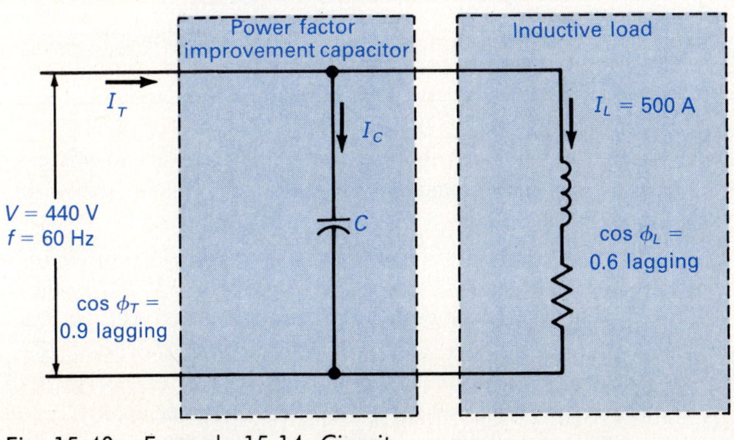

Fig. 15-40. Example 15-14: Circuit.

Consider the load alone.

The apparent power of the load is

$$S_L = VI_L = 440\text{ V} \times 500\text{ A} = 220\text{ kVA}$$

The active power of the load is

$$P_L = S_L \cos \phi_L = 220\text{ kVA} \times 0.6 = 132\text{ kW}$$

The reactive power of the load is

$$Q_L = \sqrt{S_L^2 - P_L^2} = \sqrt{(220\text{ kVA})^2 - (132\text{ kW})^2} = 176\text{ kvar}$$

Consider the load together with the power-factor-improvement capacitor.

The active power will be unchanged:

$$P_T = P_L = 132\text{ kW}$$

The apparent power of the complete circuit will be

$$S_T = \frac{P_T}{\cos \phi_T} = \frac{132\text{ kW}}{0.9} = 146.7\text{ kVA}$$

The reactive power of the complete circuit will be

$$Q_T = \sqrt{S_T^2 - P_T^2} = \sqrt{(146.7\text{ kVA})^2 - (132\text{ kW})^2} = 63.9\text{ kvar}$$

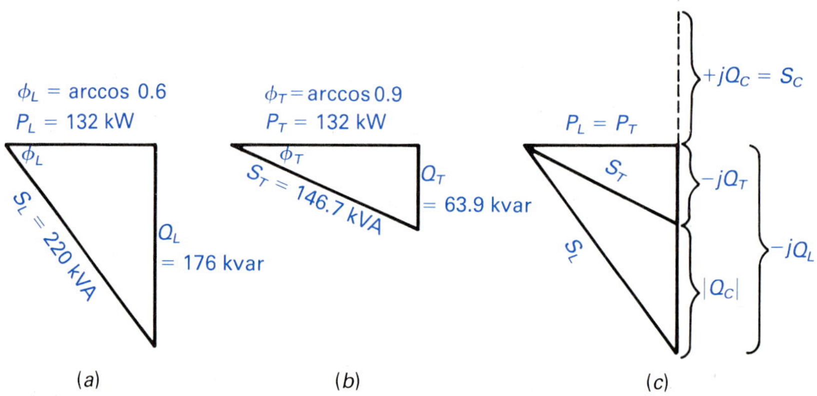

Fig. 15-41. Example 15-14: *(a)* Power triangle for the load alone; *(b)* power triangle for the complete circuit; *(c)* power triangles superimposed.

Figure 15-41 gives the power triangles relevant to this example. The power triangle for the load alone is shown in Fig. 15-41*a*, and that for the complete circuit in Fig. 15-41*b*. The power triangles in Fig. 15-41*a* and *b* are superimposed in Fig. 15-41*c*.

a. Q_L and Q_T are the reactive powers associated with the lagging power factors of the load and the complete circuit. Q_C, shown by the broken line in Fig. 15-41*c*, is the reactive power of the power-factor-improvement capacitor.

The relationship between the reactive powers is

$$Q_C = Q_L - Q_T = 176 \text{ kvar} - 63.9 \text{ kvar} = \mathbf{112.1\ kvar}$$

b. The capacitor is connected across the 440-V supply and dissipates no power:

$$S_C = Q_C = 112.1 \text{ kVA}$$
$$S_C = VI_C$$
$$I_C = \frac{S_C}{V} = \frac{112.1 \text{ kVA}}{440 \text{ V}} = \mathbf{254.8\ A}$$

c. The reactance of the capacitor is the ratio of its voltage and its current:

$$X_C = \frac{V}{I_C} = \frac{440 \text{ V}}{254.8 \text{ A}} = 1.727\ \Omega$$
$$X_C = \frac{1}{\omega C}$$
$$C = \frac{1}{\omega X_C} = \frac{1}{2\pi \times 60 \text{ Hz} \times 1.727\ \Omega} = \mathbf{1536\ \mu F}$$

It can be seen from the results for Example 15-14 that capacitors used for power-factor improvement in industrial situations are large power devices. A capacitor of 1536 μF capacitance with insulation to withstand a peak voltage of 623 V and capable of continuously passing a current of 254 A is a large, expensive device. But over a period of time the total cost saving produced by reducing the reactive power by 112.1 kvar can far exceed the cost of the capacitor.

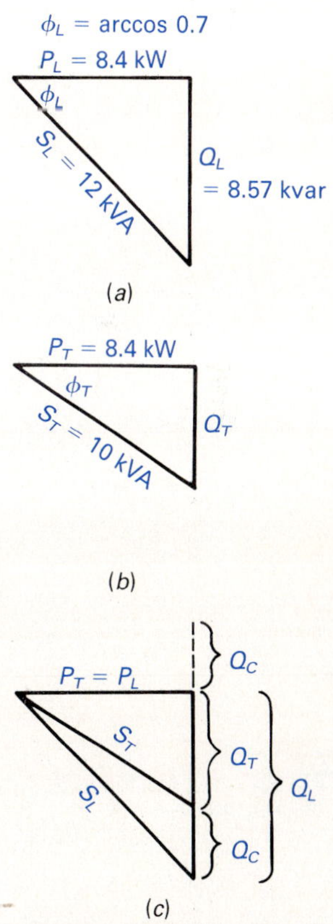

Fig. 15-42. Example 15-15: *(a)* Power triangle for the load alone; *(b)* power triangle for the complete circuit; *(c)* power triangles superimposed.

EXAMPLE 15-15 The standby power plant for a microwave radio relay station, capable of supplying 10 kVA at 220 V 60 Hz, was adequate for the station prior to the installation of additional equipment. With the new equipment installed, the load is increased to 12 kVA at a power factor of 0.7 lagging. This is too great a load for the power plant. What is the minimum capacitor rating required to enable the power plant to handle the increased load?

Solution For the relay station without the capacitor, refer to Fig. 15-42*a*:

$$S_L = 12 \text{ kVA}$$
$$P_L = S_L \cos \phi_L = 12 \text{ kVA} \times 0.7 = 8.4 \text{ kW}$$
$$Q_L = \sqrt{S_L^2 - P_L^2} = \sqrt{(12 \text{ kVA})^2 - (8.4 \text{ kW})^2} = 8.57 \text{ kvar}$$

Figure 15-42*a* shows the power triangle for the relay station load. Figure 15-42*b* gives the required power triangle with the total apparent power S_T equal to 10 kVA, the power plant maximum. The total active power P_T is the same as the load power P_L:

$$P_T = P_L = 8.4 \text{ kW}$$

The total permissible reactive power without overloading the power plant is

$$Q_T = \sqrt{S_T^2 - P_T^2} = \sqrt{(10\ \text{kVA})^2 - (8.4\ \text{kW})^2} = 5.43\ \text{kvar}$$

The capacitor must supply a correcting influence equivalent to the difference between the reactive power of the load Q_L and the total permissible reactive power Q_T:

$$Q_C = Q_L - Q_T = 8.57\ \text{kvar} - 5.43\ \text{kvar} = 3.14\ \text{kvar}$$

The current in the capacitor is

$$I_C = \frac{Q_C}{V} = \frac{3.14\ \text{kvar}}{220\ \text{V}} = 14.27\ \text{A}$$

$$X_C = \frac{V}{I_C} = \frac{220\ \text{V}}{14.27\ \text{A}} = 15.42\ \Omega$$

$$C = \frac{1}{\omega X_C} = \frac{1}{377\ \text{rad/s} \times 15.42\ \Omega} = 172\ \mu\text{F}$$

Thus the capacitor's minimum ratings are **172 μF** capacitance at a peak voltage of **311 V** and a reactive power of **3.14 kvar.**

EXERCISE PROBLEMS FOR SECTION 15-3

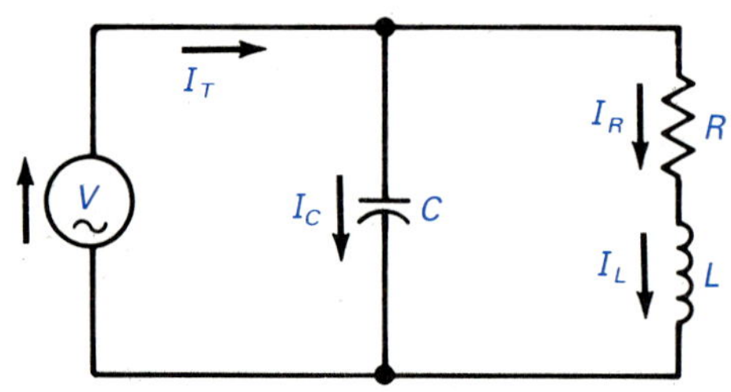

Fig. 15-43. Problems 1 to 4: Circuit.

1. Find the magnitude of I_T, I_C, I_R, I_L, and the phase angle for the circuit in Fig. 15-43 when
   a. $V = 120$ V, $R = 30\ \Omega$, $X_L = 60\ \Omega$, $X_C = 120\ \Omega$
   (*ans.* 1 A, 1 A, 1.789 A, 1.789 A, 36.87° lag)
   b. $V = 120$ V, $R = 60\ \Omega$, $X_L = 60\ \Omega$, $X_C = 120\ \Omega$
   (*ans.* 1 A, 1 A, 1.414 A, 1.414 A, 0°)
   c. $V = 1$ V, $R = 10\ \Omega$, $X_L = 200\ \Omega$, $X_C = 200\ \Omega$
   (*ans.* 249.7 μA, 5 mA, 4.994 mA, 4.994 mA, 2.86° lead)
   d. $V = 1$ V, $R = 1\ \Omega$, $X_L = 200\ \Omega$, $X_C = 200\ \Omega$
   (*ans.* 25 μA, 5 mA, 5 mA, 5 mA, 0.286° lead)
2. Find the magnitude of I_T and I_L in Fig. 15-43 when $V = 10$ mV, $R = 10\ \Omega$, $C = 400$ pF, $L = 350\ \mu$H, and
   a. $f = 400$ kHz *c.* $f = 500$ kHz
   b. $f = 450$ kHz *d.* $f = 550$ kHz

3. Find the magnitude of I_T and I_L in Fig. 15-43 when $V = 10$ V, $R = 10\ \Omega$, $L = 50\ \mu$H, $f = 2$ MHz, and C is
a. 100 pF (*ans*. 3.35 mA, 15.91 mA)
b. 120 pF (*ans*. 867 μA, 15.91 mA)
c. 140 pF (*ans*. 1.7 mA, 15.91 mA)
d. 160 pF (*ans*. 4.2 mA, 15.91 mA)

4. Find the magnitude of I_T, I_L, S, P, and the circuit phase angle in Fig. 15-43 when $V = 120$ V, $R = 24\ \Omega$, $L = 0.1$ H, $f = 60$ Hz, and C is
a. Not connected *c*. 40 μF
b. 20 μF *d*. 60 μF

5. Find the magnitude of I_T, S, and P for the circuit shown in Fig. 15-44 when $V = 120$ V, $f = 60$ Hz, and
a. $R_1 = 5\ \Omega$, $C = 100\ \mu$F, $R_2 = 10\ \Omega$, $L = 0.2$ H
(*ans*. 2.987 A, 358.5 VA, 123.7 W)
b. $R_1 = 5\ \Omega$, $C = 60\ \mu$F, $R_2 = 10\ \Omega$, $L = 0.2$ H
(*ans*. 1.228 A, 147.26 VA, 61.34 W)
c. $R_1 = 5\ \Omega$, $C = 30\ \mu$F, $R_2 = 10\ \Omega$, $L = 0.2$ H
(*ans*. 0.354 A, 42.48 VA, 34.13 W)

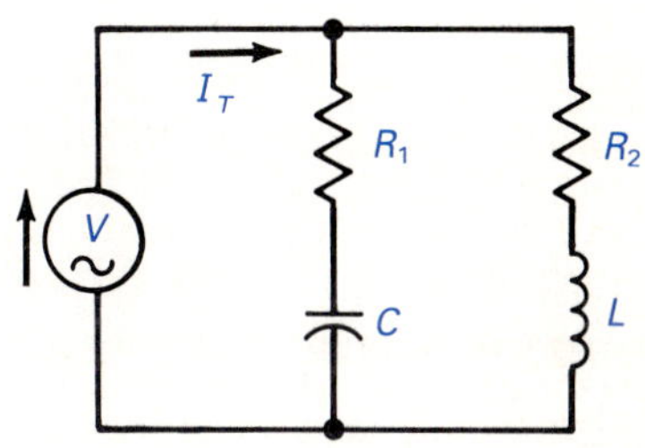

Fig. 15-44. Problem 5: Circuit.

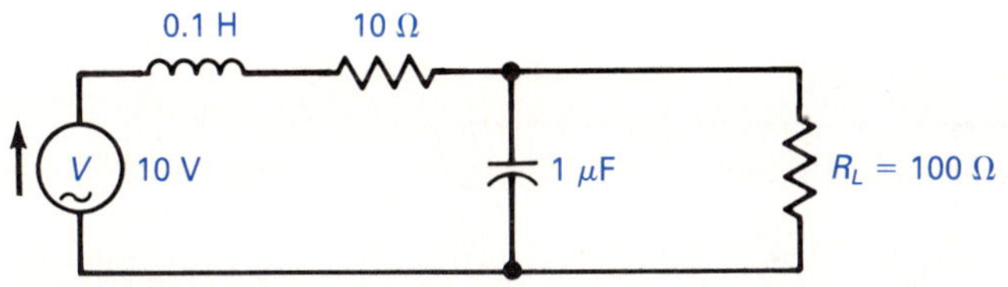

Fig. 15-45. Problem 6.

6. The *RLC* network in Fig. 15-45 is a low-pass filter. Find the voltage across R_L when the frequency of the generator is
a. 500 Hz
b. 1000 Hz
c. 1500 Hz

7. A 120-V 60-Hz electric motor has an effective resistance of 58 Ω in series with an effective inductance of 0.175 H.
a. Find the power factor, the line current, and the apparent power.
(*ans*. 0.66 lag, 1.366 A, 163.9 VA)
b. Find the power factor, the line current, and the apparent power if a capacitor of 50 μF is connected across the motor.
(*ans*. 0.589 lead, 1.53 A, 183.6 VA)

8. A 240-V 60-Hz electric motor has an output power of 1 hp, is 75 percent efficient, and has a power factor of 0.6 lagging. What value of capacitance connected across the motor will change the power factor to 0.95 lagging?

9. If the motor described in Prob. 8 has its power factor changed to 0.95 lagging by a series capacitor, find
a. The value of the capacitor. (*ans*. 126.7 μF)
b. The line current. (*ans*. 10.94 A)
c. The voltage across the capacitor. (*ans*. 229 V)
d. The voltage across the motor. (*ans*. 380 V)

10. The tank circuit of a radio-frequency amplifier contains a parallel *RLC* circuit. One branch of the circuit contains a 100-pF capacitor, the other branch contains a 150-μH inductance with an effective series resistance of 12 Ω. Find the impedance of the parallel circuit when f equals

a. 1280 kHz

b. 1305 kHz

11. If the emf across the circuit in Prob. 10 is 20 V at 1280 kHz, find

a. The line current. (*ans*. 520 μA)

b. The current in the inductor. (*ans*. 16.58 mA)

12. If a 120-kΩ resistor is shunted across the circuit in Prob. 10 when the voltage is 20 V at 1280 kHz, find

a. The line current.

b. The current in the inductor.

13. A 600-W load with a power factor of 0.65 lagging is connected in parallel with a 500-W resistive load across a 120-V 60-Hz line. Find

a. The total line current. (*ans*. 10.87 A)

b. The circuit phase angle. (*ans*. 32.53° lag)

14. If a 100-μF capacitor is connected across the circuit in Prob. 13, find

a. The new line current.

b. The new phase angle.

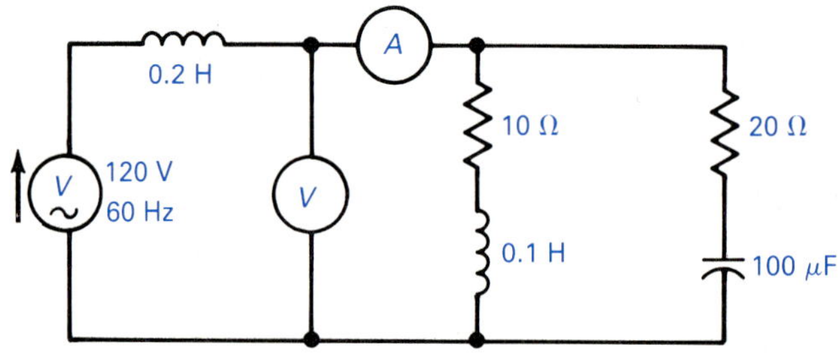

Fig. 15-46. Problems 15 and 16: Circuit.

15. Find the current and voltage which would be indicated by the meters in Fig. 15-46. (*ans*. 1.385 A, 56.07 V)

16. Find the power that would be indicated by a wattmeter if its current coil and potential coil replace the ammeter and voltmeter in Fig. 15-46.

17. A 240-V 60-Hz motor with an output of 1 hp (746 W) has an efficiency of 85 percent and a power factor of 0.75 lagging. Find the value of the capacitor which will correct the power factor to unity. (*ans*. 35.6 μF)

18. If the power-factor correction in Prob. 17 was made incorrectly by placing a capacitor in series so that unity power factor was obtained, find

a. The current through the motor.

b. The voltage across the motor.

SUMMARY OF EQUATIONS

$$V_X = jV_L - jV_C \text{ or } j(V_L - V_C) \qquad (15\text{-}1)$$

$$I_X = jI_C - jI_L \text{ or } j(I_C - I_L) \qquad (15\text{-}2)$$

CHAPTER QUESTIONS

15-1. In what major way do the phasors of a series *RLC* circuit differ from those of series *RL* or *RC* circuits?

15-2. In what major way do the phasors of a parallel *RLC* circuit differ from those of parallel *RL* or *RC* circuits?

15-3. What is the result of two antiphase phasors reacting?

15-4. What is the name given to the condition when two antiphase phasors have the same numerical value?

15-5. In a series *RLC* circuit, is the circuit phase angle leading or lagging if the inductive voltage has a greater magnitude than the capacitive voltage?

15-6. In a parallel *RLC* circuit, is the circuit phase angle leading or lagging if the inductive current has a greater magnitude than the capacitive current?

15-7. What is the effect of frequency change in a series *RLC* circuit?

15-8. What is the effect of frequency change in a parallel *RLC* circuit?

15-9. What is the effect of resistance change in a series *RLC* circuit?

15-10. What is the effect of conductance change in a parallel *RLC* circuit?

15-11. How can a circuit have its low-value lagging power factor raised?

15-12. When a component is added to a circuit to improve its power factor, why is such a component not connected in series with the circuit?

15-13. Could a circuit be modified so that its power factor changes from lagging to leading?

CHAPTER PROBLEMS

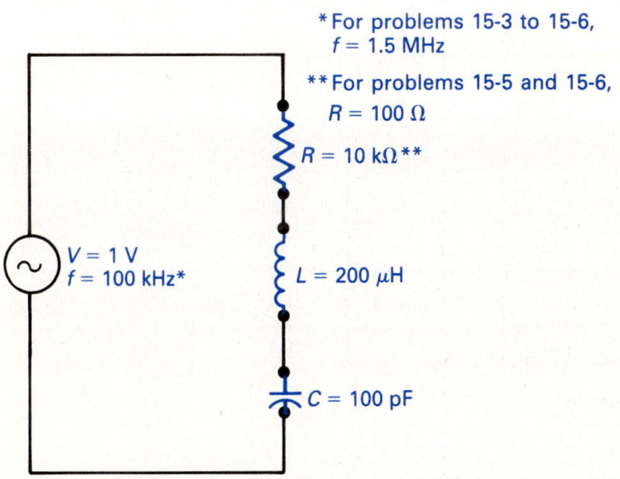

Fig. 15-47. Problems 15-1 to 15-6: Circuit.

15-1. A series ac circuit is shown in Fig. 15-47. What is the magnitude of its
a. Current? (*ans*. 53.49 μA)
b. Resistive voltage? (*ans*. 534.9 mV)

c. Inductive voltage? (*ans*. 6.724 mV)
d. Capacitive voltage? (*ans*. 851.6 mV)
e. Net reactive voltage? (*ans*. 844.8 mV)

15-2. Express each voltage in Prob. 15-1 as a percentage of the 1-V supply voltage. Identify any that exceed 1 V.

15-3. In the circuit of Fig. 15-47, the frequency is raised to 1.5 MHz. What is the magnitude of the circuit's
a. Current? (*ans*. 99.66 μA)
b. Resistive voltage? (*ans*. 996.6 mV)
c. Inductive voltage? (*ans*. 187.9 mV)
d. Capacitive voltage? (*ans*. 105.7 mV)
e. Net reactive voltage? (*ans*. 82.2 mV)

15-4. Express each voltage in Prob. 15-3 as a percentage of the 1-V supply voltage. Identify any that exceed 1 V.

15-5. The resistance of the circuit of Fig. 15-47 is reduced to 100 Ω while the frequency is at 1.5 MHz. What is the magnitude of the circuit's
a. Current? (*ans*. 1.205 mA)
b. Resistive voltage? (*ans*. 120.5 mV)
c. Inductive voltage? (*ans*. 2.271 V)
d. Capacitive voltage? (*ans*. 1.278 V)
e. Net reactive voltage? (*ans*. 993.7 mV)

15-6. Express each voltage in Prob. 15-5 as a percentage of the 1-V supply voltage. Identify any that exceed 1 V.

15-7. A parallel ac circuit is shown in Fig. 15-48. What is the magnitude of its
a. Voltage? (*ans*. 78.49 mV)
b. Resistive current? (*ans*. 784.9 μA)
c. Inductive current? (*ans*. 624.6 μA)
d. Capacitive current? (*ans*. 4.932 μA)
e. Net susceptive current? (*ans*. 619.7 μA)

15-8. Express each current in Prob. 15-7 as a percentage of the 1-mA supply current. Identify any that exceed 1 mA.

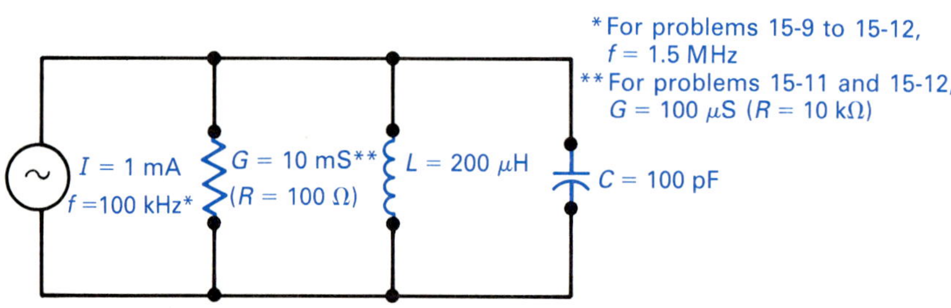

Fig. 15-48. Problems 15-7 to 15-12: Circuit.

15-9. In the circuit of Fig. 15-48, the frequency is raised to 1.5 MHz. What is the magnitude of the circuit's
a. Voltage? (*ans*. 99.9 mV)
b. Resistive current? (*ans*. 999 μA)
c. Inductive current? (*ans*. 53 μA)

d. Capacitive current? (*ans*. 94.16 μA)
e. Net susceptive current? (*ans*. 41.16 μA)

15-10. Express each current in Prob. 15-9 as a percentage of the 1-mA supply current. Identify any that exceed 1 mA.

15-11. The conductance of the circuit of Fig. 15-48 is reduced to 100 μS ($R = 10$ kΩ) while the frequency is at 1.5 MHz. What is the magnitude of the circuit's

a. Voltage? (*ans*. 2.358 V)
b. Resistive current? (*ans*. 235.8 μA)
c. Inductive current? (*ans*. 1.251 mA)
d. Capacitive current? (*ans*. 2.223 mA)
e. Net susceptive current? (*ans*. 971.7 μA)

15-12. Express each current in Prob. 15-11 as a percentage of the 1-mA supply current. Identify any that exceed 1 mA.

15-13. What are the phase angles of the circuits of Figs. 15-47 and 15-48 as applied to Probs. 15-1 to 15-11?
(*ans*. in rads: 1.006, −0.00229, −1.45, −0.66, 0.041, 1.333 or in degrees: 57.66, −4.715, −83.08, −38.29, 2.35, 76.36)

15-14. In Prob. 14-10, an electric motor was described as operating on 240 V 60 Hz with a rating of 5 hp, an efficiency of 70 percent, and a lagging power factor of 0.6. What would be required to raise the power factor of the equivalent circuit to 0.9 lagging?

16

Resonant Circuits

The *RLC* circuits dealt with in Chap. 15 all had one characteristic in common: they had net reactance. Each of those circuits could be reduced to an equivalent *RL* or *RC* network as a result of partial cancellation of reactive effects. What if the circuit values were such that the inductive and capacitive effects completely canceled? The resulting equivalent circuit would be entirely resistive in character, and the power factor would be unity. This condition of unity power factor in a circuit containing R, L, and C is called *resonance,* the circuit which results in resonance being a *resonant circuit*.

Resonance is very important since many electric and electronic systems are dependent on resonant circuits for their operation. For example, the radio-frequency spectrum is filled with countless transmissions. It is the resonant circuits that do the selecting of the wanted signals and the rejecting of the unwanted signals.

Like all electric circuit characteristics, resonance can cause problems in some circumstances. For instance, resonant effects can cause excessive voltages and currents, resulting in the breakdown of wiring and equipment.

Important resonant circuit concepts are introduced by concentrating on the cancellation of reactive effects in hypothetical, loss-free circuits. This treatment lays the foundation for dealing with practical resonant circuits.

16-1 RESONANT CIRCUITS WITHOUT RESISTANCE

Practical inductors and capacitors are imperfect; they dissipate power, and they have resistance. Therefore, practical resonant circuits, which must contain inductors and capacitors, have resistance. Considering the effects of this resistance tends to obscure some of the important basic concepts involved in resonance. An initial view of resonant circuits without resistance can be helpful, even though such circuits are not available in practice.

Series *LC* Circuit

Figure 16-1 shows a circuit in which an inductance of 100 μH and a capacitance of 250 pF are connected in series across a current source of variable frequency.

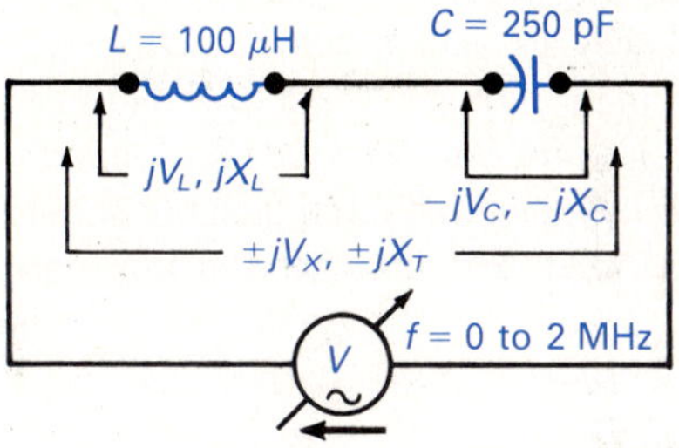

Fig. 16-1. Series *LC* circuit.

The reactance of the inductance X_L is directly proportional to the frequency, the capacitive reactance X_C being inversely proportional to the frequency. Plots of these reactances against frequency are given in Fig. 16-2*a*.

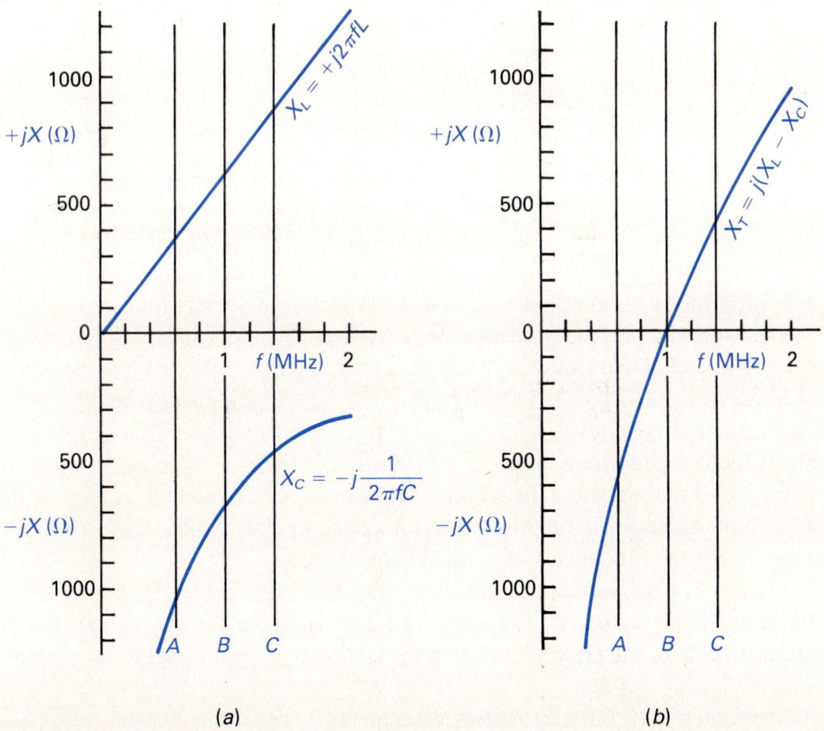

Fig. 16-2. Series *LC* circuit reactance-frequency curves. *(a)* Individual elements; *(b)* total circuit.

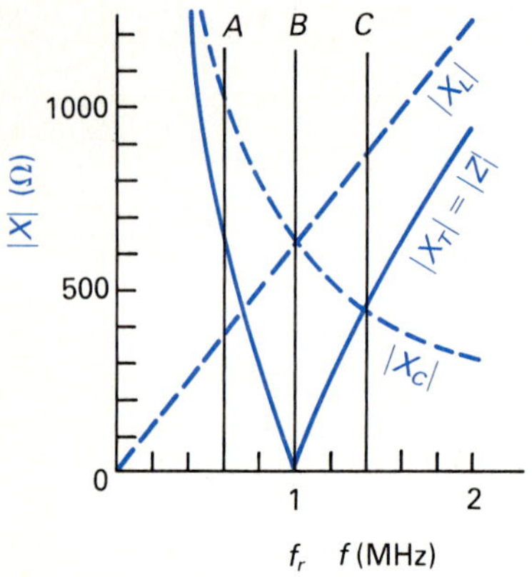

Fig. 16-3. Series *LC* circuit reactance magnitude–frequency curves.

A curve for total reactance X_T can be obtained by adding the inductive and capacitive reactance curves in Fig. 16-2*a*. The algebraic sums of the values of X_L and X_C at each frequency are plotted. At frequency *A* (0.6 MHz), $-jX_C$ is greater than $+jX_L$, so the total reactance is a $-j$ quantity of magnitude $X_C - X_L$. At frequency *B* ($\simeq$1 MHz), $-jX_C$ equals $+jX_L$, resulting in a total reactance of zero. At frequency *C* (1.4 MHz), $+jX_L$ is greater than $-jX_C$, giving a total reactance X_T which is a $+j$ quantity of magnitude $X_L - X_C$. These values of total reactance are shown in Fig. 16-2*b*, the curve for X_T for all frequencies from 0 to 2 MHz.

Figure 16-2 shows the magnitude and the phase (in terms of j sign) of the reactances X_L, X_C, and X_T. An appreciation of circuit behavior can be obtained by separating the magnitude and phase characteristics.

In Fig. 16-3 the magnitudes of the reactances are shown with respect to frequency, the broken lines depicting the individual reactances and the solid lines their total (or the total impedance Z). The values for Fig. 16-3 are taken directly from Fig. 16-2, observing the principle that the magnitude of a quantity shown graphically is represented by the distance between the curve and the zero line. This principle applies to both negative and positive quantities. It will be seen from Fig. 16-3 that the *magnitude of the total reactance X_T of a series resonant circuit decreases toward zero as the frequency increases from zero to the resonant frequency f_r. The total reactance then increases as the frequency rises above resonance.*

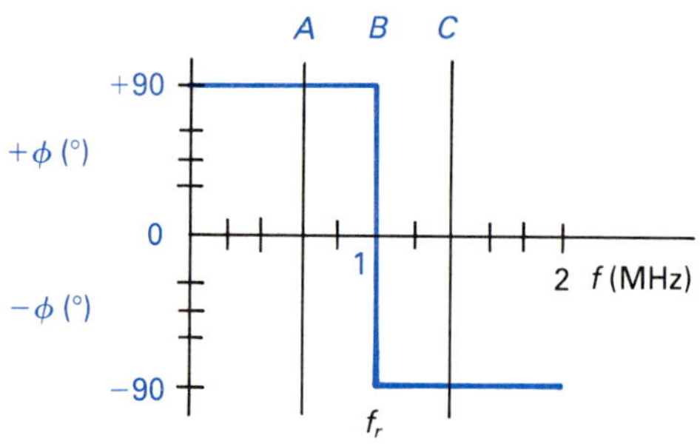

Fig. 16-4. Phase angle–frequency curve for a series *LC* circuit (no resistance).

The variation in circuit phase angle with respect to frequency for a resistanceless series resonant circuit is given in Fig. 16-4. Below the resonant frequency f_r the circuit is capacitive ($X_C > X_L$) and the phase angle is +90°. Above the resonant frequency the circuit is inductive ($X_C < X_L$), the phase angle being −90°. At the resonant frequency the circuit is in transition between being capacitive and being inductive; the phase angle is in transition between +90° and −90°.

In Fig. 16-3 the reactance magnitudes apply to all series resonant circuits, with and without resistance. But Fig. 16-4 for phase angles applies only to series resonant circuits without resistance or energy losses.

In Fig. 16-5, phasor diagrams are shown for the circuit in Fig. 16-1 in relation to the frequencies *A*, *B*, and *C* in Figs. 16-2 to 16-4. As the source supplies a constant current to the circuit, each voltage is directly proportional to the related reactance X_L, X_C, or X_T.

At frequency *A* (0.6 MHz), below resonance, $-jV_C$ is greater than $+jV_L$, so that the total circuit voltage V_T is equal to $-j(V_C - V_L)$ and lags the current by 90°. These conditions are shown in Fig. 16-5*a*.

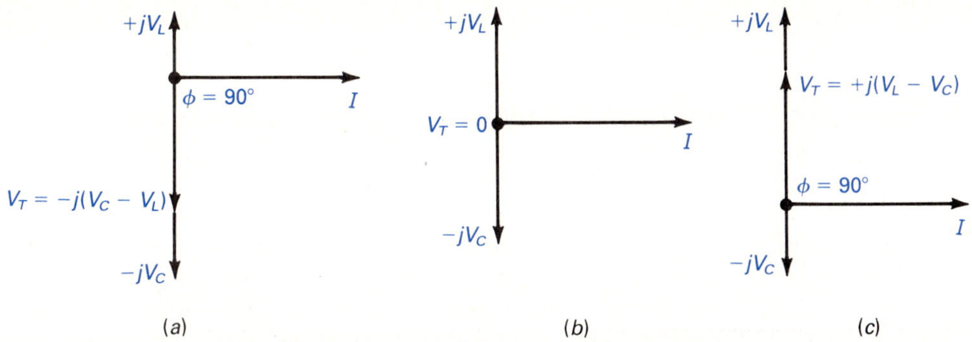

Fig. 16-5. Series *LC* circuit phasor diagrams: *(a)* Below resonance; *(b)* at resonance; *(c)* above resonance.

At frequency B (1 MHz), the resonant frequency f_r, V_L and V_C have the same magnitude. The magnitude of V_T is the difference between V_L and V_C and is thus zero at this frequency. Since there is no voltage, there is no circuit phase angle. Figure 16-5*b* shows this situation.

At frequency C (1.4 MHz), above resonance, $+jV_L$ is greater than $-jV_C$; V_T is equal to $+j(V_L - V_C)$ and leads the current by 90°, as shown in Fig. 16-5*c*.

The reason for choosing a constant-current source for the series resonant circuit in Fig. 16-1, rather than a constant-voltage source (a more usual choice for explaining series circuit action), can be seen in the descriptions of the phasor diagrams of Fig. 16-5. At resonance the total circuit voltage is zero, while at all other frequencies it has a finite value. This voltage variation renders a constant-voltage source meaningless for this convenient voltage approach to the behavior of the series resonant circuit.

The circuit conditions existing in a series *LC* circuit at the *resonant frequency* f_r can be summarized:

1. The magnitudes of the inductive reactance X_L and the capacitive reactance X_C are equal.
2. The total reactance X_T is zero.
3. There is no resistance, so that the total impedance Z equals X_T and is therefore zero; their polarities are opposite.
4. The magnitudes of the inductive voltage V_L and the capacitive voltage V_C are equal.
5. The total circuit voltage V_T is zero.
6. The ratio of either reactive voltage to the total voltage (V_L/V_T or V_C/V_T) is infinite. In other words, either of the reactive voltages is infinitely greater than the total circuit voltage.

At frequencies *below* the resonant frequency, the following conditions exist:

1. The circuit is predominantly capacitive.
2. The phase angle is +90°.

Conditions existing at frequencies *above* the resonant frequency are:

1. The circuit is predominantly inductive.
2. The phase angle is −90°.

Parallel *LC* Circuit

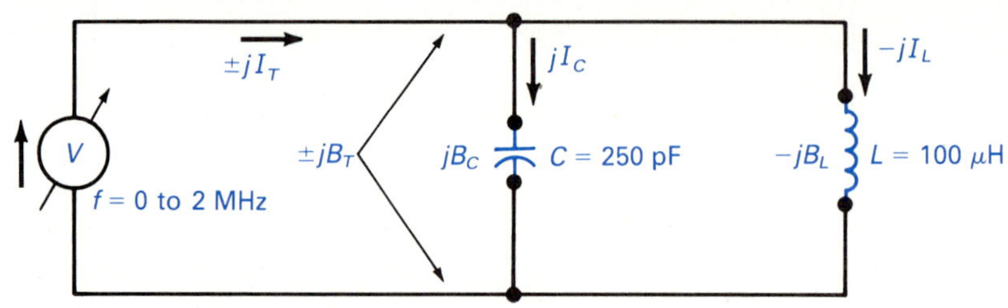

Fig. 16-6. Parallel *LC* circuit.

In Fig. 16-6 an inductor of 100 μH and a capacitor of 250 pF are shown connected in parallel across a voltage source of variable frequency. The capacitive susceptance B_C is directly proportional to the frequency, and the inductive susceptance B_L is inversely proportional to the frequency. Curves showing how these susceptances vary with respect to frequency are given in Fig. 16-7*a*.

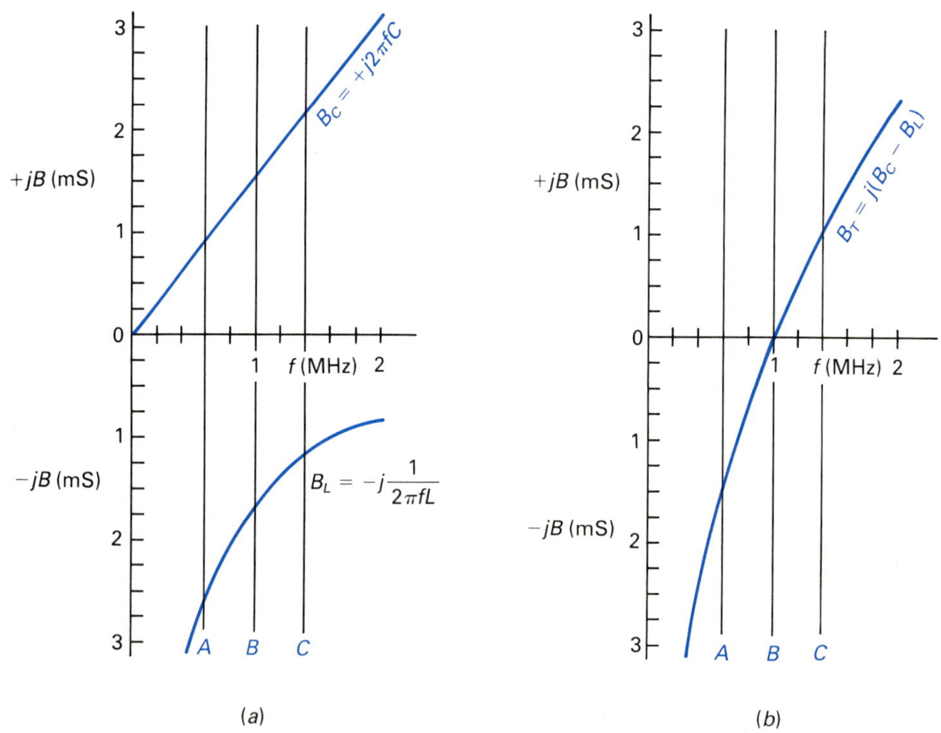

Fig. 16-7. Parallel *LC* circuit susceptance-frequency curves. *(a)* Individual elements; *(b)* total circuit.

Figure 16-7*b* shows a curve for total susceptance B_T for the parallel *LC* circuit in Fig. 16-6. This curve is the sum of the two separate curves in Fig. 16-7*a* for the individual susceptances B_L and B_C. The curve for B_T values is obtained by plotting the algebraic sums of the values of B_L and B_C for each frequency. At frequency A (0.6 MHz), $-jB_L$ is greater than $+jB_C$, and B_T is a $-j$ quantity of magnitude $B_L - B_C$. At frequency B ($\simeq$ 1 MHz), $-jB_L$ is equal to $+jB_C$, and B_T is zero. At frequency C (1.4 MHz), $+jB_C$ is greater than $-jB_L$, and B_T is a $+j$ quantity of magnitude $B_C - B_L$.

Figure 16-7 shows the variation in the magnitudes of the susceptances B_L, B_C, and B_T with frequency, together with their phase angles (in terms of j signs).

The magnitudes of the susceptances and their variation with respect to frequency are given in Fig. 16-8. The broken lines represent the individual susceptances B_L and B_C, while the solid line represents the total susceptance B_T (or total admittance Y). The magnitudes are the same as in Fig. 16-7, the j positions being ignored. *As the frequency increases from zero toward the resonant frequency f_r, the magnitude of the total susceptance B_T decreases toward zero. As the frequency then increases beyond resonance, the susceptance increases.*

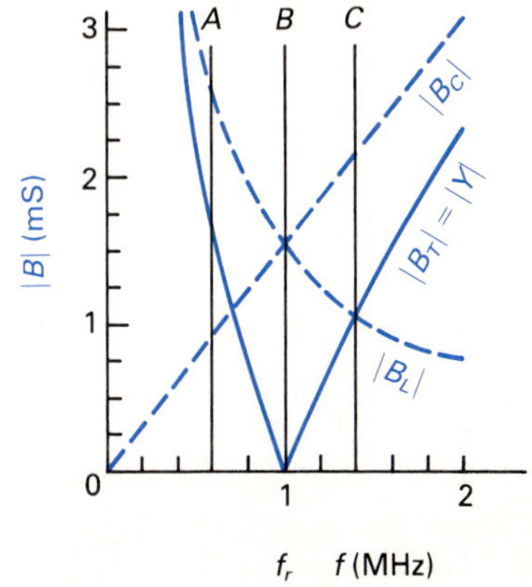

Fig. 16-8. Parallel LC circuit susceptance magnitude-frequency curves.

If you prefer to think in terms of impedance for all circuits, the statement for the variation in the total impedance magnitude of a parallel LC circuit is: *As the frequency increases from zero toward the resonant frequency f_r, the magnitude of the total impedance increases from zero. As the frequency then increases beyond resonance, the impedance decreases toward zero.*

A curve for the total impedance magnitude of the LC parallel circuit in Fig. 16-6 is given in Fig. 16-9. This curve is the reciprocal of the total admittance curve in Fig. 16-8.

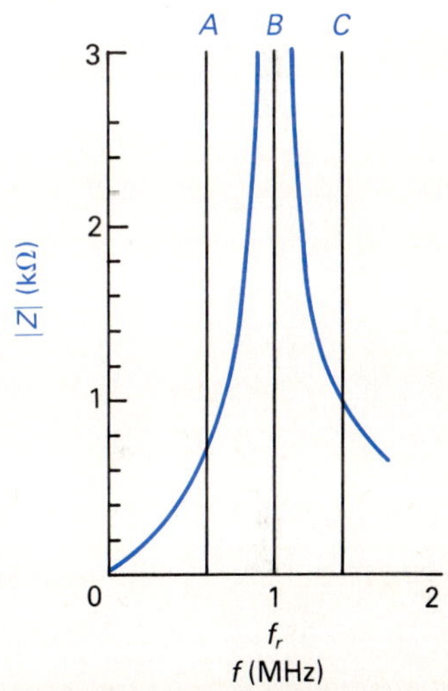

Fig. 16-9. Impedance-frequency curve for a parallel LC circuit (no resistance).

The manner in which the phase angle of a parallel LC circuit varies with respect to frequency is shown in Fig. 16-10. Below resonance the circuit is inductive ($B_L > B_C$), and the phase angle is $-90°$. Above resonance the circuit is capacitive ($B_L < B_C$), and the phase angle is $+90°$. At the resonant frequency f_r the circuit is in transition between the net inductive and net capacitive states, the phase angle being in transition between -90 and $+90°$.

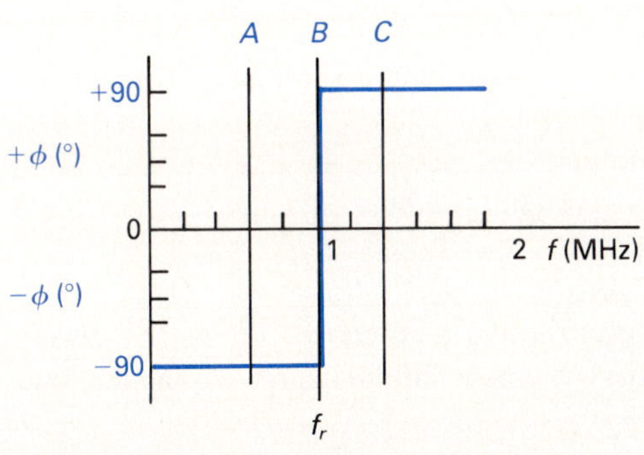

Fig. 16-10. Phase angle–frequency curve for a parallel LC circuit (no resistance).

In Fig. 16-8, the magnitudes of the susceptances apply to all parallel resonant circuits, with and without resistance. Figure 16-9 (for impedance) and Fig. 16-10 (for phase angle) apply only to parallel resonant circuits without resistance or energy losses.

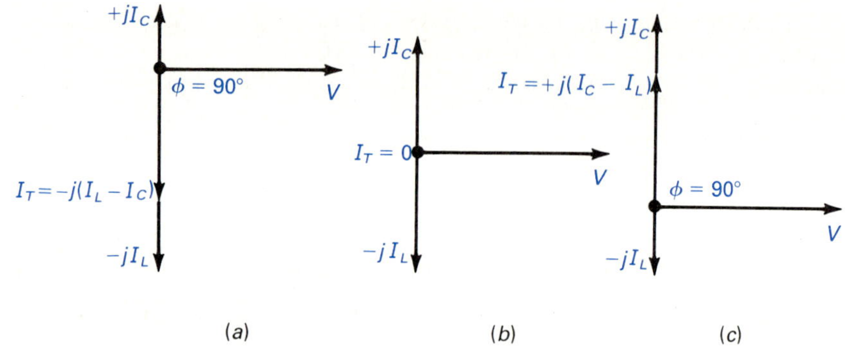

Fig. 16-11. Parallel *LC* circuit phasor diagrams: *(a)* Below resonance; *(b)* at resonance; *(c)* above resonance.

In Fig. 16-11 phasor diagrams are shown for the circuit in Fig. 16-6 in relation to frequencies *A*, *B*, and *C* in Figs. 16-7 to 16-10. The parallel *LC* network is fed with a constant voltage from the source. Each current is directly proportional to the associated susceptance.

At frequency *A* (0.6 MHz), $-jI_L$ is greater than $+jI_C$. The total current I_T is equal to $-j(I_L - I_C)$ and lags the emf *V* by 90°. These relationships are shown in Fig. 16-11*a*.

At frequency *B* (1 MHz), the resonant frequency f_r, $-jI_L$ equals $+jI_C$, the total current I_T is zero, and there is no circuit phase angle. The phasors for frequency *B* are shown in Fig. 16-11*b*.

At frequency *C* (1.4 MHz), $+jI_C$ is greater than $-jI_L$. The total current I_T equals $+j(I_C - I_L)$ and leads the emf *V* by 90°, as shown in Fig. 16-11*c*.

The circuit conditions present in a parallel *LC* circuit at the resonant frequency f_r can be summarized:

1. The magnitudes of the inductive susceptance B_L and the capacitive susceptance B_C are equal.
2. The total effective susceptance B_T is zero.
3. There is no conductance, so that the total circuit admittance *Y* equals B_T and is therefore zero.
4. The circuit impedance *Z* is infinite.
5. The magnitudes of the inductive current I_L and the capacitive current I_C are equal; their directions are opposite.
6. The total circuit current I_T is zero.
7. The ratio of either reactive current to the total current (I_L/I_T or I_C/I_T) is infinite, or either of the reactive currents is infinitely greater than the total circuit current.

Below resonance the following conditions exist in the parallel *LC* circuit:

1. The circuit is predominantly inductive.
2. The phase angle is −90°.

Above resonance the following conditions exist in the parallel *LC* circuit:

1. The circuit is predominantly capacitive.
2. The phase angle is +90°.

It is well to compare the characteristics of series and parallel *LC* circuits:

1. Compare Figs. 16-2 and 16-7 and note that the actual curves are identical. The differences between the series reactance and parallel susceptance curves lie only in the scaling and units of the vertical axis and the labeling of the curves.
2. A comparison of Figs. 16-3 and 16-8 reveals identical curves. The scaling and units of the vertical axis and the curve labels are the only differences between the series reactance and parallel susceptance scalar-value curves. As the frequency increases from zero, the values of $|X_T|$ for the series circuit and $|B_T|$ for the parallel circuit decrease from infinity toward zero at the resonant frequency, and then increase again from zero toward infinity.
3. Figures 16-4 and 16-10 show that the series and parallel *LC* circuits both change their phase angles at resonance. The series circuit goes from $+90°$ (capacitive) to $-90°$ (inductive). The parallel circuit goes from $-90°$ (inductive) to $+90°$ (capacitive).
4. In Figs. 16-5 and 16-11 the usual duality of series and parallel ac circuits can be seen. The phasor diagrams are similar if the basic rule is followed; for series circuits, voltages are shown with the current as the reference phasor; for parallel circuits, currents are shown with the voltage as the reference phasor.
5. Below resonance the series circuit is capacitive; the parallel circuit is inductive.
6. Above resonance the series circuit is inductive; the parallel circuit is capacitive.

To emphasize the duality of series and parallel circuit behavior, parallel resonance is sometimes called *antiresonance*. Where this convention is used, series resonance is merely referred to as resonance.

When, as in a practical situation, a resonant circuit contains resistance, its characteristics will be similar to those of the simple *LC* circuit but with detail modifications. By bearing in mind the characteristics of theoretically resistanceless *LC* resonant circuits and noting how the presence of resistance modifies them, the behavior of practical *RLC* resonant circuits can be visualized.

REVIEW QUESTIONS FOR SECTION 16-1

1. Why is a radio receiver dependent on resonant circuits?
2. In a series *LC* circuit at resonance, what are
   a. The comparative magnitudes of the reactances?
   b. The total reactance?
   c. The total impedance?
3. In a series *LC* circuit, when is the net reactance inductive?
4. In a parallel *LC* circuit at resonance, what are
   a. The comparative magnitudes of the susceptances?
   b. The total susceptance?
   c. The total reactance?
   d. The total impedance?
5. In a parallel *LC* circuit, when is the net reactance inductive?

16-2 SERIES *RLC* RESONANT CIRCUIT

A series *RLC* circuit is shown in Fig. 16-12. Its impedance Z is the phasor sum of the element impedances:

$$Z = R + jX_T = R + j(X_L - X_C)$$

The characteristics of the series resonant circuit are determined by this impedance equation. Initially three conditions will be considered.

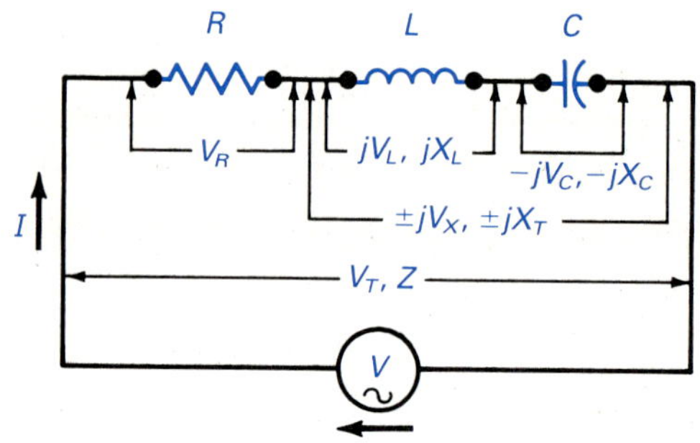

Fig. 16-12. Series *RLC* circuit.

At the Resonant Frequency f_r By definition, at the resonant frequency the power factor is unity. This implies that the circuit is purely resistive; the impedance consists only of the circuit resistance, the net reactance is zero, and the phase angle is zero. There is unity power factor ($\cos \phi = 1$).

$$Z = R$$

$$X_T = j(X_L - X_C) = 0 \qquad \text{or} \qquad X_L = X_C$$

Expanding X_L and X_C, we obtain

$$\omega_r L = \frac{1}{\omega_r C}$$

or

$$\omega_r^2 = \frac{1}{LC} \tag{16-1}$$

Expanding ω_r gives

$$(2\pi f_r)^2 = \frac{1}{LC}$$

or

$$f_r = \frac{1}{2\pi\sqrt{LC}} \tag{16-2}$$

In Eqs. (16-1) and (16-2),

ω_r = resonant angular frequency, rad/s
L = inductance, H
C = capacitance, F
f_r = resonant frequency, Hz

Equation (16-2) is one of the most important electrical relationships. Not only does it accurately give the resonant frequency of a practical series circuit, but it also gives an approximate value of the resonant frequency of a practical parallel circuit.

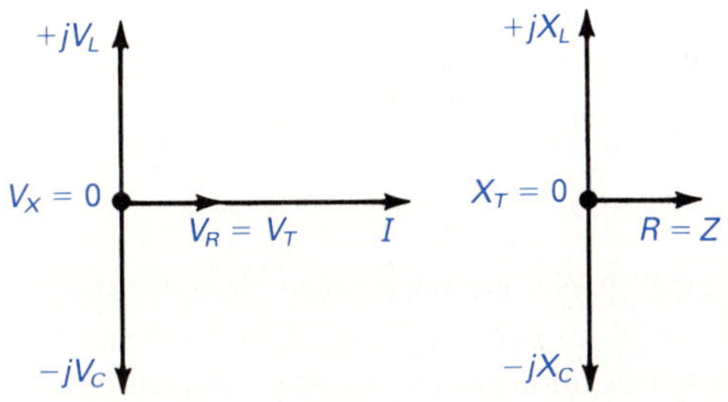

(a) (b)

Fig. 16-13. Series *RLC* circuit at resonance. *(a)* Phasor diagram; *(b)* impedance diagram.

Figure 16-13*a* shows a phasor diagram of a series *RLC* circuit at resonance. The inductive and capacitive voltages jV_L and $-jV_C$ are equal in magnitude but are in antiphase (opposite *j* signs). Thus the net reactive voltage V_X is zero. This cancellation of reactive voltages leaves the resistive voltage as the only active circuit voltage. The resistive voltage V_R equals the total circuit voltage V_T.

Figure 16-13*b* shows an impedance diagram. The reactances jX_L and $-jX_C$ cancel, leaving only the resistance *R* to constitute the total circuit impedance *Z*. It is not possible to draw an impedance triangle for the resonant condition, because there is no net reactance.

Voltage Magnification Factor

It will be seen from Fig. 16-13 that, for a given value of resistance, no matter how large the magnitude of either reactance, the total circuit impedance *Z* will equal the resistance *R*, and the total circuit voltage V_T will equal the resistive voltage V_R. This observation leads to the special significance of the quality factor *Q* in a series resonant circuit. It has been mentioned previously that *Q* is defined as the ratio of stored energy to dissipated energy and that, for a series circuit, *Q* equals the ratio of reactance to resistance. In a resonant circuit the energy is stored alternately in the inductance and the capacitance, so that at resonance *Q* becomes the ratio of either reactance to the resistance:

$$Q_r = \frac{X_{Lr}}{R} = \frac{X_{Cr}}{R}$$

or

$$X_{Lr} = X_{Cr} = Q_r R = Q_r Z \qquad \textbf{(16-3)}$$

Equation (16-3) can be written in voltage form:

$$Q_r = \frac{V_{Lr}}{V_T} = \frac{V_{Cr}}{V_T}$$

or

$$V_{Lr} = V_{Cr} = Q_r V_T \qquad \textbf{(16-4)}$$

In Eqs. (16-3) and (16-4),

Q_r = the quality factor at resonance
X_{Lr} = inductive reactance at resonance, Ω
X_{Cr} = capacitive reactance at resonance, Ω
R = equivalent resistance of circuit losses, Ω
Z = total circuit impedance, Ω
V_{Lr} = inductive voltage at resonance, V
V_{Cr} = capacitive voltage at resonance, V
V_T = total circuit voltage, V (at resonance the total voltage V_T is the same as the resistive voltage V_R)

For practical circuits designed to exhibit resonant conditions, Q_r will usually be greater than unity. Therefore in a series resonant circuit it is common for the voltage across either reactance to exceed the total circuit voltage. This fact leads to another definition of Q_r:

In a series *RLC* circuit at resonance, the quality factor Q_r is called the *voltage magnification factor*. The voltage across the entire circuit is magnified so that it appears across each reactance Q_r times as large.

At Any Frequency Other than the Resonant Frequency If the circuit is not operating at its resonant frequency, its power factor will be less than unity. The phase angle and the net reactance will both be greater than zero.

$$Z > R$$
$$X_T = j(X_L - X_C) > 0 \qquad \text{or} \qquad X_L \neq X_C$$

In Fig. 16-14*a* an impedance diagram is shown for the situation in which the frequency is below resonance. Here the capacitive reactance X_C is greater than the inductive reactance X_L, resulting in a net reactance X_T equal to $-j(X_c - X_L)$.

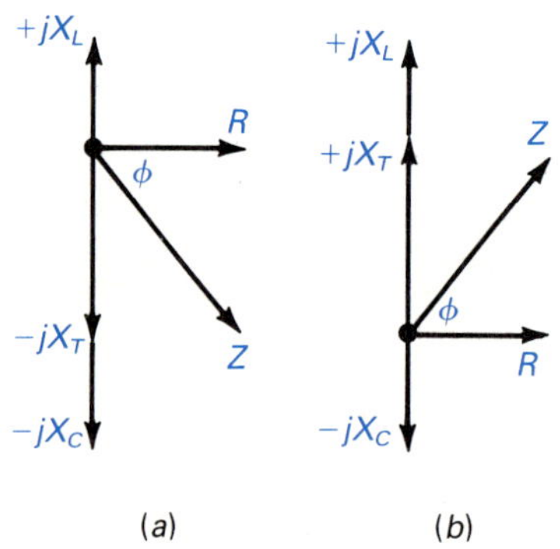

Fig. 16-14. Series *RLC* circuit impedance diagrams: *(a)* Below resonance; *(b)* above resonance.

When the frequency is above resonance, X_L is greater than X_C, and X_T equals $+j(X_L - X_C)$. This condition is shown in Fig. 16-14*b*.

For Fig. 16-14*a* and *b* the circuit impedance Z is the phasor sum of the net reactance X_T and the resistance R. The phase angle ϕ is equal to arctan X_T/R.

It follows that the voltage relationships will be

$$V_L \neq V_C$$
$$V_X = (V_L - V_C) > 0$$
$$V_T = V_R + j(V_L - V_C) > V_R$$

Figure 16-15 shows typical phasor diagrams for a series resonant circuit operated at nonresonant frequencies. In Fig. 16-15*a* the frequency is below resonance, so $V_C > V_L$, resulting in a net reactive voltage V_X equal to $-j(V_C - V_L)$.

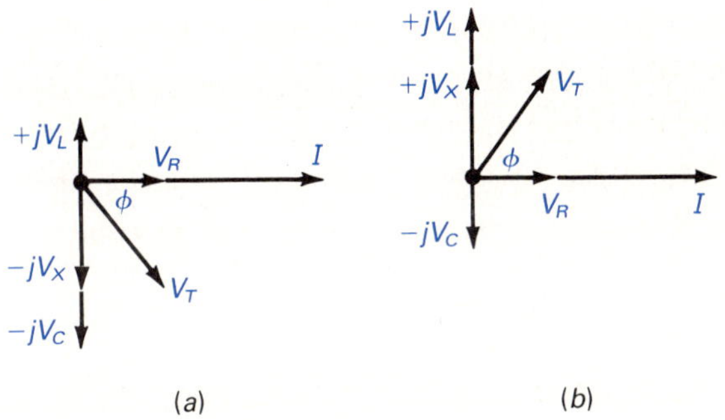

Fig. 16-15. Series *RLC* circuit phasor diagrams: *(a)* Below resonance; *(b)* above resonance.

In Fig. 16-15*b* the frequency is above resonance ($V_L > V_C$), and the net reactive voltage V_X is $+j(V_L - V_C)$. In both cases the total circuit voltage V_T is the phasor sum of the net reactive voltage V_X and the resistive voltage V_R. The phase angle ϕ is equal to arctan V_X/V_R.

As there is resistance in the *RLC* resonant circuit, impedance and voltage triangles can be drawn, as in Fig. 16-16.

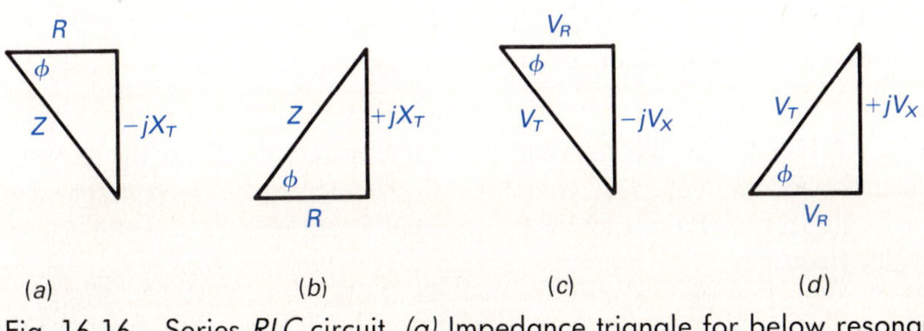

Fig. 16-16. Series *RLC* circuit. *(a)* Impedance triangle for below resonance; *(b)* impedance triangle for above resonance; *(c)* voltage triangle for below resonance; *(d)* voltage triangle for above resonance.

Figure 16-16*a* and *b* shows impedance triangles for below- and above-resonance frequencies. These triangles are derived from Fig. 16-14. In a similar manner, the voltage triangles in Fig. 16-16*c* and *d* refer to below- and above-resonance frequencies and are obtained from Fig. 16-15.

At the Frequencies Where $X_T = R$ "Bandwidth" is a term which is dealt with later in this chapter when the characteristics of resonant circuits over a range of frequencies are discussed. The term is introduced now, since its definition is related to two specific frequencies, one below and one above resonance.

Bandwidth in a series resonant circuit can be defined as the frequency difference between the two conditions at which the net reactance equals the resistance. If this definition is used, the specific frequencies are obtained from the equation:

$$X_T = R$$

At the bandwidth frequency below resonance f_1, the capacitive reactance X_{C1} will be greater than the inductive reactance X_{L1}:

$$X_{T1} = X_{C1} - X_{L1} = R \tag{16-5}$$

At the bandwidth frequency above resonance f_2, the inductive reactance X_{L2} will be greater than the capacitive reactance X_{C2}:

$$X_{T2} = X_{L2} - X_{C2} = R \tag{16-6}$$

It follows that

$$X_{T1} = X_{T2} = R$$

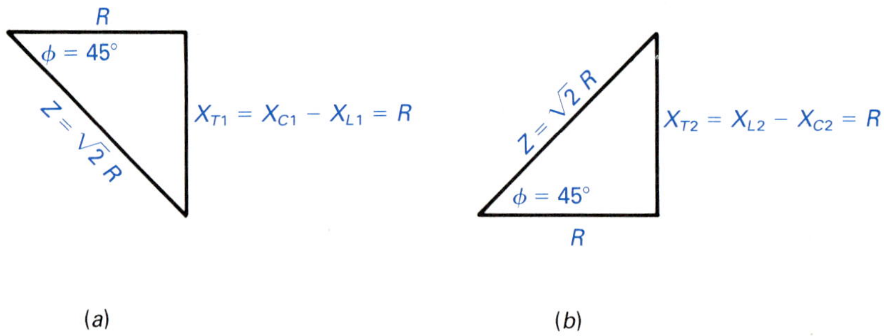

Fig. 16-17. Series *RLC* circuit, impedance triangles for the bandwidth frequencies: *(a)* Below resonance; *(b)* above resonance.

Figure 16-17*a* shows the impedance triangle which refers to the below-resonance bandwidth frequency, while Fig. 16-17*b* refers to the above-resonance bandwidth frequency. From each of the triangles it can be seen that, for either bandwidth frequency,

$$X_T = R$$
$$Z = \sqrt{R^2 + X_T^2} = \sqrt{2}\,R$$
$$\phi = \arctan \frac{X_T}{R} = \arctan 1 = 45° = \pi/4 \text{ rad}$$

Impedance and Phase Angle Curves

We have seen that, for a series resonant circuit with resistance, the practical situation

1. At resonance the impedance equals the resistance, and the phase angle is zero.

2. Off resonance the impedance is greater than the resistance, and the phase angle is greater than zero.
3. At the bandwidth frequencies the impedance is $\sqrt{2}$ times the resistance, and the phase angle is 45° ($\pi/4$ rad).

Three more statements can be added:

4. At zero frequency the capacitive reactance is infinitely high and the inductive reactance is zero. Thus the circuit in effect consists of an infinitely high capacitive reactance in series with a resistance. The impedance is therefore infinite, and the phase angle is +90° ($+\pi/2$ rad).
5. If the frequency were to become infinitely high, the inductive reactance would become infinite and the capacitive reactance zero. The effective circuit would be that of an infinite inductive reactance in series with a resistance. The impedance would be infinite, and the phase angle would be −90° ($-\pi/2$ rad).
6. Between zero frequency and resonance, and between resonance and infinite frequency, the transition of impedance and phase angle values is gradual, the rate of change depending on the circuit values involved.

Figure 16-18 shows the effect of varying frequency on the impedance magnitude and the phase angle of a series resonant circuit which has the values $L = 100\ \mu\text{H}$, $C = 250$ pF, and $R = 100\ \Omega$.

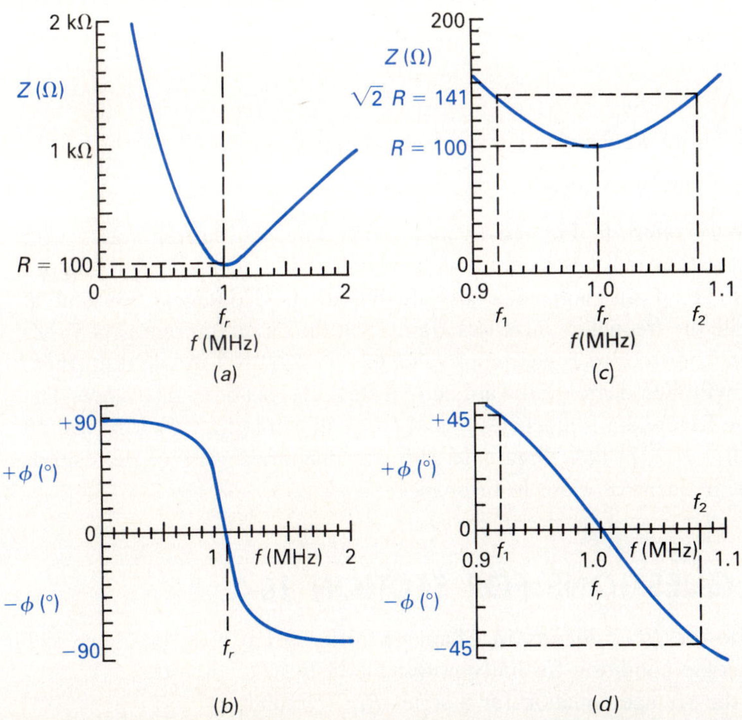

Fig. 16-18. Series *RLC* circuit curves: *(a)* Impedance-frequency; *(b)* phase angle–frequency; *(c)* impedance-frequency—expanded close to resonance; *(d)* phase angle–frequency—expanded close to resonance.

In Fig. 16-18*a* the impedance decreases from infinity to the value of the resistance ($R = 100\ \Omega$) at the resonant frequency f_r. Above f_r the impedance increases again toward infinity.

Figure 16-18*b* shows the circuit phase angle decreasing from $+90°$ (capacitive) toward zero (resistive) at the resonant frequency, and then increasing toward $-90°$ (inductive) above resonance (although the curve goes downward through zero, the change from zero toward $-90°$ is an increase in magnitude).

Figure 16-18*c* and *d* shows magnifications in horizontal and vertical scaling of the curves in Fig. 16-18*a* and *b* to show in greater detail the impedance and phase angle values close to the resonant frequency. In Fig. 16-18*c* it can be seen that the impedance is $\sqrt{2}R = 141\ \Omega$ at the bandwidth frequencies f_1 and f_2, while Fig. 16-18*d* shows the phase angle of 45° at these frequencies.

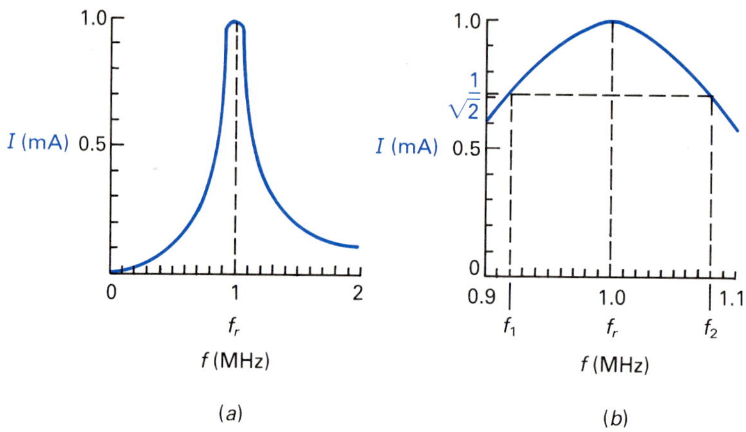

Fig. 16-19. Series *RLC* circuit current-frequency curves.

Figure 16-19 refers to Fig. 16-18 and shows how the current varies with respect to frequency in a series circuit with the values $L = 100\ \mu H$, $C = 250$ pF, and $R = 100\ \Omega$. If the voltage across the circuit is maintained constant at 100 mV while the frequency increases from zero, the circuit current ($I = V_T/Z$) will increase from zero to a maximum value of $I = V_T/R = 1$ mA at resonance. The current will then decrease toward zero at frequencies above resonance. The current at the bandwidth frequencies f_1 and f_2, as shown in Fig. 16-19*b*, will be $I = V_T/\sqrt{2}R = 0.7071$ mA. Figure 16-19*b* is a magnified view of the current variations at frequencies close to resonance.

REVIEW QUESTIONS FOR SECTION 16-2

1. How do the *RLC* circuits in Chap. 16 differ from those in Chap. 15?
2. What is the condition for unity power factor in *RLC* circuits?
3. Write the voltage equation for a series *RLC* circuit.
4. Write the equation for the frequency of a series *RLC* circuit at resonance.
5. How does the inductive voltage of a series *RLC* circuit compare with the capacitive voltage when the circuit is operated

a. At its resonant frequency?
b. Above its resonant frequency?
c. Below its resonant frequency?

16-3 BANDWIDTH OF A SERIES *RLC* CIRCUIT

Bandwidth frequencies have been referred to, but what is the significance of bandwidth? By referring to Fig. 16-19, it can be seen that there is a distinct difference between the current at the resonant frequency and the current at any other frequency. The resonant circuit is thus able to discriminate between one frequency and all others. This capability is called *selectivity*. A series circuit is highly selective if it gives a high ratio of current at the resonant frequency to current at some other stated frequency.

The degree of selectivity required of a resonant circuit depends on the application. The selectivity of a resonant circuit in an oscillator should be high, as high selectivity is required to maintain the oscillator's frequency constant at a specific value. In a radio receiver a resonant circuit used to tune to a required station must be able to pass a band of frequencies on either side of the resonant frequency. Thus a tuning circuit in a radio could be expected to have lower selectivity than a resonant circuit in an oscillator.

Bandwidth is a measure of a circuit's selectivity. High selectivity is narrow bandwidth. Low selectivity is wide or broad bandwidth.

The bandwidth of a circuit is the frequency difference between points at which the power response is one-half that of the maximum.

In a series resonant circuit supplied with a constant voltage, maximum power is dissipated at the resonant frequency. From the definition of bandwidth, the relationship between the powers at the bandwidth frequencies and the power at resonance is

$$P_1 = P_2 = \frac{P_r}{2} \qquad \textbf{(16-7)}$$

where P_1 and P_2 = power at the lower and upper bandwidth frequencies f_1 and f_2, W
P_r = power at the resonant frequency f_r, W

In any ac circuit power is dissipated in the resistance, and in a series circuit power is given by $P = I^2R$. Therefore, Eq. (16-7) can be written in the form

$$I_1^2R = I_2^2R = \frac{I_r^2R}{2}$$

$$I_1^2 = I_2^2 = \frac{I_r^2}{2}$$

or

$$I_1 = I_2 = \frac{I_r}{\sqrt{2}} = 0.7071I_r \qquad \textbf{(16-8)}$$

where I_1 = current at low bandwidth frequency f_1, A
I_2 = current at high bandwidth frequency f_2, A
I_r = resonant frequency current, A

Equation (16-8) leads to another definition of bandwidth: The bandwidth of a series *RLC* circuit is the frequency difference between conditions where the magnitudes of the currents are $1/\sqrt{2}$, or 0.7071, of the magnitude at the resonant frequency.

In the series resonant circuit to which Fig. 16-19 refers, the current at resonance is 1 mA. At frequencies f_1 and f_2 the current is 0.7071 mA, so these frequencies fit the criteria for bandwidth frequencies. At frequencies f_1 and f_2,

$$BW = f_2 - f_1 \tag{16-9}$$

where BW* = bandwidth, Hz
f_1 = lower bandwidth frequency, Hz
f_2 = upper bandwidth frequency, Hz

At frequencies f_1 and f_2 the power dissipated by the circuit is one-half that dissipated at the resonant frequency f_r.

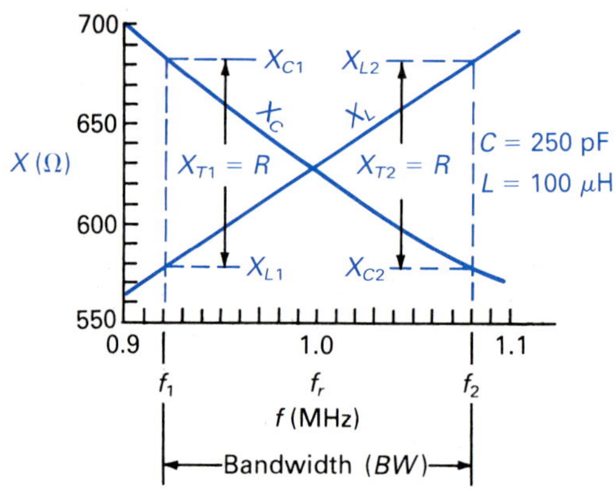

Fig. 16-20. Series resonant circuit reactance-frequency curves expanded close to resonance to show bandwidth conditions.

Curves for X_C and X_L, expanded to show the region close to resonance, are given in Fig. 16-20. Although these curves apply specifically to 250 pF and 100 μH, any other values of C and L will give curves of the same shape; only the scaling of the axes will be different. Thus the following discussion and conclusions may be considered general.

At the bandwidth frequencies the net reactance equals the resistance. Therefore, at frequency f_2,

$$X_{T2} = X_{L2} - X_{C2} = R \tag{16-6}$$

But from Fig. 16-20 it can be seen that

$$X_{C2} \simeq X_{L1}$$

Therefore,

$$\begin{aligned} X_{L2} - X_{L1} &\simeq R \\ 2\pi f_2 L - 2\pi f_1 L &\simeq R \\ 2\pi L(f_2 - f_1) &\simeq R \end{aligned}$$

*To avoid confusion with susceptance B in this chapter, BW is used for bandwidth.

Since $f_2 - f_1$ equals the bandwidth BW,

$$BW \simeq \frac{R}{2\pi L}$$

Dividing both sides of this equation by f_r gives

$$\frac{BW}{f_r} \simeq \frac{R}{2\pi f_r L} \simeq \frac{R}{X_{Lr}}$$

X_{Lr}/R is the quality factor at resonance Q_r, so

$$\frac{BW}{f_r} \simeq \frac{1}{Q_r}$$

or

$$BW \simeq \frac{f_r}{Q_r} \tag{16-10}$$

where BW = bandwidth frequency, Hz
f_r = resonant frequency, Hz
Q_r = the circuit quality factor

Equation (16-10) is used extensively for both series and parallel resonant circuits and is often memorized in statement form: *The bandwidth of a resonant circuit is approximately equal to the ratio of the resonant frequency to the circuit Q at resonance.*

In applying Eq. (16-10), the degree of approximation is dependent on the circuit values. If these values are such that Q_r is high, the degree of approximation is high. In practice it is common for Q_r to be at least 10, in which case Eq. (16-10) is accurate within 1 percent. For this reason the approximately equals sign is often omitted, treating Eq. (16-10) as a true equation rather than as an approximation. However, we will retain this sign to stress the fact that the use of Eq. (16-10) results in an approximation.

EXAMPLE 16-1 A 500-pF capacitor and a 50-μH inductor (with an effective resistance of 20 Ω) are connected in series. What is the resulting *(a)* resonant frequency f_r and *(b)* bandwidth BW?

Solution Figure 16-21 shows the circuit for this example.

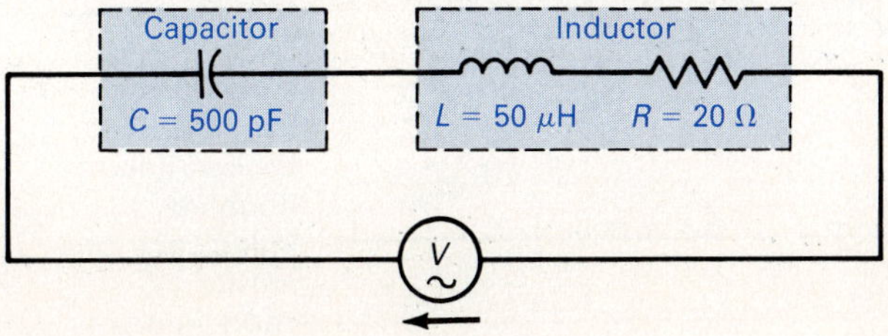

Fig. 16-21. Example 16-1: Circuit.

a. $$f_r = \frac{1}{2\pi\sqrt{LC}} \tag{16-2}$$

$$= \frac{1}{2\pi\sqrt{5 \times 10^{-10}\ \text{F} \times 5 \times 10^{-5}\ \text{H}}} = \frac{10^7}{2\pi\sqrt{2.5}}\ \text{Hz}$$

$$f_r = \mathbf{1.007\ MHz}$$

b. $$Q_r = \frac{X_{Lr}}{R} \tag{16-3}$$

$$Q_r = \frac{2\pi f_r L}{R} = \frac{2\pi \times 1.007\ \text{MHz} \times 50\ \mu\text{H}}{20\ \Omega} = 15.82$$

$$BW \simeq \frac{f_r}{Q_r} \tag{16-10}$$

$$BW \simeq \frac{1.007\ \text{MHz}}{15.82} = \mathbf{63.65\ kHz}$$

During the discussion on series resonance, reference was made to a circuit with the values $C = 250$ pF, $L = 100\ \mu$H, and $R = 100\ \Omega$, the implied resonant frequency being 1 MHz. Actually the resonant frequency for these values is 1.007 MHz, the same as that in Example 16-1. The two sets of element values result in the same resonant frequency because the product of L and C is the same in both cases:

$$250\ \text{pF} \times 100\ \mu\text{H} = 500\ \text{pF} \times 50\ \mu\text{H} = 2.5 \times 10^{-14}\ \text{F} \cdot \text{H}$$

In a series circuit, combinations of L and C which have the same product have the same resonant frequency.

In Example 16-1 the reactances at resonance are both 316.4 Ω, $Q_r = 15.82$, and $BW = 63.65$ kHz. For the values $C = 250$ pF, $L = 100\ \mu$H, and $R = 20\ \Omega$, the resonant reactances are both 632.8 Ω, $Q_r = 31.64$, and $BW = 31.83$ kHz. For any given value of resistance, the bandwidth of a series resonant circuit with an inductor of 50 μH will be wider than the bandwidth with 100 μH by a ratio of 100 μH/50 μH = 2. Figure 16-22 shows current response curves for the two sets of values with an assumed emf of 20 mV; the solid line represents the values $C = 500$ pF, $L = 50\ \mu$H, $R = 20\ \Omega$, and the broken line represents the values $C = 250$ pF, $L = 100\ \mu$H, $R = 20\ \Omega$.

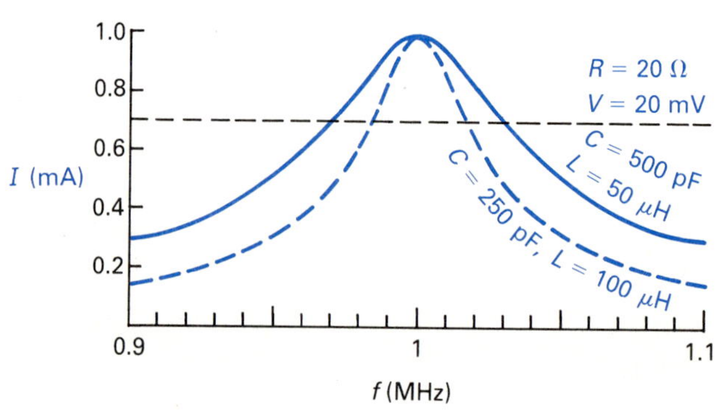

Fig. 16-22. Series *RLC* circuit current-frequency curves for two sets of values: Solid line for the values in Example 16-1; broken line for doubled inductance and halved capacitance.

It is a general principle for series circuits with the *same resonant frequency* that, for a given resistance value, *low inductance* will result in *wide bandwidth*.

If in Example 16-1 the resistance had been 40 Ω, the bandwidth would have been 127.3 kHz, twice as great as for 20 Ω. For series circuits with the *same values of L and C, high series resistance* results in *wide bandwidth*.

As the current at resonance with $R = 40\ \Omega$ will be one-half that with $R = 20\ \Omega$, the two current curves will be displaced vertically if drawn with the same current axis. This is shown in Fig. 16-23*a*.

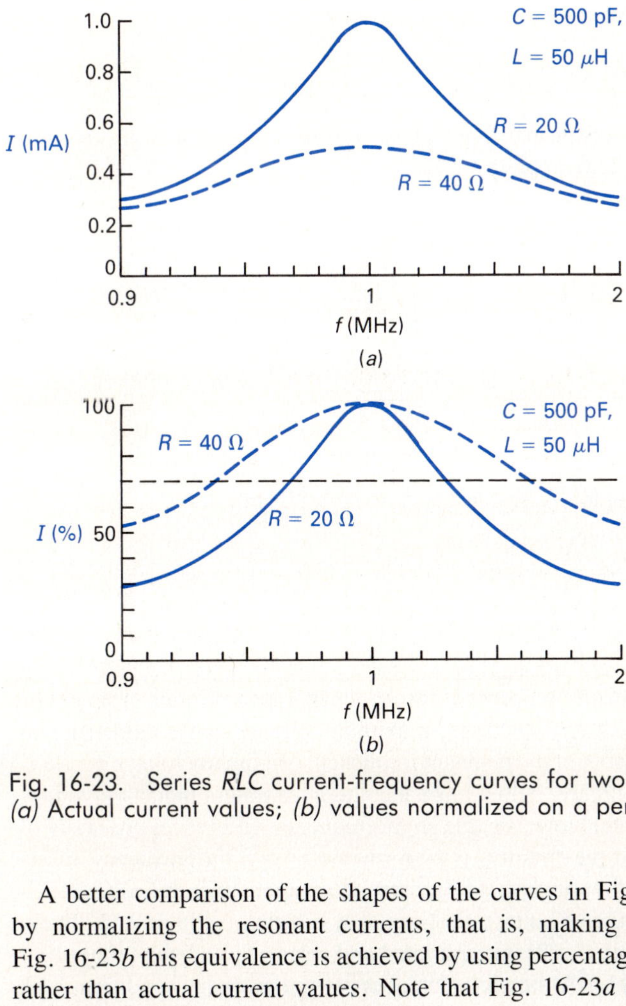

Fig. 16-23. Series *RLC* current-frequency curves for two resistance values: *(a)* Actual current values; *(b)* values normalized on a percentage basis.

A better comparison of the shapes of the curves in Fig. 16-23*a* is obtained by normalizing the resonant currents, that is, making them equivalent. In Fig. 16-23*b* this equivalence is achieved by using percentages of resonant current rather than actual current values. Note that Fig. 16-23*a* and *b* gives different presentations of the same conditions.

EXAMPLE 16-2 What are the voltages at resonance across each element in the circuit in Example 16-1 if 100 mV is applied across the complete circuit as shown in Fig. 16-24?

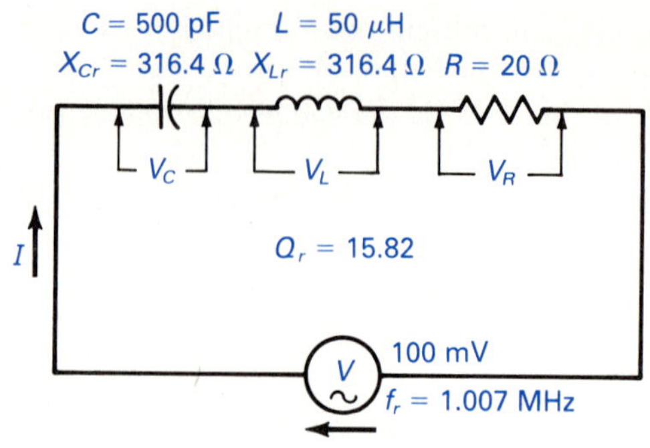

Fig. 16-24. Example 16-2: Circuit.

Solution *a.* At resonance the voltage across the resistance is the total voltage applied to the circuit.

$$V_R = V = \mathbf{100\ mV}$$

b. At resonance the voltages across the capacitance and across the inductance are Q_r times the applied voltage. From Eq. (16-4),

$$V_C = V_L = Q_rV = 15.82 \times 100\text{ mV} = V_L = \mathbf{1.582\ V}$$

The result for Example 16-2*b* can be verified by Ohm's law:

$$I_r = \frac{V}{Z_r} = \frac{V}{R} = \frac{100\text{ mV}}{20\ \Omega} = 5\text{ mA}$$

$$V_C = I_rX_{Cr} = 5\text{ mA} \times 316.4\ \Omega = 1.582\text{ V}$$

$$V_L = I_rX_{Lr} = 5\text{ mA} \times 316.4\ \Omega = 1.582\text{ V}$$

Although maximum current occurs at resonance in a series circuit, it should be pointed out here that, strictly speaking, maximum voltage across inductance or capacitance does not occur at the resonant frequency. Maximum voltage across C occurs at a frequency slightly below resonance, and maximum inductive voltage slightly above resonance. However, the slight frequency differences are rarely of practical importance. If the circuit Q is high (over 10, say), the frequency differences are usually small enough to be ignored. If an application relies on maximum capacitive or inductive voltage for frequency discrimination, it is not likely that exact values of L or C will be calculated and selected. Probably either L or C will be made variable. A simple calculation of the resonant conditions will suffice, final adjustment of maximum element voltage being made during measurement.

The voltage across either reactance is 1.582 V, 15.82 times as great as the applied emf. Thus these particular circuit values result in a voltage magnification (sometimes referred to as voltage gain) of 15.82. The circuit Q (316.4 Ω/20 Ω) is also 15.82. Although there is a voltage gain, there is no power gain; the only power dissipation takes place in the resistive component of the circuit ($P_T = VI = V_RI$).

EXAMPLE 16-3

Figure 16-25*a* shows an antenna feeding a signal to a radio receiver. The series resonant circuit, consisting of a variable capacitor and an inductor, is a wave trap installed for the purpose of rejecting an undesired signal. If the signal to be rejected originates from an aircraft navigation beacon and has a frequency of 400 kHz, to what value should the variable capacitor be set? Figure 16-25*b* is equivalent to the circuit in Fig. 16-25*a*.

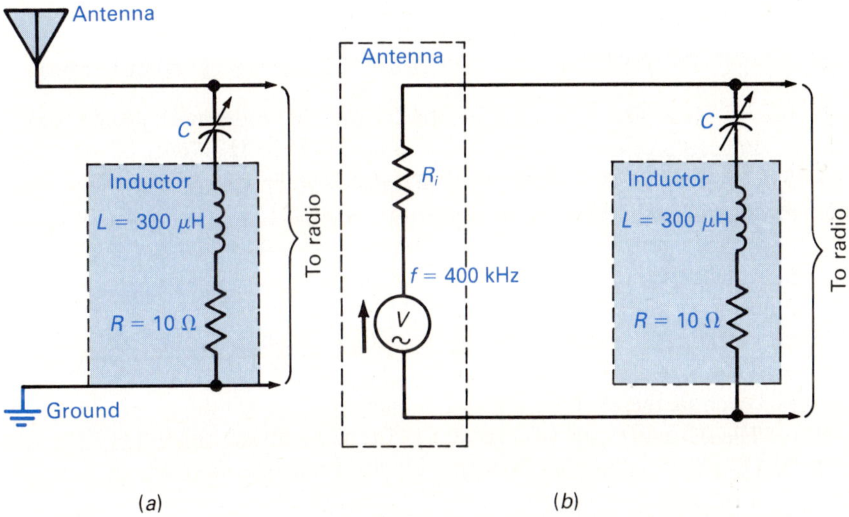

Fig. 16-25. Example 16-3: *(a)* Given circuit; *(b)* equivalent circuit.

Solution When the wave trap is resonant, its impedance will be minimum and a minimum signal will be passed on. Therefore C should be set for resonance at the undesired 400-kHz frequency.

$$\omega_r^2 = \frac{1}{LC} \qquad (16\text{-}1)$$

$$C = \frac{1}{\omega_r^2}$$

$$C = \frac{1}{(2\pi \times 4 \times 10^5\ \text{Hz})^2 \times 3 \times 10^{-4}\ \text{H}} = \frac{10^{-6}}{(2\pi \times 4)^2 \times 3}\ \text{F}$$

$$= \mathbf{527.7\ pF}$$

In Example 16-3 the level of the signal fed to the radio at 400 kHz will be $VR/(R + R_i)$, the reactances not being relevant at resonance. Thus for efficient rejection, R should be low in value compared with R_i, the effective resistance of the antenna. This requirement implies that the wave-trap inductor should have a high Q. A high Q also means narrow bandwidth, another way of viewing high ability to reject the undesired signal. The Q at resonance in the case of the inductor in Fig. 16-25 is 75, a high value for a practical inductor operating at 400 kHz.

REVIEW QUESTIONS FOR SECTION 16-3

1. How are the "bandwidth points" described?
2. In a series resonant circuit, what is the impedance at the bandwidth points?
3. In a series resonant circuit, what is the phase angle at the bandwidth points?
4. Compare the terms "selectivity" and "bandwidth."
5. What is the equation relating resonant frequency and bandwidth?
6. What is the effect of increasing the resistance of a series resonant circuit?

EXERCISE PROBLEMS FOR SECTIONS 16-2 AND 16-3

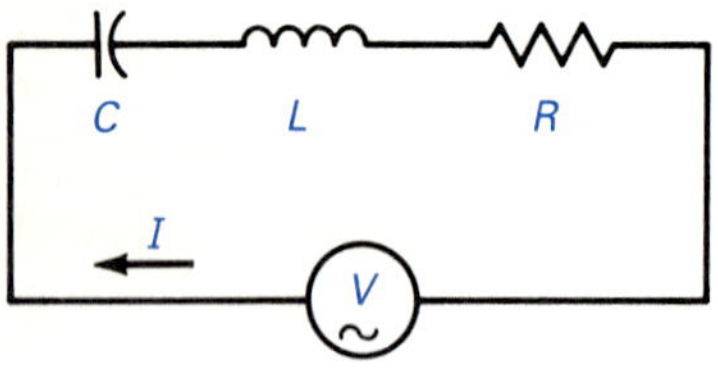

Fig. 16-26. Problems 1 to 7: Circuit.

1. For the series *RLC* circuit in Fig. 16-26, find the resonant frequency when
   a. $V = 10\ \mu V$, $C = 150$ pF, $L = 250\ \mu H$, $R = 25\ \Omega$ (*ans*. 821.9 kHz)
   b. $V = 1$ mV, $C = 350$ pF, $L = 500\ \mu H$, $R = 20\ \Omega$ (*ans*. 380.5 kHz)
   c. $V = 1$ V, $C = 0.002\ \mu F$, $L = 1$ mH, $R = 2\ \Omega$ (*ans*. 112.5 kHz)
2. Using the circuits in Prob. 1, find the reactances of L and C at their resonant frequencies.
3. Find the Q of each of the circuits in Prob. 1.
   (*ans*. *a*. 51.64; *b*. 59.75; *c*. 353.6)
4. Find the voltage across the capacitors of each of the circuits in Prob. 1.
5. For each of the circuits in Prob. 1, find
   i. The current I. (*ans*. *a*. 0.4 μA; *b*. 50 μA; *c*. 0.5 A)
   ii. IX_L. (*ans*. *a*. 516 μV; *b*. 59.75 mV; *c*. 353.6 V)
6. Find the bandwidth for each of the circuits in Prob. 1.
7. For each of the circuits in Prob. 1, find the reactances of L and C at the bandwidth frequencies and compare the difference of these two reactances with the circuit resistance.
   (*ans*. *a*. 1.304 kΩ, 1.279 kΩ, 24.9 Ω, 1.279 kΩ, 1.304 kΩ, 25.1 Ω;
   b. 1.205 kΩ, 1.185 kΩ, 19.9 Ω, 1.185 kΩ, 1.205 kΩ, 20.1 Ω;
   c. 708 Ω, 706 Ω, 2 Ω, 706 Ω, 708 Ω, 2 Ω)
8. The bandwidth of a series resonant circuit consisting of $C = 450$ pF, $L = 350\ \mu H$, and $R = 8\ \Omega$ is found to be too narrow. Find the amount of additional resistance required to increase the bandwidth to 10 kHz.
9. If the voltage across the circuit in Prob. 8 is 10 mV, find the voltage across the inductance
   a. Before the additional resistance is added. (*ans*. 1.102 V)
   b. After the additional resistance is added. (*ans*. 0.401 V)
10. It is desired to change the frequency of the circuit in Prob. 8 to 450 kHz.
    a. Find a new value of L which will obtain this frequency with C remaining at 450 pF.
    b. Find a new value of C which will obtain this frequency with L remaining at 350 μH.

16-4 PARALLEL *RLC* RESONANT CIRCUIT

A circuit consisting of resistance, capacitance, and inductance all in parallel is rarely met in practice. A practical parallel resonant circuit is usually made up of an inductor and a capacitor connected in parallel. An inductor will normally have losses which are significant and can be conveniently represented by a series resistance. The capacitor's losses will normally be negligible. Thus the practical parallel resonant circuit is usually depicted as a series-parallel network consisting of an inductance and a resistance in series and a capacitance connected across the *RL* series combination. Despite the lack of practicality of the true parallel *RLC* resonant circuit, the principles involved in its characteristics are important. Visualizing the duality of series and parallel resonant circuits is perhaps the best way to understand the behavior of all resonant circuits.

The development of series resonant circuit characteristics will be followed step by step, using parallel-circuit terms. In this manner a direct comparison between series and parallel resonance can be obtained.

R, *L*, and *C* All in Parallel

In the parallel circuit in Fig. 16-27, the total admittance Y is the phasor sum of the element admittances:

$$Y = G + j(B_C - B_L)$$

The admittance equation is the basis on which the circuit characteristics are investigated.

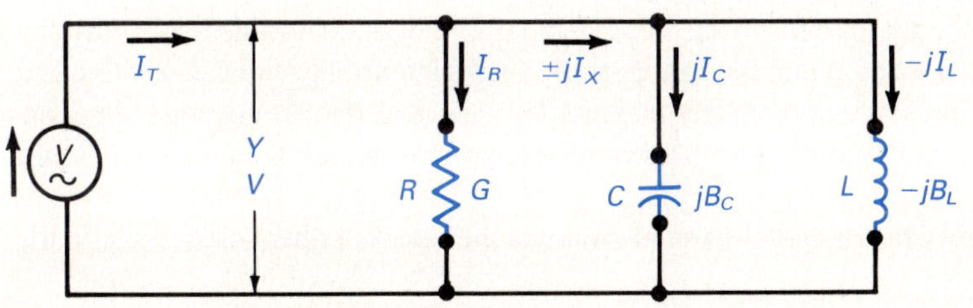

Fig. 16-27. Parallel *RLC* circuit.

At the Resonant Frequency f_r At the resonant frequency the power factor is unity, the circuit is purely resistive, the admittance consists only of the circuit conductance, the net susceptance is zero, and the phase angle is zero. There is unity power factor ($\cos\phi = 1$).

$$Y = G$$

$$B_T = j(B_C - B_L) = 0 \qquad \text{or} \qquad B_C = B_L$$

Substituting expressions for B_C and B_L gives

$$\omega_r C = \frac{1}{\omega_r L}$$

or

$$\omega_r^2 = \frac{1}{LC} \tag{16-1}$$

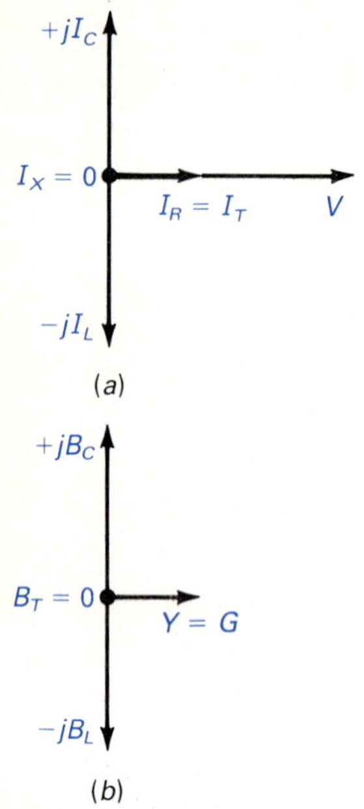

Fig. 16-28. Parallel *RLC* circuit at resonance: *(a)* Phasor diagram; *(b)* admittance diagram.

Expanding ω_r,
$$(2\pi f_r)^2 = \frac{1}{LC}$$

or
$$f_r = \frac{1}{2\pi\sqrt{LC}} \tag{16-2}$$

Thus it can be seen that the basic resonant conditions for both series and parallel circuits are the same. Although we have considered only the true parallel circuit, perhaps it should be added here that in many circumstances these conditions also apply to practical parallel resonant circuits to an adequate degree of accuracy.

A phasor diagram for the parallel *RLC* circuit at resonance is given in Fig. 16-28*a*. The capacitive current I_C is a $+j$ quantity with respect to the common voltage V. The inductive current I_L is a $-j$ quantity. Therefore, since their magnitudes are equal, the capacitive and inductive currents cancel, making the net reactive current I_X equal to zero. This cancellation of the reactive currents leaves the resistive current as the only component of the total circuit current. The resistive current I_R equals the total current I_T.

An admittance diagram for the circuit in Fig. 16-27 is given in Fig. 16-28*b*. Here are shown the equal magnitudes and opposite j signs of B_C and B_L. Cancellation of the susceptances results in the conductance of the resistive element being the total circuit admittance. Without net susceptance, an admittance triangle cannot be drawn.

Current Magnification Factor

At resonance the current in the *RLC* parallel circuit is determined solely by the resistance. Depending on the values of the susceptances, the equal capacitive and inductive currents can have any magnitude, their net effect being zero. Thus the capacitive and inductive currents can be many times greater than the total circuit current.

The quality factor Q of a parallel circuit is the ratio of either susceptance to the conductance:

$$Q_r = \frac{B_{Cr}}{G} = \frac{B_{Lr}}{G}$$

or
$$B_{Cr} = B_{Lr} = Q_r G = Q_r Y \tag{16-11}$$

Writing Eq. (16-11) in current form,

$$Q_r = \frac{I_{Cr}}{I_R} = \frac{I_{Lr}}{I_R}$$

or
$$I_{Cr} = I_{Lr} = Q_r I_R = Q_r I_T \tag{16-12}$$

In Eqs. (16-11) and (16-12),

Q_r = the quality factor at resonance
B_{Cr} = capacitive susceptance at resonance, S
B_{Lr} = inductive susceptance at resonance, S
G = equivalent conductance of the circuit losses, S
Y = total circuit admittance, S
I_{Cr} = capacitive current at resonance, A

I_{Lr} = inductive current at resonance, A
I_R = resistive current, A
I_T = total circuit current, A (at resonance the total current I_T is the same as the resistive current I_R)

In practical parallel resonant circuits, Q_r is usually greater than unity. When this is so, the capacitive and inductive currents are each greater than the total circuit current.

In a parallel *RLC* circuit the quality factor is called the *current magnification factor*. The total current in the circuit is magnified so that it appears Q_r times as large in both capacitor and inductor.

At Any Frequency Other than the Resonant Frequency When the circuit is supplied with current at a nonresonant frequency, the power factor is less than unity; the phase angle and the net susceptance are both greater than zero.

$$Y > G$$
$$B_T = j(B_C - B_L) > 0 \qquad \text{or} \qquad B_C \neq B_L$$

Figure 16-29 shows admittance diagrams for off-resonance conditions. In Fig. 16-29*a* the frequency is below resonance, and so the inductive susceptance is greater than the capacitive susceptance, making the net susceptance $B_T = -j(B_L - B_C)$.

At frequencies above the resonant frequency, B_C exceeds B_L and $B_T = +j(B_C - B_L)$, as shown in Fig. 16-29*b*.

In Fig. 16-29*a* and *b* the circuit admittance Y is the phasor sum of the net susceptance B_T and the conductance G. The phase angle ϕ is equal to arctan B_T/G.

The current relationships under off-resonance conditions will be

$$I_C \neq I_L$$
$$I_X = (I_C - I_L) > 0$$
$$I_T = I_R + j(I_C - I_L) > I_R$$

Phasor diagrams for an *RLC* parallel circuit operated at nonresonant frequencies are shown in Fig. 16-30.

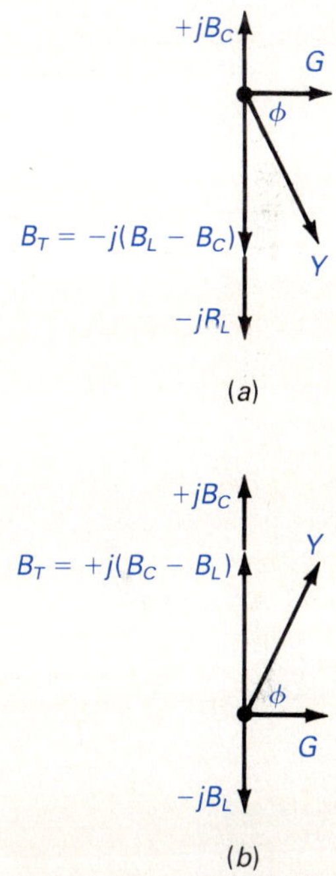

Fig. 16-29. Parallel *RLC* circuit admittance diagrams: *(a)* Below resonance; *(b)* above resonance.

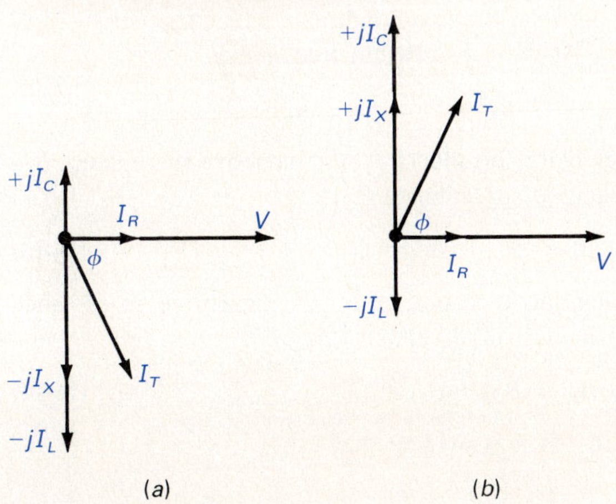

Fig. 16-30. Parallel *RLC* circuit phasor diagrams: *(a)* Below resonance; *(b)* above resonance.

Figure 16-30*a* shows the conditions below resonance, where $I_L > I_C$ and the net reactive current $I_X = -j(I_L - I_C)$. In Fig. 16-30*b* the frequency is above resonance ($I_C > I_L$), and the net reactive current is $+j(I_C - I_L)$. In both situations the total circuit current I_T is the phasor sum of the resistive current I_R and the net reactive current I_X. The phase angle ϕ is arctan I_X/I_R.

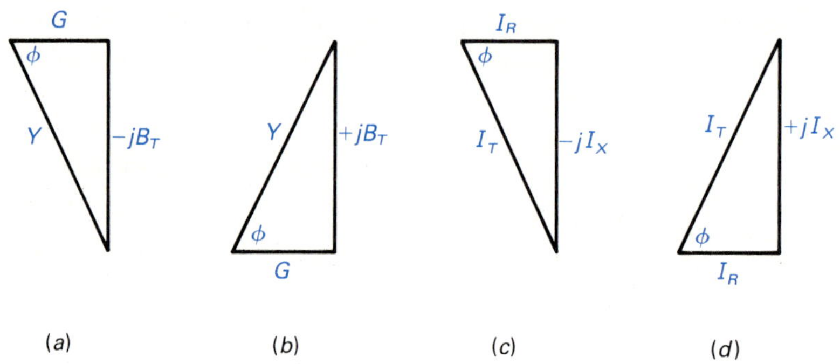

Fig. 16-31. Parallel *RLC* circuit. *(a)* Admittance triangle for below resonance; *(b)* admittance triangle for above resonance; *(c)* current triangle for below resonance; *(d)* current triangle for above resonance.

Figure 16-31*a* and *b* shows admittance triangles for frequencies below and above resonance and are derived from Fig. 16-29. The current triangles in Fig. 16-31*c* and *d* also apply to frequencies below and above resonance and are derived from Fig. 16-30.

At Frequencies Where $B_T = G$ At two frequencies, one below resonance and one above resonance, the net susceptance equals the conductance. These frequencies are the bandwidth frequencies for the parallel *RLC* circuit.

Bandwidth in a circuit consisting of *R*, *L*, and *C* all in parallel can be defined as the frequency difference between the two conditions where the net susceptance equals the conductance. At these bandwidth frequencies,

$$B_T = G$$

At the bandwidth frequency below resonance f_1, the inductive susceptance B_{L1} will be greater than the capacitive susceptance B_{C1}:

$$B_{T1} = B_{L1} - B_{C1} = G \tag{16-13}$$

At the bandwidth frequency above resonance f_2, the capacitive susceptance B_{C2} will be greater than the inductive susceptance B_{L2}:

$$B_{T2} = B_{C2} - B_{L2} = G \tag{16-14}$$

From Eqs. (16-13) and (16-14) it is evident that

$$B_{T1} = B_{T2} = G$$

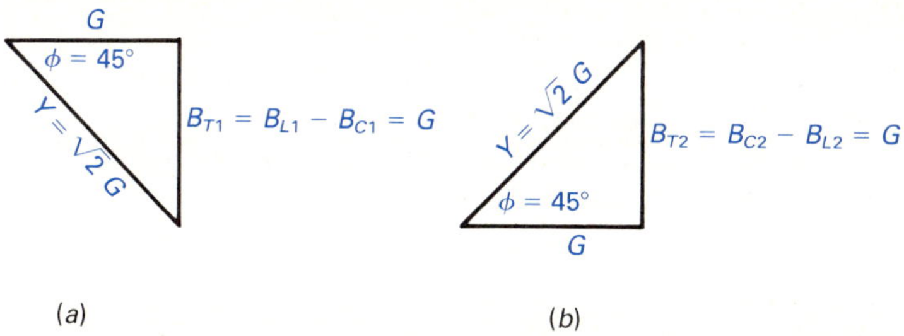

Fig. 16-32. Parallel *RLC* circuit, admittance triangles for the bandwidth frequencies: *(a)* Below resonance; *(b)* above resonance.

Figure 16-32*a* and *b* shows the admittance triangles for the bandwidth frequencies below and above resonance. For either bandwidth frequency,

$$B_T = G$$
$$Y = \sqrt{G^2 + B_T^2} = \sqrt{2}G$$

and
$$\phi = \arctan \frac{B_T}{G} = \arctan 1 = 45° = \pi/4 \text{ rad}$$

Admittance and Phase Angle Curves

From the foregoing discussion these characteristics of the parallel resonant circuit have emerged:

1. At resonance the admittance equals the conductance, and the phase angle is zero.
2. Off resonance the admittance is greater than the conductance, and the phase angle is greater than zero.
3. At the bandwidth frequencies the admittance is $\sqrt{2}$ times the conductance, and the phase angle is 45° or $\pi/4$ rad.

It can be added that:

4. At zero frequency the inductive susceptance is infinitely high, and the capacitive susceptance is zero. Thus at zero frequency the circuit consists of an infinitely high inductive susceptance in parallel with a conductance. The admittance is infinite, and the phase angle is $-90°$ or $-\pi/2$ rad.
5. If the frequency could become infinitely high, the capacitive susceptance would become infinite and the inductive susceptance zero. In this case the circuit would be in effect an infinite capacitive susceptance in parallel with a conductance. The admittance would be infinite, and the phase angle would be $+90°$ or $\pi/2$ rad.
6. Between zero frequency and resonance, and between resonance and infinite frequency, the change in admittance and phase angle is gradual, the rate of change depending on the specific circuit values.

The curves in Fig. 16-33 show the effect of varying frequency on the admittance magnitude and the phase angle of a parallel resonant circuit which has the values $L = 100\ \mu\text{H}$, $C = 250$ pF, and $R = 5\ \text{k}\Omega$ ($G = 0.2$ mS).

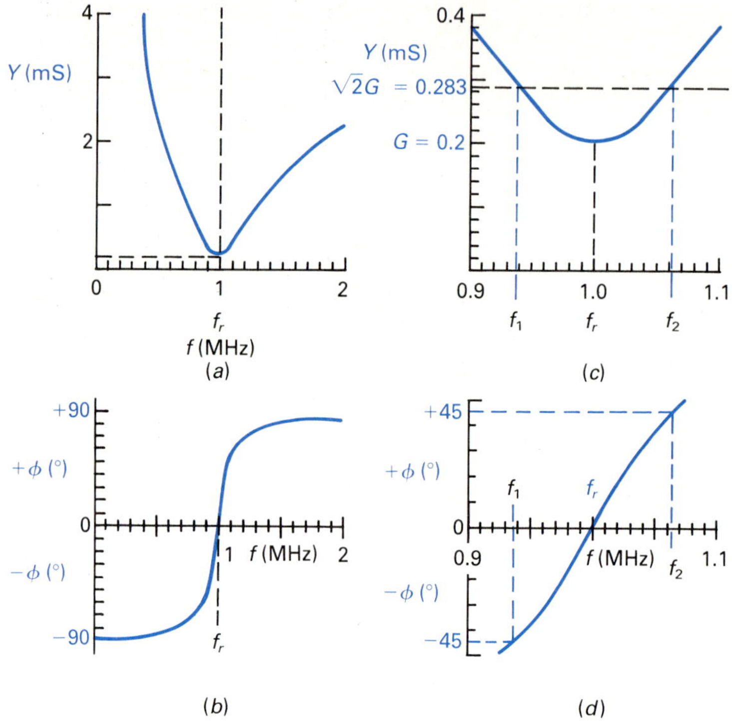

Fig. 16-33. Parallel *RLC* circuit curves. *(a)* Admittance-frequency; *(b)* phase angle–frequency; *(c)* admittance-frequency close to resonance; *(d)* phase angle–frequency close to resonance.

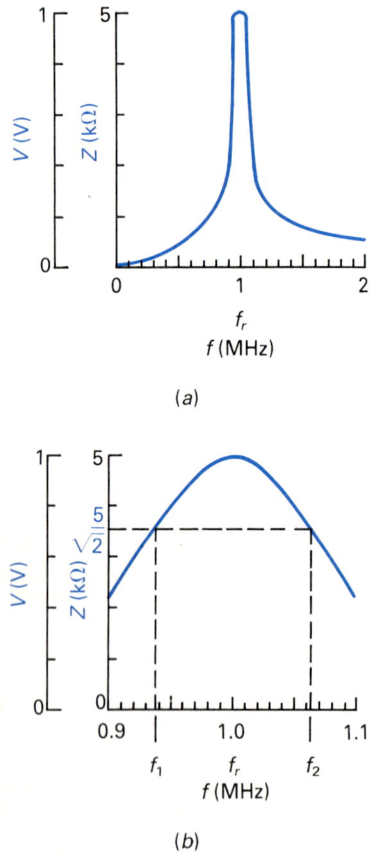

Fig. 16-34. Parallel *RLC* circuit. Curves of impedance-frequency and voltage-frequency.

In Fig. 16-33*a* the admittance decreases from infinity to the value of the conductance ($G = 0.2$ mS) at the resonant frequency f_r. Above resonance the admittance again increases toward infinity.

In Fig. 16-33*b* the circuit phase angle decreases from $-90°$ (inductive) toward zero (resistive) at the resonant frequency, and then increases toward $+90°$ (capacitive) above resonance.

Figure 16-33*c* and *d* shows in detail the variations in admittance and phase angle close to the resonant frequency. Figure 16-33*c* shows the admittance magnitude at the bandwidth frequencies f_1 and f_2 to be $\sqrt{2}G = 0.283$ mS. Figure 16-33*d* shows a phase angle of 45° at f_1 and f_2.

It can be useful to refer to the impedance of parallel circuits, particularly when series and parallel network characteristics are being compared. Figure 16-34 shows how the impedance and voltage magnitudes of a parallel resonant circuit vary with changing frequency.

In Fig. 16-34*a* the impedance of the *RLC* parallel circuit rises from zero at zero frequency to a maximum value of $R = 5$ kΩ at resonance and then drops toward zero as the frequency increases beyond resonance. The impedance curve is magnified near the region of resonance in Fig. 16-34*b*, showing the value of $Z_r/\sqrt{2} = 5\ \text{k}\Omega/\sqrt{2}$ at the bandwidth frequencies f_1 and f_2.

The voltage across a network is directly proportional to its impedance. Therefore the curves in Fig. 16-34 show how the voltage across a parallel resonant circuit varies with frequency if the total current is maintained constant.

Subsidiary scales to the left of Fig. 16-34*a* and *b* give the voltage values with a constant current of 0.2 mA.

REVIEW QUESTIONS FOR SECTION 16-4

1. Write the current equation for a parallel *RLC* circuit.
2. Write the equation for the frequency of a parallel *RLC* circuit at resonance.
3. How does the inductive current of a parallel *RLC* circuit compare with the capacitive current when the circuit is operated
   a. At its resonant frequency?
   b. Above its resonant frequency?
   c. Below its resonant frequency?

16-5 BANDWIDTH OF A PARALLEL RESONANT CIRCUIT

Bandwidth and related concepts, such as selectivity, have the same significance for all circuits. This is a convention which enables performance characteristics to be specified irrespective of circuit configuration. Thus the power definition of bandwidth for a parallel resonant circuit is the same as for a series circuit.

In a parallel resonant circuit supplied with constant current, maximum power is dissipated at the resonant frequency. From the definition of bandwidth,

$$P_1 = P_2 = \frac{P_r}{2} \tag{16-7}$$

In an ac circuit, power is dissipated only in resistive elements; in a parallel circuit it is given by $P = V^2/R$ or $P = V^2G$. Thus Eq. (16-7) can be written as

$$V_1^2G = V_2^2G = \frac{V_r^2G}{2}$$

$$V_1^2 = V_2^2 = \frac{V_r^2}{2}$$

or

$$V_1 = V_2 = \frac{V_r}{\sqrt{2}} = 0.7071\ V_r \tag{16-15}$$

where V_1 = voltage at low bandwidth frequency f_1, V
V_2 = voltage at high bandwidth frequency f_2, V
V_r = resonant frequency voltage, V

The bandwidth *BW* is given by

$$BW = f_2 - f_1 \tag{16-9}$$

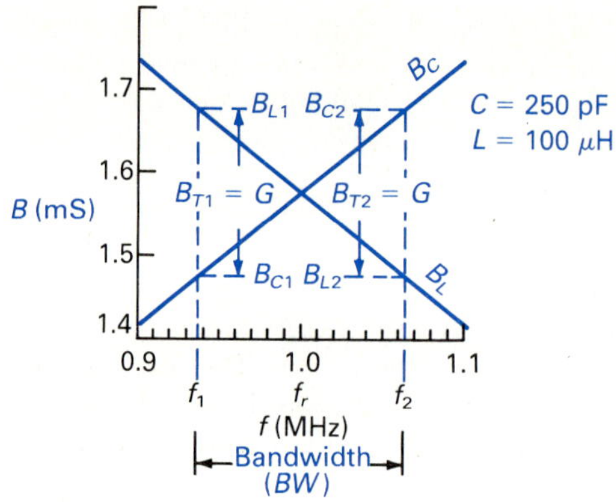

Fig. 16-35. Parallel resonant circuit susceptance-frequency curves expanded close to resonance to show bandwidth conditions.

Figure 16-35 shows, in expanded form, the curves for B_C and B_L close to the resonant frequency for the values $C = 250$ pF and $L = 100$ μH. For any other values of C and L the appropriate curves will be the same except for the axis scaling.

At the bandwidth frequencies the net susceptance equals the conductance. Therefore, at frequency f_2,

$$B_{T2} = B_{C2} - B_{L2} = G \tag{16-14}$$

By applying to Eq. (16-14) and Fig. 16-35 the same logic as was applied to Eq. (16-6) and Fig. 16-20, the bandwidth for the parallel resonant circuit is found to be the same as for the series circuit:

$$BW \simeq \frac{f_r}{Q_r} \tag{16-10}$$

EXAMPLE 16-4 A capacitor of 500 pF, an inductor of 50 μH, and a resistor of 5 kΩ are connected in parallel. What is *(a)* the resonant frequency f_r and *(b)* the bandwidth BW?

Solution The circuit applicable to this example is given in Fig. 16-36.

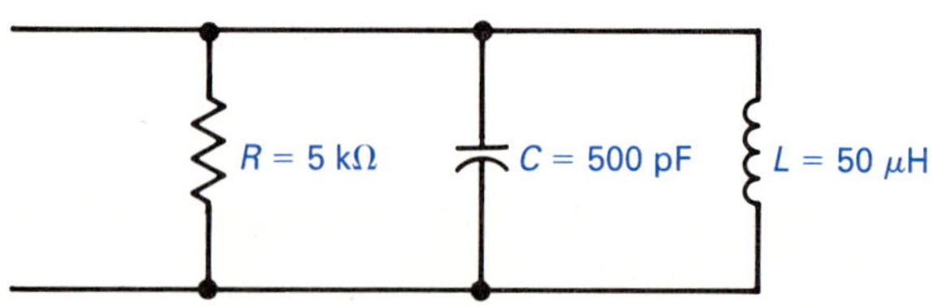

Fig. 16-36. Example 16-4: Network.

a.
$$f_r = \frac{1}{2\pi\sqrt{LC}} = \frac{1}{2\pi\sqrt{500\text{ pF} \times 50\ \mu\text{H}}} = \mathbf{1.007\ MHz}$$

$$b.\ Q_r = \frac{B_{Cr}}{G} = \frac{2\pi f_r C}{1/R} = 2\pi \times 1.007 \text{ MHz} \times 500 \text{ pF} \times 5 \text{ k}\Omega$$
$$= 15.82$$

$$BW \simeq \frac{f_r}{Q_r} \simeq \frac{1.007 \text{ MHz}}{15.82} \simeq \mathbf{63.65 \text{ kHz}}$$

The parallel circuit in Example 16-4 has the same resonant frequency, quality factor, and bandwidth as the series circuit in Example 16-1. The resonant frequency of each circuit is determined by the same relationship ($f_r = 1/2\pi\sqrt{LC}$), so, since the LC product is the same for each example, the resonant frequencies must likewise be the same.

The quality factor at resonance is the same for each example. Therefore,

$$\text{Series resonant } Q = \text{parallel resonant } Q$$
$$Q_{rs} = Q_{rp}$$
$$\frac{X_{rs}}{R_s} = \frac{B_{rp}}{G_p} \quad \text{or} \quad \frac{X_{rs}}{R_s} = \frac{R_p}{X_{rp}}$$

The two circuits have the same values of L and C, so X_r (the reactance at resonance of either L or C) is the same:

$$R_s R_p = X_{rs} X_{rp} = X_r^2$$

For series or parallel resonance ($X_{Lr} = X_{Cr}$),

$$X_r^2 = X_{Lr} X_{Cr} = \frac{\omega_r L}{\omega_r C}$$

therefore

$$R_s R_p = \frac{L}{C} \tag{16-16}$$

Substituting $R_s = 20\ \Omega$ from Example 16-1 into Eq. (16-16),

$$R_p = \frac{L}{C} \times \frac{1}{R_s} = \frac{5 \times 10^{-5} \text{ H}}{5 \times 10^{-10} \text{ F}} \times \frac{1}{20\ \Omega}$$
$$R_p = 5 \text{ k}\Omega$$

This derivation of R_p explains why 20-Ω series resistance is equivalent to 5-kΩ parallel resistance if the capacitance and inductance values are 500 pF and 50 μH.

For any combination of L and C, Eq. (16-16) gives the relationship between the series and parallel circuits which will have the same resonant frequency and bandwidth.

In Example 16-4 the reactances are both 316.4 Ω, $Q_r = 15.82$, and $BW = 63.65$ kHz. For the values $C = 250$ pF, $L = 100\ \mu$H, and $R = 5$ kΩ, the resonant reactances are both 632.8 Ω, $Q_r = 7.91$, and $BW = 127.3$ kHz. For any given value of resistance, the bandwidth of the parallel circuit with an inductance of 50 μH will be narrower than one with 100 μH by a ratio 100 μH/50 μH = 2. Figure 16-37 shows two voltage response curves (constant current); solid-line curve represents the values $C = 500$ pF, $L = 50\ \mu$H, $R = 5$ kΩ, and the broken-line curve the values $C = 250$ pF, $L = 100\ \mu$H, $R = 5$ kΩ.

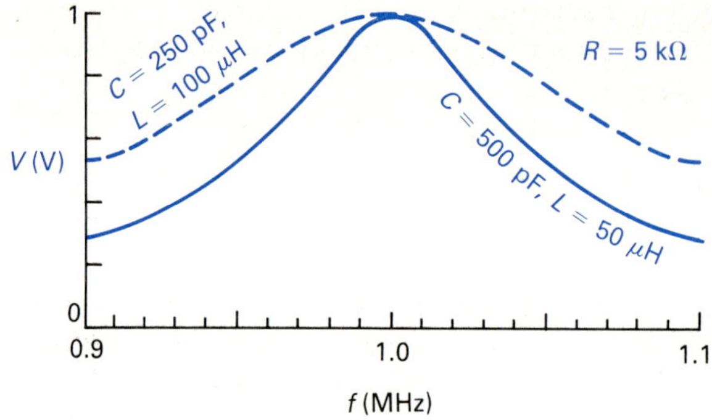

Fig. 16-37. Parallel *RLC* voltage-frequency curves for two sets of values: Solid line for the values in Example 16-4; broken line for doubled inductance and halved capacitance.

Figure 16-37 should be compared with Fig. 16-22. The solid-line curves represent parallel and series equivalent circuits. In each case the broken-line curves represent changes in capacitance and inductance by the same amount, the resistance remaining constant.

For parallel circuits with the same resonant frequency, low inductance will result in narrow bandwidth for a given resistance value.

For parallel resonant circuits with the same values of L *and* C, *high parallel resistance results in narrow bandwidth.*

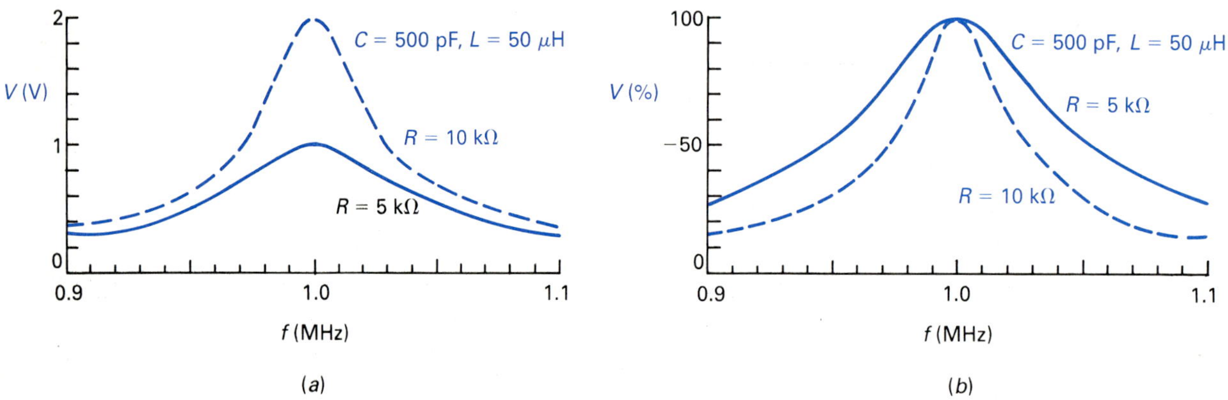

Fig. 16-38. Parallel *RLC* circuit voltage-frequency curves for two resistance values: *(a)* Actual voltage values; *(b)* values normalized on a percentage basis.

Figure 16-38*a* and *b* shows voltage curves for the L and C values in Example 16-4 with the original 5-kΩ resistance and with the resistance increased to 10 kΩ. The curves in Fig. 16-38*a* show the actual voltages with 0.2-mA current supply. The curves are normalized on a percentage basis in Fig. 16-38*b*.

Compare Fig. 16-38 with Fig. 16-23. They both represent the same values of L and C. The solid-line curves apply to parallel and series circuits which are equivalent in terms of Q_r and bandwidth. However, it can be seen from the broken-line curves that the effect of resistance change in the parallel resonant circuit is opposite that in the series circuit.

EXAMPLE 16-5 What are the currents at resonance in each element in the circuit in Fig. 16-39 if 1 mA is supplied to the complete circuit?

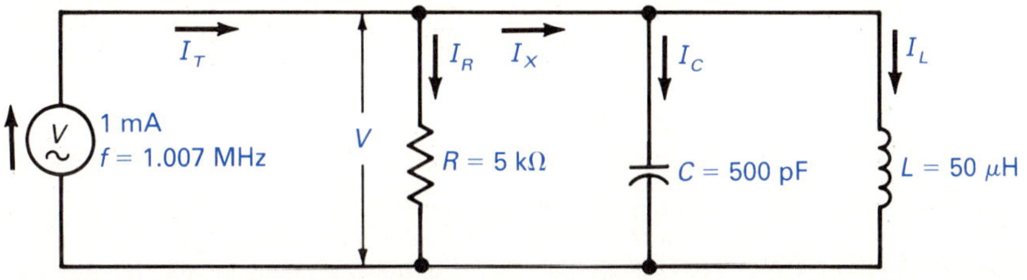

Fig. 16-39. Example 16-5: Circuit.

Solution *a.* At resonance the resistive current is the total current fed to the circuit:

$$I_R = I_T = \mathbf{1\ mA}$$

b. At resonance the currents in the capacitance and in the inductance are both Q_r times the total circuit current. The values of R, L, and C are the same as in Example 16-4, so the value of Q_r is the same.

$$I_C = I_L = Q_r I_T \qquad (16\text{-}12)$$

$$I_C = I_L = 15.82 \times 1\ \text{mA} = \mathbf{15.82\ mA}$$

Circulating Current

In Fig. 16-39 the arrows for I_C and I_L both point downward. But these two currents are in antiphase; their directions are opposite. If during one half-cycle the capacitive current is downward, the inductive current will be upward. During the next half-cycle the capacitive current will be upward and the inductive current downward. Thus a current arrow in Fig. 16-39 indicates the presence of current rather than its direction. In ac circuits, where many phase relationships can exist simultaneously, it is only feasible for arrows to indicate current presence.

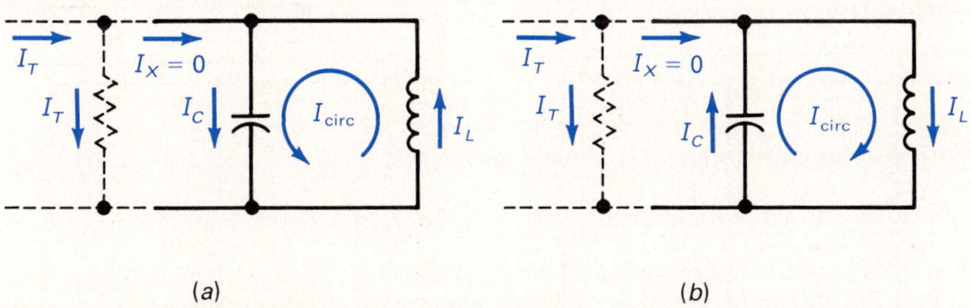

Fig. 16-40. Parallel *RLC* circuit. Instantaneous reactive current directions.

Figure 16-40 shows the two alternate current directions in capacitance and inductance. In either situation, $I_C = I_L = Q_r I_T$. Since $I_C = I_L$ and they are in the same loop direction (clockwise or counterclockwise), they may be said to form a single current which circulates within the loop.

$$I_{circ} = Q_r I_T \qquad \textbf{(16-17)}$$

This concept of a circulating current, of value $Q_r I_T$, within the capacitance-inductance loop of a parallel resonant circuit brings a new aspect to the significance of the quality factor Q_r.

The magnitude of the circulating current in a parallel resonant circuit is Q_r times the magnitude of the total current fed to the circuit.

The current in either reactance is 15.82 mA, 15.82 times as great as the current fed to the circuit. These particular circuit values therefore result in a current magnification (sometimes referred to as *current gain*) of 15.82, the same as the value of Q_r. There is no power gain, since the only dissipation occurs in the resistive component of the circuit ($P_T = I_T^2 R = I_R^2 R$).

Tank Circuit

Visualize the notion that the parallel resonant circuit contains energy which circulates back and forth within it via the circulating current. Energy is continually being interchanged between the inductance and the capacitance. The energy remains within the magnetic and electric fields. This concept leads to the view of a parallel resonant circuit as a container of electrical energy.

So evolved the term ''tank circuit'' used in radio work, a parallel resonant circuit being considered an energy storage tank. The current, and hence the energy, fed to the *RLC* parallel resonant circuit can be small, while the circulating energy is large. This notion suggests a tank of large capacity which is fed slowly, just fast enough to compensate for any loss of contents. The energy in L and C never needs replenishing because reactive elements do not dissipate power. Energy to the circuit provides for the loss incurred by the resistance.

EXAMPLE 16-6 A circuit consists of a 20-pF capacitor and a 10-kΩ resistor connected in parallel. What inductance value is needed to make the circuit resonant at 16 MHz? What is the effect of connecting a 10-μH inductance in series with the first one?

Solution

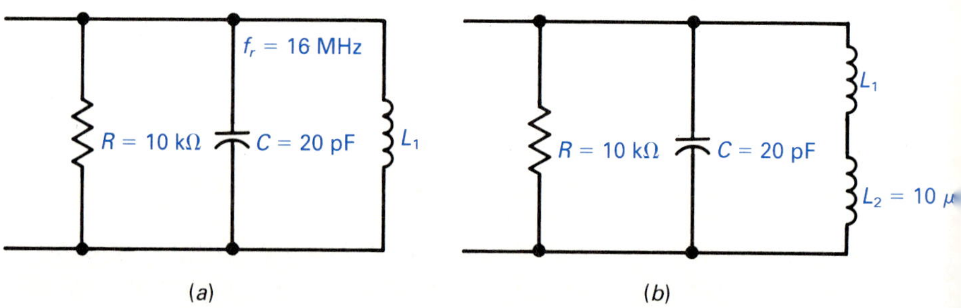

Fig. 16-41. Example 16-6: Circuits: *(a)* Original; *(b)* with added inductance.

In Fig. 16-41*a*,

$$\omega_r = 2\pi f_r = 2\pi \times 16 \text{ MHz} = 100.5 \text{ Mrad/s}$$

Applying Eq. (16-1):

$$\omega_r^2 = \frac{1}{L_1 C}$$

$$L_1 = \frac{1}{\omega_r^2 C} = \frac{1}{(100.5 \text{ Mrad/s})^2 \times 20 \text{ pF}} = \mathbf{4.95\ \mu H}$$

In Fig. 16-41*b*, L_1 and L_2 are connected in series. Assuming no mutual inductance, the total inductance is

$$L_T = L_1 + L_2 = 4.95\ \mu\text{H} + 10\ \mu\text{H} = 14.95\ \mu\text{H}$$

Apply Eq. (16-2):

$$f_r = \frac{1}{2\pi\sqrt{L_T C}}$$

$$f_r \propto \frac{1}{\sqrt{L_T}}$$

$$f_r = \sqrt{\frac{4.95\ \mu\text{H}}{14.95\ \mu\text{H}}} \times 16 \text{ MHz} = 9.207 \text{ MHz}$$

The addition of L_2 reduces the resonant frequency from **16** to **9.207 MHz.**

Figure 16-41*b* is a series-parallel circuit. However, it is not regarded as a series-parallel resonant circuit, since the total circuit inductance, capacitance, and resistance are connected in parallel.

The proportionality approach was used in the solution to Example 16-6. Not only does this technique give a simple solution, but an understanding of circuit behavior is aided by a sense of the effect of proportionality.

REVIEW QUESTIONS FOR SECTION 16-5

1. Compare the meanings of ''bandwidth'' and ''bandwidth points'' for series and parallel resonant circuits.
2. What is the effect of increasing the resistance of a parallel resonant circuit?

16-6 PRACTICAL PARALLEL RESONANT CIRCUIT

When a real inductor is connected in parallel with a real capacitor, the result is a practical parallel resonant circuit. The inductor and the capacitor will both have losses which add resistance to the inductance and capacitance. In many applications of parallel resonance it is desired to keep this resistance to a minimum, so as to achieve narrow bandwidth and/or high circulating current. An inductor's losses can be great enough to require consideration. By comparison with those of an inductor, the losses of a good capacitor can be ignored. Thus a practical parallel resonant circuit is regarded as a lossy inductor in parallel with a perfect capacitor.

The inductor can be represented by equivalent series or parallel *RL* networks. The series view is more commonly used, since many instruments for measuring the characteristics of inductors give the results in terms of series components.

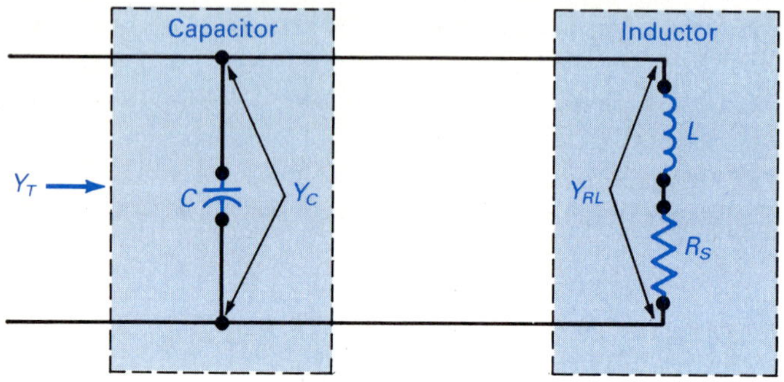

Fig. 16-42. Practical parallel resonant circuit.

A series-parallel version of the practical parallel resonant circuit is given in Fig. 16-42. To determine its resonant characteristics, the most direct method is to refer to its admittance:

$$Y_T = Y_C + Y_{RL}$$

$$= j\omega C + \frac{R_s - j\omega L}{R_s^2 + \omega^2 L^2} = j\omega C + \frac{R_s}{R_s^2 + \omega^2 L^2} - j\frac{\omega L}{R_s^2 + \omega^2 L^2}$$

$$Y_T = \frac{R_s}{R_s^2 + \omega^2 L^2} + j\left(\omega C - \frac{\omega L}{R_s^2 + \omega^2 L^2}\right) \qquad \textbf{(16-18)}$$

The circuit will be resonant at the frequency f_r, where the admittance is entirely resistive. This condition is met when the imaginary component of the admittance expression [Eq. (16-18)] becomes zero in magnitude:

$$\omega_r C - \frac{\omega_r L}{R_s^2 + \omega_r^2 L^2} = 0$$

$$\omega_r^2 = \frac{1}{LC} - \frac{R_s^2}{L^2}$$

or

$$\omega_r^2 = \frac{1}{LC}\left(1 - \frac{R_s^2 C}{L}\right) \qquad \textbf{(16-19)}$$

When Eq. (16-19) is written in the frequency form,

$$f_r = \frac{1}{2\pi\sqrt{LC}}\sqrt{1 - \frac{R_s^2 C}{L}} \qquad \textbf{(16-20)}$$

it can be seen that, theoretically, the resonant frequency of a practical parallel resonant circuit differs from that of either a series circuit or a purely parallel circuit even if all three circuits have the same values of inductance and capacitance. The degree of difference is slight in practice, since $1 - R_s^2C/L$ is usually close to unity ($R_s^2C/L << 1$). For instance, consider the elements in the series circuit in Example 16-3. If the 527.7-pF capacitor were connected in parallel with the 300-μH 10-Ω inductor, the quantity R_s^2C/L would equal 1.757×10^{-4} and the resulting resonant frequency would be almost exactly equal to the 400 kHz of the series circuit.

The frequency of a radio transmitter must be maintained to a high order of accuracy to avoid interfering with other transmissions. For this purpose and for other precision frequency needs, frequencies are often quoted to seven or more significant figures. This order of accuracy is rarely needed in the calculation of resonant frequencies of *RLC* circuits. Resonant circuits cannot be manufactured to close frequency limits, and final adjustment is achieved by the use of variable component parameters. Thus the term $\sqrt{1 - R_s^2C/L}$ is often ignored, and Eqs. (16-1) and (16-2) are applied to practical parallel resonant circuits as well as to purely parallel and series *RLC* circuits. A commonly used rule is: If the Q of the inductor at the resonant frequency is 10 or more, the resistance can be considered to have an insignificant effect on the resonant frequency.

The admittance at resonance is the real part of Eq. (16-18), the imaginary part being zero:

$$Y_r = \frac{R_s}{R_s^2 + \omega_r^2L^2}$$

From this equation we can obtain an expression for the impedance of the practical parallel resonant circuit at resonance:

$$Z_r = \frac{1}{Y_r} = \frac{R_s^2 + (\omega_rL)^2}{R_s}$$

or

$$Z_r = R_s + \frac{(\omega_rL)^2}{R_s}$$

Substituting X_{Lr} for ω_rL,

$$Z_r = R_s + \frac{X_{Lr}^2}{R_s} \tag{16-21}$$

As $X_{Lr}/R_s = Q_r$,

$$Z_r = R_s + Q_rX_{Lr} \tag{16-22}$$

where Z_r = total circuit impedance, Ω
R_s = series resistance of the inductor, Ω
Q_r = the quality factor of the inductor (also the quality factor of the complete circuit)
X_{Lr} = inductive reactance, Ω

All terms are to be taken at their resonant frequency values. This applies to the resistance R_s which represents the total losses of the inductor, these being influenced by frequency.

Equation (16-22) gives an expression for the circuit impedance which is more accurate than is usually necessary. The equation can be simplified by adopting reasonable practical assumptions:

$$X_{Lr} >> R_s$$
$$Q_r X_{Lr} >> R_s$$

Therefore,

$$Z_R \simeq Q_r X_{Lr}$$

or, since

$$X_{Lr} \simeq X_{Cr}$$
$$Z_r \simeq Q_r X_r \qquad \textbf{(16-23)}$$

where X_r = reactance of either inductor or capacitor, Ω.

Substituting X_{Cr}/R_s for $1/\omega_r C R_s$ for Q_r in Eq. (16-23),

$$Z_r \simeq \frac{\omega_r L}{\omega_r C R_s}$$

$$Z_r \simeq \frac{L}{C R_s} \qquad \textbf{(16-24)}$$

The dynamic impedance Z_r of the practical parallel resonant circuit, as given in Eq. (16-24), is important in the design of circuits for maximum voltage ($V_r = IZ_r$) or maximum power transfer (Z_r is resistive, and maximum power is transferred when Z_r equals the internal resistance of the source).

Practical parallel resonant circuits are used so extensively in commercial equipment that a summary of their characteristics is justified. You will see that a number of the items listed are the same as those for the purely parallel *RLC* circuit. Both true and practical parallel circuits behave in the same general manner; only some numerical details are different.

At resonance:

1. The power factor is unity, and the phase angle is zero.
2. The impedance is resistive and approximately of maximum value.
3. The impedance magnitude is given by $Z_r = R_s + X_{Lr}^2/R_s$, by $Z_r = R_s + Q_r X_r$, or approximately by $Z_r = L/CR_s$.
4. The resonant frequency is given by $f_r = [1/(2\pi\sqrt{LC})] \times \sqrt{1 - R_s^2 C/L}$, or approximately by $f_r = 1/2\pi\sqrt{LC}$ (the same as for a series resonant circuit).

Off resonance:

5. Below resonance the circuit is inductive.
6. Above resonance the circuit is capacitive.

At the bandwidth points:

7. The phase angle is 45°.
8. The impedance is $Z_r/\sqrt{2} \simeq L/CR_s\sqrt{2}$ or $0.7071\ L/CR_s$.
9. The bandwidth is given by $BW \simeq f_r/Q_r$.

EXAMPLE 16-7 The measured series characteristics of an inductor at 1 kHz are 200 mH inductance and 500 Ω resistance. A 0.1-μF capacitor is connected in parallel with the inductor. What is the resonant frequency of the circuit?

Solution (See Fig. 16-43.)

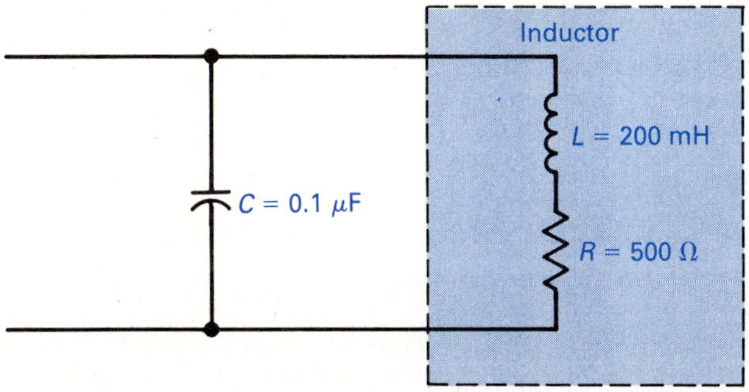

Fig. 16-43. Example 16-7: Circuit.

The Q of the coil at 1 kHz is

$$Q = \frac{X_L}{R} = \frac{2\pi fL}{R} = \frac{2\pi \times 1\text{ kHz} \times 200\text{ mH}}{500\ \Omega} = 2.513$$

This Q value is well below 10, so Eq. (16-20) should be used to determine the resonant frequency:

$$f_r = \frac{1}{2\pi\sqrt{LC}}\sqrt{1 - \frac{R^2C}{L}} \qquad (16\text{-}20)$$

$$= \frac{1}{2\pi\sqrt{200\text{ mH} \times 0.1\ \mu\text{F}}}\sqrt{1 - \frac{(500\ \Omega)^2 \times 0.1\ \mu\text{F}}{200\text{ mH}}}$$

$$= \frac{1}{2\pi\sqrt{2 \times 10^{-8}}}\sqrt{1 - 0.125} = 1125 \times \sqrt{0.875}$$

$$f_r = \mathbf{1053\ Hz}$$

As the resonant frequency in Example 16-7 is close to 1 kHz, the inductance and resistance values measured at 1 kHz were valid for use in the calculation.

If the general equation $f_r = 1/2\pi\sqrt{LC}$ had been used for this low-Q circuit, the resonant frequency would have been computed as 1.125 kHz, representing an error of 6.837 percent. With the resistance decreased to 50 Ω ($Q \simeq 25$), the error would be only 0.0626 percent.

EXAMPLE 16-8 What voltage is developed across the circuit in Example 16-7 if a current of 1 A is supplied to it *(a)* at the resonant frequency and *(b)* at 60 Hz? Assume that the inductance and resistance values of the inductor are the same at 60 Hz as at 1 kHz.

Solution *a.* At the resonant frequency, 1053 Hz, the Q at resonance will be

$$Q_r = \frac{\omega_r L}{R} = \frac{2\pi \times 1053\text{ Hz} \times 200\text{ mH}}{500} = 2.646$$

The impedance at resonance will be

$$Z_r = R + Q_r X_{Lr} \qquad (16\text{-}22)$$

$$Z_r = 500\ \Omega + 2.646 \times 2\pi \times 1053\ \text{Hz} \times 200\ \text{mH} = 4.001\ \text{k}\Omega$$

The circuit voltage at resonance is the product of the total current and the resonant impedance:

$$V_r = I_T Z_r = 1\ \text{A} \times 4.001\ \text{k}\Omega = \mathbf{4.001\ kV}$$

b. At 60 Hz,

$$\omega = 2\pi \times 60\ \text{Hz} = 377\ \text{rad/s}$$

$$X_L = \omega L = 377\ \text{rad/s} \times 200\ \text{mH} = 75.4\ \Omega$$

The impedance of the *RL* branch is

$$Z_{RL} = R + jX_L = (500 + j75.4)\ \Omega = 505.7\ \Omega\ \underline{/8.58^\circ}$$

$$X_C = -j\frac{1}{\omega C} = -j\frac{1}{377\ \text{rad/s} \times 0.1\ \mu\text{F}} = -j26.53\ \text{k}\Omega$$

The magnitude of X_C is very much greater than that of Z_{RL}. Therefore, since X_C is in parallel with Z_{RL}, its effect is negligible, and the total circuit impedance at 60 Hz is approximately the same as the impedance of the *RL* branch:

$$V = I_T Z \simeq I_T Z_{RL} \simeq 1\ \text{A} \times 505.7\ \Omega\ \underline{/8.58^\circ}$$

$$V \simeq \mathbf{505.7\ V\ \underline{/8.58^\circ}} \quad \text{or} \quad \mathbf{\underline{/0.1497\ rad}}$$

Rms values of voltage and current are assumed in Example 16-8 because no value types are specified. Thus the peak value of the circuit voltage at resonance is $4.001\sqrt{2} = 5.658$ kV. The 0.1-μF capacitor would have to be able to withstand this high voltage without sustaining damage.

EXAMPLE 16-9 The Q of a 25-μH inductor is 75 at a frequency of 3 MHz. What is *(a)* the value of a parallel capacitor required to make the inductor resonant at 3 MHz, *(b)* the resulting bandwidth, and *(c)* the resonant impedance of the circuit?

Solution As the Q of the inductor (which is also the circuit Q at resonance) is well in excess of 10, the approximate equations can be used with reasonable accuracy. In circumstances such as these, it is not usual to adhere rigidly to the approximately equals sign.

a. $\omega_r = 2\pi f_r = 2\pi \times 3\ \text{MHz} = 18.85\ \text{Mrad/s}$

$$\omega_r^2 = \frac{1}{LC} \qquad (16\text{-}1)$$

$$C = \frac{1}{\omega_r^2 \text{L}} = \frac{1}{(18.85\ \text{Mrad/s})^2 \times 25\ \mu\text{H}} = \mathbf{112.6\ pF}$$

b. $$BW = \frac{f_r}{Q_r} \qquad (16\text{-}10)$$

$$BW = \frac{3\text{ MHz}}{75} = \mathbf{40\ kHz}$$

c. $$Z_r = Q_r X_r \qquad (16\text{-}23)$$

$$Z_r = Q_r \omega_r L = 75 \times 18.85\text{ Mrad/s} \times 25\ \mu\text{H} = \mathbf{35.34\ k\Omega}$$

As noted in the solution to Example 16-9, approximately equals signs are sometimes not used even though the results are only approximate. This is because resonant circuits, particularly high-Q ones, are subject to many variables which are difficult to predict accurately. Thus it is usual to adjust these circuits by measurement, after construction. In the circuit in Example 16-9, it is likely that the capacitor, calculated as 112.6 pF, will be made variable for final adjustment. Typically, this capacitance would be made up of a 100-pF fixed capacitor with a 25-pF variable capacitor connected in parallel with it. This arrangement would give a range of capacitance values from 100 to 125 pF, or about 12.5 pF on either side of the calculated value.

EXAMPLE 16-10 The resonant network in Example 16-9 is part of an amplifier circuit. A transistor supplies 1 mA to the network at its resonant frequency of 3 MHz. What is the magnitude of the circulating current at resonance?

Solution (See Fig. 16-44.)

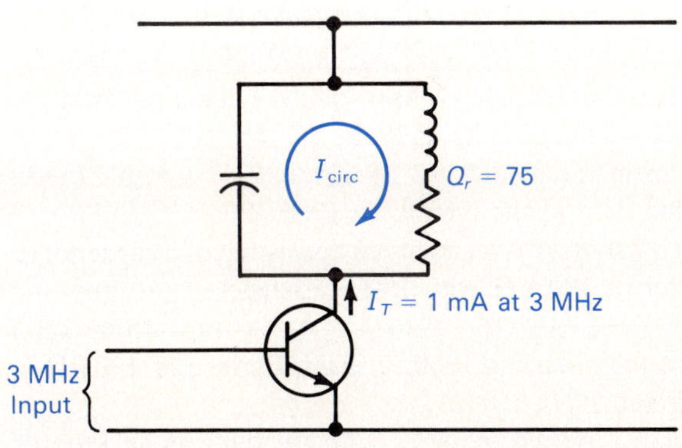

Fig. 16-44. Example 16-10: Circuit.

$$I_{circ} = Q_r I_T \qquad (16\text{-}17)$$

$$I_{circ} = 75 \times 1\text{ mA} = \mathbf{75\ mA}$$

Although the current supplied to the resonant circuit in Example 16-10 is only 1 mA, which is easily handled by a small transistor, the current in the inductor

(75 mA) is sufficiently great to pose a design and construction problem if space is limited.

REVIEW QUESTIONS FOR SECTION 16-6

1. What is meant by the term "practical parallel resonant circuit"?
2. If the Q factor of a practical parallel resonant circuit is high, what is
   a. An equation for the approximate frequency of resonance?
   b. An equation for the approximate value of impedance?
3. What is meant by
   a. Circulating current?
   b. Tank circuit?

EXERCISE PROBLEMS FOR SECTIONS 16-4 TO 16-6

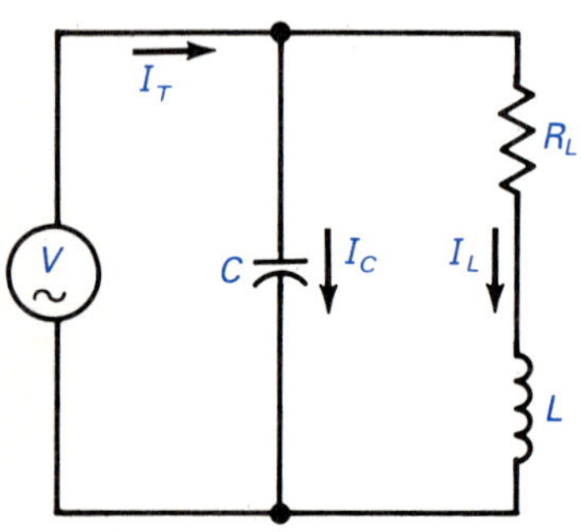

Fig. 16-45. Problems 1 to 6: Circuit.

1. For the parallel RLC circuit in Fig. 16-45, find the resonant frequency using Eq. (16-20) when
   a. $V = 10$ V, $C = 150$ pF, $L = 100$ μH, $R = 20$ Ω (*ans.* 1.299 MHz)
   b. $V = 10$ V, $C = 150$ pF, $L = 100$ μH, $R = 200$ Ω (*ans.* 1.2599 MHz)
   c. $V = 5$ V, $C = 0.022$ μF, $L = 12$ mH, $R = 10$ Ω (*ans.* 9.795 kHz)
2. Which of the frequencies of the circuits in Prob. 1 would not be appreciably changed by solving with Eq. (16-2)?
3. Using the circuits in Prob. 1, find the reactances of L and C at the resonant frequencies.
   (*ans. a.* $+j816.2$ Ω, $-j816.8$ Ω; *b.* $+j791.6$ Ω, $-j842.2$ Ω; *c.* $+j738.5$ Ω, $-j738.6$ Ω)
4. Find the Q and the total impedance of each of the circuits in Prob. 1.
5. Find I_T and I_L for each of the circuits in Prob. 1.
   (*ans. a.* 300 μA, 12.25 mA; *b.* 3 mA, 12.24 mA; *c.* 91.7 μA, 6.77 mA)
6. Find the bandwidth for each of the circuits in Prob. 1.
7. Find the bandwidth of a parallel resonant circuit consisting of a capacitor of 120 pF and an inductor of 180 μH with 18-Ω resistance. (*ans.* 15.92 kHz)
8. Find the bandwidth of the circuit in Prob. 7 if a resistance of 100 kΩ is shunted across the circuit.
9. What additional resistance placed in series with the inductor in Prob. 7 would give essentially the same results as Prob. 8? (*ans.* 15 Ω)

16-7 SOME APPLICATIONS OF RESONANCE

Resonance is a phenomenon which occurs, intentionally or otherwise, so often that only a small sampling of its many applications can be given. The following selection will serve to show the versatility of resonance and indicate how the principles we have dealt with can be applied to other resonant circuits.

Double Tuning

Arranging for a circuit to be in resonance is often called tuning, the resonant circuit then being referred to as a tuned circuit. When a circuit resonates at two different frequencies, it is called a double-tuned circuit. It should be mentioned that the term "double-tuned" is also used for a transformer which has two of its windings resonant at the same frequency.

EXAMPLE 16-11 The network in Fig. 16-46 is part of the circuit of a transistor radio which operates on either AM signals or FM signals. On AM the input to the circuit is at a frequency of 450 kHz, while on FM it is at 10.7 MHz. How does this circuit discriminate between the two frequencies? Neglect the effects of any loading on the outputs.

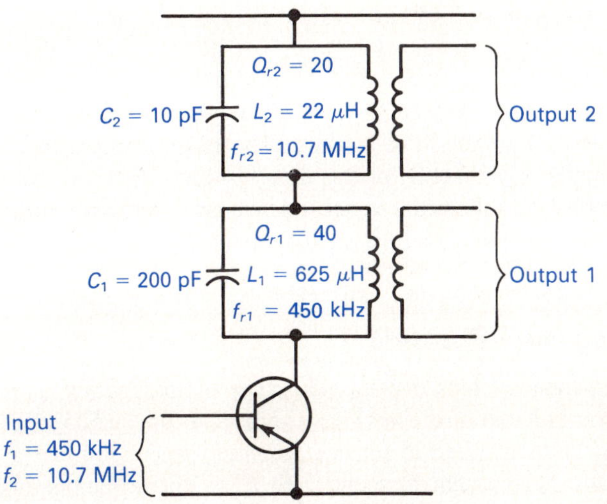

Fig. 16-46. Example 16-11: Circuit.

Solution At frequency $f_1 = 450$ kHz ($\omega_1 = 2.827$ Mrad/s):

The parallel circuit L_1C_1 resonates at 450 kHz; its impedance is

$$Z_{r1} = Q_{r1}X_{r1} \tag{16-23}$$

$$Z_{r1} = Q_{r1}\omega_1 L_1 = 40 \times 2.827 \text{ Mrad/s} \times 625\ \mu\text{H} = 70.68 \text{ k}\Omega$$

$$X_{L2} = \omega_1 L_2 = 2.827 \text{ Mrad/s} \times 22\ \mu\text{H} = 62.19\ \Omega$$

$$X_{C2} = \frac{1}{\omega_1 C_2} = \frac{1}{2.827 \text{ Mrad/s} \times 10 \text{ pF}} = 35.37 \text{ k}\Omega$$

As X_{C2} is a very high impedance in parallel with X_{L2}, a low

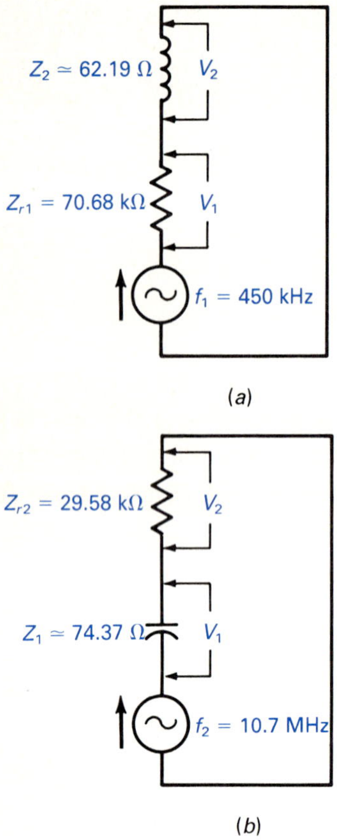

Fig. 16-47. Example 16-11: Equivalent circuits *(a)* at 450 kHz; *(b)* at 10.7 MHz

impedance, the impedance of circuit L_2C_2 at 450 kHz is approximately equal to X_{L2}:

$$Z_2 \simeq X_{L2} \simeq 62.19\ \Omega$$

L_2 and C_2 resonate at 10.7 MHz, so that the impedance of circuit L_2C_2 at this frequency is, from Eq. (16-23),

$$Z_{r2} = Q_{r2}X_{r2}$$

$$Z_{r2} = Q_{r2}\omega_2 L_2 = 20 \times 67.23\ \text{Mrad/s} \times 22\ \mu\text{H} = 29.58\ \text{k}\Omega$$

$$X_{L1} = \omega_2 L_1 = 67.23\ \text{Mrad/s} \times 625\ \mu\text{H} = 42.02\ \text{k}\Omega$$

$$X_{C1} = \frac{1}{\omega_2 C_1} = \frac{1}{67.23\ \text{Mrad/s} \times 200\ \text{pF}} = 74.37\ \Omega$$

X_{L1} is very much greater than X_{C1}. Thus the impedance of the circuit L_1C_1 at 10.7 MHz will be approximately equal to X_{C1}:

$$Z_1 \simeq X_{C1} \simeq 74.37\ \Omega$$

Using the impedance values calculated, the conditions at the two frequencies are summarized in Fig. 16-47.

Figure 16-47*a* shows the situation in which the input is $f_1 = 450$ kHz. At this frequency V_1 is approximately $70.68\ \text{k}\Omega/62.19\ \Omega = 1137$ times V_2. Therefore, in referring back to Fig. 16-46, at 450 kHz output 1 is higher than output 2.

When the input frequency is $f_2 = 10.7$ MHz, the conditions are as shown in Fig. 16-47*b*. At this frequency V_2 is approximately $29.58\ \text{k}\Omega/74.37\ \Omega = 397.8$ times V_1. Thus, at 10.7 MHz, output 2 will be considerably greater than output 1.

Summarizing the action of the circuit in Fig. 16-46, we can say that on AM stations the signal is delivered at output 1, while on FM stations the signal leaves the circuit at output 2.

Trimming, Tracking, and Padding

A radio usually receives transmissions over a wide range of frequencies. For example, the ordinary domestic radio receives signals broadcast on the 525 kHz to 1.6 MHz band. A tuned circuit used to select a required transmission has a variable capacitor to vary the resonant frequency. To allow for deviations in inductance and capacitance from their nominal values, which are the inevitable consequences of manufacturing tolerances, a trimming capacitor may be used. This trimmer is a preset variable capacitor which is connected in parallel with the main tuning capacitor. The capacitance value of the trimmer is set during the final adjustment of the tuning of the radio (called alignment).

EXAMPLE 16-12 A resonant circuit in a radio is required to tune over the band 525 kHz to 1.6 MHz. The variable tuning capacitor has a range of 30 to 300 pF. What values of inductance and trimming capacitance are needed?

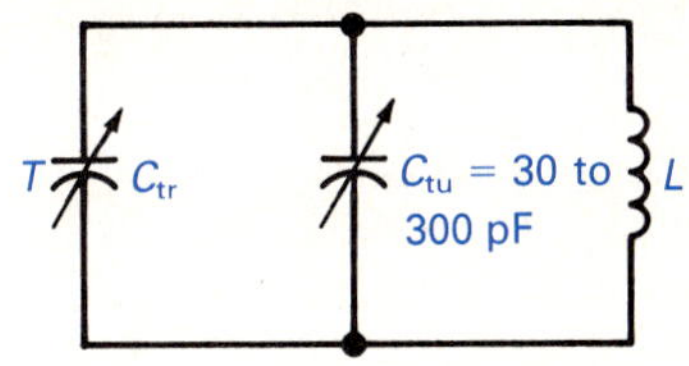

Fig. 16-48. Example 16-12: Circuit.

Solution (See Fig. 16-48.)
Apply Eq. (16-1):

$$C = \frac{1}{\omega^2 L} = \frac{1}{(2\pi f)^2 L}$$

$$C \propto \frac{1}{f^2}$$

$$\frac{C_{\text{low}}}{C_{\text{high}}} = \left(\frac{f_{\text{high}}}{f_{\text{low}}}\right)^2 = \left(\frac{1.6\text{ MHz}}{525\text{ kMz}}\right)^2 = 9.288$$

where C_{low} and C_{high} are the total circuit capacitances at the low and high ends of the frequency band. The tuning capacitor C_{tu} will have maximum capacitance at the lowest frequency. At any point in the frequency band the total capacitance is the sum of the tuning and trimming capacitances, C_{tu} and C_{tr}, respectively.

$$C_{\text{low}} = C_{\text{tu}} + C_{\text{tr}} = (300 + C_{\text{tr}})\text{ pF}$$

$$C_{\text{high}} = C_{\text{tu}} + C_{\text{tr}} = (30 + C_{\text{tr}})\text{ pF}$$

$$\frac{C_{\text{low}}}{C_{\text{high}}} = \frac{(300 + C_{\text{tr}})\text{ pF}}{(30 + C_{\text{tr}})\text{ pF}} = 9.288$$

$$C_{\text{tr}} = \mathbf{2.582\ pF}$$

The inductance can be determined at any point on the frequency band. Taking the high end, where $f = 1.6$ MHz,

$$L = \frac{1}{\omega^2 C} \qquad (16\text{-}1)$$

$$L = \frac{1}{(2\pi \times 1.6\text{ MHz})^2 \times (30 + 2.582)\text{ pF}} = \mathbf{303.7\ \mu H}$$

There are two circuits in a radio which, for any transmission, have to be resonant to two different frequencies, the difference between these frequencies being a constant for any particular radio. The relationship between the two frequencies is

$$f_{\text{osc}} - f_{\text{sig}} = f_i$$

where f_{osc} = frequency of an oscillator, Hz
f_{sig} = frequency of the desired station, Hz
f_i = intermediate frequency of the radio, Hz

The frequencies in Example 16-11 are intermediate frequencies.

The process of maintaining a constant frequency difference between the oscillator and signal frequency resonant circuits is called *tracking*. It can be accomplished by special tuning capacitor design or by padding. *Padding* is a circuit technique using a series padding capacitor, and is best illustrated by an example.

EXAMPLE 16-13 The network in Fig. 16-48 is used for the signal-frequency resonant circuit of a radio. The radio has an intermediate frequency of 450 kHz. If the capacitor which tunes the oscillator is the same type as that which tunes the signal-frequency circuit, what values of inductance and padding capacitance are needed for the oscillator circuit?

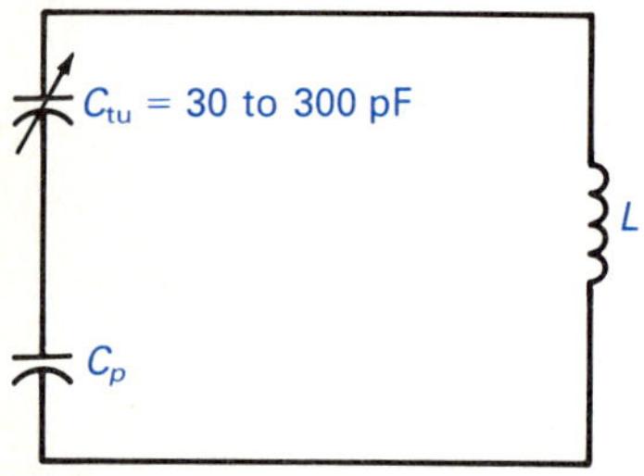

Fig. 16-49. Example 16-13: Circuit.

Solution Figure 16-49 shows the oscillator resonant circuit, C_{tu} and C_p being the tuning and padding capacitors. Values for C_p and L which will apply with reasonable accuracy over the entire frequency range of the radio can be obtained by considering each end of the band.

At the low-frequency end (f_{sig} = 525 kHz),

$$f_{osc} = f_{sig} + f_i = 525\text{ kHz} + 450\text{ kHz} = 975\text{ kHz}$$

The total capacitance, C_{tu} and C_p in series, will be given by

$$C_{T,\text{low}} = \frac{C_{tu}C_p}{C_{tu} + C_p} = \frac{300C_p}{300 + C_p}\text{ pF}$$

At the high-frequency end (f_{sig} = 1.6 MHz),

$$f_{osc} = f_{sig} + f_i = 1.6\text{ MHz} + 450\text{ kHz} = 2.05\text{ MHz}$$

The total capacitance will be

$$C_{T,\text{high}} = \frac{C_{tu}C_p}{C_{tu} + C_p} = \frac{30C_p}{30 + C_p}\text{ pF}$$

The ratio of the highest to the lowest oscillator frequency is

$$\frac{2.05\text{ MHz}}{975\text{ kHz}} = 2.103$$

The ratio of the greatest to the least total capacitance is the square of the frequency ratio:

$$\frac{C_{T,\text{low}}}{C_{T,\text{high}}} = 2.103^2 = 4.423$$

$$\frac{300C_p}{300 + C_p} \div \frac{30C_p}{30 + C_p} = 4.423$$

$$C_p = \mathbf{184.2\ pF}$$

The inductance can be found from the data for either end of the frequency range. Using the low-end data,

$$f_{osc} = 975\text{ kHz}$$

$$\omega_{osc} = 2\pi \times 975\text{ kHz} = 6.126\text{ Mrad/s}$$

$$C_{T,\text{low}} = \frac{300 \times 184.2}{300 + 184.2} = 114.1\text{ pF}$$

Apply Eq. (16-1):

$$L = \frac{1}{\omega_{osc}^2 C_{T,\text{low}}}$$

$$L = \frac{1}{(6.126\text{ Mrad/s})^2 \times 114.1\text{ pF}} = \mathbf{233.54\ \mu H}$$

REVIEW QUESTIONS FOR SECTION 16-7

1. What is a double-tuned circuit?
2. What is the difference between trimming and padding in reference to a radio receiver?

EXERCISE PROBLEM FOR SECTION 16-7

1. A variable capacitor with a tuning range of 10 to 100 pF is to be used to tune over the band of frequencies 1.6 to 4 MHz. Find the value of the trimming capacitor and the value of the inductance required. (*ans*. 7.143 pF, 92.35 μH)
2. For Prob. 1, when the circuit is resonant at 4 MHz, if the power response to a frequency of 3.975 MHz is 50 percent of the response to a 4-MHz signal, what is the resistance of the inductor?
3. A parallel resonant tank circuit in a radio transmitter contains an inductor of 350 μH and 15-Ω resistance and is tuned to 500 kHz. When the antenna is coupled to the tank circuit, its loading effect is the same as if an additional 200-Ω resistance is connected in series with the inductor. Find the resonant frequency of the circuit after the antenna is coupled. (*ans*. 490.35 kHz)
4. For Prob. 2, what shunt resistance would be required to increase the bandwidth to 70 kHz?

SUMMARY OF EQUATIONS

Series *RLC* circuit

$\omega_r^2 = 1/LC$ (16-1)

$f_r = 1/2\pi\sqrt{LC}$ (16-2)

$X_{Lr} = X_{Cr} = Q_rR = Q_rZ$ (16-3)

$V_{Lr} = V_{Cr} = Q_rV_T$ (16-4)

$X_{T1} = X_{C1} - X_{L1} = R$ (16-5)

$X_{T2} = X_{L2} - X_{C2} = R$ (16-6)

$P_1 = P_2 = P_r/2$ (16-7)

$I_1 = I_2 = I_r/\sqrt{2} = 0.7071I_r$ (16-8)

$BW = f_2 - f_1$ (16-9)

$BW \simeq f_r/Q_r$ (16-10)*

Parallel *RLC* circuit

$\omega_r^2 = 1/LC$ (16-1)

$f_r = 1/2\pi\sqrt{LC}$ (16-2)

$B_{Cr} = B_{Lr} = Q_rG = Q_rY$ (16-11)

$I_{Cr} = I_{Lr} = Q_rI_R = Q_rI_T$ (16-12)

$B_{T1} = B_{L1} - B_{C1} = G$ (16-13)

$B_{T2} = B_{C2} - B_{L2} = G$ (16-14)

*When applying this equation, note that $Q_r = X_{Lr}/R = X_{Cr}/R$ for series circuits, and $Q_r = B_{Lr}/G = B_{Cr}/G = R/X_{Lr} = R/X_{Cr}$ for parallel circuits.

$P_1 = P_2 = P_r/2$ (16-7)
$V_1 = V_2 = V_r/\sqrt{2} = 0.7071V_r$ (16-15)
$BW = f_2 - f_1$ (16-9)
$BW \simeq f_r/Q_r$ (16-10)*
$R_sR_p = L/C$ (16-16)
$I_{circ} = Q_rI_T$ (16-17)

Practical parallel resonant circuit

$Y_T = R_s/(R_s^2 + \omega^2L^2) + j[\omega C - \omega L/(R_s^2 + \omega^2L^2)]$ (16-18)
$\omega_r^2 = 1/LC - R_s^2/L^2 = (1/LC)(1 - R_s^2C/L)$ (16-19)
$f_r = [1/(2\pi\sqrt{LC})]\sqrt{1 - R_s^2C/L}$ (16-20)
$Z_r = R_s + (X_{Lr})^2/R_s$ (16-21)
$Z_r = R_s + Q_rX_{Lr}$ (16-22)
$Z_r \simeq Q_rX_r$ (16-23)
$Z_r \simeq L/CR_s$ (16-24)

CHAPTER QUESTIONS

16-1. In what ways do resonant circuits differ with and without significant energy losses?

16-2. What type of impedance is presented by an *RLC* circuit
a. At its resonant frequency?
b. At its bandwidth frequencies?

16-3. Is the impedance of a series *RLC* circuit at resonance
a. Maximum in value?
b. Minimum in value?

16-4. Is the impedance of a parallel *RLC* circuit at resonance
a. Maximum in value?
b. Minimum in value?

16-5. Is the answer to Question 16-4 influenced by whether the circuit is of true parallel form or has the series-parallel form of a ''practicial'' parallel resonant circuit?

16-6. What is the phase angle of an *RLC* circuit
a. At its resonant frequency?
b. At its bandwidth frequencies?

16-7. In what situations might resonance be undesirable?

16-8. In what circumstances can the impedance of a resonant circuit be expressed as $Z = L/CR$?

16-9. What is the connection between resonance and Q factor?

16-10. Why is it often accepted that the Q factor of a resonant circuit is the Q factor of its inductor?

16-11. How is Question 16-10 related to the concept of a practical parallel resonant circuit?

16-12. In what circumstances can the resistance of a resonant circuit be regarded as having a significant effect on its resonant frequency
a. If it is a series circuit?
b. If it is a parallel circuit?

16-13. The resonant circuits used in a radio transmitter to control its station frequency have much narrower bandwidth than those used to tune a radio receiver. Why might that be so?

16-14. In a series or true parallel *RLC* circuit, the resonant frequency occurs when the capacitive and inductive reactances are equal. In a practical resonant circuit, where the total losses are regarded as being represented by a resistance in series with the inductance, this is not quite so. Why is that?

16-15. Why is it that the response curve of a resonant circuit is not symmetrical, its shape being different below and above the resonant frequency?

16-16. Why is it that any practical circuit component, such as a resistor, inductor, capacitor, or transistor, exhibits the characteristics of a resonant frequency and a bandwidth?

16-17. In what situation is the bandwidth equation $BW \simeq f_r/Q_r$ of no use?

CHAPTER PROBLEMS

16-1. Figure 16-50 shows a series *RLC* circuit with the values of *L* and *C* the same as those in the Chap. 15 problems covered by Fig. 15-47. At what frequency is this circuit resonant? (*ans.* 1.125 MHz)

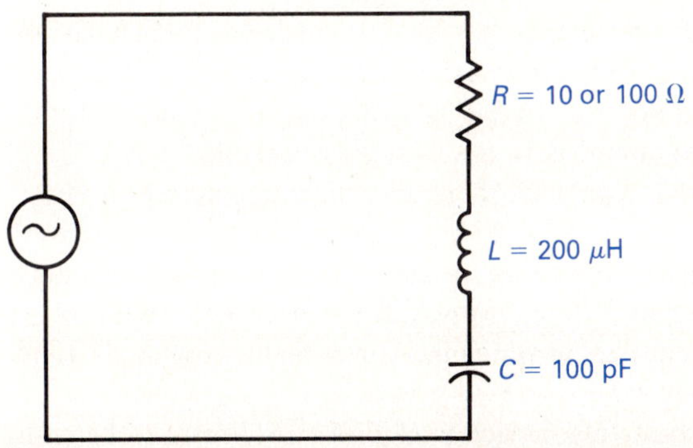

Fig. 16-50. Problems 16-1 and 16-3: Circuit.

16-2. The parallel *RLC* circuit of Fig. 16-51 has the same values of *L* and *C* as those in Fig. 15-48. At what frequency is this circuit resonant?

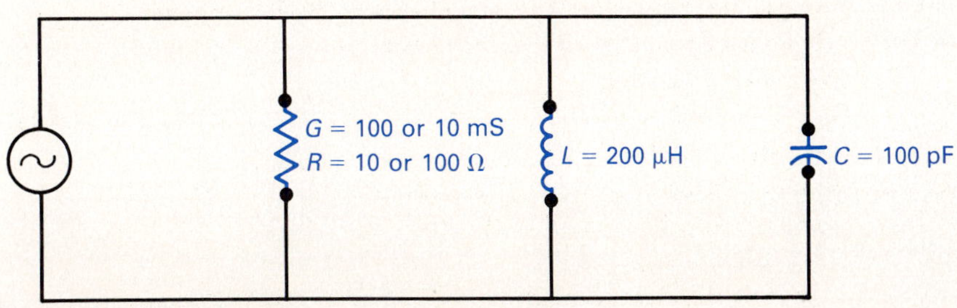

Fig. 16-51. Problem 16-2 and 16-4: Circuit.

16-3. What is the bandwidth of the circuit of Fig. 16-50 when
a. $R = 10\ \Omega$? (*ans.* 7.958 kHz)
b. $R = 100\ \Omega$? (*ans.* 79.58 kHz)

16-4. What is the bandwidth of the circuit of Fig. 16-51 when
a. $G = 100$ mS ($R = 10\ \Omega$)?
b. $G = 10$ mS ($R = 100\ \Omega$)?

16-5. In Fig. 16-52 is shown a practical parallel resonant circuit with the R, L, and C values of Fig. 16-51. What is its resonant frequency when
a. $R = 10\ \Omega$? (*ans.* 1.1249 MHz)
b. $R = 100\ \Omega$? (*ans.* 1.119 MHz)

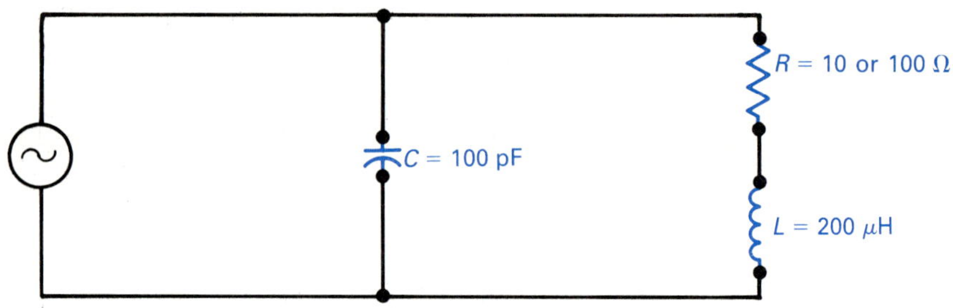

Fig. 16-52. Problem 16-5 and 16-6: Circuit.

16-6. What is the percentage difference between the resonant frequencies of the practical parallel resonant circuit of Fig. 16-52 and those of the true parallel resonant circuit of Fig. 16-51? Why does the difference *not* indicate that the practical circuit is in any way unacceptable?

16-7. Problem 15-14 referred to an electric motor with a lagging power factor of 0.6. What would be required to give the circuit a unity power factor? The motor is described as operating on 240 V 60 Hz with a 5-hp rating and 70 percent efficiency. (*ans.* 327.2-μF parallel capacitor)

16-8. What would be the frequency at which the power-factor-corrected circuit of Prob. 16-7 is resonant?

16-9. If the power-factor-correcting capacitor of Prob. 16-7 were to be connected in series with the motor, what would be the voltage across the motor with the applied circuit voltage of 240 V 60 Hz? (*ans.* 300 V)

17

AC NETWORKS

All the techniques of dc network analysis apply to ac circuits. In employing these techniques, phasor methods must be used.

The principles of network analysis were developed in Chaps. 9 and 10, and you should refer back to them if you need to refresh your memory. The only difference in applying the techniques to ac circuits lies in the phasor calculations. These can be a problem if you are not too clear on the network analysis basics.

17-1 AC NETWORK ANALYSIS

Nothing new is necessary to explain ac network analysis, and so it will be demonstrated by examples.

For simplicity, the resistance symbol -\/\/\/- is used for any type of element or group of elements when there is no need to show the details. For example, in Fig. 17-1 each of the elements has two components, a resistance and a reactance, but the circuit is easier to follow when the diagram shows six elements instead of the twelve separate components.

EXAMPLE 17-1 What is the magnitude of the voltage developed across the load Z_L in the circuit in Fig. 17-1? Use the current assumption technique (refer to Sec. 9-1).

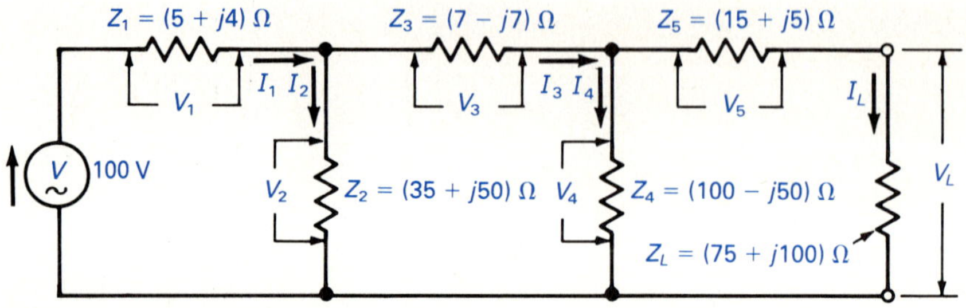

Fig. 17-1. Example 17-1: Circuit.

Solution

Assuming that the load current I_L is 1 A $\angle 0°$:

$$V_L = I_L Z_L = 1\text{ A }\angle 0° \times (75 + j100)\ \Omega = (75 + j100)\text{ V}$$
$$V_5 = I_L Z_5 = 1\text{ A }\angle 0° \times (15 + j5)\ \Omega = (15 + j5)\text{ V}$$
$$V_4 = V_L + V_5 = (75 + j100)\text{ V} + (15 + j5)\text{ V} = (90 + j105)\text{ V}$$
$$I_4 = \frac{V_4}{Z_4} = \frac{(90 + j105)\text{ V}}{(100 - j50)\ \Omega} = \frac{138.3\text{ V }\angle 49.4°}{111.8\ \Omega\ \angle -26.57°} = 1.237\text{ A }\angle 75.97°$$
$$I_3 = I_4 + I_L = 1.237\text{ A }\angle 75.97° + 1\text{ A }\angle 0°$$
$$I_3 = (0.2999 + j1.2)\text{ A} + (1 + j0)\text{ A} = (1.299 + j1.2)\text{ A}$$
$$V_3 = I_3 Z_3 = (1.299 + j1.2)\text{ A} \times (7 - j7)\ \Omega$$
$$V_3 = 1.768\text{ A }\angle 42.73° \times 9.899\ \Omega\ \angle -45° = 17.5\text{ V }\angle -2.27°$$
$$V_2 = V_3 + V_4 = 17.5\text{ V }\angle -2.27° + (90 + j105)\text{ V}$$
$$V_2 = (17.49 - j0.6932)\text{ V} + (90 + j105)\text{ V} = (107.5 + j104.3)\text{ V}$$
$$I_2 = \frac{V_2}{Z_2} = \frac{(107.5 + j104.3)\text{ V}}{(35 + j50)\ \Omega} = \frac{149.8\text{ V }\angle 44.13°}{61.03\ \Omega\ \angle 55.01°} = 2.455\text{ A }\angle -10.88°$$
$$I_1 = I_2 + I_3 = 2.455\text{ A }\angle -10.88° + (1.299 + j1.2)\text{ A}$$
$$I_1 = (2.411 - j0.4634)\text{ A} + (1.299 + j1.2)\text{ A} = (3.71 + j0.7366)\text{ A}$$
$$V_1 = I_1 Z_1 = (3.71 + j0.7366)\text{ A} \times (5 + j4)\ \Omega$$
$$V_1 = 3.782\text{ A }\angle 11.23° \times 6.403\ \Omega\ \angle 38.66° = 24.22\text{ V }\angle 49.89°$$
$$V = V_1 + V_2 = 24.22\text{ V }\angle 48.89° + (107.5 + j104.3)\text{ V}$$
$$V = (15.92 + j18.25)\text{ V} + (107.5 + j104.3)\text{ V} = (123.4 + j122.6)\text{ V}$$
$$V = 173.9\text{ V }\angle 44.81°$$

Therefore for a load current of 1 A magnitude, the generator emf would have to be of magnitude 173.9 V. The actual emf magnitude is 100 V. Thus all the calculated voltage and current values are too great by a factor of 173.9/100 = 1.739.

Using correct values,

$$|I_L| = \frac{1\text{ A}}{1.739} = 0.575\text{ A}$$
$$Z_L = (75 + j100)\ \Omega$$
$$|Z_L| = 125\ \Omega$$
$$|V_L| = |I_L| \times |Z_L| = 0.575\text{ A} \times 125\ \Omega = \mathbf{71.88\text{ V}}$$

EXAMPLE 17-2 What is the magnitude of the voltage V_L across the load in the circuit in Fig. 17-2 (same circuit as for Example 17-1)? Use nodal analysis (refer to Sec. 9-2).

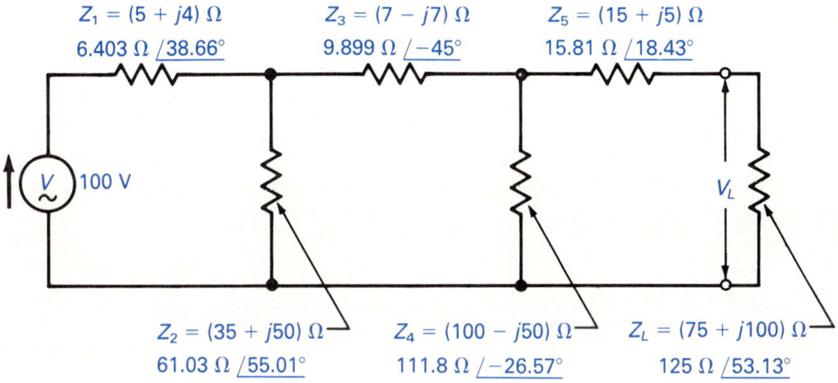

Fig. 17-2. Example 17-2: Circuit.

Solution

Redrawing Fig. 17-2 with the voltage source converted to a current source and the impedance values converted to admittance values gives Fig. 17-3.

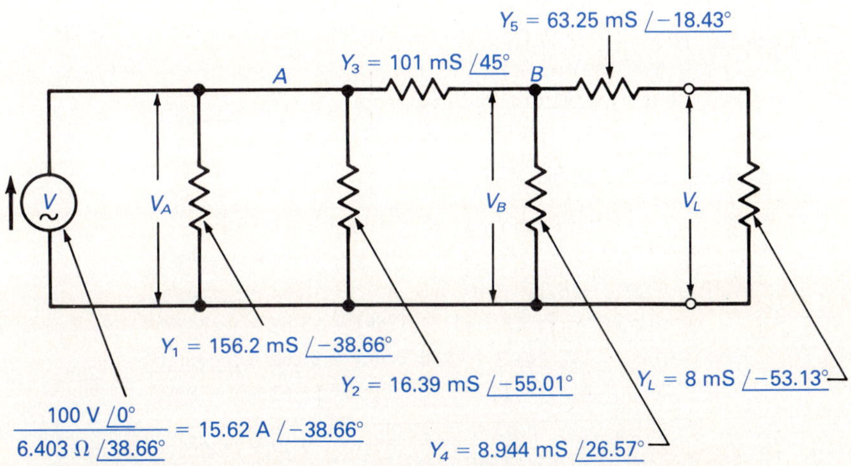

Fig. 17-3. Example 17-2: The voltage source in Fig. 17-2 has been converted to a current source and the impedances have been converted to admittances.

In Fig. 17-3 two independent nodes A and B are shown. The reference node is selected to be the common connection of the generator and the elements Y_1, Y_2, Y_4, and Y_L. The nodal voltages V_A and V_B are the voltages between the independent nodes and the reference node.

At Node *A*

$$V_A(Y_1 + Y_2 + Y_3) - V_B Y_3 = I$$

$$V_A(156.2\ \underline{/-38.66^\circ} + 16.39\ \underline{/-55.01^\circ} + 101\ \underline{/45^\circ})\ \text{mS} - V_B(101\ \text{mS}\ \underline{/45^\circ} = 15.62\ \text{A}\ \underline{/-38.66^\circ}$$

$$V_A[(122 - j97.58) + (9.399 - j13.43) + (71.42 + j71.42)]\ \text{mS} - V_B(101\ \text{mS}\ \underline{/45^\circ}) = 15.62\ \text{A}\ \underline{/-38.66^\circ}$$

$$V_A(202.8 - j39.59)\ \text{mS} - V_B(101\ \text{mS}\ \underline{/45^\circ} = 15.62\ \text{A}\ \underline{/-38.66^\circ}$$

$$V_A(206.6\ \text{mS}\ \underline{/-11.05^\circ}) - V_B(101\ \text{mS}\ \underline{/45^\circ}) = 15.62\ \text{A}\ \underline{/-38.66^\circ} \qquad (A)$$

At Node *B* The series combination of admittances Y_5 and Y_L is given by

$$Y_{5L} = \frac{1}{Z_5 + Z_L} = \frac{1}{(15 + j5)\ \Omega + (75 + j100)\ \Omega} = \frac{1}{(90 + j105)\ \Omega}$$

$$Y_{5L} = \frac{1}{138.3\ \Omega\ \underline{/49.4^\circ}} = 7.231\ \text{mS}\ \underline{/-49.4^\circ}$$

$$V_B(Y_3 + Y_4 + Y_{5L}) - V_A Y_3 = 0$$

$$V_B(101\ \underline{/45^\circ} + 8.944\ \underline{/26.57^\circ} + 7.231\ \underline{/-49.4^\circ})\ \text{mS} - V_A(101\ \text{mS}\ \underline{/45^\circ}) = 0$$

$$V_B[(71.42 + j71.42) + (7.999 + j4) + (4.706 - j5.49)]\ \text{mS} - V_A(101\ \text{mS}\ \underline{/45^\circ}) = 0$$

$$V_B(84.13 + j69.93)\ \text{mS} - V_A(101\ \text{mS}\ \underline{/45^\circ}) = 0$$

$$V_B(109.4\ \text{mS}\ \underline{/39.73^\circ}) - V_A(101\ \text{mS}\ \underline{/45^\circ}) = 0 \qquad (B)$$

From Eq. *(B)*,

$$V_A = \frac{V_B(109.4\ \text{mS}\ \underline{/39.73^\circ})}{101\ \text{mS}\ \underline{/45^\circ}} = V_B(1.083\ \underline{/-5.27^\circ})$$

Substituting for V_A in Eq. *(A)*,

$$V_B(1.083\ \underline{/-5.27^\circ}) \times (206.6\ \text{mS}\ \underline{/-11.05^\circ}) - V_B(101\ \text{mS}\ \underline{/45^\circ}) = 15.62\ \text{A}\ \underline{/-38.66^\circ}$$

$$V_B(223.7\ \text{mS}\ \underline{/-16.32^\circ} - 101\ \text{mS}\ \underline{/45^\circ}) = 15.62\ \text{A}\ \underline{/-38.66^\circ}$$

$$V_B[(214.7 - j62.86)\ \text{mS} - (71.42 + j71.42)\ \text{mS}] = 15.62\ \text{A}\ \underline{/-38.66^\circ}$$

$$V_B(143.3 - j134.3)\ \text{mS} = 15.62\ \text{A}\ \underline{/-38.66^\circ}$$

$$V_B(196.4\ \text{mS}\ \underline{/-43.14^\circ}) = 15.62\ \text{A}\ \underline{/-38.66^\circ}$$

$$V_B = \frac{15.62\ \text{A}\ \underline{/-38.66^\circ}}{196.4\ \text{mS}\ \underline{/-43.14^\circ}} = 79.53\ \text{V}\ \underline{/4.48^\circ}$$

V_B is the voltage across Y_5 and Y_L in series. Using impedance values from Fig. 17-2,

$$V_L = V_B \frac{Z_L}{Z_5 + Z_L} = 79.53\ \text{V}\ \underline{/4.48^\circ}\ \frac{125\ \Omega\ \underline{/53.13^\circ}}{(90 + j105)\ \Omega}$$

$$= 79.53\ \text{V}\ \underline{/4.48^\circ}\ \frac{125\ \Omega\ \underline{/53.13^\circ}}{138.3\ \Omega\ \underline{/49.4^\circ}} = 71.88\ \text{V}\ \underline{/8.21^\circ}$$

$|V_L| = \mathbf{71.88\ V}$

EXAMPLE 17-3 What is the value of the inductive current I_L in the circuit in Fig. 17-4? Use loop (mesh) analysis (refer to Sec. 9-3).

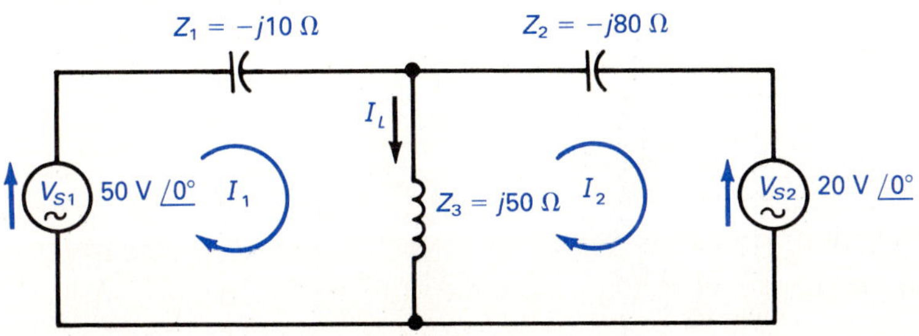

Fig. 17-4. Example 17-3: Circuit.

Solution The loop equations are

$$(Z_1 + Z_3)I_1 - Z_3 I_2 = V_{S1} \qquad (A)$$
$$-Z_3 I_1 + (Z_2 + Z_3)I_2 = -V_{S2} \qquad (B)$$

Substituting values into Eqs. *(A)* and *(B)*,

$$(-j10\ \Omega + j50\ \Omega)I_1 - j50\ \Omega\, I_2 = 50\ \text{V}\ \underline{/0^\circ}$$
$$j40\ \Omega\, I_1 - j50\ \Omega\, I_2 = 50\ \text{V}\ \underline{/0^\circ} \qquad (A)$$
$$-j50\ \Omega\, I_1 + (-j80\ \Omega + j50\ \Omega)I_2 = -20\ \text{V}\ \underline{/0^\circ}$$
$$-j50\ \Omega I_1 - j30\ \Omega\, I_2 = -20\ \text{V}\ \underline{/0^\circ} \qquad (B)$$

Putting Eqs. *(A)* and *(B)* into matrix form for solution by determinants (see appendix):

$$\begin{bmatrix} 40\ \underline{/90^\circ} & 50\ \underline{/-90^\circ} \\ 50\ \underline{/-90^\circ} & 30\ \underline{/-90^\circ} \end{bmatrix} \begin{bmatrix} I_1 \\ I_2 \end{bmatrix} = \begin{bmatrix} 50\ \underline{/0^\circ} \\ -20\ \underline{/0^\circ} \end{bmatrix}$$

The solutions are

$$I_1 = 0.6757\ \text{A}\ \underline{/-90^\circ} \qquad \text{and} \qquad I_2 = 0.4595\ \text{A}\ \underline{/90^\circ}$$

The current I_L is the difference between I_1 and I_2:

$$I_L = I_1 - I_2 = -j0.6757\ \text{A} - j0.4595\ \text{A} = -j1.135\ \text{A}$$

Thus the current I_L with its phase relative to the phase of the two voltage sources is

$$I_L = \mathbf{1.135\ A}\ \underline{\mathbf{/-90^\circ}} \qquad \text{or} \qquad \underline{\mathbf{/-\pi/2\ rad}}$$

In Example 17-3 the magnitude of the voltage across the inductor ($I_L X_L = 56.75$ V) is greater than the emf of either source. This is another reminder that voltages (or currents) in circuits containing *both* inductance and capacitance can be greater than the source values.

EXAMPLE 17-4 What is the value of the current I_L in the circuit in Fig. 17-5 (a repeat of Fig. 17-4)? Apply the superposition principle (refer to Sec. 9-4).

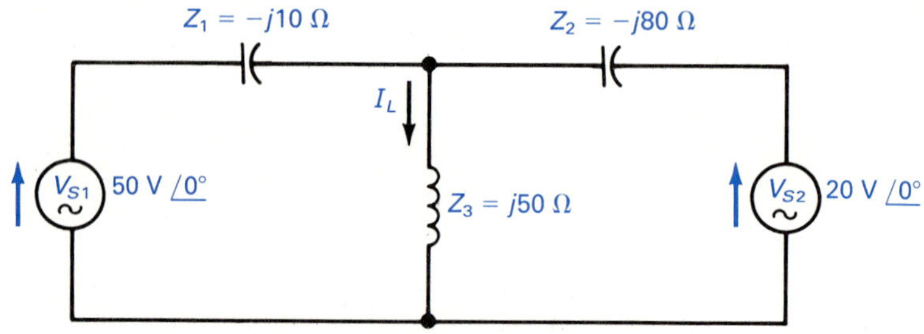

Fig. 17-5. Example 17-4: Given circuit.

Solution

In the application of the superposition principle, the circuit in Fig. 17-5 is reduced by the elimination of sources, so that each voltage source is alternately the only source in the circuit. Figure 17-6*a* and *b* shows the resulting reduced circuits and the currents I_{La} and I_{Lb}, which are components of the current I_L due to each of the two voltage sources.

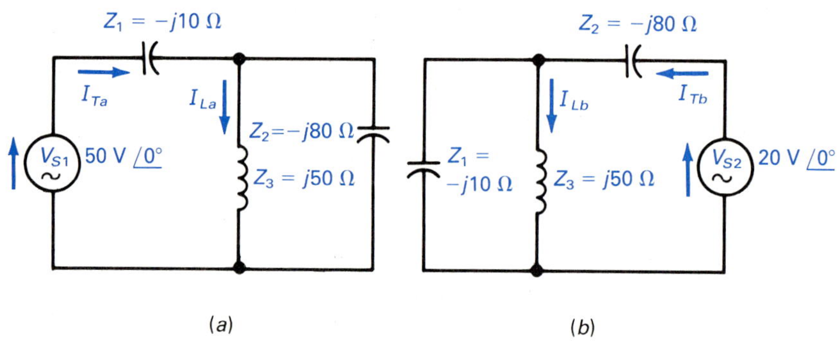

Fig. 17-6. Example 17-4: *(a)* Circuit with V_{S2} removed; *(b)* circuit with V_{S1} removed.

From Fig. 17-6*a*,

$$Z_{Ta} = Z_1 + \frac{Z_2 Z_3}{Z_2 + Z_3} = -j10\ \Omega + \frac{(-j80\ \Omega)(j50\ \Omega)}{(-j80\ \Omega) + (j50\ \Omega)}$$

$$Z_{Ta} = (-j10 + j133.3)\ \Omega = j123.3\ \Omega = 123.3\ \Omega\ \underline{/90^\circ}$$

$$I_{Ta} = \frac{V_{S1}}{Z_{Ta}} = \frac{50\ \text{V}\ \underline{/0^\circ}}{123.3\ \Omega\ \underline{/90^\circ}} = 0.4055\ \text{A}\ \underline{/-90^\circ}$$

$$I_{La} = I_{Ta}\frac{Z_2}{Z_2 + Z_3} = 0.4055\ \text{A}\ \underline{/-90^\circ}\,\frac{80\ \Omega\ \underline{/-90^\circ}}{30\ \Omega\ \underline{/-90^\circ}} = 1.081\ \text{A}\ \underline{/-90^\circ}$$

From Fig. 17-6*b*,

$$Z_{Tb} = Z_2 + \frac{Z_1 Z_3}{Z_1 + Z_3} = -j80\ \Omega + \frac{(-j10\ \Omega)(j50\ \Omega)}{(-j10 + j50)\ \Omega}$$

$$Z_{Tb} = (-j80 - j12.5)\ \Omega = -j92.5\ \Omega = 92.5\ \Omega\ \underline{/-90^\circ}$$

$$I_{Tb} = \frac{V_2}{Z_{Tb}} = \frac{20\ \text{V}\ \underline{/0^\circ}}{92.5\ \Omega\ \underline{/-90^\circ}} = 0.2162\ \text{A}\ \underline{/90^\circ}$$

$$I_{Lb} = I_{Tb}\frac{Z_1}{Z_1 + Z_3} = 0.2162\ \text{A}\ \underline{/90^\circ}\ \frac{10\ \Omega\ \underline{/-90^\circ}}{40\ \Omega\ \underline{/90^\circ}} = 0.0541\ \text{A}\ \underline{/-90^\circ}$$

As each voltage source tends to produce currents through Z_3 in the same direction, the total current I_L is the sum of the components I_{La} and I_{Lb}:

$$I_L = I_{La} + I_{Lb} = 1.081\ \text{A}\ \underline{/-90^\circ} + 0.0541\ \text{A}\ \underline{/-90^\circ} = -j(1.081 + 0.0541)\ \text{A}$$

$$I_L = \mathbf{1.135\ A}\ \underline{/\mathbf{-90^\circ}} \quad \text{or} \quad \underline{/\mathbf{-\pi/2\ rad}}$$

EXAMPLE 17-5 What impedance values are necessary for the T network in Fig. 17-7*b* to be equivalent to the π network in Fig. 17-7*a*? Apply π-to-T conversion (refer to Sec. 10-1).

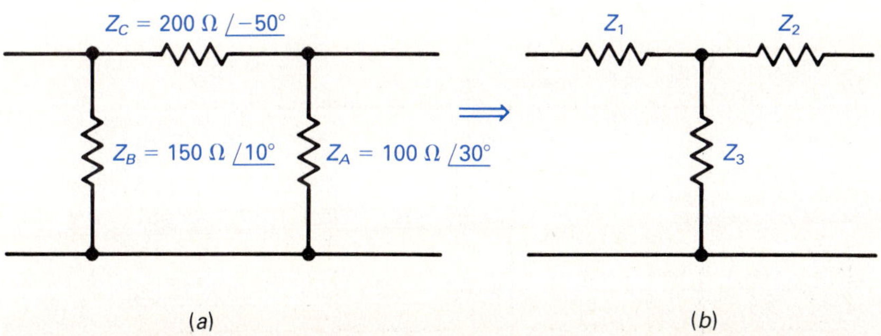

Fig. 17-7. Example 17-5: *(a)* Given π network; *(b)* equivalent T network.

Solution

$$\Sigma Z_\nabla = Z_A + Z_B + Z_C = 100\ \Omega\ \underline{/30^\circ} + 150\ \Omega\ \underline{/10^\circ} + 200\ \Omega\ \underline{/-50^\circ}$$

$$= (86.6 + j50)\ \Omega + (147.7 + j26.04)\ \Omega + (128.6 - j153.2)\ \Omega$$

$$\Sigma Z_\nabla = (362.9 - j77.16)\ \Omega = 371.1\ \Omega\ \underline{/-12^\circ}$$

$$Z_1 = \frac{Z_B Z_C}{\Sigma Z_\nabla} = \frac{150\ \Omega\ \underline{/10^\circ} \times 200\ \Omega\ \underline{/-50^\circ}}{371.1\ \Omega\ \underline{/-12^\circ}} = \mathbf{80.84\ \Omega}\ \underline{/\mathbf{-28^\circ}} \text{ or } \underline{/\mathbf{-0.4887\ rad}}$$

$$Z_2 = \frac{Z_C Z_A}{\Sigma Z_\nabla} = \frac{200\ \Omega\ \underline{/-50^\circ} \times 100\ \Omega\ \underline{/30^\circ}}{371.1\ \Omega\ \underline{/-12^\circ}} = \mathbf{53.89\ \Omega}\ \underline{/\mathbf{-8^\circ}} \text{ or } \underline{/\mathbf{-0.1396\ rad}}$$

$$Z_3 = \frac{Z_A Z_B}{\Sigma Z_\nabla} = \frac{100\ \Omega\ \underline{/30^\circ} \times 150\ \Omega\ \underline{/10^\circ}}{371.1\ \Omega\ \underline{/-12^\circ}} = \mathbf{40.42\ \Omega}\ \underline{/\mathbf{52^\circ}} \quad \text{or} \quad \underline{/\mathbf{0.9075\ rad}}$$

EXAMPLE 17-6 Obtain the original values of the π network in Example 17-5 from those calculated for the equivalent T network. Apply T-to-π conversion (refer to Sec. 10-1).

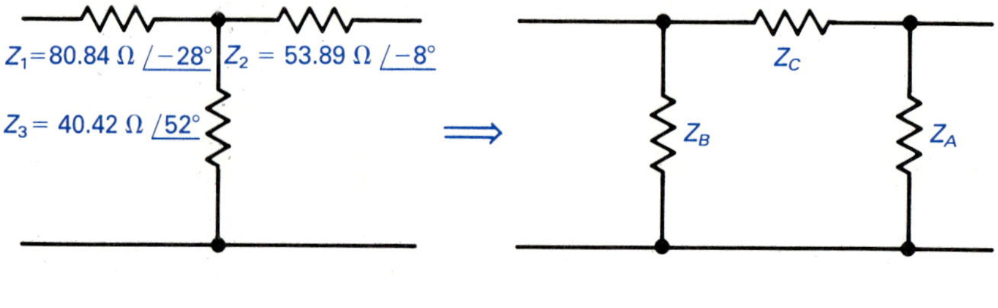

(a) (b)

Fig. 17-8. Example 17-6: *(a)* Given T network; *(b)* equivalent π network.

Solution

Figure 17-8*a* is the same as Fig. 17-7*b* but with the values from Example 17-5 inserted.

$$\begin{aligned}
\Sigma Z_Y &= Z_1Z_2 + Z_2Z_3 + Z_3Z_1 \\
&= 80.84\ \Omega\ \angle{-28^\circ} \times 53.89\ \Omega\ \angle{-8^\circ} + 53.89\ \Omega\ \angle{-8^\circ} \times 40.42\ \Omega\ \angle{52^\circ} \\
&\qquad + 40.42\ \Omega\ \angle{52^\circ} \times 80.84\ \Omega\ \angle{-28^\circ} \\
&= 4356\ \Omega^2\ \angle{-36^\circ} + 2178\ \Omega^2\ \angle{44^\circ} + 3267\ \Omega^2\ \angle{24^\circ} \\
&= (3524 - j2560)\ \Omega^2 + (1567 + j1513)\ \Omega^2 + (2984 + j1329)\ \Omega^2 \\
\Sigma Z_Y &= (8075 + j282)\ \Omega^2 = 8080\ \Omega^2\ \angle{2^\circ}
\end{aligned}$$

$$Z_A = \frac{\Sigma Z_Y}{Z_1} = \frac{8080\ \Omega^2\ \angle{2^\circ}}{80.84\ \Omega\ \angle{-28^\circ}} = \mathbf{99.95\ \Omega\ \angle{30^\circ}} \quad \text{or} \quad \mathbf{\angle{0.5236\ rad}}$$

$$Z_B = \frac{\Sigma Z_Y}{Z_2} = \frac{8080\ \Omega^2\ \angle{2^\circ}}{53.89\ \Omega\ \angle{-8^\circ}} = \mathbf{149.9\ \Omega\ \angle{10^\circ}} \quad \text{or} \quad \mathbf{\angle{0.1745\ rad}}$$

$$Z_C = \frac{\Sigma Z_Y}{Z_3} = \frac{8080\ \Omega^2\ \angle{2^\circ}}{40.42\ \Omega\ \angle{52^\circ}} = \mathbf{199.9\ \Omega\ \angle{-50^\circ}} \quad \text{or} \quad \mathbf{\angle{0.8726\ rad}}$$

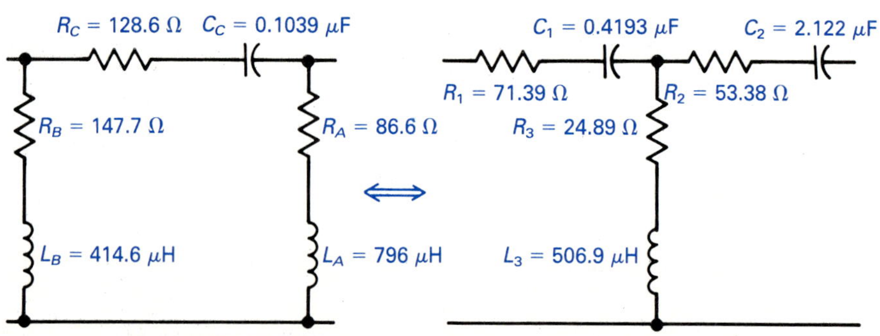

(a) (b)

Fig. 17-9. Networks of Examples 17-5 and 17-6 with individual element values for a frequency of 10 kHz.

The results for Example 17-6 all agree with the original values of Example 17-5 within 0.1 percent. That this accuracy is achieved after so many arithmetic operations is justification for trust in the principles of T $\leftrightarrow$ π conversions.

To bring into focus the results of Examples 17-5 and 17-6, Fig. 17-9 shows the equivalent π and T networks with individual element values for a frequency of 10 kHz (ω = 62.83 krad/s).

EXAMPLE 17-7 The load resistance R_L in the circuit in Fig. 17-10 is variable over the range 500 Ω to 5 kΩ. What is the load voltage magnitude range? Employ the Thevenin principle (refer to Sec. 10-2).

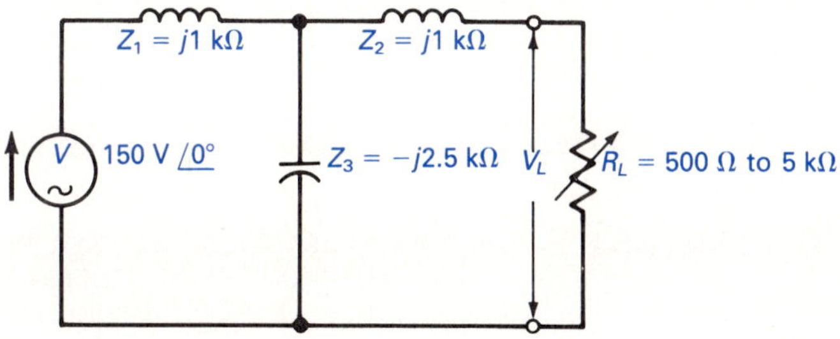

Fig. 17-10. Example 17-7: Given circuit.

Solution

The Thevenin equivalent of the circuit of Fig. 17-10 is given in Fig. 17-11.

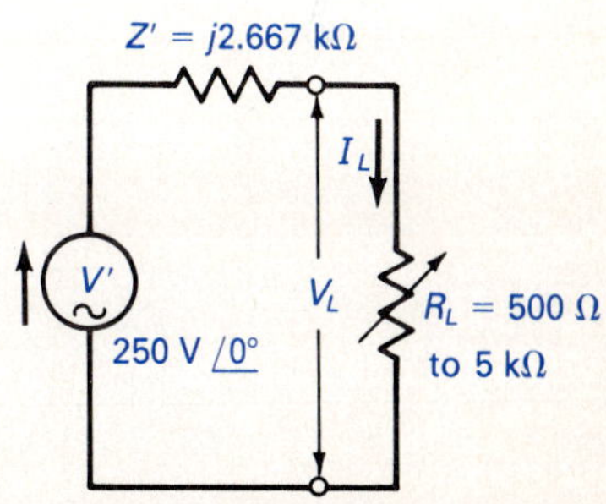

Fig. 17-11. Example 17-7: Thevenin equivalent circuit.

Obtain the Thevenin voltage V':

$$V' = V\frac{Z_3}{Z_1 + Z_3} = 150 \text{ V} \angle 0° \frac{2.5 \text{ k}\Omega \angle -90°}{j1 \text{ k}\Omega - j2.5 \text{ k}\Omega} = 150 \text{ V} \angle 0° \frac{2.5 \text{ k}\Omega \angle -90°}{1.5 \text{ k}\Omega \angle -90°}$$

$$V' = 250 \text{ V} \angle 0°$$

Obtain the Thevenin impedance Z':

$$Z' = Z_2 + \frac{Z_1 Z_3}{Z_1 + Z_3} = j1 \text{ k}\Omega + \frac{(j1 \text{ k}\Omega)(-j2.5 \text{ k}\Omega)}{j1 \text{ k}\Omega - j2.5 \text{ k}\Omega} = j1 \text{ k}\Omega + \frac{2.5 \text{ (k}\Omega)^2}{-j1.5 \text{ k}\Omega}$$

$$Z' = j1 \text{ k}\Omega + j1.667 \text{ k}\Omega = j2.667 \text{ k}\Omega$$

The load current I_L is given by

$$I_L = \frac{V'}{R_L + Z'} = \frac{250 \text{ V} \angle 0°}{R_L + j2.667 \text{ k}\Omega}$$

When R_L = 500 Ω (0.5 kΩ),

$$R_L + j2.667 \text{ k}\Omega = (0.5 + j2.667) \text{ k}\Omega = 2.713 \text{ k}\Omega \angle 79.38°$$

$$I_L = \frac{250 \text{ V } \angle 0^\circ}{2.713 \text{ k}\Omega \angle 79.38^\circ} = 92.15 \text{ mA } \angle -79.38^\circ$$

$$|V_L| = |I_L| R_L = 92.15 \text{ mA} \times 0.5 \text{ k}\Omega = \mathbf{46.07 \text{ V}}$$

When $R_L = 5 \text{ k}\Omega$,

$$R_L + j2.667 \text{ k}\Omega = (5 + j2.667) \text{ k}\Omega = 5.667 \text{ k}\Omega \angle 28.08^\circ$$

$$I_L = \frac{250 \text{ V } \angle 0^\circ}{5.667 \text{ k}\Omega \angle 28.08^\circ} = 44.12 \text{ mA } \angle -28.08^\circ$$

$$|V_L| = |I_L| R_L = 44.12 \text{ mA} \times 5 \text{ k}\Omega = \mathbf{220.6 \text{ V}}$$

The load voltage magnitude range is **46.07** to **220.6 V.**

Again in Example 17-7 we see the voltage magnification effect in an *LC* circuit. The magnitude of the Thevenin voltage is 250 V. When the load resistance is 5 kΩ, the load voltage magnitude is 220.6 V. Both of these voltages are higher than the 150-V applied emf.

EXAMPLE 17-8 What load impedance would result in maximum load power in the circuit in Fig. 17-12*a*? Refer to Sec. 10-3 for the conditions for maximum power transfer in a circuit containing steady direct current.

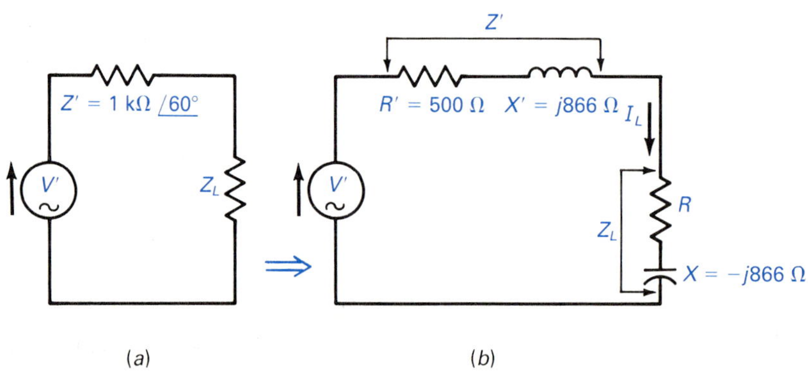

Fig. 17-12. Example 17-8: *(a)* Given circuit; *(b)* circuit showing the components of the impedances.

Solution

Figure 17-12*a* is the Thevenin equivalent of some unspecified circuit. Figure 17-12*b* is the same as Fig. 17-12*a*, but with the Thevenin impedance shown as a series equivalent *RL* network [$Z' = 1 \text{ k}\Omega \angle 60^\circ = (500 + j866) \ \Omega$].

The circuit current I_L, and hence the load power, for any value of load resistance will be maximized if the circuit is arranged to be in series resonance. This can be achieved by including in the load impedance a capacitive reactance equal in magnitude to the inductive reactance component of the Thevenin impedance. Thus Fig. 17-12*b* shows the load impedance as $(R - j866)\ \Omega$.

After the circuit is made entirely resistive by resonance, the dc maximum power transfer conditions will apply: The load resistance should equal the Thevenin resistance. Therefore the load impedance for maximum power transfer will be

$$Z_L = R - jX = \mathbf{(500 - j866)\ \Omega = 1\ k\Omega\ \underline{/-60^\circ}} \quad \text{or} \quad \mathbf{\underline{/-1.047\ rad}}$$

The conditions for maximum power transfer determined for Example 17-8 are general for ac circuits:

1. The load impedance should contain a reactance equal in magnitude and opposite in type to the reactive component of the Thevenin equivalent impedance of the remainder of the circuit.
2. The resistive components of the load and Thevenin impedances should be equal.

These conditions can be summarized by the statement: *Maximum power is transferred from an ac network to a load if the load impedance is the conjugate of the Thevenin equivalent impedance of the network.*

The above statement is written in equation form as

$$\text{When } Z' = R' + jX',\ Z_L = R_L - jX_L \qquad |R'| = |R_L|$$
$$\text{When } Z' = R' - jX',\ Z_L = R_L + jX_L \qquad |X'| = |X_L|$$

where R_L and jX_L are the resistive and reactive components of the load impedance, and R' and X' are the resistive and reactive components of the Thevenin equivalent impedance of the network feeding the load.

EXAMPLE 17-9 Find the range of load current values in the circuit in Fig. 17-13*a*. The load resistance is variable over the range 500 Ω to 5 kΩ (conductance range 2 to 0.2 mS). Apply the Norton principle (refer to Sec. 10-4).

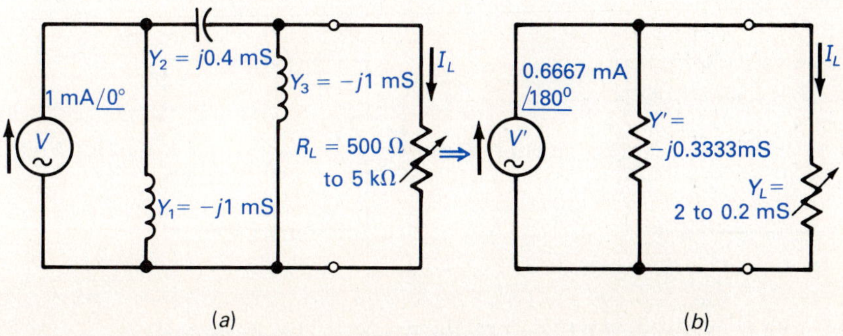

Fig. 17-13. Example 17-9: *(a)* Given circuit; *(b)* Norton equivalent.

Solution

Figure 17-13*b* is the Norton equivalent of Fig. 17-13*a*.

Obtain the Norton equivalent current I':

$$I' = I\frac{Y_2}{Y_1 + Y_2} = 1 \text{ mA } \underline{/0^\circ}\,\frac{j0.4 \text{ mS}}{-j1 \text{ mS} + j0.4 \text{ mS}}$$

$$I' = 0.6667 \text{ mA } \underline{/180^\circ}$$

Obtain the Norton equivalent admittance Y':

$$Y' = Y_3 + \frac{Y_1Y_2}{Y_1 + Y_2} = -j1 \text{ mS} + \frac{(-j1 \text{ mS})(j0.4 \text{ mS})}{-j1 \text{ mS} + j0.4 \text{ mS}}$$

$$Y' = -j1 \text{ mS} + j0.6667 \text{ mS} = -j0.3333 \text{ mS}$$

$$I_L = I'\frac{Y_L}{Y_L + Y'} = 0.6667 \text{ mA } \underline{/180^\circ} \times \frac{Y_L}{(Y_L - j0.3333) \text{ mS}}$$

When $Y_L = 2$ mS ($R_L = 500\ \Omega$),

$$I_L = \frac{0.6667 \text{ mA } \underline{/180^\circ} \times 2 \text{ mS } \underline{/0^\circ}}{(2 - j0.3333) \text{ mS}} = \frac{0.6667 \text{ mA } \underline{/180^\circ} \times 2 \text{ mS } \underline{/0^\circ}}{2.028 \text{ mS } \underline{/-9.46^\circ}}$$

$$I_L = 0.6575 \text{ mA } \underline{/189.5^\circ} = 0.6575 \text{ mA } \underline{/-170.5^\circ} \quad \text{or} \quad \underline{/-2.976 \text{ rad}}$$

When $Y_L = 0.2$ mS ($R_L = 5$ kΩ),

$$I_L = \frac{0.6667 \text{ mA } \underline{/180^\circ} \times 0.2 \text{ mS } \underline{/0^\circ}}{(0.2 - j0.3333) \text{ mS}} = \frac{0.6667 \text{ mA } \underline{/180^\circ} \times 0.2 \text{ mS } \underline{/0^\circ}}{0.3887 \text{ mS } \underline{/-59.04^\circ}}$$

$$I_L = 0.343 \text{ mA } \underline{/239^\circ} = 0.343 \text{ mA } \underline{/-121^\circ} \quad \text{or} \quad \underline{/-2.112 \text{ rad}}$$

The range of load current values is **0.343 mA $\underline{/-121^\circ}$** to **0.6575 mA $\underline{/-170.5^\circ}$**, the phase angles being with respect to the source current.

EXAMPLE 17-10 Obtain the Norton equivalent of the Thevenin circuit in Fig. 17-11 (refer to Sec. 10-5).

Solution Figure 17-14*a* is a repeat of Fig. 17-11. Figure 17-14*b* is the Norton equivalent of Fig. 17-14*a*.

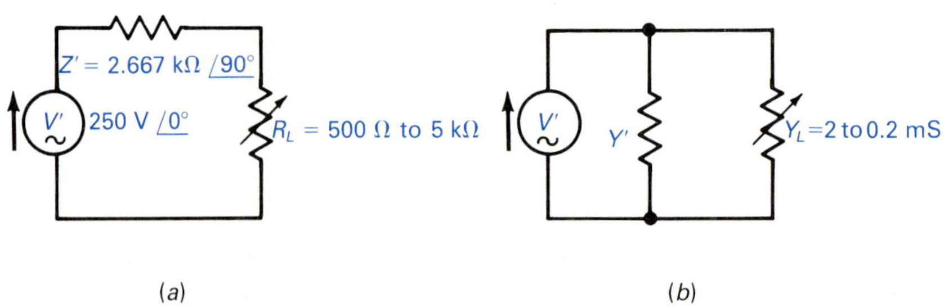

Fig. 17-14. Example 17-10: *(a)* Given Thevenin circuit; *(b)* Norton equivalent.

$$I' = \frac{V'}{Z'} = \frac{250\text{ V}\ \underline{/0°}}{2.667\text{ k}\Omega\ \underline{/90°}} = \mathbf{93.74\text{ mA}\ \underline{/-90°}\text{ or }\underline{/-\pi/2\text{ rad}}}$$

$$Y' = \frac{1}{Z'} = \frac{1}{2.667\text{ k}\Omega\ \underline{/90°}} = \mathbf{0.375\text{ mS}\ \underline{/-90°}\text{ or }\underline{/-\pi/2\text{ rad}}}$$

EXAMPLE 17-11 Obtain the Thevenin equivalent of the Norton circuit in Fig. 17-13*b* (refer to Sec. 10-5).

Solution Figure 17-15*a* is the same as Fig. 17-13*b*. Figure 17-15*b* is the Thevenin equivalent of Fig. 17-15*a*.

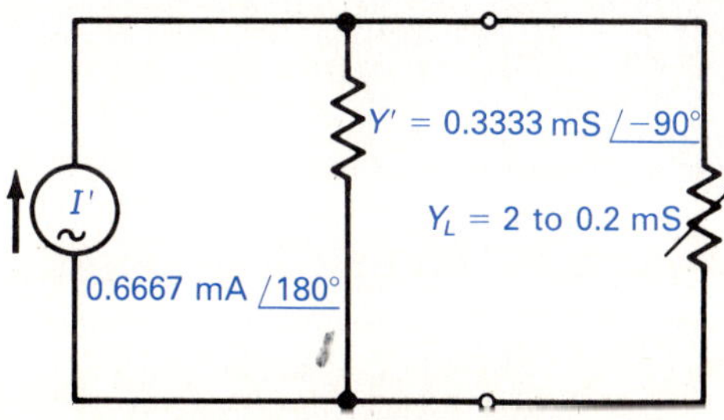

(*a*)

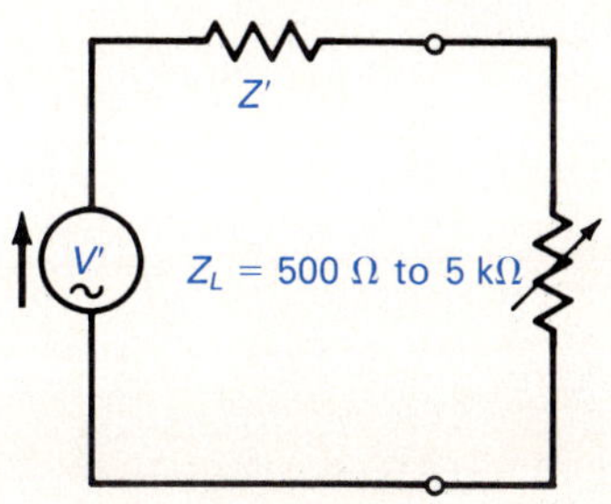

(*b*)

Fig. 17-15. Example 17-11: *(a)* Given Norton circuit; *(b)* Thevenin equivalent.

$$V' = \frac{I'}{Y'} = \frac{0.6667\text{ mA}\ \underline{/180°}}{0.3333\text{ mS}\ \underline{/-90°}} = \mathbf{2\text{ V}\ \underline{/270°} = 2\text{ V}\ \underline{/-90°}\text{ or }\underline{/-\pi/2\text{ rad}}}$$

$$Z' = \frac{1}{Y'} = \frac{1}{0.3333\text{ mS}\ \underline{/-90°}} = \mathbf{3\text{ k}\Omega\ \underline{/90°}\quad\text{or}\quad\underline{/\pi/2\text{ rad}}}$$

EXAMPLE 17-12 What is the magnitude of the load voltage in the circuit in Fig. 17-16*a* (a repeat of the circuit in Fig. 17-1)? Use the consecutive Thevenin approach (refer to Sec. 10-6).

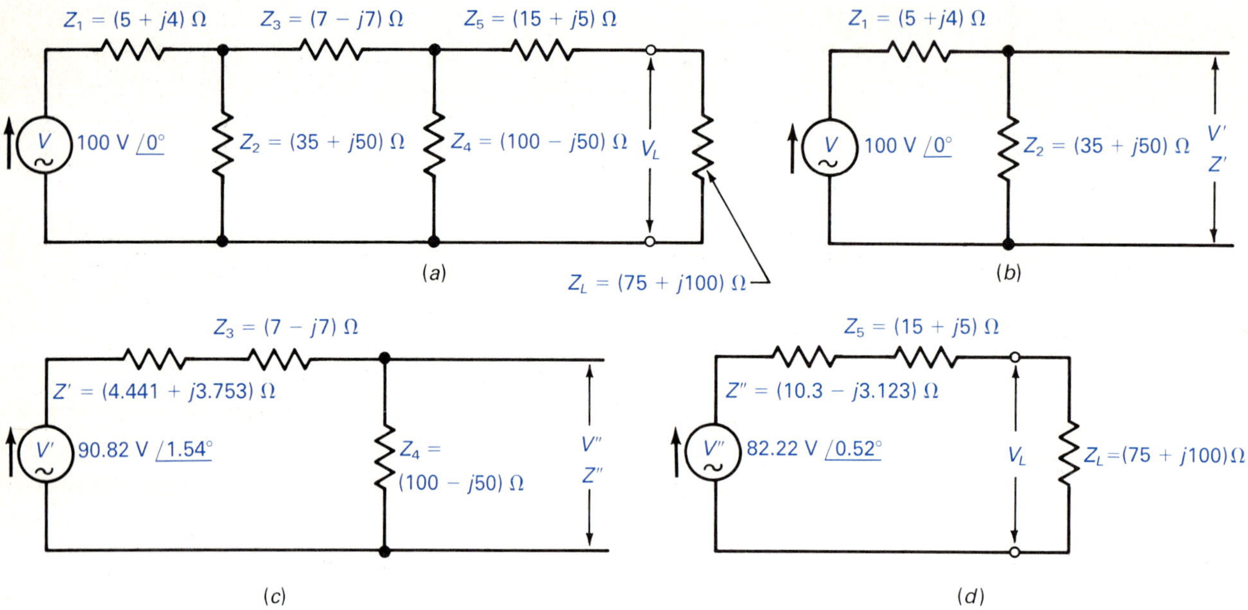

Fig. 17-16. Example 17-12: *(a)* Given circuit; *(b)* to *(d)* steps in the solution.

Solution

Refer to Fig. 17-16*b* to obtain the values of the first Thevenin emf V', and impedance Z':

$$V' = V\frac{Z_2}{Z_1 + Z_2} = 100\ \text{V}\ \underline{/0^\circ}\ \frac{(35 + j50)\ \Omega}{(5 + j4)\ \Omega + (35 + j50)\ \Omega}$$

$$= 100\ \text{V}\ \underline{/0^\circ}\ \frac{(35 + j50)\ \Omega}{(40 + j54)\ \Omega}$$

$$V' = 100\ \text{V}\ \underline{/0^\circ}\ \frac{61.03\ \Omega\ \underline{/55.01^\circ}}{67.2\ \Omega\ \underline{/53.47^\circ}} = 90.82\ \text{V}\ \underline{/1.54^\circ}$$

$$Z' = \frac{Z_1 Z_2}{Z_1 + Z_2} = \frac{(5 + j4)\ \Omega(35 + j50)\ \Omega}{(5 + j4)\ \Omega + (35 + j50)\ \Omega}$$

$$= \frac{6.403\ \Omega\ \underline{/38.66^\circ} \times 61.03\ \Omega\ \underline{/55.01^\circ}}{67.2\ \Omega\ \underline{/53.47^\circ}}$$

$$Z' = 5.815\ \Omega\ \underline{/40.2^\circ} = (4.441 + j3.753)\ \Omega$$

Refer to Fig. 17-16*c* to obtain the values of the second Thevenin emf V'' and impedance Z'':

$$V'' = V'\frac{Z_4}{Z' + Z_3 + Z_4}$$

$$= 90.82\ \text{V}\ \underline{/1.54^\circ}\ \frac{(100 - j50)\ \Omega}{(4.441 + j3.753)\ \Omega + (7 - j7)\ \Omega + (100 - j50)\ \Omega}$$

$$= 90.82\ \text{V}\ \underline{/1.54^\circ}\ \frac{111.8\ \Omega\ \underline{/-26.57^\circ}}{(111.4 - j53.25)\ \Omega}$$

$$V'' = 90.82 \text{ V } \angle 1.54° \frac{111.8 \ \Omega \ \angle -26.57°}{123.5 \ \Omega \ \angle -25.55°}$$

$$V'' = 82.22 \text{ V } \angle 0.52°$$

$$Z'' = \frac{Z_4(Z' + Z_3)}{Z' + Z_3 + Z_4} = \frac{(100 - j50) \ \Omega[(4.441 + j3.753) \ \Omega + (7 - j7) \ \Omega]}{(4.441 + j3.753) \ \Omega + (7 - j7) \ \Omega + (100 - j50) \ \Omega}$$

$$= \frac{111.8 \ \Omega \ \angle -26.57°(11.44 - j3.247) \ \Omega}{(111.4 - j53.25) \ \Omega}$$

$$= \frac{111.8 \ \Omega \ \angle -26.57° \times 11.89 \ \Omega \ \angle -15.85°}{123.5 \ \Omega \ \angle -25.55°} = 10.76 \ \Omega \ \angle -16.87°$$

$$Z'' = (10.3 - j3.123) \ \Omega$$

The magnitude of the load voltage V_L can now be found from Fig. 17-16*d*:

$$V_L = V'' \frac{Z_L}{Z'' + Z_5 + Z_L}$$

$$= 82.22 \text{ V } \angle 0.52° \frac{(75 + j100) \ \Omega}{(10.3 - j3.123) \ \Omega + (15 + j5) \ \Omega + (75 + j100) \ \Omega}$$

$$V_L = 82.22 \text{ V } \angle 0.52° \frac{125 \ \Omega \ \angle 53.13°}{143 \ \Omega \ \angle 45.45°} = 71.88 \text{ V } \angle 8.2°$$

Therefore the magnitude of the load voltage V_L is **71.88 V.**

The same problem was posed in Examples 17-1, 17-2, and 17-12. By three completely different analysis techniques the same answer was obtained, thus validating the methods.

EXERCISE PROBLEMS FOR SECTION 17-1

1. Find the power dissipated in Z_L in Fig. 17-17, using current assumption, when
   a. $V = 120$ V, $Z_1 = 15 \ \Omega \ \angle +25°$, $Z_2 = 10 \ \Omega \ \angle -30°$, $Z_3 = 20 \ \Omega \ \angle +40°$, $Z_L = 10 \ \Omega \ \angle +45°$ (*ans.* 58.58 W)
   b. $V = 5$ V, $Z_1 = 350 \ \Omega \ \angle +50°$, $Z_2 = 400 \ \Omega \ \angle +50°$, $Z_3 = 450 \ \Omega \ \angle -60°$, $Z_L = 600 \ \Omega \ \angle +30°$ (*ans.* 7.36 mW)

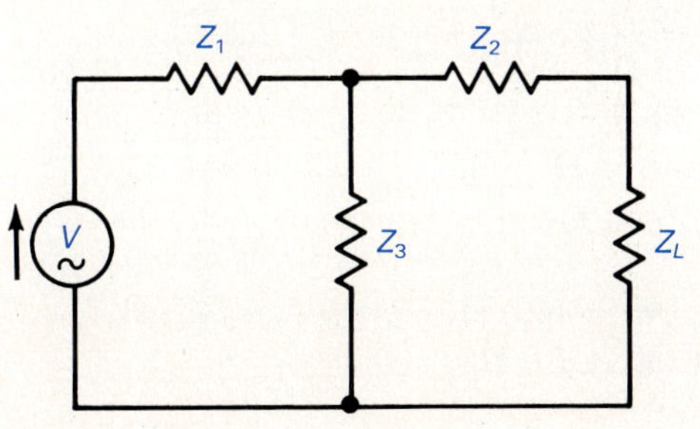

Fig. 17-17. Problems 1 to 4 and 14: Circuits.

2. Find the voltage across Z_L in Fig. 17-17, using nodal analysis, when
   a. $V = 50$ V, $Z_1 = 30\ \Omega\ \underline{/+30^\circ}$, $Z_2 = 20\ \Omega\ \underline{/-25^\circ}$, $Z_3 = 30\ \Omega\ \underline{/+35^\circ}$, $Z_L = 40\ \Omega\ \underline{/-40^\circ}$
   b. $V = 30$ V, $Z_1 = 850\ \Omega\ \underline{/-50^\circ}$, $Z_2 = 1000\ \Omega\ \underline{/+40^\circ}$, $Z_3 = 650\ \Omega\ \underline{/+30^\circ}$, $Z_L = 750\ \Omega\ \underline{/-20^\circ}$
3. Find the current in Z_L in Fig. 17-17, using loop analysis, when
   a. $V = 1$ V, $Z_1 = 1500\ \Omega\ \underline{/+45^\circ}$, $Z_2 = 1800\ \Omega\ \underline{/-60^\circ}$, $Z_3 = 2000\ \Omega\ \underline{/+55^\circ}$, $Z_L = 1200\ \Omega\ \underline{/+60^\circ}$ (*ans*. 278 μA)
   b. $V = 50$ V, $Z_1 = 10\ \Omega\ \underline{/-75^\circ}$, $Z_2 = 6\ \Omega\ \underline{/-65^\circ}$, $Z_3 = 5\ \Omega\ \underline{/-45^\circ}$, $Z_L = 12\ \Omega\ \underline{/-50^\circ}$ (*ans*. 0.81 A)
4. Find the current in Z_L in Fig. 17-17, using Thevenin's theorem, when
   a. $V = 20$ V, $Z_1 = (8 - j8)\ \Omega$, $Z_2 = (11 + j15)\ \Omega$, $Z_3 = (15 + j12)\ \Omega$, $Z_L = (10 + j10)\ \Omega$
   b. $V = 100$ V, $Z_1 = (100 + j100)\ \Omega$, $Z_2 = (50 - j50)\ \Omega$, $Z_3 = (75 + j75)\ \Omega$, $Z_L = (25 - j25)\ \Omega$

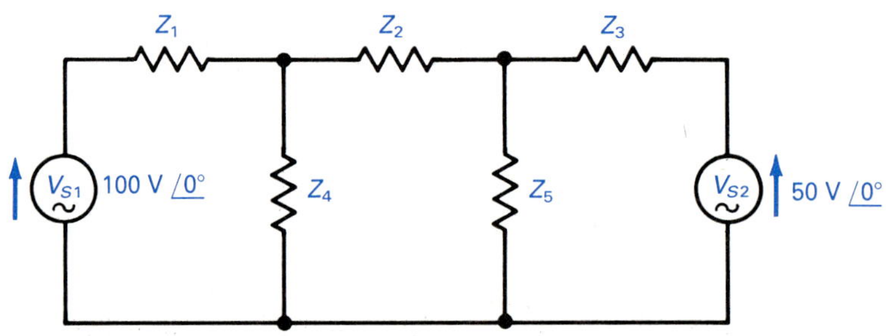

Fig. 17-18. Problems 5 to 8 and 15: Circuit.

5. Find the voltage across Z_5 in Fig. 17-18, using nodal analysis, when
   a. $Z_1 = 15\ \Omega\ \underline{/+18^\circ}$, $Z_2 = 24\ \Omega\ \underline{/-24^\circ}$, $Z_3 = 22\ \Omega\ \underline{/+30^\circ}$, $Z_4 = 40\ \Omega\ \underline{/-30^\circ}$, $Z_5 = 30\ \Omega\ \underline{/+40^\circ}$ (*ans*. 43.5 V $\underline{/+7.6^\circ}$)
   b. $Z_1 = 200\ \Omega\ \underline{/-15^\circ}$, $Z_2 = 240\ \Omega\ \underline{/-30^\circ}$, $Z_3 = 200\ \Omega\ \underline{/-15^\circ}$, $Z_4 = 300\ \Omega\ \underline{/+20^\circ}$, $Z_5 = 300\ \Omega\ \underline{/+20^\circ}$ (*ans*. 39.5 V $\underline{/+16.8^\circ}$)
6. Find the current in Z_5 in Fig. 17-18, using loop analysis, when
   a. $Z_1 = (30 + j40)\ \Omega$, $Z_2 = (40 + j20)\ \Omega$, $Z_3 = (30 - j20)\ \Omega$, $Z_4 = (50 - j40)\ \Omega$, $Z_5 = (20 + j20)\ \Omega$
   b. $Z_1 = (90 - j150)\ \Omega$, $Z_2 = (200 + j100)\ \Omega$, $Z_3 = (60 + j50)\ \Omega$, $Z_4 = (60 - j100)\ \Omega$, $Z_5 = (90 + j75)\ \Omega$
7. Find the current in Z_5 in Fig. 17-18, using the superposition theorem, when
   a. $Z_1 = 25\ \Omega\ \underline{/-30^\circ}$, $Z_2 = 25\ \Omega\ \underline{/+45^\circ}$, $Z_3 = 30\ \Omega\ \underline{/+30^\circ}$, $Z_4 = 15\ \Omega\ \underline{/-30^\circ}$, $Z_5 = 15\ \Omega\ \underline{/+60^\circ}$ (*ans*. 1.526 A $\underline{/-44.56^\circ}$)
   b. $Z_1 = 20\ \Omega\ \underline{/-45^\circ}$, $Z_2 = 30\ \Omega\ \underline{/-45^\circ}$, $Z_3 = 40\ \Omega\ \underline{/-45^\circ}$, $Z_4 = 25\ \Omega\ \underline{/-45^\circ}$, $Z_5 = 35\ \Omega\ \underline{/-45^\circ}$ (*ans*. 0.954 A $\underline{/-45^\circ}$)
8. Find the power dissipated by Z_5 in Fig. 17-18, using Norton's theorem, when
   a. $Z_1 = (50 + j75)\ \Omega$, $Z_2 = (60 - j40)\ \Omega$, $Z_3 = (45 + j45)\ \Omega$, $Z_4 = (50 + j75)\ \Omega$, $Z_5 = (50 + j0)\ \Omega$

b. $Z_1 = (50 + j75)\ \Omega$, $Z_2 = (60 - j40)\ \Omega$, $Z_3 = (45 + j45)\ \Omega$, $Z_4 = (50 + j75)\ \Omega$, $Z_5 = (50 + j100)\ \Omega$

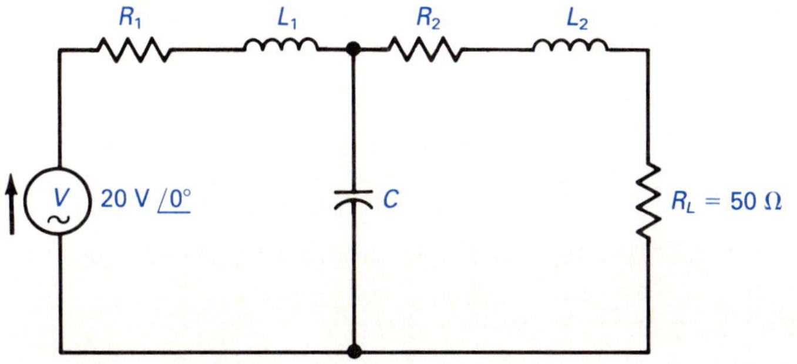

Fig. 17-19. Problems 9 and 10: Circuit.

9. In the low-pass filter circuit in Fig. 17-19, find the voltage across R_L when $R_1 = 0\ \Omega$, $R_2 = 0\ \Omega$, $L_1 = 100$ mH, $L_2 = 100$ mH, $C = 300\ \mu$F, and
a. $f = 30$ Hz (*ans.* 55.84 V $\angle +100.7°$)
b. $f = 120$ Hz (*ans.* 0.719 V $\angle +125.3°$)

10. In the low-pass filter circuit in Fig. 17-19, find the voltage across R_L when $R_1 = 10\ \Omega$, $R_2 = 10\ \Omega$, $L_1 = 100$ mH, $L_2 = 100$ mH, $C = 300\ \mu$F, and
a. $f = 30$ Hz *b.* $f = 120$ Hz *c.* $f = 0$ Hz

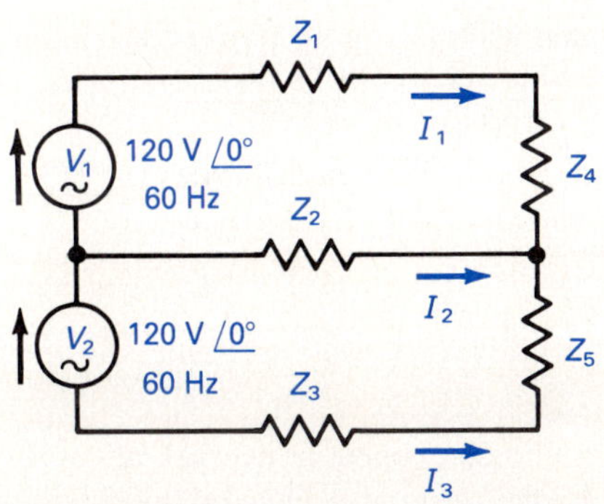

Fig. 17-20. Problems 11 and 12: Circuit.

11. In the three-wire, single-phase circuit in Fig. 17-20, find I_1, I_2, and I_3 when $Z_1 = Z_2 = Z_3 = 0\ \Omega$ and
a. $Z_4 = 8\ \Omega \angle +20°$, $Z_5 = 10\ \Omega \angle +35°$
(*ans.* 15 A $\angle -20°$, 4.61 A $\angle +22.3°$, 12 A $\angle -35°$)
b. $Z_4 = 12\ \Omega \angle -25°$, $Z_5 = 10\ \Omega \angle +30°$
(*ans.* 10 A $\angle +25°$, 10.31 A $\angle -82.6°$, 12 A $\angle -30°$)

12. In the three-wire, single-phase circuit in Fig. 17-20, find I_1, I_2, and I_3 when $Z_1 = Z_2 = Z_3 = 2\ \Omega$ and
a. $Z_4 = 8\ \Omega \angle +20°$, $Z_5 = 10\ \Omega \angle +35°$
b. $Z_4 = 12\ \Omega \angle -25°$, $Z_5 = 10\ \Omega \angle +30°$

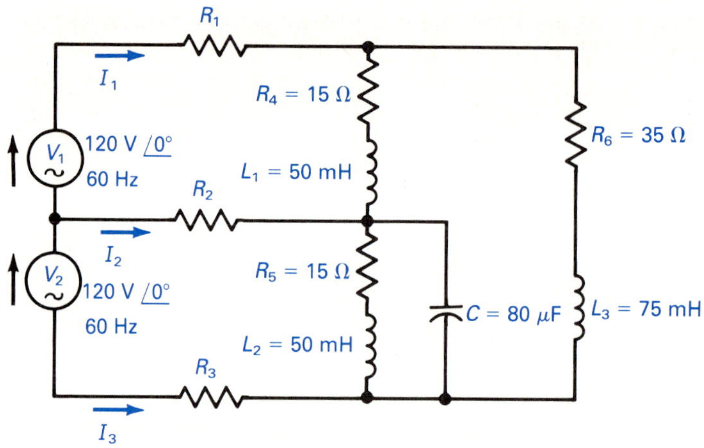

Fig. 17-21. Problem 13: Circuit.

13. In the three-wire, single-phase circuit in Fig. 17-21, find I_1, I_2, and I_3 when
a. $R_1 = R_2 = R_3 = 0\ \Omega$
(*ans.* 10.25 A $\angle -45°$, 3.618 A $\angle -90°$, 8.11 A $\angle -26.6°$)
b. $R_1 = R_2 = R_3 = 3\ \Omega$
(*ans.* 9.4 A $\angle -35.7°$, 3.11 A $\angle -67.2°$, 6.935 A $\angle -22.2°$)

14. For Fig. 17-17, find the value of Z_L that will result in its power dissipation being maximum when
a. $Z_1 = 15\ \Omega\ \angle +25°$, $Z_2 = 10\ \Omega\ \angle -30°$, $Z_3 = 20\ \Omega\ \angle +40°$
b. $Z_1 = 350\ \Omega\ \angle +50°$, $Z_2 = 400\ \Omega\ \angle +50°$, $Z_3 = 450\ \Omega\ \angle -60°$

15. For Fig. 17-18, find the value of Z_5 that will result in its power dissipation being maximum when $Z_1 = 15\ \Omega\ \angle +18°$, $Z_2 = 24\ \Omega\ \angle -24°$, $Z_3 = 22\ \Omega\ \angle +30°$, and $Z_4 = 40\ \Omega\ \angle -30°$.
(*ans.* 14.48 $\Omega\ \angle +13°$)

17-2 AC BRIDGES

Measurement of some of the characteristics of a component or network under ac conditions may be carried out using an ac bridge. A number of ac bridge circuits appear in commercial measuring instruments. Each of these different circuits has its own field of usefulness, and all are ac adaptations of the dc Wheatstone bridge typified in Example 9-12.

The basic ac bridge circuit is shown in Fig. 17-22.

When the bridge arms Z_1 to Z_4 in the circuit in Fig. 17-22 have a definite relationship to each other, the bridge will be balanced. At balance there will be zero voltage across the detector. Thus the balance condition, a zero detector indication, can be used as a means for showing when a certain relationship exists between the arms. The impedance or admittance of three of the arms, at balance, can be used to determine the impedance or admittance of the remaining arm. A practical bridge will have three of its arms of known (fixed or variable) value, the fourth arm being the impedance or admittance to be measured and called the unknown.

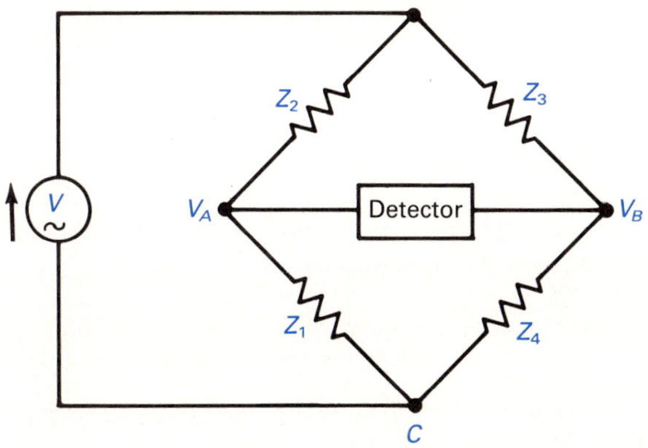

Fig. 17-22. Basic ac bridge circuit.

EXAMPLE 17-13 Derive an equation for determining the balance condition for the ac bridge circuit in Fig. 17-22.

Solution At balance there will be zero voltage across the detector. When this is so, the node voltages V_A and V_B, with respect to the reference node C, will be equal. There will be zero detector current, so that Z_1 and Z_2 will each carry the same current and thus be in series. Similarly, Z_3 and Z_4 will be in series. The voltage-divider principle can be used.

$$V_A = V_B$$

$$V\frac{Z_1}{Z_1 + Z_2} = V\frac{Z_4}{Z_3 + Z_4}$$

$$Z_1Z_3 + Z_1Z_4 = Z_1Z_4 + Z_2Z_4$$

$$Z_1Z_3 = Z_2Z_4$$

The impedance relationship for ac bridge balance in the specific case of Fig. 17-22 can be couched in general terms applicable to any ac impedance bridge by the statement: *A bridge is balanced when the products of the impedances of the opposite bridge arms are equal.*

The statement for impedance bridge balance applies for any supply frequency, including zero frequency (steady dc).

EXAMPLE 17-14 Express the resistance R_4 and inductance L in the Hay bridge circuit of Fig. 17-23 in terms of R_1, R_2, R_3, and C when the bridge is balanced.

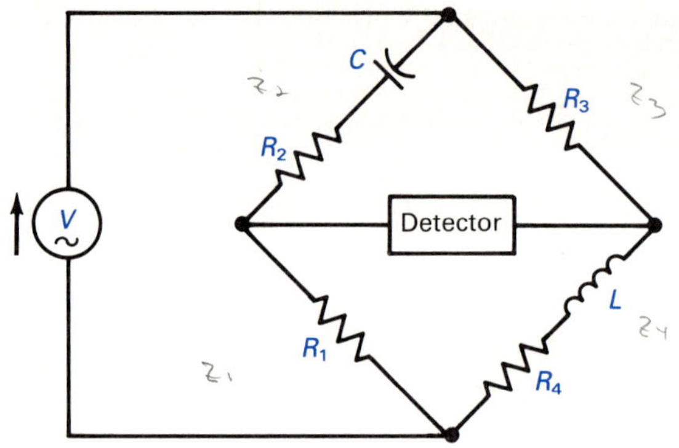

Fig. 17-23. Example 17-14: Hay bridge circuit.

Solution From the conclusion for Example 17-13, the bridge circuit in Fig. 17-23 will be balanced if

$$Z_1Z_3 = Z_2Z_4$$
$$R_1R_3 = (R_2 - jX_C)(R_4 + jX_L)$$
$$R_1R_3 = (R_2R_4 + X_LX_C) + j(R_2X_L - R_4X_C)$$

This equation states that the impedance expression on the left (R_1R_3) equals the impedance expression on the right $[(R_2R_4 + X_LX_C) + j(R_2X_L - R_4X_C)]$. This equality will be true only if real (in-phase) components are equal and the imaginary (quadrature) components are also equal.

Equating real components,

$$R_1R_3 = R_2R_4 + X_LX_C$$
$$R_4 = \frac{R_1R_3 - X_LX_C}{R_2} \quad \text{or} \quad X_L = \frac{R_1R_3 - R_2R_4}{X_C} \qquad (A)$$

Equating imaginary components,

$$0 = j(R_2X_L - R_4X_C)$$
$$R_4X_C = R_2X_L$$
$$R_4 = \frac{R_2X_L}{X_C} \quad \text{or} \quad X_L = \frac{R_4X_C}{R_2} \qquad (B)$$

Substituting for R_4 from Eq. *(B)* in Eq. *(A)*,

$$\frac{R_2X_L}{X_C} = \frac{R_1R_3 - X_LX_C}{R_2}$$
$$R_2^2X_L = R_1R_3X_C - X_LX_C^2$$
$$X_L(R_2^2 + X_C^2) = R_1R_3X_C$$
$$X_L = R_1R_3\frac{X_C}{R_2^2 + X_C^2}$$
$$\omega L = R_1R_3\frac{1/\omega C}{R_2^2 + 1/(\omega C)^2} = R_1R_3\frac{\omega C}{(\omega CR_2)^2 + 1}$$
$$\boldsymbol{L = C\frac{R_1R_3}{(\omega CR_2)^2 + 1}}$$

Substituting for X_L from Eq. *(B)* in Eq. *(A)*,

$$\frac{R_4X_C}{R_2} = \frac{R_1R_3 - R_2R_4}{X_C}$$

$$R_4X_C^2 = R_1R_2R_3 - R_2^2R_4$$

$$R_4(R_2^2 + X_C^2) = R_1R_2R_3$$

$$R_4 = R_1R_3\frac{R_2}{R_2^2 + X_C^2} = R_1R_3\frac{R_2}{R_2^2 + 1/(\omega C)^2}$$

$$= R_1R_3\frac{(\omega C)^2R_2}{(\omega CR_2)^2 + 1}$$

$$\mathbf{R_4 = (\omega C)^2R_2\frac{R_1R_3}{(\omega CR_2)^2 + 1}}$$

The circuit in Fig. 17-23 is called a Hay bridge. It is one of several circuit configurations used in commercially produced instruments for measuring the characteristics of electric and electronic components. The Hay bridge is used for measuring the characteristics of inductors. Typically, the values of R_1, R_2, R_3, and C in Fig. 17-23 are switched or varied to obtain a balance condition. Calibrated dials associated with these elements indicate the measured inductor values. The inductance value is usually indicated directly. As the Q value of an inductor gives a more graphic impression of its merit than does its resistance, it is usual for commercial instruments to read Q directly, using a relationship derived as follows:

Using the symbols in Example 17-14 (Fig. 17-23),

$$Q = \frac{X_L}{R_4} = \omega C\frac{R_1R_3}{(\omega CR_2)^2 + 1} \div (\omega C)^2R_2\frac{R_1R_3}{(\omega CR_2)^2 + 1} = \frac{1}{\omega CR_2}$$

EXAMPLE 17-15 What are the values of inductance and Q of the inductor in Fig. 17-24? The Hay bridge circuit is balanced with the values given for R_1, R_2, R_3, and C.

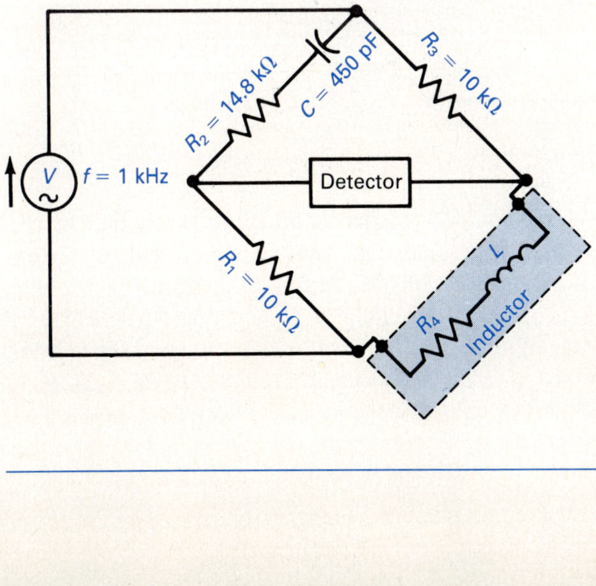

Fig. 17-24. Example 17-15: Hay bridge circuit.

Solution The term ωCR_2 occurs in the equations for both L and Q.

$$\omega CR_2 = 2\pi \times 1 \text{ kHz} \times 450 \text{ pF} \times 14.8 \text{ k}\Omega = 4.185 \times 10^{-2}$$

$$L = C\frac{R_1R_2}{(\omega CR_2)^2 + 1} = 450 \text{ pF}\frac{10 \text{ k}\Omega \times 10 \text{ k}\Omega}{(4.185 \times 10^{-2})^2 + 1} \simeq \mathbf{45 \text{ mH}}$$

$$Q = \frac{1}{\omega CR_2} = \frac{1}{4.185 \times 10^{-2}} = \mathbf{23.89}$$

REVIEW QUESTIONS FOR SECTION 17-2

1. Compare the circuits of ac bridges with those intended for use with steady dc.
2. Compare the balance conditions of ac bridges with those intended for use with steady dc.
3. In an ac bridge circuit, what is the objection to having opposite a reactive arm, one of the same type (both capacitive or both inductive)?
4. In a bridge circuit, ac or dc, the balance condition does not depend on the supply voltage. Why may it depend on the supply frequency in an ac bridge? What is an exception to this limitation?

EXERCISE PROBLEMS FOR SECTION 17-2

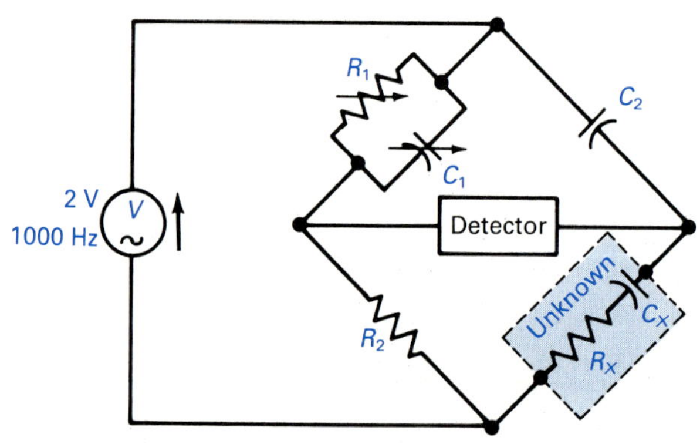

Fig. 17-25. Problem 1: Circuit.

1. Derive the balance equations for the Schering bridge shown in Fig. 17-25. For the unknown capacitor, find the values of R_x and C_x when $R_1 = 2300\ \Omega$, $R_2 = 1\ \text{k}\Omega$, $C_1 = 800$ pF, and $C_2 = 0.005\ \mu$F.
   (*ans.* $C_x = C_2R_1/R_2$, $R_x = R_2C_1/C_2$, 160 Ω, 0.0115 μF)
2. Derive the balance equations for the Maxwell bridge shown in Fig. 17-26. Find the values of R_x, L_x, and Q when $R_1 = 8300\ \Omega$, $R_2 = 2420\ \Omega$, $R_3 = 100\ \Omega$, and $C_1 = 0.1\ \mu$F.

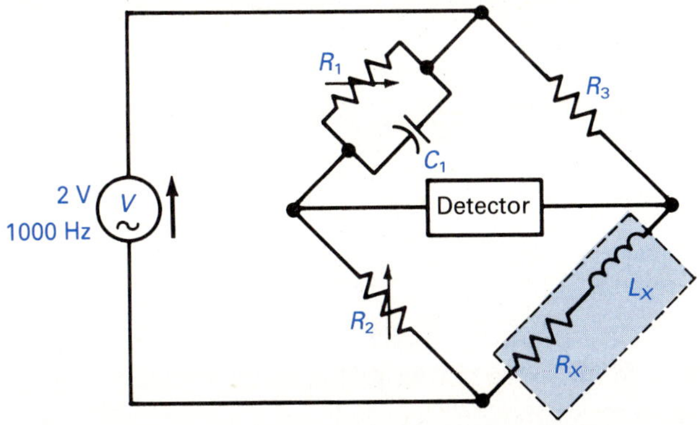

Fig. 17-26. Problem 2: Circuit.

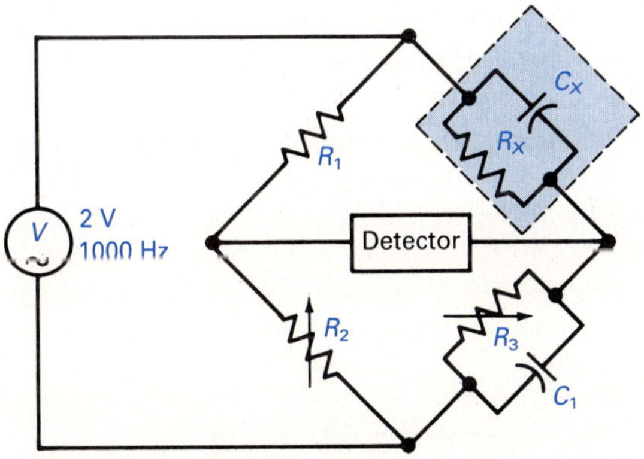

Fig. 17-27. Problem 3: Circuit.

3. Derive the balance equations for the capacitance comparison bridge shown in Fig. 17-27. Find the values of R_x and C_x when $R_1 = 10\ \text{k}\Omega$, $R_2 = 1800\ \Omega$, $R_3 = 5.5\ \text{k}\Omega$, and $C_1 = 0.1\ \mu\text{F}$.

(*ans.* $R_x = R_3R_1/R_2$, $C_x = C_1R_2/R_1$, 30.56 kΩ, 0.018 μF)

17-3 THREE-PHASE POWER

Single-Phase and Two-Phase

Most of the ac circuits dealt with to this point in the text were single-phase. A *single-phase circuit* can be described as one in which there is either (1) one energy source, or (2) a number of sources all acting in phase with each other in supplying energy to the load system.

A single-phase circuit with one energy source is called *two-wire,* because there are just two conductors connecting the source to its load. A *three-wire,* single-phase circuit has in effect two energy sources connected in series and acting in phase across the load system.

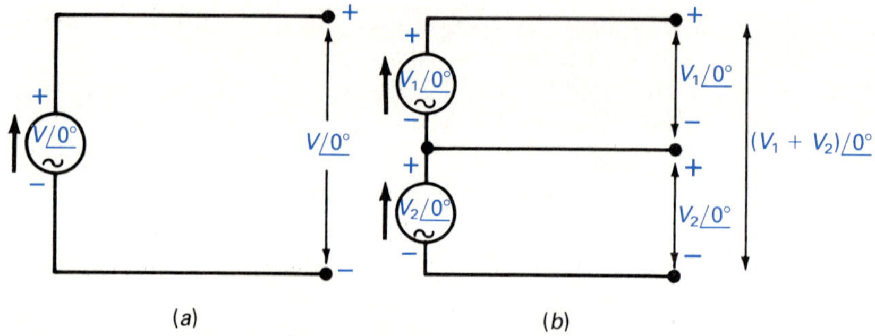

Fig. 17-28. *(a)* Single-phase, two-wire circuit; *(b)* single-phase, three-wire circuit.

Figure 17-28 illustrates the essential similarity of single-phase circuits.

In the two-wire circuit in Fig. 17-28*a*, an instantaneous polarity is shown for the single generator, its emf being indicated as the reference phasor $V \underline{/0^\circ}$.

The three-wire circuit in Fig. 17-28*b* also shows instantaneous polarities, with each source producing in-phase emfs $V_1 \underline{/0^\circ}$ and $V_2 \underline{/0^\circ}$, both of reference phase. The two emfs, acting in series, result in a third emf $(V_1 + V_2) \underline{/0^\circ}$, also of reference phase. This three-wire circuit forms the basis for domestic electric supplies, giving two 120-V sources and one of 240 V, all in phase with each other.

Compare the three-wire, single-phase circuit in Fig. 17-28*b* with the circuit in Examples 17-3 and 17-4. The latter circuit is redrawn in Fig. 17-29*a*.

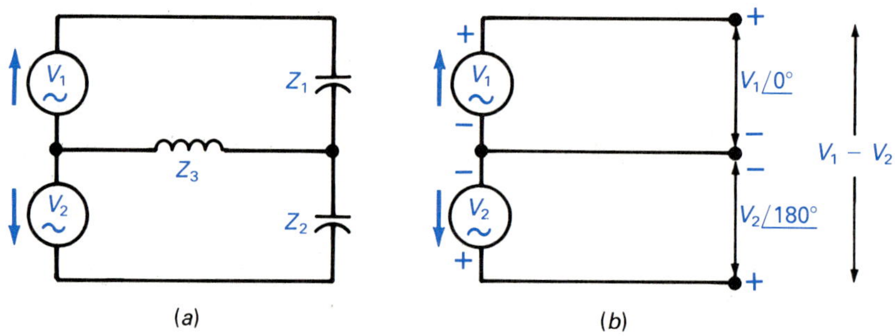

Fig. 17-29. *(a)* Circuit in Examples 17-3 and 17-4 redrawn; *(b)* the emfs of the circuit.

Figure 17-29*a* has two generators and three wires feeding the load system, as for a three-wire, single-phase circuit. But its two sources are connected in series opposition. The generators alone are shown in Fig. 17-29*b* for direct comparison with Fig. 17-28*b*. In assigning the reference phase to V_1 in Fig. 17-29*b*, the three circuit emfs are $V_1 \underline{/0^\circ}$, $V_2 \underline{/180^\circ}$, and $(V_1 \underline{/0^\circ} + V_2 \underline{/180^\circ}) = V_1 - V_2$. If the magnitudes of V_1 and V_2 are equal, the third emf will be of zero magnitude.

The circuit in Fig. 17-29 is *two-phase*. A two-phase circuit with emfs in antiphase has little practical use, but a two-phase circuit with emfs in quadrature is found in some industrial control systems.

Single-phase and two-phase circuits are usually restricted to low-power applications, such as electronic equipment and domestic installations.

For the generation, transmission, and distribution of electrical energy, and its industrial use, there are technical and economic advantages in using a three-phase electric system. As some of the details of these considerations involve equipment rather than circuits, they are not appropriate to this text. Three-phase circuits, however, raise some ac network issues which warrant our attention.

Three-Phase Conventions

In our treatment of three-phase circuits, it will be assumed that two basic conventions of commercial electric systems apply:

1. The three phases originate within the circuit as three emfs of equal magnitude.
2. The three emfs have mutual phase differences of 120° or $2\pi/3$ rad.

The equal emf magnitudes imply a balanced three-phase energy source. The term $2\pi/3$ indicates that every cycle (2π rad) is effectively divided into three equal angle or time segments.

The three emfs required for operating three-phase (often symbolically abbreviated 3ϕ) have the waveform and phasor relationships given in Fig. 17-30.

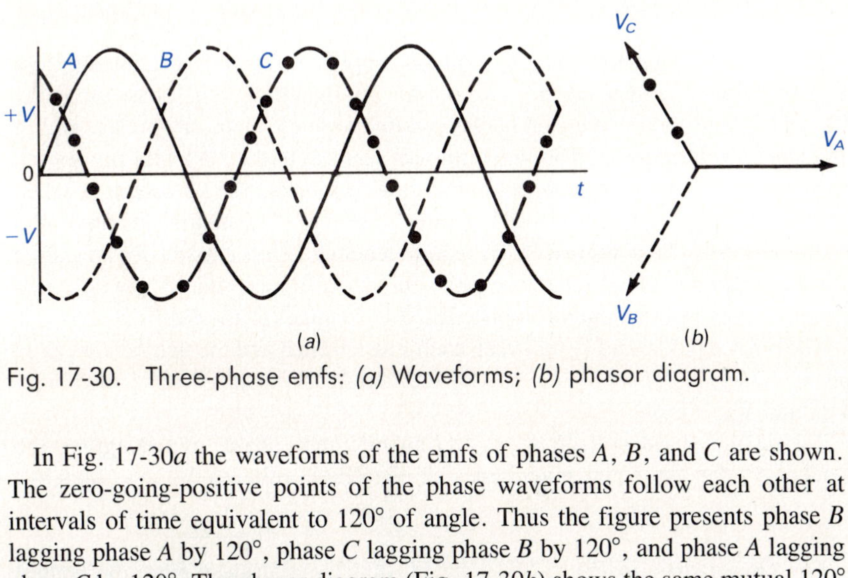

Fig. 17-30. Three-phase emfs: *(a)* Waveforms; *(b)* phasor diagram.

In Fig. 17-30*a* the waveforms of the emfs of phases *A*, *B*, and *C* are shown. The zero-going-positive points of the phase waveforms follow each other at intervals of time equivalent to 120° of angle. Thus the figure presents phase *B* lagging phase *A* by 120°, phase *C* lagging phase *B* by 120°, and phase *A* lagging phase *C* by 120°. The phasor diagram (Fig. 17-30*b*) shows the same mutual 120° lag between adjacent phases.

Source Connections in 3ϕ Circuits

The energy sources originate at an electric generating station. Typically, in a hydroelectric station energy is converted from the potential energy of elevated water to electrical energy by a turbine-generator combination. The actual output of the generator is unlikely to be fed to the electric system. Rather, it will be fed through a transformer. Furthermore, it is often desired to refer to circuits remote from the generating station, where again the effective energy source is likely to be a transformer. Despite the fact that it may be more realistic to show transformers as virtual energy sources, it simplifies the understanding of 3ϕ circuits if sources are represented by generator symbols.

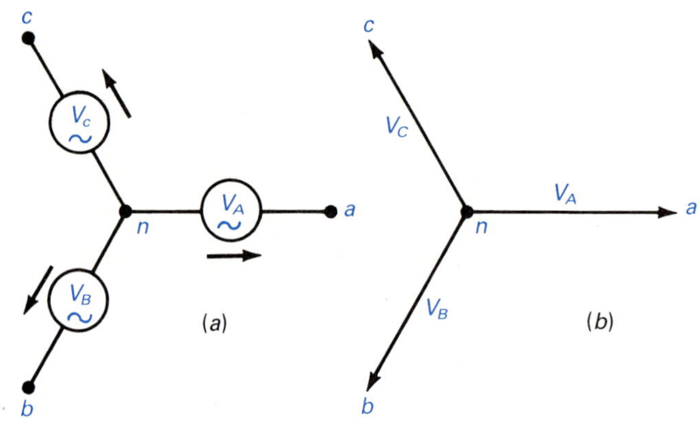

Fig. 17-31. Wye-connected three-phase supply source: *(a)* Connections; *(b)* phasor diagram.

In Fig. 17-31*a* the emfs of a three-phase supply are shown as originating from three generators connected in a *wye* or *star* configuration. The phase emfs V_A, V_B, and V_C are each connected between a *line* terminal a, b, or c, respectively, and a *neutral* terminal n. The line terminals connect to the 3ϕ load. The neutral terminal may or may not connect to the load; if it does, the 3ϕ source is called four-wire wye; if not, it is classed as three-wire wye. The neutral terminal is so called because (1) it represents an equipotential reference point for the three emfs, and (2) if the load is balanced (the only condition we shall deal with), zero current exists in a conductor connecting this terminal to the load.

The emfs V_A, V_B, and V_C between each line terminal and the neutral are called phase voltages. Between each pair of line terminals exists a line voltage, V_{ab}, V_{bc}, and V_{ca}.

Taking the phase voltage V_A as having reference phase, and assigning the symbol V_p for the equal magnitudes of the phase voltages V_A, V_B, and V_C,

$$V_A = V_p \underline{/0^\circ} \qquad V_B = V_p \underline{/-120^\circ} \qquad V_C = V_p \underline{/-240^\circ}$$

The phasor lines in Fig. 17-31*b* are all directionally away from the neutral terminal. Thus, tracing from the end of one line to the end of another, the path goes through two opposing arrow directions, indicating that adjacent phase voltages are in phasor opposition. The line voltages, therefore, are the phasor differences between adjacent phase voltages.

$$\begin{aligned} V_{ab} &= V_A - V_B = V_p \underline{/0^\circ} - V_p \underline{/-120^\circ} \\ &= V_p - (-0.5 - j0.866)V_p = (1.5 + j0.866)V_p \\ V_{ab} &= \sqrt{3}V_p \underline{/30^\circ} \end{aligned}$$

$$V_{bc} = V_B - V_C = V_p \underline{/-120^\circ} - V_p \underline{/-240^\circ}$$
$$= (-0.5 - j0.866)V_p - (-0.5 + j0.866)V_p = -j1.732V_p$$
$$V_{bc} = \sqrt{3}V_p \underline{/-90^\circ}$$
$$V_{ca} = V_C - V_A = V_p \underline{/-240^\circ} - V_p \underline{/0^\circ}$$
$$= (-0.5 + j0.866)V_p - V_p = (-1.5 + j0.866)V_p$$
$$V_{ca} = \sqrt{3}V_p \underline{/-210^\circ}$$

All three line voltages are of magnitude equal to $\sqrt{3}$ times a phase voltage. Using the symbol V_l for the line voltage magnitude,

$$V_l = \sqrt{3}V_p$$

The three phase voltages V_A, V_B, and V_C are shown in the phasor diagram in Fig. 17-32*a*, together with the three line voltages V_{ab}, V_{bc}, and V_{ca}. There is a mutual lagging phase angle of 120° between adjacent phase voltages and between adjacent line voltages, but a phase voltage lags by 30° the line voltage existing between that phase and the next adjacent phase; for example, V_A lags V_{ab} by 30°.

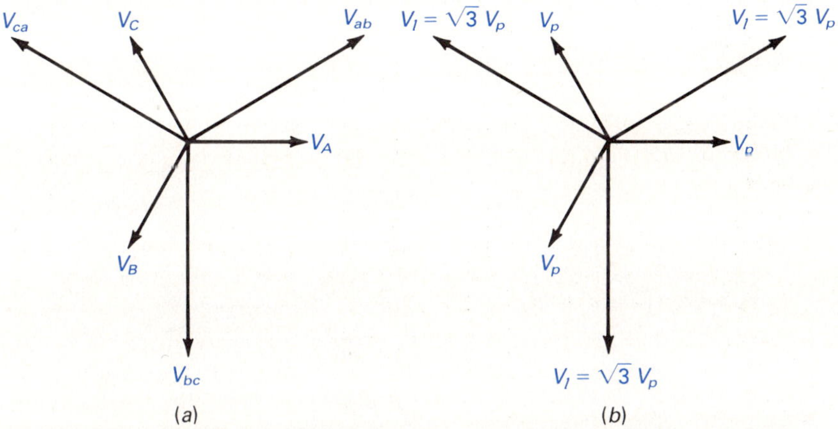

Fig. 17-32. Wye source phasor diagrams: *(a)* Terminal voltages; *(b)* line and phase voltage.

Figure 17-32*b* is the same as Fig. 17-32*a* except that it is labeled in terms of the common phase voltage V_p and the common line voltage V_l.

A *delta* three-phase source has the format of Fig. 17-33*a*. This type of source connection is less common than the wye. No neutral line is available, although it is possible to refer to a *virtual* or *floating* neutral. As the phasors in Fig. 17-33*b* and *c* have the same geometry as those of the line voltages of the wye connection in Fig. 17-32*a* and *b*, in the case of a balanced system the virtual neutral of a delta source will have the same relative potential as the true neutral of the wye source; the phase voltage from each line terminal to the virtual neutral will be $1/\sqrt{3}$ times that of the line voltage.

In a delta source the generator emfs are the line voltages. The potentials between each line terminal and the virtual neutral are the phase voltages. The relationship between the line and phase voltages is the same for the delta as for the wye connection:

$$V_l = \sqrt{3}V_p$$

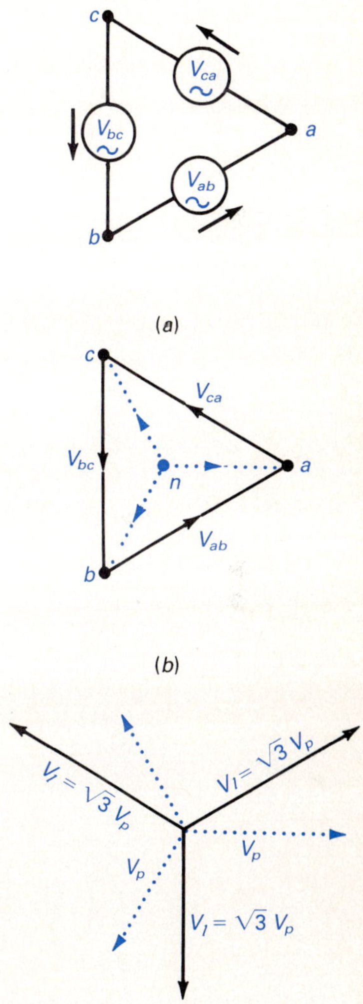

Fig. 17-33. Delta-connected three-phase supply source: *(a)* Connections; *(b)* terminal voltages; *(c)* line and phase voltages.

The phasor diagram in Fig. 17-33*b* is shown in *funicular* form (each phasor line drawn from the end of the adjacent line), as this clearly relates to the formation of the delta circuit. To be consistent with all our other phasor diagrams, we will maintain the principle that phasor lines radiate from a common origin, as in Fig. 17-33*c*. This convention has the added advantages of presenting all 3ϕ diagrams as basically the same, and of simplification where both currents and voltages are to be shown.

Load Connections in 3ϕ Circuits

As for sources, loads may be connected in wye or delta configurations. In either case the load will consist of three elements, each of which can have any impedance value. This treatment will be confined to balanced loads in which the three elements are of the same impedance value.

A schematic diagram for a wye-connected load is given in Fig. 17-34*a*. The diagram is oriented in the same manner as the source diagrams in Figs. 17-31*a* and 17-33*a*; the symbols are also basically the same for direct comparison between a source and its load.

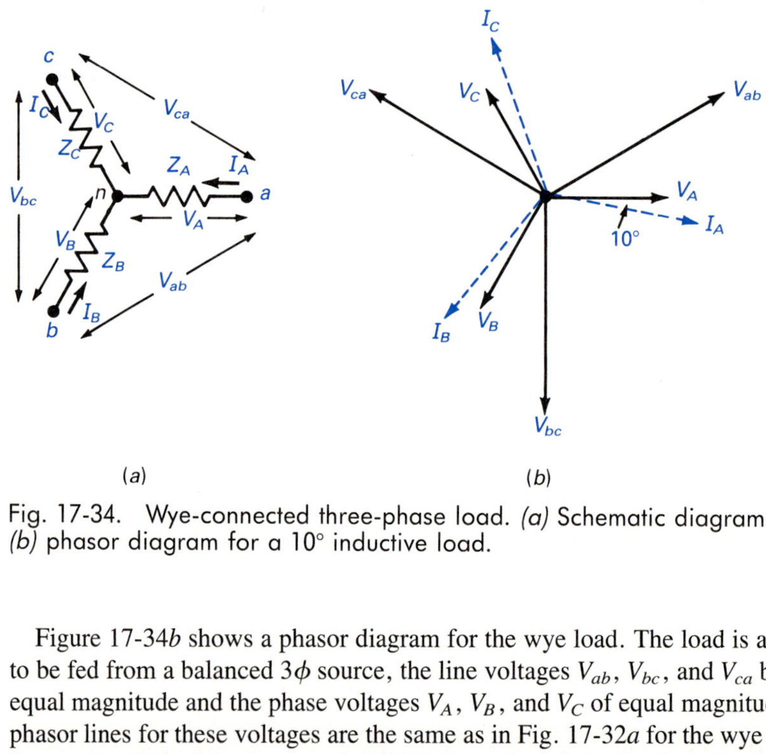

Fig. 17-34. Wye-connected three-phase load. *(a)* Schematic diagram; *(b)* phasor diagram for a 10° inductive load.

Figure 17-34*b* shows a phasor diagram for the wye load. The load is assumed to be fed from a balanced 3ϕ source, the line voltages V_{ab}, V_{bc}, and V_{ca} being of equal magnitude and the phase voltages V_A, V_B, and V_C of equal magnitude. The phasor lines for these voltages are the same as in Fig. 17-32*a* for the wye source. A common inductive phase angle of 10° was selected for the three impedances. The voltage across each impedance is a phase voltage, so that the related current lags this voltage by 10°. Any other value of impedance phase angle, common or individual, will result in appropriately related current phase angles. The current in each element of a wye load is both phase current and line current.

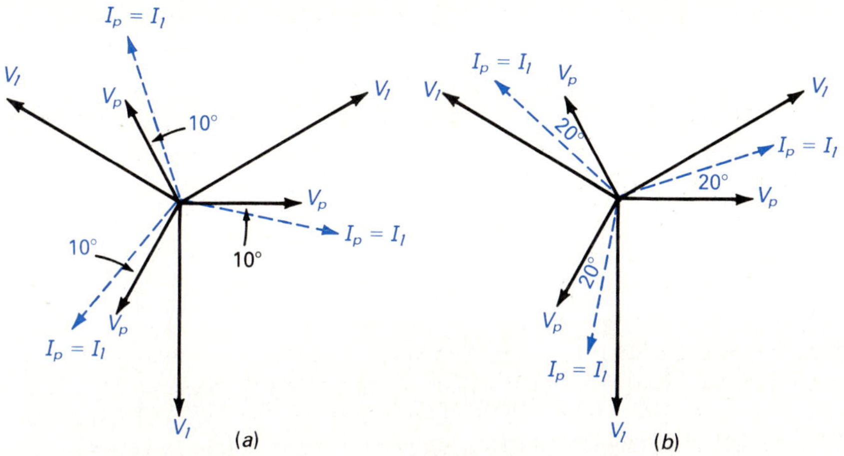

Fig. 17-35. Phasor diagrams of the line and phase voltages and currents for a wye load *(a)* with a 10° inductive load; *(b)* with a 20° capacitive load.

The phasor diagrams in Fig. 17-35 refer to a wye load, the phasors being expressed in terms of phase and line voltages V_p and V_l, and phase and line currents I_p and I_l. The common load impedances Z for Fig. 17-35*a* are inductive with a 10° phase angle, while those in Fig. 17-35*b* are capacitive with a 20° phase angle. In Fig. 17-35,

$$V_l = \sqrt{3}V_p \qquad \text{and} \qquad I_l = I_p = \frac{V_p}{Z}$$

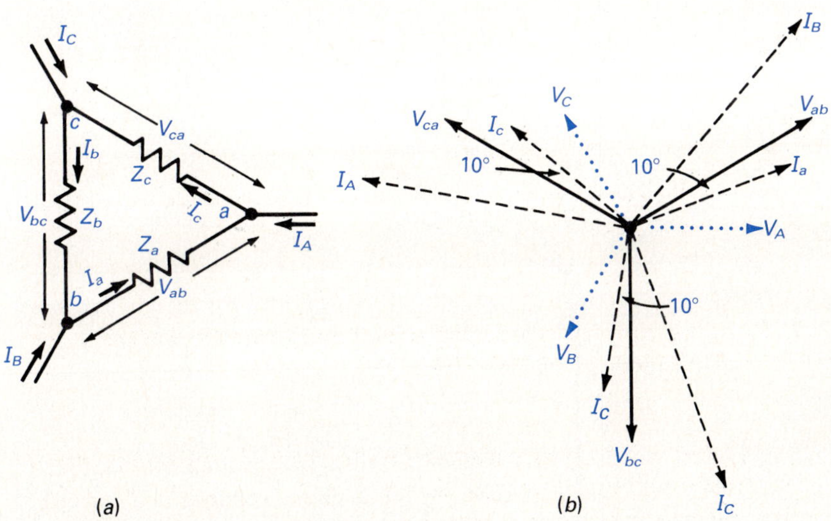

Fig. 17-36. Delta-connected three-phase load. *(a)* Schematic diagram; *(b)* phasor diagram for a 10° inductive load.

Figure 17-36*a* gives a schematic diagram of a delta load. The voltages V_{ab}, V_{bc}, and V_{ca} are line voltages which, if the load is supplied from a balanced source, are equal in magnitude with a mutual 120° phase difference. Furthermore, if the load is balanced, each of the three impedances will be equal and their currents will be

$$I_a = \frac{V_{ab}}{Z_a} = \frac{V_l \angle 30^\circ}{Z}$$

$$I_b = \frac{V_{bc}}{Z_b} = \frac{V_l \angle -90^\circ}{Z}$$

$$I_c = \frac{V_{ca}}{Z_c} = \frac{V_l \angle -210^\circ}{Z}$$

At each of the line terminals, a, b, and c the line current is the phasor difference between currents in two impedances:

$$I_A = I_c - I_a = \frac{V_l \angle -210^\circ}{Z} - \frac{V_l \angle 30^\circ}{Z} = \sqrt{3}\,\frac{V_l \angle 180^\circ}{Z}$$

Similarly, $\quad I_B = \sqrt{3}\,\dfrac{V_l \angle 60^\circ}{Z} \quad$ and $\quad I_C = \sqrt{3}\,\dfrac{V_l \angle -60^\circ}{Z}$

To illustrate a typical case, in Fig. 17-36b the impedances are assumed to be inductive with a 10° phase angle, so their currents lag the line voltages by 10°:

$$I_a = \frac{V_l}{Z} \angle 20^\circ \qquad I_b = \frac{V_l}{Z} \angle -100^\circ \qquad I_c = \frac{V_l}{Z} \angle -220^\circ$$

$$I_A = \sqrt{3}\,\frac{V_l}{Z} \angle 170^\circ \qquad I_B = \sqrt{3}\,\frac{V_l}{Z} \angle 50^\circ \qquad I_C = \sqrt{3}\,\frac{V_l}{Z} \angle -70^\circ$$

In Fig. 17-36b the phase voltages V_A, V_B, and V_C between the line terminals and the virtual neutral are shown as dotted lines.

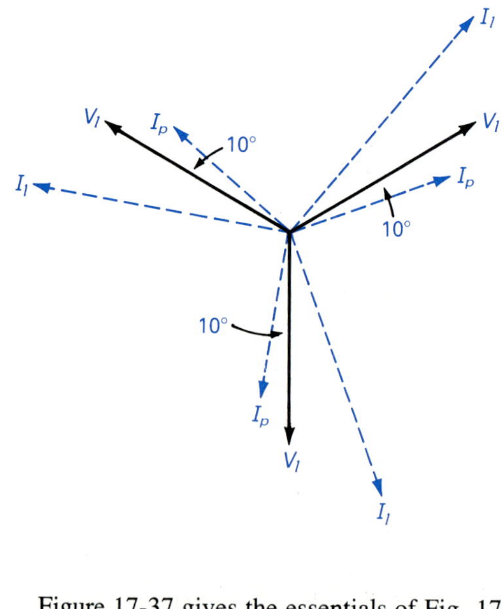

Fig. 17-37. Phasor diagram of the line and phase voltages and currents for a 10° inductive delta load.

Figure 17-37 gives the essentials of Fig. 17-36b with the line voltages shown as V_l, the line currents as I_l, and the currents in the individual impedances as I_p.

Summary of the Main 3ϕ Data

The following summary of voltage and current magnitude data can be of help in understanding balanced 3ϕ networks.

1. Wye source: $V_l = \sqrt{3}V_p$
   $I_l = I_p$
2. Delta source: $V_l = \sqrt{3}V_p$
   $I_l = \sqrt{3}I_p$
3. Wye load: $V_l = \sqrt{3}V_p$
   $I_l = I_p = \dfrac{V_p}{Z}$
4. Delta load: $V_l = \sqrt{3}V_p$
   $I_l = \sqrt{3}I_p = \sqrt{3}\dfrac{V_l}{Z}$

In the above relationships the symbols represent phasor terms:

V_l = source line voltage, voltage between a pair of source terminals, V
V_p = source phase voltage, voltage between a line terminal and the neutral terminal or the virtual neutral, V
I_l = line current, current in a line conductor from a source or to a load, A
I_p = phase current, current in an element connected between a line terminal and a neutral terminal or between two line terminals, A
V_l = load line voltage, V
V_p = load phase voltage, V

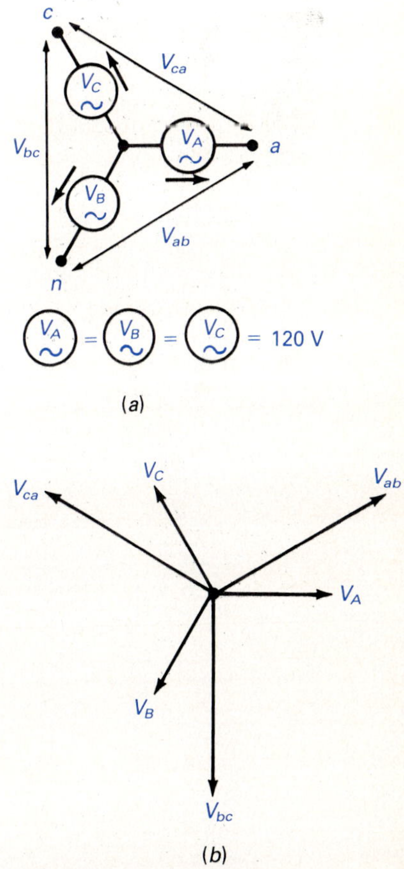

Fig. 17-38. Example 17-16: *(a)* Source connections; *(b)* phasor diagram.

EXAMPLE 17-16 A 3ϕ source consists of three 120-V transformer windings connected in wye formation. What are the phase and line voltages?

Solution Figure 17-38*a* is a schematic diagram of the wye 3ϕ source in which the transformer windings are represented as generators.

By definition, if V_A is made the reference phasor, the phase voltages are V_A = **120 V $\underline{/0°}$**, V_B = **120 V $\underline{/-120°}$**, and V_C = **120 V $\underline{/-240°}$**.

Each line voltage is the phasor difference between two phase voltages:

$V_{ab} = V_A - V_B = 120\text{ V }\underline{/0°} - 120\text{ V }\underline{/-120°} = \sqrt{3} \times 120\text{ V }\underline{/30°}$
V_{ab} = **207.8 V $\underline{/30°}$** or **$\underline{/0.5236\text{ rad}}$**
$V_{bc} = V_B - V_C = 120\text{ V }\underline{/-120°} - 120\text{ V }\underline{/-240°} = \sqrt{3} \times 120\text{ V }\underline{/-90°}$
V_{bc} = **207.8 V $\underline{/-90°}$** or **$\underline{/-\pi/2\text{ rad}}$**
$V_{ca} = V_C - V_A = 120\text{ V }\underline{/-240°} - 120\text{ V }\underline{/0°} = \sqrt{3} \times 120\text{ V }\underline{/-210°}$
V_{ca} = **207.8 V $\underline{/-210°}$** or **$\underline{/-3.665\text{ rad}}$**

The combination of phase and line voltages in Example 17-16 is commonly used in 3ϕ electric systems. In such systems it is usual to regard the line voltage magnitude as 208 V, rather than the true value of 207.8 V, resulting in the so-called 120/208-V system.

EXAMPLE 17-17 Figure 17-39 shows a balanced delta load connected to a three-phase supply of 208-V line voltage. What are the magnitudes of the line currents if each load element has an apparent power of 1 kVA?

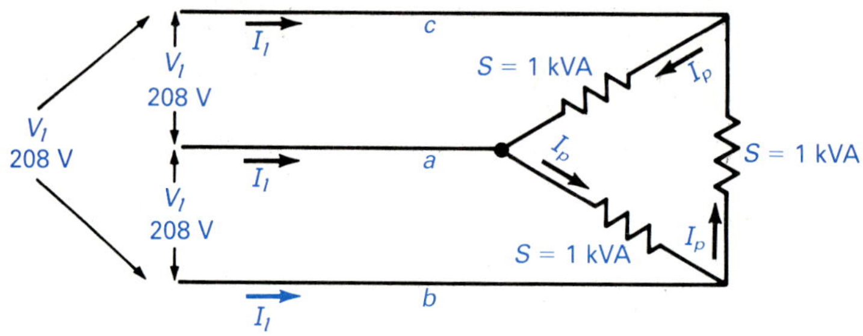

Fig. 17-39. Example 17-17: Network.

Solution Each load element is directly connected across a 208-V line voltage.

For each load element,

$$S = V_l I_p$$

$$I_p = \frac{S}{V_l} = \frac{1 \text{ kVA}}{208 \text{ V}} = 4.808 \text{ A}$$

The magnitude of each line current is the same in this balanced circuit:

$$I_l = \sqrt{3} I_p = \sqrt{3} \times 4.808 \text{ A} = \mathbf{8.327\ A}$$

For a single-phase 208-V circuit with the same 3-kVA total load, the line current would be 3×4.808 A = 14.42 A in each of two conductors. Thus if these two conductors were each of 1 Ω resistance, there would be individual line voltages of 14.42 V and the source emf would have to be 236.8 V to deliver 208 V to the load. With the 3ϕ circuit in the example, there are three conductors, each carrying 8.327 A. A conductor resistance of 1 Ω means an individual line voltage of 8.327 V. Two conductors connect each load element to the source, and so the source emf need be only 224.7 V to develop 208 V at the load. The lower 3ϕ source voltage relative to the single-phase source voltage required to deliver the same voltage to the same total load implies a greater efficiency for the

3ϕ circuit. The greater efficiency can be viewed in several ways:

1. For a given conductor resistance, the load voltage regulation is better.
2. For a given load voltage regulation, line conductors can be of smaller gauge.
3. For a given conductor resistance and load, line losses are less.

From a circuit viewpoint, as distinct from an equipment viewpoint (for example, in electric motors), the advantages of three-phase operation are usually cited as greater efficiency and economy.

REVIEW QUESTIONS FOR SECTION 17-3

1. How do single-phase, two-phase, and three-phase systems differ?
2. What is the angle between phases of a three-phase alternator, or source?
3. If one phase of a 3ϕ source can be described in phasor form as 208 V $\underline{/45^\circ}$, how would the other phases be shown?
4. Compare a *wye* source with a *delta* source.
5. In a three-phase system, what is the difference between balanced and unbalanced loads?

EXERCISE PROBLEMS FOR SECTION 17-3

1. What are the line-to-line voltages on a
   a. 120-V single-phase system?
   b. 120-V two-phase system?
2. What are the line-to-line voltages at the output of a three-phase alternator that gives 120 V line-to-neutral?
3. Express in phasor form the line currents when a 208-V wye-connected 3ϕ system feeds balanced loads of 3 kVA each at 0.75 power factor.

17-4 ELECTRONICS

The superposition principle was introduced in Chap. 9 to find the total effect on a circuit of a number of separate *dc* sources. Earlier in this present chapter, the use of this principle was extended to circuits with more than one *ac* source. Another application of superposition, particularly important in electronics, is for the separate treatment of the effects of *dc and ac* sources acting in the same circuit.

Often the sole purpose of an electronic circuit is to produce an ac output from an ac input source. To do this, the circuit may need a dc energy source. For example, the low-power *ac* from a microphone may be the input that is amplified by the circuit to deliver a high-power *ac* output to loudspeakers. To increase the effective *ac* power, *dc* energy is supplied to the circuit, with the dc source also used to set the conditions within which the ac operates.

The separate dc view of electronics, ignoring any ac operation, was the subject of Examples 9-7, 9-13, 9-17, 10-10, and 10-18, which you may wish to review. We will now look at some electronic circuits from the ac viewpoint alone. If both the dc and the ac conditions are put together, the complete circuit situation may be seen.

Equivalent ac circuits can be handled by the various analysis methods. However, this seems an appropriate point to introduce a subject particularly suited to electronics. It is convenient to approach many electronic circuits using *input-output,* or *two-port,* equivalence. A number of sets of parameters may be developed this way, each having preferred areas of use. A two-port equivalent circuit may be constructed from any of these parameter sets, giving a suitable basis for solving problems. Three such parameter sets will be discussed: z and y parameters show the basic principles involved while h parameters demonstrate how other parameter sets may be obtained by applying the basic principles.

Two-Port Circuit

Figure 17-40 shows a generalized two-port circuit with standard conventions. Two generators are shown. At a given instant each generator feeds a current i_1 or i_2 into the network via the upper terminal of one of its two ports. At the same instant there is a voltage v_{11} or v_{22} between each pair of port terminals, the voltage polarity being such that the upper terminal is positive with respect to the lower terminal.

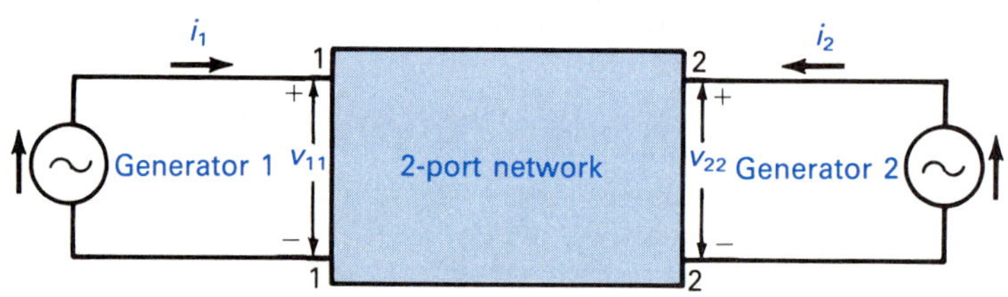

Fig. 17-40. Two-port circuit conventions.

In Fig. 17-40 lowercase symbols i and v are used to denote ac or varying currents and voltages.

Input-Output Approach

The conventions in Fig. 17-40 are deliberately arranged to avoid differentiating functionally between ports 11 and 22, which is desirable from a completely general viewpoint. But many applications of two-port principles relate to the transmission of energy through a network in one dominant direction, permitting the use of an input-output approach. This view involves simpler and more obvious symbols than the rigorous general approach, and so we will adopt it in this introduction to two-port network parameters and their use in ac network analysis.

In the diagram in Fig. 17-41, energy is transmitted from the generator to the load through the two-port network. The energy enters the network at the input

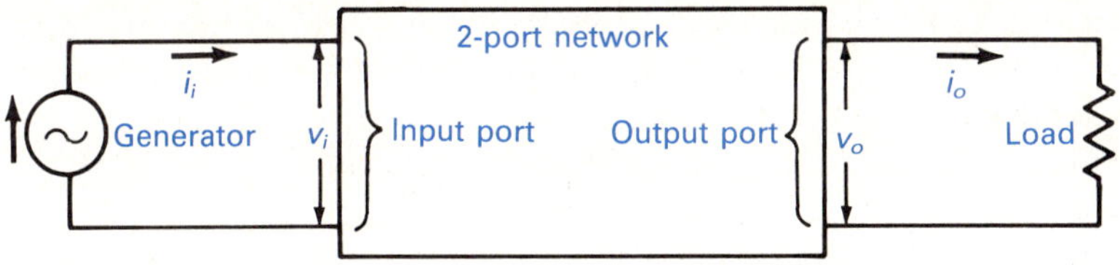

Fig. 17-41. Input-output form of a two-port circuit.

port and leaves at the output port. The input current i_i and the input voltage v_i are associated with the generator and the input port. The output current i_o and the output voltage v_o are associated with the output port and the load.

Input, Output, Forward, and Reverse Parameters

The electrical characteristics of a two-port network can be described by four parameters, shown pictorially in Fig. 17-42. These parameters are:

1. The input parameter, describing the characteristics determined at the input port.
2. The output parameter, describing the characteristics determined at the output port.
3. The forward parameter, describing the effect at the output port of a voltage or current applied at the input port. This effect is called *feedforward*.
4. The reverse parameter, describing the effect at the input port of a voltage or current applied at the output port. This effect is called *feedback*.

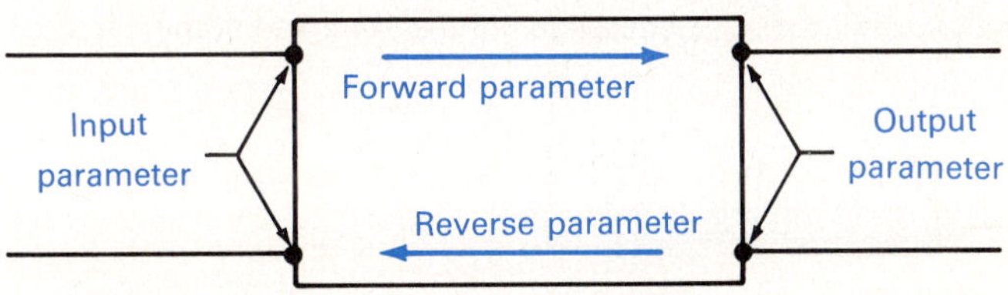

Fig. 17-42. Two-port network parameters.

Parameter Sets

Various ways of specifying the nature of network currents and voltages and the relationships between them are used for different circuit needs. This text will refer to impedance *(z)*, admittance *(y)*, and hybrid *(h)* parameters. These are three parameter sets, each consisting of four parameters (input, output, forward, and reverse). Each parameter is a specifically defined ratio of two quantities. The quantities we shall use are sine-wave voltages and currents, again adopting the convention of associating lowercase symbols v and i with ac voltages and currents of similar value, rms, peak-to-peak, etc.

It is important to realize that two-port parameters refer only to the characteristics of the two-port network, not to the characteristics of a complete circuit of which the network is part. The parameters are defined in terms of the reaction of the network to externally applied energy sources.

The two-port networks with which we shall concern ourselves will consist of basic resistances, capacitances, and inductances, and will thus be linear passive two-port networks.

Impedance *(z)* Parameters

Consider the situation shown in Fig. 17-43. Generators simultaneously apply voltages v_i and v_o to the input and output ports of the network. Currents i_i and i_o enter the input and output ports at their upper terminals and are shown as loop currents. This is the basic setup for determining z parameters.

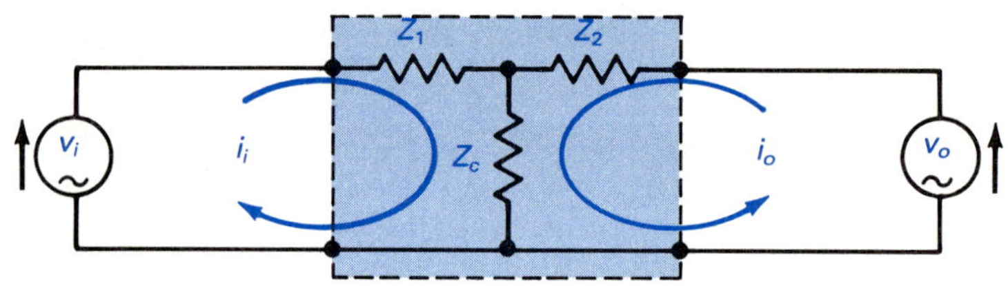

Fig. 17-43. Determination of z parameters.

Any linear passive two-port network can be reduced to an equivalent T network; thus the network in Fig. 17-43 can be regarded as representative of all linear passive two-port networks. The T network gives *common impedance coupling* between the input and output circuits, Z_c being the coupling impedance.

Applying loop analysis methods, the equations for the input and output circuits in Fig. 17-43 are

Input port

$$(Z_1 + Z_c)i_i + Z_c i_o = v_i \tag{17-1}$$

Output port

$$Z_c i_i + (Z_2 + Z_c)i_o = v_o \tag{17-2}$$

The impedance (z) parameters are described as follows:

1. The *input impedance parameter* z_i is the impedance of the network at the input port with the output port open-circuit (disconnected from the external circuit).

$$z_i = Z_1 + Z_c \tag{17-3}$$

2. The *output impedance parameter* z_o is the impedance of the network at the output port with the input port open-circuit.

$$z_o = Z_2 + Z_c \tag{17-4}$$

3. In Eq. (17-1), $Z_c i_o$ is a component of the input port voltage and represents a voltage fed back from the output to the input. Thus, in this context, Z_c can be regarded as an impedance controlling feedback. As such, Z_c is called the *reverse impedance parameter* z_r.

$$z_r = Z_c \tag{17-5}$$

4. In Eq. (17-2), $Z_c i_i$ is a component of the output port voltage, representing a voltage fed forward from the input to the output. Therefore, in this context, Z_c can be regarded as an impedance controlling feedforward. From an input-to-output viewpoint, Z_c is called the *forward impedance parameter* z_f.

$$z_f = Z_c \tag{17-6}$$

It can be seen from Eqs. (17-1) and (17-2) that Z_c, the coupling impedance, is common to both input and output circuits. The effect of Z_c in controlling transmission in the forward and reverse directions (from input to output and from output and input) is indicated by the parameters z_f and z_r.

Equations (17-1) and (17-2) can be rewritten using the symbols z_i, z_o, z_f, and z_r:

$$z_i i_i + z_r i_o = v_i \tag{17-7}$$

$$z_f i_i + z_o i_o = v_o \tag{17-8}$$

The two-port impedance parameters are defined in open-circuit terms (for example, the input impedance parameter is the impedance at the input port with the output port open-circuit). Applying these terms to Eqs. (17-7) and (17-8) results in expressions for each of the four parameters.

With the output port open-circuit ($i_o = 0$), Eq. (17-7) becomes

$$z_i i_i = v_i \qquad \text{or} \qquad z_i = \frac{v_i}{i_i}\bigg|_{i_o = 0} \tag{17-9}$$

and Eq. (17-8) becomes

$$z_f i_i = v_o \qquad \text{or} \qquad z_f = \frac{v_o}{i_i}\bigg|_{i_o = 0} \tag{17-10}$$

With the input port open-circuit ($i_i = 0$), Eq. (17-7) becomes

$$z_r i_o = v_i \qquad \text{or} \qquad z_r = \frac{v_i}{i_o}\bigg|_{i_i = 0} \tag{17-11}$$

and Eq. (17-8) becomes

$$z_o i_o = v_o \qquad \text{or} \qquad z_o = \frac{v_o}{i_o}\bigg|_{i_i = 0} \tag{17-12}$$

Equations (17-9) to (17-12) correctly express the impedance parameters as ratios of voltage to current.

z-Parameter Definitions The two-port, open-circuit impedance parameters are defined as follows:

The *open-circuit input impedance parameter* z_i is the ratio of input voltage to input current with the output port open-circuit.

The *open-circuit output impedance parameter* z_o is the ratio of output voltage to output current with the input port open-circuit.

The *open-circuit forward impedance parameter* z_f is the ratio of output voltage to input current with the output port open-circuit.
The *open-circuit reverse impedance parameter* z_r is the ratio of input voltage to output current with the input port open-circuit.

Each z parameter has the unit ohms.

Each of the open-circuit impedance parameters may be said to represent a *local* impedance (input or output) or *transfer* impedance (forward or reverse) which controls the relative magnitudes of a voltage and a current.

z-Parameter Equivalent Circuit Equations (17-7) to (17-12) can be used to produce a general two-port network with the four z parameters, as shown in Fig. 17-44.

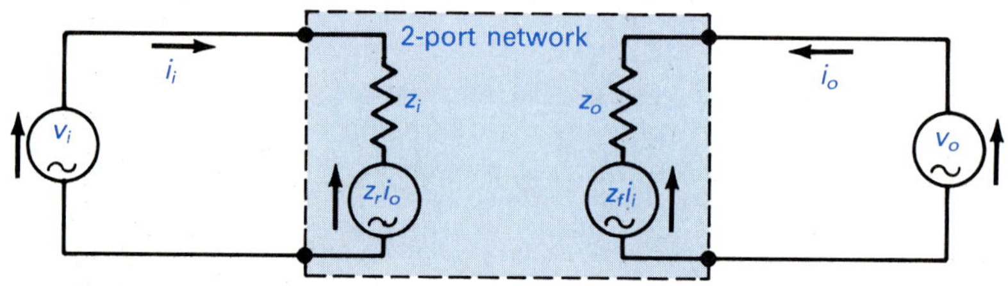

Fig. 17-44. z-parameter circuit.

In the network in Fig. 17-44 two elements are associated with each port:

1. A local impedance (z_i or z_o) whose voltage is a function of the local current (i_i or i_o)
2. A source of voltage equal to the product of a transfer impedance (z_r or z_f) and the other port current (i_o or i_i)

EXAMPLE 17-18 What are the values of the input and output currents (i_i and i_o) in the circuit in Fig. 17-45*a*?

Solution Figure 17-45*b* is a z-parameter equivalent of Fig. 17-45*a*.
The z parameters are, from Eq. (17-3),

$$z_i = R_1 + R_c = 11\ \text{k}\Omega$$

from Eq. (17-4),

$$z_o = R_2 + R_c = 15\ \text{k}\Omega$$

and from Eqs. (17-5) and (17-6),

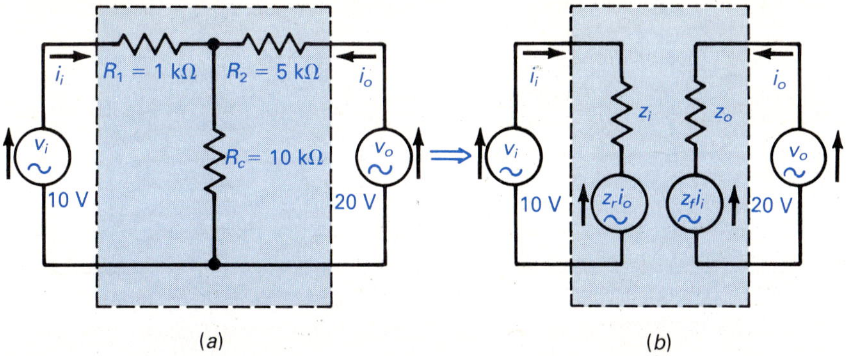

Fig. 17-45. Example 17-18: *(a)* Given circuit; *(b)* *z*-parameter equivalent circuit.

$$z_f = z_r = 10\ \text{k}\Omega$$

The input-port loop equation is

$$z_i i_i + z_r i_o = v_i \tag{17-7}$$

$$11\ \text{k}\Omega\ i_i + 10\ \text{k}\Omega\ i_o = 10\ \text{V} \tag{A}$$

The output-port loop equation is

$$z_f i_i + z_o i_o = v_o \tag{17-8}$$

$$10\ \text{k}\Omega\ i_i + 15\ \text{k}\Omega\ i_o = 20\ \text{V} \tag{B}$$

From Eq. *(A)*,

$$i_o = \frac{10\ \text{V} - 11\ \text{k}\Omega\ i_i}{10\ \text{k}\Omega} = (1 - 1.1 i_i)\ \text{mA} \tag{C}$$

Substituting i_o from Eq. *(C)* into Eq. *(B)*,

$$10\ \text{k}\Omega\ i_i + 15\ \text{k}\Omega \times (1 - 1.1 i_i)\ \text{mA} = 20\ \text{V}$$

$$10\ \text{k}\Omega\ i_i + 15\ \text{V} - 16.5\ \text{k}\Omega\ i_i = 20\ \text{V}$$

$$(10 - 16.5)\ \text{k}\Omega\ i_i = 5\ \text{V}$$

$$i_i = -\frac{5\ \text{V}}{6.5\ \text{k}\Omega} = \mathbf{-0.7692\ mA}$$

Substituting for i_i in Eq. *(C)*,

$$i_o = (1 - 1.1 i_i)\ \text{mA} = (1 + 0.8462)\ \text{mA} = \mathbf{1.846\ mA}$$

Therefore i_i = 0.7692 mA with a direction opposite that shown in Fig. 17-45, and i_o = 1.846 mA with a direction as shown in Fig. 17-45.

EXAMPLE 17-19 The T network in Fig. 17-46 acts as an attenuator, so that the load voltage v_L is of smaller magnitude than the input voltage v_i. What is the ratio of the load voltage to the input voltage?

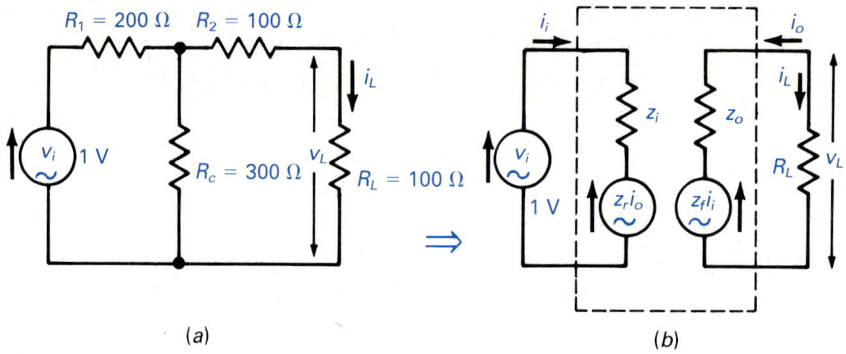

Fig. 17-46. Example 17-19: *(a)* Given circuit; *(b)* z-parameter equivalent circuit.

Solution Figure 17-46*b* is the *z*-parameter equivalent of Fig. 17-46*a*. In Fig. 17-46*b* the current i_o is shown in the conventional direction entering the output port, the direction used in developing the parameters. The load current direction is actually opposite this, and so $i_L = -i_o$.

The *z* parameters are, from Eq. (17-3),

$$z_i = R_1 + R_c = 500\ \Omega$$

from Eq. (17-4),

$$z_o = R_2 + R_c = 400\ \Omega$$

and from Eqs. (17-5) and (17-6),

$$z_f = z_r = 300\ \Omega$$

The input-port loop equation is

$$z_i i_i + z_r i_o = v_i \qquad (17\text{-}7)$$

$$500\ \Omega\ i_i + 300\ \Omega\ i_o = 1\ \text{V} \qquad (A)$$

The output-port equation is

$$z_f i_i + z_o i_o = v_o \qquad (17\text{-}8)$$

The output-port voltage is $v_o = R_L i_L = R_L(-i_o)$. Therefore the output-port loop equation becomes

$$z_f i_i + z_o i_o = -R_L i_o$$

$$300\ \Omega\ i_i + 400\ \Omega\ i_o = -100\ \Omega\ i_o$$

$$300\ \Omega\ i_i + 500\ \Omega\ i_o = 0 \qquad (B)$$

The matrix form of Eqs. *(A)* and *(B)* is

$$\begin{bmatrix} 500 & 300 \\ 300 & 500 \end{bmatrix} \begin{bmatrix} i_i \\ i_o \end{bmatrix} = \begin{bmatrix} 1 \\ 0 \end{bmatrix} \qquad i_i \text{ and } i_o \text{ in amperes}$$

The solutions for i_i and i_o are

$$i_i = 3.125\ \text{mA} \qquad \text{and} \qquad i_o = -1.875\ \text{mA}$$

$$i_L = -i_o = 1.875 \text{ mA} \qquad \text{direction as shown in Fig. 17-46}$$
$$v_L = R_L i_L = 100\ \Omega \times 1.875 \text{ mA} = 0.1875 \text{ V}$$

Therefore the ratio v_L/v_i = **0.1875.**

EXAMPLE 17-20 A linear electronic network of unknown circuit design has the measured two-port z parameters $z_i = 500\ \Omega$, $z_o = 2\ \text{k}\Omega$, $z_f = 1\ \text{k}\Omega$, and $z_r = 5\ \text{k}\Omega$. What is the magnitude of the voltage developed across a 10-kΩ load resistor if the input voltage is 100 mV?

Solution Figure 17-47a shows the electronic network as an unspecified type of unit. The two-port equivalent network in Fig. 17-47b is based on the measured parameters.

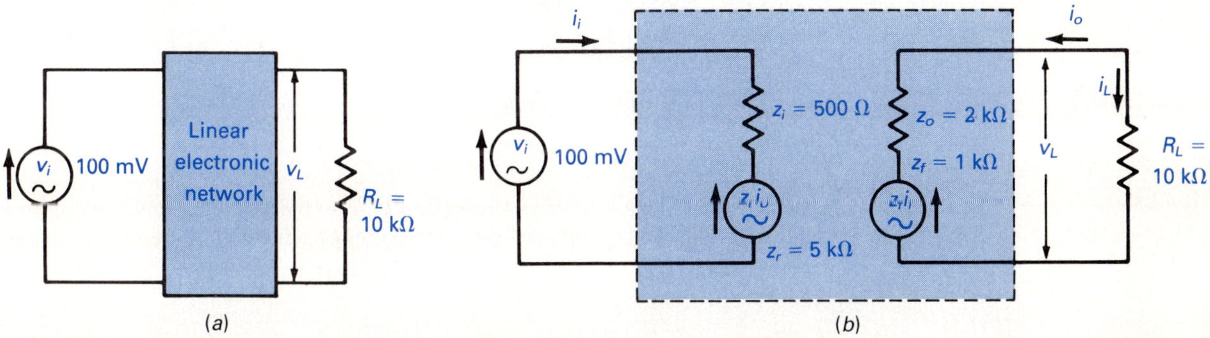

Fig. 17-47. Example 17-20: *(a)* Block diagram; *(b)* z-parameter equivalent circuit.

The input-loop equation is

$$z_i i_i + z_r i_o = v_i \qquad (17\text{-}7)$$
$$0.5\ \text{k}\Omega\ i_i + 5\ \text{k}\Omega\ i_o = 100 \text{ mV} \qquad (A)$$

The output-loop equation is, from Eq. (17-8),

$$z_f i_i + z_o i_o = -R_L i_o$$
$$1\ \text{k}\Omega\ i_i + 2\ \text{k}\Omega\ i_o = -10\ \text{k}\Omega\ i_o$$
$$1\ \text{k}\Omega\ i_i + 12\ \text{k}\Omega\ i_o = 0 \qquad (B)$$

In matrix form the loop equations are

$$\begin{bmatrix} 0.5 & 5 \\ 1 & 12 \end{bmatrix} \begin{bmatrix} i_i \\ i_o \end{bmatrix} = \begin{bmatrix} 100 \\ 0 \end{bmatrix} \qquad i_i \text{ and } i_o \text{ in microamperes}$$

The solution for i_o is

$$i_o = -100\ \mu\text{A}$$

The magnitude of i_L is the same as that of i_o, the directions being opposite.

$$i_L = 100\ \mu\text{A} \qquad \text{direction as shown in Fig. 17-47}$$
$$v_L = R_L i_L = 10\ \text{k}\Omega \times 100\ \mu\text{A} = \mathbf{1\ V}$$

Admittance (y) Parameters

A set of four admittance (y) parameters can be developed in a manner similar to that adopted for the z parameters. A circuit consisting of a two-port π network fed by current sources applied to each port lends itself to a nodal equation derivation for the y parameters. As the method is essentially the same as for the z parameters, it will not be detailed here. A y-parameter equivalent of generalized two-port circuit, expressed in input-output terms, is given in Fig. 17-48.

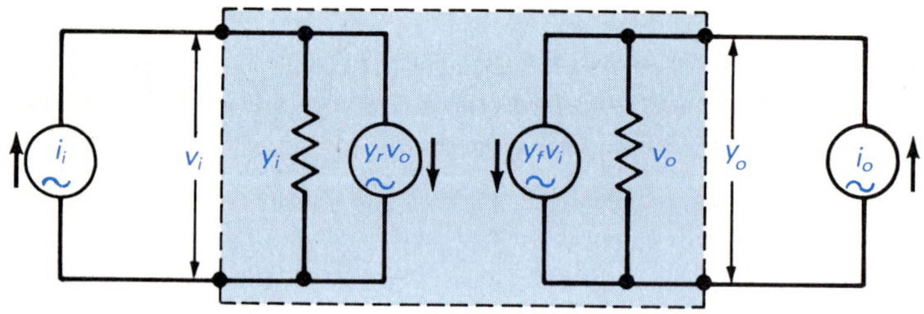

Fig. 17-48. y-parameter circuit.

In Fig. 17-48 two current sources supply currents i_i and i_o to the input and output ports. These currents each have two components within the network, one local $(y_i v_i$ and $y_o v_o)$ and one transfer $(y_r v_o$ and $y_f v_i)$. The input- and output-port nodal equations are

$$y_i v_i + y_r v_o = i_i \tag{17-13}$$
$$y_f v_i + y_o v_o = i_o \tag{17-14}$$

y_i, y_o, y_f, and y_r are described as short-circuit admittance parameters. For the determination of a parameter it is assumed that one source is removed from the circuit and replaced by its internal admittance, resulting in a short circuit. The remaining source thus supplies current to a network terminated by a short-circuited port.

With the output port short-circuit ($v_o = 0$), Eq. (17-13) becomes

$$y_i v_i = i_i \qquad \text{or} \qquad y_i = \left.\frac{i_i}{v_i}\right|_{v_o = 0} \tag{17-15}$$

and Eq. (17-14) becomes

$$y_f v_i = i_o \qquad \text{or} \qquad y_f = \left.\frac{i_o}{v_i}\right|_{v_o = 0} \tag{17-16}$$

With the input port short-circuit ($v_i = 0$), Eq. (17-13) becomes

$$y_r v_o = i_i \qquad \text{or} \qquad y_r = \left.\frac{i_i}{v_o}\right|_{v_i = 0} \tag{17-17}$$

and Eq. (17-14) becomes

$$y_o v_o = i_o \qquad \text{or} \qquad y_o = \left.\frac{i_o}{v_o}\right|_{v_i = 0} \tag{17-18}$$

Equations (17-15) to (17-18) correctly express the admittance parameters as ratios of current to voltage.

y-Parameter Definitions The two-port, short-circuit admittance parameters are defined as follows:

The *short-circuit input admittance parameter* y_i is the ratio of input current to input voltage with the output port short-circuited.
The *short-circuit output admittance parameter* y_o is the ratio of output current to output voltage with the input port short-circuited.
The *short-circuit forward admittance parameter* y_f is the ratio of output current to input voltage with the output port short-circuited.
The *short-circuit reverse admittance parameter* y_r is the ratio of input current to output voltage with the input port short-circuited.

Each y parameter has the unit siemens.

It may be said that each of the short-circuit admittance parameters represents a local admittance (input or output) or transfer admittance (forward or reverse) which controls the relative magnitudes of a current and a voltage.

EXAMPLE 17-21 What are the values of the input and output voltages (v_i and v_o) in the circuit in Fig. 17-49*a*?

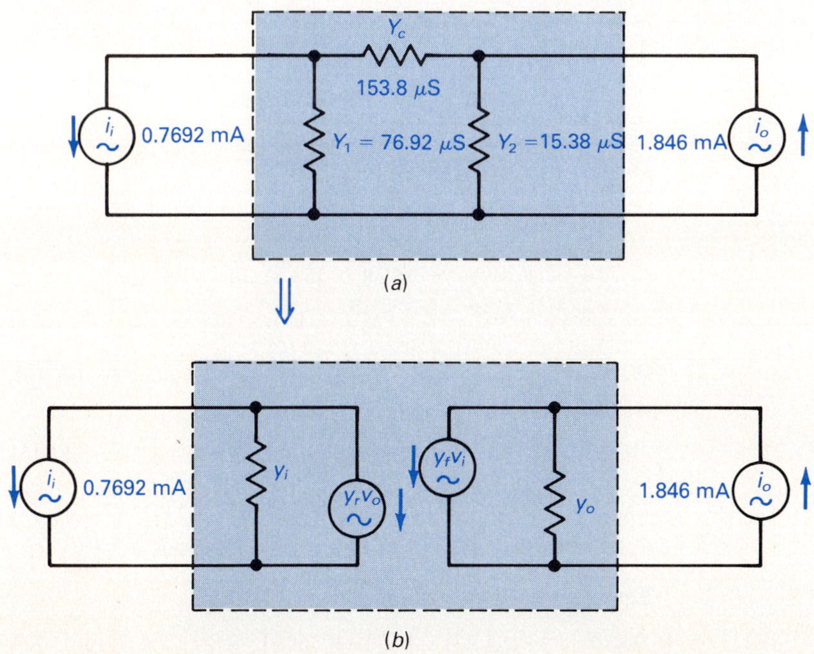

Fig. 17-49. Example 17-21: *(a)* Given circuit; *(b)* y-parameter equivalent circuit.

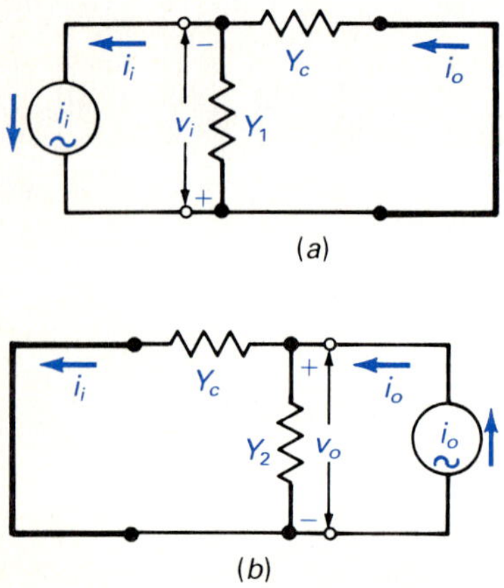

Fig. 17-50. Example 17-21: Figure 17-49*a* (*a*) with output short circuit; (*b*) with input short circuit.

Solution Figure 17-49*b* is the *y*-parameter equivalent of Fig. 17-49*a*. The sense of each *y* parameter can be determined by reference to Fig. 17-49*a*. With the output port short-circuited, the input source and the network will appear as in Fig. 17-50*a*. From the viewpoint of two-port circuit conventions, in Fig. 17-50*a* the input current i_i has the wrong direction, the input voltage v_i has the wrong polarity, and the output current i_o has the correct direction.

$$y_i = \left.\frac{i_i}{v_i}\right|_{v_o = 0} \tag{17-15}$$

$$y_i = \frac{-}{-} = + \qquad y_i \text{ is positive}$$

$$y_f = \left.\frac{i_o}{v_i}\right|_{v_o = 0} \tag{17-16}$$

$$y_f = \frac{+}{-} = - \qquad y_f \text{ is negative}$$

With the input port short-circuited, the output source and the network will be as shown in Fig. 17-50*b*. From the viewpoint of two-port circuit conventions, in Fig. 17-50*b* the output current i_o has the correct direction, the output voltage v_o has the correct polarity, and the input current i_i has the wrong direction. Thus,

$$y_r = \left.\frac{i_i}{v_o}\right|_{v_i = 0} \tag{17-17}$$

$$y_r = \frac{-}{+} = - \qquad y_r \text{ is negative}$$

$$y_o = \left.\frac{i_o}{v_o}\right|_{v_i = 0} \tag{17-18}$$

$$y_o = \frac{+}{+} = + \qquad y_o \text{ is positive}$$

The values of the *y* parameters are

$$\begin{aligned}
y_i &= Y_1 + Y_c = (76.92 + 153.8)\ \mu\text{S} = 230.7\ \mu\text{S} \\
y_f &= -Y_c = -153.8\ \mu\text{S} \\
y_o &= Y_2 + Y_c = (15.38 + 153.8)\ \mu\text{S} = 169.2\ \mu\text{S} \\
y_r &= -Y_c = -153.8\ \mu\text{S}
\end{aligned}$$

With reference to Fig. 17-49*b*, the nodal equations for the input and output ports are

$$y_i v_i + y_r v_o = i_i \tag{17-13}$$

$$230.7\ \mu\text{S}\ v_i - 153.8\ \mu\text{S}\ v_o = -0.7692\ \text{mA} \tag{A}$$

$$y_f v_i + y_o v_o = i_o \tag{17-14}$$

$$-153.8\ \mu\text{S}\ v_i + 169.2\ \mu\text{S}\ v_o = 1.846\ \text{mA} \tag{B}$$

Equations (*A*) and (*B*) in matrix form are:

$$\begin{bmatrix} 230.7 & -153.8 \\ -153.8 & 169.2 \end{bmatrix} \begin{bmatrix} v_i \\ v_o \end{bmatrix} = \begin{bmatrix} -0.7692 \\ 1.846 \end{bmatrix} \qquad v_i \text{ and } v_o \text{ in kilovolts}$$

The determinant solutions for v_i and v_o are

$$v_i = \mathbf{10\ V}$$
$$v_o = \mathbf{20\ V}$$

The values for Example 17-21 were chosen to be the same as those for Example 17-18; the π network in Example 17-21 is the equivalent of the T network in Example 17-18.

The current sources in Example 17-21 produce the same input and output currents as were obtained in the solutions to Example 17-18. The solutions to Example 17-21 result in the same voltages (value and sign) as do the voltage sources in Example 17-18.

EXAMPLE 17-22

Figure 17-51*a* shows the circuit of a power-supply filter. The input to the filter may be regarded as an 85-V steady dc with 60-V rms, 120-Hz ac superimposed on it. This combined input has an effective internal resistance of 100 Ω. The purpose of the filter is to reduce to a low value the magnitude of the ac component of the output voltage. What is the magnitude of this voltage?

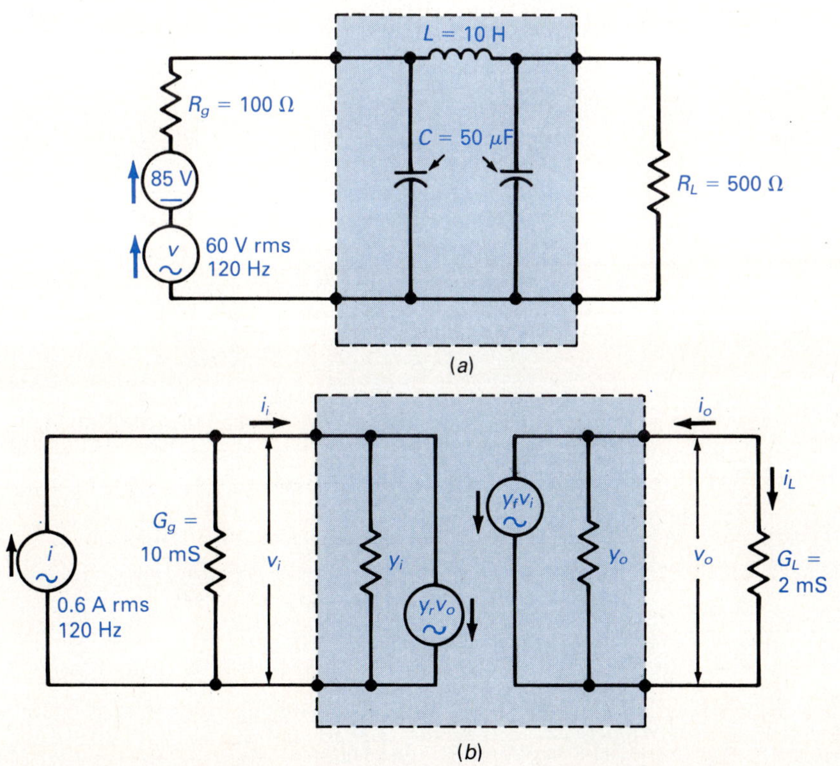

Fig. 17-51. Example 17-22: *(a)* Given circuit; *(b)* *y*-parameter equivalent.

Solution Applying the superposition principle, we may consider the ac components of the circuit voltages and currents separately from the dc components. Figure 17-51*b* shows an ac equivalent circuit in admittance parameter terms. The ac voltage v has been converted to a source current $i = v/R_g$. The current-source internal conductance is $G_g = 1/R_g$.

$$\omega = 2\pi f = 2\pi \times 120 \text{ Hz} = 754 \text{ rad/s}$$

$$B_L = -j\frac{1}{\omega L} = -j0.1326 \text{ mS}$$

$$B_C = j\omega C = j37.7 \text{ mS}$$

$$y_i = y_o = B_L + B_C = j(37.7 - 0.1326) \text{ mS} = j37.57 \text{ mS}$$

$$y_r = y_f = +jB_L = +j0.1326 \text{ mS}$$

the sign of $-jB_L$ is changed because, in a conductive network, a transfer admittance parameter is the negative of the coupling admittance

The equation for the input-port node is

$$y_i v_i + y_r v_o = i_i \tag{17-13}$$

where i_i = total current entering the input-port upper terminal,

$$i_i = i - G_g v_i$$

Substituting for i_i, the input nodal equation becomes

$$y_i v_i + y_r v_o = i - G_g v_i$$

$$j37.57 \text{ mS } v_i + j0.1326 \text{ mS } v_o = 0.6 \text{ A} - 10 \text{ mS } v_i$$

$$(10 + j37.57) \text{ mS } v_i + j0.1326 \text{ mS } v_o = 0.6 \text{ A}$$

The equation for the output-port node is

$$y_f v_i + y_o v_o = i_o \tag{17-14}$$

where i_o = current entering the output-port upper terminal,

$$v_o = \frac{i_L}{G_L} \qquad \text{or} \qquad i_L = G_L v_o$$

But i_L *leaves* the output-port upper terminal, so $i_o = -i_L = -G_L v_o$.

The output nodal equation becomes

$$y_f v_i + y_o v_o = -G_L v_o$$

$$j0.1326 \text{ mS } v_i + j37.57 \text{ mS } v_o = -2 \text{ mS } v_o$$

$$j0.1326 \text{ mS } v_i + (2 + j37.57) \text{ mS } v_o = 0$$

The nodal equations in matrix form are

$$\begin{bmatrix} 38.88 \angle 75.1^\circ & 0.1326 \angle 90^\circ \\ 0.1326 \angle 90^\circ & 37.62 \angle 87^\circ \end{bmatrix} \begin{bmatrix} v_i \\ v_o \end{bmatrix} = \begin{bmatrix} 0.6 \\ 0 \end{bmatrix} \qquad v_i \text{ and } v_o \text{ in kilovolts}$$

The solution for v_o is

$$v_o = 54.42 \text{ mV} \angle -72.1^\circ \qquad \text{or} \qquad \angle -1.258 \text{ rad}$$

Therefore the magnitude of the ac component of the output voltage is **54.42 mV rms.**

Hybrid *(h)* Parameters

Two-port parameters are particularly useful when the characteristics of a network can be readily determined only by measurement at its ports. The transistor is an important example. However, with the transistor it is often difficult to make reliable measurements of y and z parameters, because of its markedly different input and output characteristics. There are other networks which also do not lend themselves to satisfactory z- and y-parameter treatment.

It may be roughly generalized that y and z parameters are suitable for networks in which the input and output characteristics are comparable. If input and output ports differ greatly in characteristics, h (hybrid) parameters may be more suitable.

h parameters are developed by combining the concepts involved in developing z and y parameters. The four h parameters are an impedance, an admittance, a current ratio, and a voltage ratio; hence the name hybrid.

Figure 17-52 shows the equivalent circuit of a two-port network drawn in conventional h-parameter form. As the circuits associated with the input and output ports in h-parameter terms are quite different, it is important to stress that the input-output approach has been adopted for convenience. For a particular network it might be desirable to reverse the format, making the input a parallel circuit and the output a series circuit. The generalized 1,1–2,2 approach, without specifying input and output, could be adopted to reduce confusion. But, since you will probably use h parameters mainly in transistor work and Fig. 17-52 applies to most transistors, the convenience of this configuration makes its choice reasonable. Although it should not be regarded as an invariable rule, the circuit in Fig. 17-52 may be considered to apply to networks which have low input impedance and high output impedance. Transistors usually fall into this category.

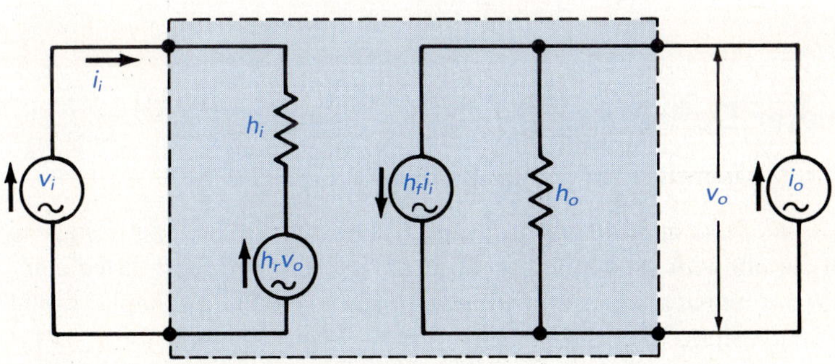

Fig. 17-52. *h*-parameter circuit.

The input-port elements within the network in Fig. 17-52 are:

1. An impedance h_i which, with the input current i_i, forms a voltage of $h_i i_i$ volts.

2. A voltage source $h_r v_o$ formed by a voltage ratio h_r and the output voltage v_o. The voltage $h_r v_o$ is fed back from the output to the input.

The output-port elements within the network are:

1. An admittance h_o which, with the output voltage v_o, forms a current of $h_o v_o$ amperes.
2. A current source $h_f i_i$ formed by a current ratio h_f and the input current i_i. The current $h_f i_i$ is fed forward from the input to the output.

The input loop equation and the output nodal equation together define the four h parameters:

$$h_i i_i + h_r v_o = v_i \qquad \textbf{(17-19)}$$

$$h_f i_i + h_o v_o = i_o \qquad \textbf{(17-20)}$$

Expressions for the four h parameters are obtained by alternately considering the output port to be short-circuit and the input port to be open-circuit.

With the output port short-circuit ($v_o = 0$), Eq. (17-19) becomes

$$h_i i_i = v_i \qquad \text{or} \qquad h_i = \left.\frac{v_i}{i_i}\right|_{v_o = 0} \qquad \textbf{(17-21)}$$

and Eq. (17-20) becomes

$$h_f i_i = i_o \qquad \text{or} \qquad h_f = \left.\frac{i_o}{i_i}\right|_{v_o = 0} \qquad \textbf{(17-22)}$$

With the input port open-circuit ($i_i = 0$), Eq. (17-19) becomes

$$h_r v_o = v_i \qquad \text{or} \qquad h_r = \left.\frac{v_i}{v_o}\right|_{i_i = 0} \qquad \textbf{(17-23)}$$

and Eq. (17-20) becomes

$$h_o v_o = i_o \qquad \text{or} \qquad h_o = \left.\frac{i_o}{v_o}\right|_{i_i = 0} \qquad \textbf{(17-24)}$$

h-Parameter Definitions

The two-port h parameters are defined as follows:

The *short-circuit input impedance parameter* h_i is the ratio of the input voltage to the input current with the output port short-circuit. The unit for h_i is the ohm.

The *open-circuit output admittance parameter* h_o is the ratio of the output current to the output voltage with the input port open-circuit. The unit for h_o is the siemens.

The *forward current gain* or *forward current parameter* h_f is the ratio of output current to input current with the output port short-circuit.

The *reverse voltage gain* or *reverse voltage parameter* h_r is the ratio of input voltage to output voltage with the input port open-circuit.

EXAMPLE 17-23

What are the values of the input current i_i and output voltage v_o in the circuit in Fig. 17-53*a*?

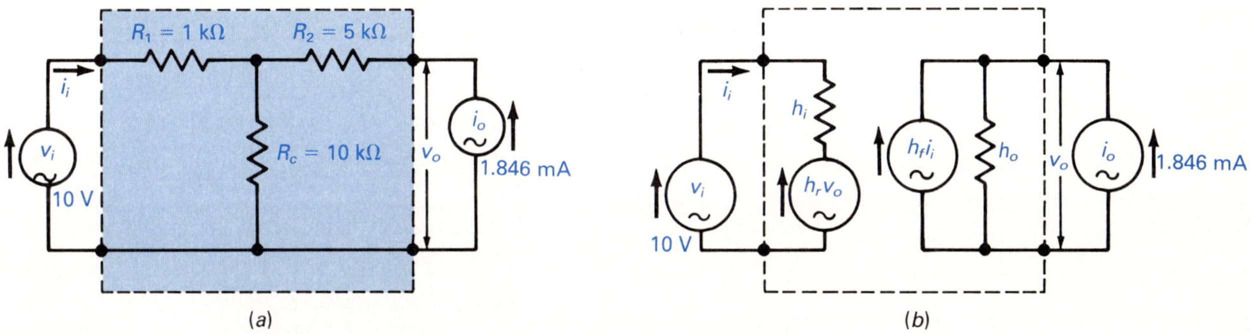

Fig. 17-53. Example 17-23: *(a)* Given circuit; *(b)* *h*-parameter equivalent.

Solution Figure 17-53*b* is an *h*-parameter equivalent of Fig. 17-53*a*.

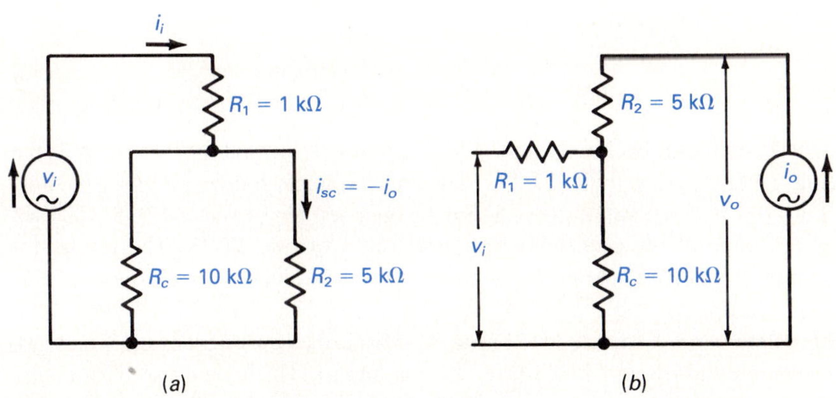

Fig. 17-54. Example 17-23: *(a)* Output-port short circuit; *(b)* input-port open circuit.

Figure 17-54*a* and *b* shows the short-circuit output and open-circuit input conditions.

With the output short-circuit (Fig. 17-54*a*),

$$h_i = \frac{v_i}{i_i} = R_1 + \frac{R_2 \times R_c}{R_2 + R_c} = 1\text{ k}\Omega + \frac{5\text{ k}\Omega \times 10\text{ k}\Omega}{5\text{ k}\Omega + 10\text{ k}\Omega} = 4.333\text{ k}\Omega$$

Applying the current-divider principle,

$$h_f = \frac{-i_o}{i_i} = -\frac{R_c}{R_2 + R_c} = -\frac{10\text{ k}\Omega}{5\text{ k}\Omega + 10\text{ k}\Omega} = -0.6667$$

With the input open-circuit (Fig. 17-54*b*),

$$h_o = \frac{i_o}{v_o} = \frac{1}{Z_o}$$

where Z_o is the impedance seen by the source i_o:

$$h_o = \frac{1}{R_2 + R_c} = \frac{1}{15\ \text{k}\Omega} = 0.06667\ \text{mS}$$

Applying the voltage-divider principle,

$$h_r = \frac{v_i}{v_o} = \frac{R_c}{R_2 + R_c} = \frac{10\ \text{k}\Omega}{5\ \text{k}\Omega + 10\ \text{k}\Omega} = 0.6667$$

The input loop and output nodal equations are

$$h_i i_i + h_r v_o = v_i \qquad (17\text{-}19)$$

$$4.333\ \text{k}\Omega\ i_i + 0.6667\ v_o = 10\ \text{V}$$

$$h_f i_i + h_o v_o = i_o \qquad (17\text{-}20)$$

$$-0.6667\ i_i + 0.06667\ \text{mS}\ v_o = 1.846\ \text{mA}$$

The matrix form of the equations is

$$\begin{bmatrix} 4.333 & 0.6667 \\ -0.6667 & 0.06667 \end{bmatrix} \begin{bmatrix} i_i \\ v_o \end{bmatrix} = \begin{bmatrix} 10 \\ 1.846 \end{bmatrix} \qquad \begin{matrix} i_i \text{ in milliamperes;} \\ v_o \text{ in volts} \end{matrix}$$

The solutions for i_i and v_o are

$$i_i = \mathbf{-0.7692\ mA}$$

$$v_o = \mathbf{20\ V}$$

If you will refer back to Example 17-18, you will see that the T network is the same as that in Example 17-23. The input voltage source is also the same ($v_i = 10$ V). The result ($i_o = 1.846$ mA) obtained for Example 17-18 was used for the value of the output current source in Example 17-23. The results for Example 17-23 complete the input and output port values as for Example 17-18.

The six diagrams in Fig. 17-55 compare Examples 17-18, 17-21, and 17-23. The π network in Example 17-21 (Fig. 17-55*c*) is the equivalent of the T network in Examples 17-18 and 17-23 (Fig. 17-55*a* and *e*). The *z*-, *y*-, and *h*-parameter configurations used in Examples 17-18, 17-21, and 17-23 are given beside the original circuit schematics as Fig. 17-55*b*, *d*, and *f*. All six diagrams are equivalent to each other.

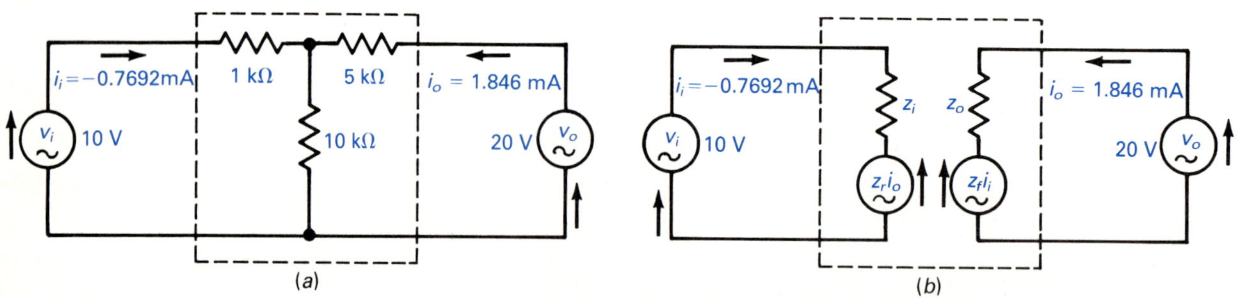

Fig. 17-55. Comparison of circuits. *(a)* and *(b)* Example 17-18.

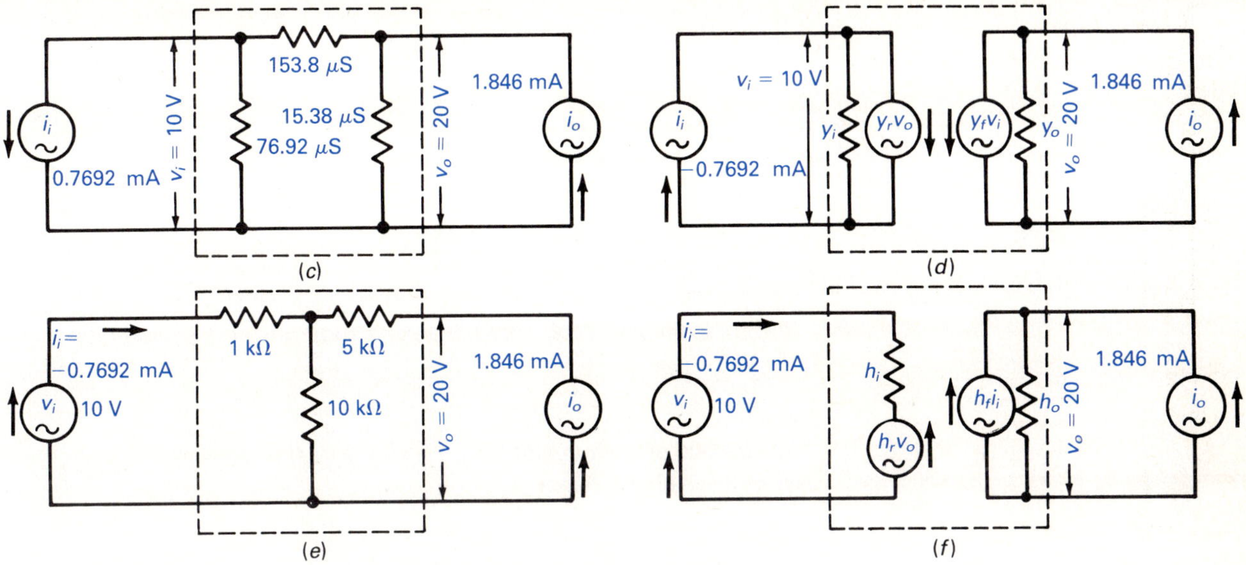

Fig. 17-55 *(Continued)*. *(c)* and *(d)* Example 17-21; *(e)* and *(f)* Example 17-23.

EXAMPLE 17-24

What are the h parameters for the π network in Fig. 17-56*a* at a frequency of 1 MHz?

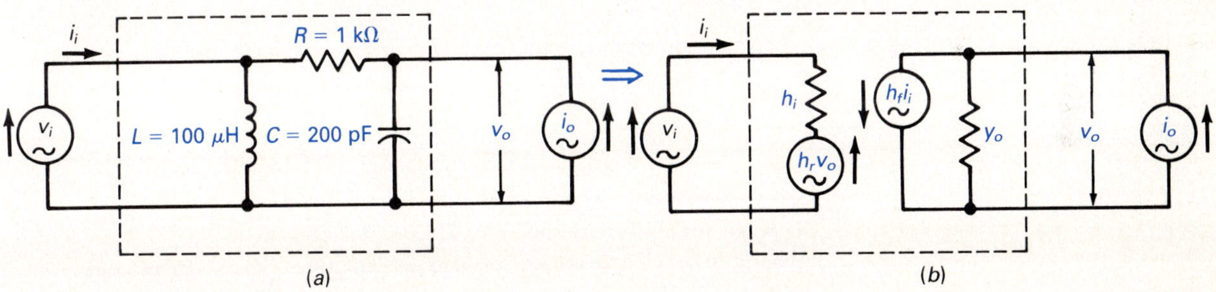

Fig. 17-56. Example 17-24: *(a)* Given circuit; *(b)* h-parameter equivalent.

Solution

Figure 17-56*b* is the h-parameter equivalent of Fig. 17-56*a*.

$$f = 1\text{ MHz} \qquad \omega = 6.283\text{ Mrad/s}$$

$$X_L = \omega L = 6.283\text{ Mrad/s} \times 100\ \mu\text{H} = 628.3\ \Omega$$

$$B_C = \omega C = 6.283\text{ Mrad/s} \times 200\text{ pF} = 1.257\text{ mS}$$

With the output short-circuit, Fig. 17-56*a* becomes as shown in Fig. 17-57*a*. With the input open-circuit, Fig. 17-56*a* becomes as shown in Fig. 17-57*b*.

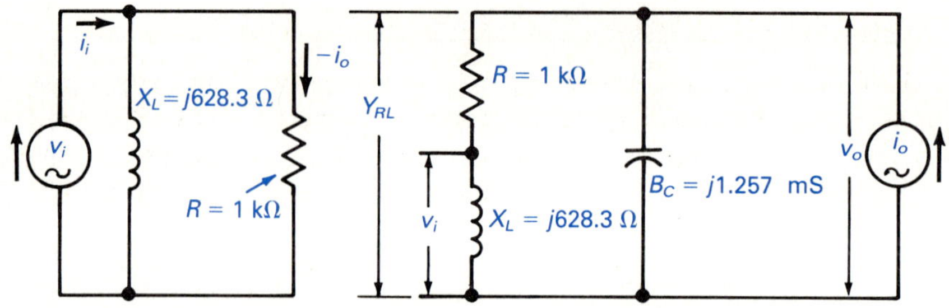

Fig. 17-57. Example 17-24: *(a)* Output-port short circuit; *(b)* input-port open circuit.

For the output short-circuit (Fig. 17-57*a*), h_i is the impedance seen by the voltage source v_i:

$$h_i = \frac{R \times jX_L}{R + jX_L} = \frac{1\ k\Omega \times j628.3\ \Omega}{1\ k\Omega + j628.3\ \Omega} = \frac{628.3\ k\Omega^2\ \underline{/90°}}{1.181\ k\Omega\ \underline{/32.14°}}$$

$$h_i = \mathbf{532\ \Omega\ \underline{/57.86°}} \quad \text{or} \quad \mathbf{\underline{/1.01\ rad}}$$

By current division,

$$h_f = \frac{-i_o}{i_i} = -\frac{jX_L}{R + jX_L} = -\frac{j628.3\ \Omega}{1\ k\Omega + j628.3\ \Omega}$$

$$h_f = -\frac{628.3\ \Omega\ \underline{/90°}}{1.181\ k\Omega\ \underline{/32.14°}}$$

$$= \mathbf{-0.532\ \underline{/57.86°}} \quad \text{or} \quad \mathbf{0.532\ \underline{/-122.14°}} \quad \text{or} \quad \mathbf{\underline{/-2.132\ rad}}$$

For the input open-circuit (Fig. 17-57*b*), h_o is the admittance seen by current source i_o:

$$Y_{RL} = \frac{1}{Z_{RL}} = \frac{1}{R + jX_L} = \frac{1}{1\ k\Omega + j628.3\ \Omega} = \frac{1}{1.181\ k\Omega\ \underline{/32.14°}}$$

$$Y_{RL} = 0.8467\ mS\ \underline{/-32.14°} = (0.7169 - j0.4505)\ mS$$

$$h_o = B_C + Y_{RL} = j1.257\ mS + (0.7169 - j0.4505)\ mS$$

$$h_o = (0.7169 + j0.8065)\ mS = \mathbf{1.079\ mS\ \underline{/48.36°}} \quad \text{or} \quad \mathbf{\underline{/0.844\ rad}}$$

By voltage division,

$$h_r = \frac{v_i}{v_o} = \frac{jX_L}{R + jX_L} = -h_f = \mathbf{0.532\ \underline{/57.86°}} \quad \text{or} \quad \mathbf{\underline{/1.01\ rad}}$$

In the solutions for Example 17-24 the value of h_r is given as $-h_f$, [$h_r = jX_L/(R + jX_L)$ and $h_f = -jX_L/(R + jX_L)$]. This equality is not a coincidence. As a check back to each of the z, y, and h parameters indicates, it is a general principle that, for a bilateral two-port network (a network which operates equally well for transmission in either direction), the transfer parameters have the same magnitude. For conductive bilateral networks the y and z transfer parameters have the same sense ($y_f = y_r$ and $z_f = z_r$), but the h transfer parameters have opposing sense ($h_f = -h_r$).

Transistor as Two-Port Network

A transistor is a semiconductor device used as an amplifier (magnifier) of direct and alternating voltages, currents, and powers. The transistor is a nonlinear device and in some applications must be considered as such. There are many applications, however, in which the transistor's action is *quasi-linear,* so near to linear that it can be so regarded.

As an example of the practical use of h parameters, the transistor will be viewed here as a linear amplifier of ac voltages.

The internal functioning of a transistor is the concern of electronics or physics textbooks. However, the following treatment will show how transistor circuits can be regarded as electric circuits if the transistor parameters are obtained by measurement or by manufacturer's specification, both of which are usual in industry.

To obtain the h parameters of a transistor by measurement, a setup such as that depicted in Fig. 17-58 may be used.

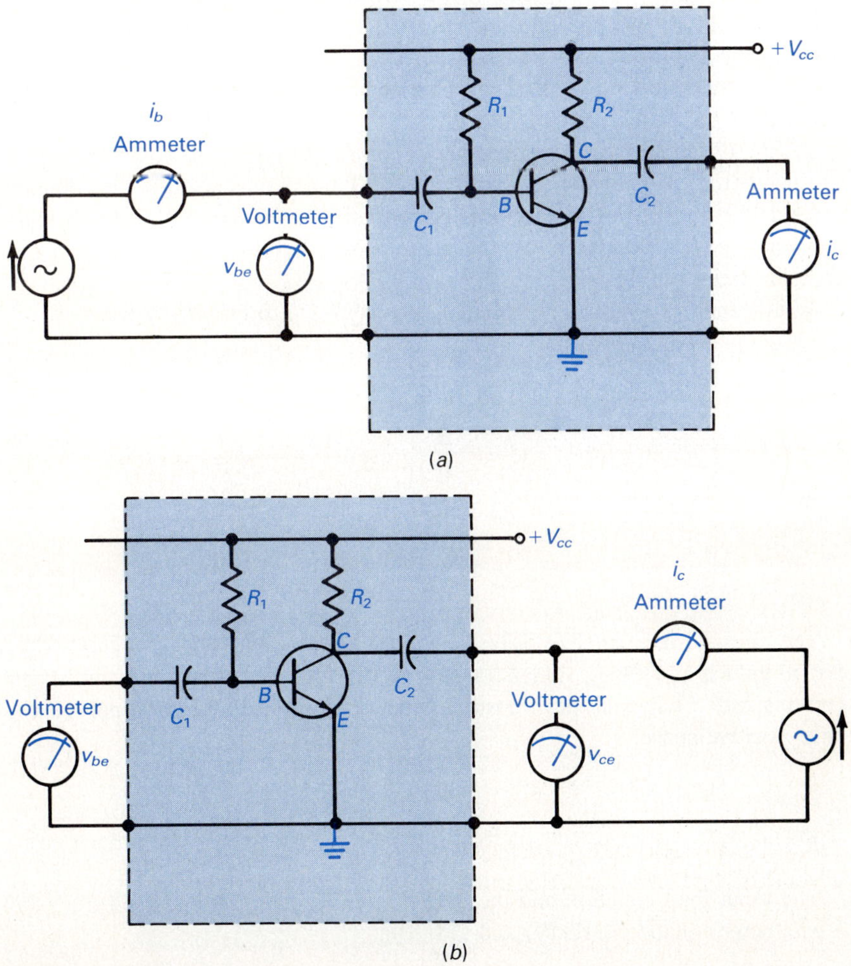

Fig. 17-58. Measurement of transistor parameters: *(a)* h_{ie} and h_{fe}; *(b)* h_{oe} and h_{re}.

In Fig. 17-58 an NPN transistor is shown connected to a dc supply V_{cc} which, via resistors R_1 and R_2, provides the correct dc values of voltage and current to set the transistor operating conditions. If the characteristics of a PNP transistor are to be measured, the polarity of V_{cc} is reversed.

The ammeters and voltmeters are connected to the transistor by capacitors and thus measure ac values only.

The values of resistors R_1 and R_2 are selected to be high enough to avoid significant interference with the ac quantities being measured.

The meters measure the ac values of base-to-emitter voltage v_{be}, base current i_b, collector-to-emitter voltage v_{ce}, and collector current i_c.

The hybrid parameters h_i and h_f are defined under output short-circuit conditions. These conditions may be simulated for a transistor by connecting a large-value capacitor between the collector and the emitter (C_2 in Fig. 17-58*a*). The low reactance of C_2, together with the low resistance of the ammeter, is an effective short circuit to ac but does not affect the dc operation of the transistor. The source supplies ac to the transistor input via capacitor C_1.

The hybrid parameters h_o and h_r are defined under open-circuit input conditions. In Fig. 17-58*b* the base connects to V_{cc} through R_1 and to the emitter through the voltmeter. As both R_1 and the voltmeter are high resistances, the input is virtually open-circuit. The source supplies ac to the output port through capacitor C_2.

In Fig. 17-58 the transistor is arranged for measurement of its common-emitter characteristics, as is evidenced by the emitter *e* being connected to the common ground line. In these circumstances, the hybrid parameters are given an extra subscript, *e*. Manufacturers often give transistor hybrid characteristics in this form in their specifications.

With transistor symbols substituted, Eqs. (17-21) to (17-24) become

$$h_{ie} = \left.\frac{v_{be}}{i_b}\right|_{v_{ce}=0} \tag{17-25}$$

$$h_{fe} = \left.\frac{i_c}{i_b}\right|_{v_{ce}=0} \tag{17-26}$$

$$h_{re} = \left.\frac{v_{be}}{v_{ce}}\right|_{i_b=0} \tag{17-27}$$

$$h_{oe} = \left.\frac{i_c}{v_{ce}}\right|_{i_b=0} \tag{17-28}$$

The transistor is neither an ordinary conductive network, such as a T network, nor an inductive network, like a transformer. In Fig. 17-58 the input and output currents and voltages all have positive sense, and so all four hybrid parameters have positive sense.

Transistor Equivalent Circuit A hybrid-parameter equivalent circuit of a transistor is shown in Fig. 17-59.

The input loop equation and the output nodal equation are adapted from the *h*-parameter equations (17-19) and (17-20):

$$h_{ie}i_b + h_{re}v_{ce} = v_{be} \tag{17-29}$$

$$h_{fe}i_b + h_{oe}v_{ce} = i_c \tag{17-30}$$

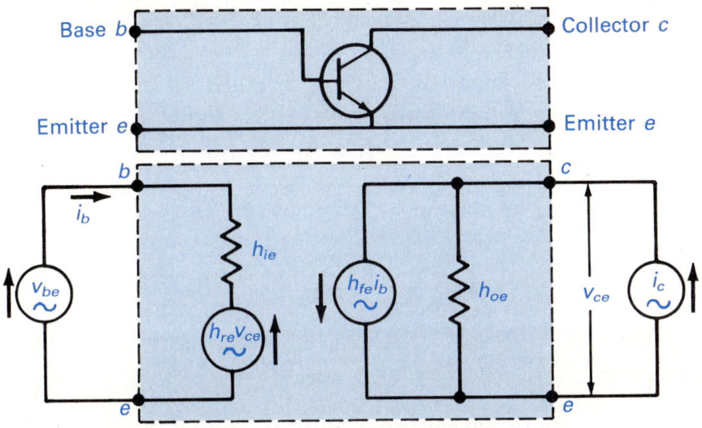

Fig. 17-59. Transistor *h*-parameter equivalent circuit.

EXAMPLE 17-25

Figure 17-60 shows the circuit of a transistor linear amplifier.

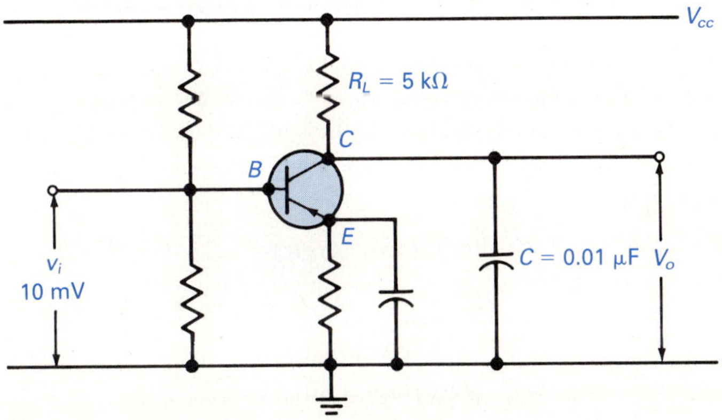

Fig. 17-60. Example 17-25: Given circuit.

Figure 17-61 gives an ac equivalent circuit of the amplifier.

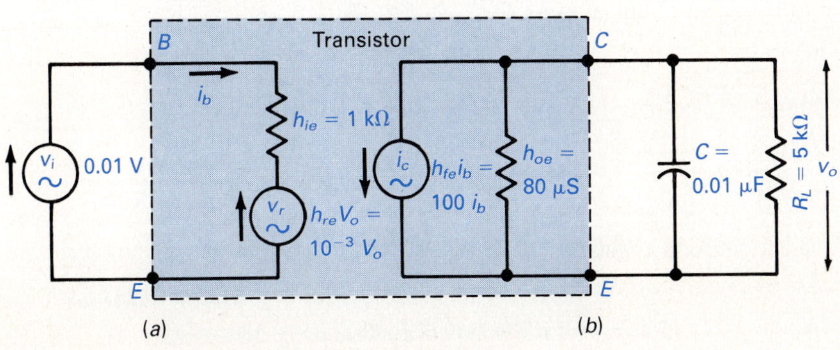

Fig. 17-61. Example 17-25: Equivalent circuit.

Figure 17-61 is shown as being in effect two separate circuits; Fig. 17-61*a* is the input circuit and Fig. 17-61*b* the output circuit. If the ac input voltage to the amplifier v_i is 10 mV rms, what is the output voltage v_o if the input frequency is *(a)* 100 Hz, *(b)* 1 kHz, *(c)* 10 kHz, and *(d)* any value, but the capacitor C is removed from the circuit?

The values of the transistor parameters are:

$$h_{ie} \text{ (input impedance)} = 1 \text{ k}\Omega$$
$$h_{re} \text{ (reverse voltage function)} = 10^{-3}$$
$$h_{fe} \text{ (forward current gain)} = 100$$
$$h_{oe} \text{ (output admittance)} = 80\ \mu\text{S}$$

Solution For the input circuit (refer to Fig. 17-61*a*), $v_r = h_{re}v_o$ is a voltage fed back from the output circuit to the input circuit, and i_b is the input current.

The input loop equation is

$$h_{ie}i_b = v_i - v_r$$
$$h_{ie}i_b = v_i - h_{re}v_o$$
$$1 \text{ k}\Omega\ i_b = 0.01 \text{ V} - 10^{-3}v_o \qquad i_b \text{ in milliamperes; } v_o \text{ in volts} \quad (A)$$

For the output circuit (refer to Fig. 17-61*b*), $i_c = h_{fe}i_b$ is a current fed forward from the input circuit to the output circuit.

With C (the collector) as an independent node and E (the emitter) as the reference node, the node voltage is v_o. The nodal equation is

$$(h_{oe} + G_L + jB_C)v_o = -i_c$$
$$\left(h_{oe} + \frac{1}{R_L} + jB_C\right)v_o = -h_{fe}i_b$$
$$(80\ \mu\text{S} + 200\ \mu\text{S} + jB_C)v_o = -100i_b$$
$$[(280 + jB_C)\ \mu\text{S}]v_o = -100i_b$$
$$i_b \text{ in milliamperes; } v_o \text{ in volts} \quad (B)$$

From Eq. *(A)*,

$$i_b = \frac{0.01 \text{ V} - 10^{-3}v_o}{1 \text{ k}\Omega} = 0.01 \text{ mA} - 1\ \mu\text{S}\ v_o$$

Substituting for i_b in Eq. *(B)*,

$$[(280 + jB_C)\ \mu\text{S}]v_o = -100(0.01 \text{ mA} - 1\ \mu\text{S}\ v_o)$$
$$= -1 \text{ mA} + 100\ \mu\text{S}\ v_o$$
$$[(280 - 100 + jB_C)\ \mu\text{S}]v_o = -1 \text{ mA}$$
$$v_o = -\frac{1 \text{ mA}}{(180 + jB_C)\ \mu\text{S}}$$

a. When $f = 100$ Hz, $\omega = 628.3$ rad/s, and

$$B_C = \omega C = 628.3 \text{ rad/s} \times 0.01\ \mu\text{F} = 6.283\ \mu\text{S}$$
$$(180 + jB_C)\ \mu\text{S} = (180 + j6.283)\ \mu\text{S} \simeq 180\ \mu\text{S}\ \underline{/0^\circ}$$

$$v_o \simeq -\frac{1\text{ mA}}{180\ \mu\text{S}\ \angle 0°} \simeq \mathbf{-5.556\ V\ \angle 0°}\text{ or }\mathbf{\angle 0\ rad}$$

b. When $f = 1$ kHz, $\omega = 6.283$ krad/s, and $B_C = 62.83\ \mu$S,

$$(180 + jB_C)\ \mu\text{S} = (180 + j62.83)\ \mu\text{S} = 190.7\ \mu\text{S}\ \angle 19.24°$$

$$v_o = -\frac{1\text{ mA}}{190.7\ \mu\text{S}\ \angle 19.24°}$$
$$= \mathbf{-5.244\ V\ \angle -19.24°} \quad\text{or}\quad \mathbf{\angle -0.3358\ rad}$$

c. When $f = 10$ kHz, $\omega = 62.83$ krad/s, and $B_C = 628.3\ \mu$S,

$$(180 + j628.3)\ \mu\text{S} = 653.6\ \mu\text{S}\ \angle 74.02°$$

$$v_o = \frac{1\text{ mA}}{653.6\ \mu\text{S}\ \angle 74.02°}$$
$$= \mathbf{-1.53\ V\ \angle -74.02°} \quad\text{or}\quad \mathbf{\angle -1.292\ rad}$$

d. If C is removed from the circuit, $B_C = \omega C = 0$,

$$(180 + jB_C)\ \mu\text{S} = 180\ \mu\text{S}\ \angle 0°$$

$$v_o = -\frac{1\text{ mA}}{180\ \mu\text{S}\ \angle 0°} = \mathbf{-5.556\ V\ \angle 0°}\text{ or }\mathbf{\angle 0\ rad}$$

Example 17-25 shows the effect of shunt capacitance. At low frequencies (100 Hz in the example), the effect is negligible, and so the output voltage with the capacitor in circuit is approximately the same as without the capacitor. As the frequency increases, the output voltage drops. At 1 kHz in the example, the effect on the voltage is not great: v_o = 94 percent of the value without C. At 10 kHz, v_o is only 28 percent of the value without C. When the frequency is low, the phase angle between the input and output voltages is small, but it becomes significant at higher frequencies.

In the example lowercase letters i and v were used for the current and voltage symbols although they represented rms values. This use of lowercase symbols is common in electronic circuit work to distinguish between ac signal components (lowercase) and dc components (capital).

The values of the transistor parameters h_{ie}, h_{re}, h_{fe}, and h_{oe} were given in Example 17-25 with no attempt to explain how they might be obtained.

EXAMPLE 17-26 A transistor has these parameters: $h_{ie} = 1.5\ \text{k}\Omega$, $h_{oe} = 40\ \mu$S, $h_{re} = 0.0005$, and $h_{fe} = 100$. What is the voltage gain v_{ce}/v_{be} of the amplifier in Fig. 17-62*a*, in which this transistor operates into a 4.7-kΩ load? Assume that the circuit elements R_1, R_2, R_3, C_1, C_2, and C_3 have negligible effect on the transistor's ac characteristics.

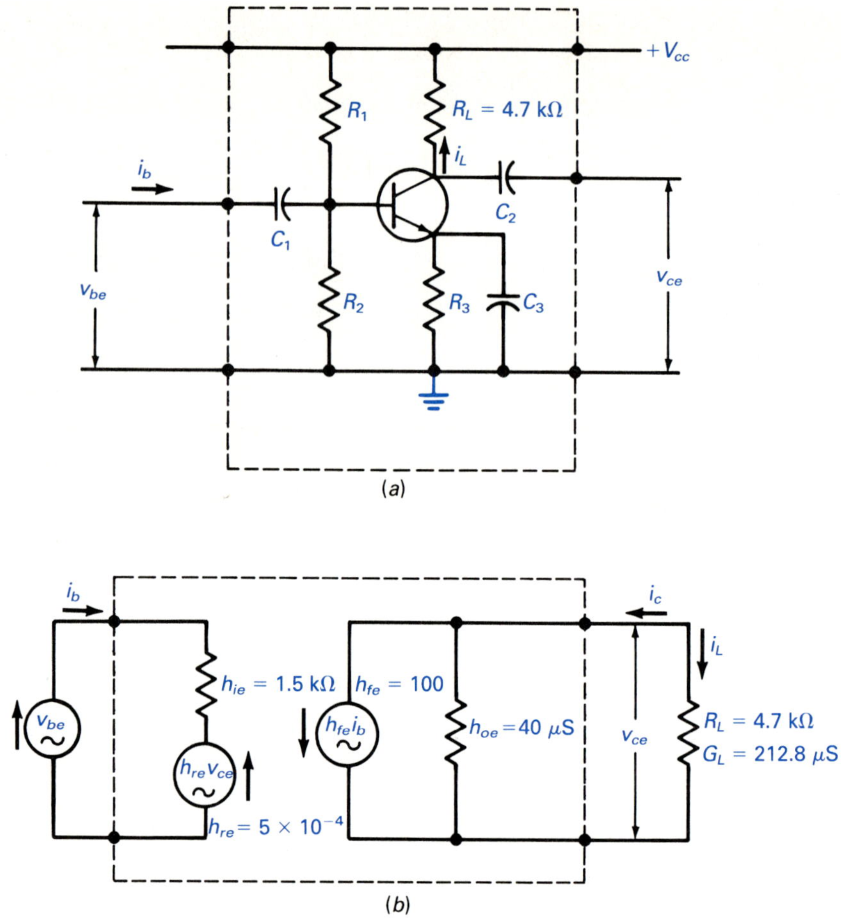

Fig. 17-62. Example 17-26: *(a)* Given circuit; *(b)* *h*-parameter equivalent.

Solution Figure 17-62*b* shows an *h*-parameter equivalent to the circuit in Fig. 17-62*a*. The input loop and output nodal equations are

$$h_{ie}i_b + h_{re}v_{ce} = v_{be} \tag{17-29}$$

$$1.5\ \text{k}\Omega\ i_b + 0.0005\ v_{ce} = v_{be}$$

From Eq. (17-30),

$$h_{fe}i_b + h_{oe}v_{ce} = i_c = -\ i_L = -\ G_L v_{ce}$$

$$100 i_b + (40 + 212.8) \times 10^{-6}\ \text{S}\ v_{ce} = 0$$

The equations in matrix form are

$$\begin{bmatrix} 1.5 \times 10^3 & 5 \times 10^{-4} \\ 10^2 & 2.528 \times 10^{-4} \end{bmatrix} \begin{bmatrix} i_b \\ v_{ce} \end{bmatrix} = \begin{bmatrix} v_{be} \\ 0 \end{bmatrix} \qquad i_b \text{ in amperes; } v_{ce} \text{ and } v_{be} \text{ in volts}$$

The solution for v_{ce} is

$$v_{ce} = -303.8 v_{be}$$

or

$$\frac{v_{ce}}{v_{be}} = \mathbf{-303.8}$$

The negative result for Example 17-26 indicates that there is a phase reversal between input and output. This phase reversal is a true characteristic of a common-emitter transistor amplifier.

EXAMPLE 17-27 What is the input impedance of the transistor amplifier in Example 17-26?

Solution The input impedance of a network is the ratio of input voltage to input current. In the case of the common-emitter transistor amplifier, this is $Z_{\text{in}} = v_{be}/i_b$.

The solution for i_b, from the matrix in Example 17-26, is

$$i_b = (7.679 \times 10^{-4}\ \text{S})v_{be}$$

Therefore, $$Z_{\text{in}} = \frac{v_{be}}{i_b} = \frac{v_{be}}{(7.679 \times 10^{-4}\ \text{S})\ v_{be}} = \mathbf{1.302\ k\Omega}$$

EXAMPLE 17-28 A small power transistor, such as might be used in a car radio, has the following parameters: $h_{ie} = 250\ \Omega$, $h_{oe} = 4.5$ mS, $h_{fe} = 80$, and $h_{re} = 1.5 \times 10^{-3}$. What is the power gain of an amplifier which uses this transistor and operates into a 100-Ω resistive load?

Solution A typical common-emitter power amplifier circuit is shown in Fig. 17-63*a*. In this circuit the output transformer connects the transistor to the loudspeaker load. If the loudspeaker is regarded as a resistive impedance of 3 Ω, the transformer (if assumed to be perfect) will have a primary-to-secondary turns ratio of 5.77, resulting in a collector load of 100 Ω. The hybrid-parameter equivalent circuit is given in Fig. 17-63*b*.

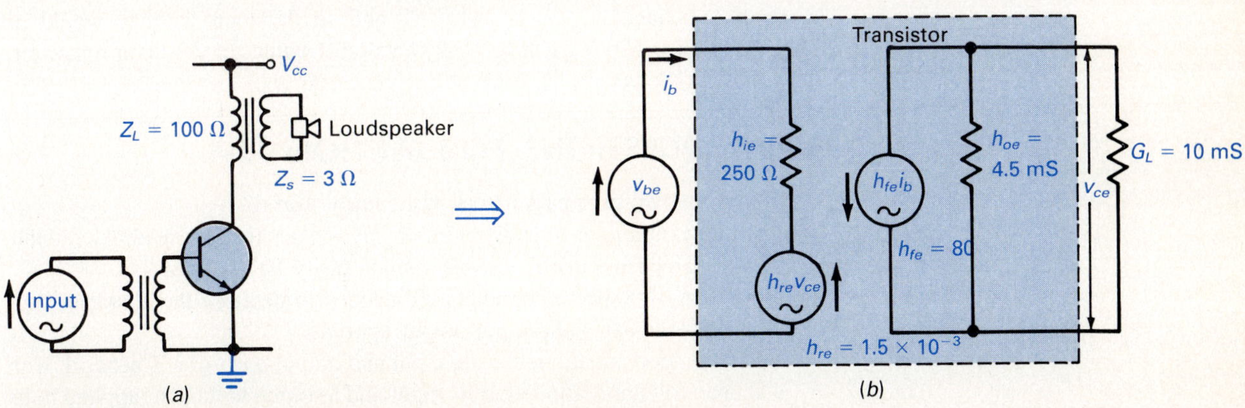

Fig. 17-63. Example 17-28: Circuits: *(a)* Typical common-emitter power amplifier; *(b)* *h*-parameter equivalent.

The input loop and output nodal equations are

$$h_{ie}i_b + h_{re}v_{ce} = v_{be} \qquad (17\text{-}29)$$
$$250\ \Omega\ i_b + 1.5 \times 10^{-3}\ v_{ce} = v_{be}$$

From Eq. (17-30),

$$h_{fe}i_b + (h_{oe} + G_L)v_{ce} = 0$$
$$80\ i_b + (4.5 + 10) \times 10^{-3}\ \text{S}\ v_{ce} = 0$$

The matrix form for these equations is

$$\begin{bmatrix} 250 & 1.5 \times 10^{-3} \\ 80 & 1.45 \times 10^{-2} \end{bmatrix} \begin{bmatrix} i_b \\ v_{ce} \end{bmatrix} \begin{bmatrix} v_{be} \\ 0 \end{bmatrix} \qquad i_b \text{ in amperes; } v_{ce} \text{ and } v_{be} \text{ in volts}$$

The solutions for i_b and v_{ce} are

$$i_b = 4.137 \times 10^{-3}\ \text{S}\ v_{be}$$
$$v_{ce} = -22.82v_{be}$$

The input and output powers are

$$P_{\text{in}} = i_b v_{be} = 4.137 \times 10^{-3}\ \text{S}(v_{be})^2\ \text{W}$$
$$P_{\text{out}} = (v_{ce})^2 G_L = (-22.82v_{be})^2 \times 10\ \text{mS} = 5.207\ \text{S}(v_{be})^2\ \text{W}$$

The power gain A_P is given by

$$A_P = \frac{P_{\text{out}}}{P_{\text{in}}} = \frac{5.207\ \text{S}(v_{be})^2\ \text{W}}{4.137 \times 10^{-3}\ \text{S}(v_{be})^2\ \text{W}}$$
$$A_P = \mathbf{1259}$$

Some feeling for the quantities involved in the amplifier in Example 17-28 can be obtained. It is typical for the peak-to-peak voltage across the load impedance to be of the order of 12 V. For a sine wave, this would be 4.243 V rms. If it is assumed that this is indeed the load voltage, then the output power is

$$P_{\text{out}} = (V_L)^2 G_L = (4.243\ \text{V})^2 \times 10\ \text{mS} = 180\ \text{mW}$$

For a power gain of 1259 the input power would thus need to be 0.143 mW. The input impedance v_{be}/i_b is 241.7 Ω, and so the input voltage would be 0.186 V rms or 0.526 V peak to peak, a suitable value for a silicon transistor.

REVIEW QUESTIONS FOR SECTION 17-4

1. When are two-port parameters particularly useful?
2. Give an example of a device with characteristics that are readily available but for which the circuit details cannot easily be obtained.
3. How many two-port parameters are needed to describe the characteristics of a network using the input-output approach?
4. Write the loop equations for the input and output ports of a T network with a source voltage v_i applied to its input and a source voltage v_o applied to its output.
5. Write each of the z parameters as a ratio of voltage and current.

6. Write the nodal equations for a y-parameter circuit that has a source current i_i applied to its input and a source current i_o applied to its output.
7. Write each of the y parameters as a ratio of voltage and current.
8. What distinguishes an h-parameter circuit from equivalent z- and y-parameter circuits?
9. Write input- and output-port equations in terms of h parameters.
10. Write each of the h parameters as a ratio of voltage and current.

EXERCISE PROBLEMS FOR SECTION 17-4

1. Using the input-output approach, find i_i and i_o in Fig. 17-64 when
   a. $R_1 = 1.2\ \text{k}\Omega$, $R_2 = 1.5\ \text{k}\Omega$, $R_c = 2.2\ \text{k}\Omega$, $v_i = 10$ V, $v_o = 10$ V
   (*ans.* 1.938 mA, 1.55 mA)
   b. $R_1 = 3.3\ \text{k}\Omega$, $R_2 = 6.8\ \text{k}\Omega$, $R_c = 4.7\ \text{k}\Omega$, $v_i = 20$ V, $v_o = 5$ V
   (*ans.* 2.954 mA, −0.7727 mA)
   c. $R_1 = 220\ \Omega$, $R_2 = 470\ \Omega$, $R_c = 1\ \text{k}\Omega$, $v_i = 10$ V, $v_o = 20$ V
   (*ans.* −6.68 mA, 18.15 mA)
   d. $R_1 = 10\ \text{k}\Omega$, $R_2 = 15\ \text{k}\Omega$, $R_c = 18\ \text{k}\Omega$, $v_i = 10$ V, $v_o = -10$ V
   (*ans.* 850.1 μA, −766.7 μA)

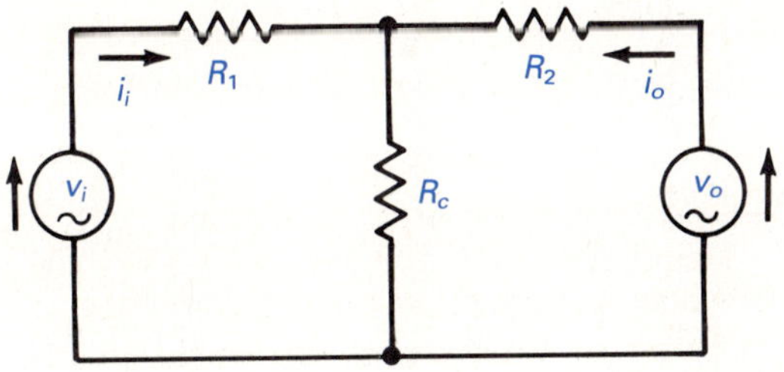

Fig. 17-64. Problem 1: Circuit.

2. Using the input-output approach, find i_L and v_L in Fig. 17-65 when
   a. $v_i = 10$ V, $R_1 = 1.2\ \text{k}\Omega$, $R_2 = 1.5\ \text{k}\Omega$, $R_c = 2.2\ \text{k}\Omega$, $R_L = 1\ \text{k}\Omega$
   b. $v_i = 50$ mV, $R_1 = 960\ \Omega$, $R_2 = 360\ \Omega$, $R_c = 320\ \Omega$, $R_L = 600\ \Omega$

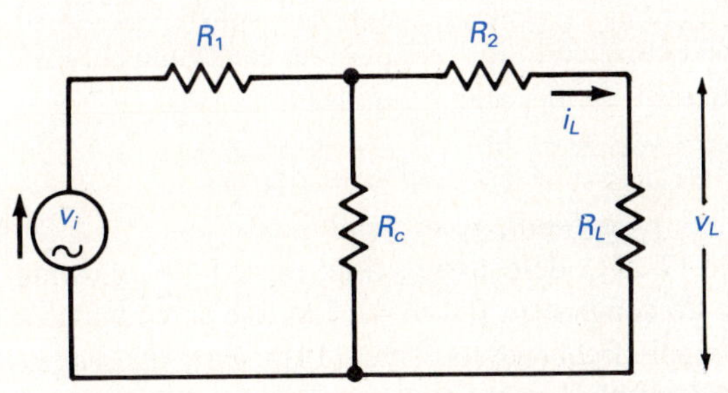

Fig. 17-65. Problem 2: Circuit.

3. Find v_L across a 1-kΩ load attached to the output of an electronic network which has an input of 5 V and z parameters of
   a. $z_i = 8\ \text{k}\Omega$, $z_o = 8\ \text{k}\Omega$, $z_f = 5\ \text{k}\Omega$, $z_r = 6\ \text{k}\Omega$ (*ans.* 0.595 V)

b. $z_i = 1\ \text{k}\Omega,\ z_o = 8\ \text{k}\Omega,\ z_f = 5\ \text{k}\Omega,\ z_r = 6\ \text{k}\Omega$ (*ans.* −1.19 V)
c. $z_i = 2\ \text{k}\Omega,\ z_o = 5\ \text{k}\Omega,\ z_f = 25\ \text{k}\Omega,\ z_r = 2\ \text{k}\Omega$ (*ans.* −3.29 V)

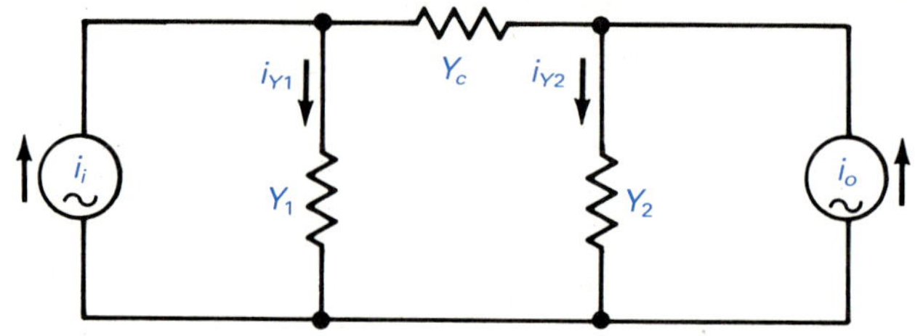

Fig. 17-66. Problem 4: Circuit.

4. Using the input-output approach and y parameters, find i_{Y1} and i_{Y2} in Fig. 17-66 when
a. $Y_1 = 1$ mS, $Y_2 = 2$ mS, $Y_c = 3$ mS, $i_i = 10$ mA, $i_o = 10$ mA
b. $Y_1 = 1$ mS, $Y_2 = 2$ mS, $Y_c = 3$ mS, $i_i = 10$ mA, $i_o = -10$ mA
c. $Y_1 = 15$ mS, $Y_2 = 5$ mS, $Y_c = 1$ mS, $i_i = -5$ mA, $i_o = -15$ mA

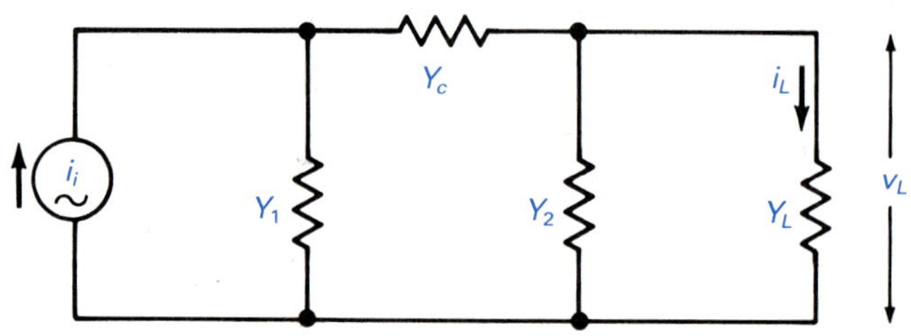

Fig. 17-67. Problem 5: Circuit.

5. Using the input-output approach and y parameters, find i_L and v_L in Fig. 17-67 when
a. $i_i = 50$ mA, $Y_1 = 1$ mS, $Y_2 = 5$ mS, $Y_c = 4$ mS, $Y_L = 10$ mS
(*ans.* 25.32 mA, 2.532 V)
b. $i_i = 25$ mA, $Y_1 = 2.5$ mS, $Y_2 = 2.5$ mS, $Y_c = 4$ mS, $Y_L = 1.667$ mS
(*ans.* 4.495 mA, 2.697 V)
c. $i_i = 100\ \mu$A, $Y_1 = 50\ \mu$S, $Y_2 = 250\ \mu$S, $Y_c = 250\ \mu$S, $Y_L = 50\ \mu$S
(*ans.* 12.2 μA, 0.2439 V)

6. Find v_L across a 1-mS load attached to the output of an electronic network which has an input current of 20 mA and y parameters:
a. $y_i = 250\ \mu$S, $y_o = 500\ \mu$S, $y_f = 150\ \mu$S, $y_r = 200\ \mu$S
b. $y_i = 200\ \mu$S, $y_o = 500\ \mu$S, $y_f = 150\ \mu$S, $y_r = 250\ \mu$S
c. $y_i = 2$ mS, $y_o = 2$ mS, $y_f = 5$ mS, $y_r = 3$ mS

7. In the circuit for Example 17-22, a defective rectifier caused the following conditions to occur. The dc component fell to 42.5 V, the ac component fell to 51.42 V, and the ripple frequency became 60 Hz. Find the magnitude of the output ripple ("ripple" refers to the ac component).
(*ans.* 345 mV $\angle 124.4°$)

8. Find i_i and v_o for the circuit in Fig. 17-68 when $v_i = 10$ V, $i_o = 5$ mA, and
a. $h_i = 1\ \text{k}\Omega$, $h_o = 1$ mS, $h_f = 10$, $h_r = 0.2$

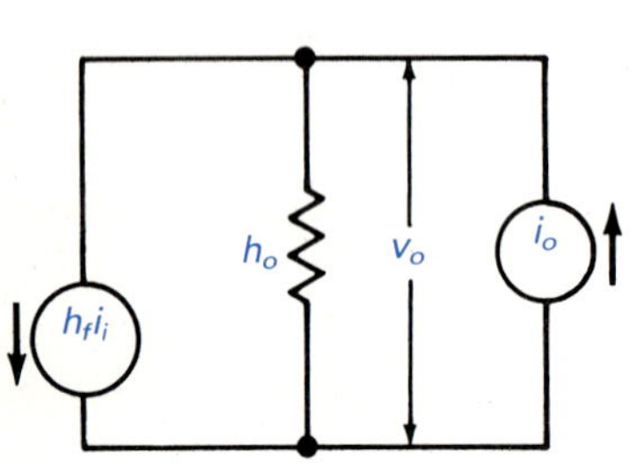

Fig. 17-68. Problem 8: Circuit.

b. $h_i = 1\ \text{k}\Omega$, $h_o = 1\ \text{mS}$, $h_f = 10$, $h_r = -0.2$
c. $h_i = 1\ \text{k}\Omega$, $h_o = 1\ \text{mS}$, $h_f = -10$, $h_r = 0.2$
d. $h_i = 1\ \text{k}\Omega$, $h_o = 1\ \text{mS}$, $h_f = -10$, $h_r = -0.2$

9. Find v_o for the circuit in Fig. 17-69 when $v_i = 1$ V and
a. $h_i = 500\ \Omega$, $h_o = 100\ \mu\text{S}$, $h_f = 15$, $h_r = 0.01$, $Z_L = 2\ \text{k}\Omega$ (*ans.* -100 V)
b. $h_i = 1\ \text{k}\Omega$, $h_o = 50\ \mu\text{S}$, $h_f = 25$, $h_r = 0.001$, $Z_L = 1\ \text{k}\Omega$ (*ans.* -24.39 V)
c. $h_i = 1\ \text{k}\Omega$, $h_o = 50\ \mu\text{S}$, $h_f = 25$, $h_r = 0.001$, $Z_L = 500\ \Omega$ (*ans.* -12.35 V)

10. A common-emitter amplifier, using a 2N1305 transistor, has an equivalent circuit as shown in Fig. 17-70. The characteristics of the transistor under the given operating conditions are $h_{ie} = 800\ \Omega$, $h_{oe} = 230\ \mu\text{S}$, $h_{fe} = 95$, and $h_{re} = 13 \times 10^{-4}$. If $v_i = 0.5$ V and $R_s = 1.5\ \text{k}\Omega$, find v_o when Z_L is
a. 560 Ω
b. 1000 Ω
c. 1500 Ω

11. A common-base amplifier, using a 2N1305 transistor, has an equivalent circuit as shown in Fig. 17-70. The characteristics of the transistor under the given operating conditions are $h_{ib} = 20\ \Omega$, $h_{ob} = 1.3 \times 10^{-6}$ S, $h_{fb} = -0.99$, and $h_{rb} = 7 \times 10^{-4}$. If $v_i = 0.5$ V and $R_s = 100\ \Omega$, find v_o when Z_L is
a. 560 Ω (*ans.* 2.3 V)
b. 1000 Ω (*ans.* 4.096 V)
c. 1500 Ω (*ans.* 6.123 V)

12. A common-emitter power amplifier, with an equivalent circuit as shown in Fig. 17-70, uses an MJE2801 power transistor with the following characteristics: $h_{ie} = 170\ \Omega$, $h_{oe} = 320\ \mu\text{S}$, $h_{fe} = 160$, and $h_{re} = 9 \times 10^{-4}$. If $v_i = 0.2$ V and $R_s = 220\ \Omega$, find the power output to the load when Z_L is
a. 200 Ω
b. 300 Ω
c. 400 Ω

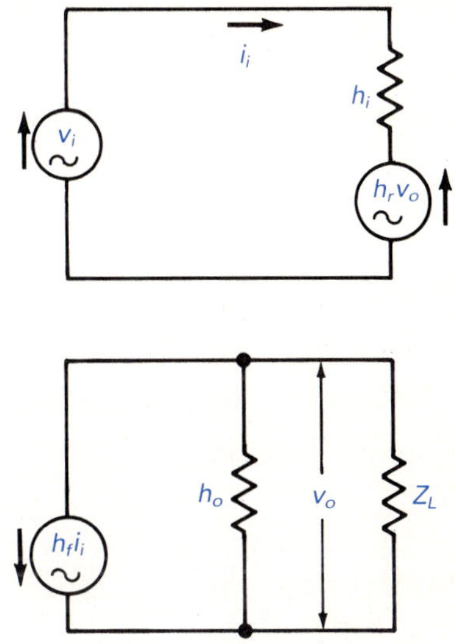

Fig. 17-69. Problem 9: Circuit.

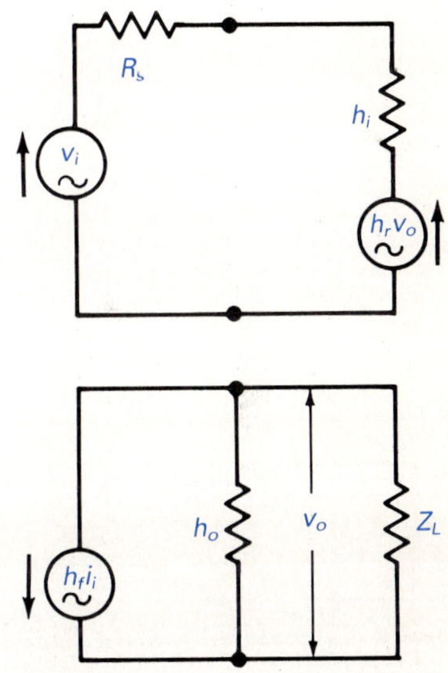

Fig. 17-70. Problems 10 to 13: Circuit.

SUMMARY OF EQUATIONS

z parameters

$(Z_1 + Z_c)i_i + Z_c i_o = v_i$ (17-1)
$Z_c i_i + (Z_2 + Z_c)i_o = v_o$ (17-2)
$z_i = Z_1 + Z_c$ (17-3)
$z_o = Z_2 + Z_c$ (17-4)
$z_r = Z_c$ (17-5)
$z_f = Z_c$ (17-6)
$z_i i_i + z_r i_o = v_i$ (17-7)
$z_f i_i + z_o i_o = v_o$ (17-8)
$z_i = v_i/i_i \,|i_o = 0$ (17-9)
$z_f = v_o/i_i \,|i_o = 0$ (17-10)

$z_r = v_i/i_o\,|i_i = 0$ (17-11)

$z_o = v_o/i_o\,|i_i = 0$ (17-12)

y parameters

$y_iv_i + y_rv_o = i_i$ (17-13)

$y_fv_i + y_ov_o = i_o$ (17-14)

$y_i = i_i/v_i\,|v_o = 0$ (17-15)

$y_f = i_o/v_i\,|v_o = 0$ (17-16)

$y_r = i_i/v_o\,|v_i = 0$ (17-17)

$y_o = i_o/v_o\,|v_i = 0$ (17-18)

h parameters

$h_ii_i + h_rv_o = v_i$ (17-19)

$h_fi_i + h_ov_o = i_o$ (17-20)

$h_i = v_i/i_i\,|v_o = 0$ (17-21)

$h_f = i_o/i_i\,|v_o = 0$ (17-22)

$h_r = v_i/v_o\,|i_i = 0$ (17-23)

$h_o = i_o/v_o\,|i_i = 0$ (17-24)

Transistors

$h_{ie} = v_{be}/i_b\,|v_{ce} = 0$ (17-25)

$h_{fe} = i_c/i_b\,|v_{ce} = 0$ (17-26)

$h_{re} = v_{be}/v_{ce}\,|i_b = 0$ (17-27)

$h_{oe} = i_c/v_{ce}\,|i_b = 0$ (17-28)

$h_{ie}i_b + h_{re}v_{ce} = v_{be}$ (17-29)

$h_{fe}i_b + h_{oe}v_{ce} = i_c$ (17-30)

CHAPTER QUESTIONS

17-1. Compare a *single-phase three-wire* system with a *three-phase* system.

17-2. Draw phasor diagrams of a
a. Single-phase two-wire source.
b. Single-phase three-wire source.
c. Three-phase three-wire source.
d. Three-phase four-wire source.

17-3. What is meant when a *wye* source is said to be the equivalent of a *delta* source?

17-4. Which of the dc network analysis methods may be used with ac circuits?

17-5. How may the details of applying analysis to ac circuits differ from those for dc circuits?

17-6. What is a *two-port* circuit?

17-7. What is a set of two-port parameters?

17-8. The two-port parameters dealt with in this book are z, y, and h parameters.
a. For what do the letters z, y, and h stand?
b. What justifies the use of these terms?

17-9. How would knowledge of a parameter set's values give information on the behavior of a circuit?

17-10. If a two-port parameter is negative in value, what is known about the circuit if
a. It is a forward parameter?
b. It is a reverse parameter?

17-11. Compare *short-circuit* and *open-circuit* parameters.

17-12. How are short-circuit and open-circuit conditions simulated when measuring transistor *h* parameters?

17-13. In what manner may *h* parameters be useful in dealing with transistor circuits?

17-14. How might the two-port parameter technique be applied to dc circuits?

CHAPTER PROBLEMS

17-1. Find the current in the load Z_L of the circuit in Fig. 17-71. First find the total circuit impedance, then the total current, etc.
(*ans.* 18.98 μA, −1.084 rad or −62.11°)

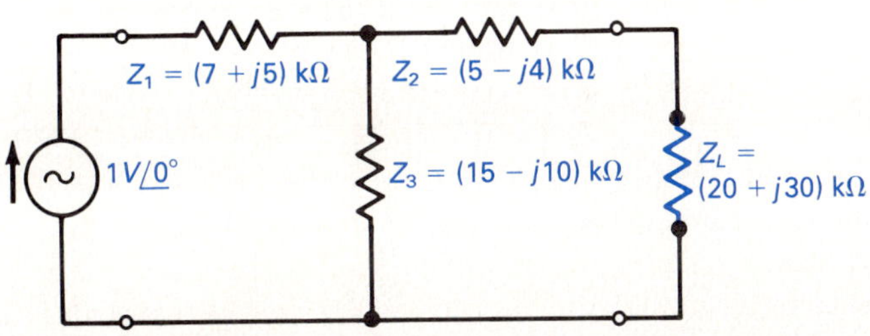

Fig. 17-71. Problems 17-1 to 17-5, 17-11, and 17-12: Circuit.

17-2. Repeat Prob. 17-1 using the assumed current method.

17-3. Check the answer to Prob. 17-1, using the Thevenin approach.

17-4. What would be the load impedance for maximum power transfer in the circuit of Fig. 17-71?

17-5. How much power is transferred to the load Z_L in the circuit of Fig. 17-71? How much power would be transferred to the load if its impedance were correct for maximum load power?
(*ans.* 7.205 μW, 13.71 μW)

17-6. Using loop analysis, find the current in Z_3 of the circuit of Fig. 17-72.

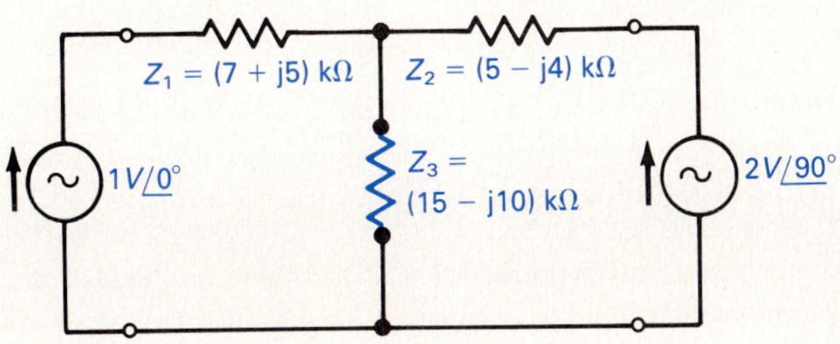

Fig. 17-72. Problems 17-6 to 17-10: Circuit.

17-7. Repeat Prob. 17-6 using nodal analysis.

(*ans*. 41.72 μA, 2.45 rad or 140.4°)

17-8. Repeat Prob. 17-6 using the superposition method.

17-9. Repeat Prob. 17-6

a. Using Thevenin analysis.

b. Converting the Thevenin equivalent circuit to a Norton equivalent.

(*ans*. See Prob. 17-7)

17-10. Repeat Prob. 17-9*b* using the Norton equivalent circuit developed from the Norton principles.

17-11. Refer to the circuit of Fig. 17-71. Draw two-port-parameter equivalent circuits with values for

a. z parameters.

b. y parameters.

c. h parameters.

(*ans*. $z_i = 22.56\ \text{k}\Omega\ \underline{/-12.8^\circ}$,
$z_o = 24.41\ \text{k}\Omega\ \underline{/-34.99^\circ}$,
$z_f = z_r = 18.03\ \text{k}\Omega\ \underline{/-33.69^\circ}$;
$y_i = 0.02697\ \text{mS}\ \underline{/13.01^\circ}$,
$y_o = 0.1562\ \text{mS}\ \underline{/38.66^\circ}$,
$y_f = y_r = -0.1562\ \text{mS}\ \underline{/38.66^\circ}$;
$h_i = 10.97\ \text{k}\Omega\ \underline{/11.2^\circ}$,
$h_o = 0.04097\ \text{mS}\ \underline{/34.99^\circ}$,
$h_f = -0.7386\ \underline{/1.3^\circ}$, $h_r = 0.7386\ \underline{/1.3^\circ}$)

17-12. With the two-port parameters for the circuit of Fig. 17-71, check the value of the load current. Use

a. z parameters.

b. y parameters.

c. h parameters.

17-13. A three-phase 240-V (phase to neutral) wye-connected 60-Hz power system supplies an electric motor. The motor takes a total of 15 kVA, balanced between phases. If the motor's power factor is 0.8 lagging and its windings form a delta load, what is the current in each winding? How much power does each winding take from the supply?

(*ans*. 12.03 A, 4 kW)

17-14. Refer to Example 17-28 on page 707. Make a graph of the power gain with respect to the load impedance at the values 100 Ω, 200 Ω, 250 Ω, and 500 Ω. What do you conclude from this graph?

APPENDIX

DETERMINANTS

The solution of simultaneous equations is important in network analysis. However, methods of obtaining these solutions have not been discussed in the text for several reasons: (1) the subject rightly belongs in a mathematics book, (2) although there are various methods of manual solution, people tend to prefer one of them, and (3) low-cost calculators are available with programs for automatic solution of simultaneous equations.

The determinant method of solving simultaneous equations is useful with or without a calculator. In addition, determinants relate directly to matrices, which you will need to know if you are to pursue electrical and electronic circuit theory beyond the level of this book. So you may welcome the following brief introduction to determinants. If you want a more thorough treatment of determinants and matrices in general, consult a mathematics textbook.

Equations with Two Unknowns

Two general simultaneous equations are

$$ax + by = c$$
$$dx + ey = f$$

The unknowns are x and y; a, b, d, and e are coefficients of the unknowns; c and f are constants. The procedure for solving these equations by determinants is as follows.

1. Write the equations in *matrix* form:

$$\begin{bmatrix} a & b \\ d & e \end{bmatrix} \begin{bmatrix} x \\ y \end{bmatrix} = \begin{bmatrix} c \\ f \end{bmatrix}$$

   The solutions will initially appear as fractions.

2. To obtain the common denominator D of these fractions,

   a. Write the section of the matrix containing the coefficients as a determinant (between parallel lines):

$$\begin{vmatrix} a & b \\ d & e \end{vmatrix}$$

   b. Cross-multiply the coefficients:

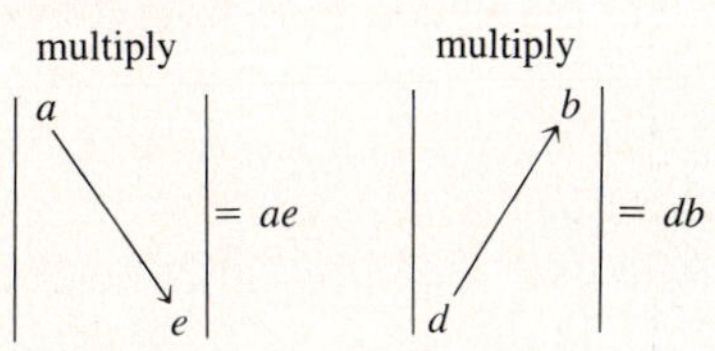

   c. Give signs to the products:

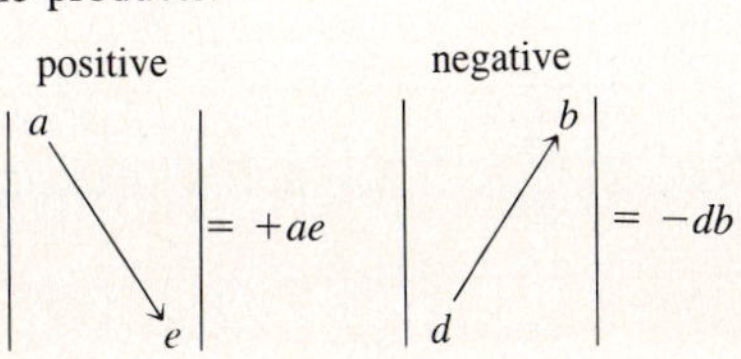

d. Write the denominator:

$$D = ae - db$$

3. To obtain the numerator N_x for the unknown x:

a. Write a determinant consisting of the coefficients of y in their original matrix locations and the coefficients of x replaced by the constants:

$$\begin{vmatrix} c & b \\ f & e \end{vmatrix}$$

b. Apply operations 2*b* to 2*d* to this determinant:

$$N_x = ce - fb$$

4. To obtain the numerator N_y for the unknown y, repeat operations similar to 3*a* and 3*b*:

$$N_y = \begin{vmatrix} a & c \\ d & f \end{vmatrix} = af - dc$$

The solutions for the pair of equations are

$$x = \frac{ce - fb}{ae - db} \quad \text{and} \quad y = \frac{af - dc}{ae - db}$$

EXAMPLE 1

$$3x + 2y = 10$$
$$5x + 6y = 15$$
$$\begin{bmatrix} 3 & 2 \\ 5 & 6 \end{bmatrix}\begin{bmatrix} x \\ y \end{bmatrix} = \begin{bmatrix} 10 \\ 15 \end{bmatrix}$$
$$D = \begin{vmatrix} 3 & 2 \\ 5 & 6 \end{vmatrix} = 3 \times 6 - 5 \times 2 = 8$$
$$N_x = \begin{vmatrix} 10 & 2 \\ 15 & 6 \end{vmatrix} = 10 \times 6 - 15 \times 2 = 30$$
$$N_y = \begin{vmatrix} 3 & 10 \\ 5 & 15 \end{vmatrix} = 3 \times 15 - 5 \times 10 = -5$$

Solution

$$x = \tfrac{30}{8} = \mathbf{3.75} \quad \text{and} \quad y = -\tfrac{5}{8} = \mathbf{-0.625}$$

EXAMPLE 2

$$3x - 2y = 10$$
$$-5x + 6y = 15$$
$$\begin{bmatrix} 3 & -2 \\ -5 & 6 \end{bmatrix}\begin{bmatrix} x \\ y \end{bmatrix} = \begin{bmatrix} 10 \\ 15 \end{bmatrix}$$
$$D = \begin{vmatrix} 3 & -2 \\ -5 & 6 \end{vmatrix} = 3 \times 6 - [(-5)(-2)] = 8$$
$$N_x = \begin{vmatrix} 10 & -2 \\ 15 & 6 \end{vmatrix} = 10 \times 6 - [15(-2)] = 90$$
$$N_y = \begin{vmatrix} 3 & 10 \\ -5 & 15 \end{vmatrix} = 3 \times 15 - [(-5) \times 10] = 95$$

Solution

$$x = \tfrac{90}{8} = \mathbf{11.25} \qquad \text{and} \qquad y = \tfrac{95}{8} = \mathbf{11.88}$$

Equations with Three Unknowns

The procedure is an extension of that used for two unknowns.

$$\begin{aligned} ax + by + cz &= d \\ ex + fy + gz &= h \\ jx + ky + lz &= m \end{aligned}$$

1. Write the equations in matrix form:

$$\begin{bmatrix} a & b & c \\ e & f & g \\ j & k & l \end{bmatrix} \begin{bmatrix} x \\ y \\ z \end{bmatrix} = \begin{bmatrix} d \\ h \\ m \end{bmatrix}$$

2. To obtain the common denominator for the solutions:

a. Write the coefficient section of the matrix as a determinant:

$$\begin{vmatrix} a & b & c \\ e & f & g \\ j & k & l \end{vmatrix}$$

b. To simplify the cross-multiplication, repeat the first two columns to the right of the determinant:

$$\left|\begin{array}{ccc|cc} a & b & c & a & b \\ e & f & g & e & f \\ j & k & l & j & k \end{array}\right.$$

c. Cross-multiply the coefficients and give signs:

$$\left|\begin{array}{ccc|cc} a & b & c & a & b \\ e & f & g & e & f \\ j & k & l & j & k \end{array}\right. = +(afl + bgj + cek)$$

$$\left|\begin{array}{ccc|cc} a & b & c & a & b \\ e & f & g & e & f \\ j & k & l & j & k \end{array}\right. = (jfc + kga + leb)$$

d. Write the common denominator D:

$$D = \begin{vmatrix} a & b & c \\ e & f & g \\ j & k & l \end{vmatrix} = (afl + bgj + cek) - (jfc + kga + leb)$$

3. Obtain the numerator N_x for x by applying methods 2*a* to 2*d*, but with the coefficients of x replaced by the constants:

$$N_x = \begin{vmatrix} d & b & c \\ h & f & g \\ m & k & l \end{vmatrix} = (dfl + bgm + chk) - (mfc + kgd + lhb)$$

4. Repeat operation 3 for N_y:

$$N_y = \begin{vmatrix} a & d & c \\ e & h & g \\ j & m & l \end{vmatrix} = (ahl + dgj + cem) - (jhc + mga + led)$$

5. Repeat operation 3 for N_z:

$$N_z = \begin{vmatrix} a & b & d \\ e & f & h \\ j & k & m \end{vmatrix} = (afm + bhj + dek) - (jfd + kha + meb)$$

The solutions for the equations are

$$x = \frac{N_x}{D} = \frac{(dfl + bgm + chk) - (mfc + kgd + lhb)}{(afl + bgj + cek) - (jfc + kga + leb)}$$

$$y = \frac{N_y}{D} = \frac{(ahl + dgj + cem) - (jhc + mga + led)}{(afl + bgj + cek) - (jfc + kga + leb)}$$

$$z = \frac{N_z}{D} = \frac{(afm + bhj + dek) - (jfd + kha + meb)}{(afl + bgj + cek) - (jfc + kga + leb)}$$

EXAMPLE 3

$$x + 3y + 4z = 14$$
$$x + 2y + z = 7$$
$$2x + y + 2z = 2$$

$$\begin{bmatrix} 1 & 3 & 4 \\ 1 & 2 & 1 \\ 2 & 1 & 2 \end{bmatrix} \begin{bmatrix} x \\ y \\ z \end{bmatrix} = \begin{bmatrix} 14 \\ 7 \\ 2 \end{bmatrix}$$

$$D = \begin{vmatrix} 1 & 3 & 4 \\ 1 & 2 & 1 \\ 2 & 1 & 2 \end{vmatrix} \begin{matrix} 1 & 3 \\ 1 & 2 \\ 2 & 1 \end{matrix} = (4 + 6 + 4) - (16 + 1 + 6) = -9$$

$$N_x = \begin{vmatrix} 14 & 3 & 4 \\ 7 & 2 & 1 \\ 2 & 1 & 2 \end{vmatrix} \begin{matrix} 14 & 3 \\ 7 & 2 \\ 2 & 1 \end{matrix} = (56 + 6 + 28) - (16 + 14 + 42) = 18$$

$$N_y = \begin{vmatrix} 1 & 14 & 4 \\ 1 & 7 & 1 \\ 2 & 2 & 2 \end{vmatrix} \begin{matrix} 1 & 14 \\ 1 & 7 \\ 2 & 2 \end{matrix} = (14 + 28 + 8) - (56 + 2 + 28) = -36$$

$$N_z = \begin{vmatrix} 1 & 3 & 14 \\ 1 & 2 & 7 \\ 2 & 1 & 2 \end{vmatrix} \begin{matrix} 1 & 3 \\ 1 & 2 \\ 2 & 1 \end{matrix} = (4 + 42 + 14) - (56 + 7 + 6) = -9$$

Solution

$$x = \frac{18}{-9} = \mathbf{-2} \qquad y = \frac{-36}{-9} = \mathbf{4} \qquad z = \frac{-9}{-9} = \mathbf{1}$$

EXAMPLE 4 The following equations are from Example 9-10.

$$30\text{ k}\Omega \times I_1 - 10\text{ k}\Omega \times I_2 - 10\text{ k}\Omega \times I_3 = 100\text{ V}$$
$$-10\text{ k}\Omega \times I_1 + 20\text{ k}\Omega \times I_2 - 10\text{ k}\Omega \times I_3 = -50\text{ V}$$
$$-10\text{ k}\Omega \times I_1 - 10\text{ k}\Omega \times I_2 + 30\text{ k}\Omega \times I_3 = 0$$

$$\begin{bmatrix} 30 & -10 & -10 \\ -10 & 20 & -10 \\ -10 & -10 & 30 \end{bmatrix} \begin{bmatrix} I_1 \\ I_2 \\ I_3 \end{bmatrix} = \begin{bmatrix} 100 \\ -50 \\ 0 \end{bmatrix} \qquad I_1,\ I_2,\text{ and } I_3 \text{ in milliamperes}$$

$$D = \begin{vmatrix} 30 & -10 & -10 \\ -10 & 20 & -10 \\ -10 & -10 & 30 \end{vmatrix} = (18\,000 - 1000 - 1000) - (2000 + 3000 + 3000) = 8000$$

$$N_1 = \begin{vmatrix} 100 & -10 & -10 \\ -50 & 20 & -10 \\ 0 & -10 & 30 \end{vmatrix} = (60\,000 + 0 - 5000) - (0 + 10\,000 + 15\,000) = 30\,000$$

$$N_2 = \begin{vmatrix} 30 & 100 & -10 \\ -10 & -50 & -10 \\ -10 & 0 & 30 \end{vmatrix} = (-45\,000 + 10\,000 + 0) - (-5000 + 0 - 30\,000) = 0$$

$$N_3 = \begin{vmatrix} 30 & -10 & 100 \\ -10 & 20 & -50 \\ -10 & -10 & 0 \end{vmatrix} = (0 - 5000 + 10\,000) - (-20\,000 + 15\,000 + 0) = 10\,000$$

Solution

$$I_1 = \frac{30\,000}{8000} = \mathbf{3.75\ mA} \qquad I_2 = \frac{0}{8000} = \mathbf{0} \qquad I_3 = \frac{10\,000}{8000} = \mathbf{1.25\ mA}$$

EXAMPLE 5 These equations are from Example 17-3.

$$j40\ \Omega\ I_1 - j50\ \Omega\ I_2 = 50\text{ V}\ \angle 0^\circ$$
$$-j50\ \Omega\ I_1 - j30\ \Omega\ I_2 = -20\text{ V}\ \angle 0^\circ$$

$$\begin{bmatrix} 40\ \angle 90^\circ & 50\ \angle -90^\circ \\ 50\ \angle -90^\circ & 30\ \angle -90^\circ \end{bmatrix} \begin{bmatrix} I_1 \\ I_2 \end{bmatrix} = \begin{bmatrix} 50\ \angle 0^\circ \\ -20\ \angle 0^\circ \end{bmatrix}$$

$$D = \begin{vmatrix} 40\ \angle 90^\circ & 50\ \angle -90^\circ \\ 50\ \angle -90^\circ & 30\ \angle -90^\circ \end{vmatrix} = 1200\ \angle 0^\circ - 2500\ \angle -180^\circ = 3700\ \angle 0^\circ \qquad \text{note that } \angle -180^\circ = \angle +180^\circ$$

$$N_1 = \begin{vmatrix} 50\ \angle 0^\circ & 50\ \angle -90^\circ \\ -20\ \angle 0^\circ & 30\ \angle -90^\circ \end{vmatrix} = 1500\ \angle -90^\circ - (-1000\ \angle -90^\circ) = 2500\ \angle -90^\circ$$

$$N_2 = \begin{vmatrix} 40\ \angle 90^\circ & 50\ \angle 0^\circ \\ 50\ \angle -90^\circ & -20\ \angle 0^\circ \end{vmatrix} = -800\ \angle 90^\circ - 2500\ \angle -90^\circ = -j800 + j2500 = 1700\ \angle 90^\circ$$

Solution

$$I_1 = \frac{2500\ \angle -90^\circ}{3700\ \angle 0^\circ} = \mathbf{0.6757\ A}\ \angle \mathbf{-90^\circ} \qquad \text{and} \qquad I_2 = \frac{1700\ \angle 90^\circ}{3700\ \angle 0^\circ} = \mathbf{0.4595\ A}\ \angle \mathbf{90^\circ}$$

SIGNIFICANT FIGURES

Consider the number 12.34. This number has four digits, ordered as to their significance. The digit corresponding to the highest power of 10 has the greatest significance. Thus the 1 has the greatest significance, as it corresponds to 10^1. Conversely, the 4 has the least significance, as it relates to 10^{-2}.

For any number, the number of digits from the highest significant nonzero digit to, and including, the least significant digit considered important is the number of significant digits or figures.

In our example there are four significant figures. Written as 0012.34, the number still has four significant figures; the two zeros to the left of the highest significant digit add nothing to our knowledge of the precision to which the number is known. But if it is written 12.3400, the number has six significant figures, the two zeros to the right of the least significant digit indicating that the number is known accurately to four decimal places (the number is neither 12.3401 nor 12.3399).

The number of significant figures can be reduced by rounding off. This means dropping the least significant digits until the desired number of digits remains. The new least significant digit may have to be changed as determined by the following rules.

If the highest significant digit dropped is

1. Less than 5, the new least significant digit is not changed.
2. Greater than 5, the new least significant digit is increased by 1.
3. 5, the new least significant digit is not changed if it is even.
4. 5, the new least significant digit is increased by 1 if it is odd.

These rules result in a rounding-off technique which, on the average of a large number of operations, gives the most consistent reliability.

Examples are:

1. 12 340 has five significant figures.
2. 12 340 written as 1.234×10^4 has four significant figures.
3. 0.001 234 has four significant figures.
4. 12.344 to four significant figures is 12.34.
5. 12.346 to four significant figures is 12.35.
6. 12.345 to four significant figures is 12.34.
7. 12.335 to four significant figures is 12.34.

INDEX